T0310287

STATISTICAL LEARNING FOR BIG DEPENDENT DATA

WILEY SERIES IN PROBABILITY AND STATISTICS
Established by *Walter A. Shewhart and Samuel S. Wilks*

Editors: *David J. Balding, Noel A. C. Cressie, Garrett M. Fitzmaurice, Geof H. Givens, Harvey Goldstein, Geert Molenberghs, David W. Scott, Adrian F. M. Smith, Ruey S. Tsay*

Editors Emeriti: *J. Stuart Hunter, Iain M. Johnstone, Joseph B. Kadane, Jozef L. Teugels*

The ***Wiley Series in Probability and Statistics*** is well established and authoritative. It covers many topics of current research interest in both pure and applied statistics and probability theory. Written by leading statisticians and institutions, the titles span both state-of-the-art developments in the field and classical methods.

Reflecting the wide range of current research in statistics, the series encompasses applied, methodological and theoretical statistics, ranging from applications and new techniques made possible by advances in computerized practice to rigorous treatment of theoretical approaches.
This series provides essential and invaluable reading for all statisticians, whether in academia, industry, government, or research.

A complete list of titles in this series can be found at
http://www.wiley.com/go/wsps

STATISTICAL LEARNING FOR BIG DEPENDENT DATA

DANIEL PEÑA
Universidad Carlos III de Madrid
Madrid, Spain

RUEY S. TSAY
University of Chicago
Chicago, United States

This first edition first published 2021
© 2021 John Wiley and Sons, Inc.

All rights reserved. No part of this publication may be reproduced, stored in a retrieval system, or transmitted, in any form or by any means, electronic, mechanical, photocopying, recording or otherwise, except as permitted by law. Advice on how to obtain permission to reuse material from this title is available at http://www.wiley.com/go/permissions.

The right of Daniel Peña and Ruey S. Tsay to be identified as the authors of this work has been asserted in accordance with law.

Registered Office
John Wiley & Sons, Inc., 111 River Street, Hoboken, NJ 07030, USA

Editorial Office
111 River Street, Hoboken, NJ 07030, USA

For details of our global editorial offices, customer services, and more information about Wiley products visit us at www.wiley.com.

Wiley also publishes its books in a variety of electronic formats and by print-on-demand. Some content that appears in standard print versions of this book may not be available in other formats.

Limit of Liability/Disclaimer of Warranty
While the publisher and authors have used their best efforts in preparing this work, they make no representations or warranties with respect to the accuracy or completeness of the contents of this work and specifically disclaim all warranties, including without limitation any implied warranties of merchantability or fitness for a particular purpose. No warranty may be created or extended by sales representatives, written sales materials or promotional statements for this work. The fact that an organization, website, or product is referred to in this work as a citation and/or potential source of further information does not mean that the publisher and authors endorse the information or services the organization, website, or product may provide or recommendations it may make. This work is sold with the understanding that the publisher is not engaged in rendering professional services. The advice and strategies contained herein may not be suitable for your situation. You should consult with a specialist where appropriate. Further, readers should be aware that websites listed in this work may have changed or disappeared between when this work was written and when it is read. Neither the publisher nor authors shall be liable for any loss of profit or any other commercial damages, including but not limited to special, incidental, consequential, or other damages.

Library of Congress Cataloging-in-Publication Data

Names: Peña, Daniel, 1948- author. | Tsay, Ruey S., 1951- author.
Title: Statistical learning for big dependent data / Daniel Peña, Ruey S. Tsay.
Description: First edition. | Hoboken, NJ : Wiley, 2021. | Series: Wiley series in probability and statistics | Includes bibliographical references and index.
Identifiers: LCCN 2020026630 (print) | LCCN 2020026631 (ebook) | ISBN 9781119417385 (cloth) | ISBN 9781119417392 (adobe pdf) | ISBN 9781119417415 (epub) | ISBN 9781119417408 (obook)
Subjects: LCSH: Big data–Mathematics. | Time-series analysis. | Data mining–Statistical methods. | Forecasting–Statistical methods.
Classification: LCC QA76.9.B45 P45 2021 (print) | LCC QA76.9.B45 (ebook) | DDC 005.7–dc23
LC record available at https://lccn.loc.gov/2020026630
LC ebook record available at https://lccn.loc.gov/2020026631

Cover Design: Wiley
Cover Images: Colorful graph Courtesy of Daniel Peña and Ruey S.Tsay, Abstract background © duncan1890/Getty Images

Set in 10/12pt TimesTenLTStd by SPi Global, Chennai, India

SKY10025220_031221

To our good friend & mentor George C. Tiao (DP & RST)

To my wife, Jette Bohsen, with love (DP)

To Teresa, Julie, Richard, and Vicki (RST)

CONTENTS

PREFACE

For the first time in human history, we are collecting data everywhere and every second. These data grow exponentially, are produced and stored at minimum costs, and are changing the way we learn things and control our activities, including monitoring our health and using our leisure time. Statistics, as a scientific discipline, was created in a different environment, where data were scarce, and focused mainly on obtaining maximum efficiency of available information from small-structured data sets. New methods are, therefore, needed to extract useful information from big data sets, which are heterogeneous and unstructured. These methods are being developed in statistics, computer science, machine learning, operation research, artificial intelligence, and other fields. They constitute what is usually called data science. Most advances in big data analysis so far have assumed that the data are collected from independent subjects, so that they can be treated as independent observations. On the other hand, empirical data are often generated over time or in space, and, hence, they have dynamic and/or spatial dependence.

The main goal of this book is to learn from big dependent data. Data with temporal dynamics have been studied in statistics as time series analysis, whereas spatial dependence belongs to the newer area of spatial statistics. Several key ideas that formed the core of recent developments in big data methods are from time series analysis. For instance, the first model selection criterion was proposed by Akaike for selecting the order of an autoregressive process, an iterative model building procedure with nonlinear estimation was proposed by Box and Jenkins for autoregressive integrated moving-average (ARIMA) models, and methods for combining forecasts, which open the way to model averaging and ensemble methods in machine learning, can be traced back to Bayesian forecasting. Today, analyzing big data with temporal and spatial dependence is needed in many scientific fields, ranging from economics and business to environmental and health science, and to engineering and computer vision. It is our belief that modeling and processing big dependent data is a key emerging area of data science.

Several excellent books have been written for statistical learning with independent data, but much less work has been done for dependent data. This book tries to stimulate the use of available methods in forecasting and classification, to promote further research in statistical methods for extracting useful information from big dependent data, and to point out the potential weaknesses when the data dependence is overlooked. We start with brief reviews of time series analysis, both univariate and multivariate, followed by methods for handling heterogeneity when analyzing many time series. We then discuss methods for classification and clustering many time series and introduce dynamic factor models for modeling multivariate or high-dimensional time series. This is followed by a thorough discussion on forecasting in a data-rich environment, including nowcasting. We then turn to deep learning and analysis of spatio-temporal data.

The book is applied oriented, but it also provides some basic theory to help readers understand better the methods and procedures used. Due to the high-dimensional nature of the problems discussed, we include some technical derivations in a few sections, but for most parts, we refer readers interested in rigorous mathematical proofs to the available literature. Real examples are used throughout the book to demonstrate the analysis and applications. For empirical data analysis, we use R extensively and provide the necessary instructions and R scripts so that readers can reproduce the results shown in the book.

The book is organized as follows. Chapter 1 provides some examples of the data sets considered in the book and introduces the basic ideas of temporal dependency, stochastic process, and time series. It also discusses principal component analysis and illustrates the weaknesses of traditional statistical inference when the data dependence is overlooked. Chapter 2 starts with new ways to visualize large sets of time series, summarizes the main properties of ARIMA and state space models, including spectral analysis and Kalman filter, and presents a methodology for building univariate models for large sets of time series. Chapter 3 reviews the available methods for building multivariate vector ARMA models and their limitations with high-dimensional time series. It also discusses cointegration and ways to handle unit-root multivariate nonstationary series. Real time series data are typically heterogeneous, with clustering and certain common features, contain missing values and outliers, and may encounter structural breaks. We address these issues in the next two chapters. Chapter 4 presents missing value estimation and some new methods for detecting and modeling outliers in univariate and multivariate time series. The outliers considered include additive outliers, innovative outliers, level shifts, transitory shifts, and parameter changes. Chapter 5 analyzes clustering, or unsupervised classification methods, for time series, including recent procedures for clustering time series based on their dependency or for selecting the number of clusters. The chapter also considers time series discrimination, or supervised classification, and covers approaches developed not only in statistics, but also in machine learning such as support vector machines. Chapter 6 focuses on an active research topic for modeling large sets of time series, namely dynamic factor models and other types of factor models. The chapter describes in detail the developments of the topic and discusses the pros and cons of several models available in the literature, including those for the high-dimensional setting. Chapter 7 concentrates on the key problem of forecasting large sets of time series. The application of Lasso regression to time series is investigated, as well as procedures developed mostly in the statistic and econometric literature for forecasting high-dimensional time series. It also studies the method of nowcasting and demonstrates its usefulness with real examples. Chapter 8 considers machine learning for forecasting and classification, including neural networks and deep learning. It also studies methods developed in the interface between Statistics and Machine Learning, commonly known as Data Science, including Classification and Regression Trees (CART) and Random Forests. Finally, Chapter 9 studies spatio-temporal data and their applications, including various kriging methods for spatial predictions.

Dependent data cover a wide range of topics and methods, especially in the presence of high-dimensional data. It is too much to expect that a book can cover all these important topics. This book is no exception. We need to make decision on the material to include. Like other books, our choices depend heavily on our experience, research areas, and preference. There are some important subjects, especially those under rapid development, that we do not cover, for instance, functional data analysis for dependent data. We hope to include this and others relevant topics in a future edition.

The book takes advantages of many available R packages. Yet, there remain some methods discussed in the book that cannot be implemented with any existing package. We have developed some R scripts to perform such analyses. For instance, we have developed a new automatic modeling procedure for a large set of time series, which is used and demonstrated in Chapter 2. We have included, with the great help of Angela Caro and Antonio Elías, these R scripts into a new package for the book and refer to it as the **SLBDD** package. In addition, the data sets used, except for those subject to copyright protection are also included. Some R scripts are provided in the web page of the book at https://www.rueytsay.com/slbdd.

We are grateful to many people who have contributed to our research in general and to this book in particular. We have dedicated the book to Professor George C. Tiao, a giant in time series analysis and a pioneer of many procedures presented here. He is a generous mentor and a good friend of us. We have also learned a lot from our coauthors on topics covered in the book. In particular, we like to acknowledge the contributions from Andrés Alonso, Stevenson Bolívar, Jorge Caiado, Angeles Carnero, Rong Chen, Nuno Crato, Chaoxing Dai, Soudeep Deb, Pedro Galeano, Zhaoxing Gao, Victor Guerrero, Yuefeng Han, Nan-Jung Hsu, Hsin-Cheng Huang, Ching-Kang Ing, Ana Justel, Mladen Kolar, Tengyuan Liang, Agustin Maravall, Fabio Nieto, Pilar Poncela, Javier Prieto, Veronika Rockova, Julio Rodríguez, Rosario Romera, Juan Romo, Esther Ruiz, Ismael Sánchez, Maria Jesús Sánchez, Ezequiel Smucler, Victor Yohai, and Rubén Zamar. Some of the programs used in our analyses were written by Angela Caro, Pedro Galeano, Carolina Gamboa, Yuefeng Han, and Javier Prieto. We sincerely thank them for their wonderful works. Chaoxing Dai helps us in many ways with the `keras` package for deep learning. Without his help, we would not be able to demonstrate its applications.

Finally, we have always had the constant support of our families. DP wants to thank his wife, Jette Bohsen, for her constant love and encouragement in all his projects, and, in particular, for her generous continued help during the years he was writing this book. RST likes to thank the love and care of his wife and the support from his children. Our families are the source of our energy and inspiration. Without their unlimited and unselfish support, it would not have been possible for us to complete this book, especially in this challenging time of COVID-19 pandemic during which the book was finished. We hope that the methods presented in this book can help, by learning from big complex dependent data, to prevent and manage the challenging problems we will face in the future.

April 2020

D.P. MADRID, SP
R.S.T. CHICAGO, USA

CHAPTER 1

INTRODUCTION TO BIG DEPENDENT DATA

Big data are common nowadays everywhere. Statistical methods capable of extracting useful information embedded in those data, including machine learning and artificial intelligence, have attracted much interest among researchers and practitioners in recent years. Most of the available statistical methods for analyzing big data were developed under the assumption that the observations are from independent samples. See, for instance, most methods discussed in Bühlmann and van de Geer (2011). Observations of big data, however, are dependent in many applications. The dependence may occur in the order by which the data were taken (such as time series data) or in space by which the sampling units reside (such as spatial data). Monthly civilian unemployment rates (16 years and older) of the 50 states in the United States is an example. Unemployment rates tend to be sticky over time and geographically neighboring states may share similar industries and, hence, have similar unemployment patterns. For dependent data, the spatial and/or temporal dependence is often the focus of statistical analysis. Consequently, there is a need to study analysis of big dependent data.

The main focus of this book is to provide readers a comprehensive treatment of statistical methods that can be used to analyze big dependent data. We start with some examples and simple methods for their descriptive analysis. More sophisticated methods will be introduced in other chapters. Keep in mind, however, that the analysis of big dependent data is gaining more attention as time advances. The methods discussed in this book reflect, to a large degree, our personal experience and preferences. We hope that the book can be helpful to many researchers and practitioners

Statistical Learning for Big Dependent Data, First Edition. Daniel Peña and Ruey S. Tsay.
© 2021 John Wiley & Sons, Inc. Published 2021 by John Wiley & Sons, Inc.

in different scientific fields in their data analysis. Also, it could attract more attention to and motivate further research in the challenging problem of analyzing big dependent data.

1.1 EXAMPLES OF DEPENDENT DATA

In this section, we present some examples of big dependent data considered in the book, introduce some statistical methods for describing such data, and provide a framework for making statistical inference. Details of the methods introduced will be given in later chapters. Our goal is to demonstrate the data we intend to analyze and to highlight the consequences of ignoring their dependency. Whenever possible, we also illustrate the difficulties statistical methods developed for independent observations are facing when they are applied to big dependent data.

By dependent data, we mean observations of the variables of interest were taken over time or across space or both. Consequently, in these data the order in time or space matters. By big data, we mean the number of variables, k, is large or the number of data points, T, is large or both, and it could be that $k > T$. Figure 1.1 shows a data set of three time series representing the relative average temperature of November in Europe, North America, and South America from 1910 to 2014. The data are in the file `Temperatures.csv` of the book web site. The measurement is relative to the average temperature of November in the twentieth century. These data are dependent because observations of consecutive years tend to be closer to each other than observations that are many years apart. Furthermore, the relative temperature in South America shows a growing trend since 1980, whereas the trend can only be seen in Europe starting from 2000, yet it does not appear in the North America data.

Statistical analysis of these monthly temperature series is relatively easy, as the number of series is only 3 and the sample size is small. It belongs to the multivariate

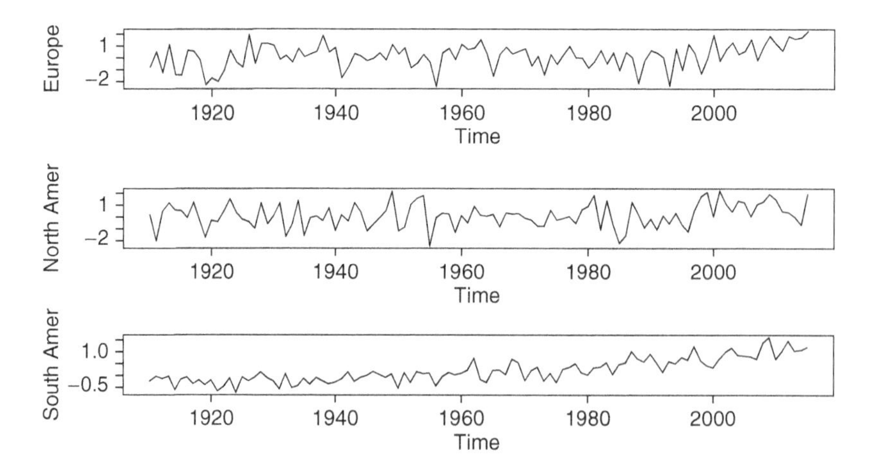

Figure 1.1 Relative average temperatures of November in Europe, North America, and South America from 1910 to 2014. The measurement is relative to the average temperature of November in the twentieth century.

time series analysis in statistics. See, for instance, Tsay (2014). The main concern of this book is statistical analysis when the number of time series is large or the time intervals between observations are small, resulting in high-frequency data with large sample sizes. Modeling big temperature data is important in many applications. As a matter of fact, daily (even hourly) temperature is routinely taken at many locations around the world. These temperatures affect the demands of electricity and heating oil, are highly related to air pollution measurements, such as the particulate matter $PM_{2.5}$ and ozone concentration, and play an important role in commodity prices around the world.

Figure 1.2 shows the standardized daily stock indexes of the 99 most important financial markets around the world, from 3 January 2000 to 16 December 2015 for 4163 observations. The data are in the file `Stock_indexes_99world.csv`. The magnitudes of stock index vary markedly from one market to another so that we standardize the indexes in the plot. Specifically, each standardized index series has mean zero and variance one. In applications, we often analyze the returns of stock indexes. Figures 1.3 and 1.4 show the time plots of the first six indexes and their log returns, respectively. The log returns are obtained by taking the first difference of the logarithm of stock index.

These time plots demonstrate several features of multiple time series. First, they show the overall evolution and cross dependence of the financial markets. All markets exhibited a trough in 2003, had a steady increase reaching a peak around early 2008, then showed a dramatic drop caused by the 2008 financial crisis. Yet some markets experienced a gradual recovery after 2009. Second, the variabilities of the world financial markets appear to be higher when the markets were down; see Figure 1.4. This is not surprising as the fear factor, i.e. market volatility, is likely to be dominant during a bear market. On the other hand, the ranges of world market indexes at a given time seem to be smaller when the market were down; see Figure 1.2.

Figures 1.5 and 1.6 show the time plots of the daily financial indexes for 22 Asian and 47 European markets, respectively, for the same time period, but with x-axis now

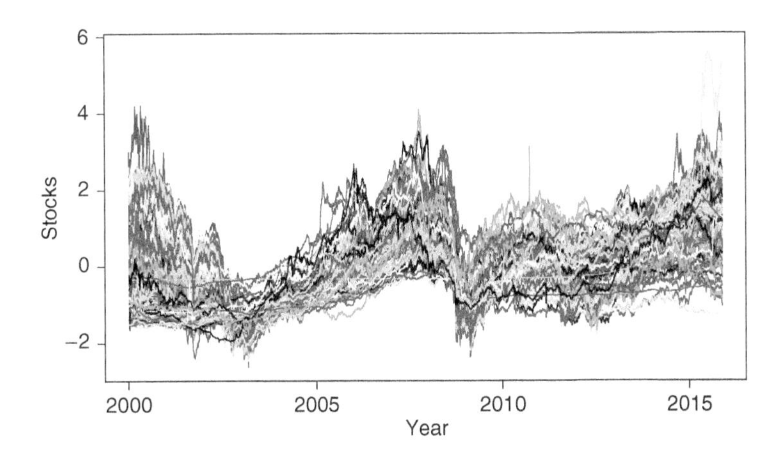

Figure 1.2 Time plots of standardized daily stock market indexes of the 99 most important financial markets around the world from 3 January 2000 to 16 December 2015.

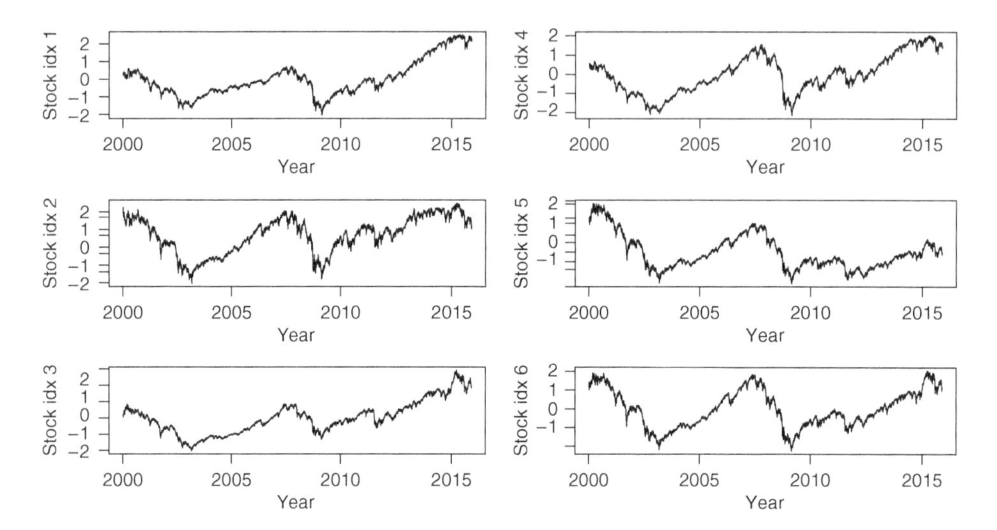

Figure 1.3 Time plots of the first six standardized daily stock market indexes: 3 January 2000 to 16 December 2015.

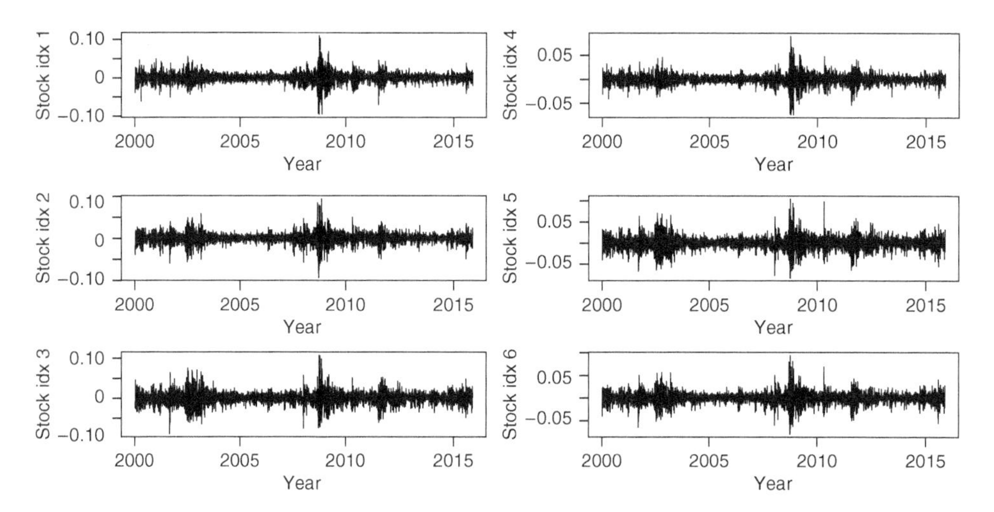

Figure 1.4 Time plots of the log returns of the first six daily stock market indexes: 3 January 2000 to 16 December 2015.

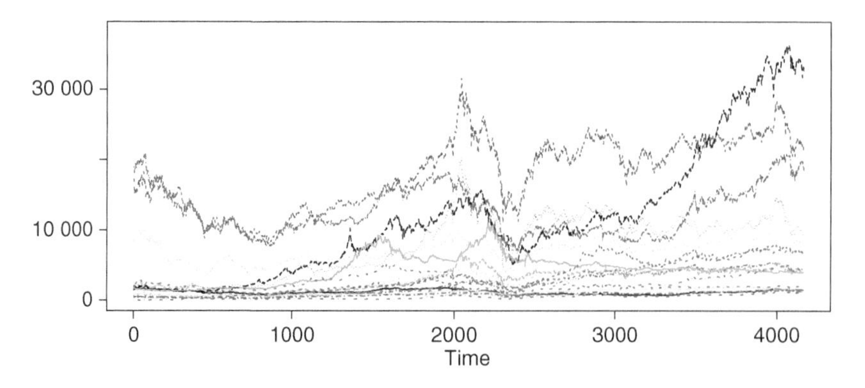

Figure 1.5 Time plots of daily stock market indexes of 22 Asian financial markets from 3 January 2000 to 16 December 2015.

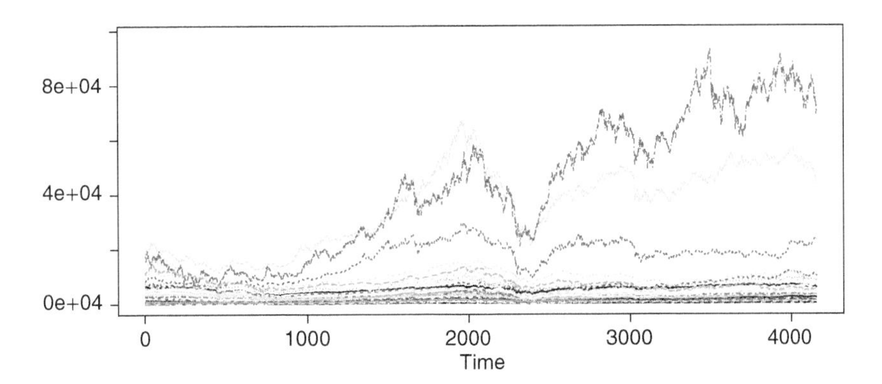

Figure 1.6 Time plots of daily stock market indexes of 47 European financial markets from 3 January 2000 to 16 December 2015.

being in time index. The difference between Asian and European financial markets is easily seen, implying that care must be exercised when analyzing similar time series across different continents.

As a third example, Figure 1.7 shows 33 monthly consumer price indexes of European countries from January 2000 to October 2015. The data are in the file `CPIEurope2000-15.csv`. In the plot, the series have been standardized to have zero mean and unit variance. As expected, the plot shows an upward trend of the price indexes, but it also demonstrates sufficient differences between the series. For example, some series show a clear seasonal pattern, but others do not.

Large sets of data are available in marketing research. As an example, Figure 1.8 shows the time plots of daily sales, in natural logarithms, of a clothing brand in 25 provinces in China from 1 January 2008 to 9 December 2012 for 1805 observations. The data are from Chang et al. (2014) and are given in the file `clothing.csv`. The plots exhibit certain annual pattern, referring to as seasonal behavior in time

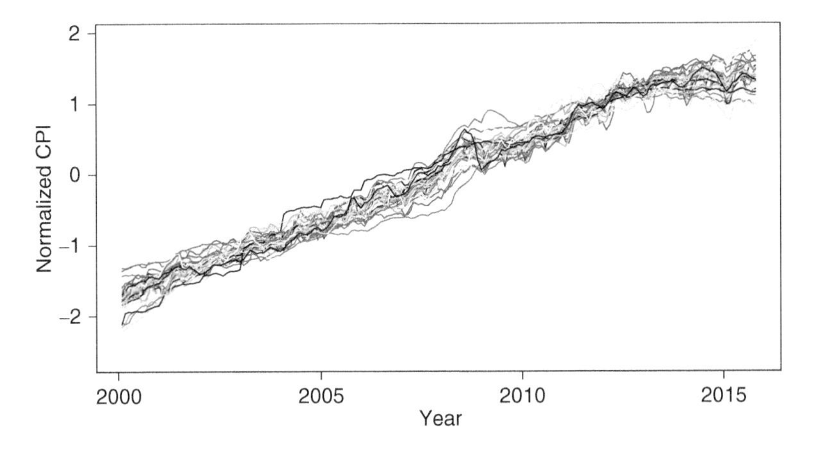

Figure 1.7 Time plots of 33 monthly price indexes of European countries from January 2000 to October 2015.

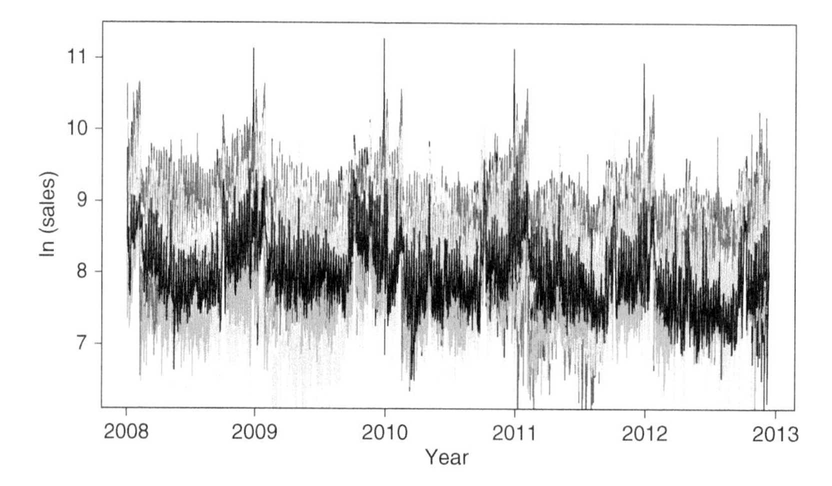

Figure 1.8 Time plots of daily sales in natural logarithms of a clothing brand in 25 provinces in China from 1 January 2008 to 9 December 2012.

series analysis. To gain further insight, Figure 1.9 shows the same time plots for the first eight provinces. They are Beijing, Fujian, Guangdong, Guangxi, Hainan, Hebei, Henan, and Hubei. The levels of the plots are adjusted so that there is no overlapping in the figure. A special characteristic of the plots is that local peaks occur irregularly in the early part of each year, followed by certain drops in sales. This is caused by the Chinese New Year holidays that vary from year to year. In addition, the peaks do not occur for all provinces; see the second plot (from the top) of Figure 1.9. Analyzing these series jointly requires modeling the common features as well as the variations from one province to another.

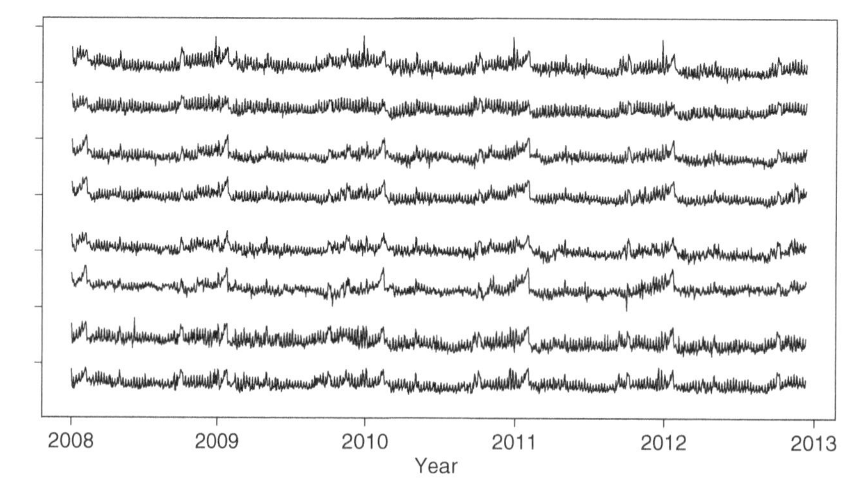

Figure 1.9 Time plots of daily sales in natural logarithms of a clothing brand in eight provinces in China from 1 January 2008 to 9 December 2012.

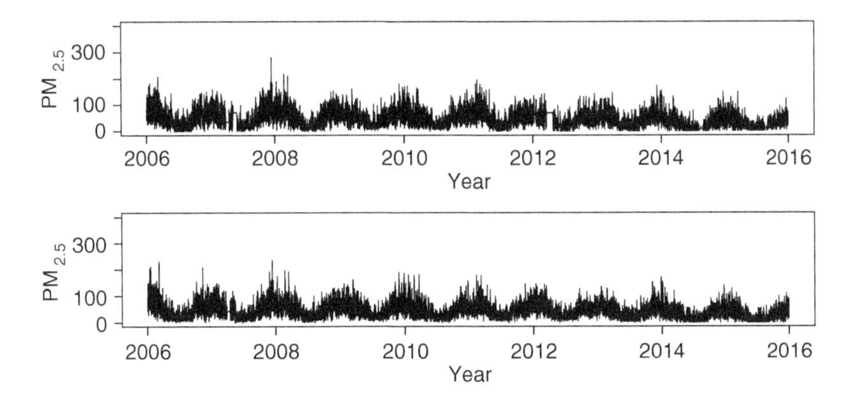

Figure 1.10 Time plots of hourly observations of the particulate matter $PM_{2.5}$ from two monitoring stations in southern Taiwan from 1 January 2006 to 31 December 2015.

Figure 1.10 shows the time plots of hourly observations of the particulate matter $PM_{2.5}$ from two monitoring stations in southern Taiwan from 1 January 2006 to 31 December 2015. We drop the data of February 29 for simplicity, resulting in a sample size 87 600. The data are given in the file `TaiwanPM25.csv`. The two series are parts of many monitoring stations throughout the island. Also available at each station are measurements of temperature, dew points, and other air pollution indexes. Thus, this is a simple example of big dependent data. The time plots exhibit certain strong annual patterns and seem to indicate a minor decreasing trend. Figure 1.11 shows the first 30 000 sample autocorrelations of the $PM_{2.5}$ series of Station 1. The annual cycle with periodicity $s = 24 \times 365 = 8760$ is clearly seen from the autocorrelations. Figure 1.12, on the other hand, shows the first 120 sample autocorrelations of the same time series. The plot shows that there also exists a daily cyclic pattern in the

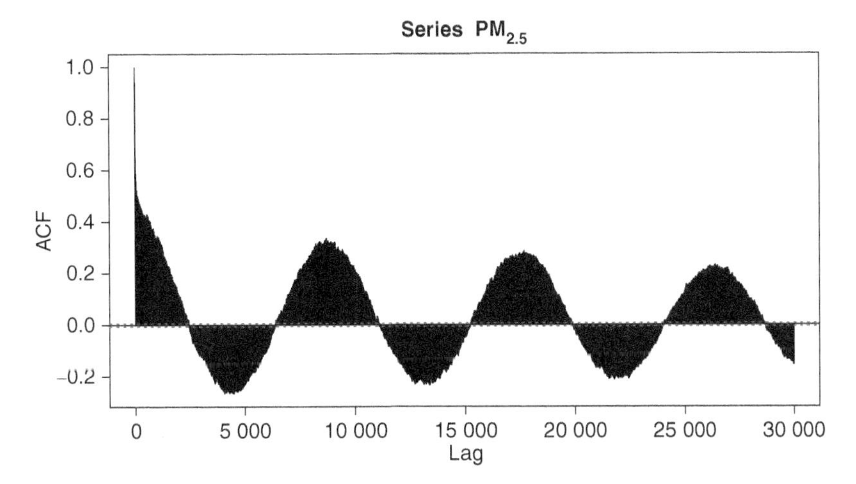

Figure 1.11 Sample autocorrelations of $PM_{2.5}$ measurement of Station 1 in southern Taiwan. The first 30 000 autocorrelations are shown. The annual cycle with periodicity 8760 is clearly seen.

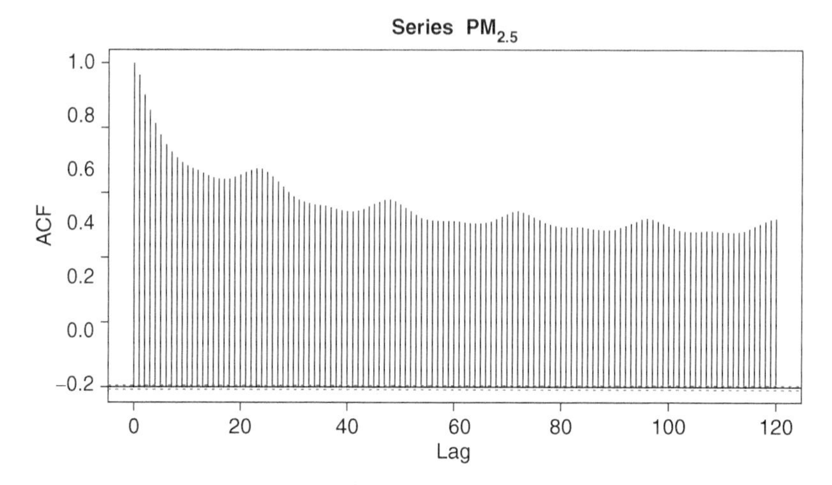

Figure 1.12 Sample autocorrelations of $PM_{2.5}$ measurement of Station 1 in southern Taiwan. The first 120 autocorrelations are shown. The daily cyclical pattern with periodicity 24 is clearly seen.

$PM_{2.5}$ series with a periodicity $\lambda = 24$. Consequently, analysis of such a big dependent data should consider not only the overall trend, but also various cyclic patterns with different frequencies.

Finally, Figure 1.13 plots the monitoring stations (in circle) and the ozone levels (in color-coded squares) of US Midwestern states on 20 June 1987. The data set is available from the R package **fields**. The states are added to the plot so that readers can understand the geographical features of ozone levels in the Midwest. Clearly, on this particular day, high ozone readings appeared in the southwest of Illinois and south of Indiana. Also, ozone levels of neighboring monitoring stations are similar. This is an example of spatial process discussed in Chapter 9.

The six data sets presented are examples of multivariate spatio-temporal series, where we observe the values of several variables of interest over time and space.

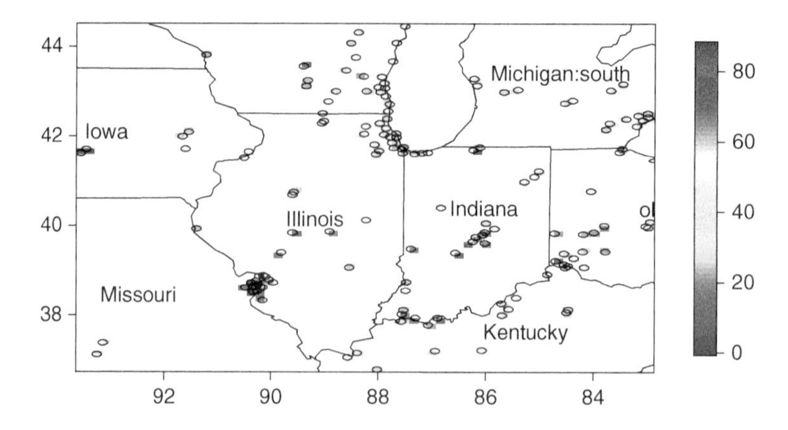

Figure 1.13 Locations and ozone levels at various monitoring stations in US Midwestern states. The measurements are 8-hour averages (9 a.m. to 4 p.m.) on 20 June 1987.

The time sequence introduces dynamic dependency in a given series, which may be time varying. The data also show cross dependence between series and the strength of cross dependence may, in turn, depend on geographical locations or economic relations among countries. One of the main focuses of the book is to investigate both the dynamic and cross dependence between multiple time series.

The objectives of analyzing big dependent data include (a) to study the relationships between the series, (b) to explore the dynamic dependence within a series and across series, and (c) to predict the future values of the series. For spatial processes, one might be interested in predicting the variable of interest at a location where no observations are available. The first two objectives are referred to as data analysis and modeling, whereas the third objective involves statistical inference. Both modeling and inference require not only methods for handling data, but also the theory on which proper inference can be drawn. To this end, we need a framework for understanding the characteristics of the time series under study and for making sound predictions. We discuss this framework by starting with stochastic processes in the next section.

1.2 STOCHASTIC PROCESSES

Theory and properties of stochastic processes form the foundation for statistical analysis and inference of dependent data.

1.2.1 Scalar Processes

A scalar stochastic process is a sequence of random variables $\{z_t\}$, where the subscript t takes values in a certain set C. In most cases, the elements of C are ordered and often correspond to certain calendar time (days, months, years, etc.). The resulting stochastic process z_t is called a time series. Unless specifically mentioned, we assume that the time index t is equally spaced in the book. For a spatial process, C consists of locations of certain geographical regions. The locations may denote the latitude and longitude of an observation station or a city. For each value t in C (e.g. each point in time), z_t is a real-valued random variable defined on a common measurable space. The observed values, also denoted by $\{z_t\}$ for simplicity, form a realization of the stochastic process. We shall denote the realization by $\boldsymbol{z}_1^T = (z_1, \dots, z_t, \dots, z_T)'$, where T is the sample size and $\boldsymbol{A}'$ denotes the transpose of the matrix or vector $\boldsymbol{A}$. The sample size T denotes the number of observations for a time series or the number of locations for a pure spatial process.

Statistical properties of a stochastic process z_t are characterized by its probability distribution. More specifically, for a given T (finite and fixed), properties of $\boldsymbol{z}_1^T$ are determined by the joint probability distribution of its elements $\{z_t|t = 1, \dots, T\}$. The marginal distribution of any non-empty subset of $\boldsymbol{z}_1^T$ can be obtained from the joint distribution by integrating out elements not in the subset. In particular, the probability distribution of the individual element z_t is called the marginal distribution of z_t. Thus, similar to the general finite-dimensional random variable, we say that the probabilistic structure of $\boldsymbol{z}_1^T$ is known when we know its joint probability distribution.

In this book, we assume that a probability distribution has a well-defined density function. Let $f_t(z)$ be the marginal probability density function of z_t. Define the *expected value* or *mean* of z_t as

$$\mu_t = E(z_t) = \int_R z f_t(z) dz,$$

provided that the integral exists, where R denotes the real line. The sequence $\boldsymbol{\mu}_1^T = (\mu_1, \mu_2, \ldots, \mu_T)$ is the *mean function* of $\boldsymbol{z}_1^T$. Define the *variance* of z_t as

$$\sigma_t^2 = E(z_t - \mu_t)^2 = \int_R (z - \mu_t)^2 f_t(z) dz$$

provided that the integral exists. The sequence $\{\sigma_t^2\}_1^T = (\sigma_1^2, \ldots, \sigma_T^2)$ is the *variance function* of $\boldsymbol{z}_1^T$. Higher order moments can be defined similarly, assuming that they exist.

A special case of interest is that all elements of $\boldsymbol{z}_1^T$ have the same mean. In this case, $\boldsymbol{\mu}_1^T$ is a constant function. If all elements of $\boldsymbol{z}_1^T$ have the same variance, then $\{\sigma_t^2\}_1^T$ is a constant function. If both $\boldsymbol{\mu}_1^T$ and $\{\sigma_t^2\}_1^T$ are constants for all finite T, then the stochastic process z_t is stable and we say that it is a homogeneous process. In some applications, $\boldsymbol{\mu}_1^T$ or $\{\sigma_t^2\}_1^T$ may not exist.

In what follows, we assume that σ_t^2 exists for all t. The linear dynamic dependence of a stochastic process z_t is determined by its *autocovariance* or *autocorrelation* function (ACF). For z_t and z_{t-j}, where j is a given integer, we define their autocovariance as

$$\gamma(t, t-j) = \text{Cov}(z_t, z_{t-j}) = E[(z_t - \mu_t)(z_{t-j} - \mu_{t-j})]. \tag{1.1}$$

When $j = 0$, $\gamma(t, t) = \text{Var}(z_t) = \sigma_t^2$. Also, from the definition, it is clear that $\gamma(t, t-j) = \gamma(t-j, t)$. The *autocorrelation* between z_t and z_{t-j} is defined as

$$\rho(t, t-j) = \frac{\gamma(t, t-j)}{\sigma_t \sigma_{t-j}}. \tag{1.2}$$

Of particular importance in real applications is the case in which the autocovariance and, hence, the autocorrrelation between z_t and z_{t-j} is a function of the lag j, and not a function of t. In this case, the autocovariance and autocorrelation are time-invariant and we write $\gamma(t, t-j) = \gamma_j$ and $\rho(t, t-j) = \rho_j$.

1.2.1.1 *Stationarity* A stochastic process z_t is *strictly stationary* if the joint distributions of the m-dimensional random vectors $(z_{t_1}, z_{t_2}, \ldots, z_{t_m})$ and $(z_{t_1+h}, z_{t_2+h}, \ldots, z_{t_m+h})$ are the same, where m is an arbitrary positive integer, $\{t_1, \ldots, t_m\}$ are arbitrary m time indexes, and h is an arbitrary integer. In other words, the joint distribution of arbitrary m random variables is time-invariant. In particular, the marginal probability density $f_t(z)$ of a strictly stationary process z_t does not depend on t, and we simply denote it by $f(z)$. A stochastic process z_t is *weakly stationary* if its mean and autocovariance functions exist and are time-invariant. Specifically, a process z_t is stationary in the weak sense if (1) $\mu_t = \mu$ for all t; (2) $\sigma_t^2 = \sigma^2$ for all t; and (3) $\gamma(t, t-j) = \gamma_j$ for all t, where j is a given integer. The first two conditions say that the mean and variance are time-invariant. The third condition states that the autocovariance and, hence, the autocorrelation between two variables depend only on their time separation j. Unless specifically stated, we use stationarity to denote weak stationarity of a stochastic process. For this process the autocovariance and autocorrelation functions depend only on the difference of lags and, in particular, we have $\gamma_0 = \sigma^2$, $\gamma_j = \gamma_{-j}$ and $\rho_j = \rho_{-j}$, where j is an arbitrary integer.

Assume that z_t is stationary. For a given positive integer m, the covariance matrix of $(z_t, z_{t-1}, \ldots, z_{t-m+1})'$ is defined as

$$\boldsymbol{M}_z(m) = E\left\{\begin{bmatrix} z_t - \mu \\ z_{t-1} - \mu \\ \vdots \\ z_{t-m+1} - \mu \end{bmatrix} [(z_t - \mu), (z_{t-1} - \mu), \ldots, (z_{t-m+1} - \mu)]\right\}$$

$$= \begin{bmatrix} \gamma_0 & \gamma_1 & \cdots & \gamma_{m-1} \\ \gamma_1 & \gamma_0 & \cdots & \gamma_{m-2} \\ \vdots & \vdots & \ddots & \vdots \\ \gamma_{m-1} & \gamma_{m-2} & \cdots & \gamma_0 \end{bmatrix}. \tag{1.3}$$

This covariance matrix has the special pattern that (a) all diagonal elements are the same, (b) all elements above and below the main diagonal are the same, and so on. Such a matrix is called a *Toeplitz* matrix. The corresponding *autocorrelation matrix* is

$$\boldsymbol{R}_z(m) = \begin{bmatrix} 1 & \rho_1 & \cdots & \rho_{m-1} \\ \rho_1 & 1 & \cdots & \rho_{m-2} \\ \vdots & \vdots & \ddots & \vdots \\ \rho_{m-1} & \rho_{m-2} & \cdots & 1 \end{bmatrix}, \tag{1.4}$$

which is also a Toeplitz matrix.

An important property of a stationary process is that any non-trivial finite linear combination of its elements remains stationary. In other words, a process obtained by any non-zero linear combination (with finite elements) of a stationary process is also stationary. For example, if z_t is stationary, the process w_t defined by $w_t = z_t - z_{t-1}$ is also stationary. A consequence of this property is that the autocovariances of z_t must satisfy certain conditions that are useful for studying stochastic processes. In particular, we can consider a non-trivial linear combination of a stationary process z_t and its lagged values, namely

$$y_t = c_1 z_t + c_2 z_{t-1} + \cdots + c_m z_{t-m+1} = \boldsymbol{c}' \boldsymbol{z}_{t,m}$$

where $\boldsymbol{c} = (c_1, \ldots, c_m)' \neq \boldsymbol{0}$ and $\boldsymbol{z}_{t,m} = (z_t, z_{t-1}, \ldots, z_{t-m+1})'$. Since y_t is stationary, its variance exists and is given by $\boldsymbol{c}'\boldsymbol{M}_z(m)\boldsymbol{c} \geq 0$, where $\boldsymbol{M}_z(m)$ is defined in Eq. (1.3). Consequently, $\boldsymbol{M}_z(m)$ must be a nonnegative definite matrix. Similarly, the correlation matrix $\boldsymbol{R}_z(m)$ must also be nonnegative definite.

Strict stationarity implies weak stationarity provided that the first two moments of the process exist. But weak stationarity does not guarantee strict stationarity of a stochastic process. Consider the following simple example. Let $\{z_t\}$ be a sequence of independent and identically distributed standard normal random variables, i.e. $z_t \sim_{iid} N(0, 1)$. Define a stochastic process x_t by

$$x_t = \begin{cases} 0.5(z_t^2 - 1), & \text{if } t \text{ is odd} \\ z_t, & \text{if } t \text{ is even.} \end{cases}$$

It is easy to show that (a) $E(x_t) = 0$, (b) $\text{Var}(x_t) = 1$, and (c) $E(x_t x_{t-j}) = 0$ for $j \neq 0$. Therefore, $\{x_t\}$ is a weakly stationary process. On the other hand, x_t is Gaussian if t is even and it is a shifted χ^2 with one degree of freedom when t is odd. The process $\{x_t\}$ is not identically distributed and, hence, is not strictly stationary. If we assume that $\boldsymbol{z}_1^T$ follows a multivariate normal distribution for all T, then weak stationarity is equivalent to strict stationarity, because a normal (or Gaussian) distribution is determined by its first two moments.

1.2.1.2 *White Noise Process* An important stationary process is the *white noise process*. A stochastic process z_t is a white noise process if it satisfies the following conditions: (1) $E(z_t) = 0$; (2) $\text{Var}(z_t) = \sigma^2 < \infty$; and (3) $\text{Cov}(z_t, z_{t-j}) = 0$ for all $j \neq 0$. In other words, z_t is a white noise series if and only if it has a zero mean and finite variance, and is not serially correlated. Thus, a white noise is not necessarily strictly stationary. If we impose the additional condition that z_t and z_{t-j} are independent and have the same distribution, where $j \neq 0$, then z_t is a *strict white noise* process.

1.2.1.3 *Conditional Distribution* In addition to marginal distributions, conditional distributions also play an important role in studying stochastic processes. Let $\boldsymbol{z}_\ell^k = (z_\ell, z_{\ell+1}, \ldots, z_k)'$, where ℓ and k are positive integers and $\ell \leq k$. In prediction, we are interested in the conditional distribution of $\boldsymbol{z}_{T+1}^{T+h}$ given $\boldsymbol{z}_1^T$, where $h \geq 1$ is referred to as the *forecast horizon*. In estimation, we express the joint probability density function of $\boldsymbol{z}_1^T$ as a product of the conditional density functions of z_{t+1} given $\boldsymbol{z}_1^t$ for $t = T-1, T-2, \ldots, 1$. More details are given in later chapters.

If the stochastic process z_t has the following property

$$f(z_{t+1}|\boldsymbol{z}_1^t) = f(z_{t+1}|z_t), \qquad t = 1, 2, \ldots,$$

then z_t is a first-order *Markov* process. The aforementioned property is referred to as the memoryless of a Markov process as the conditional distribution only depends on the variable one period before.

1.2.2 Vector Processes

A stochastic vector process is a sequence of random vectors $\{\boldsymbol{z}_t\}$, where the index t assumes values in a certain set C and $\boldsymbol{z}_t = (z_{1t}, \ldots, z_{kt})'$ is a k-dimensional vector. Each component z_{it} follows a scalar stochastic process, and the vector process is characterized by the joint probability distribution of the random variables $(\boldsymbol{z}_1, \ldots, \boldsymbol{z}_T)$. Similar to the scalar case, we define the mean vector of $\boldsymbol{z}_t$ as

$$E(\boldsymbol{z}_t) = [E(z_{1t}), E(z_{2t}), \ldots, E(z_{kt})]' = (\mu_{1t}, \ldots, \mu_{kt})' \equiv \boldsymbol{\mu}_t,$$

provided that the expectations exist, and the lag-ℓ autocovariance function as

$$\boldsymbol{\Gamma}(t, t-\ell) = \text{Cov}(\boldsymbol{z}_t, \boldsymbol{z}_{t-\ell}) = E[(\boldsymbol{z}_t - \boldsymbol{\mu}_t)(\boldsymbol{z}_{t-\ell} - \boldsymbol{\mu}_{t-\ell})'] \tag{1.5}$$

provided that the covariances involved all exist, where ℓ is an integer. We say that the vector process $\{z_t\}$ is weakly stationary if (a) $\mu_t = \mu$, a constant vector, and (b) $\Gamma(t, t-\ell)$ depends only on ℓ. In other words, a k-dimensional stochastic process z_t is weakly stationary if its first two moments are time-invariant. In this case, we write $\Gamma(t, t-\ell) = E[(z_t - \mu)(z_{t-\ell} - \mu)'] = \Gamma(\ell) = [\gamma_{ij}(\ell)]$, where $1 \leq i, j \leq k$. In particular, the lag-0 autocovariance matrix of a weakly stationary k-dimensional stochastic process z_t is given by

$$\Gamma(0) = \begin{bmatrix} \gamma_{11}(0) & \gamma_{12}(0) & \cdots & \gamma_{1k}(0) \\ \gamma_{21}(0) & \gamma_{22}(0) & \cdots & \gamma_{2k}(0) \\ \vdots & \vdots & \ddots & \vdots \\ \gamma_{k1}(0) & \gamma_{k2}(0) & \cdots & \gamma_{kk}(0) \end{bmatrix} \tag{1.6}$$

which is symmetric, because $\gamma_{ij}(0) = \gamma_{ji}(0)$. Note that the diagonal element $\gamma_{ii}(0) = \sigma_i^2 = \text{Var}(z_{it})$.

For a stationary vector process z_t, its lag-ℓ autocovariance matrix $\Gamma(\ell) = [\gamma_{ij}(\ell)]$ measures the linear dependence among the component series z_{it} and their lagged variables $z_{j,t-\ell}$. It pays to study the exact meaning of individual element $\gamma_{ij}(\ell)$ of $\Gamma(\ell)$. First, the diagonal element $\gamma_{ii}(\ell)$ is the lag-ℓ autocovariance of the scalar process z_{it} $(i = 1, \ldots, k)$. Second, the off-diagonal element $\gamma_{ij}(\ell)$ is

$$\gamma_{ij}(\ell) = E[(z_{it} - \mu_i)(z_{j,t-\ell} - \mu_j)],$$

which measures the linear dependence of z_{it} on the lagged value $z_{j,t-\ell}$ for $\ell > 0$. Thus, $\gamma_{ij}(\ell)$ quantifies the linear dependence of z_{it} on the ℓth past value of z_{jt}. Third, in general, $\Gamma(\ell)$ is not symmetric if $\ell > 0$. As a matter of fact, $\gamma_{ij}(\ell) \neq \gamma_{ji}(\ell)$ for most stochastic process z_t because, as stated before, $\gamma_{ij}(\ell)$ denotes the linear dependence of z_{it} on $z_{j,t-\ell}$, whereas $\gamma_{ji}(\ell)$ quantifies the linear dependence of z_{jt} on $z_{i,t-\ell}$. Finally, we have

$$\begin{aligned} \gamma_{ij}(-\ell) &= E[(z_{it} - \mu_i)(z_{j,t+\ell} - \mu_j)] \\ &= E[(z_{j,v} - \mu_j)(z_{i,v-\ell} - \mu_i)] \quad (v = t + \ell) \\ &= \gamma_{ji}(\ell). \quad \text{(by stationarity)} \end{aligned}$$

Therefore, we have $\gamma_{ij}(-\ell) = \gamma_{ji}(\ell)$ for all ℓ. Consequently, the autocovariance matrices of a stationary vector process z_t satisfy

$$\Gamma(-\ell) = [\Gamma(\ell)]',$$

and, hence, it suffices to consider $\Gamma(\ell)$ for $\ell \geq 0$ in practice. If the dimension of $\Gamma(\ell)$ is needed to avoid any confusion, we write $\Gamma(\ell) \equiv \Gamma_k(\ell)$ with the subscript k denoting the dimension of the underlying stochastic process z_t.

A global measure of the overall variability of the process z_t is given by its generalized variance, which is the determinant of the lag-0 covariance matrix $\Gamma(0)$, i.e. $|\Gamma(0)|$. The standardized, by the dimension, measure is the effective variance, defined by

$$\text{EV}(z_t) = |\Gamma(0)|^{1/k}. \tag{1.7}$$

For instance, if z_t is two dimensional, then $\text{EV}(z_t) = [\gamma_{11}(0)\gamma_{22}(0)\{1 - \rho_{12}^2(0)\}]^{1/2}$, where $\rho_{12}(0)$ is the contemporaneous correlation coefficient between z_{1t} and z_{2t}.

Clearly, $\text{EV}(\boldsymbol{z}_t)$ assumes its largest value when the two series are uncorrelated, i.e. $\rho_{12}(0) = 0$. For a k-dimensional process $\boldsymbol{z}_t$, as the determinant of a square-matrix is the product of its eigenvalues, the effective variance is the geometrical mean of the eigenvalues of $\boldsymbol{\Gamma}(0)$.

We can summarize the properties of a stationary vector process $\boldsymbol{z}_t$ by using linear combinations of its components. Consider, for example, the linear process

$$y_t = \boldsymbol{c}'\boldsymbol{z}_t = \sum_{i=1}^{k} c_i z_{it},$$

where $\boldsymbol{c} = (c_1, \ldots, c_k)' \neq \boldsymbol{0}$. This process is stationary because it is a finite linear combination of the stationary components z_{it} $(i = 1, \ldots, k)$. More specifically, the expectation is $E(y_t) = c_1 E(z_{1t}) + \cdots + c_k E(z_{kt}) = \boldsymbol{c}' E(\boldsymbol{z}_t) = \boldsymbol{c}'\boldsymbol{\mu}$, which is time-invariant. Similarly, the variance of y_t is given by $\text{Var}(y_t) = \text{Var}(\boldsymbol{c}'\boldsymbol{z}_t) = \boldsymbol{c}'\text{Var}(\boldsymbol{z}_t)\boldsymbol{c} = \boldsymbol{c}'\boldsymbol{\Gamma}(0)\boldsymbol{c}$, which is also time-invariant. In general, we have $\text{Cov}(y_t, y_{t-\ell}) = \boldsymbol{c}'\boldsymbol{\Gamma}(\ell)\boldsymbol{c}$, which depends only on ℓ. Thus, y_t is stationary. In addition, since $\text{Var}(y_t)$ is non-negative and $\boldsymbol{c}$ is an arbitrary non-zero vector, we conclude that $\boldsymbol{\Gamma}(0)$ is nonnegative definite.

Similar to the scalar case, the autocovariance matrices in Eq. (1.5) of a stationary vector process $\boldsymbol{z}_t$ must satisfy certain conditions. To see this, we consider a non-zero linear combination of $\boldsymbol{z}_t$ and its m lagged values, say,

$$y_t = \sum_{i=1}^{m} \boldsymbol{b}_i'\boldsymbol{z}_{t-i+1} = \boldsymbol{b}_1'\boldsymbol{z}_t + \boldsymbol{b}_2'\boldsymbol{z}_{t-1} + \cdots + \boldsymbol{b}_m'\boldsymbol{z}_{t-m+1}$$

where $\boldsymbol{b}_j = (b_{j1}, \ldots, b_{jk})'$ is a k-dimensional real-valued vector for $j = 1, \ldots, m$. Defining $\boldsymbol{b} = (\boldsymbol{b}_1', \boldsymbol{b}_2', \ldots, \boldsymbol{b}_m')'$, which is a non-zero km-dimensional constant vector, and $\boldsymbol{z}_{t,m} = (\boldsymbol{z}_t', \boldsymbol{z}_{t-1}', \ldots, \boldsymbol{z}_{t-m+1}')'$, we have $\text{Var}(y_t) = \boldsymbol{b}'\text{Var}(\boldsymbol{z}_{t,m})\boldsymbol{b} = \boldsymbol{b}'\boldsymbol{M}_{\boldsymbol{z}}(m)\boldsymbol{b}$, where

$$\boldsymbol{M}_{\boldsymbol{z}}(m) = \begin{bmatrix} \boldsymbol{\Gamma}(0) & \boldsymbol{\Gamma}(1) & \cdots & \boldsymbol{\Gamma}(m-1) \\ \boldsymbol{\Gamma}(-1) & \boldsymbol{\Gamma}(0) & \cdots & \boldsymbol{\Gamma}(m-2) \\ \vdots & \vdots & \ddots & \vdots \\ \boldsymbol{\Gamma}(-m+1) & \boldsymbol{\Gamma}(-m+2) & \cdots & \boldsymbol{\Gamma}(0) \end{bmatrix} \tag{1.8}$$

is a nonnegative block Toeplitz matrix. This $\boldsymbol{M}_{\boldsymbol{z}}(m)$ matrix is the generalization of $\boldsymbol{M}_z(m)$ in Eq. (1.3) for the scalar series z_t.

The cross-correlation matrices (CCMs) of $\boldsymbol{z}_t$ are defined as the autocovariances of the standardized process $\boldsymbol{x}_t = \boldsymbol{D}^{-1/2}(\boldsymbol{z}_t - \boldsymbol{\mu})$, where the matrix $\boldsymbol{D}$ is defined as $\text{diag}(\gamma_{11}(0), \ldots, \gamma_{kk}(0))$, consisting of the variances of the components of $\boldsymbol{z}_t$. Specifically, the lag-ℓ CCM of $\boldsymbol{z}_t$ is

$$\begin{aligned} \mathbf{R}_{\boldsymbol{z}}(\ell) &= \text{Cov}(\boldsymbol{x}_t, \boldsymbol{x}_{t-\ell}) = \boldsymbol{D}^{-1/2}\text{Cov}(\boldsymbol{z}_t, \boldsymbol{z}_{t-\ell})\boldsymbol{D}^{-1/2} \\ &= \begin{bmatrix} \rho_{11}(\ell) & \rho_{12}(\ell) & \cdots & \rho_{1k}(\ell) \\ \rho_{21}(\ell) & \rho_{22}(\ell) & \cdots & \rho_{2k}(\ell) \\ \vdots & \vdots & \ddots & \vdots \\ \rho_{k1}(\ell) & \rho_{k2}(\ell) & \cdots & \rho_{kk}(\ell) \end{bmatrix}. \end{aligned} \tag{1.9}$$

From the definition, we have $\rho_{ii}(0) = 1$ and

$$\rho_{ij}(\ell) = \frac{\gamma_{ij}(\ell)}{\sqrt{\gamma_{ii}(0)\gamma_{jj}(0)}}.$$

Similarly to the autocovariance matrices, we have $\boldsymbol{M}_{\mathbf{z}}(\ell) = [\boldsymbol{M}_{\mathbf{z}}(-\ell)]'$.

The *effective correlation* of z_t is given by

$$\text{ER}_{\mathbf{z}} = 1 - |\mathbf{R}_{\mathbf{z}}(0)|^{1/k} \tag{1.10}$$

and it assumes a value between zero and one. For instance, for $k=2$, $\text{ER}_{\mathbf{z}} = 1 - [1 - \rho_{12}(0)]^{1/2}$. See Peña and Rodriguez (2003) for more properties.

1.2.2.1 *Vector White Noises* A k-dimensional vector process z_t is a white noise process if (a) $E(z_t) = \mathbf{0}$, (b) $\text{Var}(z_t) = \boldsymbol{\Gamma}(0)$ is positive-definite, and (c) $\boldsymbol{\Gamma}(\ell) = \mathbf{0}$ for all $\ell \neq 0$. This is a direct generalization of the scalar white noise. Thus, a vector white noise process consists of random vectors that have zero means, positive-definite covariance, and no lagged serial or cross correlations.

1.2.2.2 *Invertibility* In many applications, we use linear regression type of models to describe the dynamic dependence of a vector process. The model can be written as

$$z_t = \boldsymbol{\pi}_0 + \sum_{i=1}^{\infty} \boldsymbol{\pi}_i z_{t-i} + \boldsymbol{a}_t \tag{1.11}$$

where $\{\boldsymbol{a}_t\}$ is a stationary vector white noise process, $\boldsymbol{\pi}_0$ is a k-dimensional vector of constants, and $\boldsymbol{\pi}_i$ are $k \times k$ real-valued matrices. For Eq. (1.11) to be meaningful, the coefficient matrices must satisfy the condition

$$\sum_{i=1}^{\infty} \|\boldsymbol{\pi}_i\| < \infty \tag{1.12}$$

where $\|\boldsymbol{A}\|$ denotes a norm of the matrix $\boldsymbol{A}$, e.g. $\|\boldsymbol{A}\| = \left(\sum_{i=1}^{k}\sum_{j=1}^{k} a_{ij}^2\right)^{1/2}$, which is the *Frobenius* norm. A vector process z_t is *invertible* if it can assume the model in (1.11) that satisfies the condition of Eq. (1.12). If the summation in Eq. (1.11) is truncated at a finite integer p, then z_t is invertible and the process follows a vector autoregressive (VAR) model.

Invertibility is another important concept in time series analysis. A necessary condition for invertible model is that $\|\boldsymbol{\pi}_i\| \to 0$ as $i \to \infty$. This means that the added contribution of $z_{t-\ell}$ to z_t conditioned on the available information $\{z_{t-1}, z_{t-2}, \ldots, z_{t-\ell+1}\}$ is diminishing as ℓ increases. Consider the scalar process z_t, any AR(p) process with finite p is invertible. On the other hand, the process $z_t = a_t - a_{t-1}$, where $\{a_t\}$ is a white noise, is stationary, but non-invertible.

1.3 SAMPLE MOMENTS OF STATIONARY VECTOR PROCESS

In real applications, we often observe a sequence of realizations, say $\{z_1, \ldots, z_T\}$, of a stationary vector process z_t. Our goal is to investigate properties of z_t based on the observed data. To this end, certain exploratory data analysis is useful. In particular, much can be learned about z_t by studying its sample moments.

1.3.1 Sample Mean

An unbiased estimator of the population mean of the scalar process z_{it} is the sample mean

$$\bar{z}_i = \frac{1}{T}\sum_{t=1}^{T} z_{it}. \tag{1.13}$$

Putting them together, the sample mean of the vector process z_t is

$$\bar{z} = (\bar{z}_1, \ldots, \bar{z}_k)' = \frac{1}{T}\sum_{t=1}^{T} z_t. \tag{1.14}$$

For an independent and identically distributed random sample, it is easy to see that $E(\bar{z}_i) = E(z_{it}) = \mu_i$, and the variance of $\bar{z}_i$ is σ_i^2/T, where $\sigma_i^2 = \text{Var}(z_{it})$. Therefore, the sample mean in this case is a consistent estimator of the population mean μ_i as $T \to \infty$.

Note that for a stationary process, the consistency of the sample mean does not necessarily hold. For example, let z_{11} be a random sample from a normal distribution with mean zero and variance σ_1^2. Let $z_{1t} = z_{11}$ for $t = 2, \ldots, T$. This special process $\{z_{1t}\}$ is stationary because (1) $E(z_{1t}) = 0$, which is time-invariant, (2) $\text{Var}(z_{1t}) = \sigma_1^2$, which is also time-invariant, and $\text{Cov}(z_{1t}, z_{1,t-j}) = \sigma_1^2$, which is time-invariant for all j. However, in this particular case we have $\bar{z}_1 = z_{11}$ and $\text{Var}(\bar{z}_1) = \sigma_1^2$, no matter what the sample size T is. Therefore, the sample mean is not a consistent estimator for this special process as $T \to \infty$.

The condition under which the sample mean $\bar{z}_i$ is a consistent estimator of the population mean μ_i is referred to as the *ergodic* condition of a stochastic process. The ergodic theory is concerned with conditions under which the time average of a long realization (sample mean) converges to the space average (average of many random samples at a given time index) of a dynamic system. For a scalar weakly stationary z_t, the ergodic condition for $\bar{z}$ to converge to $\mu = E(z_t)$ is that $\sum_{i=0}^{\infty} |\gamma_i| < \infty$. We assume this condition in the book from now on.

For a weakly stationary vector process z_t, the covariance matrix of the sample mean $\bar{z}$ is given by

$$\begin{aligned}\text{Var}(\bar{z}) &= E\left[(\bar{\mathbf{z}} - \boldsymbol{\mu})(\bar{\mathbf{z}} - \boldsymbol{\mu})'\right] = E\left[\left\{\frac{1}{T}\sum_{t=1}^{T}(\mathbf{z}_t - \boldsymbol{\mu})\right\}\left\{\frac{1}{T}\sum_{j=1}^{T}(\mathbf{z}_j - \boldsymbol{\mu})'\right\}\right] \\ &= \frac{1}{T^2}\sum_{t=1}^{T} E\left[(\mathbf{z}_t - \boldsymbol{\mu})\left\{\sum_{j=1}^{T}(\mathbf{z}_j - \boldsymbol{\mu})'\right\}\right] = \frac{1}{T^2}\left[\sum_{j=-(T-1)}^{T-1}(T - |j|)\boldsymbol{\Gamma}(j)\right].\end{aligned} \tag{1.15}$$

This implies that the asymptotic variance of the scalar sample mean $\bar{z}_i$ is

$$T \times \text{Var}(\bar{z}_i) \to \gamma_{ii}(0) + 2\sum_{\ell=1}^{\infty}\gamma_{ii}(\ell), \quad \text{as} \quad T \to \infty. \tag{1.16}$$

where, again, $\gamma_{ii}(\ell)$ is the lag-ℓ autocovariance of z_{it}. From Equation (1.2), if $\{z_{it}\}$ is white noise the variance of the sample mean $\bar{z}_i$ is σ_i^2/T. In general, we have

$$\text{Var}(\bar{z}_i) \approx \frac{1}{T}\left[\gamma_{ii}(0) + 2\sum_{l=1}^{T}\gamma_{ii}(l)\right], \quad \text{as} \quad T \to \infty.$$

Therefore, $\text{Var}(\bar{z}_i)$ could be large if the sum of the autocovariances of z_{it} is large. A necessary (although not sufficient) condition for the sum of autocovariances of z_{it} to converge is

$$\lim_{l \to \infty} \gamma_{ii}(l) \to 0,$$

which implies that the dependence of z_{it} on its past values $z_{i,t-l}$ vanishes as l increases.

1.3.2 Sample Covariance and Correlation Matrices

For a vector process $\boldsymbol{z}_t$, the lag-ℓ sample autocovariance matrix is

$$\hat{\boldsymbol{\Gamma}}(\ell) = \frac{1}{T} \sum_{t=\ell+1}^{T} (\boldsymbol{z}_t - \bar{\mathbf{z}})(\boldsymbol{z}_{t-\ell} - \bar{\mathbf{z}})'. \tag{1.17}$$

In particular, the sample covariance matrix is

$$\hat{\boldsymbol{\Gamma}}(0) = \frac{1}{T} \sum_{t=1}^{T} (\boldsymbol{z}_t - \bar{\mathbf{z}})(\boldsymbol{z}_t - \bar{\mathbf{z}})'.$$

The lag-ℓ sample CCM of $\boldsymbol{z}_t$ is then

$$\hat{\boldsymbol{R}}(\ell) = \hat{\boldsymbol{D}}^{-1/2} \hat{\boldsymbol{\Gamma}}(\ell) \hat{\boldsymbol{D}}^{-1/2}, \tag{1.18}$$

where $\hat{\boldsymbol{D}} = \text{diag}\{\hat{\gamma}_{11}(0), \ldots, \hat{\gamma}_{kk}(0)\}$. From Eq. (1.18), we have

$$\hat{\boldsymbol{R}}(\ell) = [\hat{\rho}_{ij}(\ell)] \quad \text{with} \quad \hat{\rho}_{ij}(\ell) = \frac{\hat{\gamma}_{ij}(\ell)}{\sqrt{\hat{\gamma}_{ii}(0)\hat{\gamma}_{jj}(0)}}.$$

Under some mild conditions, $\hat{\boldsymbol{R}}(\ell)$ is a consistent estimator of $\boldsymbol{R}(\ell)$ for a weakly stationary process $\boldsymbol{z}_t$. The conditions are met if $\boldsymbol{z}_t$ is a stationary Gaussian process. The asymptotic properties of $\hat{\boldsymbol{R}}(\ell)$, however, are rather complicated. They depend on the dynamic dependence of the process $\boldsymbol{z}_t$. However, the results for some special cases are easier to understand. Suppose the vector time series $\boldsymbol{z}_t$ is a white noise process so that $\boldsymbol{R}(\ell) = \boldsymbol{0}$ for $\ell \neq 0$. Then, we have

$$\text{Var}[\hat{\rho}_{ij}(\ell)] \approx \frac{1}{T}, \quad \ell \neq 0.$$

and, for $i \neq j$,

$$\text{Var}[\hat{\rho}_{ij}(0)] \approx \frac{[1 - \rho_{ij}^2(0)]^2}{T}. \tag{1.19}$$

For further details, see, for instance, Reinsel (1993, section 4.1.2). These simple results are useful in exploratory data analysis of vector time series. This is particularly so when the dimension k is large. An important feature in analyzing time series data is to explore the dynamic dependence of the observed time series. For a k-dimensional series, each sample CCM $\hat{\boldsymbol{R}}(\ell)$ is a $k \times k$ matrix. If $k = 10$, then each CCM contains 100 sample correlations. It would not be easy to decipher the pattern embedded in the sample CCM $\hat{\boldsymbol{R}}(\ell)$ simultaneously for $\ell = 1, \ldots, m$, where m is a given positive integer. To aid the reading of sample CCM, Tiao and Box (1981) device a simplified approach to inspect the features of sample CCM. Specifically, consider $\hat{\boldsymbol{R}}(\ell)$ for $\ell > 0$. Tiao and Box define a corresponding simplified matrix $\boldsymbol{S}(\ell) = [S_{ij}(\ell)]$, where

$$S_{ij}(\ell) = \begin{cases} + & \text{if } \hat{\rho}_{ij}(\ell) \geq 2/\sqrt{T}, \\ \cdot & \text{if } |\hat{\rho}_{ij}(\ell)| < 2/\sqrt{T}, \\ - & \text{if } \hat{\rho}_{ij}(\ell) \leq -2/\sqrt{T}, \end{cases}$$

where it is understood that $1/\sqrt{T}$ is the sample standard error of elements of $\hat{\boldsymbol{R}}(\ell)$ provided that $\boldsymbol{z}_t$ is a Gaussian white noise. It is much easier to comprehend the dynamic dependence of $\boldsymbol{z}_t$ by inspecting the simplified CCM $\boldsymbol{S}(\ell)$ for $\ell = 1, \ldots, m$. We demonstrate the effectiveness of the simplified CCM via an example.

Example 1.1

Consider the three temperature series of Figure 1.1. The sample means of the three series are 0.15, 0.14, and 0.12, respectively, for Europe, North America, and South America. The sample standard deviations are 1.02, 1.06, and 0.47, respectively. The average temperatures of November for South America appear to be lower and relatively less variable compared with those of Europe and North America. Turn to sample CCMs. In this particular instance, $k = 3$ so that each CCM contains 9 real numbers. Suppose we like to examine the dynamic dependence of the three series using the first 12 lags of sample CCM. This would require us to decipher 108 numbers simultaneously. On the other hand, the dynamic dependence pattern is relatively easy to comprehend by examining the simplified CCMs, which are given below.

```
Simplified matrix:
CCM at lag: 1
. + +
. . +
. . +
CCM at lag: 2
. . +
. . +
+ . +
CCM at lag: 3
. . +
. . .
. + +
CCM at lag: 4
. . .
. . .
. + +
CCM at lag: 5
. . .
. . +
. + +
```

```
CCM at lag: 6
. . +
. . .
. . +
CCM at lag: 7
. . .
. . +
. . +
CCM at lag: 8
. . .
. . .
. . +
CCM at lag: 9
. . +
. . .
. . +
CCM at lag: 10
. . .
. . .
. . +
CCM at lag: 11
. . .
. . +
. . +
CCM at lag: 12
. . .
. . +
. . +
```

From the simplified CCM, we make the following observations. First, the dynamic dependence of the temperature series of South America appears to be persistent as all of its sample ACFs are large (showing by the + sign). This is not surprising as the time plot of the series in Figure 1.1 shows an upward trend. Second, the temperature series of Europe and North America are dynamically correlated with that of South America because there exist several + signs at the (1,3)th and (2,3)th elements of the CCMs. Finally, there seems to have no strong dynamic dependence in November temperatures between Europe and North America, because most sample cross-correlations at the (1,2)th and (2,1)th positions are small. ■

Remarks. Some remarks are in order.

- The demonstration of Example 1.1 is carried out by the **MTS** package in R. The command used is `ccm` with default options.

```
da <- read.table("temperatures.txt",header=TRUE)
da <- da[,-1] # remove time index
require(MTS)
ccm(da)
```

- If the dimension k is close to or larger than the sample size T, then the aforementioned properties of sample CCMs do not hold. We shall discuss the situation in later chapters, e.g. Chapter 6.
- When the dimension k is large, even the simplified CCM $\boldsymbol{S}(\ell)$ becomes hard to comprehend and some summary statistics must be used. In this book, we provide three summary statistics to extract helpful information embedded in the sample CCMs. They are given below:
  1. P-value plot: A scatterplot of the p-values of the null hypothesis $H_0 : \boldsymbol{R}(\ell) = \boldsymbol{0}$ versus lag ℓ. The test statistic used asymptotically follows a $\chi^2_{k^2}$ distribution under the null hypothesis that there are no serial or cross correlations in z_t for $\ell > 0$. Details of the test statistic is given in Chapter 3.
  2. Diagonal element plot: A scatterplot of the fraction of significant diagonal elements of $\widehat{\boldsymbol{R}}(\ell)$ versus ℓ. Here an element of $\widehat{\boldsymbol{R}}(\ell)$ is said to be significant if it is greater than $2/\sqrt{T}$ in absolute value.
  3. Off-diagonal element plot: A scatterplot of the fraction of significant off-diagonal elements of $\widehat{\boldsymbol{R}}(\ell)$ versus ℓ. Again, the significance of an element of $\widehat{\boldsymbol{R}}(\ell)$ is with respect to $2/\sqrt{T}$.

The R command for the three summary plots is `Summaryccm` of the **SLBDD** package, which is a package associated with this book.

Example 1.2

Consider the daily log returns of the 99 financial market indexes of the world in Figure 1.2. Here $k = 99$ and it would be hard to comprehend any 99-by-99 cross correlation matrix. Figure 1.14 shows the three summary statistics of the daily log returns. The upper plot is the scatterplot of p-values. From the plot, all p-values are close to zero indicating that there are serial and cross-correlations in the 99-dimensional log returns for lags 1 to 12. The middle plot shows the fraction of significant diagonal elements of $\widehat{\boldsymbol{R}}(\ell)$. This plot shows that there are significant serial correlations in the daily log returns of 99 world financial indexes. But, as expected, the serial correlations decay quickly so that the fraction approaches zero as ℓ increases. The bottom plot shows the fraction of significant off-diagonal elements of $\widehat{\boldsymbol{R}}(\ell)$ versus ℓ. The plot suggests that there is significant cross dependence among the daily log returns of world financial market indexes when ℓ is small. As expected, the cross dependence also decays to zero quickly as ℓ increases.

R command used for the summary CCM plots:

```
require(SLBDD)
da <- read.csv2("Stock_indexes_99world.csv",header=FALSE)
da <- as.matrix(da[,-1]) # remove the time index
rt <- diff(log(da)) # log returns
Summaryccm(rt)
```

■

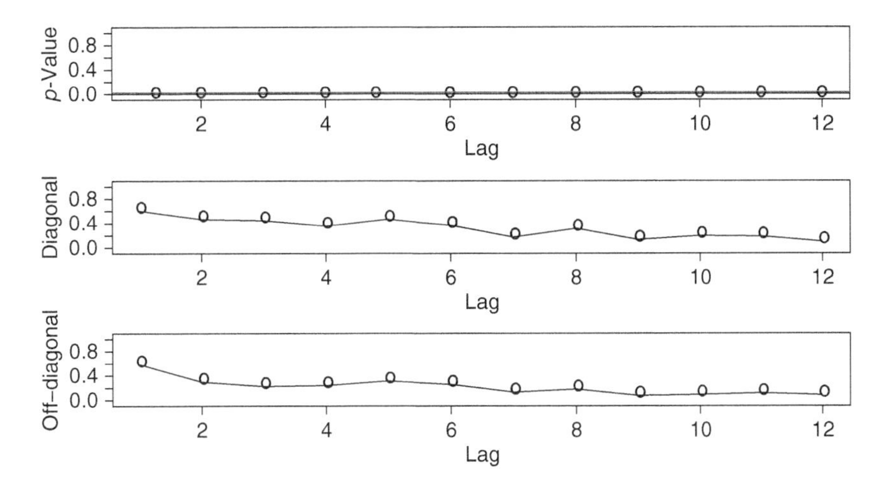

Figure 1.14 Scatterplots of three summary statistics of lagged CCMs of the daily log returns of 99 world financial market indexes. The summary statistics are (*a*) *p*-values of testing $\boldsymbol{R}(\ell) = \boldsymbol{0}$, (*b*) fractions of significant diagonal elements of $\widehat{\boldsymbol{R}}(\ell)$, and (*c*) fractions of significant off-diagonal elements of $\widehat{\boldsymbol{R}}(\ell)$.

1.4 NONSTATIONARY PROCESSES

A vector process $\boldsymbol{z}_t$ is nonstationary if it fails to meet the stationarity conditions. Simply put, if some joint distributions of $\boldsymbol{z}_t$ change over time, then $\boldsymbol{z}_t$ is nonstationary. In this case, the moments of $\boldsymbol{z}_t$ may depend on time. Since we typically have a single observation at a given time index t, further assumptions are needed to render estimation possible for a nonstationary process. The most commonly employed nonstationary processes are the *integrated processes*. Such processes are commonly referred to as unit-root nonstationary series.

A scalar time series z_t is called an integrated process of order 1 if its increment $w_t = z_t - z_{t-1}$ is a stationary and invertible process. In the econometric literature, a stationary and invertible process is called an $I(0)$ process, and an integrated process of order 1 is called an $I(1)$ process. A random-walk series is an example of an $I(1)$ process. Let B denote the back-shift operator, such that $Bz_t = z_{t-1}$. Then, z_t is an $I(d)$ process if $w_t = (1 - B)^d z_t$ is an $I(0)$ process. Following the convention, we may also use the notation $\nabla = (1 - B)$ in describing integrated processes. The operator $(1 - B)$ is called the first-difference operator, and the process $w_t = (1 - B)z_t$ is the first-differenced series of z_t.

In a similar way, a vector process $\boldsymbol{z}_t$ is called an $I(d)$ process if $\boldsymbol{w}_t = (1 - B)^d \boldsymbol{z}_t$ is a stationary and invertible series, where $(1 - B)^d$ applies to each and every component of $\boldsymbol{z}_t$. Of course, there is no particular reason to expect that all components of $\boldsymbol{z}_t$ are integrated of the same order d. Therefore, let $\boldsymbol{d} = (d_1, \dots, d_k)'$, where d_i are nonnegative integers and $\max\{d_i\} > 0$. We call the vector process $\boldsymbol{z}_t$ an $I(\boldsymbol{d})$ process if $\boldsymbol{w}_t = (w_{1t}, \dots, w_{kt})'$, where $w_{it} = (1 - B)^{d_i} z_{it}$, is a stationary and invertible process. In this case, we may define $\boldsymbol{D} = \text{diag}\{\nabla^{d_1}, \dots, \nabla^{d_k}\}$ and write $\boldsymbol{w}_t = \boldsymbol{D}\boldsymbol{z}_t$.

An important concept of the integrated vector process $\boldsymbol{z}_t$ is *co-integration*. Suppose that the components z_{it} are $I(1)$ processes. If there exists a

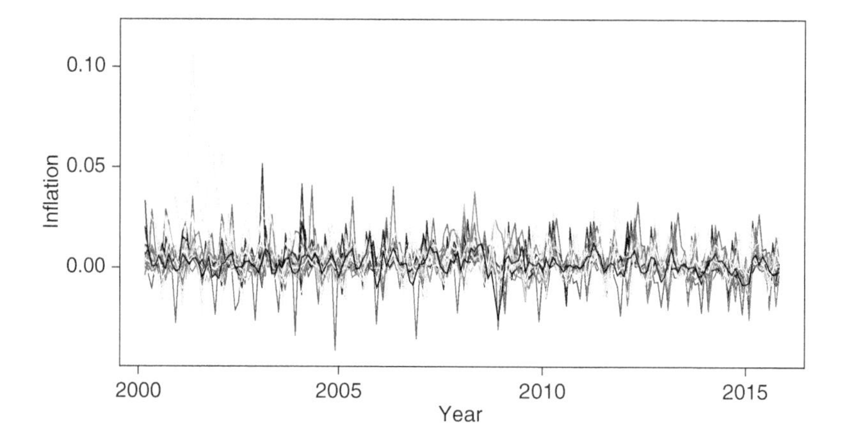

Figure 1.15 Time plots of the first-differenced series of the 33 log monthly price indexes of European countries from January 2000 to October 2015.

k-dimensional vector $\boldsymbol{c} = (c_1, \ldots, c_k)' \neq \boldsymbol{0}$ such that $w_t = \boldsymbol{c}'\boldsymbol{z}_t$ is an $I(0)$ process, then $\boldsymbol{z}_t$ is called a co-integrated process and $\boldsymbol{c}$ is the co-integrating vector. The $I(0)$ process w_t is the co-integrated series, which is stationary and invertible. In general, one can have more than one co-integrated process for a k-dimensional unit-root process $\boldsymbol{z}_t$. We shall discuss co-integration and its associated properties later.

An important class of nonstationary processes consists of seasonal time series. For a seasonal time series, the mean function is not constant but varies over time accordingly to certain periodic pattern. This pattern is repeated every season in a given year such as every quarter, month, or week. The number of seasons in a year is referred to as *periodicity*. For monthly series, the pattern is often repeated year after year and we have a periodicity $s = 12$. For quarterly data, we have $s = 4$. Consider the 33 monthly price indexes of Figure 1.7. All of the series exhibit an upward trend so that they are not stationary. Figure 1.15 shows the time plots of the first-differenced series of the log price indexes. From the plots, as expected, the trends disappear and the series (i.e. inflation rate) now fluctuate around zero. Therefore, there is evidence that the log price indexes are $I(1)$ processes. The cyclic patterns shown in Figure 1.15 indicate that some months have higher values than others so that the processes belong to the class of seasonal time series. We shall discuss seasonal time series later. For now, it suffices to introduce the idea of *seasonal difference*. The seasonal difference is given by $w_t = (1 - B^s)z_t = z_t - z_{t-s}$, which denotes the annual increment. For convenience, we refer to $1 - B$ as the regular difference.

Figure 1.16 shows the time plots of the regularly and seasonally differenced series of the 40 price indexes of Figure 1.7. Specifically, the time series shown are $w_t = (1 - B)(1 - B^{12})z_t$. From the plots, the seasonal patterns are largely removed. A seasonal time series that requires seasonal difference to achieve stationarity yet remains invertible is referred to as a seasonally integrated process. The idea of co-integration also generalizes to seasonal co-integration. Again, details are given in Chapter 3.

Another type of nonstationarity that attracts much attention is the locally stationary process. A stochastic process is locally stationary if its mean and variance functions evolve slowly over time. A formal definition of such processes can be found in Dahlhaus (2012). Different from the conventional time series analysis, statistical inference on locally stationary processes is based on in-filled asymptotics. To illustrate, consider the time-varying parameter autoregressive process

$$z_t = \phi_t z_{t-1} + \sigma_t \epsilon_t, \tag{1.20}$$

where $\{\epsilon_t\}$ is a sequence of standard normal random variates, and ϕ_t and σ_t are rescaled smooth functions such that $\phi_t \equiv \phi(t/T) : [0,1] \to (-1,1)$ and $\sigma_t \equiv \sigma(t/T) : [0,1] \to (0,\infty)$, where T is the sample size. When T increases, the observations on $\phi(t/T)$ and $\sigma(t/T)$ become denser in their domain [0,1] so that we can obtain better estimates of the time-varying parameters. Figure 1.17 shows the time plot of a realization with 300 observations from the time-varying parameter AR(1) model of Eq. (1.20), where $\phi(t/T) = 0.2 + 0.4(t/T) - 0.2(t/T)^2$ and $\sigma(t/T) = \exp(t/T)$ with $T = 300$. As expected, the series fluctuates around zero with an increasing variability. The process is locally stationary because it can be approximated by a stationary series within a small time interval. For instance, in this particular example, the first 120 observations of Figure 1.17 appears to be weakly stationary.

1.5 PRINCIPAL COMPONENT ANALYSIS

Curse of dimensionality is a well-known difficulty in analyzing big dependent data. This is so because it is hard to decipher useful information embedded in high-dimensional time series. Some exploratory data analyses are helpful in this situation. A powerful tool to explore the dynamic dependence in a vector time series is the *principal component analysis* (PCA). Even though PCA was developed primarily for independent data, it has been shown to be useful in time series analysis too. See, for instance, Peña and Box (1987), Tsay (2014, chapter 6) and the references therein. See also Taniguchi and Krishnaiah (1987) and Chang et al. (2014) for some theoretical developments. However, there are differences between applying PCA to independent data and to time series data. The main difference is that PCA is

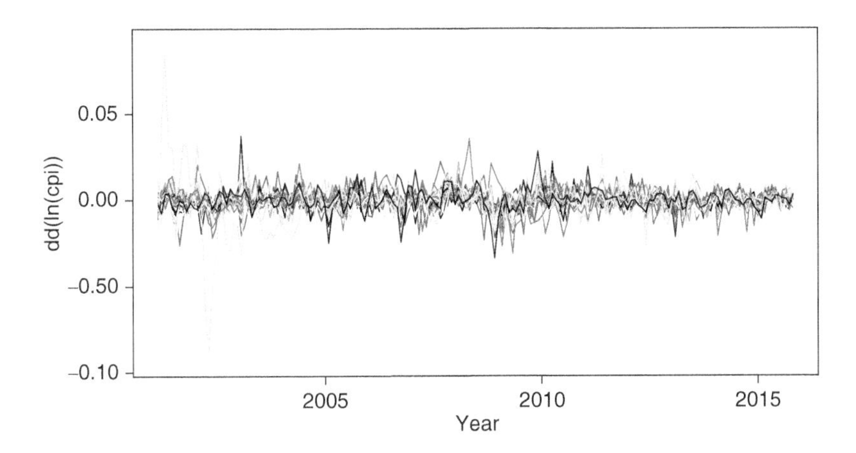

Figure 1.16 Time plots of the first and seasonally differenced series of the 30 monthly price indexes of European countries from January 2000 to October 2015.

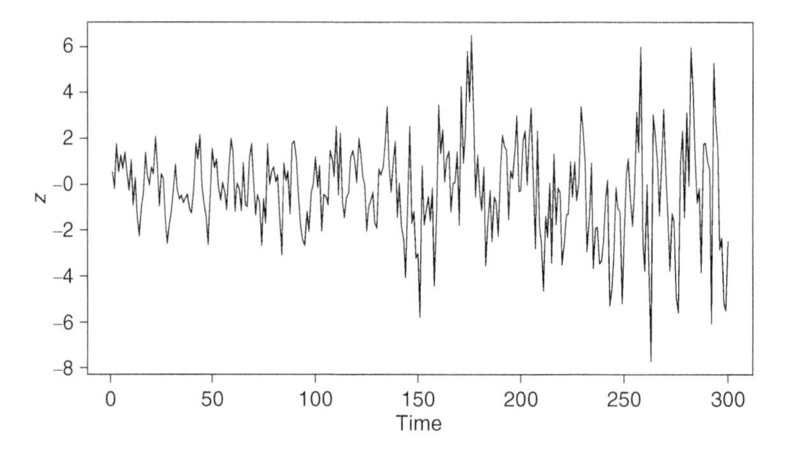

Figure 1.17 Time plot of 300 realizations from a time-varying parameter autoregressive model of order 1 in Eq. (1.20), where $\phi(t/T) = 0.2 + 0.4(t/T) - 0.2(t/T)^2$ and $\sigma(t/T) = \exp(t/T)$.

mainly concerned with the covariance matrix, not the lag covariance matrices, which are important in time series analysis. Therefore, the limiting properties of the eigenvalues and eigenvectors of PCA are different for time series data. See a recent paper by Zhang and Tong (2020) and some discussions in Section 1.6. Also, we present in Chapter 6 a more general procedure, called the dynamic principal component analysis (DPCA) for time series data, that takes into account the lag dependence of a vector time series.

There are two main ways to introduce PCA. The first way is variance decomposition, and the second one optimal reconstruction or interpolation. We start with the variance decomposition of independent data at the population level. Let $\boldsymbol{z} = (z_1, \ldots, z_k)'$ be a k-dimensional random vector with mean zero and positive-definite covariance $\boldsymbol{\Gamma}(0)$. The first population principal component (PC) y_1 of $\boldsymbol{z}$ is a linear combination $y_1 = \boldsymbol{c}_1'\boldsymbol{z}$ satisfying $\boldsymbol{c}_1'\boldsymbol{c}_1 = 1$ such that $\text{Var}(y_1)$ attends the maximum among all possible linear combinations of $\boldsymbol{z}$ built with vectors of unit length. The second PC of $\boldsymbol{z}$ is defined as a linear combination of $\boldsymbol{z}$, say $y_2 = \boldsymbol{c}_2'\boldsymbol{z}$ satisfying (i) $\boldsymbol{c}_2'\boldsymbol{c}_2 = 1$ and (ii) $\boldsymbol{c}_2'\boldsymbol{c}_1 = 0$ such that $\text{Var}(y_2)$ assumes the maximum variance among all linear combinations $\boldsymbol{z}$ built with vectors of unit length and such that the second PC is uncorrelated with the first PC. In general, the ith PC of $\boldsymbol{z}$ is a non-zero linear combination of $\boldsymbol{z}$, say $y_i = \boldsymbol{c}_i'\boldsymbol{z}$ using a vector of unit length that has the maximum variance among all linear combinations of $\boldsymbol{z}$ and that is uncorrelated to all the previous PCs, that is, $\boldsymbol{c}_i$ satisfies (iii) $\boldsymbol{c}_i'\boldsymbol{c}_i = 1$ and (iv) $\boldsymbol{c}_i'\boldsymbol{c}_j = 0$ for $j = i - 1, \ldots, 1$. The unit length normalization, $\boldsymbol{c}_i'\boldsymbol{c}_i = 1$, is needed to control the scaling effect; otherwise the maximum has no meaning.

Using properties of positive-definite matrices given in the Appendix, the PCA of a stationary process $\boldsymbol{z}_t$ with positive-definite covariance matrix $\boldsymbol{\Gamma}(0)$ can be obtained from the spectral decomposition of $\boldsymbol{\Gamma}(0)$. Specifically, let $(\lambda_i, \boldsymbol{e}_i)$ be the ith eigenvalue-eigenvector pair of $\boldsymbol{\Gamma}(0)$, where the eigenvalues satisfy $\lambda_1 \geq \lambda_2 \geq \cdots \geq \lambda_k$. The spectral decomposition of $\boldsymbol{\Gamma}(0)$ is

$$\boldsymbol{\Gamma}(0) = \lambda_1 \boldsymbol{e}_1 \boldsymbol{e}_1' + \cdots + \lambda_k \boldsymbol{e}_k \boldsymbol{e}_k'.$$

Then, the ith PCA is $y_{it} = \boldsymbol{e}_i'\boldsymbol{z}_t$. From the definition, $\text{Var}(y_{it}) = \text{Var}(\boldsymbol{e}_i'\boldsymbol{z}) = \boldsymbol{e}_i'\boldsymbol{\Gamma}(0)\boldsymbol{e}_i = \lambda_i\boldsymbol{e}_i'\boldsymbol{e}_i = \lambda_i$. Since $\text{tr}[\boldsymbol{\Gamma}(0)] = \sum_{i=1}^k \sigma_{ii} = \sum_{i=1}^k \lambda_i$, the proportion of variability of $\boldsymbol{z}$ explained by its ith PC is $\lambda_i/(\sum_{j=1}^k \lambda_j)$. Similarly, the proportion of variability of $\boldsymbol{z}_t$ explained by the first m PCs is $(\sum_{i=1}^m \lambda_i)/(\sum_{j=1}^k \lambda_j)$. If a small m can be found such that the first m PCs explain a large proportion of the variability in $\boldsymbol{z}$, then one can focus analysis on the first m PCs.

Suppose now that, instead of independent variables, we have a random realization $\{\boldsymbol{z}_1, \ldots, \boldsymbol{z}_T\}$ of a stationary vector time series $\boldsymbol{z}_t$. For simplicity, we assume that the process is mean-adjusted, i.e. $\sum_{t=1}^T \boldsymbol{z}_t/T = \boldsymbol{0}$. Define the $T \times k$ matrix $\boldsymbol{Z} = [\boldsymbol{z}_{(1)}, \ldots, \boldsymbol{z}_{(k)}]$, where $\boldsymbol{z}_{(j)} = (z_{j1}, \ldots, z_{jT})'$ is the vector of observations of the jth component z_{jt}. Then, the covariance matrix of $\boldsymbol{z}_t$ is estimated by the sample covariance matrix $\widehat{\boldsymbol{\Gamma}}(0) = \boldsymbol{Z}'\boldsymbol{Z}/T$ and the eigenvalue-eigenvector pairs $(\lambda_i, \boldsymbol{c}_i)$ of this matrix can be used to construct PCs. The first PC is given by $\boldsymbol{y}_1 = \boldsymbol{Z}\boldsymbol{c}_1$, where

$$\frac{1}{T}\boldsymbol{Z}'\boldsymbol{Z}\boldsymbol{c}_1 = \lambda_1\boldsymbol{c}_1, \quad \boldsymbol{c}_1'\boldsymbol{c}_1 = 1.$$

Pre-multiplying the above equation by $\boldsymbol{Z}$, we have

$$\frac{1}{T}\boldsymbol{Z}\boldsymbol{Z}'\boldsymbol{y}_1 = \lambda_1\boldsymbol{y}_1. \tag{1.21}$$

Therefore, the first PC $\boldsymbol{y}_1$ of $\boldsymbol{z}_t$ is proportional to the eigenvector corresponding to the largest eigenvalue of the matrix $\boldsymbol{Z}\boldsymbol{Z}'$. Other PCs can be obtained in a similar manner.

An alternative approach to introduce PCA is from an optimal reconstruction or interpolation of the data. For simplicity, we consider the sample version with the mean-adjusted stationary time series data as before, i.e. $\{\boldsymbol{z}_1, \ldots, \boldsymbol{z}_T\}$. We want to obtain a new time series, f_{1t}, defined as a linear combination of $\boldsymbol{z}_t$ that minimizes the error in the reconstruction of the observed data. More formally, let $f_{1t} = \boldsymbol{z}_t'\boldsymbol{\beta}_1$ and $\boldsymbol{f}_1 = \boldsymbol{Z}\boldsymbol{\beta}_1$, where $\boldsymbol{\beta}_1 = (\beta_{11}, \ldots, \beta_{k1})'$ and $\boldsymbol{f}_1 = (f_{11}, \ldots, f_{1T})'$. The coefficients β_{i1} are obtained by minimizing the objective function

$$L(\boldsymbol{\beta}_1, \boldsymbol{f}_1) = \sum_{t=1}^{T}\sum_{i=1}^{k}(z_{it} - \beta_{i1}f_{1t})^2. \tag{1.22}$$

Since both f_{1t} and $\boldsymbol{\beta}_1$ are unknown, the solution to Eq. (1.22) is not well-defined, because $\beta_{i1}f_{1t} = (\beta_{i1}/h)(hf_{1t}) = \beta_{i1}^*f_{1t}^*$, where h is an arbitrary non-zero real number. To overcome the difficulty of non-uniqueness, we assume that f_{1t} satisfies $\sum_{t=1}^T f_{1t}^2 = 1$. The scale of f_{1t} is then identified, and we can take the partial derivative of L with respective to β_{i1}. By equating the partial derivatives to zero, we have

$$\hat{\beta}_{i1} = \frac{\sum_{t=1}^T z_{it}f_{1t}}{\sum_{t=1}^T f_{1t}^2} = \sum_{t=1}^{T} z_{it}f_{1t}, \quad i = 1, \ldots, k, \tag{1.23}$$

or, in vector form, $\widehat{\boldsymbol{\beta}}_1 = \boldsymbol{Z}'\boldsymbol{f}_1$. Using $\boldsymbol{f}_1 = \boldsymbol{Z}\boldsymbol{\beta}_1$, we obtain

$$\widehat{\boldsymbol{\beta}}_1 = \boldsymbol{Z}'\boldsymbol{Z}\widehat{\boldsymbol{\beta}}_1,$$

and, hence, $\hat{\beta}_1$ is an un-normalized eigenvector of the matrix $Z'Z$.

Similarly, we can take the partial derivative of L with respective to f_{1t} and obtain

$$f_{1t} = \frac{\sum_{i=1}^{k} z_{it}\beta_{i1}}{\sum_{i=1}^{k} \beta_{i1}^2}, \quad t = 1, \ldots, T. \tag{1.24}$$

The numerator of Eq. (1.24) can be written as $z_t'(Z'f_1)$. Letting $c = \sum_{i=1}^{k} \beta_{i1}^2$, plugging Eq. (1.23) into Eq. (1.24), and using matrix notation, we have

$$ZZ'f_1 = cf_1. \tag{1.25}$$

This implies that f_1 is an eigenvector of the matrix ZZ' associated with eigenvalue c.

The function in Eq. (1.22) is an objective function of least squares estimation, and the fitted value is given by $ZZ'f_1$. Therefore, the value of objective function evaluated at the estimates becomes

$$\begin{aligned}\hat{L} &= \sum_{t=1}^{T}\sum_{i=1}^{k} z_{it}^2 - [ZZ'f_1]'(ZZ'f_1)\\ &= \sum_{t=1}^{T}\sum_{i=1}^{k} z_{it}^2 - f_1'ZZ'(cf_1)\\ &= \sum_{t=1}^{T}\sum_{i=1}^{k} z_{it}^2 - c^2,\end{aligned}$$

where we have used $f_1'f_1 = \sum_{t=1}^{T} f_{1t}^2 = 1$. Consequently, to achieve the minimum value of the objective function $\hat{L}$, we choose c to be the largest eigenvalue of ZZ'. The vector f_1 is not equal to the first PC y_1, because it has variance 1, whereas the PC has variance λ_1, but apart from this scaling effects they are identical, as both are proportional to the eigenvalues of the matrix ZZ', as shown in (1.21) and (1.25).

For the second sample PC, we can follow a similar procedure by replacing the observed data z_t by the residual $\epsilon_t = z_t - \hat{\beta}_1 f_{1t}$. This process can be repeated until all sample PCs are introduced.

In summary, both approaches led to the same result with a difference in scale. In the first approach, the linear combinations of maximum variance are defined by unit norm vectors that are the eigenvectors of the covariance matrix and the PCs have different variances. In the second approach, the optimal interpolations are defined by vectors that are proportional to the eigenvalues of the covariance matrix and the linear combinations are standardized to have unit variance.

1.5.1 Discussion

We have introduced PCA using both the variance decomposition and the reconstruction approaches. However, there is a subtle difference between the two versions. In the first case we assume stationarity. On the other hand, for a given data set, the reconstruction approach can be applied to any time series.

1.5.2 Properties of the PCs

From the introduction, PCs $\boldsymbol{y}_t = (y_{1t}, \ldots, y_{kt})'$ of a stationary vector process $\boldsymbol{z}_t$ are linear transforms of the observed data, and they are ordered according to their variabilities such that y_{1t} has the largest variance. In particular, $\text{Var}(y_{it}) = \lambda_i$, which is the ith largest eigenvalue of the covariance matrix of $\boldsymbol{z}_t$. Since $\sum_{i=1}^{k} \text{Var}(z_{it}) = \sum_{i=1}^{k} \gamma_{ii}(0) = \text{tr}[\boldsymbol{\Gamma}(0)]$, where $\text{tr}[\boldsymbol{\Gamma}(0)]$ denotes the trace of the matrix $\boldsymbol{\Gamma}(0)$, and $\text{tr}[\boldsymbol{\Gamma}(0)] = \sum_{i=1}^{k} \lambda_i$, we see that PCs retain the total variability of the vector process $\boldsymbol{z}_t$.

Next, let $\boldsymbol{e}_i$ be the eigenvector associated with eigenvalue λ_i of $\boldsymbol{\Gamma}(0)$ such that $\boldsymbol{e}_i'\boldsymbol{e}_i = 1$. We have $\boldsymbol{y}_t = \boldsymbol{P}'\boldsymbol{z}_t$, where $\boldsymbol{P} = [\boldsymbol{e}_1, \ldots, \boldsymbol{e}_k]$ is the matrix of eigenvectors. This transformation matrix is an orthonormal matrix so that $\boldsymbol{PP}' = \boldsymbol{P}'\boldsymbol{P} = \boldsymbol{I}$. Consequently, we have $1 = |\boldsymbol{PP}'| = |\boldsymbol{P}||\boldsymbol{P}'|$. From the transformation, $\text{Cov}(\boldsymbol{y}_t) = \boldsymbol{P}'\text{Cov}(\boldsymbol{z}_t)\boldsymbol{P}$. Using $|\boldsymbol{P}||\boldsymbol{P}'| = 1$, we have $|\text{Cov}(\boldsymbol{y}_t)| = |\text{Cov}(\boldsymbol{z}_t)|$, implying that the generalized variance of $\boldsymbol{y}_t$ is the same as that of $\boldsymbol{z}_t$.

An important application of PCA is to reduce the dimension of a vector process. A scalar stochastic process would be close to a constant series if its variance is close to zero. Therefore, a PC associated with an eigenvalue $\lambda_i \approx 0$ would carry little information about the underlying process $\boldsymbol{z}_t$. As discussed before, the proportion of variability in $\boldsymbol{z}_t$ explained by the first m PCs is $(\sum_{i=1}^{m} \lambda_i)/(\sum_{i=1}^{k} \lambda_k)$. In practice, one can select an m such that the first m PCs, $\{y_{it}|i = 1, \ldots, m\}$, explain a substantial portion of the variability of $\boldsymbol{z}_t$. In the literature, some methods have been proposed to aid the choice of m. A commonly used, yet informal, method is the *scree plot*, which is simply the scatterplot of λ_i versus i. To choose m, one looks for an elbow (bend) in the scree plot.

Example 1.3

To demonstrate PCA, we consider, again, the 33 monthly price indexes shown in Figure 1.7. In this example, we employ the logarithms of the price indexes. Let $\boldsymbol{z}_t$ be the log series of the 33 monthly price indexes from January 2000 to October 2015 for 190 observations. A key feature of the price series shown in Figure 1.7 is the upward trends. These trends do not move in unison as different price series may increase at different paces. We apply PCA to $\boldsymbol{z}_t$ and obtain the results in Table 1.1.

From Table 1.1, we see that (a) the eigenvalue λ_i decays quickly as i increases, (b) the first PC alone explains about 95.8% of the variability in the data, and (c) the second PC explains about 3.3% of the total variability. As a matter of fact, the first

TABLE 1.1 The Results of PCA Applied to the Log Series of Monthly Price Indexes of 33 European Countries. Only the Results of the First Six PCs Are Shown, Where λ_i Is the ith Largest Eigenvalue

Component	1	2	3	4	5	6
$\sqrt{\lambda_i}$	0.968	0.176	0.075	0.042	0.031	0.020
Proportion	0.958	0.033	0.006	0.002	0.001	0.001
Cumulative	0.958	0.990	0.996	0.997	0.998	0.999

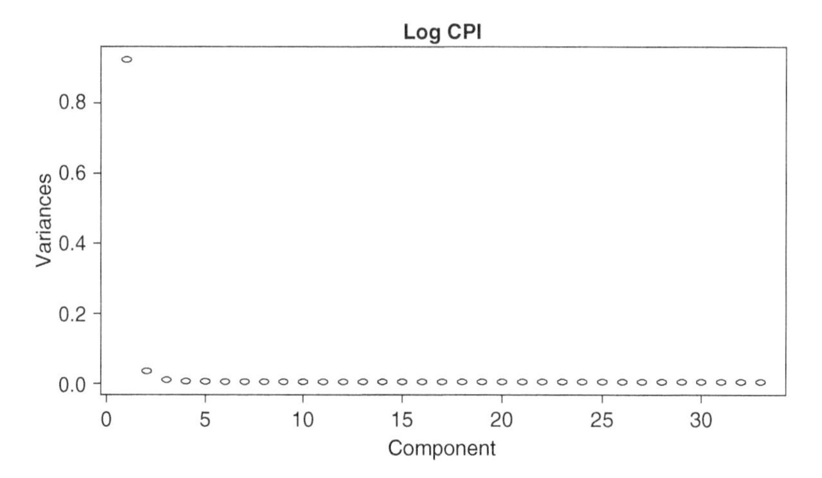

Figure 1.18 Scree plot for the log series of 33 monthly price indexes of European countries from January 2000 to October 2015.

TABLE 1.2 The Results of PCA Applied to the First-Differenced Log Series of Monthly Price Indexes. Only the Results of the First Six PCs Are Shown, Where λ_i Is the *i*th Largest Eigenvalue

Component	1	2	3	4	5	6
$\sqrt{\lambda_i}$	0.0217	0.0165	0.0124	0.0120	0.0108	0.0075
Proportion	0.3011	0.1725	0.0975	0.0909	0.0736	0.0360
Cumulative	0.3011	0.4736	0.5711	0.6620	0.7356	0.7716

six PCs explain more than 99.9% of the sample variability in z_t. Figure 1.18 shows the scree plot of the PCA of z_t and it confirms the findings that the first PC is the dominating factor and the first few PCs capture most of the variabilities in the log price indexes. Figure 1.19 shows the time plots of the first six PCs. From the plots, we see that the first PC captures the overall upward trend of the log price indexes. The second and third PCs also exhibit certain trending behavior, but they indicate that the trends do not necessarily move in a monotone manner. The fourth and sixth PCs show some periodic patterns of the price indexes, signifying the seasonal behavior of price indexes. Figure 1.20 shows the time plots of the 7th to 12th PCs. Even though those plots show certain variabilities, they are in a smaller scale compared with those of Figure 1.19.

To demonstrate that PCA are simply for variance decomposition, we also consider the first-differenced series $\boldsymbol{x}_t = (1 - B)z_t$. The component x_{it} of the series $\boldsymbol{x}_t$ represents the inflation rates (with respective to the previous month) of the ith price index. Table 1.2 gives the results of applying PCA to $\boldsymbol{x}_t$. The eigenvalues are much smaller compared with those of Table 1.1. This is understandable as trends are the dominating factors of the monthly price indexes. The inflation rates, on the other hand, are in a much smaller scale. For the inflation rates, the first six PCs explain

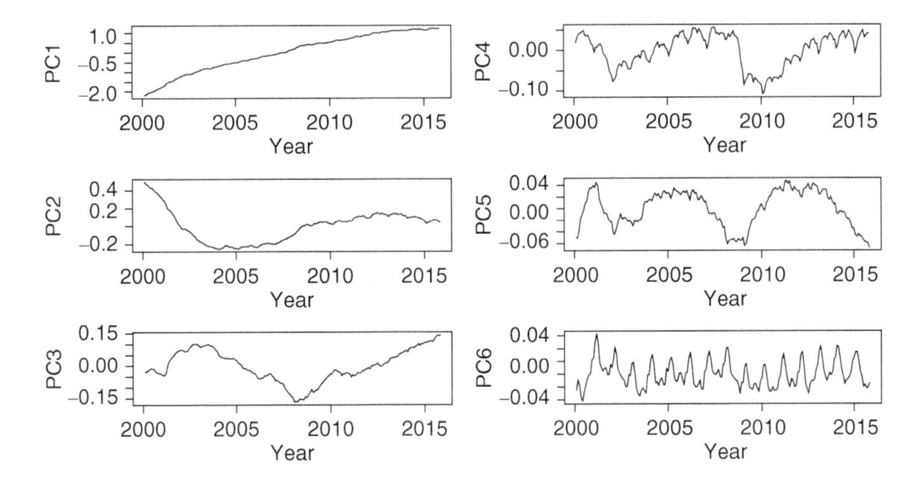

Figure 1.19 Time plots of the first six PCs of the log series of 33 monthly price indexes of European countries.

about 77.16% of the data variabilities. Figure 1.21 shows the corresponding scree plot. From the plot, the first five PCs seem to distance themselves from the others and the decay of the eigenvalues λ_i appears to be slow for $i \geq 6$. Finally, Figure 1.22 shows the time plots of the first six PCs of $\boldsymbol{x}_t$. It is interesting to see that the seasonal pattern of the monthly price indexes is now clearly captured by the first PC. In addition, as expected, there is no upward trend in the PCs.

This simple example shows that PCA can be applied to unit-root nonstationary time series. It also demonstrates that, as a tool for variance decomposition, results of PCA depend on the scale. The method is useful in capturing the dominating factors of variability for a given data set, but care must be exercised in interpreting the results of PCA. If one thought that the first two PCs of the log price indexes, which explain about 99.0% of the variabilities, are sufficient for the data, she would miss the important characteristics of seasonality.

Main R commands used in Example 1.3

```
da <- read.csv2("CPIEurope2000-15.csv",header=TRUE)
x <- log(as.matrix(da))
m1 <- princomp(x)
m1
names(m1)
#[1] "sdev"    "loadings" "center"   "scale"    "n.obs" "scores"    "call"
# sdev: the square-root of eigenvalues
# loadings: the eigenvectors
# scores: the principal components
plot(1:33,m1$sdev^2,xlab="component",ylab="variance", pch="o",
main="log CPI") # Figure 1.18.
```

■

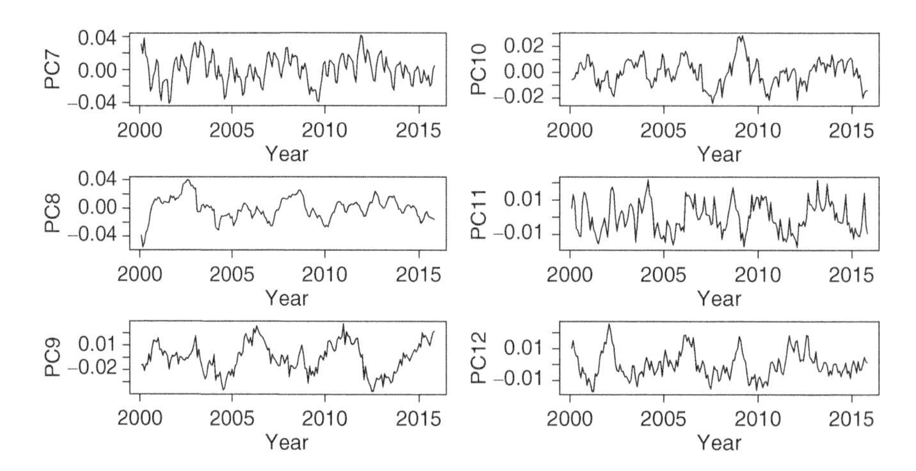

Figure 1.20 Time plots of the 7th to 12th PCs of the log series of 33 monthly price indexes from European countries.

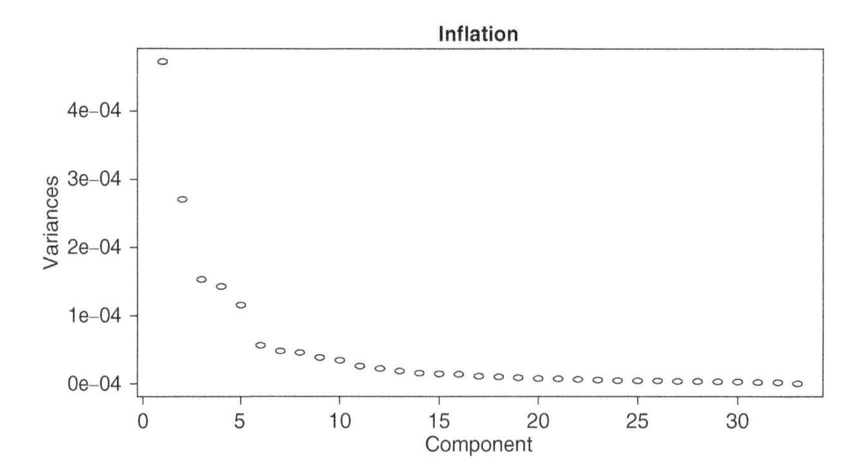

Figure 1.21 Scree plot for the first-differenced log series of 33 monthly price indexes of European countries from January 2000 to October 2015.

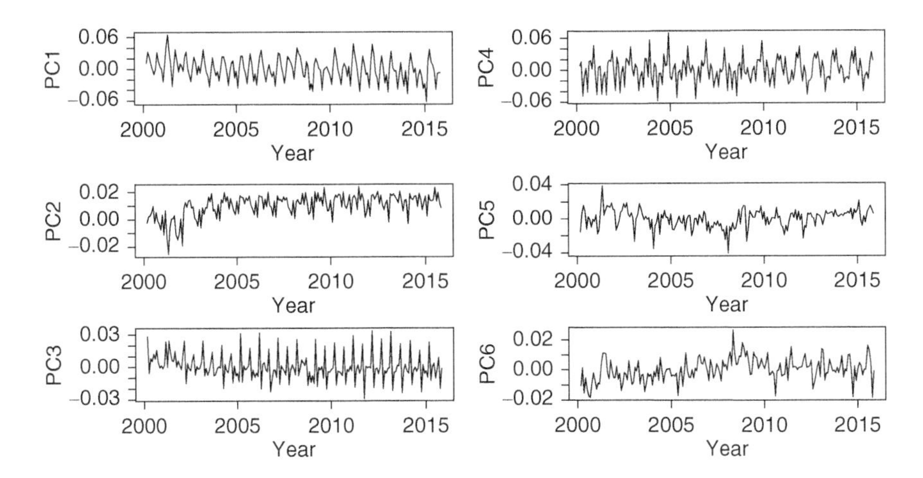

Figure 1.22 Time plots of the first six PCs of the first-differenced log series of 33 monthly price indexes of European countries.

1.6 EFFECTS OF SERIAL DEPENDENCE

In the previous sections, we have introduced some examples of big dependent data and some basic concepts of stochastic processes. In this section, we use some simple examples to illustrate that statistical methods developed for analysis of independent data may encounter difficulties when they are applied to dependent data.

As stated before, PCA was proposed originally for independent data and the sample version of PCA is applicable to dependent data. Let $\boldsymbol{y}_t = \boldsymbol{P}'\boldsymbol{z}_t$ be the PCs of a vector process $\boldsymbol{z}_t$, where $\boldsymbol{P}$ is the orthonormal transformation matrix consisting of the eigenvectors of the sample covariance matrix of $\boldsymbol{z}_t$. From the orthogonality, we have $\text{Cov}(y_{it}, y_{jt}) = 0$ for $i \neq j$. That is, the PCs are contemporaneously uncorrelated. For independent data, $\text{Cov}(\boldsymbol{z}_t, \boldsymbol{z}_{t-j}) = \boldsymbol{0}$ provided $j \neq 0$. Consequently, for independent data, the PCs $\boldsymbol{y}_t$ are contemporaneously and serially uncorrelated. Therefore, we may analyze, in this case, each component y_{it} separately if the analysis focuses on the first two moments of the data. In particular, under the Gaussian assumption, PCs are mutually independent.

Turn to dependent data. It is true that the PCs $\boldsymbol{y}_t$ are contemporaneously uncorrelated, but they can be dynamically (or serially) correlated. To demonstrate, Figure 1.23 shows the plots of autocorrelations and cross-correlations of y_{2t} and y_{3t} of the growth rates of the 40 price indexes of Figure 1.7. These two PCs are shown in Figure 1.22. From the correlation plots, y_{2t} and y_{3t} are serially correlated and, hence, they must be analyzed jointly. Consequently, PCA for dependent data may differ markedly from that for independent data. As a matter of fact, Zhang and Tong (2020) show that (a) the eigenvalues of the sample covariance matrix of a stationary and serially correlated vector time series are asymptotically correlated and (b) those eigenvalues are asymptotically dependent on all eigenvectors. On the other hand, for independent data, (a) eigenvalues and eigenvectors are asymptotically independent and (b) eigenvalues are asymptotically uncorrelated under certain conditions.

Turn to statistical methods proposed for big independent data. One of the most commonly used test statistics is the one-sample t-test for testing that the mean of a scalar random variable is zero. For an observed data set $\{z_1, \ldots, z_T\}$, one computes the test statistic

$$t = \frac{\bar{z}}{\sqrt{s^2/T}}, \tag{1.26}$$

where $\bar{z} = \sum_{t=1}^{T} z_t/T$ is the sample mean and $s^2 = \sum_{t=1}^{T} (z_t - \bar{z})^2/(T-1)$ is the sample variance. For i.i.d. sample, the t statistic of Eq. (1.26) follows asymptotically the $N(0,1)$ distribution so that statistical inference can be made. Suppose, on the other hand, z_t follows an AR(1) model, say $z_t = \phi_0 + \phi_1 z_{t-1} + a_t$, where ϕ_0 is a constant, $\phi_1 \neq 0$, and $\{a_t\}$ is a scalar white noise series. Then, the test statistic in Eq. (1.26) does not follow asymptotically the $N(0,1)$ distribution, because it is easy to verify that $\mathrm{Var}(\sqrt{T}\bar{z})$ converges to $\mathrm{Var}(z_t) \times \frac{1+\phi}{1-\phi}$ as $T \to \infty$. Therefore, the proper test statistic to use in this case is

$$t_{\text{dep}} = \frac{\bar{z}}{\sqrt{s^2(1+\phi)/[T(1-\phi)]}}. \tag{1.27}$$

From Eqs. (1.26) and (1.27), we have $t = t_{\text{dep}} \times \sqrt{(1+\phi)/(1-\phi)}$, so that t and t_{dep} can differ markedly, indicating that overlooking the serial dependence can lead to erroneous inference for the one-sample t-test. For example, if $\phi = 0.9$, then $t = t_{\text{dep}} \times \sqrt{19}$, and for $\phi = -0.9$, $t = t_{\text{dep}}/\sqrt{19}$. As another example, a widely used statistical method for analyzing big data in recent years is the Lasso regression of Tibshirani (1996). The method emphasizes on sparsity, and is applicable even when the sample size is smaller than the number of variables. Details of the method and its extensions will be given in Chapter 7. The Lasso was developed for independent data. Here we use a simple simulated example to demonstrate that Lasso regression may fail when it is applied to serially dependent data.

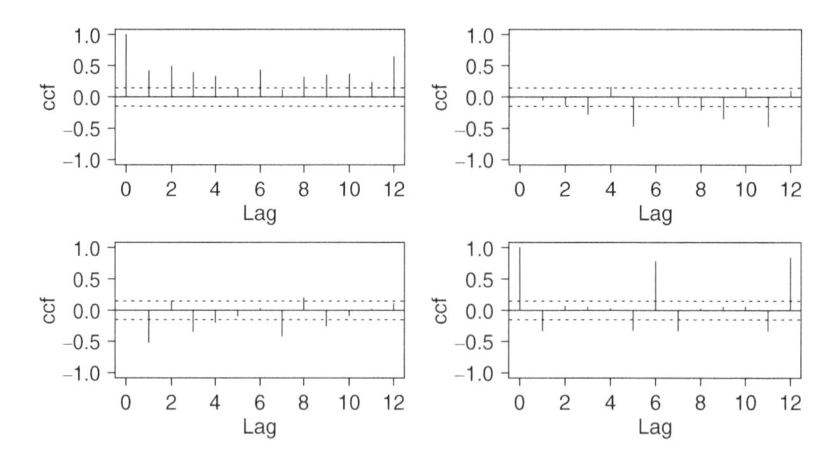

Figure 1.23 The sample autocorrelations and cross-correlations of the second and third PCs of the inflation rates of 33 monthly price indexes. The diagonal plots are sample autocorrelations and the off-diagonal plots are sample cross-correlations.

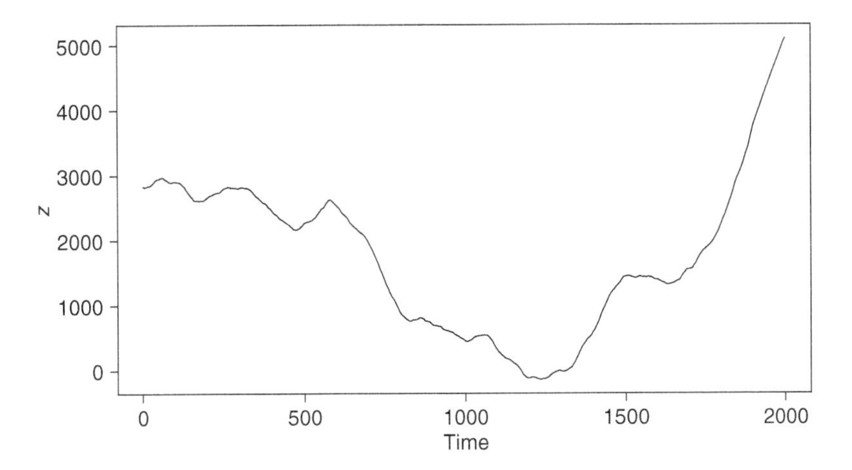

Figure 1.24 Time plot of a simulated series with 2000 observations based on the model in Eq. (1.28).

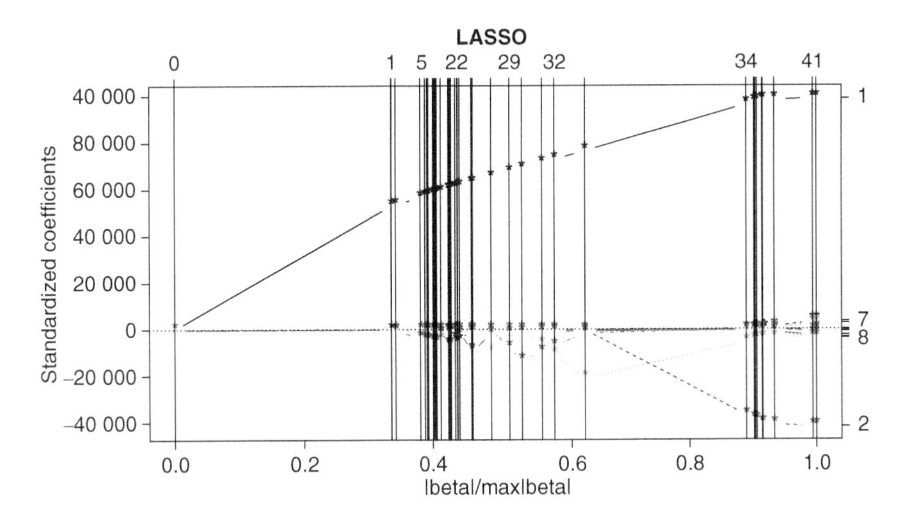

Figure 1.25 Result of Lasso regression for the data in Example 1.4. The plot is obtained by the `lars` package of R.

Example 1.4

We generated 2000 observations using the model

$$x_t = 1.9x_{t-1} - 0.8x_{t-2} - 0.1x_{t-3} + \epsilon_t, \quad t = 1, \ldots, 2000, \tag{1.28}$$

where ϵ_t are random draws from the standard normal distribution, i.e. $\epsilon_t \sim_{iid} N(0, 1)$. The model in Eq. (1.28) is a Gaussian ARIMA(1,2,0) model. The series is shown in Figure 1.24. We then consider a linear regression

$$y_t = \boldsymbol{x}_t'\boldsymbol{\beta} + e_t, \quad t = 11, \ldots, 2000,$$

where the design vector is given by

$$\boldsymbol{x}_t = (x_{t-1}, x_{t-2}, \ldots, x_{t-10}, z_{1t}, \ldots, z_{10,t})',$$

where $z_{it} \sim_{iid} N(0,1)$ for all i and t. Thus, we have a linear regression model with $p = 20$ and $T = 1990$. The data generating model is sparse with three non-zero coefficients. Our goal is to estimate the regression model and to identify the non-zero coefficients. In this particular case, a proper statistical method would identify x_{t-1}, x_{t-2} and x_{t-3} as the only predictors with non-zero coefficients. Figure 1.25 shows the result of the Lasso estimation. The plot is called a coefficient profile plot, which is a scatterplot of standardized coefficient estimates versus a properly scaled norm of the coefficient estimates. The numbers on top of the plot indicate the number of non-zero coefficients. The numbers on the right side of the plot identify the key predictors. Details and use of the coefficient profile plot are given in Chapter 7. It suffices now to note that large non-zero coefficient estimates are shown in the plot. From the plot, it is clear that the Lasso can easily identify x_{t-1} and x_{t-2} as important explanatory variables, but it fails to pin down x_{t-3}. This is not surprising because the AR(3) model in Eq. (1.28) is an $I(2)$ process which has strong serial dependence, leading to the difficulty of multicollinearity in linear regression analysis. In this particular case, the explanatory variables x_{t-1} to x_{t-10} are strongly correlated with sample correlations being close to 1. In addition, estimation of Lasso regression often requires normalization of each column of the design matrix. For an $I(2)$ series, the sample standard deviation grows at a higher order than the sample size. The normalization can easily lead to focusing purely on the unit-root dependence. This, in turn, may overlook the serial dependence of the stationary part of the process. Details can be justified by using the theory of unit-root processes. See, for instance, Tsay (2014) and the references therein. It suffices here to say that Lasso regression may fail for dependent data when the serial dependence is strong. We shall discuss ways to overcome such a difficulty in Chapter 7. Consequently, there is a need to develop statistical methods for analyzing big dependent data. ■

APPENDIX 1.A: SOME MATRIX THEORY

In this appendix, we provide some properties of matrix useful in multivariate statistical analysis. Let $\boldsymbol{A}$ be a $k \times k$ real-valued matrix. The solutions of the determinant equation $|\boldsymbol{A} - \lambda \boldsymbol{I}| = 0$ are the eigenvalues of $\boldsymbol{A}$. A k-dimension vector $\boldsymbol{x} = (x_1, \ldots, x_k)'$ is an eigenvector associated with an eigenvalue λ of $\boldsymbol{A}$ if and only if $\boldsymbol{A}\boldsymbol{x} = \lambda\boldsymbol{x}$. In general, eigenvalues and eigenvectors of $\boldsymbol{A}$ may assume complex values.

The matrix $\boldsymbol{A}$ is positive-definite if (a) $\boldsymbol{A} = \boldsymbol{A}'$, i.e. $\boldsymbol{A}$ is symmetric, and (b) for any non-zero k-dimensional vector $\boldsymbol{x} = (x_1, \ldots, x_k)'$, $\boldsymbol{x}'\boldsymbol{A}\boldsymbol{x} > 0$. Positive-definite matrices play an important role in many statistical applications and it pays to study their properties. It is easy to see that all eigenvalues of a positive-definite matrix $\boldsymbol{A}$ are positive.

Spectral decomposition: Assume that $\boldsymbol{A}$ is a $k \times k$ real-valued positive-definite matrix. Let $\lambda_1 \geq \lambda_2 \geq \cdots \geq \lambda_k$ be the eigenvalues of $\boldsymbol{A}$. Let $\boldsymbol{e}_i$ be an eigenvector of $\boldsymbol{A}$ associated with eigenvalue λ_i such that $\boldsymbol{e}_i'\boldsymbol{e}_1 = 1$. That is the Euclidean norm of $\boldsymbol{e}_i$ is 1. Let $\boldsymbol{P} = [\boldsymbol{e}_1, \ldots, \boldsymbol{e}_k]$ be the $k \times k$ matrix of eigenvectors. Then, we have (1)

$\boldsymbol{AP} = \boldsymbol{P\Lambda}$, where $\boldsymbol{\Lambda} = \text{diag}\{\lambda_1, \ldots, \lambda_k\}$; (2) $\boldsymbol{P}'\boldsymbol{P} = \boldsymbol{PP}' = \boldsymbol{I}$, the k-dimensional identity matrix; (3)

$$\boldsymbol{A} = \boldsymbol{P\Lambda P}' = \sum_{i=1}^{k} \lambda_i \boldsymbol{e}_i \boldsymbol{e}_i'.$$

Property (3) is referred to as the spectral decomposition of the positive-definite matrix $\boldsymbol{A}$. Using the spectral decomposition and properties of eigenvalues and eigenvectors of $\boldsymbol{A}$, we have the following properties.

Property I: Assume that $\boldsymbol{A}$ is a positive-definite $k \times k$ real-valued matrix with spectral decomposition given by eigenvalue-eigenvector pairs $(\lambda_i, \boldsymbol{e}_i)$. Then,

$$\max_{x \neq \mathbf{0}} \frac{\boldsymbol{x}'\boldsymbol{A}\boldsymbol{x}}{\boldsymbol{x}'\boldsymbol{x}} = \lambda_1$$

$$\min_{x \neq \mathbf{0}} \frac{\boldsymbol{x}'\boldsymbol{A}\boldsymbol{x}}{\boldsymbol{x}'\boldsymbol{x}} = \lambda_k,$$

where the maximum and minimum are attended with $\boldsymbol{x} = \boldsymbol{e}_1$ and $\boldsymbol{x} = \boldsymbol{e}_k$, respectively. Furthermore,

$$\max_{\boldsymbol{x} \neq \mathbf{0}, \boldsymbol{x}'\boldsymbol{e}_j = 0; j=1,\ldots,i} \frac{\boldsymbol{x}'\boldsymbol{A}\boldsymbol{x}}{\boldsymbol{x}'\boldsymbol{x}} = \lambda_{i+1}$$

where the maximum is attended when $\boldsymbol{x} = \boldsymbol{e}_{i+1}$, where $i = 2, \ldots, k$.

EXERCISES

1. Consider the 99 world financial market indexes. Compute the log returns of the indexes. Obtain a time plot of all series and perform a PCA of the log returns. Summarize the results of PCA, including scree plot and time plots of the first six PCs. [An R command for PCA is `princomp`].
2. Consider the clothing data set of Figure 1.8. Perform PCA on the sales data and summarize the results. Obtain time plots of the first 12 PCs with 6 series on one page.
3. Consider, again, the clothing data set. Obtain the three summary plots of the sample cross-correlations for lags 1 to 21.
4. Consider the temperature data of Figure 1.1. (a) Obtain the sample mean and sample covariance matrix of the data. (b) Obtain the lag-1 to lag-10 sample CCMs of the data.
5. Consider the hourly $PM_{2.5}$ measurements at 15 monitoring stations in the southern Taiwan; columns 4 to 18 of the file `TaiwanPM25.csv`. (a) Compute the sample mean and sample covariance matrix of the 15 time series. (b) Obtain a time plot of the 15 time series.
6. Prove Eq. (1.16) and discuss the increase in the variance of sample mean with respect to the independent case when the series has a non-zero first-order autocorrelation coefficient and zero autocorrelations for all higher lags. That is, the series follows an MA(1) model.

REFERENCES

Bühlmann, P. and van de Geer, S. (2011). *Statistics for High-Dimensional Data*. Springer, New York, NY.

Chang, J., Guo, B., and Yao, Q. (2014). Segmenting multiple time series by contemporaneous linear transformation: PCA for time series. Working paper, London School of Economics.

Dahlhaus, R. (2012). Locally stationary processes. In *Handbook of statistics*, **30**: 351–413.

Peña, D. and Box, G. (1987). Identifying a simplifying structure in time series. *Journal of the American Statistical Association*, **82**: 836–843.

Peña, D. and Rodriguez, J. (2003). Descriptive measures of multivariate scatter and linear dependence. *Journal of Multivariate Analysis*, **85**: 361–374.

Reinsel, G. (1993). *Elements of Multivariate Time Series Analysis*. Springer-Verlag, New York, NY.

Taniguchi, M. and Krishnaiah, P. (1987). Asymptotic distributions of functions of the eigenvalues of sample covariance matrix and canonical correlation matrix in multivariate time series. *Journal of Multivariate Analysis*, **22**: 156–176.

Tiao, G. C. and Box, G. E. P. (1981). Modeling multivariate time series with applications. *Journal of American Statistical Association*, **72**: 802–816.

Tibshirani, R. (1996). Regression shrinkage and selection via Lasso. *Journal of the Royal Statistical Society, Series B*, **58**: 267–288.

Tsay, R. S. (2014). *Multivariate Time Series Analysis with R and Financial Applications*. John Wiley & Sons, Hoboken, NJ.

Zhang, X. and Tong, H. (2020). Some cautionary comments on PCA for time series data. Working paper, London School of Economics and Political Science, UK.

CHAPTER 2

LINEAR UNIVARIATE TIME SERIES

In this chapter, we study methods for modeling and forecasting univariate time series. Our study is brief and focuses on the automatic analysis of large sets of time series. Interested readers can consult any of the many time series textbooks for further details. See, for instance, Peña et al. (2001), Box et al. (2015), Brockwell and Davis (2013), Cryer and Chan (2008), Shumway and Stoffer (2017), and Tsay (2010, 2014), among others.

Given a set of time series, the first step of the analysis is to visualize the data by plotting the series in order to understand their basic properties. One would also try to detect gross measurement errors that are quite common with data recorded in an automatic way. These errors may happen at some specific times in many series, or be concentrated in a few series making them different from the others. The graphical representations we present are designed to identify both cases of measurement errors. Once the set of time series has gone through some cleaning, we present statistical methods to model and forecast the series. The models most commonly used for stationary time series are the autoregressive moving-average (ARMA) models, that have proved to be useful in many scientific fields. Particular members of this family are the autoregressive (AR) models, where the series depends only on its first p lagged values with p being a positive integer. AR models are characterized by having infinitely many non-zero autocorrelation coefficients which decay exponentially to zero as the lag increases. Thus, AR processes have a relatively long memory, since the current value of a series is correlated with many previous ones, although with vanishing coefficients. Scalar processes with finite-memory property are the moving-average (MA) models, that are formed by weighted averages of a finite number of past innovations (or noises). Combining the AR and MA structures

Statistical Learning for Big Dependent Data, First Edition. Daniel Peña and Ruey S. Tsay.
© 2021 John Wiley & Sons, Inc. Published 2021 by John Wiley & Sons, Inc.

together, we obtain the ARMA models, that provide a broad and flexible family of stationary stochastic processes useful in representing many empirical time series.

Some stationary time series can be thought of as a mixture of several cyclical processes. This is the motivation for the spectral analysis in which the variability of the series is explained by harmonic processes with different frequencies and variabilities. The spectrum of a stationary time series is a linear combination of its autocovariances of all lags, and each ARMA process implies a given spectrum and vice versa. This approach, referred to as the frequency domain analysis, provides complementary information to the time domain analysis using ARMA models.

A time series can be nonstationary in the mean, the variance, the autocorrelations, or in other characteristics of its distribution. As mentioned in Chapter 1, the most important nonstationary processes are the integrated processes, especially the I(1) processes. The basic feature of an I(1) process is that its mean changes over time, but its increments form a stationary process. Therefore, the series can become a stationary one by differencing. The family of autoregressive integrated moving-average (ARIMA) models applies to time series that become stationary ARMA process by differencing. An important property, that distinguishes integrated processes from stationary ones, is the way by which the serial dependence behaves. For stationary processes the autocorrelations decay to zero exponentially with the increase in lag, whereas those of an integrated time series do not converge to zero as the lag increases.

Another important class of nonstationary processes consists of seasonal processes, for which the mean of the series is not constant, but exhibits a cyclical pattern. For example, the series of monthly temperature at a geographical location often shows strong seasonality with higher mean values in the summer. If this seasonal pattern is persistent, then one would need a seasonal difference to transform the series into stationarity. ARMA models with seasonal differencing are called seasonal ARIMA (SARIMA) models.

An alternative way to model nonstationary time series is to treat it as an additive model of trend, seasonality, and irregular components. This approach is called the *structural time series modeling*. It is a special case of a general procedure to describe the evolution of a dynamic system, called the *state space representation*. One advantage of the state-space model (SSM) is that it provides an effective recursive way for estimating the model parameters and generating forecasts. Also, this model generalizes straightforwardly to multivariate time series, which we discuss in Chapter 3.

The second part of this chapter studies how to build ARIMA models for a large set of time series, with emphasis in automatic procedures. The fitting of a model to each time series is carried out in three steps. In the first step, we examine whether a transformation (power family) is required for the series and then classify the resulting series into one of the following categories: stationary, unit-root nonstationary, or seasonal series. This step can be regarded as a *classification step*. The second step is *model selection* in which multiple models are estimated and an information criterion is used to select the best one among the entertained models. In the third step, *diagnostic checking* is performed and the fitted model is refined, if necessary. The selected model for each series can then be used for forecasting. Keep in mind that in some applications several models may fit a given time series well and one can then use model average to produce combined forecasts.

2.1 VISUALIZING A LARGE SET OF TIME SERIES

The visualization of a large set of time series is important to detect gross measurement errors, to identify series that are different from the others, to show the existence of possible grouping, and to understand the overall structure of the series under study. Putting the k time series with T observations into a two-way table, with each column containing an individual time series and each row all the series at a time point, we focus on two types of plot that show properties of rows and columns, respectively. Specifically, they exhibit properties of (1) All of the series jointly over time, plotting some properties of each row over time; (2) Each series in the complete period, plotting some properties of the marginal distribution of each series (column).

The first type of plots is called *dynamic plots*, because they show the evolution over time of selected summaries of all series. The second type of plots is referred to as *static plots*, because they show some summaries of the behavior of each series (columns) over the whole observation period. When the order of the columns (series) is non-informative, as it is usually the case, we can make scatterplots of the selected summaries of the series or boxplots of selected quantiles of the distribution of the columns (series). If a meaningful ordering of the columns can be found, for instance with spatial data discussed in Chapter 9, then we can also plot some selected summaries of the rows with respect to this order. In what follows, we briefly explain and demonstrate these two types of plots.

2.1.1 Dynamic Plots

A useful way to summarize the distribution of a set of data is to use the quantiles. Given a random sample $\{x_i\}_{i=1}^k$ of a scalar random variable X with empirical cumulative distribution function (CDF) $\widehat{F}_k(x)$, the empirical pth quantile $q^{*(p)}$ is defined as

$$q^{*(p)} = \inf_{x \in R} \{x | \widehat{F}_k(x) \geq p\}, \tag{2.1}$$

and it is well-known, see for instance Ferguson (1967), that this quantile can be computed by

$$q^{*(p)} = \arg\min_{y \in R} \left[p \sum_{x_i \geq y} |x_i - y| + (1-p) \sum_{x_i < y} |x_i - y| \right]. \tag{2.2}$$

Consider a large data set of k-dimensional time series $\{\boldsymbol{z}_t | t = 1, \ldots, T\}$. If $\boldsymbol{z}_t$ is strictly stationary, then the distribution of $\boldsymbol{z}_t$ is time invariant. We can compute at every time t some selected empirical quantiles of the observations at this point, and form with these values a time series of timewise quantiles (TWQ). The quantile at each point $t = v$ is computed by (2.2), where the values x_i are now the $z_{i,v}$. When the series are stationary these quantiles should be similar over time.

Consequently, by plotting the time series formed by selected timewise quantiles one can examine the stationarity of the underlying time series.

To demonstrate, consider the daily log prices of the world financial market indexes employed in Chapter 1. Figure 2.1 shows the time plots of timewise quantiles with probabilities 0.025, 0.5, and, 0.975, respectively. That is, the plot shows the series of 2.5% quantiles, median, and 97.5% quantiles of each row of the two-way data table.

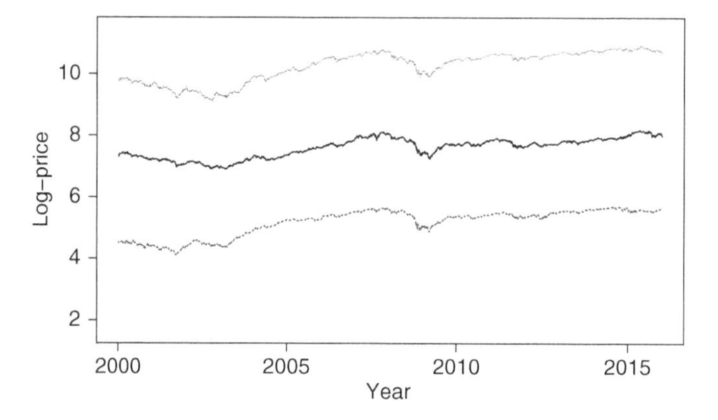

Figure 2.1 Timewise quantiles of daily log prices of the 99 world financial market indexes from 4 January 2000 to 16 December 2015. The quantiles are 2.5, 50, and 97.5% quantiles of returns for each trading day.

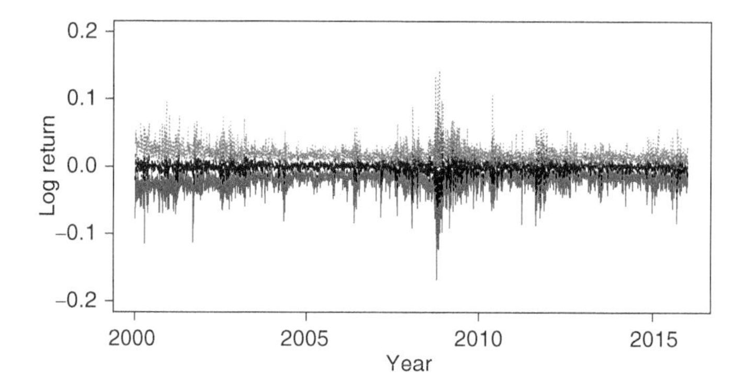

Figure 2.2 Timewise quantiles of daily log returns of the 99 world financial market indexes from 4 January 2000 to 16 December 2015. The quantiles are 2.5, 50, and 97.5% quantiles of returns for each trading day.

In the plot, the black (solid) line is the time series plot of daily median log returns, the red (dashed) line is the time plot of 2.5% quantiles, and the blue (dotted) line is the time plot of 97.5% quantiles. These quantiles show that the series are, as expected, nonstationary, because all three quantiles show an upward trend. Figure 2.2 shows the time plots of the corresponding quantiles for the return series, that is, the first log difference of the WSI series. From the plots, it is easily seen that (a) the period of 2008 financial crisis has the highest variability (volatility) in the world financial markets, and (b) the inter-quantile ranges [97.5–2.5%] of the world financial markets appear to be larger in early 2000 than the recent period, indicating that the financial markets around the world are more integrated than ever before. All these changes indicate that the return series are also nonstationary over the whole period considered, although in some subperiods they appear to be stationary.

These timewise quantile plots convey the visual impression about the stationarity of the data. However, they do not represent well the dynamic dependence of the data, because their dependency need not be the same as that of the individual time series.

For instance, a set of stationary time series may have quantiles that form almost parallel lines even though the individual series may have strong autocorrelations. For series with weak dynamics, as the series of asset returns that have only a few lags of serial correlations, this is not a limitation, but for series with marked autocorrelation structure their dynamic may not be shown by the timewise quantiles. To mitigate this problem, Peña et al. (2019) propose a dynamic quantile plot designed to preserve the dynamic dependence of the series.

Suppose that the set of time series are generated by a multivariate stochastic process, i.e. we have k possibly dependent and different scalar stochastic processes. Focusing on a given time point $t = v$, we have a vector of data $\boldsymbol{z}_v = (z_{1v}, \dots, z_{kv})'$ from an k-dimensional random vector $(Z_{1v}, \dots, Z_{kv})'$ with distribution $\boldsymbol{F}_v(\boldsymbol{z})$ that represents a sample from some stochastic process depending on two indexes. For instance, it could be a spatio-temporal process or any other process with one index being time and the other a set of well-defined elements. It can be shown (see Peña et al., 2019, and the references therein) that the empirical CDF of $\boldsymbol{z}_v$ converges to a univariate limiting distribution under some general conditions (for instance, stationary processes) as k increases to infinity. Assuming that the set of time series verifies these conditions, the marginal distribution function $\boldsymbol{F}_v(\boldsymbol{z})$ converges to a well-defined distribution function for every time point v as k increases to infinity.

For ease in reference, let $\mathbb{C}_k = \{z_{it} | 1 \leq i \leq k,\ 1 \leq t \leq T\}$ be the observed set of time series. Based on the aforementioned properties of the empirical CDF of $\boldsymbol{z}_t$, we define the pth empirical dynamic quantile (EDQ) as the series $\{q_t^{(p)}\}$ in $\mathbb{C}_k$ that satisfies the optimization problem:

$$\{q_t^{(p)}\} = \text{argmin}_{\{y_t\} \in \mathbb{C}_k} \left[\sum_{t=1}^{T} \left(\sum_{z_{it} \geq y_t} p|z_{it} - y_t| + \sum_{z_{it} \leq y_t} (1-p)|z_{it} - y_t| \right) \right]. \tag{2.3}$$

Note that this optimization problem is different from the one solved in (2.2) to find the timewise quantiles. The TWQ are computed by solving the optimization problem at every time point, whereas the EDQ are found solving it over the whole period. In the TWQ the solution obtained at every time point can be any real number, whereas the EDQ time series that solves the problem is constrained to be one of the observed time series. The direct computation of the EDQ requires $O(k^2)$ operations to calculate the distances between two series and $k \log(k)$ sorting operations. For moderate k, let say no more than 1000 time series, they can be computed directly in a few minutes with nowadays desktop computers. However, for large k, the direct estimation can be slow. Peña et al. (2019) proposed an algorithm that gives a good approximation to the dynamic quantiles with $O(k)$ computations. Thus, it can be applied to a collection of high-dimensional time series.

To demonstrate the EDQ, we consider a data set of hourly $PM_{2.5}$ measurements taken in March 2017 in Taiwan. Those measurements (in micrograms per cubic meters, $\mu g/m^2$) are obtained by a new device, called AirBox,[1] which is portable and widely available. The data set contains 516 series each with 744 observations (i.e. $k = 516$ and $T = 744$). Figures 2.3 and 2.4 show the timewise and EDQs of the $PM_{2.5}$ with probabilities 0.025, 0.5, 0.975. It can be seen that the results are different. The EDQ shows the real dynamic of the series that is stronger than the one shown in the TWQ.

[1] https://en.smartcity.org.tw

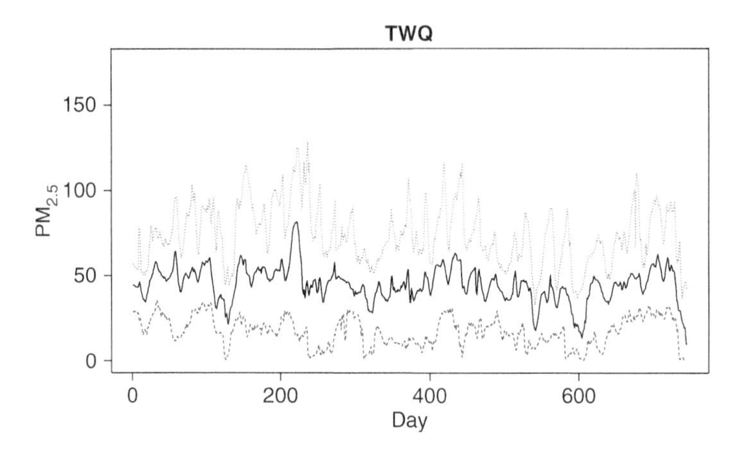

Figure 2.3 Timewise quantile plot for the AirBox $PM_{2.5}$ data. The TWQ 0.025 (red), 0.5 (black), and 0.975 (blue) are shown.

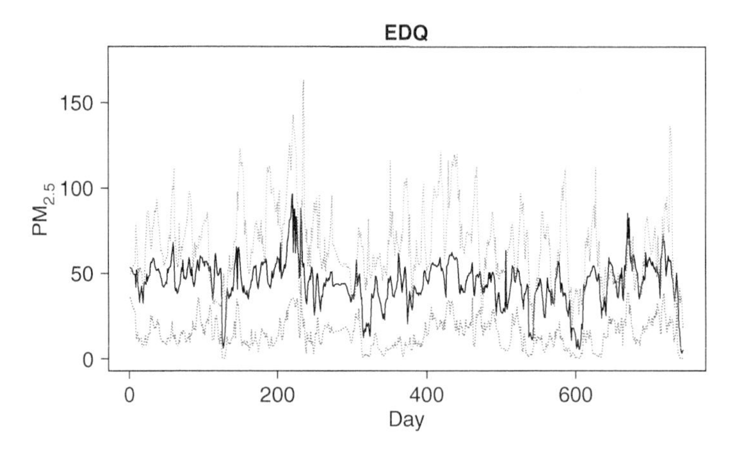

Figure 2.4 EDQ plot for the AirBox $PM_{2.5}$ data. The EDQ 0.025 (red), 0.5 (black), and 0.975 (blue) are shown.

Instead of using quantiles of the marginal distribution of all the series at each time point, we can compute other summary statistics such as the mean and the standard deviation, and plot them in a bivariate plot, or over time. Figure 2.5 shows the scatterplot of the mean and variance of the return series of world financial indexes for each trading day. From the plot, it is easy to see that there are three days in which the stock returns have very high variances. These three days are marked in the plot and they are 1 and 3 January 2001 and 15 October 2008. The plot also indicates that there was a particular day on which the volatility of the financial indexes was low, yet the stock dropped more than 6% (the far-left point). It turns out that this occurred on 6 October 2008, the beginning of the world financial crisis. These findings seem to be reasonable in view of (a) the drops in US market in December 2000 and the rally of US market in January 2001, and (b) the 2008 world financial crisis. Overall, the scatterplot indicates that the variances of daily log returns of the 99 financial indexes

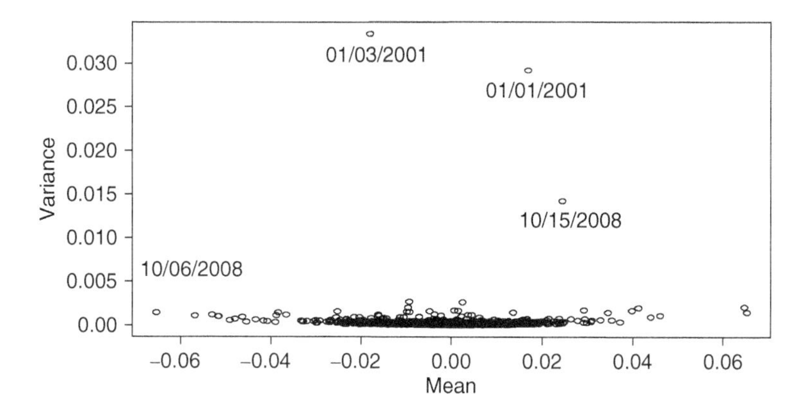

Figure 2.5 Scatterplot of cross-sectional variances versus means for the daily log returns of 99 world financial market indexes. The time period is from 4 January 2000 to 16 December 2015.

are not large, but the range of the means of daily long returns could be substantial, ranging from −0.06 to 0.06. We conclude that the plots of summary statistics of the rows provide useful information concerning the joint evolution of the k time series under study over time.

R commands and packages used:

```
> require(MTS)
> stock <- read.csv2("Stock_indexes_99world.csv",header=FALSE)
> stock <- stock[,-1]  # remove observation index
> stock <- as.matrix(stock)# #plot twq stock return
> rtn <- diffM(log(stock))
> q <- apply(rtn,1,quantile,c(0.025,0.5,0.975))
> tdx <- c(1:4162)/260+2000
> plot(tdx,q[1,],type="l",xlab="year",ylab="log-return",
                                          ylim=c(-0.2,0.2),col="red")
> lines(tdx,q[2,],col="black",lty=2)
> lines(tdx,q[3,],col="blue",lty=3)
#plot twq  logstock
>x=log(stock)
>tdx <- c(1:4163)/260+2000
>twq <-apply(x,1,quantile,c(0.025,0.5,0.975))
plot(tdx,twq[1,],main="TWQ",type="l",xlab="year",ylab="log-price",
                                          ylim=c(min(x),max(x)),col="red")
lines(tdx,twq[2,],col="black",lty=2)
lines(tdx,twq[3,],col="blue",lty=3)
#plot edq  logstock
>q50=edqts(x,p=0.5,h=30)
>q025=edqts(x,p=0.025,h=30)
>q975=edqts(x,p=0.975,h=30)
>plot(tdx,x[,q50],main="EDQ",type="l",xlab="year",ylab="log-price",
      ylim=c(min(x),max(x)),col="black")
>lines(tdx,x[,q025],col="red",lty=3)
>lines(tdx,x[,q975],col="blue",lty=3)
> ave <- apply(rtn,1,mean)
```

```
> v <- apply(rtn,1,var)
> plot(ave,v,xlab="mean",ylab="variance")
> text(locator(1),"10/15/2008")
> text(locator(1),"01/03/2001")
> text(locator(1),"01/01/2001")
> text(locator(1),"10/06/2008")
```

2.1.2 Static Plots

Static plots are graphical tools to summarize the behavior over the sample period of each of the k time series under study. In these plots, we examine some properties (or summaries) of the marginal distribution of each column and plot those summaries together for all series. We refer to such plots as *static plots*, because they show a summary of each time series over the observed period and thus the time dynamics is not revealed. The first plot we propose here is to make traditional boxplots of selected quantiles of the distribution of each column (series), including the minimum and maximum of each series. This plot is referred to as a *quantile-box* plot.

Figure 2.6 shows the quantile-box plot of the AirBox measurements of $PM_{2.5}$. The selected quantiles are minimum, 25, 50, 75% quantiles, and the maximum of each series. From the plot, it is clear that two series of the $PM_{2.5}$ measurements have median values close to zero. As a matter of fact, there exist two low points at 25, 50, and 75% quantiles. Since $PM_{2.5}$ measurements cannot assume any negative value, this plot immediately identifies two potentially outlying series in the data set. Figure 2.7 shows the time plots of three series, which have the smallest sample variances. The series are 29th, 70th, and 348th. From the time plots, it is easily seen that Series 29 and 70 assume lots of zero measurements, indicating that these two series have more than 50% missing values and, hence, should be discarded from further modeling.

Two important summaries of each observed time series are its central location and variability, and we recommend a scatterplot of the standard deviation (or variance) versus mean of each series. This plot is capable of identifying series that have higher variability or unusual sample means. In other words, the plot can spot certain heterogeneity in the data set. Figure 2.8 shows the scatterplot of sample variances versus sample means of the 99 daily log returns of world financial market indexes. From the

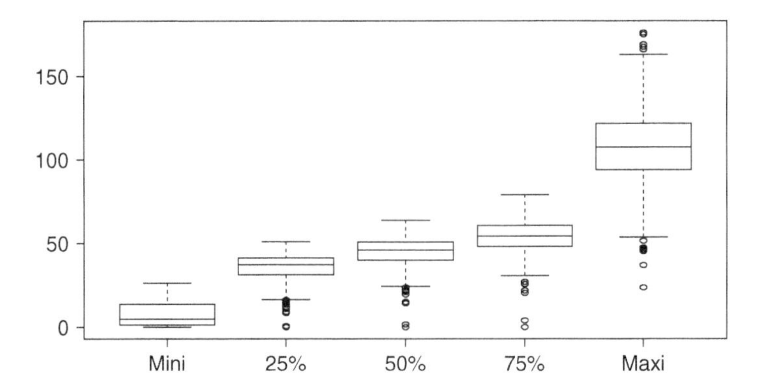

Figure 2.6 Quantile-box plot for the AirBox measurements of $PM_{2.5}$ data set. The quantiles selected are the minimum, 25, 50, 75% quantiles, and the maximum.

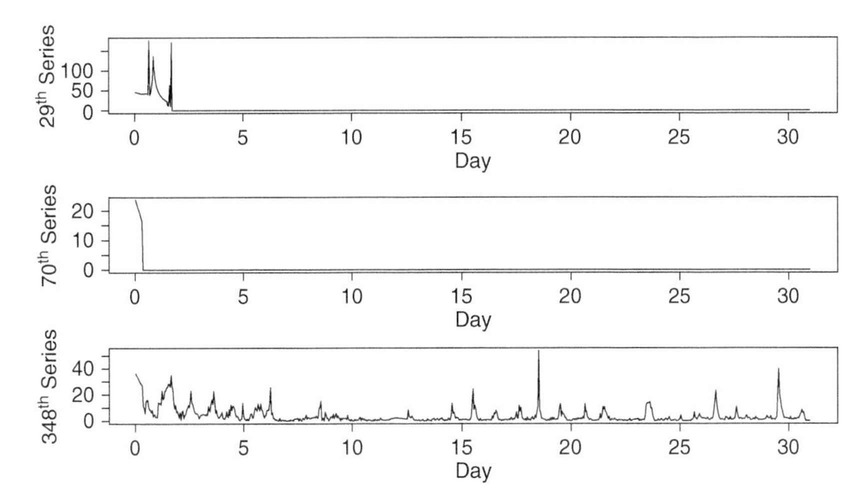

Figure 2.7 Time plots of three extreme series of the hourly AirBox $PM_{2.5}$ measurements in Taiwan, in March 2017. The x-axis denotes day in March.

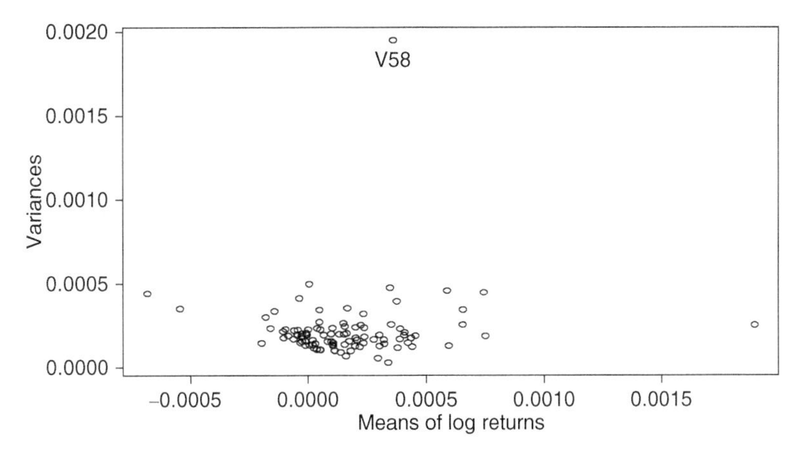

Figure 2.8 Sample variances versus sample means of 99 daily log returns of world financial market indexes from January 2000 to December 2015.

plot, an isolated point with relatively large variance is easily seen. It turns out that it is Series 58 (which corresponds to Russia Stock Market). Also, there is a series with a very large mean return (series 85, Venezuela Stock Market) and two series with very low return (56 and 79, Netherlands and Greece Stock Market). Figure 2.9 shows the time plot of the series 58 and 85. From the plot, it is clear that series 58 contains two huge outliers with daily returns exceed 150 percentage in absolute value. This example demonstrates the need to clean the data when analyzing large data sets of time series. Series 85 shows also high variability around the 2008 world financial crisis and in the last few years.

The third static plot we propose is the scatterplot of lag-1 sample autocorrelation coefficients versus lag-2 sample autocorrelation coefficients for all the series. This

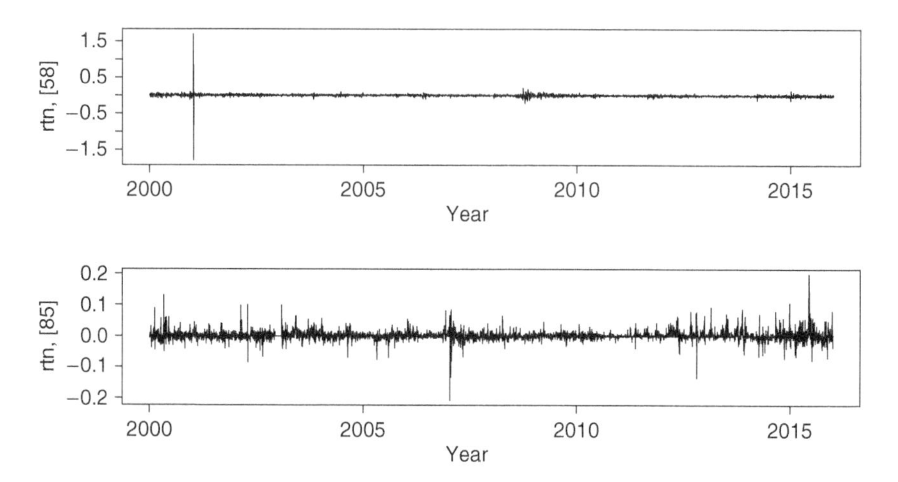

Figure 2.9 Daily log returns of two series of the financial market indexes. Series 58 (Russia), upper figure, and 85 (Venezuela), lower figure.

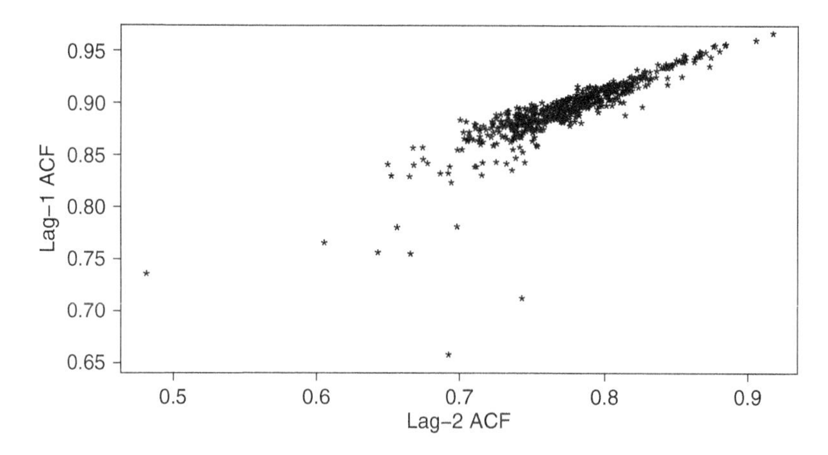

Figure 2.10 Scatterplot of the lag-1 versus lag-2 sample autocorrelation coefficients of the hourly AirBox $PM_{2.5}$ measurements in Taiwan, in March 2017.

simple plot can provide information concerning the dynamic dependence of individual series. If the series share a similar dynamic dependence, then the scatterplot should show a cluster of points along a straight line, otherwise, different types of dependency clusters of points will be found. Figure 2.10 shows the scatterplot of lag-1 ACF versus lag-2 ACF of the AirBox hourly $PM_{2.5}$ series. From the plot, we make the following observations. First, as expected, there exists certain clustering behavior along a straight line. Second, there exists a cluster of 5 points with lag-1 ACF around 0.78 and lag-2 ACF around 0.68. Finally, the plot also points out 3 isolated points with a different pattern of dynamic dependence. It turns out that the minimum lag-1 ACF occurs in Series 29 whereas the two smallest lag-2 ACFs occur at Series 19 and 213. The identification of these isolated series is given in the attached R output.

It is interesting to see the Series 70, which only has eight non-zero observations, does not show up as an isolated point in Figure 2.10. This is likely due to the fact that the eight observations share similar dynamic dependence with other series. Of course, Series 70 would become an isolated point if we consider scatterplot of lag-1 ACF versus lag-ℓ ACF with $\ell > 7$.

R demonstration of cross-section plots:

```
### Cross section plots
air <- read.csv('TaiwanAirBox032017.csv',header=FALSE)
air <- as.matrix(air)[,-1]
x <- air
### Quantile Box plot
require(SLBDD)
quantileBox(air)
q50 <- apply(air,2,quantile,0.5)
s1 <- sort(q50,index.return=TRUE)
s1$ix[1:3]

stock <- read.csv2('Stock_indexes_99world.csv')
stock <- stock[,-1]
stock <- as.matrix(stock)
rtn <- diffM(log(stock))
ave <- apply(rtn,2,mean)
v1 <- apply(rtn,2,var)
plot(ave,v1,xlab=''Sample means'',ylab=''Sample variance'')
which.max(v1)
tdx <- c(1:4163)/260+2000
plot(tdx[-1],rtn[,58],xlab=''year'',ylab=''log-return'',
         main=''Series 58'',type=''l'')
## scatterplot of ACFs
scatterACF(air)
```

Example 2.1

We analyze the electricity price data in the file `PElectricity1344.csv`, that contains the hourly electricity prices in the ISO New England electricity market from January 2004 to December 2016. The data are weekly series of the electricity price each hour of each day during T =678 weeks, in the eight regions in New England. We have k =1344 series, corresponding to each one of the eight regions, for one of the 24 hours of the day, and one of the seven days of the week ($8 \times 24 \times 7$=1344). The first series corresponds to the price in the first region at 1 a.m. Eastern Time of Thursday, 1 January 2004, the second to the price at 2 a.m. in the same region on the same day, and so on. Thus, the first 192 series (8 regions $\times$ 24 hours) are the electricity prices of all the hours of Thursday for the eight regions, the next 192 series are the prices for Friday, and so on. These time series were analyzed and corrected for the missing values at days of day-light saving time in Alonso and Peña (2019). We use corrected

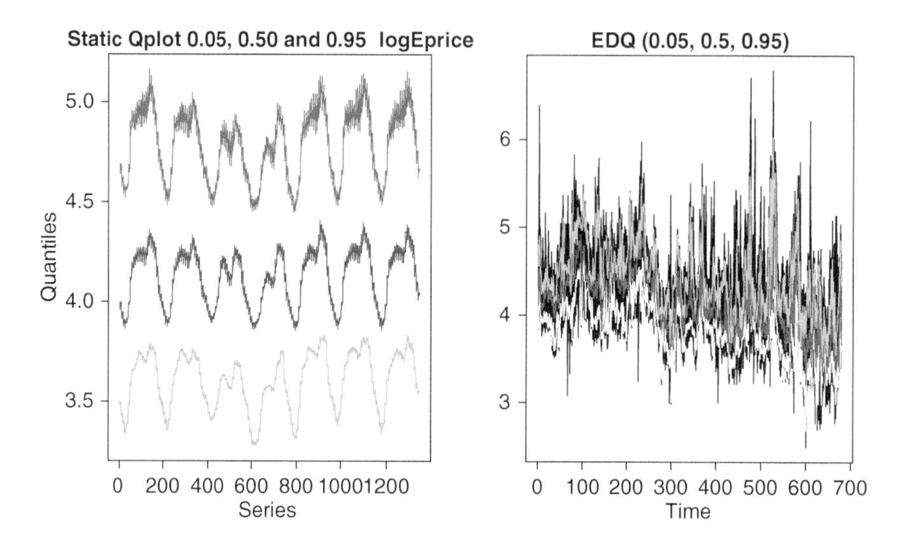

Figure 2.11 Static quantiles of each time series (left panel) and the EDQs with probabilities 0.05, 0.5, and 0.95 (right panel) for the hourly electricity price in New England.

data in this example. Still a few values of the series have negative values and we have added 15 to all the series in order to take the log transformation. The plots are for the transformed data.

The left plot of Figure 2.11 shows the static quantiles of each time series plotted with respect to the number of the series. As in these data the series are ordered by blocks of the same day of the week, the plot shows seven cycles indicating the evolution of the quantiles of the electricity price for each hour and region. The right plot shows three EDQ of the prices over the 678 weeks. They have a decreasing trend and a variability that seems to be increasing over time. The R commands for this figure are given below.

```
x0 <- read.csv("PElectricity1344.csv", header=FALSE)
 x=as.matrix(x0)
> x = log(x+15)
> sq = apply(x,2,quantile,c(0.05,0.50,0.95))
## sq is a 3-by-1344 matrix. The first row is the 0.25 quantile
## of the first time series. The second row is the 0.5 quantile
## of the second time series, etc.
> par(mfcol=c(1,2))
#Plot of sq in the original order of the data
>  q05=sq[1,]
>  q50=sq[2,]
>  q95=sq[3,]
>  Ma=max(q05,q50,q95)
>  Mi=min(q05,q50,q95)
> ts.plot(q50,col=2,main="Static Qplot .05,.50 and .95 logEprice",
          ylim=c(Mi,Ma),xlab="series",ylab="Quantiles")
> lines(q05,col=3)
> lines(q95,col=4)
```

```
> #Dynamic Plots of the series
> sal=edqplot(x,prob=c(0.05,0.5,0.95),h=30,loc=NULL,
                color=c("yellow","red","green","blue"))
```

■

2.2 STATIONARY ARMA MODELS

Stationary time series are useful and widely used in practice, and many nonstationary time series can be transformed to stationary ones via differencing. The general representation of a stationary process was obtained by Wold (1938). He proved that any weakly stationary stochastic process, z_t, with a finite mean that does not contain any deterministic component, can be written as a linear function of uncorrelated random variables $\{a_t\}$. That is,

$$z_t = \mu + a_t + \psi_1 a_{t-1} + \psi_2 a_{t-2} + \cdots = \mu + \sum_{i=0}^{\infty} \psi_i a_{t-i} \qquad (\psi_0 = 1), \tag{2.4}$$

where $E(z_t) = \mu$, $E(a_t) = 0$, $\text{Var}(a_t) = \sigma^2$, and $E(a_t a_{t-h}) = 0$, for $h > 0$. Letting $\tilde{z}_t = z_t - \mu$, and using the lag operator B, defined as $Bz_t \equiv z_{t-1}$ and, therefore, $B^h z_t = z_{t-h}$, we can write Eq. (2.4) as

$$\tilde{z}_t = \psi(B) a_t, \tag{2.5}$$

with $\psi(B) = 1 + \psi_1 B + \psi_2 B^2 + \cdots$ being an infinite polynomial in the lag operator B such that $\sum_{i=0}^{\infty} \psi_i^2 < \infty$. We denote Eq. (2.5) as the general linear representation of a nondeterministic stationary process. It is often referred to as the *Wold decomposition* of a stationary stochastic process. The representation is important, because it guarantees that any stationary process can be approximated by a linear function of certain white noise process a_t and all its lags. In this chapter, we search for adequate approximations to this general representation, but using a small number of unknown parameters. In certain specific cases, the process z_t can be written as a linear function of normal independent variables $\{a_t\}$. Then the variable $\tilde{z}_t$ assumes a normal distribution, and weak stationarity and strict stationarity coincide.

The representation in Eq. (2.5) has the problem of requiring an infinite number of coefficients. In the next sections, we study particular cases with a finite number of parameters, namely, the AR and the MA models.

The variance of z_t in Eq. (2.4) is

$$\text{Var}(z_t) = \gamma_0 = \sigma^2 \sum_{i=0}^{\infty} \psi_i^2, \tag{2.6}$$

and the process has a finite variance, because we assume that the series $\{\psi_i^2\}$ is convergent. The autocovariances of z_t are obtained by

$$\gamma_h = E(\tilde{z}_t \tilde{z}_{t-h}) = \sigma^2 \sum_{i=0}^{\infty} \psi_i \psi_{i+h},$$

and the autocorrelation coefficients are given by

$$\rho_h = \frac{\sum_{i=0}^{\infty} \psi_i \psi_{i+h}}{\sum_{i=0}^{\infty} \psi_i^2}. \tag{2.7}$$

This autocorrelation coefficient of z_t depends only on h. In fact, properties of a stationary process are described by its autocorrelation function (ACF). We discuss some features of ACF for some selected processes later.

It was proved by Koopmans (1974) that any stationary process also admits an autoregressive representation of infinite order provided that it is invertible. This condition is general and will be explained later. The AR representation is the inverse of the Wold decomposition, and is given by

$$\tilde{z}_t = \pi_1 \tilde{z}_{t-1} + \pi_2 \tilde{z}_{t-2} + \cdots + a_t,$$

which, in the back-shift operator B, is reduced to

$$\pi(B)\tilde{z}_t = a_t, \tag{2.8}$$

where $\pi(B) = 1 - \pi_1 B - \pi_2 B^2 - \cdots$. This is the dual representation of Eq. (2.5). Inserting the last equation in (2.5), we have

$$\pi(B)\psi(B)a_t = a_t,$$

which implies that

$$\pi(B)\psi(B) = 1, \tag{2.9}$$

and operator $\psi(B)$ is the inverse of $\pi(B)$ and vice versa. Therefore, we write $\pi^{-1}(B) = \psi(B)$. In general, by equating coefficients of the powers of B in the product of Eq. (2.9) to zero, we can obtain the coefficients of one representation from those of another. In the next sections, we present some particular cases of the general representations in Eqs. (2.5) and (2.8).

2.2.1 The Autoregressive Process

Suppose that z_t is a time series representing the amount of water in a reservoir in a given day t. The desired stable objective with respect to the amount of water to maintain is μ and when $z_t > \mu$ some fraction $1 - \phi_1$ of the deviation from μ, namely $\tilde{z}_t = z_t - \mu$, is released and some fraction of the deviation, $\phi_1 \tilde{z}_{t-1}$, is kept, where $|\phi_1| < 1$. On the other hand, when $z_t < \mu$ some fraction, $1 - \phi_1$, of the deviation $\tilde{z}_t = z_t - \mu$ is added from other connected reservoir. In addition, some additional water may be added to the reservoir every day, due to possible rain, or may disappear, because water simply evaporates. Denoting the random amount by a_t, and assuming that has a zero expected value, that is, an average null effect in the long run, the amount of daily water in the reservoir would follow the equation

$$\tilde{z}_t = \phi_1 \tilde{z}_{t-1} + a_t. \tag{2.10}$$

In this case, z_t follows a stationary AR model of order 1, i.e. AR(1), with mean μ, parameter $|\phi_1| < 1$, and innovations a_t that follow a white noise process with zero mean and finite variance σ^2. Using the lag operator B, the equation of an AR(1) model is

$$(1 - \phi_1 B)\tilde{z}_t = a_t.$$

By either repeated substitutions or inverting the operator $(1 - \phi_1 B)$ with

$$(1 - \phi_1 B)^{-1} = 1 + \phi_1 B + \phi_1^2 B^2 + \cdots, \tag{2.11}$$

the process can be written as

$$\tilde{z}_t = a_t + \sum_{i=1}^{\infty} \phi_1^i a_{t-i}. \tag{2.12}$$

In the water reservoir example, if the amount of water is determined by a policy that depends on the deviations of the previous p periods, we have an AR(p) model with the general equation

$$\tilde{z}_t = \phi_1 \tilde{z}_{t-1} + \cdots + \phi_p \tilde{z}_{t-p} + a_t. \tag{2.13}$$

The coefficients ϕ_i must satisfy certain conditions, explained next, for the process to be stationary. Using the lag operator, B, the equation for an AR(p) model is

$$(1 - \phi_1 B - \cdots - \phi_p B^p)\tilde{z}_t = a_t, \tag{2.14}$$

and letting $\phi_p(B) = 1 - \phi_1 B - \cdots - \phi_p B^p$ be the polynomial of degree p in the lag operator, we have

$$\phi_p(B)\tilde{z}_t = a_t, \tag{2.15}$$

which is the general expression of an autoregressive process. The representation in Eq. (2.8) is called the AR(∞) form of a linear stationary process.

The *characteristic equation* of the AR(p) process is defined as

$$\phi_p(B) = 0, \tag{2.16}$$

considered as a function of B. This characteristic equation has p roots $G_1^{-1}, \ldots, G_p^{-1}$, which can be complex and are generally distinct in practice (see Appendix 2.A), and we have

$$\phi_p(B) = \prod_{i=1}^{p} (1 - G_i B),$$

where the components $(1 - G_i B)$ are the factors of the characteristic equation. It can be proved that the process is stationary if $|G_i| < 1$ for all i. In particular, for an AR(1) model, if $|G_1| = |\phi_1| < 1$, then the process is stationary.

2.2.1.1 *Autocorrelation Functions* Consider an AR(p) process. Multiplying (2.13) by $\tilde{z}_{t-h}$ $(h > 0)$, taking expectations, and dividing the result by the variance of the process, we find that the autocorrelation coefficients of an AR(p) verify the following difference equation:

$$\rho_h = \phi_1 \rho_{h-1} + \cdots + \phi_p \rho_{h-p}, \quad h > 0.$$

Therefore, ACFs of the AR(p) process satisfy the same difference equation as the process itself,

$$\phi_p(B)\rho_h = 0 \qquad h > 0. \tag{2.17}$$

The general solution to the difference equation is (see Appendix 2.A)

$$\rho_h = \sum_{i=1}^{p} A_i G_i^h, \tag{2.18}$$

where A_i are constants to be determined by the initial conditions and G_i are the inverses of the roots of the characteristic equation. For z_t to be stationary, the modulus of G_i must be less than 1, implying that the roots of the characteristic equation (2.16) must be greater than 1 in modulus. To prove this, we observe that the condition $|\rho_h| < 1$ requires that there cannot be any $|G_i|$ greater than 1 in Eq. (2.18), otherwise $|G_i^h|$ will increase without any limit as h increases. Furthermore, for the process to be stationary, there cannot be a root G_i^{-1} equal to 1, since then its component G_i^h would not decrease as h increases and the coefficients ρ_h would not tend to zero for any lag.

It is shown in the Appendix that the ACFs of an AR(p) process is a mixture of exponentials, due to the terms with real roots, and sinusoids, due to the complex conjugates. As a result, their structure can be complex. Consequently, it would be difficult to determine the order of an AR process from examining its ACFs. However, suppose that we fit a family of consecutive autoregressions of increasing order:

$$\begin{aligned}
\tilde{z}_t &= \alpha_{11}\tilde{z}_{t-1} + \eta_{1t}, \\
\tilde{z}_t &= \alpha_{21}\tilde{z}_{t-1} + \alpha_{22}\tilde{z}_{t-2} + \eta_{2t}, \\
\vdots &= \vdots \\
\tilde{z}_t &= \alpha_{k1}\tilde{z}_{t-1} + \cdots + \alpha_{kk}\tilde{z}_{t-k} + \eta_{kt},
\end{aligned}$$

and refer to the sequence of coefficients α_{ii} as the *partial autocorrelation function* (PACF) of z_t. Then, from the definition it is clear that the lag-p PACF α_{pp} should be non-zero for an AR(p) process. Therefore, the largest p for each α_{pp} is non-zero can be used to determine the order of an AR(p) process. The sample version of this PACF property is widely used in practice to identify the order of an AR process.

Consider jointly the AR moment equation in Eq. (2.17) for $h = 1, \ldots, p$. We have p equations for the AR coefficients ϕ_i, $(i = 1, \ldots, p)$. This system of linear equations is referred to as the *Yule–Walker equation* of an AR(p) model, which can be used to solve for ϕ_i once the ACFs are given.

2.2.2 The Moving Average Process

Similar to the way that we truncate the AR(∞) representation in Eq. (2.8) to form a finite-order AR(p) process, we can truncate the MA representation in Eq. (2.5) to obtain a finite-order MA process. The simplest MA process is the MA(1) model given by

$$\tilde{z}_t = (1 - \theta_1 B)a_t,$$

where the minus sign of θ_1 is used for simplicity in order to have the same polynomial operator as that of an AR(1) process. One can use $(1 + \theta B)$ if preferred; see, for instance, the model form used by the `arima` command in R. An MA(1) process is always stationary, but is called invertible if $|\theta_1| < 1$. In this case the MA(1) operator can be inverted, as in (2.11), and we obtain

$$\tilde{z}_t = -\sum_{i=1}^{\infty} \theta_1^i \tilde{z}_{t-i} + a_t, \tag{2.19}$$

which is an AR(∞) model but with coefficients decaying to zero exponentially. Note that if $|\theta_1| \geq 1$, then $|\theta_1^i|$ would not converge to zero as i increases, implying that $\tilde{z}_t$ would depend on its remote past values, which is not realistic. The general, MA(q) process assumes the form

$$\tilde{z}_t = a_t - \theta_1 a_{t-1} - \theta_2 a_{t-2} - \cdots - \theta_q a_{t-q}. \tag{2.20}$$

Using the B operator, we can write the model as

$$\tilde{z}_t = (1 - \theta_1 B - \theta_2 B^2 - \cdots - \theta_q B^q) a_t = \theta_q(B) a_t.$$

An MA(q) process with $q < \infty$ is always stationary, as it is a linear combination of stationary processes. We say that the process is *invertible* if the roots of the operator $\theta_q(B) = 0$ are greater than 1 in modulus.

The autocovariances and autocorrelations of an MA(q) process can be obtained by multiplying Eq. (2.20) by $\tilde{z}_{t-h}$, for $h \geq 0$, and taking expectations. Then, we have

$$\begin{aligned}
\gamma_0 &= (1 + \theta_1^2 + \cdots + \theta_q^2)\sigma^2, \\
\gamma_h &= (-\theta_h + \theta_1\theta_{h+1} + \cdots + \theta_{q-h}\theta_q)\sigma^2, \qquad h = 1, \ldots, q, \\
\gamma_h &= 0, \qquad\qquad\qquad\qquad\qquad\qquad\qquad h > q.
\end{aligned}$$

In particular, this result shows that the lag-q autocovariance $\gamma_q = -\theta_q\sigma^2$, which is non-zero, but all autocovariances are zero if the lag is greater than q. Dividing the covariances by γ_0 and utilizing a more compact notation, the ACFs of an MA(q) model are

$$\rho_h = \frac{\sum_{i=0}^{i=q} \theta_i \theta_{h+i}}{\sum_{i=0}^{i=q} \theta_i^2}, \qquad h = 1, \ldots, q, \tag{2.21}$$

$$\rho_h = 0, \qquad h > q,$$

where $\theta_0 = -1$, and $\theta_h = 0$ for $h > q$. Note that the MA(1) process only has the lag-1 autocorrelation coefficient different from zero and its value is $\rho_1 = -\theta_1/(1 + \theta_1^2)$. With $|\theta_1| < 1$, we have $|\rho_1| < 0..5$.

Thus, the absolute value of the first-order autocorrelation coefficient must be smaller than 0.5 for an MA(1), whereas for an AR(1) it has the standard bound of being smaller than 1.. From the ACF properties of an MA model, one can use sample ACFs to identify its order. Finally, as we have shown in Eq. (2.19), an MA(q) process can be written as an AR(∞) process. It is then clear that PACFs of an MA model are different from zero. Thus, there is a duality between the properties of AR and MA models.

2.2.3 The ARMA Process

Combining properties of AR and MA processes together, we obtain an ARMA process that has proven to be widely applicable and useful in real applications. The simplest ARMA process is the ARMA(1,1) model given by

$$(1 - \phi_1 B)\tilde{z}_t = (1 - \theta_1 B)a_t,$$

where, again, $\tilde{z}_t = z_t - E(z_t)$ is the deviation of z_t from its mean and $\{a_t\}$ is a white noise process with mean zero and variance $\sigma^2 < \infty$. The ARMA(1,1) process is stationary if $|\phi_1| < 1$, and it is invertible if $|\theta_1| < 1$. Also, we assume $\phi_1 \neq \theta_1$ to avoid the cancellation of the AR and MA polynomials.

In general, the ARMA(p, q) process is defined as

$$(1 - \phi_1 B - \cdots - \phi_p B^p)\tilde{z}_t = (1 - \theta_1 B - \cdots - \theta_q B^q)a_t, \tag{2.22}$$

or, in the compact notation with back-shift operator,

$$\phi_p(B)\tilde{z}_t = \theta_q(B)a_t,$$

where we assume that no common roots exist between the AR operator $\phi(B)$ and the MA operator $\theta(B)$. The process z_t is stationary if all of the roots of $\phi_p(B) = 0$ are outside the unit circle, and invertible if all of the roots of $\theta_q(B) = 0$ are outside the unit circle.

The MA representation of z_t can be obtained using

$$\tilde{z}_t = \phi_p(B)^{-1}\theta_q(B)a_t = \psi(B)a_t,$$

where ψ_i can be obtained by equating the coefficients of B^i in the identity $\psi(B)\phi_p(B) = \theta_q(B)$. For instance, an ARMA(1,1) has $\psi(B) = 1 + (\phi_1 - \theta_1)B + \phi_1(\phi_1 - \theta_1)B^2 + \phi_1^2(\phi_1 - \theta_1)B^3 + \cdots$. Analogously, we can obtain the AR representation of an ARMA(p, q) process via

$$\pi(B)\tilde{z}_t = \theta_q^{-1}(B)\phi_p(B)\tilde{z}_t = a_t,$$

where $\pi(B) = 1 - \sum_{i=1}^{\infty} \pi_i B^i$ with π_i being obtained by equating the coefficients of B^i in the identity $\phi_p(B) = \theta_q(B)\pi(B)$.

To calculate the autocovariances of z_t, we multiply (2.22) by $\tilde{z}_{t-h}$ and take expectations, resulting in

$$\gamma_h - \phi_1\gamma_{h-1} - \cdots - \phi_p\gamma_{h-p} = E[a_t\tilde{z}_{t-h}] - \sum_{i=1}^{q} \theta_i E[a_{t-i}\tilde{z}_{t-h}]. \tag{2.23}$$

For $h = q$, we have $\gamma_q - \phi_1\gamma_{q-1} - \cdots - \phi_p\gamma_{q-p} = -\theta_q\sigma^2 \neq 0$, but, for $h > q$, all of the expectations on the right side of Eq. (2.23) are zero and $\gamma_h - \phi_1\gamma_{h-1} - \cdots - \phi_p\gamma_{h-p} = 0$. Dividing by γ_0, we obtain

$$\rho_h - \phi_1\rho_{h-1} - \cdots - \phi_p\rho_{h-p} = 0, \quad h > q. \tag{2.24}$$

This is the moment equation of a stationary ARMA model. In particular, considering jointly Eq. (2.24) for $h = q + 1, \ldots, q + p$, we have a system of p linear equations for the AR parameters ϕ_i. This system of equations is referred to as the *generalized Yuel–Walker equations* of a stationary ARMA model.

Eq. (2.23) can be rewritten, by the back-shift operator, as

$$\phi_p(B)\rho_h = 0 \quad h > q, \tag{2.25}$$

from which, we conclude that, for $h > q$, the ACFs of z_t would behave in a similar manner as those of an AR(p) model. Therefore, the decay of the ACFs of an ARMA model is only determined by the autoregressive part $\phi(B)$ of the model. However, the first q ACFs of z_t depend on both the AR and MA parameters. For a stationary ARMA(p, q) model, ACFs decay to zero eventually according to Eq. (2.25), but the ACFs depend on some initial values. For $p \leq q$, the ACFs ρ_i $(i = q - p + 1, \ldots, q)$ provide the initial values whereas for $p > q$, the ACFs ρ_i $(i = p - 1, \ldots, 0)$ provide necessary initial values.

2.2.4 Linear Combinations of ARMA Processes

ARMA processes are frequently found in practice because a non-zero linear combination of pure AR processes leads to an ARMA model. In fact, the class of ARMA models is closed under linear combinations. In particular, adding a finite number of ARMA processes together results in a new ARMA process. To illustrate, we start with the simplest case of adding a white noise to a zero-mean AR(1) process, say,

$$z_t = y_t + v_t, \tag{2.26}$$

where $y_t = \phi y_{t-1} + a_t$ and v_t is white noise independent of a_t, and thus of y_t. Process z_t can be interpreted as the result of observing an AR(1) process with measurement error. The variance of z_t is

$$\gamma_z(0) = E(z_t^2) = E[(y_t^2 + v_t^2 + 2y_t v_t)] = \gamma_y(0) + \sigma_v^2. \tag{2.27}$$

As the autocovariances of y_t verify $\gamma_y(h) = \phi^h \gamma_y(0)$ and those of v_t for $h \geq 1$ are zero, we have

$$\gamma_z(h) = E(z_t z_{t-h}) = E[(y_t + v_t)(y_{t-h} + v_{t-h})] = \gamma_y(h) = \phi^h \gamma_y(0).$$

Specifically, for $h = 1$, Eq. (2.27) implies that

$$\gamma_z(1) = \phi\gamma_y(0) - \phi\gamma_z(0) - \phi\sigma_v^2, \tag{2.28}$$

and for $h \geq 2$,

$$\gamma_z(h) = \phi\gamma_z(h - 1). \tag{2.29}$$

We conclude that process z_t follows an ARMA(1,1) model with an AR parameter equal to ϕ. The MA parameter and the variance of the innovations of the resulting ARMA(1,1) process depend on the parameters of the y_t and v_t.

The above results can be generalized to any AR(p) process. It can be proved (see Granger and Morris, 1976) that, under rather weak conditions,

$$\text{AR}(p) + \text{AR}(q) = \text{ARMA}(p + q, \max(p, q)),$$

and

$$\text{MA}(q_1) + \text{MA}(q_2) = \text{MA}(\max(q_1, q_2)).$$

For ARMA processes, it can also be proved that, under the same weak conditions,

$$\text{ARMA}(p_1, q_1) + \text{ARMA}(p_2, q_2) = \text{ARMA}(a, b),$$

where

$$a \le p_1 + p_2, \qquad b \le \max(p_1 + q_1, p_2 + q_2).$$

These results suggest that whenever we observe a process that is the sum of other processes some of which follows an AR model, we expect to have an ARMA process. The weak conditions used include that the innovations to the individual time series either are independent or have no lagged cross-correlations.

A practical implication of the aforementioned results is that ARMA(p, q) models are often used in real applications as the observed time series may be a linear combination of several underlying processes. For instance, the gross domestic product of an economy typically consists of several components measuring the output of that economy. Another practical implication is that ARMA(p, p) models are sometimes found in real applications. As illustrated by the above case of AR(1) model, this phenomenon might be caused by measurement errors.

Example 2.2

The R command `arima.sim` generates stationary ARMA processes. Some time series are simulated following low orders ARMA models and their sample ACFs are computed.

```
 #simulates ARMA models and plot the ACF
> y1=arima.sim(n=100,list(ar=c(.8)))
> y2=arima.sim(n=100,list(ar=c(-.7)))
> y3=arima.sim(n=100,list(ar=c(1.2,-.9)))
> y4=arima.sim(n=100,list(ma=c(.6)))
> y5=arima.sim(n=100,list(ma=c(-.7)))
> y6=arima.sim(n=100,list(ar=c(.8),ma=c(-.4)))
> par(mfrow=c(3,2))
> ts.plot(y1)
.......
> ts.plot(y6)
> par(mfrow=c(3,2))
.......
> acf(y6,lag.max=15)
```

Figure 2.12 shows some simulated time series and Figure 2.13 the corresponding sample ACFs. Note that an AR(2) model with complex roots generates pseudo-cyclical behavior. ■

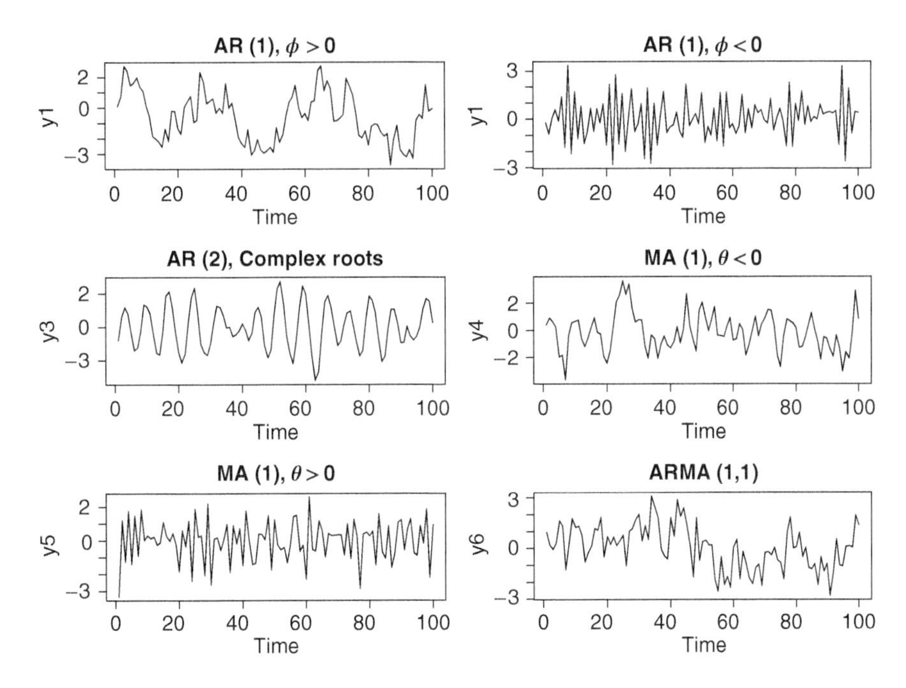

Figure 2.12 Plots of some simple ARMA models.

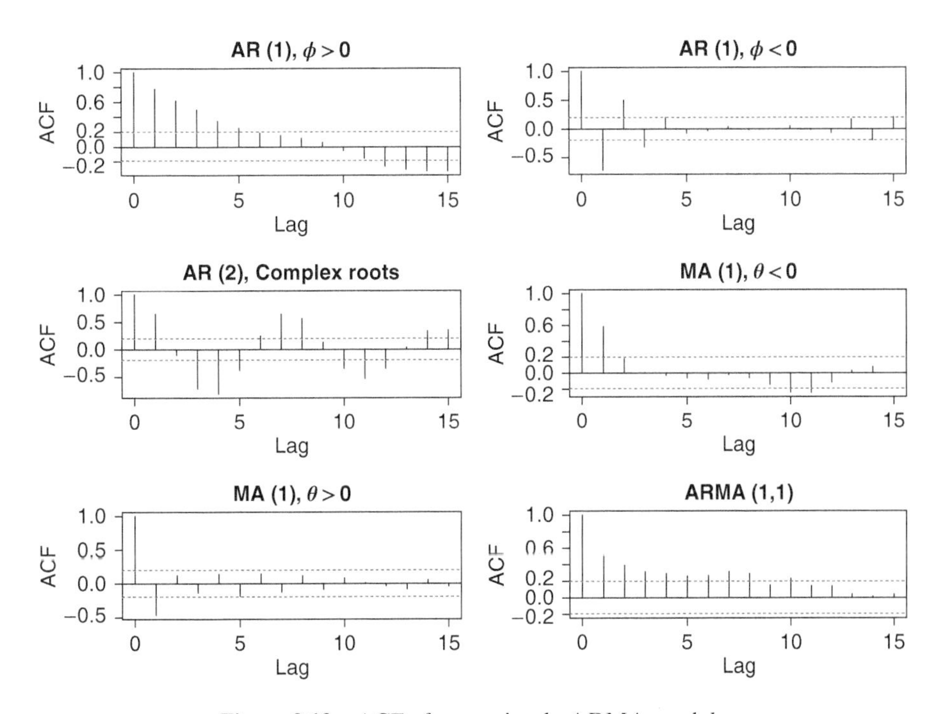

Figure 2.13 ACF of some simple ARMA models.

2.3 SPECTRAL ANALYSIS OF STATIONARY PROCESSES

2.3.1 Fitting Harmonic Functions to a Time Series

An alternative approach to analyzing stationary time series is to explore the cyclical properties of the data. Fourier, a French mathematician, proved at the beginning of the nineteenth century that any periodic function can be represented as a sum of sinusoidal functions of different amplitudes and frequencies. This approach was later extended to represent any continuous function in an interval. The harmonic or sinusoidal functions of time we use here are the sine and the cosine, given by $\sin(2\pi t/P)$ and $\cos(2\pi t/P)$ for $t = 1, \ldots, P$, where P is called the period and defines the properties of the harmonic function. The sine and cosine functions complete a cycle every P periods, because $\sin(2\pi(t+P)/P) = \sin(2\pi t/P + 2\pi) = \sin(2\pi t/P)$. An equivalent way to define these harmonic functions is by using the corresponding frequency, that is the inverse of the period $f = 1/P$. This frequency indicates the fraction of a full cycle that is observed between two consecutive units of time. For example, for a monthly seasonal series ($P = 12$), the frequency is $1/12 = 0.0833$, indicating that a month represents 8.33% of the seasonal cycle of 12 months. Instead of the frequency, we can also use the angular frequency, $w = 2\pi f = 2\pi/P$, that indicates, the angle in radians covered in a unit of time. Note that the complete cycle of 2π radians is covered in the period P. For example, a quarterly series with $P = 4$ has an angular frequency of $w = 2\pi/4 = \pi/2$, indicating that in one observation (quarter) an angle of $\pi/2$ is covered with respect to the full cycle of 2π.

Suppose that we fit a sinusoidal function of a given angular frequency w to a stationary time series observed at times $t = 1, \ldots, T$. That is, we entertain the model

$$z_t = \mu + R\sin(wt + \theta) + a_t, \tag{2.30}$$

where μ denotes the mean of z_t, R is the amplitude of the sine wave, and θ is the phase or distance from the origin at $t = 0$ of the sine function. Introducing the phase θ we do not need to decide between the functions sine and cosine. In fact, we use both. To see this, using $\sin(a + b) = \sin(a)\sin(b) + \cos(a)\cos(b)$, we can write (2.30) as

$$z_t = \mu + R\sin(wt)\sin(\theta) + R\cos(wt)\cos(\theta) + a_t,$$

and denoting $A = R\sin\theta$ and $B = R\cos\theta$, we have

$$z_t = \mu + A\sin(wt) + B\cos(wt) + a_t. \tag{2.31}$$

This expression is better for estimation than Eq. (2.30) because it represents the series as the sum of two sinusoidal functions of the same angular frequency but different coefficients, A and B. The original unknown parameters of the sine wave, θ and R in Eq. (2.30), have been transformed to the coefficients, A and B, which can easily be estimated from the data: Eq. (2.31) is linear in the three unknown parameters, μ, A, and B, and can be fitted by *least squares* (LS). As the explanatory variables $\sin(wt)$ and $\cos(wt)$ have zero mean, variance 1/2 and are orthogonal, their LS estimates are $\widehat{\mu} = \sum_{t=1}^{T} z_t/T$ and

$$\widehat{A} = \frac{2}{T}\sum_{t=1}^{T} z_t \sin(wt), \tag{2.32}$$

$$\widehat{B} = \frac{2}{T}\sum_{t=1}^{T} z_t \cos(wt). \tag{2.33}$$

The original amplitude, R, is then estimated by

$$\widehat{R}^2 = \widehat{A}^2 + \widehat{B}^2. \tag{2.34}$$

The residuals of the fitted model are calculated using

$$\widehat{a}_t = z_t - \widehat{\mu} - \widehat{A}\sin(wt) - \widehat{B}\cos(wt),$$

and, by construction, they have zero mean and variance $\widehat{\sigma}^2 = \sum_{t=1}^{T} \widehat{a}_t^2/T$. The fitted model performs a decomposition of the variability of the data into two parts; the first part, explained by the sinusoidal explanatory variables and the second part, due to the residuals. Since the variance of the two sinusoidal components is 1/2 and they are uncorrelated, it is easy to obtain that

$$s_z^2 = \frac{1}{T}\sum_{t=1}^{T}(z_j - \widehat{\mu})^2 = \frac{\widehat{A}^2}{2} + \frac{\widehat{B}^2}{2} + \widehat{\sigma}^2 = \frac{\widehat{R}^2}{2} + \widehat{\sigma}^2, \tag{2.35}$$

where s_z^2 is the variance of z_t. Equation (2.35) provides a decomposition of the variance of the series into two orthogonal components, namely, the signal, represented by the square of amplitude of the wave divided by two, and the variance of the noises.

2.3.2 The Periodogram

Given a real-valued time series, the value of the angular frequency w that provides the best fit is unknown, and we may try different values to see which one fits the data best. In practice, it would be difficult to estimate any cycle with period P larger than the sample size, T, or any cycle with period P smaller than 2. Therefore, the angular frequency w of the sinusoidal waves entertained in practice satisfies $2\pi/T \leq w \leq \pi$. In addition, instead of using a single-wave function, a better approach is to assume that the series is generated by a sum of waves with different amplitudes and frequencies, and estimate the relative importance of each wave in explaining the variability of the data. With T observations, where to simplify the exposition in this section we assume that T is even, we need to estimate two amplitudes for each wave and the mean of z_t so that the number of angular frequencies that can be fitted, n_f must verify $2n_f + 1 \leq T$. Consequently, the maximum number of angular frequencies that can be considered is $n_f = T/2$.

We define the *basic* or *Fourier periods* as those that are exact fractions of the sample size, that is, $P_j = T/j$ with j being a positive integer. The maximum value of the basic period is $P_{\max} = T$, the sample size, that is $j = 1$, and within this period we only observe the sinusoidal function once. The minimum value for the period is $P_{\min} = 2$, because we cannot observe periods that last fewer than two observations, and, therefore, $j = T/2$. The *Fourier frequencies* are defined as the inverses of these basic periods, $w_j = 2\pi j/T$, for $j = 1, 2, \ldots, T/2$, and the frequencies we can observe are $2\pi/T \leq w \leq \pi$, as indicated before. In this way, we can obtain a general representation of a stationary time series as the sum of waves associated with all the basic frequencies:

$$z_t = \mu + \sum_{j=1}^{T/2} A_j \sin(w_j t) + \sum_{j=1}^{T/2} B_j \cos(w_j t). \quad (2.36)$$

Equation (2.36) allows us to decompose exactly an observed time series into the sum of harmonic components because we have T observations and T parameters. Note that the coefficient $A_{T/2}$ is set to zero, because it corresponds to the wave $\sin(w_{T/2}t) = \sin(\pi t) = 0$.

According to Eq. (2.35), the contribution of each wave to the variance of the series is the square of its amplitude divided by 2. As the importance of a wave depends on its amplitude, we select the important ones by calculating the parameters A_j and B_j for all the basic frequencies (with the exemption of $A_{T/2}$, that is set to zero). These computations are easily carried out because the explanatory variables $\sin(w_j t)$ and $\cos(w_j t)$ are orthogonal and the coefficients A_j and B_j are computed by Eqs. (2.32) and (2.33), with $w_j = 2\pi j/T$. Given the estimated coefficients $\hat{A}_j$ and $\hat{B}_j$ for each frequency w_j, we calculate $\hat{R}_j = \hat{A}_j^2 + \hat{B}_j^2$. Via Eq. (2.35), we decompose the variance of the time series into components associated with each one of the harmonic functions. Letting s_z^2 be the sample variance of the series, we write

$$Ts_z^2 = \sum_{t=1}^{T} (z_t - \mu)^2 = \sum_{j=1}^{T/2} \frac{T}{2} \hat{R}_j^2. \quad (2.37)$$

The *periodogram* represents the contribution, $T\hat{R}_j^2/2$, of each frequency to the variance of the data as a function of the angular frequency $w_j = 2\pi j/T$. Specifically, the periodogram of a stationary time series z_t at frequency w_j is defined as

$$I(w_j) = \frac{T\hat{R}_j^2}{2}, \quad \text{with} \quad 2\pi/T \leq w_j \leq \pi. \quad (2.38)$$

We observe that, with the above representation, the sum of the periodogram ordinates is, by Eq. (2.37), equal to the total variability, Ts_z^2, of the series, and the average value of the periodogram ordinates is the variance of the series.

The above analysis assumes that we are interested only in the basic frequencies. Such an assumption is fairly nonrestrictive if the sample size is large so that the number of basic frequencies is large. In this case, there always exists some basic frequency close to the one that is of interest to us in an application. For example, suppose that we have $T = 140$ monthly data points and wish to estimate the amplitude with period 12. In this case, the basic period for $j = 11$ is $P_{11} = 140/11 = 12.7$ and that for $j = 12$, $P_{12} = 140/12 = 11.66$. Therefore, we can obtain an approximate estimate of the amplitude of the wave with period 12 from the amplitudes calculated with $j = 11$ and 12. Naturally, we can fit a model for any period but then, by losing the symmetry, the formulas shown for the estimators are only approximate, although we can always calculate the coefficients exactly by multiple regression.

The periodogram can be seen as a tool for detecting deterministic cycles or, in general, important cycles in a time series. For example, in a monthly seasonal series we expect to find a high periodogram at the angular frequency $w = 2\pi/12$, but we can also find high values at $w = 2\pi j/12$, for $j = 1, 2, \ldots$, which are harmonics of the seasonal frequency. On the other hand, the series may have other cycles that are not necessarily tied to the seasonal period, and the periodogram is a useful tool for detecting these additional components. The usefulness of the periodogram increases by noting that when we estimate the amplitude of a wave for a given frequency we are, in fact, calculating an average amplitude of all the possible cycles with frequencies close to the one being estimated. Since the distances between basic frequencies are $2\pi(j+1)/T - 2\pi j/T = 2\pi/T$, we can consider the amplitude calculated for the frequency w_j as an average of the amplitudes that exist in the frequencies in the interval $w_j \pm 2\pi/2T$. As a result, instead of representing bars in the basic frequencies with heights $I(w_j)$, we can construct a *smoothed periodogram* using rectangles with area proportional to the periodogram, as we do in the histogram. Thus, we build rectangles with bases in $w_j \pm 2\pi/2T$ (center at w_j and side $2\pi/T$) and with height $I(w_j) = T\widehat{R}_j^2/2$, so that its area is equal to $\widehat{R}_j^2/2$. In this way, we distribute the estimated variability over the whole range of frequencies. This smoothed periodogram is now defined for all the frequencies in the interval $\pi \geq w_j \geq 0$ and the total area enclosed within this function is the variance of the series. We can go one step further and instead of representing a function in steps, such as the one just described, by treating it as a continuous function. To this end, we smooth the abovementioned amplitude estimation by weighting the adjacent estimators in the same way as we smooth a histogram to obtain an estimation of the density function. The resulting smoothed periodogram, that is usually called the sample spectrum, has ordinate at each point calculated by

$$I(w) = \sum_{w_i - a}^{w_i + a} p_i I(w_i), \quad \text{with } 0 \leq w \leq \pi, \tag{2.39}$$

where a represents the window width used and the p_i are symmetric weights that add up to one.

2.3.3 The Spectral Density Function and Its Estimation

The previous analysis can be applied to any stationary process. Consider a zero-mean stationary process, $\{z_t | t = 1, \ldots, T\}$ It can be shown (see, for instance, chapter 2 of Box et al., 2015) that the periodogram is related to the estimated covariance function by

$$I(w_j) = 2\left[\widehat{\gamma}_z(0) + 2\sum_{k=1}^{T-1} \widehat{\gamma}_z(k)\cos(w_j k)\right], \qquad 0 \leq w_j \leq \pi,$$

and the normalized periodogram is defined as

$$NI(w_j) = I(w_j)/\widehat{\gamma}_z(0) = 2\left[1 + 2\sum_{k=1}^{T-1} \widehat{\rho}_z(k)\cos(w_j k)\right]. \qquad 0 \leq w_j \leq \pi.$$

Since the estimated autocovariances converge to the theoretical ones as $T \to \infty$, we have that the periodogram converges to the power spectrum of z_t defined by

$$p(w) = 2\left[\gamma_z(0) + 2\sum_{k=1}^{\infty}\gamma_z(k)\cos(wk)\right]. \qquad 0 \le w \le \pi. \tag{2.40}$$

The spectrum of a time series shows how its variance is distributed among the frequencies because

$$\gamma_z(0) = \int_0^{\pi} p(w)dw.$$

The spectrum can be standardized so that its area integrates to unity, similar to a density function. The standardized spectrum is called *the spectral density function* and it is the asymptotic limit of the normalized periodrogram. Specifically, the spectral density function of z_t is given by

$$g(w) = p(w)/\gamma_z(0) = 2\left[1 + 2\sum_{j=1}^{\infty}\rho_z(j)\cos(2\pi j/T)\right], \qquad 0 \le w_j \le \pi. \tag{2.41}$$

The spectral density can be estimated by standardizing the sample spectrum. Alternatively, we can fit an ARMA model to the stationary time series and use the estimated parameters to estimate the spectral density. It can be shown, (see, for instance, chapter 3 of Box et al., 2015) that an ARMA(p, q) model $\phi(B)\tilde{z}_t = \theta(B)a_t$ has spectral density given by

$$g(w) = 2\frac{|\theta_q(e^{-iw})|^2}{|\phi_p(e^{-iw})|^2} = 2\frac{|1 - \theta_1 e^{-iw} - \cdots - \theta_q e^{-iw}|^2}{|1 - \phi_1 e^{-iw} - \cdots - \phi_p e^{-iw}|^2}, \tag{2.42}$$

and, plugging in the estimated parameters, we obtain an estimate of the spectral density function.

The representation of a time series as the sum of orthogonal harmonic functions with different frequencies is useful for stationary series, but is less useful for irregular time series with no periodic behavior. This type of decomposition using orthogonal (Fourier) functions has been generalized by using wavelets, which are designed to approximate locally continuous functions over a short period of time by using a set of orthogonal functions more flexible than sines and cosines. Wavelets were introduced with this objective and can be useful for certain nonstationary and nonlinear time series. Interested readers are referred to Tsay and Chen (2018) for an introduction and to Percival and Walden (2000) for a broader discussion.

Example 2.3

In Figure 2.14, the smoothed periodograms, or sample spectrums, of the ARMA time series simulated in Example 2.2 are shown. The positive autocorrelations of an AR(1) model produces a higher concentration of power in low frequencies (large periods) and the opposite is true when the AR(1) model has a negative parameter. The pseudo cyclical behavior of an AR(2) model with complex roots is also shown. Figure 2.15 shows the plot of an AR(1) process and the sum of this process with a cyclical time series of period $P = 4$. The smoothed periodograms of both series are

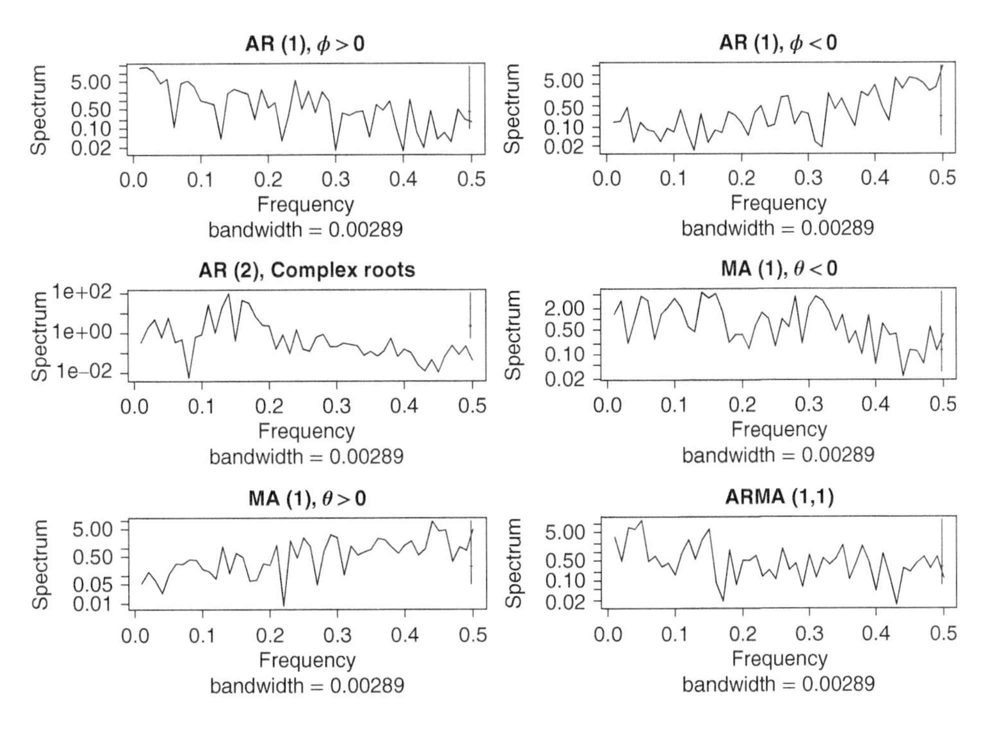

Figure 2.14 Sample spectrum of the ARMA time series simulated in Example 2.2.

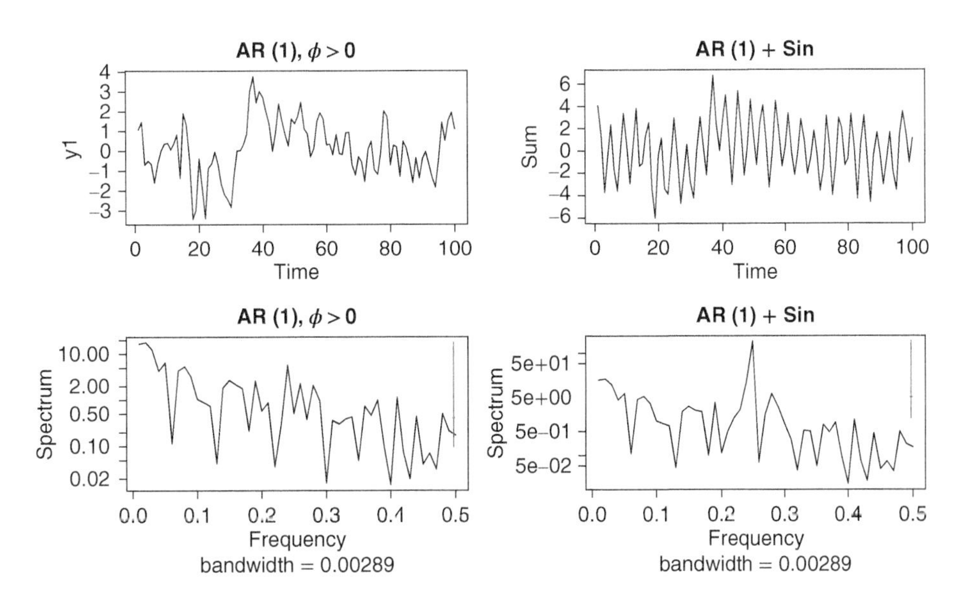

Figure 2.15 Upper part, time series plots, and lower part, sample spectrums, of an AR(1) and the sum of this process and a sine wave of frequency 0.25.

given and the peak at frequency 0.25, or period $P = 4$, is clearly seen in the second series with this cycle added. The R commands used to generate these two series are given below.

```
> y1=arima.sim(n=100,list(ar=c(0.6)))
> par(mfrow=c(2,2))
> t=seq(1,100,1)
> z=3*sin(2*pi*t/4)
> s=y1+z
> plot(y1,main="AR(1), fi>0")
> plot(s, main="AR(1)+Sin")
> spectrum(y1,main="AR(1), fi>0")
> spectrum(s,main="AR(1)+Sin")
```

■

2.4 INTEGRATED PROCESSES

Most time series observed in practice are nonstationary and show certain trends over time. However, often the series of changes (or increments), $y_t = z_t - z_{t-1} = \nabla z_t$, or the changes in the increments, $c_t = y_t - y_{t-1} = z_t - 2z_{t-1} + z_{t-2} = \nabla^2 z_t$ are stationary. Processes of this type, which become stationary after differencing, are called *integrated processes*. Here the name *integration* denotes the inverse of differencing, signifying that the time series $z_t = y_t + y_{t-1} + \cdots$ is a sum of the past values of the stationary series y_t. Note that if time is continuous, then sum becomes integration and differencing becomes differentiation. We discuss some integrated processes next.

2.4.1 The Random Walk Process

We have seen that finite-order MA processes are always stationary and the stationarity of an AR process requires that all roots of the AR polynomial equation, $\phi(B) = 0$, lie outside the unit circle. For instance, consider an AR(1) model $z_t = c + \phi z_{t-1} + a_t$. If $|\phi| < 1$, the process is stationary. On the other hand, if $|\phi| > 1$, then the process is explosive as $|z_t| \to \infty$ as t increases. An interesting case is when $|\phi| = 1$ when the process is not stationary and belongs to the class of first-order integrated processes. Indeed, its first difference becomes

$$w_t = \nabla z_t = z_t - z_{t-1} = c + a_t, \tag{2.43}$$

which is a stationary process. Such a z_t process is called a *random walk* with drift c. An important characteristic that distinguishes stationary from nonstationary processes is the role played by the constant c. In a stationary process, the constant is related to the expected value of z_t, which can easily be estimated by the sample mean. One can then subtract the mean from z_t and analyzes the mean-adjusted process. Statistically speaking, higher-order properties of a stationary time series, such as the moment equations, do not depend on its mean. On the contrary, for a nonstationary process, the constant, if exists, plays an important role. For instance, consider the random walk with drift, (2.43). By repeated substitutions of $z_j = c + z_{j-1} + a_j$, for $1 \leq j \leq t$, we obtain

$$z_t = ct + \sum_{t=1}^{t} a_t. \tag{2.44}$$

where z_0, the initial value of the process, is assumed to be zero. If $c \neq 0$, we see that, from Eq. (2.44), $E(z_t) = ct$, indicating that the mean of z_t is time-varying with slope c. Furthermore,

$$\text{Var}(z_t) = E(a_t + a_{t-1} + a_{t-2} + \cdots + a_1)^2 = \sigma^2 t, \tag{2.45}$$

which converges to infinity as t increases, indicating that the uncertainty of z_t increases with $\sqrt{t}$, which is unbounded. Finally, using

$$z_{t+h} = c(t+h) + a_{t+h} + a_{t+h-1} + \cdots + a_1,$$

and $\text{Cov}(t, t+h)=\text{Cov}(z_t, z_{t+h})$ for $h > 0$, we have

$$\text{Cov}(t, t+h) = E[(z_t - ct)(z_{t+h} - c(t+h))] = \sigma^2 t, \tag{2.46}$$

which is also time-varying. As a matter of fact, one can easily see that

$$\text{Cov}(t, s) = \text{Cov}(z_t, z_s) = E\left[\left(\sum_{i=1}^{t} a_i\right)\left(\sum_{j=1}^{s} a_j\right)\right] = \sigma^2 \wedge (t, s),$$

where $\wedge(t, s) = \min\{t, s\}$. Consequently, for $h > 0$, we have

$$\text{Cov}(t, t-h) = \sigma^2(t-h) \neq \text{Cov}(t, t+h).$$

The aforementioned properties show that the random walk process z_t is nonstationary. Furthermore, the autocorrelation coefficients of z_t are, $h > 0$,

$$\rho(t, t-h) = \frac{t-h}{\sqrt{t(t-h)}} = \sqrt{1 - \frac{h}{t}}. \tag{2.47}$$

The above expression indicates that, for large t, the autocorrelation coefficients of z_t would approach 1 and they decay approximately in a linear manner as h increases. Indeed, if we assume that the process starts at the remote past so that h/t is small, then the function $(1 - h/t)^{1/2}$ can be approximated, using a first-order Taylor expansion, by

$$\rho(t, t-h) \approx 1 - \frac{2h}{t},$$

indicating that, assuming fixed t, the autocorrelation coefficients is a function of h and follows roughly a straight line with slope $(-1/2t)$. The property shown in Eq. (2.47) continues to hold for integrated processes, even though the derivation is a bit more complicated. We provide further details in the next section.

2.4.2 ARIMA Models

This idea of transforming a nonstationary series into a stationary one by differencing can be generalized to any ARMA process, leading to the ARIMA(p, d, q) processes, where *ARIMA* stands for autoregressive integrated moving-average. Here p is the order of the AR part, d is the number of unit roots (or the order of integration, or the number of differences required for stationarity), and q is the MA order. For instance, an ARIMA(0,1,1) process has no AR part, requires the first difference, and has an MA(1) component. Its model assumes the form

$$(1 - B)z_t = (1 - \theta B)a_t, \tag{2.48}$$

where $|\theta| < 1$ for the process to be invertible. Inverting the MA operator, the process can be written as

$$z_t = (1 - \theta)[z_{t-1} + \theta z_{t-2} + \theta^2 z_{t-3} + \cdots] + a_t, \tag{2.49}$$

so that each value z_t is a weighted average of previous values plus a random noise. Note that the weights sum to 1 and decay exponentially to zero as lag increases. An alternative way to express Eq. (2.49) is as follows: at the forecast origin $t - 1$, the 1-step ahead forecast of z_t is

$$\hat{z}_{t-1}(1) = (1 - \theta) \sum_{i=1}^{\infty} \theta^{i-1} z_{t-i}.$$

Here $\theta > 0$ is called the discounting rate, and this approach to forecasting is called the *simple exponential smoothing*, which is widely used in practice, especially in predicting sales or inventories.

The general ARIMA(p, d, q) process is given by

$$(1 - \phi_1 B - \cdots - \phi_p B^p)(1 - B)^d z_t = c + (1 - \theta_1 B - \cdots - \theta_q B^q)a_t,$$

or

$$\phi_p(B)\nabla^d z_t = c + \theta_q(B)a_t. \tag{2.50}$$

Let $\omega_t = \nabla^d z_t$ be the stationary process. The observed process z_t is obtained as a sum (integration) of the stationary process ω_t. If $d = 1$, then $z_t = z_{t-1} + \omega_t$ and, by repeated substitutions, we have

$$z_t = \sum_{j=0}^{t} \omega_t,$$

where, for simplicity, we assume the process starts at $t = 0$ with $z_0 = 0$. If $\omega_t = a_t$, a white noise, then z_t is a random-walk process equivalent to an ARIMA(0,1,0) process. We have shown in the previous section that the autocorrelation coefficients of a random walk are close to 1 and decay slowly as lag increases. Here we briefly prove that the same property continues to hold for all nonstationary ARIMA(p, d, q) processes with $d > 0$. Recall that the correlation coefficients of an ARMA(p, q) process satisfies the difference equation for $h > q$,

$$\phi_p(B)\rho_h = 0, \quad h > q,$$

the solution of which can be written as

$$\rho_h = \sum_{1}^{p} A_i G_i^h,$$

where G_i^{-1} are the roots of $\phi_p(B) = 0$ and $|G_i| < 1$. If one of these roots G_i is very close to 1, say $G_i = 1 - \varepsilon$ with a small deviation ε, and all other roots are away from 1, then for large h, the terms $A_j G_j^h$ $(j \neq i)$ converge to zero quickly and we have

$$\rho_h \approx A_i(1 - \varepsilon)^h \approx A_i(1 - h\varepsilon),$$

for a large h. Consequently, the ACFs of z_t would assume a value close to A_i and decay slowly and linearly with the lag h. One would expect to see this feature of integrated process when the sample size T is large. Note that the value A_i depends on the AR coefficients ϕ_j, but the characteristic of slowly decaying ACF signifies the unit-root nonstationarity.

2.4.3 Seasonal ARIMA Models

We say that a series z_t is seasonal if its expected values follow a nonconstant cyclical pattern. Specifically, if $E(z_t) = E(z_{t+s})$, where $s > 1$ is the smallest positive integer, then z_t is seasonal with a *seasonality s* periods. For example, consider a monthly time series z_t, if its expected values depend on the month within a year, but are the same for the same month each year, then z_t is a seasonal time series with seasonality 12. Mathematically, denote the time index as $t = m \times 12 + j$, where m is the number of years and $j = 1, \ldots, 12$, if $E(z_t) = E(z_{t+12}) = \mu_j$ and $\mu_j \neq \mu_i$ for some $i \neq j$, then z_t is seasonal with seasonality 12. Thus, a seasonal series is nonstationary. The seasonality s denotes the number of observations within a seasonal cycle. Typically, we have $s = 12$ for monthly time series and $s = 4$ for quarterly series. Here we assume that s is fixed and known. It is possible that a time series may have multiple seasonal periods. For example, as shown in Chapter 1, the hourly $PM_{2.5}$ measurements at a monitoring station may exhibit both the annual and diurnal cycle patterns. Also, it is possible that, in an application, the seasonal period s only holds approximately. Consider, for instance, the daily average temperature at a monitoring station. This series would exhibit strong seasonal patterns as temperature is seasonal in most part of the world. However, in this particular instance, the seasonal period is neither 365 nor 366 precisely due to the leap years. Similarly, the series may exhibit certain monthly pattern, but the length of each month differs. In the time series literature, there are ways to address this issue. For instance, for business and economics data, one may adjust for the number of working days within a month to better represent the seasonality.

There are several ways available to modeling seasonal time series. The choice often depends on the seasonality s and the sample size T. For simplicity in discussion, we assume that $T = ns$ and write $t = ms + j$, where $m = 0, \ldots, n-1$ and $j = 1, \ldots, s$. The observed data can then be arranged in the following two-way table:

	1	2	3	$\cdots$	$s-2$	$s-1$	s
1	z_1	z_2	z_3	$\cdots$	z_{s-2}	z_{s-1}	z_s
2	z_{s+1}	z_{s+2}	z_{s+3}	$\cdots$	z_{2s-2}	z_{2s-1}	z_{2s}
3	z_{2s+1}	z_{2s+2}	z_{2s+3}	$\cdots$	z_{3s-2}	z_{3s-1}	z_{3s}
$\vdots$	$\vdots$	$\vdots$	$\vdots$		$\vdots$	$\vdots$	$\vdots$
$n-1$	$z_{(n-1)s+1}$	$z_{(n-1)s+2}$	$z_{(n-1)s+3}$	$\cdots$	z_{ns-2}	z_{ns-1}	z_{ns}

If s is not large, such as the monthly time series with $s = 12$, then one can treat each column of the above two-way table as an individual time series and entertain an ARIMA model for each series of n observations, namely

$$\Phi_j(B^s)(1 - B^s)^{d_j} z_{ms+j} = \Theta_j(B^s) b_{ms+j}, \quad j = 1, \ldots, s, \tag{2.51}$$

where $\Phi_j(B^s)$ and $\Theta_j(B^s)$ are, respectively, the AR and MA polynomials for z_{ms+j}, d_j is a nonnegative integer, and $\{b_{ms+j} | m = 0, \ldots, n-1\}$ is the innovations for column j with mean zero and variance σ_j^2. Here we use the back-shift operator B^s because $B^s z_{ms+j} = z_{(m-1)s+j}$ gives the previous value in the same column of the two-way table. This approach to modeling seasonal time series is relatively simple, but it has several weaknesses. For instance, it fails to consider the interdependence between consecutive observations, e.g. the dependence between columns. It may also result in using too many parameters for the data under study, especially when n is not large and s is large.

An alternative approach is to assume that the column-wise model in Eq. (2.51) are the same for all the columns, resulting in the parsimonious model

$$\Phi(B^s)(1 - B^s)^D z_t = \Theta(B^s) b_t, \quad t = 1, \ldots, T, \tag{2.52}$$

where the order of $\Phi(B^s)$ and $\Theta(B^s)$ are P and Q, respectively, and D is a nonnegative integer denoting seasonal differencing. As before, we assume that there is no common factor between $\Phi(B^s)$ and $\Theta(B^s)$.

While Model (2.52) is parsimonious, it still fails to take into account the column dependence of the aforementioned two-way table. One possible modification is to assume that the serial dependence between the columns is homogeneous, i.e. the same dependence applies to each row of the two-way table. Such a modification can be achieved by postulating an ARIMA(p, d, q) model for the b_t series in Eq. (2.52), namely

$$\phi(B)(1 - B)^d b_t = \theta(B) a_t, \tag{2.53}$$

where $\phi(B) = 1 - \phi_1 B - \cdots - \phi_p B^p$ and $\theta(B) = 1 - \theta_1 B - \cdots - \theta_q B^q$ are the AR and MA polynomials, respectively, and d is a nonnegative integer denoting the order of regular differencing. This model considers the serial dependence between consecutive data points of the series just like the ordinary case.

By putting Eqs. (2.52) and (2.53) together, one obtains the well-known multiplicative seasonal ARIMA$(p, d, q)(P, D, Q)_s$ model,

$$\Phi_P(B^s)\phi_p(B)\nabla_s^D \nabla^d z_t = \theta_q(B)\Theta_Q(B^s) a_t, \tag{2.54}$$

where $\Phi_P(B^s) = (1 - \Phi_1 B^s - \cdots - \Phi_P B^{sP})$ is the seasonal AR operator of order P, $\phi_p = (1 - \phi_1 B - \cdots - \phi_p B^P)$ is the regular AR operator of order p, $\nabla_s^D = (1 - B^s)^D$ represents the seasonal difference, and $\nabla^d = (1 - B)^d$ the regular difference, $\Theta_Q(B^s) = (1 - \Theta_1 B^s - \cdots - \Theta_Q B^{sQ})$ is the seasonal moving average operator of order Q, $\theta_q(B) = (1 - \theta_1 B - \cdots - \theta_q B^q)$ is the regular MA operator of order q and a_t is a white noise process. This class of models, introduced by Box and Jenkins (1976), offers a good representation of many empirical time series. Note that the multiplicative nature of the model in Eq. (2.54) signifies that the row and column dependences of the data, shown in the aforementioned two-way table, are orthogonal.

Based on the discussion of multiplicative seasonal ARIMA models, it is clear that the models are derived under some relatively strong assumptions. For instance, the serial dependence of each column in the two-way table is the same for all columns and the serial dependence of each row is also the same for all rows. In addition, the row and column dependences are orthogonal. While the model is parsimonious in parameterization and has been found useful in many applications, one should keep in mind that there exist situations in which such highly structured models may not be adequate. Care need to be exercised in analyzing seasonal time series.

2.4.3.1 ***The Airline Model*** The simplest multiplicative seasonal ARIMA model is the ARIMA$(0,1,1)_s \times (0,1,1)$, which assumes the form

$$\nabla_s \nabla z_t = (1 - \theta_1 B)(1 - \Theta_1 B^s) a_t, \tag{2.55}$$

where $-1 < \theta_1, \Theta_1 < 1$. This model is referred to as the *Airline model*, because it was first proposed by Box and Jenkins to model monthly number of airlines passengers. See Box et al. (2015). It has been found to be widely applicable in modeling and forecasting seasonal time series. One possible explanation is that the model, in effect, consists simultaneously of a regular and a seasonal exponential smoothing. For simplicity, assume monthly series with $s = 12$. In this case, the seasonal part of the model in Eq. (2.55) can be written as

$$(1 - B^{12}) u_t = (1 - \Theta_1 B^{12}) a_t, \tag{2.56}$$

where $u_t = [(1 - B)/(1 - \theta_1 B)] z_t$. By repeated substitutions of Eq. (2.56), we see that

$$u_t = (1 - \Theta_1)(u_{t-12} + \Theta_1 u_{t-24} + \Theta_1^2 u_{t-36} + \cdots) + a_t,$$

from which the 1-step ahead prediction of u_t at the forecast origin $t - 1$ is

$$\hat{u}_{t-1}(1) = (1 - \Theta_1) \sum_{i=1}^{\infty} \Theta_1^{i-1} u_{t-12i},$$

indicating that u_t is a weighted average of its past lagged values at lags $t - 12$, $t - 24, \ldots$. Therefore, u_t follows a seasonal exponential smoothing model with discounting rate Θ_1. Next, from $u_t = [(1 - B)/(1 - \theta_1 B)] z_t$, we have $(1 - B) z_t = (1 - \theta_1 B) u_t$, which shows that z_t follows an exponential smoothing model with discounting rate θ_1 and innovation u_t. Consequently, the Airline model in Eq. (2.55) can be thought of as applying a regular exponential smoothing model with discounting rate θ_1 on top of a seasonal exponential smoothing model with discounting rate Θ_1. The model is a double-exponential smoothing model.

It is also of interest to study the ACF of the stationary part of the Airline model. Let $w_t = (1 - B^s)(1 - B) z_t$ be the differenced series. Then, $w_t = (1 - \theta_1 B)(1 - \Theta_1 B^s) a_t$. It is easy to see that the ACFs of w_t are given by

$$\rho_\ell = \begin{cases} \dfrac{-\theta_1^2}{1 + \theta_1^2}, & \text{if } \ell = 1, \\ \dfrac{-\Theta_1}{1 + \Theta_1^2}, & \text{if } \ell = \mathrm{s}, \\ \rho_1 \rho_s, & \text{if } \ell = \mathrm{s} - 1 \text{ or } s + 1, \\ 0, & \text{otherwise.} \end{cases}$$

Therefore, the lag-1 ACF is determined by the regular part of the model, the lag-s ACF is determined by the seasonal part of the model, and the ACFs at lags $s-1$ and $s+1$ can be thought of as interactions between the regular and the seasonal parts of the model. The ACFs also signify the parsimonious parameterization of the Airline model as it employs two coefficients θ_1 and Θ_1 yet it has four non-zero ACFs.

Example 2.4

We illustrate seasonal models with data of consumer price indexes in Europe from file `CPIEurope2000-15.csv`. One of the seasonal time series is the ninth series that corresponds to Spain, and Figure 2.16 presents some plots of the series illustrating the need to apply a regular and a seasonal difference to transform the series into stationarity. The program `sarimaSpec` indicates that this series is seasonal and can well be modeled by an Airline model. The R commands are given next.

```
> x0 <- read.csv2("CPIEurope2000-15.csv")
> x1=log(x0[,9])
> y1=diff(x1)
> par(mfrow=c(3,2))
> ts.plot(x1, main="log CPI in Spain")
> ts.plot(y1, main="first difference of log CPI Spain")
> acf(x1, main="ACF log CPI in Spain")
> acf(y1, main="ACF first difference of log CPI Spain")
```

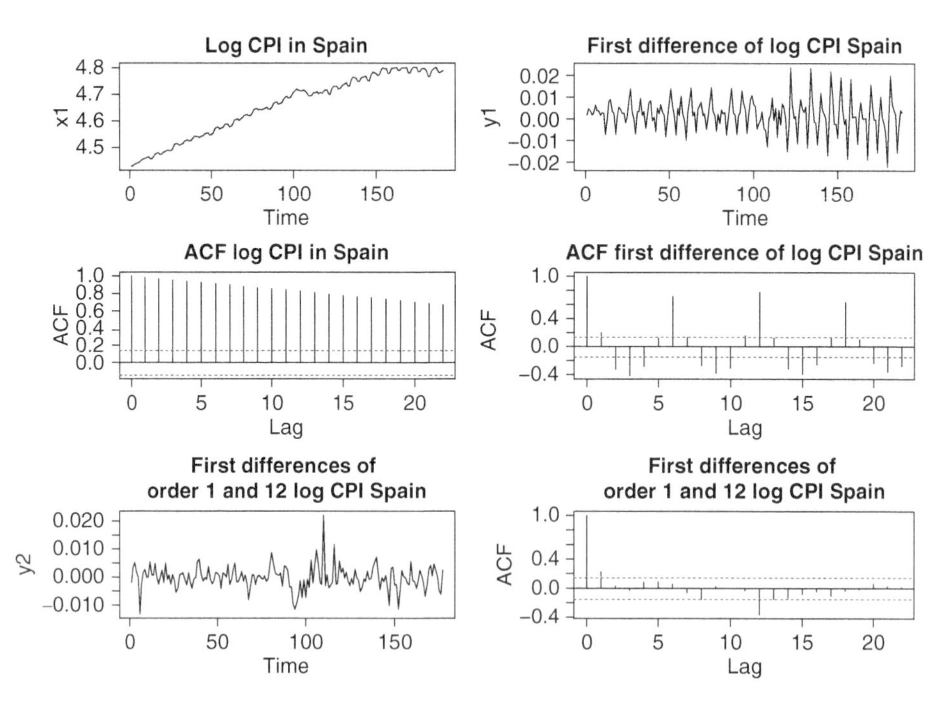

Figure 2.16 Plots and ACF of different transformations of the series of Consumer Price Index in Spain.

```
> y2=diff(y1,lag=12)
> ts.plot(y2, main="first differences of order 1 and 12 log CPI Spain")
> acf(y2, main="first differences of order 1 and 12 log CPI Spain")
> sarimaSpec(x1)
Selected differencing order (d,D): 1 1
Selected order (p,d,q,P,D,Q): 0 1 1 0 1 1
minimum criterion: -1457.766
```

■

2.5 STRUCTURAL AND STATE SPACE MODELS

2.5.1 Structural Time Series Models

An alternative approach to modeling nonstationary linear time series is to postulate that the series is the sum of a deterministic component, a cyclical component and an irregular component. The deterministic component may contain a time trend, which is often assumed to be linear. The cyclical component consists of some harmonic functions or dummy variables, and the irregular component assumes the form of an ARMA model. This approach is referred to as the *structural approach* and the entertained model *the structural model*. See Harrison and Stevens (1976), West and Harrison (2006), and Harvey (1990), among others. A structural time series model assumes the form

$$z_t = \mu_t + S_t + I_t, \tag{2.57}$$

where μ_t represents the trend of z_t, S_t is the cyclical (or seasonal) component and I_t is the irregular component, which may be a white noise. The functional form of each component of z_t in Eq. (2.57) depends on the applications, but some general functions are available. A typical formulation of the trend is

$$\begin{aligned} \mu_t &= \mu_{t-1} + \beta_{t-1} + v_t, \\ \beta_t &= \beta_{t-1} + u_t, \end{aligned} \tag{2.58}$$

where μ_t is the trend level, β_t denotes the trend growth rates, and u_t and v_t are two white noise series, which are often assumed to be independent. If the variances of u_t and v_t are zero, the trend has a constant growth rate, because β_t is then a constant. In general, the model in Eq. (2.58) represents a stochastic quadratic trend, which is sufficiently flexible in most applications. For instance, if $\beta_t = 0$ for all t and the variance of v_t is positive, then the trend of z_t follows a random walk.

The cyclical or seasonal component is usually modeled using a stochastic seasonal summation function. Let S_t be the seasonal component at time index t and assume that the seasonality is s, then S_t follows the model

$$\sum_{j=0}^{s-1} S_{t-j} = S_s(B)S_t = \epsilon_t, \tag{2.59}$$

where

$$S_s(B) = 1 + B + \cdots + B^{s-1},$$

is the seasonal sum operator of the order s. The variance of the noise ϵ_t, σ_ϵ^2, governs the variability of the seasonal coefficients and the seasonal pattern is deterministic if $\sigma_\epsilon^2 = 0$. The irregular component I_t follows a regular ARMA model with innovation a_t,

$$\phi(B)I_t = \theta(B)a_t. \tag{2.60}$$

The innovations v_t, u_t, ϵ_t, and a_t are independent white noise processes. Equations (2.57)–(2.60) jointly form the structural time series representation. In the particular case of $I_t = a_t$, the structural model contains only four parameters, which are the variances of the four innovations.

It is easy to see that any structural time series model has an implied ARIMA representation. For instance, assume $I_t = a_t$ for simplicity. Taking the second difference of (2.57) and using Eqs. (2.58)–(2.60), we have

$$\nabla^2 z_t = u_t + \nabla v_t + \nabla^2 S_t + \nabla^2 a_t. \tag{2.61}$$

Noting that

$$\nabla_s \equiv 1 - B^s = (1 - B)(1 + B + \cdots + B^{s-1}) = (1 - B)S_s(B),$$

and applying $S_s(B)$ to Eq. (2.61), we further obtain

$$\nabla\nabla_s z_t = S_s(B)u_t + \nabla_s v_t + \nabla^2 \epsilon_t + \nabla\nabla_s a_t,$$

which says that $\nabla\nabla_s z_t$ is the sum of four finite-order MA processes with the maximum order $s + 1$ and, hence, can be written as an MA($s + 1$) model. In fact, under certain conditions, the structural model series z_t would follow an airline model, i.e. an ARIMA$(0, 1, 1)_s(0, 1, 1)$ model. See Proietti (2000) for an analysis of the relation between seasonal structural patterns and seasonal ARIMA models.

2.5.2 State-Space Models

A general framework for linear time series models is the State-Space Model (SSM). Indeed, the structural time series models are special cases of the SSM and an ARMA model can also be transformed into an SSM. Assume that we observe a vector of time series $\boldsymbol{z}_t$ that can be represented by means of an observation equation:

$$\boldsymbol{z}_t = \boldsymbol{H}_t\boldsymbol{\alpha}_t + \boldsymbol{\epsilon}_t, \tag{2.62}$$

where $\boldsymbol{z}_t$ is a $k \times 1$ vector of observations, $\boldsymbol{H}_t$ is a $k \times p$ matrix assumed to be known for all t, $\boldsymbol{\alpha}_t$ is an unobserved $p \times 1$ state vector and $\boldsymbol{\epsilon}_t$ is a white noise process that, for simplicity, follows the multivariate distribution $N(\boldsymbol{0}, \boldsymbol{V}_t)$, where $\boldsymbol{V}_t$ denotes the covariance matrix. The state vector $\boldsymbol{\alpha}_t$ is assumed to evolve over time according to the following state transition equation

$$\boldsymbol{\alpha}_t = \boldsymbol{\Omega}_t\boldsymbol{\alpha}_{t-1} + \boldsymbol{u}_t, \tag{2.63}$$

where $\boldsymbol{\Omega}_t$ is a known $p \times p$ transition matrix and $\boldsymbol{u}_t$ is another white noise process with distribution $N_p(\boldsymbol{0}, \boldsymbol{R}_t)$ and independent of $\boldsymbol{\epsilon}_t$.

Jointly Eqs. (2.62) and (2.63) are referred to as an SSM. The concept of state might be foreign at the first glance, but it will become clear and meaningful once we see some specific examples later. It suffices to say now that the state vector $\boldsymbol{\alpha}_t$ contains information needed to predict $\boldsymbol{z}_{t+1}$ at the forecast origin t. Consider a simple zero-mean AR(2) model,

$$z_t = \phi_1 z_{t-1} + \phi_2 z_{t-2} + a_t.$$

The 1-step ahead prediction of z_{t+1} at the forecast origin t is $\hat{z}_t(1) = \phi_1 z_t + \phi_2 z_{t-1}$. This quantity can be computed at the forecast origin because we know the model and have z_t and z_{t-1}. Thus, given the model, to compute the 1-step ahead prediction of an AR(2) model at the forecast origin t, we need two quantities z_t and z_{t-1}. Therefore, we can define the state vector as $\boldsymbol{\alpha}_t = (z_t, z_{t-1})'$. Since z_t is the first element of $\boldsymbol{\alpha}_t$, the observation equation is simply

$$z_t = [1, 0]\boldsymbol{\alpha}_t,$$

implying that the variance of $\boldsymbol{\varepsilon}_t$ is zero. The state equation for the AR(2) model is also easy to construct as

$$\boldsymbol{\alpha}_t = \begin{bmatrix} z_t \\ z_{t-1} \end{bmatrix} = \begin{bmatrix} \phi_1 & \phi_2 \\ 1 & 0 \end{bmatrix} \begin{bmatrix} z_{t-1} \\ z_{t-2} \end{bmatrix} + \begin{bmatrix} a_t \\ 0 \end{bmatrix}.$$

For the SSM in Eqs. (2.62) and (2.63), we assume that the matrices $\boldsymbol{H}_t$ and $\boldsymbol{\Omega}_t$ are known for all t. In real applications, their parameters can consistently be estimated from the data under certain conditions. We discuss estimation later. For now, we provide further information concerning the generality of the SSM. Assume that we observe the position of a satellite in space at each time t by measuring the variable $\boldsymbol{z}_t$, which is a three-dimensional vector (e.g. longitude, latitude, and height) and may contains some measurement errors, which are assumed to be independent of the actual location of the satellite. We also assume that satellite's position depends on a set of state variables, $\boldsymbol{\alpha}_t$, which are not observable such as the gravity of earth and the speed and acceleration of the satellite. Our observed data $\boldsymbol{z}_t$ is related to $\boldsymbol{\alpha}_t$ via some aerodynamic force approximated by $\boldsymbol{H}_t$, i.e.

$$\boldsymbol{z}_t = \boldsymbol{H}_t \boldsymbol{\alpha}_t + \boldsymbol{\varepsilon}_t,$$

where $\boldsymbol{\varepsilon}_t$ denotes the measurement errors. Since the satellite would move in certain smooth manner, the state variable $\boldsymbol{\alpha}_t$ should depend on its past value at time $t-1$ and some additional disturbance at time t. It is then reasonable to assume that $\boldsymbol{\alpha}_t$ can be modeled by the state transition equation in (2.63). Consequently, the SSM in Eqs. (2.62) and (2.63) could be used to study the process $\boldsymbol{z}_t$. In this particular example, we do not provide a precise definition of the state vector $\boldsymbol{\alpha}_t$. In practice, we often seek a state vector $\boldsymbol{\alpha}_t$ that attains the minimum dimension in the sense that any state vector with dimension less than p cannot adequately represent the system under study. The aforementioned simple AR(2) example serves as an example as no state vector of dimension 1 can adequately represent the z_t process.

Structural time series models can be put into a state-space form. For instance, consider a monthly time series with periodicity $s = 12$ and irregular component $I_t = a_t$, we can define the state variable as

$$\boldsymbol{\alpha}_t = (\mu_t, \beta_t, S_t, S_{t-1}, \dots, S_{t-10})'$$

which includes μ_t, β_t, and 11 seasonal coefficients. Then, the state transition equation becomes

$$\begin{bmatrix} \mu_t \\ \beta_t \\ S_t \\ S_{t-1} \\ \vdots \\ S_{t-10} \end{bmatrix} = \begin{bmatrix} 1 & 1 & 0 & \cdots & 0 & 0 \\ 0 & 1 & 0 & \cdots & 0 & 0 \\ 0 & 0 & -1 & \cdots & -1 & -1 \\ 0 & 0 & 1 & \cdots & 0 & 0 \\ \vdots & \vdots & \vdots & \ddots & \vdots & \vdots \\ 0 & 0 & 0 & \cdots & 1 & 0 \end{bmatrix} \begin{bmatrix} \mu_{t-1} \\ \beta_{t-1} \\ S_{t-1} \\ S_{t-2} \\ \vdots \\ S_{t-11} \end{bmatrix} + \begin{bmatrix} v_t \\ u_t \\ \epsilon_t \\ 0 \\ \vdots \\ 0 \end{bmatrix},$$

or

$$\boldsymbol{\alpha}_t = \boldsymbol{\Omega}\alpha_{t-1} + \boldsymbol{u}_t, \tag{2.64}$$

where, in this case, the matrix $\boldsymbol{\Omega}_t = \boldsymbol{\Omega}$ is a constant matrix and the vector $\boldsymbol{u}_t$ has only three non-zero components. The observation equation can be written as

$$z_t = \boldsymbol{h}'\boldsymbol{\alpha}_t + a_t, \tag{2.65}$$

where $\boldsymbol{h}' = (1, 0, 1, 0, \ldots, 0)$ is a constant vector of the same dimension as that the state vector $\boldsymbol{\alpha}_t$.

An ARMA(p, q) model can be written in multiple forms of SSM, depending on the objective of the data analysis. Suppose that we are interested in on-line estimation of an AR(p) process. Then, we can write the observation equation as

$$z_t = \boldsymbol{h}_t'\boldsymbol{\alpha}_t + a_t,$$

with $\boldsymbol{h}_t = (z_{t-1}, \ldots, z_{t-p})'$ and $\boldsymbol{\alpha}_t = (\phi_1, \ldots, \phi_p)'$, which is a constant vector. The state transition equation is simply $\boldsymbol{\alpha}_t = \boldsymbol{\alpha}_{t-1}$. As will be seen in the next session, the state vector, which consists of the AR parameters, can be updated via the Kalman filter algorithm as the new observation becomes available.

In general, we can put an ARMA(p, q) model into a state-space form by the following procedure. Let $m = \max(p, q+1)$ and define the state vector as $\boldsymbol{\alpha}_t = (\alpha_{1,t}, \alpha_{2,t}, \ldots, \alpha_{m,t})'$, where $\alpha_{1t} = z_t$ and α_{jt} for $j > 1$ will be defined shortly. By the definition of $\boldsymbol{\alpha}_t$, the observation equation is simply

$$z_t = \boldsymbol{h}'\boldsymbol{\alpha}_t,$$

where $\boldsymbol{h}' = (1, 0, \ldots, 0)$. There is no measurement error.

For convenience, it is understood that $\phi_i = 0$ for $i > p$ and $\theta_j = 0$ for $j > q$. To derive the other elements of the state vector $\boldsymbol{\alpha}_t$ and the state transition equation, we start with $\alpha_{1t} = z_t$ and rewrite the model as

$$\begin{aligned} \alpha_{1t} &= \phi_1 z_{t-1} + \sum_{i=2}^{m} \phi_i z_{t-i} - \sum_{j=1}^{m} \theta_j a_{t-j} + a_t \\ &= \phi_1 \alpha_{1,t-1} + \alpha_{2,t-1} + a_t, \end{aligned} \tag{2.66}$$

where $\alpha_{2,t-1} = \sum_{i=2}^{m} \phi_i z_{t-i} - \sum_{j=1}^{m} \theta_j a_{t-j}$.

Next, consider

$$
\begin{aligned}
\alpha_{2t} &= \sum_{i=2}^{m} \phi_i z_{t+1-i} - \sum_{j=1}^{m} \theta_j a_{t+1-j} \\
&= \phi_2 z_{t-1} + \sum_{i=3}^{m} \phi_i z_{t+1-i} - \sum_{j=2}^{m} \theta_j a_{t+1-j} - \theta_1 a_t \\
&= \phi_2 \alpha_{1,t-1} + \alpha_{3,t-1} - \theta_1 a_t, \qquad (2.67)
\end{aligned}
$$

where $\alpha_{3,t-1} = \sum_{i=3}^{m} \phi_i z_{t+1-i} - \sum_{j=2}^{m} \theta_j a_{t+1-j}$. One can repeat the same procedure to define $\alpha_{4,t-1}$ and find the equation $\alpha_{3t} = \phi_3 \alpha_{1,t-1} + \alpha_{4,t-1} - \theta_2 a_t$. Since m is finite, the aforementioned procedure comes to the end with $\alpha_{mt} = \phi_m \alpha_{1,t-1} - \theta_m a_t$. Putting the recursive definitions of α_{it} and its associated equation together, we obtain the following state transition equation

$$
\begin{bmatrix} \alpha_{1,t} \\ \alpha_{2,t} \\ \alpha_{3t} \\ \vdots \\ \alpha_{m,t} \end{bmatrix} = \begin{bmatrix} \phi_1 & 1 & 0 & \cdots & 0 \\ \phi_2 & 0 & 1 & \cdots & 0 \\ \phi_3 & 0 & 0 & \cdots & 0 \\ \vdots & \vdots & \vdots & \ddots & 1 \\ \phi_m & 0 & 0 & \cdots & 0 \end{bmatrix} \begin{bmatrix} \alpha_{1,t-1} \\ \alpha_{2,t-1} \\ \alpha_{3,t-1} \\ \vdots \\ \alpha_{m,t-1} \end{bmatrix} + \begin{bmatrix} 1 \\ -\theta_1 \\ -\theta_2 \\ \vdots \\ -\theta_m \end{bmatrix} a_t.
$$

Define the constant state space matrix $\mathbf{\Omega}$ as

$$
\mathbf{\Omega}_t = \begin{bmatrix} \boldsymbol{\phi}_{m-1} & \boldsymbol{I} \\ \phi_m & \mathbf{0}' \end{bmatrix},
$$

where $\boldsymbol{\phi}_{m-1}$ is an $(m-1)$-dimensional column vector, $\boldsymbol{I}$ is the $(m-1) \times (m-1)$ identity matrix and $\mathbf{0}'$ is a $(m-1)$-dimensional vector of zeros. Also, define the vector of innovations as

$$
\boldsymbol{u}_t = \boldsymbol{\theta} a_t,
$$

where $\boldsymbol{\theta} = (1, -\theta_1, \ldots, -\theta_m)'$, and denote the covariance matrix of $\boldsymbol{u}_t$ as

$$
\boldsymbol{R}_t = \boldsymbol{\theta}\boldsymbol{\theta}'\sigma^2.
$$

We see that the state transition equation for the ARMA model is

$$
\boldsymbol{\alpha}_t = \mathbf{\Omega}\boldsymbol{\alpha}_{t-1} + \boldsymbol{u}_t,
$$

where the random vector $\boldsymbol{u}_t$ has mean zero and covariance matrix $\boldsymbol{R}_t$, which in this particular case is a constant matrix.

There are other ways to put an ARMA(p, q) model into an SSM. Interested readers are referred to Durbin and Koopman (2001), Brockwell and Davies (2013), Shumway and Stoffer (2017), and Tsay (2010), among others. In general, for an ARMA(p, q) model, the dimension of the state vector $\boldsymbol{\alpha}_t$ is $\max\{p, q+1\}$.

Based on the SSM in Eqs. (2.62) and (2.63), there are three types of statistical inference available. The first type of inference is *prediction*, which makes use of observed data $F_t = \{\ldots, \boldsymbol{z}_{t-2}, \boldsymbol{z}_t\}$ to predict $\boldsymbol{z}_{t+1}$. The second type of inference is called *filtering*, which uses the available information F_t to estimate $\boldsymbol{\alpha}_t$, which may not be directly observable. The third type of inference is *smoothing*, which makes use of information available in F_T to estimate $\boldsymbol{\alpha}_t$, where $T > t$. The Kalman filter algorithm discussed next is a useful tool to make those statistical inferences. The algorithm can also be used to evaluate efficiently the log likelihood function in parameter estimation.

2.5.3 The Kalman Filter

The Kalman filter is a fast-recursive algorithm for forecasting and parameter estimation of a linear system presented in an SSM. The filter has a straightforward Bayesian interpretation, and it can be derived easily by properties of the multivariate normal distribution. The algorithm operates in three steps. First, we predict the future state using available information about the current state. Second, we forecast a new data point using the observation equation. Third, once a new data is observed the state vector is updated using information contained in the new data. These three steps are iterated to form the Kalman filter algorithm.

Assume that we have observed the data $F_{t-1} = \{z_1, \ldots, z_{t-1}\}$ and from which we have computed an estimator of the state vector, $\widehat{\boldsymbol{\alpha}}_{t-1}$. We wish to predict the next state vector, say $\widehat{\boldsymbol{\alpha}}_{t|t-1}$, using the available information in F_{t-1}. The prediction, under the mean squared error loss, is the conditional expectation of $\boldsymbol{\alpha}_t$ in (2.63) given F_{t-1}. That is,

$$\widehat{\boldsymbol{\alpha}}_{t|t-1} = \boldsymbol{\Omega}_t \widehat{\boldsymbol{\alpha}}_{t-1}, \tag{2.68}$$

where it is understood that $\widehat{\boldsymbol{\alpha}}_{t-1|t-1} = \widehat{\boldsymbol{\alpha}}_{t-1}$. Let $\boldsymbol{S}_{t|t-1}$ denote the conditional covariance matrix of $\widehat{\boldsymbol{\alpha}}_{t|t-1}$ given by

$$\boldsymbol{S}_{t|t-1} = E[(\boldsymbol{\alpha}_t - \widehat{\boldsymbol{\alpha}}_{t|t-1})(\boldsymbol{\alpha}_t - \widehat{\boldsymbol{\alpha}}_{t|t-1})' | F_{t-1}].$$

By (2.68) and (2.63),

$$\boldsymbol{\alpha}_t - \widehat{\boldsymbol{\alpha}}_{t|t-1} = \boldsymbol{\Omega}_t(\boldsymbol{\alpha}_{t-1} - \widehat{\boldsymbol{\alpha}}_{t-1}) + \boldsymbol{u}_t.$$

Plugging the above expression into the definition of $\boldsymbol{S}_{t|t-1}$ and using $\boldsymbol{S}_{t-1} = \boldsymbol{S}_{t-1|t-1}$, we obtain

$$\boldsymbol{S}_{t|t-1} = \boldsymbol{\Omega}_t \boldsymbol{S}_{t-1} \boldsymbol{\Omega}_t' + \boldsymbol{R}_t. \tag{2.69}$$

This equation has a clear intuitive interpretation: the uncertainty in predicting a new state with available information F_{t-1} is the sum of the uncertainty that we had with respect to the previous state using the same information, measured by $\boldsymbol{S}_{t-1}$, and the uncertainty of the noise of the state equation, $\boldsymbol{R}_t$. The matrix $\boldsymbol{\Omega}_t$ appears in order to relate the components of the state vectors at time $t-1$ and time t. If this matrix was the identity, $\boldsymbol{\Omega}_t = \boldsymbol{I}$, the state vector would evolve like a random walk, and the uncertainty in state estimation would increase continuously through the sum of matrices $\boldsymbol{R}_t$. Since $\boldsymbol{\Omega}_t$ is generally not the identity matrix the increase in uncertainty depends on its structure. For example, if we assume an AR(1) model, then the state vector is scalar, and $\Omega_t = \phi < 1$. The variance of the estimation follows the process $s_{t|t-1} = \phi^2 s_{t-1} + \sigma^2$ and only a portion of the uncertainty at $t-1$ is transferred to time t.

The second step of the Kalman filter is to predict the new observation z_t given F_{t-1}. This prediction is calculated again using the conditional expectation given F_{t-1} and we obtain, by the observation equation,

$$\widehat{z}_{t|t-1} = E(z_t | F_{t-1}) = \boldsymbol{H}_t \widehat{\boldsymbol{\alpha}}_{t|t-1}. \tag{2.70}$$

The uncertainty of this prediction depends on the covariance matrix of the prediction errors, namely,

$$\boldsymbol{e}_t = \boldsymbol{z}_t - \hat{\boldsymbol{z}}_{t|t-1}.$$

Define

$$\boldsymbol{P}_{t|t-1} = E[\boldsymbol{e}_t \boldsymbol{e}_t'], \tag{2.71}$$

which can be obtained as follows. Subtracting the prediction (2.70) from the observation equation (2.62), we have

$$\boldsymbol{e}_t = \boldsymbol{z}_t - \hat{\boldsymbol{z}}_{t|t-1} = \boldsymbol{H}_t(\boldsymbol{\alpha}_t - \hat{\boldsymbol{\alpha}}_{t|t-1}) + \boldsymbol{\epsilon}_t. \tag{2.72}$$

Plugging this expression into the definition of $\boldsymbol{P}_{t|t-1}$, we obtain

$$\boldsymbol{P}_{t|t-1} = \boldsymbol{H}_t \boldsymbol{S}_{t|t-1} \boldsymbol{H}_t' + \boldsymbol{V}_t. \tag{2.73}$$

Equation (2.73) indicates that the uncertainty of the prediction $\hat{\boldsymbol{z}}_{t|t-1}$ is the sum of the uncertainty in the state vector and the measurement error. The prediction error that comes from the state estimation is modulated depending on the matrix $\boldsymbol{H}_t$. If $\boldsymbol{H}_t$ is the identity matrix, meaning that the observation $\boldsymbol{z}_t$ is the measurement of the state vector plus a random error, the measurement error of the observation is added to the error of the state variable.

The third and final step in the filter is to update the state estimation in light of the new information. Assume that $\boldsymbol{z}_t$ has been observed and the available information becomes $F_t = (F_{t-1}, \boldsymbol{z}_t)$. The new state estimation, $\hat{\boldsymbol{\alpha}}_t = \hat{\boldsymbol{\alpha}}_{t|t} = E(\boldsymbol{\alpha}_t|F_t)$, is calculated by regression (or by properties of the multivariate normal) with

$$\begin{aligned} E(\boldsymbol{\alpha}_t|F_{t-1}, \boldsymbol{z}_t) &= E(\boldsymbol{\alpha}_t|F_{t-1}) \\ &\quad + \mathrm{Cov}(\boldsymbol{\alpha}_t, \boldsymbol{z}_t|F_{t-1})\mathrm{Var}(\boldsymbol{z}_t|F_{t-1})^{-1}(\boldsymbol{z}_t - E(\boldsymbol{z}_t|F_{t-1})). \end{aligned} \tag{2.74}$$

In the above equation, the expectations $E(\boldsymbol{\alpha}_t|F_{t-1}) = \hat{\boldsymbol{\alpha}}_{t|t-1}$ and $E(\boldsymbol{z}_t|F_{t-1}) = \hat{\boldsymbol{z}}_{t|t-1}$ are known, as is the matrix $\mathrm{Var}(\boldsymbol{z}_t|F_{t-1}) = \boldsymbol{P}_{t|t-1}$. All that remains to be calculated is the covariance between the state and the new observation, which is given by

$$\mathrm{Cov}(\boldsymbol{\alpha}_t, \boldsymbol{z}_t|F_{t-1}) = E[(\boldsymbol{\alpha}_t - \hat{\boldsymbol{\alpha}}_{t|t-1})(\boldsymbol{z}_t - \hat{\boldsymbol{z}}_{t|t-1})'] = E[(\boldsymbol{\alpha}_t - \hat{\boldsymbol{\alpha}}_{t|t-1})\boldsymbol{e}_t'],$$

and by Eq. (2.72)

$$\mathrm{Cov}(\boldsymbol{\alpha}_t, \boldsymbol{z}_t|F_{t-1}) = E[(\boldsymbol{\alpha}_t - \hat{\boldsymbol{\alpha}}_{t|t-1})((\boldsymbol{\alpha}_t - \hat{\boldsymbol{\alpha}}_{t|t-1})'\boldsymbol{H}_t' + \boldsymbol{\epsilon}_t')] = \mathbf{S}_{t|t-1}\mathbf{H}_t'. \tag{2.75}$$

Equation (2.75) holds because the observation error $\boldsymbol{\epsilon}_t'$ is white noise and independent of $\boldsymbol{\alpha}_t - \hat{\boldsymbol{\alpha}}_{t|t-1}$. Plugging this covariance into (2.74), we can write:

$$\hat{\boldsymbol{\alpha}}_t = \hat{\boldsymbol{\alpha}}_{t|t-1} + \boldsymbol{K}_t(\boldsymbol{z}_t - \hat{\boldsymbol{z}}_{t|t-1}), \tag{2.76}$$

where $\boldsymbol{K}_t$ is the matrix of regression coefficients, which is called the *Kalman gain* of the filter, and is given by

$$\boldsymbol{K}_t = \boldsymbol{S}_{t|t-1}\boldsymbol{H}_t'\boldsymbol{P}_{t|t-1}^{-1}.$$

Equation (2.76) indicates that the revision we make on the previous state estimation depends on the prediction error, $\boldsymbol{e}_t = \boldsymbol{z}_t - \hat{\boldsymbol{z}}_{t|t-1}$. If this error is zero, we do not modify the estimation because no new information is available. Otherwise, we make a modification in the state estimation that depends on the quotient of the uncertainty in the state estimation, $\boldsymbol{S}_{t|t-1}$, and the precision of the prediction error, $\boldsymbol{P}_{t|t-1}^{-1}$. The matrix $\boldsymbol{H}_t'$ allows us to compare these two matrices. An equivalent way of writing Eq. (2.76) is

$$\hat{\boldsymbol{\alpha}}_t = (\boldsymbol{I} - \boldsymbol{K}_t\boldsymbol{H}_t)\hat{\boldsymbol{\alpha}}_{t|t-1} + \boldsymbol{K}_t z_t,$$

which indicates that the state estimation is a linear combination of the two sources of information that are available to us. On the one hand, the prior estimation, $\hat{\boldsymbol{\alpha}}_{t|t-1}$, and, on the other, the observation z_t. It can be proved that the weights of the two information sources are equal to their relative precision. The conditional covariance matrix of this estimation is

$$\boldsymbol{S}_t = E[(\boldsymbol{\alpha}_t - \hat{\boldsymbol{\alpha}}_t)(\boldsymbol{\alpha}_t - \hat{\boldsymbol{\alpha}}_t)'|F_t].$$

Replacing $\hat{\boldsymbol{\alpha}}_t$ with its expression in Eq. (2.76), we have

$$\boldsymbol{S}_t = E[(\boldsymbol{\alpha}_t - \hat{\boldsymbol{\alpha}}_{t|t-1} - \boldsymbol{K}_t\boldsymbol{e}_t)(\boldsymbol{\alpha}_t - \hat{\boldsymbol{\alpha}}_{t|t-1} - \boldsymbol{K}_t\boldsymbol{e}_t)'|F_t],$$

and using (2.71) and (2.75) we finally obtain

$$\boldsymbol{S}_t = \boldsymbol{S}_{t|t-1} - \boldsymbol{S}_{t|t-1}\boldsymbol{H}_t'\boldsymbol{P}_{t|t-1}^{-1}\boldsymbol{H}_t\boldsymbol{S}_{t|t-1} = (\boldsymbol{I} - \boldsymbol{K}_t\boldsymbol{H}_t)\boldsymbol{S}_{t|t-1}. \tag{2.77}$$

The system formed by Eqs. (2.68)–(2.70), (2.76), and (2.77) is referred to as the *Kalman filter* algorithm. To start the algorithm, an initial value must be given for the state variables, $\boldsymbol{\alpha}_0$, and for the covariance matrix $\boldsymbol{S}_0$. These initial values are not crucial, because the filter does not depend heavily on the initial conditions. The reader interested in the implementation of the algorithm can consult Young (1984), Harvey (1990), and Gómez and Maravall (1994). Corresponding to the Kalman filter algorithm, there are recursive algorithms for filtering and smoothing too. Interested readers are referred to Tsay (2010, chapter 11), among other textbooks on SSMs.

2.6 FORECASTING WITH LINEAR MODELS

2.6.1 Computing Optimal Predictors

Given a realization of a scalar time series z_t with T equally spaced observations, say $\boldsymbol{z}_T = (z_1, \dots, z_T)$, our goal is to predict z_{T+h}, where h is a positive integer referring to as the *forecast horizon* and T is called the *forecast origin*. In this section, we focus on linear predictors, which are linear combinations of the available data, and employ the minimum squared prediction error (MSPE) criterion. That is, we use a square loss function. This is mainly for simplicity and is also commonly used in the literature. Denote the optimal point forecast by $\hat{z}_T(h)$. It is easy to see that, under the MSPE criterion, we have $\hat{z}_T(h) = E(z_{T+h}|\boldsymbol{z}_T)$ and the prediction error, $e_T(h)$, is given by

$$e_T(h) = z_{T+h} - \hat{z}_T(h). \tag{2.78}$$

Assume that the data are generated by an ARIMA(p, d, q) process, $\phi_p(B)\nabla^d z_t = c + \theta_q(B)a_t$, with known parameters, including the variance of a_t. Let $\varphi_m(B) = \phi_p(B)\nabla^d$ be the AR operator of order $m = p + d$ obtained by multiplying the stationary AR(p) operator and the difference operator $(1-B)^d$. Then, we have

$$z_{T+h} = c + \sum_{i=1}^{m} \varphi_i z_{t+h-i} + a_{T+h} - \sum_{j=1}^{q} \theta_j a_{T+h-j}. \tag{2.79}$$

Taking conditional expectation of Eq. (2.79) given $\boldsymbol{z}_T$ and calling $\hat{a}_T(j) = E(a_{t+j}|\boldsymbol{z}_T)$, we obtain

$$\hat{z}_T(h) = c + \sum_{i=1}^{m} \varphi_i \hat{z}_T(h-i) - \sum_{j=1}^{q} \theta_j \hat{a}_T(h-j), \quad h > 0, \tag{2.80}$$

where it is understood that $\hat{z}_T(j) = z_{T+j}$ if $j \leq 0$, $\hat{a}_T(j) = 0$ if $j > 0$ and $\hat{a}_T(j) = a_{T+j}$ of $j \leq 0$. In practice, a_t denotes the residuals that are available for $t = 1, \dots, T$, using the estimated parameters and some initial values. The initial values for a_t are assumed to be zero, i.e. $a_j = 0$ for $j \leq 0$, and those for z_t can be set to zero or the sample mean of z_t provided that the series is stationary. For sufficiently large forecast origin T, the effect of initial values is immaterial. The predictions $\hat{z}_T(j)$ for $j > 0$ can be computed recursively via Eq. (2.80).

To illustrate, consider a stationary ARMA(1,1) model. The innovations, $a_2, \dots, a_T$, are computed recursively via

$$a_t = z_t - c - \phi z_{t-1} + \theta a_{t-1}, \qquad t = 2, \dots, T$$

and for $t = 1$ we have

$$a_1 = z_1 - c - \phi z_0 + \theta a_0,$$

where neither z_0 nor a_0 is known. To obtain a_1, we can set both z_0 and a_0 to their unconditional expectations. Specifically, we set $a_0 = 0$, because $E(a_0) = E(a_t) = 0$, and $z_0 = c/(1-\phi)$, because $E(z_0) = E(z_t) = c/(1-\phi)$ under stationarity.

In summary, given an ARIMA model and some initial values of a_t and z_t, we can compute the innovations $a_2, \dots, a_T$ recursively. Consequently, the point forecasts $\hat{z}_T(j)$ can also be computed recursively via Eq. (2.80) for $j = 1, \dots, h$. In addition, we can also compute the corresponding forecast errors accordingly via Eq. (2.78). In particular, we have

$$e_T(1) = z_{T+1} - \hat{z}_T(1) = a_{T+1},$$

implying that the innovation a_{T+1} can be interpreted as 1-step ahead prediction error provided that the model parameters are known. The variance of the 1-step ahead prediction of z_t is then σ^2.

Equation (2.80) indicates that, for $h > q$, the point forecasts $\hat{z}_T(h)$ does not explicitly depend on the MA part of the ARIMA model. The point forecasts would then determine exclusively by the AR part of the model. Indeed, for $h > q$, the point forecasts $\hat{z}_T(h)$ satisfy the difference equation

$$\hat{z}_T(h) = c + \varphi_1 \hat{z}_T(h-1) + \cdots + \varphi_m \hat{z}_T(h-m). \tag{2.81}$$

Since in an application the forecast origin T is fixed, we can apply the backshift operator to the forecast horizon. Specifically, define

$$B\hat{z}_T(h) = \hat{z}_T(h-1),$$

Then, Eq. (2.81) is written as

$$\phi(B)\nabla^d \hat{z}_T(h) = c, \quad h > q. \tag{2.82}$$

This equation is called the *eventual forecast equation* because it governs the behavior of the point forecasts $\hat{z}_T(h)$ as h increases and is greater than q. The solution of this difference equation assumes the form (see Appendix 2.A),

$$\hat{z}_T(h) = A_0^{(T)} + A_1^{(T)}h + \cdots + A_{d-1}^{(T)}h^{d-1} + \sum_{i=1}^{p} B_i^{(T)}G_i^h,$$

where $A_{d-1}^{(T)} = c$, $A_i^{(T)}$ for $i = 0, \ldots, d-2$ are constants which depend on the forecast origin T, and $B_i^{(T)}$ and G_i^h are solutions of the difference equation $\phi(B) = 0$ discussed before. Specifically, the solution consists of two parts; the first part is generated by the difference operator ∇^d and the second part is generated by $\phi(B)$ representing a mixture of exponential decays (for the real roots) and sinusoidal functions (for the complex conjugate roots). In particular, if $d = 2$, the long-run forecasts would contain a linear time trend with fixed slope c. For seasonal models the eventual forecast equation is

$$\Phi(B^s)\phi(B)\nabla_s\nabla^d\hat{z}_T(h) = c, \quad h > M,$$

where $M = q + Qs$ is the order of the MA polynomial. Note that the solutions of $\nabla_s x_t = 0$ satisfy $x_t = x_{t-s}$, which are the seasonal coefficients that repeat themselves in s periods. Thus, the long-run predictions can be written as

$$\hat{z}_T(h) = T_T(h) + S_T(h) + N_T(h),$$

where $T_T(h) = A_0^{(T)} + A_1^{(T)}h + \cdots + ch^{d-1}$ is a polynomial trend, $S_T(h)$ consists of seasonal components corresponding to the period $T + h$ and the irregular component $N_T(h)$ is a mixture of exponential decays and sinusoidal functions that decay to zero as h increases. The structure of the model would determine how the seasonal coefficients change over time.

2.6.2 Variances of the Predictions

The variances of predictions are easy to compute for an ARIMA model. Let $z_t = c + \psi(B)a_t$ be the MA(∞) representation of the model, where c denotes the mean if z_t is stationary and is a constant otherwise. Then, we have

$$z_{T+h} = c + a_{T+h} + \psi_1 a_{T+h-1} + \cdots + \psi_{h-1}a_{T+1} + \sum_{0}^{\infty} \psi_{h+i}a_{T-i}. \tag{2.83}$$

Taking conditional expectation of Eq. (2.83) given $\mathbf{z}_T$, we see that

$$\hat{z}_T(h) = c + \sum_{i=h}^{\infty} \psi_i a_{T+h-i}, \tag{2.84}$$

because $E(a_{T+j}) = 0$ for $j > 0$. Consequently, the forecast error is

$$e_T(h) = z_{T+h} - \hat{z}_T(h) = a_{T+h} + \psi_1 a_{T+h-1} + \cdots + \psi_{h-1}a_{T+1},$$

and the variance of forecast error is

$$\mathrm{Var}[e_T(h)] = \sigma^2(1 + \psi_1^2 + \cdots + \psi_{h-1}^2). \tag{2.85}$$

From Eq. (2.85), it is clear that the uncertainty in point forecast $\hat{z}_T(h)$ is a nondecreasing function of h. This makes sense for a linear model because we would be more uncertain about the remote than the near future. The equation also shows that the uncertainty in prediction is rather different between stationary and nonstationary models. For a stationary model $\psi_h \to 0$ as $h \to \infty$, and the long-term variance of the prediction converges at a constant, which is the marginal variance of the process. Recall that for a stationary series z_t, $\mathrm{Var}(z_t) = \sigma^2 \sum_{i=0}^{\infty} \psi_i^2$, where $\psi_0 = 1$.

Also, from Eq. (2.84), we see that, for a stationary process, $\hat{z}_T(h) \to c$ as $h \to \infty$, where $c = E(z_t)$ is the mean of the series. This phenomenon is referred to as *mean reverting* of a stationary series in the finance literature.

Finally, if we further assume that the process z_t follows a normal distribution, then we can use Eqs. (2.84) and (2.85) to construct interval forecast of z_{T+h} at the forecast origin T. Specifically, a $100(1-\alpha)$ percent interval forecast is

$$\hat{z}_T(h) \pm \lambda_{1-\alpha/2}\sqrt{\mathrm{Var}[e_T(h)]},$$

where λ_γ denotes the $100(1-\gamma)$ quantile of the standard normal distribution.

2.6.3 Measuring Predictability

A stationary ARMA process, z_t, can be decomposed as

$$z_t = \hat{z}_{t-1}(1) + a_t,$$

where $\hat{z}_{t-1}(1)$ is the 1-step ahead prediction, knowing the past values and the parameters of the model, and a_t is the innovation, which is independent of the past values of the series. Therefore, we have

$$\sigma_z^2 = \sigma_{\hat{z}}^2 + \sigma^2,$$

which decomposes the variance of the series, σ_z^2, into two independent sources of variability: (1) that of the predictable part, $\sigma_{\hat{z}}^2$, and (2) that of the unpredictable part, σ^2. Box and Tiao (1977) proposed measuring the predictability of a stationary series using the quotient between the variance of the predictable part and the total variance, namely

$$P = \frac{\sigma_{\hat{z}}^2}{\sigma_z^2} = 1 - \frac{\sigma^2}{\sigma_z^2}. \tag{2.86}$$

Clearly, $0 \leq P \leq 1$, and it is similar to the determination coefficient in linear regression analysis. Indeed, it indicates the proportion of variability of the series that can be explained by its history. For an ARMA process $\sigma_z^2 = \sigma^2 \sum \psi_i^2$, and the coefficient P can be written as

$$P = 1 - \left(\sum \psi_i^2\right)^{-1}.$$

For example, for an AR(1) process, since $\sigma_z^2 = \sigma^2/(1-\phi^2)$, we have

$$P = 1 - (1 - \phi^2) = \phi^2.$$

If the parameter ϕ is near zero, the process is close to being a white noise and its predictability should be close to zero. If ϕ is near one, the process approaches a random walk, and the coefficient P is close to 1.

The measure P is not helpful for a nonstationary ARIMA process because then the marginal variance approaches infinity and the value of P is always 1. To generalize the predictability measure, one can examine carefully the quantities used in Eq. (2.86). Consider the variance ratio σ^2/σ_z^2. The numerator is the variance of 1-step ahead forecast error, whereas the denominator is the variance of the forecast error with infinite forecast horizon. Thus, P can also be interpreted as the relative reduction in prediction variability when going from the long run prediction (infinite forecast horizon) to the short run prediction (1-step ahead forecast). Therefore, one possible generalization of predictability measure is to extend the forecast horizon of the numerator yet shorten the forecast horizon of the denominator.

Diebold and Kilian (2001) define the predictability of a time series with forecast horizon h as follows:

$$P_h(a) = 1 - \frac{\text{Var}[e_t(h)]}{\text{Var}[e_t(h+a)]},$$

where both h and a are positive integers. The predictability of Eq. (2.86) is the limiting case of $h = 1$ and $a \to \infty$. The statistic $P_h(a)$ measures the increase in the variabilities of forecast errors when one increases the forecast horizon from h to $h + a$.

The measure $P_h(a)$ is well-defined for ARIMA model so long as a and h are finite. Consider the variance of h-step ahead forecast error given in Eq. (2.85). It is easy to see that

$$P_h(a) = 1 - \frac{\sum_{i=0}^{h} \psi_i^2}{\sum_{i=0}^{h+a} \psi_i^2} = \frac{\sum_{i=h+1}^{h+a} \psi_i^2}{\sum_{i=0}^{h+a} \psi_i^2},$$

where $\psi_0 = 1$.

2.7 MODELING A SET OF TIME SERIES

Consider a set of k scalar time series $\{z_{it} | i = 1, \ldots, k; t = 1, \ldots, T\}$. The goal of this section is to explore the basic features of the data, to model individual series, and to summarize the results. In real applications, the k series may have different sample sizes or were observed at different time spans. In this case, some pre-screening is needed to divide the k series into subgroups such that each subgroup contains series that were observed at the same time span and have approximately the same sample sizes.

The first step of the proposed analysis is to visualize the set of series using plots discussed in Section 2.1. As illustrated before, those plots can reveal series with gross measurement errors and other sources of heterogeneity. If some series have a small proportion of random missing values or outlying data points, say <10%, then it is recommended to apply interpolation to clean those aberrant data. For a missing value z_v, the simplest interpolation is to use the average $(z_{v-1} + z_{v+1})/2$, provided that $v \neq 1$

or T. It will be seen in Chapter 4 that this solution turns out to be the optimal one if the time series z_t follows a random walk. In some cases, interpolation may require a few iterations to adequately handle a batch of missing values. If the proportion of missing values or outlying observations is high, say > 25%, then special care is needed to handle the aberrant data. See, Chapter 4 for statistical methods that can handle missing values and outliers.

If the number of time series k is not large, one can perform a detailed analysis of each individual series and summarize the resulting models accordingly. On the other hand, when k is large, it would not be feasible, or be too time-consuming, to model individual series by hand and an automatic modeling procedure is needed. Several procedures have been proposed for automatic ARIMA modeling, such as the TRAMO program, developed by Gómez and Maravall (1994), that was a pioneer approach widely used for analyzing economic time series. The automatic modeling procedure proposed in the next section is based on this and other previous approaches. It focuses on dealing with large data sets with many time series and consists of three steps. The first step, called *identification or classification*, explores the need of data transformation of individual series. It also investigates the degree of differencing for achieving weak stationarity. The step further includes the specification of a finite set of ARIMA models to be entertained for the series under study. The specified set is typically carried out by choosing the maximum values of p, d, and q of ARIMA models.

The second step is called *estimation and model selection*. The parameters of all specified models are estimated and the best model is selected. The estimation is carried out by a quasi-maximum likelihood (ML) method with a Gaussian likelihood function and the best model is selected via an information criterion. The third step is *diagnosis checking or validation*. The selected model for each series is checked for its adequacy using several test statistics. The goal is to verify that the selected model fulfills the basic assumptions and has no serious inadequacy. If some deficiencies of a selected model are found, the model is refined accordingly. Finally, the best selected model, or a set of models, is used for forecasting. In what follows, we describe the three steps of this procedure in more details.

Keep in mind, however, that there exists no best automatic modeling procedure for all applications. In fact, one is likely to pay a price for using an automatic program, because the complexity of real data can exceed the capability of a given automatic modeling procedure. For instance, most automatic procedures, including the one presented next, only allow for a single seasonal period, yet in reality a seasonal time series can have multiple seasonal periods. Consider the hourly measurements of $PM_{2.5}$ discussed in Chapter 1. Experience indicates that such a time series may have both diurnal and annual cycles with seasonal periods 24 and 8760 (365×24), respectively.

2.7.1 Data Transformation

The first decision to make for the proposed automatic procedure is whether to transform the observed series with the objective of working with a series that has a constant variance over time. For many empirical time series, the variability tends to increase with the level of the series in such a way that the standard deviation of the series at a given time, σ_t, is a function of the mean μ_t. That is,

$$\sigma_t \propto \mu_t^{\alpha}, \tag{2.87}$$

where α is a real number. In this case, it can be shown (Box and Cox, 1964) that the following power transformation

$$y_t = \frac{x_t^{1-\alpha} - 1}{1-\alpha}, \tag{2.88}$$

can result in y_t having approximately a constant variability. The family of *Box–Cox transformations* defined by Eq. (2.88) includes the powers of the variable and, when $\alpha \to 1$, $y_t = \ln(z_t)$, where, for simplicity, we assume $z_t > 0$. The parameter α can be estimated by making consecutive homogeneous groups of the observations, calculating the standard deviation, say s_i, and the mean, say $\bar{x}_i$, of each group, and fitting the linear regression

$$\log s_i = \beta_0 + \widehat{\beta}_1 \log \bar{x}_i, \quad i = 1, \ldots, g, \tag{2.89}$$

where g is the number of groups, via the ordinary LS method. The α of Eq. (2.87) can then be estimated by $\widehat{\beta}_1$ of Eq. (2.89).

Some remarks of the transformation are in order. First, if one is concerned with the impact of outliers or measurement errors on the decision of transformation, he/she can use some robust estimators for the scale and location of each group in lieu of the standard deviation and mean. For instance, one can make use of the group MAD, defined as

$$\text{MAD}(x_1, \ldots, x_n) = \text{median}\{|x_i - \text{median}_j(x_j)|\},$$

and choose the scaling parameter as 0.6745×MAD and the group median as the location parameter. Second, if the number of series k is large, it might be too time-consuming to run the regression (2.89) for all series. To simplify the computation, we only consider four possible values of α in the proposed procedure. They are $\alpha = 0$ (no transformation), $\alpha = 0.5$ (square root), $\alpha = 1$ (logarithm), and $\alpha = 2$ (inverse transformation). These four transformations are easier to interpret in practice. Third, with the four possible values of α, we use the following procedure to determine the transformation. Consider the hypotheses $H_{10} : \beta_1 = 0$; $H_{20} : \beta_1 = 1$; $H_{30} : \beta_1 < 1$; and $H_{40} : \beta_1 > 1$. Starting with H_{10}. If the hypothesis is not rejected, we take no transformation and stop the procedure. If H_{10} is rejected, we consider H_{20}. If this hypothesis is not rejected, we take the log transformation and terminate the procedure. If H_{20} is rejected, we consider H_{30}. If the hypothesis is not rejected, we take the square root transformation; otherwise, we take the inverse transformation. Fourth, it is advisable that the grouping of data be carried out so that each group is as homogeneous as possible. For a seasonal series with periodicity s the size of each group should be a multiple of s so that at least a complete season is included. Also, the number of groups g should be at least 7 to have enough data points to estimate the regression in Eq. (2.89). For nonseasonal series we take $n_g \geq 6$ contiguous observations to form a group and require $g \geq 7$ so that we assume that the sample size T is at least 42 for the consideration of checking data transformation.

2.7.2 Testing for White Noise

The second decision to make in the proposed procedure is to test whether a given time series is a white noise. To this end, we apply the rank-based Ljung–Box statistics $Q(m)$ and reject the white noise hypothesis if $Q(m) > \chi_m(0.05)$, where $\chi_m(0.05)$ denotes the 95% quantile of a chi-square distribution with m degrees of freedom. For a given time series $\{z_t | t = 1, \ldots, T\}$, the rank-based ACF is the ACFs of its corresponding rank time series, obtained by replacing z_t with its rank in the sample. The rank-based ACFs have been found to be robust to outliers and resulting Ljung–Box statistics do not require the existence of any moment of z_t. See Dufour and Roy (1985), Tsay (2010) and the references therein. We use $m = 12$ in the procedure. This step is important because no further modeling is needed for a detected white noise series.

2.7.3 Determination of the Difference Order

The third decision to make for a non-white noise series is to determine the order of differencing to achieve weak stationarity. That is, the selection of d of an ARIMA model for a nonseasonal series and both d and D for a seasonal series with a single periodicity s. To this end, we employ the well-known augmented Dickey–Fuller unit-root test for nonseasonal time series. See Dickey and Fuller (1979). Specifically, consider the model

$$\nabla z_t = c + \alpha z_{t-1} + \sum_{j=1}^{\ell} \phi_j \nabla z_{t-j} + a_t,$$

where ℓ is a nonnegative integer, and test the null hypothesis $H_0 : \alpha = 0$ versus $H_a : \alpha < 0$. The test statistic used is the t-ratio of the LS estimate of α, namely $t = \hat{\alpha}/s(\hat{\alpha})$. Under H_0, the t-ratio follows asymptotically a function of the standard Brownian motion (or Wiener process), and its critical values have been tabulated in the literature. Again, see Dickey and Fuller (1979). The null hypothesis H_0 is rejected for a large t-ratio. In the proposed procedure, we reject H_0 if $t < -3.43$. This corresponds roughly to using a type-I error of 1%. This choice of type-I error is based on the consideration that, for the purpose of forecasting a scalar time series, over-differencing often fares better than under-differencing. Type-I errors 0.05 and 0.025 are also available. Since ℓ is unknown, we use an automatic AR order selection to determine its value. An alternative is to fix ℓ *a priori* such as $\ell = 2$, resulting in the model

$$\nabla z_t = c + \alpha z_{t-1} + \beta_1 \nabla z_{t-1} + \beta_2 \nabla z_{t-2} + a_t, \tag{2.90}$$

to make inference in selecting d. If the null hypothesis H_0 is rejected, no difference is needed so that $d = 0$. On the other hand, if H_0 is not rejected, then $d \geq 1$, and we repeat the testing process using ∇z_t as the observed series. This second stage of unit-root test enables us to determine $d = 1$ or $d = 2$. Experience indicates that $d = 1$ is common in business and economic series, but $d \geq 2$ is rare. The second-stage of testing can be dropped in most applications.

For a seasonal time series z_t with seasonal period $s > 1$, let $\nabla z_t = (1 - B)z_t$, $\nabla_s z_t = (1 - B^s)z_t$, and $y_t = \nabla_s \nabla z_t = (1 - B)(1 - B^s)z_t$. We employ the model

$$y_t = c + \alpha_s \nabla z_{t-s} + \alpha_r z_{t-1} + \sum_{j=1}^{\ell} \phi_j y_{t-j} + a_t, \quad \ell \geq 0, \tag{2.91}$$

and consider the t-ratios of the LS estimates of α_1 and α_2 as unit-root test statistics. Specifically, let $t_s = \hat{\alpha}_s/\text{std}(\hat{\alpha}_s)$ and $t_r = \hat{\alpha}_r/\text{std}(\hat{\alpha}_r)$. If $t_s \geq -3.43$, seasonal difference is needed so that $D = 1$, otherwise, $D = 0$. Similarly, if $t_r \geq -3.43$, then $d = 1$, otherwise $d = 0$. If $d = 1$ and $D = 0$, then we test for the need of a second regular difference using $(1 - B)z_t$ as the observed series. We do not check for the second regular difference if $d = D = 1$, because the series already contains the difference factor $(1 - B)^2$.

Example 2.5

To illustrate the proposed methods for data transformation and differencing, we consider the data set of 99 series formed by daily World Stock Indexes in the file `Stock_indexes_99world.csv`. See Chapter 1 for further information on the data. The time plots of these 99 series are available there. The program for checking possible transformation is called `chktrans` in the R program developed for the book. Details of the commands used and the results are given below:

```
>WStock <- read.csv2("Stock_indexes_99world.csv", header=FALSE)
>x=WStock
>x=x[,-1] ## Remove the time index
> out=chktrans(x, output=F)
Transformed series are in the output.
> names(out)
> out$Summary
[1] 47 12 40
 > out$lnTran ## Series that require log-transformation
 [1]  3  9 10 11 14 21 25 26 29 32 33 34 35 36 40 42 45 47 53 54
 55 57 58 59 60 62 64 65 67 69 73 75 81 83 84 85 86 87 89 90
 92 93 94 95 97 98 99
> length(out$lnTran)
[1] 47
> out$sqrtTran ## Series that require square-root transformation
 [1] 20 41 48 49 56 61 66 70 71 79 80 96
> out$noTran ## Series that require no transformation.
 [1]  1  2  4  5  6  7  8 12 13 15 16 17 18 19 22 23 24 27 28 30
 31 37 38 39 43 44 46 50 51 52 63 68 72 74 76 77 78 82 88 91
> length(out$noTran)
[1] 40
```

The Summary indicates that 47 series require the log transformation, 12 series need the square root and 40 do not require any transformation. The output of the program also identifies the series in each group.

The R command used to classify stationary or unit-root nonstationary series is called `SummaryModel` in the R program developed for the book. In fact, the command also identifies the ARIMA model for each individual series. Here we use it only for the classification purpose. Consider the order of differencing for the 99 stock index series. For simplicity, we take the log transformation of each stock index series, because return series are typically used in financial study. The output of the command is given before. From which, we see that all the series require the first difference to be stationary.

```
>lx=log(x)
>out1=SummaryModel(lx) All series are nonstationary
Number of first-differenced series: 99
```
■

2.7.4 Model Identification

This is an important step of the proposed automatic modeling procedure. Its goal is to specify the order (p, q) for nonseasonal series and the orders (p, q) and (P, Q) for seasonal series. Ideally, we would like to try as many candidate models as possible, but due to various issues such as the difficulty of over-fitting and computational time constraint, some restrictions are needed to narrow down the entertained models. To this end, we specify the maximum values of p and q, say $p_{\max}$ and $q_{\max}$, respectively, and estimate all possible models with $0 \leq p \leq p_{\max}$ and $0 \leq q \leq q_{\max}$. Then, we use an information criterion such as Bayesian information criterion (BIC) to perform model selection. For ease in computation, especially for large k, we set $(p_{\max}, q_{\max}) = (5,3)$. For seasonal time series, we also specify the maximum values of P and Q. More specifically, for seasonal time series, we set $(p_{\max}, d, q_{\max}) = (2, 1, 3)$ and $(P_{\max}, D, Q_{\max}) = (1, 1, 1)$ in our procedure. These maximum orders are default values, and they can be changed if needed.

2.8 ESTIMATION AND INFORMATION CRITERIA

2.8.1 Conditional Likelihood

Estimation of a specified ARIMA model is often carried out via the conditional ML method under the normality assumption. For AR models, the LS method can be used and is faster. Here we briefly outline the likelihood function approach. For simplicity, we define $\boldsymbol{z}_s^n = (z_s, z_{s+1}, \ldots, z_n)'$ as the vector of data collection from time index s to n of a time series z_t. Suppose we want to estimate a zero-mean AR(p) model for a stationary Gaussian series $\boldsymbol{\omega}_1^T = (\omega_1, \ldots, \omega_T)'$. In practice, ω_t might be a differenced series of z_t as $\omega_t = \nabla^d z_t$. For an AR(p) model, we have that, for $t > p$, the conditional density function $f(\omega_t|\boldsymbol{\omega}_1^{t-1})$ is normal with mean $\boldsymbol{\phi}_p'\boldsymbol{\omega}_{t-1}^{t-p}$ and variance σ^2, where $\boldsymbol{\phi}_p = (\phi_1, \ldots, \phi_p)'$. Therefore,

$$f(\boldsymbol{\omega}_1^T) = f(\omega_1, \ldots, \omega_p) \prod_{t=p+1}^{T} f(\omega_t|\boldsymbol{\omega}_{t-1}^{t-p}).$$

Since p is fixed, the likelihood function would be dominated by the second term of the above equation for large T. One can then focus on the conditional likelihood function, namely $f(\boldsymbol{\omega}_{p+1}^T|\boldsymbol{\omega}_1^p)$ to perform estimation. The log likelihood function then becomes

$$L_C(\boldsymbol{\phi}_p, \sigma^2) = -\frac{(T-p)}{2} \ln \sigma^2 - \frac{1}{2} \sum_{t=p+1}^{T} \frac{(\omega_t - \boldsymbol{\phi}_p'\boldsymbol{\omega}_{t-1}^{t-p})^2}{\sigma^2}.$$

Define $\boldsymbol{W} = [\boldsymbol{\omega}_{p+1}^T, \ldots, \boldsymbol{\omega}_1^{T-p}]$. We can express the conditional log likelihood function in a matrix form as

$$L_C(\boldsymbol{\phi}_p, \sigma^2) = -\frac{(T-p)}{2} \ln \sigma^2 - \frac{1}{2\sigma^2}(\boldsymbol{\omega}_{p+1}^T - \boldsymbol{W}\boldsymbol{\phi}_p)'(\boldsymbol{\omega}_{p+1}^T - \boldsymbol{W}\boldsymbol{\phi}_p).$$

Taking the derivative with respect to $\boldsymbol{\phi}_p$ and setting them to zero, we have

$$\frac{\partial L_C(\boldsymbol{\phi}_p, \sigma^2)}{\partial \boldsymbol{\phi}_p} = 0 = -\boldsymbol{W}'\boldsymbol{\omega}_{p+1}^T + \boldsymbol{W}'\boldsymbol{W}\widehat{\boldsymbol{\phi}}_p, \tag{2.92}$$

and, therefore,

$$\widehat{\boldsymbol{\phi}}_p = (\boldsymbol{W}'\boldsymbol{W})^{-1}\boldsymbol{W}'\boldsymbol{\omega}_{p+1}^T, \tag{2.93}$$

which is the LS estimate of $\boldsymbol{\phi}_p$ for an AR(p) model and can be computed easily and efficiently. The covariance matrix of the estimate is given by

$$\text{Cov}(\widehat{\boldsymbol{\phi}}_p) = (\boldsymbol{W}'\boldsymbol{W})^{-1}\widehat{\sigma}_a^2, \tag{2.94}$$

where

$$\widehat{\sigma}_a^2 = \frac{1}{T-p}\sum_{t=p+1}^{T}(\omega_t - \widehat{\boldsymbol{\phi}}_p'\boldsymbol{\omega}_{t-1}^{t-p})^2. \tag{2.95}$$

The estimates $\widehat{\boldsymbol{\phi}}_p$ and $\widehat{\sigma}_a^2$ are consistent and they are close to their exact likelihood counterparts provided that p/T is small.

For ARMA(p, q) models, the likelihood function is nonlinear with respect to the MA parameters, but a fast LS estimation has been proposed by Hannan and Rissanen (1982). This is an approximation method and involves two steps. In the first step, a high-order AR(p_o) model is fitted to obtain the residuals. Ideally, $p_o > p + q$ and the residuals should be serially uncorrelated. Denoted the residuals by $\widehat{b}_t$, which is used as a proxy for the innovation a_t. In the second step, one considers the model

$$w_t = \phi_1 w_{t-1} + \cdots + \phi_p w_{t-p} + \alpha_1 \widehat{b}_{t-1} + \cdots + \alpha_q \widehat{b}_{t-q} + a_t, \tag{2.96}$$

and applies the LS estimates. The coefficients $\widehat{\phi}_i$ and $-\widehat{\alpha}_i$ serve as estimators of the ϕ_i and θ_i parameters, respectively. With p_o increases with the sample size T, one expects that the approach would produce good estimates of a stationary ARMA(p, q) model. In practice, this approach can be used to provide good initial parameter estimates for the ML estimation, especially for the vector ARMA models. See Tsay (2014).

In practice, the high-order p_o can be selected by an information criterion. One can then use the AR(p_o) residuals as a proxy for the innovation a_t and use Eq. (2.96) to estimate low-order ARMA(p, q) models. The estimation results can then be used to compute information criteria for ARMA model selection. Such an approximation method could be useful when the number of time series k is large.

2.8.2 On-line Estimation

Once an ARMA model has been estimated, we can perform a fast updating estimation via the conditional likelihood approach when a new observation becomes available. Such an on-line estimation is useful in practice. Suppose that we have estimated the parameters with the sample ω_1^T and observe the new data ω_{T+1}. Instead of re-estimating the model, we can update the parameters using the Kalman filter. We present the case of an AR(p) model first. Using the equations of Section 2.4.3, the observation equation is now $\omega_t = \boldsymbol{h}_t'\boldsymbol{\alpha}_t + a_t$, where $\boldsymbol{h}_t' = (\omega_{t-1}, \ldots, \omega_{t-p})$ and $\boldsymbol{\alpha}_t = (\phi_{1,t}, \ldots, \phi_{p,t})'$ and the state transition equation is $\boldsymbol{\alpha}_t = \boldsymbol{\alpha}_{t-1}$. In the general state representation of Section 2.4.2, we have $\boldsymbol{\Omega}_t = \boldsymbol{I}$, $\boldsymbol{R}_t = 0$, $\boldsymbol{H}_t = \boldsymbol{h}'_t$ and $\boldsymbol{V}_t = \sigma^2$. The estimation of the parameters, or the state vector, at time T is given by Eq. (2.93). In other words, we have

$$\boldsymbol{\alpha}_T = \hat{\boldsymbol{\phi}}_{p,T} = (\boldsymbol{W}'_T\boldsymbol{W}_T)^{-1}\boldsymbol{W}'_T\boldsymbol{\omega}^T_{p+1},$$

where the subscript T is used to denote the data $\boldsymbol{w}^T_1$, and the associated covariance matrix is

$$\boldsymbol{S}_{T+1|T} = \boldsymbol{S}_T = \text{Cov}(\hat{\boldsymbol{\phi}}_{p,T}) = (\boldsymbol{W}'_T\boldsymbol{W}_T)^{-1}\hat{\sigma}^2_{a,T}.$$

The prediction error of the new observation ω_{T+1} at the forecast origin T is

$$e_{T+1} = \hat{a}_{T+1|T} = \omega_{T+1} - \hat{\boldsymbol{\phi}}_{p,T}\boldsymbol{\omega}^T_{T-p+1}, \tag{2.97}$$

with variance (see Eq. (2.73))

$$\begin{aligned} p_{T+1|T} = E(e^2_{T+1}) &= [(\boldsymbol{\omega}^T_{T-p+1})'(\boldsymbol{W}'_T\boldsymbol{W}_T)^{-1}(\boldsymbol{\omega}^T_{T-p+1}) + 1]\hat{\sigma}^2_{a,T} \\ &= (1 + h_{T+1|T})\hat{\sigma}^2_{a,T}, \end{aligned} \tag{2.98}$$

where

$$h_{T+1|T} = (\boldsymbol{\omega}^T_{T-p+1})'(\boldsymbol{W}'_T\boldsymbol{W}_T)^{-1}(\boldsymbol{\omega}^T_{T-p+1}),$$

and the new estimate of the parameters, given ω_{T+1}, is

$$\boldsymbol{\alpha}_{T+1} = \hat{\boldsymbol{\phi}}_{p,T+1} = \hat{\boldsymbol{\phi}}_{p,T} + (\boldsymbol{W}'_T\boldsymbol{W}_T)^{-1}\boldsymbol{\omega}^T_{T-p+1}e^{T+1}/(1 + h_{T+1|T}). \tag{2.99}$$

Note that the changes in the parameters are proportional to the prediction error e_{T+1}, and if this prediction error is zero the parameters are not updated. Otherwise, the parameter updates depend on the vector of regressors $\boldsymbol{\omega}^T_{T-p+1}$. The new covariance matrix of the parameters is given by

$$\boldsymbol{S}_{T+1} = (\boldsymbol{W}'_{T+1}\boldsymbol{W}_{T+1})^{-1}\hat{\sigma}^2_{a,T+1}, \tag{2.100}$$

that can be written

$$\boldsymbol{S}_{T+1} = (\boldsymbol{W}'_T\boldsymbol{W}_T)^{-1}\hat{\sigma}^2_{a,T}[\boldsymbol{I} - \boldsymbol{\omega}^T_{T-p+1}(\boldsymbol{\omega}^T_{T-p+1})'(\boldsymbol{W}'_T\boldsymbol{W}_T)^{-1}/(1 + h_{T+1|T})]. \tag{2.101}$$

Since $\hat{\sigma}^2_a(T) = \sum^T_{t=p+1}(\omega_t - \hat{\boldsymbol{\phi}}'_p(T)\boldsymbol{\omega}^{t-p}_{t-1})^2/(T-p)$ and $\hat{\sigma}^2_a(T+1) = \sum^{T+1}_{t=p+1}(\omega_t - \hat{\boldsymbol{\phi}}'_p(T+1)\boldsymbol{\omega}^{t-p}_{t-1})^2/(T+1-p)$ are known, we can compute $(\boldsymbol{W}'_{T+1}\boldsymbol{W}_{T+1})^{-1}$ by Eqs. (2.100) and (2.101) without the need to compute the inverse matrix directly.

For ARMA models an approximate updating scheme can also be derived by using the linear model representation Eq. (2.96). In fact, the updating algorithm given by Eqs. (2.97) to (2.99) was obtained by Plackett (1950) for any linear model of the form $y_t = \boldsymbol{x}'_t\boldsymbol{\beta}_t + u_t$. Calling $e_{T+1} = y_{T+1} - \boldsymbol{x}'_{T+1}\hat{\boldsymbol{\beta}}_T$, the updating is

$$\hat{\boldsymbol{\beta}}_{T+1} = \hat{\boldsymbol{\beta}}_T + (\boldsymbol{X}'_T\boldsymbol{X}_T)^{-1}\boldsymbol{x}_{T+1}e_{T+1}/(1 + h_{T+1|T}), \tag{2.102}$$

where $h_{T+1|T} = \boldsymbol{x}'_{T+1}(\boldsymbol{X}'_T\boldsymbol{X}_T)^{-1}\boldsymbol{x}_{T+1}$, and

$$(\boldsymbol{X}'_{T+1}\boldsymbol{X}_{T+1})^{-1} = (\boldsymbol{X}'_T\boldsymbol{X}_T)^{-1}[\boldsymbol{I} - \boldsymbol{x}_{T+1}\boldsymbol{x}'_{T+1}(\boldsymbol{X}'_T\boldsymbol{X}_T)^{-1}/(1 + h_{T+1|T})]\frac{\hat{\sigma}^2_{a,T}}{\hat{\sigma}^2_{a,T+1}}, \quad (2.103)$$

that are similar to Eqs. (2.99) and (2.101).

2.8.3 Maximum Likelihood (ML) Estimation

The ML estimation of an ARMA model for a stationary series ω_t is carried out by decomposing the joint probability density function of the data as products of conditional and marginal density functions and maximizing it with respect to the parameters. The data are treated as fixed. The joint density function of $\boldsymbol{\omega}_1^T$ is

$$f(\boldsymbol{\omega}_1^T) = f(\omega_1)f(\omega_2|\omega_1)f(\omega_3|\omega_2, \omega_1)\cdots f(\omega_T|\omega_{T-1}, \ldots, \omega_1).$$

and, under joint normality, all the conditional distributions involved are normal and their expectations are the 1-step ahead predictions

$$E(\omega_t|\omega_{t-1}, \ldots, \omega_1) = \hat{\omega}_{t-1}(1) = \omega_{t|t-1}.$$

Let $e_{t|t-1}$ be the 1-step ahead prediction error of ω_t using the information of $\boldsymbol{\omega}_1^{t-1}$ and giving the parameters of the process. Thus,

$$e_{t|t-1} = \omega_t - \omega_{t|t-1}.$$

These prediction errors are highly related to the innovations a_t of the process, but they are not identical because they depend on the initial values. Calling $\sigma^2 v_{t|t-1}$ the variance of these prediction errors, the joint density function of the data for an ARMA process can be written as

$$f(\boldsymbol{\omega}_T) = \prod_{t=1}^{T} \sigma^{-1} v_{t|t-1}^{-1/2} (2\pi)^{-1/2} \exp\left\{-\frac{1}{2\sigma^2}\sum_{t=1}^{T}\frac{(\omega_t - \omega_{t|t-1})^2}{v_{t|t-1}}\right\},$$

where $v_{1|0}$ and $\omega_{1|0}$ are initial values. Taking natural logarithm and calling $\boldsymbol{\beta} = (\mu, \sigma^2, \phi_1, \ldots, \phi_p, \theta_1, \ldots, \theta_q)$ the vector of parameters to be estimated, the log likelihood function of the ARMA model to be maximized is

$$L(\boldsymbol{\beta}) = -\frac{1}{2}T\ln\sigma^2 - \frac{1}{2}\sum_{t=1}^{T}\ln v_{t|t-1} - \frac{1}{2}\sum_{t=1}^{T}\frac{e^2_{t|t-1}}{\sigma^2 v_{t|t-1}}. \quad (2.104)$$

In practice, the likelihood function can be efficiently computed for given parameter values by putting the model in the state-space form and using the Kalman filter. Readers are referred to the algorithm discussed in Section 2.4, which provides a fast-recursive way to compute $\omega_{t|t-1}$ and $\sigma^2 v_{t|t-1}$. The maximization of the exact likelihood function is carried out using a nonlinear optimization algorithm (see Box at al., 2015; Shumway and Stoffer, 2017).

2.8.4 Model Selection

Suppose that we have estimated via the ML method a set of ARMA(p, q) models, where $0 \leq p \leq p_{\max}$ and $0 \leq q \leq q_{\max}$. Denote the estimated models by $M_1, \ldots, M_m$, and we wish to select the model which explains best the observed series or provides the best forecasts. Keep in mind that the model with the best in sample fit or the largest likelihood does not necessarily yield the best out of sample predictions, because a model with more parameters tend to have a smaller sum of squares errors. In order to select a suitable model, we can use model selection criteria or out-of-sample prediction. Model selection can be carried out either from a classical or from a Bayesian point of view. Beginning with the classical approach, we can select the model with better expected out of sample forecasting performance. This approach leads to the Akaike information criterion (AIC). From the Bayesian approach we may put prior probabilities on the estimated models, say $P(M_i)$ for $i = 1, \ldots, m$, and select the one with maximum posterior probability given the data. That is, we calculate for $i = 1, \ldots, m$,

$$P(M_i|\boldsymbol{\omega}_T) = \frac{P(\boldsymbol{\omega}_1^T|M_i)P(M_i)}{\sum_{i=1}^{m} P(\boldsymbol{\omega}_1^T|M_j)P(M_j)},$$

and choose the most probable model in view of the data. This approach does not require the series to be stationary, so it can be used to compare models with different number of differences and, hence, can serve as an alternative to the unit-root tests. If we assume that the prior probabilities of all the models are equal, this approach leads to the BIC. Details are given below. An alternative approach for selecting the best model is by means of out-of-sample forecasting performance, which is similar to using cross-validation. We describe the three approaches to model selection below.

Finally, instead of selecting a single model, we can work with a mixture of models and generate forecast using all the models involved. This approach to forecasting is called *model averaging* and has proven to be useful in applications.

2.8.4.1 *The Akaike Information Criterion (AIC)* It can be proved that if we estimate the parameters of a model via the ML method and then compute the expectation of the likelihood with respect to future observations, we obtain

$$\text{AIC}(h) = \frac{1}{T}E(-2L(\boldsymbol{\beta})) = \ln \hat{\sigma}^2_{\text{MV}} + \frac{2h}{T}, \tag{2.105}$$

where T is the sample size used to estimate the model, $\hat{\sigma}^2_{\text{MV}}$ is the ML estimator of the variance of the innovations, and h is the number of estimated parameters in calculating the 1-step ahead predictions. Therefore, selecting the model with maximum expected likelihood is equivalent to choosing the one that minimizes the AIC. This criterion is due to Akaike (1973) and, under certain conditions, is asymptotically efficient as it selects the model with smaller prediction error. However, if there exists a true model, then AIC is not asymptotically consistent, as it does not select the correct model with probability one when the sample grows.

In Eq. (2.105), the first term $\ln \hat{\sigma}^2_{\text{MV}}$ measures the goodness of fit of the model and the second term $2h/T$ is called the penalty. The criterion seeks a model that provides a good fit to the data, but prefers a model that is parsimonious in parameterization. All information criteria share this decomposition but different criteria use different penalties. Note that if h is fixed, the penalty goes to zero with the sample size, and when both h and T grow, but h/T goes to c, the penalty goes to $2c$.

If we compare models by AIC, it is important that T, the effective number of observations used to estimate the model, be the same for all of them. The number of stationary data points is equal to the original data minus $d + sD$, where d denotes the number of regular differences, s is the seasonal period and D is the number of seasonal differences. If we estimate the model by the exact ML we can calculate residuals for all data points and, hence, the effective number of data points is $T - d - sD$. If we consider models with different numbers of differences and let d_{max} and D_{max} denote the highest degrees of differences of the models involved, then the number of effective data points is

$$T_0 = T - d_{max} - sD_{max}. \tag{2.106}$$

2.8.4.2 *The Bayesian Information Criterion (BIC)* An alternative criterion was proposed by Schwarz (1978) using a Bayesian approach. The criterion is to maximize the posterior probability of the model, $P(M_i|\boldsymbol{\omega}_1^T)$, assuming that the prior probabilities are the same for all the models. Since $P(M_i|\boldsymbol{\omega}_1^T)$ is proportional to $P(\boldsymbol{\omega}_1^T|M_i)P(M_i)$, if the prior probabilities are the same, the posterior probability of the model is proportional to $P(\boldsymbol{\omega}_1^T|M_i)$. Selecting the model that maximizes this probability is equivalent to selecting the model that minimizes $-\frac{2}{T}\ln P(\boldsymbol{\omega}_1^T|M_i)$. It can be proved that replacing the parameters with their ML estimates the model that asymptotically minimizes this quantity is the one that minimizes the criterion

$$\text{BIC}(h) = \ln \hat{\sigma}^2_{\text{MV}} + \frac{h \ln T}{T}, \tag{2.107}$$

where, as before, T is the sample size, $\hat{\sigma}^2_{\text{MV}}$ is the ML estimator of the innovation variance and h is the number of parameters. This criterion is asymptotically consistent and selects the true model, if exists, with probability one when $T \to \infty$.

If we compare Eqs. (2.107) and (2.105), we see that the BIC penalizes the introduction of new parameters heavier than the AIC does; hence, it tends to choose a more parsimonious model. In both criteria if h is fixed the penalty goes to zero with the sample size. However, when both h and T grow but h/T goes to c the AIC penalty tends to $2c$ and the BIC to $c\ln(T)$ that is larger than the AIC penalty for $\ln(T) > 2$. To compare BIC of various models fitted to a time series, as with the AIC, the size of T must be calculated using Eq. (2.106). In simulations the BIC criterion seems to fare better in selecting the correct orders of ARIMA models (see for instance Lütkepohl, 1985). Keep in mind that in real applications, there exists no true model so that the consistency might not be a good property to use in comparing AIC and BIC.

2.8.4.3 *Other Criteria* Several other information criteria have been proposed in the literature. Some are modifications of the existing ones. For instance, AIC does not work well when the sample size is small so that Hurvich and Tsai (1989) proposed a modification to improve its performance in small samples. The resulting criterion is referred to as the AIC_c. See also Bengtsson and Cavanaugh (2006).

Hannan and Quinn (1979) proposed a criterion with a penalty larger than AIC, but smaller than BIC. The Hannan–Quinn (HQ) criterion is

$$\text{HQ}(h) = \ln \hat{\sigma}^2_{\text{MV}} + \frac{2h \ln(\ln T)}{T}.$$

The penalty is larger than AIC for $T > 15$, but much smaller than BIC when T is large. This criterion is asymptotically consistent and has high efficiency. In fact, the penalty $\ln(\ln T)$ is the minimum penalty rate needed for an information criterion to be consistent.

2.8.4.4 Cross-Validation An alternative approach to model selection is to split the data in two parts, estimate the model in the first and compute the forecast errors in the second. Suppose we start with an estimation or training subsample, $\{\omega_1, \ldots, \omega_{T_1}\}$, and a validation of testing part, $\{\omega_{T_1+1}, \ldots, \omega_T)\}$. Starting with the forecast origin T_1, one estimates each model with the first T_1 observations in the training sample and compute the forecasting errors for several forecast horizons h, say $e_{T_1}^{M_i}(\ell)$, where M_i denotes the model used and ℓ is the forecast horizon with $\ell = 1, \ldots, h$. Then, one increases the forecast origin by 1, i.e. $T_1 + 1$, and re-estimate the model using the first $T_1 + 1$ observations to compute ℓ-step ahead forecast errors for $\ell = 1, \ldots, h$. Denote the forecasting errors by $e_{T_1+1}^{M_i}(\ell)$ for $\ell = 1, \ldots, h$. Repeat this estimation-forecasting process until reaching the last available forecast origin, $T - 1$, and compute the 1-step ahead forecast error for the last observation $e_{T-1}^{M_i}(1)$.

In general, for a given forecast horizon h, we have out-of-sample forecast errors $\{e_{T_1+j}^{M_i}(h)\}$ for $j = 0, 1 \ldots, T - T_1 - h$. The h-step mean squared forecast errors (MSFEs) can then be computed by

$$\mathrm{Var}[e^{M_i}(h)] = \frac{1}{(T - T_1 - h + 1)} \sum_{j=1}^{T-T_1-h} [e_{T_1+j}^{M_i}(h)]^2. \tag{2.108}$$

If the h-step ahead forecasts are of interest, one can then select the model with the smallest $\mathrm{Var}[e^{M_i}(h)]$. Note that usually $h = 1$ is used, but depending upon the objective other values can be chosen. If the data were indeed generated by an underlying true model the prediction errors should behave consistently by providing the minimum MSFE at the same model for all forecast horizons. Nevertheless, in real applications, if there is no true model, the prediction errors can select different models for different horizons (see, Tiao and Xu, 1993). Clearly, the MSFE may depend on the choice of the forecasting origin T_1. There is no obvious choice that works for all cases. In practice, T_1 often satisfies $T/2 \leq T_1 \leq 3T/4$ with $T_1 = 2T/3$ being commonly used. It can be proved that, in large samples, there is an equivalence between the cross-validation method and model selection criteria.

Example 2.6

We illustrate model selection using the set of time series of world stock indexes. This is carried out using one of three R commands developed for this book. The first command is `arimaSpec`, which specifies automatically a model for each individual series of a set of nonseasonal time series. The command implements the modeling procedure discussed in the previous sections. The second command is `sarimaSpec`, which uses the procedures discussed in the book to perform model selection for seasonal time series. The third command is `arimaID`, which checks whether or not a given time series is seasonal with a given seasonality. Finally, the command `SummaryModel` developed for the book is used to provide a summary of model specification for all 99 time series. This command summarizes the results of the command `arimaSpec`.

These three commands make use of the R built-in command `arima` for time series estimation, which estimates a specified ARIMA model for a given time series. Therefore, we briefly explain the `arima` command of R. The command performs ML estimation of an ARIMA model. The order of the model must be given. For a regular ARIMA model, the order is specified by the subcommand `order=c(p,d,q)`. For a seasonal ARIMA model, one also needs to specify the seasonal part of the model with the following subcommand:

```
seasonal=list(order=c(P,D,Q),period=s).
```

Three important issues to mention for the `arima` command. First, the command uses a slightly different model formulation. For instance, an ARIMA(1,0,1) model is written as

$$(1 - \phi_1 B)(z_t - \mu) = (1 + \theta_1 B)a_t.$$

Therefore, the output of an MA parameter of the command is negative of the MA coefficient used in the book. That is, the MA coefficient involves a sign change. Second, the mean μ is called *intercept* in the output of the `arima` command. Third, if $d = 1$ or $D = 1$, the `arima` command automatically sets $\mu = 0$. Thus, care must be exercised in using the `arima` command when there exists a drift in the data.

In what follows, we use the first series of the world stock indexes of Example 2.5 to demonstrate the proposed automatic modeling procedure. The R commands are as follows:

```
> lx=log(x)
> out1=arimaSpec(lx[,1])
include.mean:   FALSE
Selected order (p,d,q):   1 1 1
minimum criterion:   -24707.47
> out2=arima(lx[,1],order=c(1,1,1)) ## R estimation
> out2
 Call: arima(x = lx[, 1], order = c(1, 1, 1))
 Coefficients:
         ar1       ma1
      0.4847   -0.5686
s.e.  0.1430    0.1352

sigma^2 estimated as 0.0001534: log likelihood=12370.39,aic=-24734.78
>  out3=arima(lx[,1],order=c(3,1,2)) ## Try another model
> out3
Call:
arima(x = lx[, 1], order = c(3, 1, 2))

Coefficients:
         ar1      ar2      ar3       ma1       ma2
      0.3428   0.2127   0.0291   -0.4277   -0.2305
s.e.  0.3831   0.2038   0.0217    0.3831    0.2255

sigma^2 estimated as 0.0001533:  log likelihood=12371.53,aic=-24731.06
> p1 = c(1,-c(out3$coef[1:3])) ## AR polynomial
> q1 = c(1,-c(out3$coef[4:5])) ## MA polynomial
```

```
> polyroot(p1)
[1]  1.420945+0.000000i -4.359908+2.268511i -4.359908-2.268511i
> polyroot(q1)
[1] -0.927846+1.865029i -0.927846-1.865029i
##The best model selected by arimaSelec is ARIMA(1,1,1). The fitted model is
## (1-0.48B)(1-B)z(t) = (1-0.57B)a(t).  Also, a more complicated
## ARIMA(3,1,2) is estimated, the coefficients are not significant at
## the 5% level, indicating over fitting. The roots of the AR and MA
## polynomials are shown. Next, the output of the program SummaryModel is
## presented
> out1=SummaryModel(lx)

All series are nonstationary
Number of first-differenced series:  99
d=1
    q=0 q=1 q=2 q=3
p=0  18   4   0   0
p=1  18   3   2   0
p=2  14   2   3   0
p=3   6   0   2   0
p=4   1   0   2   0
p=5  19   1   3   1
```

The result indicates that all the series of log stock indexes require a first difference and most of the differenced series either are white noise (18/99=18%) or assume a low order model. The specific information about the model for each time series is given as an output in matrix M2 of the command `SummaryModel`. This information is not printed, but we can find the series that follow each of the models with `SelectedSeries` that takes the information of the matrix M2 in the output of the `SummaryModel` and give the series for each model as shown next.

```
> SelectedSeries(out1$M2,c(1,1,0))
There are  18  of order:  1 1 0
Names of the series:
 [1] "x21" "x33" "x41" "x47" "x49" "x50" "x51" "x52" "x59" "x60"
     "x62" "x64" "x67" "x71" "x79" "x89" "x93" "x98"
> SelectedSeries(out1$M2,c(0,1,1))
There are  4  of order:  0 1 1
Names of the series:
[1] "x56" "x87" "x92" "x99"
> SelectedSeries(out1$M2,c(5,1,2))
There are  3  of order:  5 1 2
Names of the series:
[1] "x23" "x38" "x77"
```

■

2.9 DIAGNOSTIC CHECKING

Diagnostic checking is to perform certain residual analysis of a fitted time series model. The goal is to verify that there are no indications that the assumptions made in the modeling process are wrong. It also provides an opportunity for model refinement or improvement if any deficiency is found. The checking focuses not only on the estimated parameters, but also on the residuals of the fitted model. Ideally, parameter estimates should be statistically significant, the residuals should form a white noise series, and the fitted model is meaningful for the case under study. For a fitted ARIMA model, there should be no commonly factors between the AR and MA polynomials. In addition, the AR and MA polynomial should contain no roots inside the unit cycle. We discuss some useful tools commonly used in diagnostic checking.

2.9.1 Residual Plot

Time plot of the standardized residuals, $\hat{a}_t/\hat{\sigma}_a^2$, is important in model checking. The plot can be used for (a) identifying outlying observations such as standardized residuals with magnitude greater than 3.5, (b) checking for structural breaks, (c) verifying constant variance, and (d) inspecting serial correlations. Ideally, the standardized residuals should be random with zero mean and constant variance. They should also be broadly between −3.5 and 3.5. In Chapter 4, we will study how to analyze the residuals to identify outliers and structural breaks and correct the model from their effects.

2.9.2 Portmanteau Test for Residual Serial Correlations

The residuals of a well-specified ARIMA model should form a white noise series. To this end, the most widely used model checking statistics is the portmanteau test using residual ACFs. Let h be a positive integer, the null hypothesis of interest is $H_0 : \rho_1 = \cdots = \rho_h = 0$ versus $H_a : \rho_i \neq 0$ for some $i \in \{1, \ldots, h\}$, where ρ_i is the lag-i ACF of the residuals. Let $\hat{r}_j$ be the lag-j sample ACF of the residuals. If the residuals have no serial correlations, then $\hat{r}_j$ is asymptotically normal with mean zero and variance $(T-j)/T(T+2)$, where T is the sample size. Therefore, the portmanteau test statistic for H_0 is

$$Q(h) = T(T+2)\sum_{j=1}^{h} \frac{\hat{r}_j^2}{T-j}, \tag{2.109}$$

which, under H_0, is asymptotically distributed as χ^2_{h-m} distribution, where m is the number of parameters used in the model. Therefore, one rejects the null hypothesis H_0 if $Q(h) > \chi^2_{h-m}(\alpha)$, where $\chi^2_{h-m}(\alpha)$ is the $100(1-\alpha)$ quantile of the chi-square distribution with $h-m$ degrees of freedom. Typically, $\alpha = 0.05$ or 001 is used. A robust version of the portmanteau test statistics is to replace $\hat{r}_j$ by its rank-based counterpart. See, Tsay (2010) and the references therein.

This test statistic in Eq. (2.109) has a drawback of treating all residual ACFs equally. In practice, one may argue that certain lags of ACFs are more important than the others. For instance, in a seasonal model, $\hat{r}_s$ is likely to be more important than $\hat{r}_2$ if $s > 2$. Peña and Rodríguez (2002, 2006) proposed a more powerful test statistic to mitigate this weakness. The test statistic is based on the autocorrelation matrix of the residuals:

$$\boldsymbol{R}_m = \begin{bmatrix} 1 & \hat{r}_1 & \cdots & \hat{r}_{m-1} \\ \hat{r}_1 & 1 & \cdots & \hat{r}_{m-2} \\ \vdots & \vdots & \ddots & \vdots \\ \hat{r}_{m-1} & \hat{r}_{m-2} & \cdots & 1 \end{bmatrix},$$

and is defined as

$$D_m = -\frac{T}{m+1} \ln |\hat{\boldsymbol{R}}_m|,$$

which follows, asymptotically, a gamma distribution with parameters

$$\alpha = \frac{3(m+1)\{m-2(p+q)\}^2}{2\{2m(2m+1)-12(m+1)(p+q)\}},$$

$$\beta = \frac{3(m+1)\{m-2(p+q)\}}{2m(2m+1)-12(m+1)(p+q)}.$$

This distribution has mean $\alpha/\beta = (m+1)/2 - (p+q)$ and variance:

$$\alpha/\beta^2 = (m+1)(2m+1)/3m - 2(p+q).$$

The percentiles of the test statistic D_m are easily obtained by calculating the parameters of the gamma distribution with the above formulas. Alternatively, the variable

$$ND_m^* = (\alpha/\beta)^{-1/4}(4/\sqrt{\alpha})\left[(D_m)^{1/4} - (\alpha/\beta)^{1/4}\left(1-\frac{1}{6\alpha}\right)\right], \tag{2.110}$$

follows approximately a standard normal distribution.

It can be shown that the D_m statistic can be written as

$$D_m = T\sum_{i=1}^{m} \frac{(m+1-i)}{(m+1)}\hat{\pi}_i^2, \tag{2.111}$$

where $\hat{\pi}_i$ is the lag-i partial autocorrelation coefficient of the residuals. Thus, the test can be seen as a modified Ljung–Box test, where instead of utilizing the sample ACFs of the residuals we use the residual PACFs with weights $\frac{(m+1-i)}{m}$. These weights decrease linearly with the lag, such that $\hat{\pi}_1^2$ has weight 1 and $\hat{\pi}_m^2$ weight $1/m$.

2.9.3 Homoscedastic Tests

To test for constant variance, one can divide the residuals into two parts and apply the traditional F-test to resulting two variances. Specifically,

$$F = \frac{\sum_{t=1}^{n_1} \hat{a}_t^2/n_1}{\sum_{t=n_1+1}^{T} \hat{a}_t^2/(T-n_1)} = \frac{s_1^2}{s_2^2},$$

where n_1 is approximately $T/2$. Under the null hypothesis of constant variance, the F-statistic follows asymptotically a F-distribution with degrees of freedom n_1 and $T-n_1$. In a similar way, one can divide the residuals into h subsamples with sizes $n_1, \dots, n_h$ and compute the test statistics

$$\lambda = T\log\hat{\sigma}^2 - \sum_{i=1}^{h} n_i \log s_i^2,$$

where $\hat{\sigma}^2$ is the residual variance of the entire sample and s_i^2 is the variance of the ith subsample with n_i observations. Under the hypothesis that the variances are the same for all subsamples, the test statistic λ follows asymptotically a chi-square distribution with $h-1$ degrees of freedom. There are several tests available for testing conditional heteroscedasticity. See, for instance, Tsay (2010, chapter 3).

2.9.4 Normality Tests

A normality test for the residuals can be made by computing the coefficients of skewness and kurtosis to construct the test statistic

$$X = \frac{T\alpha_1^2}{6} + \frac{T(\alpha_2 - 3)^2}{24},$$

where α_1 and α_2 are the third and fourth sample moments of the residuals, respectively. Under the normality assumption, X is asymptotically a chi-square random variable with 2 degrees of freedom. This test statistic is commonly known as the Jarque–Bera test in the financial econometrics literature. See, Jarque and Bera (1987).

2.9.5 Checking for Deterministic Components

If an estimated model contains some difference operators and some MA polynomials, it is advisable to check the unit roots of the MA polynomial. This is particularly so for a seasonal model. The fact that the MA polynomial contains a factor that is near cancellation with the difference operator is indicative that the time series under study may contain some deterministic components. For example, the model

$$\nabla z_t = \beta_1 + (1 - \theta B)a_t \text{ with } \theta \simeq 1$$

is approximately equivalent to

$$z_t = \beta_0 + \beta_1 t + a_t,$$

which contains a deterministic linear trend. In general, the model

$$\nabla^d z_t = a_t,$$

is equivalent to the following model with a deterministic polynomial trend

$$z_t = \beta_0 + \beta_1 t + \cdots + \beta_d t^{d-1} + a_t.$$

Canceling a seasonal MA operator with $(1 - B^s)$ produces a deterministic seasonal component. For instance, the model

$$(1 - B^s)z_t = (1 - \Theta B^s)a_t \text{ with } \Theta \simeq 1$$

is approximately equivalent to

$$z_t = S_t^{(s)} + a_t,$$

where $S_t^{(s)} = S_{t-s}^{(s)}$ is a periodic function. If a fitted model contains regular and seasonal differences and the MA terms of both types that are close to $(1 - B)(1 - B^s)$, then the underlying series contains certain time trends and deterministic seasonality. For example, the airline passenger model with MA coefficients near 1, say,

$$\nabla\nabla_{12} z_t = (1 - \theta B)(1 - \Theta B)a_t, \qquad \theta \approx \Theta \approx 1 \tag{2.112}$$

is equivalent to the model

$$z_t = \beta_0 + \beta_1 t + S_t^{(12)} + a_t, \tag{2.113}$$

which contains a deterministic linear trend and a seasonality that is also deterministic.

Example 2.7

The R script `tsdiag` provides diagnostic checking of a fitted ARIMA model. The R command and its output are as follows:

```
>out1=arima(lx[,23],order=c(5,1,3))
>out2=tsdiag(out1)
> t.test(out1$residuals)
        One Sample t-test
data: m1$residuals
t = -0.51816, df = 4161, p-value = 0.6044
alternative hypothesis: true mean is not equal to 0
95 percent confidence interval:
 -0.0005044595  0.0002935480
```

Figure 2.17 provides model checking of the fitted ARIMA(5,1,3) model for the logarithm of the 23th world stock index. From the middle and bottom plots, there

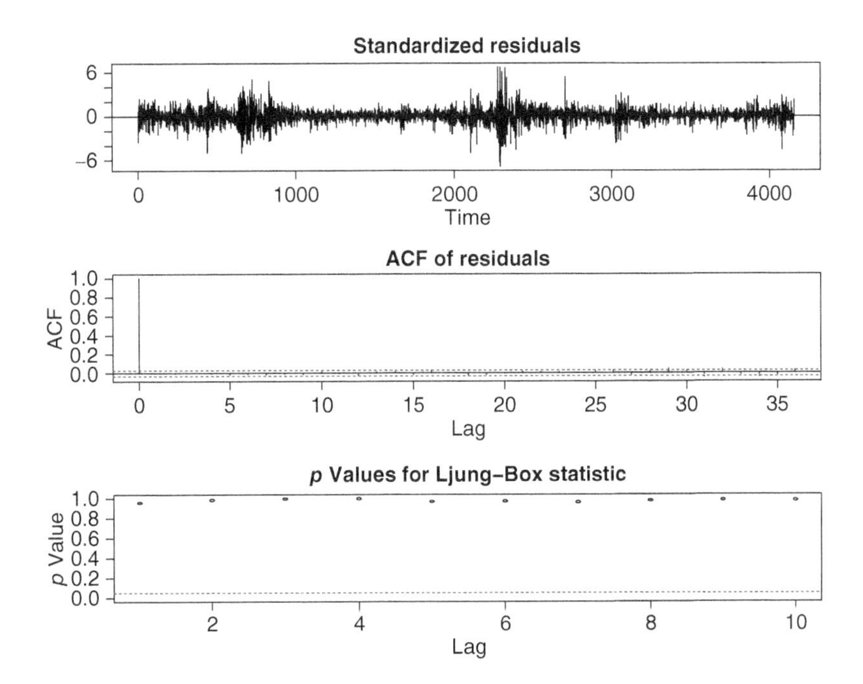

Figure 2.17 Model checking of a fitted ARIMA(5,1,3) model to the log series of a world stock index (series 23). The top figure is the time plot of standardized residuals, the middle plot is the residual sample autocorrelations with blue line showing approximate 95% pointwise interval, and the bottom plot shows the p-values of Ljung–Box Q statistics for the residual autocorrelations.

are no significant serial correlations in the residuals, indicating that the model is adequate in describing the serial dependence of the data. But the plot of standardized residual indicates that the assumption of constant variance is questionable. Indeed, most financial time series show certain degree of conditional heteroscedasticity.

Finally, as mentioned before, the `arima` command of R automatically drops the intercept of a model when a difference operator is used, it is then important to check that the mean of residuals is zero. This can easily be done using the conventional one-sample t-test, i.e. `t.test` in R. For the stock index series used, the test statistic is -0.518 with p-value 0.60 so that one cannot reject the null hypothesis that the mean of residuals is zero. ■

2.10 FORECASTING

2.10.1 Out-of-Sample Forecasts

Out-of-sample forecasts can be used for model comparison and for model checking. Suppose that we are interested in forecasts with horizon h. If the fitted model is adequate, we can apply the results of Section 2.5 to compute the forecasts and expect that the variance of 1-step ahead forecast errors would equal to the residual variance, $\hat{\sigma}_a^2$, and the variance of h-step ahead forecasts is equal to $\hat{\sigma}_a^2 \sum_{i=0}^{h} \hat{\psi}_i^2$ with $\hat{\psi}_0 = 1$. These properties serve as additional tools for model checking. One way to check these properties is to use a procedure similar to that of cross-validation discussed before. Specifically, one divides the data into an estimation (or training) subsample and a testing (or validation) subsample. Details of cross-validation is given in Section 2.8.4.4. The MSFE of Eq. (2.108) can be used to compute the out-of-sample forecast variances. For model checking, we can consider the test statistics

$$p_h = \frac{1}{n_h} \sum_{j=1}^{n_h} \frac{e_{T_1+j}^2(h)}{\hat{\sigma}_{T_1}^2(h)},$$

where $\hat{\sigma}_{T_1}^2$ is the residual variance of the training subsample and $e_{T_1+j}^2(h)$ is h-step ahead forecast error at forecast origin $T_1 + j$. Assuming that the model is correct and for large sample size T, the statistic p_h follows asymptotically a chi-square distribution divided by its degrees of freedom (χ_g^2/g). In particular, for $h = 1$, the statistic p_1 follows asymptotically a $\chi_{n_h}^2/n_h$ distribution.

We remark that there is substantial literature on using rolling forecasts for checking a fitted time series model or for model comparison. See, for instance, Diebold and Mariano (2002), Clark and McCracken (2009), and Rossi and Inoue (2012). For a discussion on the Box–Cox transformation in forecasting time series, see Proietti and Lütkepohl (2013).

2.10.2 Forecasting with Model Averaging

Instead of selecting the best model and disregarding the others, we can generate forecasts by using all the fitted models with different weights. This could be useful when we have several models that fit the data well. Assume that m models have been estimated for a time series. Recall that the BIC of model M_i can be written as

$$\mathrm{BIC}_i = -\frac{2}{T} \ln P(M_i|\mathbf{z}_1^T) + c,$$

where c is a constant, and this value is related to the posterior probabilities of model M_i by

$$P(M_i|\mathbf{z}_1^T) = c_1 e^{-\frac{T}{2}\mathrm{BIC}_i},$$

where $c_1 = \exp(c)$. To calculate this constant, c_1, we make use of the fact that sum of the probabilities of all the models is 1. Therefore,

$$\sum_{i=1}^{m} P(M_i|\mathbf{z}_1^T) = 1 = c_1 \sum_{i=1}^{m} e^{-\frac{T}{2}\mathrm{BIC}_i}.$$

With these results, we can transform BIC values into posterior probabilities of different models by means of

$$P(M_i|\mathbf{z}_1^T) = \frac{e^{-\frac{T}{2}\mathrm{BIC}_i}}{\sum_{j=1}^{k} e^{-\frac{T}{2}\mathrm{BIC}_j}}.$$

Therefore, we may conclude that the probability distribution $f(z_f)$ of a new future observation z_f would follow the mixed distribution

$$f(z_f) = \sum_{i=1}^{m} f(z_f|M_i)P(M_i|\mathbf{z}_1^T), \tag{2.114}$$

where $f(z_f|M_i)$ is the density function of the new observation in accordance with model M_i. This distribution has a mean value that is the prediction generated by the model and a variance equal to that of the corresponding prediction error. For example, suppose that we generate 1-step ahead predictions with each model and let $\hat{z}_T^{(M_i)}(1)$ be the prediction for the period $T+1$ with model M_i. The expectation of the distribution (2.114), denoted by $\hat{z}_T(1)$, is

$$\hat{z}_T(1) = \sum_{i=1}^{m} \hat{z}_T^{(M_i)}(1)P(M_i|\mathbf{z}_1^T),$$

which is a combination of all the predictions weighted by their probabilities. This approach of calculating predictions is known as *Bayesian Model Averaging*. In general, the combined prediction is more accurate than that generated by a single model. Moreover, as it has a better assessment of the uncertainty, it allows us to construct more realistic prediction intervals than those obtained by ignoring the model uncertainty. Letting $\hat{\sigma}_i^2$ denote the variance of the innovations of model M_i, which coincides with the variance of 1-step ahead prediction error of the model, the variance of the combination of forecast is

$$\mathrm{Var}[\hat{z}_T(1)] = E[(z_{T+1} - \hat{z}_T(1))^2],$$

where the expectation is calculated with respect to the mixed distribution in Eq. (2.114). It can be proved that the variance of the mixed distribution is

$$\mathrm{Var}[\hat{z}_T(1)] = \sum_{i=1}^{k} \hat{\sigma}_i^2 P(M_i|z_1^T) + \sum_{i=1}^{k} [\hat{z}_T^{(M_i)}(1) - \hat{z}_T(1)]^2 P(M_i|z_1^T),$$

which allows a construction of a more realistic interval prediction. The results can be generalized to any prediction interval.

2.10.3 Forecasting with Shrinkage Estimators

In some applications, we are interested in forecasting many time series that share certain common characteristics such as the monthly sales of products in a given category. In this case, there is empirical evidence, e.g. Garcia-Ferrer et al. (1987), that we can improve the forecasts of individual series, say $\hat{z}_{i,T}(h)$, by incorporating information of an aggregate series of the whole data. Specifically, let y_t be the simple average series (or a weighted average series) of all time series under study, and $\hat{y}_T(h)$ be the h-step ahead forecast of the aggregate series at the forecast origin T. The shrinkage forecast of the ith series, defined as

$$\hat{z}_{i,T}^S(h) = (1-\alpha)\hat{z}_{i,T}(h) + \alpha\hat{y}_T(h),$$

where $0 \le \alpha \le 1$, usually fares better than $\hat{z}_{i,T}(h)$. This is not surprising if the series under study are highly related and have strong correlation, because the aggregate series pools information together so that the shrinkage forecasts make use of more information than the individual forecast does. It will be shown in Chapter 6 that this kind of forecasts can be justified when the series have common factors.

Example 2.8

In this example, we use the time series of 23rd world financial index to compare model selection via AIC and a rolling out-of-sample forecast. The sample size is $T = 4162$. The rolling forecasts are computed by the command `backtest` of the R package **MTS**. Our goal is to demonstrate the comparison between the best model selected by AIC with a higher order and a simple ARIMA(1,1,0) model. To this end, we compute the 1-step ahead forecasts out-of-sample in two ways. In the first case, the starting forecast origin is $t = 4000$ whereas in the second case the starting forecast origin is $t = 3000$. Also, we use the root mean squared forecast error (RMSE) as the criterion in our comparison. The R commands used are given below.

```
>out1=arima(lx[,23],order=c(5,1,3))
>out2=arima(lx[,23],order=c(1,1,0))
 ##computes the residual MSE of c(5,1,3) model
>sqrt(out1$sigma2)
[1] 0.01314981
##computes the residual MSE of c(1,1,0) model
>sqrt(out2$sigma2)
[1] 0.01326138
```

```
## The difference in residual variances is rather small.
>vv=backtest(out1,lx[,23],4000,h=1)
[1] "RMSE of out-of-sample forecasts"
[1] 0.01606062
##The out-of-sample RMSE is about 23% larger than in sample
>vv=backtest(out2,lx[,23],4000,h=1) #1-step ahead forecasts
[1] "RMSE of out-of-sample forecasts"
[1] 0.01609272
>vv=backtest(out1,lx[,23],3000,h=1)
[1] "RMSE of out-of-sample forecasts"
[1] 0.01246
>vv=backtest(out2,lx[,23],3000,h=1)
[1] "RMSE of out-of-sample forecasts"
[1] 0.01244233
##By enlarging the number of out of sample forecasts,
#  the difference in RMSE remains small, indicating there
#  is no big difference between the two models.
```

From the output, we see that the two models fit the data similarly with residual standard error 0.01315 versus 0.01326 slightly in favor of the ARIMA(5,1,3) model. For rolling out-of-sample forecasts, there is no big difference between the two models. For forecasting origin $t = 4000$, the RMSE are 0.01606 versus 0.01609, and for forecasting origin $t = 3000$, the RMSE are 0.01246 versus 0.01244. ∎

APPENDIX 2.A: DIFFERENCE EQUATIONS

The term *difference equation* is referred to equation in the form

$$x_t - \phi_1 x_{t-1} - \phi_2 x_{t-2} - \cdots - \phi_h x_{t-h} = c, \quad \phi_h \neq 0,$$

where the coefficients ϕ_i are constant. The equation is of order h, which is the maximum lag of the variable in the equation. We say that the difference equation is homogeneous if $c = 0$. The solution to this equation is a sequence x_t that verifies the equation for all t. Obtaining the solutions of the equation is simplified by introducing the lag operator and writing

$$\phi(B)x_t = c.$$

We call the polynomial $\phi(B)$ the characteristic polynomial of a difference equation, and the equation $\phi(B) = 0$, considered as a function of the variable B, is the *characteristic equation*. For example, the characteristic equation of a first-order difference equation is

$$1 - \phi B = 0,$$

and it has as solution $B = 1/\phi$. In this appendix, we prove the following properties of the solutions of a difference equation:

1. The solution to a first-order homogeneous equation is AG^t, where A is a constant that depends on the initial conditions and G^{-1} is the solution to the characteristic equation.

2. The solution to a homogeneous equation whose polynomial characteristic can be written as a product of first-degree polynomials with different roots is the sum of the solutions associated with each of the first-order characteristic equations of the polynomials.
3. The solution to a second-degree homogeneous equation with complex conjugate roots is a sinusoidal function.
4. The solution to a second-degree homogeneous equation with a double root, G^{-1}, is $(A_1 + A_2 t)G^t$.
5. The general solution to a difference equation is the sum of the solution to the homogeneous equation plus a particular solution to the equation.

We now prove these five properties. The first one is arrived at by considering the first-order difference equation:

$$x_t - \phi x_{t-1} = (1 - \phi B)x_t = 0.$$

This equation produces from an initial value x_0, the sequence of values $x_1 = \phi x_0$, $x_2 = \phi^2 x_0$, and, in general:

$$x_t = \phi^t x_0.$$

This proves the first property: the solution is the product of a constant, which is the initial value, multiplied by the inverse of the solution to the characteristic equation (ϕ^{-1} in this case) raised to t. Note that if $\phi = 1$ the solution is $x_t = x_0$, a constant.

To prove the second property, we will consider a second-degree equation, $x_t - \phi_1 x_{t-1} - \phi_2 x_{t-2} = 0$, which we write as:

$$(1 - \phi_1 B - \phi_2 B^2)x_t = 0, \tag{2.115}$$

and if $G_1^{-1}$and G_2^{-1} are the roots of the characteristic equation, we can express this equation as:

$$(1 - G_1 B)(1 - G_2 B)x_t = 0. \tag{2.116}$$

Assuming that $G_1 \neq G_2$ we will show that the solution to this equation has the form

$$x_t = S_{1t} + S_{2t}, \tag{2.117}$$

where S_{1t} verifies $(1 - G_1 B)S_{1t} = 0$ and S_{2t} verifies $(1 - G_2 B)S_{2t} = 0$. Indeed, substituting Eq. (2.117) in Eq. (2.116), we have

$$(1 - G_1 B)(1 - G_2 B)S_{1t} + (1 - G_1 B)(1 - G_2 B)S_{2t} = 0.$$

Since the solution to $(1 - G_i B)S_{it} = 0$ must have the form $S_{it} = A_i G_i^t$, where the A_i are constants that depend on the initial conditions, we find that the general solution to Eq. (2.116) must have the form:

$$x_t = A_1 G_1^t + A_2 G_2^t. \tag{2.118}$$

For example, let us assume that we have the sequence of autocorrelations of an AR(2) with $\rho_0 = 1$ and $\rho_1 = \phi_1/(1 - \phi_2)$. Thus, according to (2.118) the general solution is:

$$\rho_h = A_1 G_1^h + A_2 G_2^h, \tag{2.119}$$

where, because G_1 and G_2 are factors of the equation it can be verified that $\phi_1 = G_1 + G_2$ and $\phi_2 = -G_1G_2$. To obtain the constants A_1, A_2, specifying for $h = 0$ we have $\rho_0 = A_1 + A_2 = 1$ and for $h = 1$ we obtain $\rho_1 = A_1G_1 + A_2G_2$. From these two equations we arrive at

$$A_1 = \frac{\rho_0 G_1 - \rho_1}{(G_2 - G_1)}; \quad A_2 = \frac{\rho_1 - \rho_0 G_1}{(G_2 - G_1)}. \tag{2.120}$$

If the two roots of Eq. (2.116) are real the solution will be the sum of two geometric decays.

The proof of this second property for general polynomials as

$$(1 - \phi_1 B - \phi_2 B^2 - \cdots - \phi_h B^h)x_t = (1 - G_1 B)(1 - G_2 B)\cdots(1 - G_h B)x_t = 0,$$

where the roots, G_i, are assumed to be different is a straightaway extension of the case of two factors, and the general solution is

$$x_t = A_1 G_1^t + \cdots + A_h G_h^t, \tag{2.121}$$

where $A_1, \ldots, A_h$ are constants that depend on the initial conditions and $G_1^{-1}, \ldots, G_h^{-1}$ are the roots of the characteristic equation. Moreover, we observe that for x_t going to zero when $t \to \infty$, G_i^t must go to zero, which requires that the modulus of all the solutions be less than the unit.

Now we prove property 3: when two roots are complex the solution implied by these two roots has sinusoidal behavior. Consider again Eqs. (2.115) and (2.116). When G_1 and G_2 are complex conjugates they can be written as $G_1 = R\exp(i\omega)$ and $G_2 = R\exp(-i\omega)$, and

$$G_1 G_2 = R^2 = -\phi_2,$$

$$R = \sqrt{-\phi_2}, \tag{2.122}$$

and we see that for the solution to be complex ϕ_2 must be negative. Also,

$$G_1 + G_2 = \phi_1 = R(\exp(i\omega) + \exp(-i\omega)) = 2R\cos\omega,$$

and by Eq. (2.122) ω can be obtained by

$$\cos\omega = \frac{\phi_1}{2\sqrt{-\phi_2}}. \tag{2.123}$$

The values of A_1, A_2 are given by Eq. (2.120) with $\rho_0 = x_0$ and $\rho_1 = x_1$ as starting values. Then inserting these values and those of G_1 and G_2 in Eq. (2.119) we have

$$x_t = R^t \frac{x_1(e^{i\omega t} - e^{\ i\omega t}) - x_0(e^{i\omega(t-1)} - e^{-i\omega(t-1)})}{(e^{i\omega} - e^{-i\omega})},$$

that can be written as

$$x_t = \frac{R^t}{\sin\omega}(x_1 \sin\omega t - x_0 \sin\omega(t-1))$$

and the solution is a sinusoidal function.

We prove next the fourth property about multiple roots in the characteristic equation. To begin, we show that when two roots are equal the two corresponding terms $A_i G_i^t$ can be written as $(A_1 + A_2 t)G^t$. Let us assume Eq. (2.115), whose characteristic equation is

$$(1 - \phi_1 B - \phi_2 B^2) = 0. \tag{2.124}$$

If there is a double root in this equation, G_0^{-1}, it is verified:

$$(1 - G_0 B)^2 = (1 - \phi_1 B - \phi_2 B^2),$$

which implies $\phi_2 = -G_0^2$ and $\phi_1 = 2G_0$, thus

$$\phi_1 G_0 + 2\phi_2 = 0. \tag{2.125}$$

We check that if G_0^t is a solution to Eq. (2.124), tG_0^t is also a solution. Substituting

$$(1 - \phi_1 B - \phi_2 B^2)tG_0^t = tG_0^t - \phi_1(t-1)G_0^{t-1} - \phi_2(t-2)G_0^{t-2},$$

and taking out common factors, we have:

$$tG_0^{t-2}(G_0^2 - \phi_1 G_0 - \phi_2) + G_0^{t-2}(\phi_1 G_0 + 2\phi_2) = 0.$$

The first term is zero, because G_0 is a solution, and the second as well, according to Eq. (2.125). Hence, we conclude that the general solution is

$$x_t = (A_1 + A_2 t)G_0^t.$$

These results can be extended to equations of any order and with roots that are equal or different. We assume a homogeneous linear difference equation

$$(1 - \phi_1 B - \phi_2 B^2 - \cdots - \phi_h B^h)x_t = 0. \tag{2.126}$$

If there are roots repeated r times, that is, a term exists with the form $(1 - G_0 B)^r$, it is easy to prove, as in the case $r = 2$, that the solution associated with this term is

$$x_t = (A_1 + A_2 t + \cdots + A_r t^{r-1})G_0^t. \tag{2.127}$$

Therefore, the general solution of Eq. (2.126) contains: (1) a term of type $A_i G_i^t$ for each non-repeated real root; (2) a polynomial term of order $(r - 1)$ that multiplies G_0^t for each real root repeated r times; (3) a sinusoidal term for each pair of complex conjugate roots.

Finally, we prove the fifth property. Suppose the equation $\phi(B)x_t = c$. We are going to prove that if y_t is a solution then so is $y_t + q_t$ where q_t is a solution to the homogeneous equation, $\phi(B)q_t = 0$. This is straightforward to show, since

$$\phi(B)(y_t + q_t) = \phi(B)y_t = c.$$

Validity of a Solution and Initial Values

A variable can verify a difference equation from a defined point in time. Thus, the solution to the equation provides values of the variable that verify the equation not only from this defined point, but also for the previous times that are necessary for calculating the solution. That is, if the equation is valid for $t \geq t_0$ and it needs h initial values to solve it, the solution will be valid for values from $t \geq t_0 - h$, that is, from the origin of the equation minus the number of initial values required to calculate the solution.

To illustrate this result, we start with the first-order equation:

$$x_t - \phi x_{t-1} = 0, \qquad t \geq t_0.$$

If the equation is valid for $t = t_0$, we have $x_{t_0} = \phi x_{t_0-1}, x_{t_0+1} = \phi x_{t_0} = \phi^2 x_{t_0-1}$, and, in general,

$$x_t = \phi^{t-t_0+1} x_{t_0-1}.$$

Note that this solution is valid for $t \geq t_0 - 1$. Indeed, for $t = t_0 - 1$ it provides the result $x_{t_0-1} = x_{t_0-1}$, which is trivially true. Therefore, we conclude that the solution is valid for the interval of definition of the equation, minus the number of initial values needed to initiate the sequence.

We check this rule in a second-degree equation:

$$x_t - \phi_1 x_{t-1} - \phi_2 x_{t-2} = 0, \qquad t \geq 1.$$

From the above results we know that the solution is:

$$x_t = A_1 G_1^t + A_2 G_2^t,$$

where $G_1^{-1}$and G_2^{-1} are the roots of the characteristic equation. Let us check that this solution is valid for $t \geq -1$. For $t = 0$ we obtain

$$x_0 = A_1 + A_2,$$

and for $t = -1$, we have

$$x_{-1} = A_1 G_1^{-1} + A_2 G_2^{-1}.$$

We now prove that the three values x_1, x_0 and x_{-1} verify the equation. Substituting in the second-degree equation

$$\begin{aligned} x_1 &= A_1 G_1 + A_2 G_2 = \phi_1 x_0 + \phi_2 x_{-1} \\ &= \phi_1 (A_1 + A_2) + \phi_2 (A_1 G_1^{-1} + A_2 G_2^{-1}), \end{aligned}$$

and grouping terms

$$A_1 G_1 + A_2 G_2 = A_1(\phi_1 + \phi_2 G_1^{-1}) + A_2(\phi_1 + \phi_2 G_2^{-1}),$$

which is always true, as is proved by using the relationship between the parameters and the roots, $\phi_1 = G_1 + G_2$ and $\phi_2 = -G_1 G_2$, which was obtained by solving the second-degree equation.

Generalizing, if we have a difference equation of order h valid for $t \geq t_0$ we need h initial values to begin the sequence. Since these initial values are obtained with the equation that provides the solution, the solution provides the valid values of the sequence for $t \geq t_0 - h$.

EXERCISES

1. Compute the variance and the ACF (lag-1 to lag-4) of the following ARMA models with $Var(a_t) = 1$: (a) $z_t = 0.7z_{t-1} + a_t$; (b) $z_t = 0.4a_{t-1} + a_t$; (c) $z_t = 0.7z_{t-1} + 0.4a_{t-1} + a_t$.
2. Simulate the three ARIMA models of Exercise 1 with the command `arima.sim` and compare the theoretical ACFs with the sample ACFs.
3. Compare the EDQ and the TWQ for probabilities (0.05,0.5,0.95) of the logs of CPI series in file `CPIEurope2000-15.csv`. Compute these quantiles in levels and first differences. Note the extreme behavior of Series 30 and 34, that correspond to Iceland and Turkey in this period. Find which series are seasonal and which are not with the command `chksea`.
4. Compute the ACF of the process $z_t = y_t + v_t$, where $y_t = 0.4a_{t-1} + a_t$ with $Var(a_t) = 4$ and v_t a white noise process with $Var(v_t) = 2.0$ Simulate the series with the command `arima.sim` and compare the theoretical ACFs with the sample ACFs (first four lags).
5. Find the roots of the characteristic equation of the following ARMA models: (a) $(1 - 6B)z_t = a_t$; (b) $(1 - 1.4B + 0.8B^2)z_t = a_t$; (c) $(1 - 0.6B + 1.2B^2)z_t = (1 - 0.5B)a_t$.
6. Compute the periodogram of the first three series identified as seasonal in Exercise 2 (Series 4th, 5th, and 9th) with the transformation $\nabla \log(\text{CPI}_t)$ and compare the ACF of these series (first 4 lags). Note the peaks in the spectrum at frequencies 1/12 and 1/6. Then look at the spectrum and the ACF of the transformation $\nabla\nabla_{12} \log(\text{CPI}_t)$. Which of these three series has stronger seasonality?
7. Write the Kalman filter equations for an AR(1) process written is state space form with $H_t = 1, \alpha_t = z_t$ and $V_t = 0, \Omega_t = \phi, R_t = \sigma_a^2$. Show that the Kalman gain is one in this case.
8. Compare the 1-step and 2-step ahead forecast error variances of the ARMA models of Exercise 1.

REFERENCES

Akaike, H. (1973). Information theory and an extension of the maximum likelihood principle. In *Selected Papers of Hirotugu Akaike*. Springer, New York, NY.

Alonso, A. M. and Peña, D. (2019). Clustering time series by linear dependency. *Statistics and Computing*, **29**: 655–676.

Bengtsson, T. and Cavanaugh, J. E. (2006). An improved Akaike information criterion for state-space model selection. *Computational Statistics & Data Analysis*, **50**: 2635–2654.

Box, G. E. P. and Cox, D. R. (1964). An analysis of transformations. *Journal of the Royal Statistical Society, Series B*, **26**: 211–243.

Box, G. E. P. and Jenkins, G. M. (1976). Time Series Analysis: Forecasting and Control. Holden Day-Inc., San Francisco, CA.

Box, G. E. P., Jenkins, G. M., Reinsel, G. C., and Ljung, G. M. (2015). Time Series Analysis: Forecasting and Control, 5th Edition. John Wiley & Sons, Hoboken, NJ.

Box, G. E. P. and Tiao, G. C. (1977). A canonical analysis of multiple time series. *Biometrika*, **64**: 355–365.

Brockwell, P. J. and Davis, R. A. (2013). *Time Series: Theory and Methods*. Springer Science & Business Media, New York, NY.

Clark, T. E. and McCracken, M. W. (2009). Improving forecast accuracy by combining recursive and rolling forecasts. *International Economic Review*, **50**: 363–395.

Cryer, J. D. and Chan, K. S. (2008). Time Series Analysis with Applications in R, 2nd Edition. Springer, New York, NY.

Dickey, D. A. and Fuller, W. A. (1979). Distribution of the estimators for autoregressive time series with a unit root. *Journal of the American Statistical Association*, **74**: 427–431.

Diebold, F. X. and Kilian, L. (2001). Measuring predictability: Theory and macroeconomic applications. *Journal of Applied Econometrics*, **16**: 657–669.

Diebold, F. X. and Mariano, R. S. (2002). Comparing predictive accuracy. *Journal of Business and Economic Statistics*, **20**: 134–144.

Dufour, J. M. and Roy, R. (1985). Some robust exact results on sample autocorrelations and tests of randomness. *Journal of Econometrics*, **29**: 257–273.

Durbin, J. and Koopman, S. K. (2001). *Time Series Analysis by State Space Models*. Oxford University Press, Oxford.

Ferguson, T. S. (1967). *Mathematical Statistics: A Decision Theoretic Approach*. Academic Press, New York, NY.

Garcia-Ferrer, A., Highfield, R. A., Palm, F., and Zellner, A. (1987). Macroeconomic forecasting using pooled international data. *Journal of Business and Economic Statistics*, **5**: 53–67.

Gómez, V. and Maravall, A. (1994). Estimation, prediction, and interpolation for nonstationary series with the Kalman filter. *Journal of the American Statistical Association*, **89**: 611–624.

Gómez, V. and Maravall, A. (2001). Automatic modeling mehods for univariate series. In D. Peña, G. C. Tiao, and R. S. Tsay (eds.). *A Course in Time Series Analyisis*. John Wiley & Sons, Hoboken, NJ.

Granger, C. W. J. and Morris, M. J. (1976). Time series modelling and interpretation. *Journal of the Royal Statistical Society: Series A*, **139**: 246–257.

Hannan, E. J. and Rissanen, J. (1982). Recursive estimation of mixed autoregressive-moving average order. *Biometrika*, **69**: 81–94.

Hannan, E. J. and Quinn, B. G. (1979). The determination of the order of an autoregression. *Journal of the Royal Statistical Society B*, **41**: 190–195.

Harrison, P. J. and Stevens, C. F. (1976). Bayesian forecasting. *Journal of the Royal Statistical Society B*, **38**: 205–228.

Harvey, A. C. (1990). *Forecasting, Structural Time Series Models and the Kalman Filter*. Cambridge University Press, Cambridge, UK.

Hurvich, C. M. and Tsai, C. L. (1989). Regression and time series model selection in small samples. *Biometrika*, **76**: 297–307.

Jarque, C. M. and Bera, A. K. (1987). A test of normality of observations and residuals. *International Statistical Review*, **55**: 163–172.

Koopmans, L. H. (1974). *The Spectral Analysis of Time Series*. Academic Press, New York, NY.

Lütkepohl, H. (1985). Comparison of criteria for estimating the order of a vector autoregressive process. *Journal of Time Series Analysis*, **6**: 35–52.

Peña, D. and Rodriguez, J. (2002). A powerful portmanteau test of lack of fit for time series. *Journal of the American Statistical Association*, **97**: 601–610.

Peña, D. and Rodríguez, J. (2006). The log of the determinant of the autocorrelation matrix for testing goodness of fit in time series. *Journal of Statistical Planning and Inference*, **136**: 2706–2718.

Peña, D., Tiao, G. C., and Tsay, R. S. (2001). *A Course in Time Series Analysis*. John Wiley & Sons, Hoboken, NJ.

Peña, D., Tsay, R. S., and Zamar, R. (2019). Empirical dynamic quantiles for visualization of high-dimensional time series. *Technometrics*, **61**: 429–444.

Percival, D. B. and Walden, A. T. (2000). *Wavelet Methods for Time Series Analysis*. Cambridge University Press, Cambridge, UK.

Plackett, R. L. (1950). Some theorems in least squares. *Biometrika*, **37**: 149–157.

Proietti, T. (2000). Comparing seasonal components for structural time series models. *International Journal of Forecasting*, **16**: 247–260.

Proietti, T. and Lütkepohl, H. (2013). Does the Box–Cox transformation help in forecasting macroeconomic time series? *International Journal of Forecasting*, **29**: 88–99.

Rossi, B. and Inoue, A. (2012). Out-of-sample forecast tests robust to the choice of window size. *Journal of Business & Economic Statistics*, **30**: 432–453.

Shumway, R. H. and Stoffer, D. S. (2017). *Time Series Analysis and Its Applications: With R Examples*. Springer, New York, NY.

Schwarz, G. (1978). Estimating the dimension of a model. *The Annals of Statistics*, **6**: 461–464.

Tiao, G. C. and Xu, D. (1993). Robustness of maximum likelihood estimates for multi-step predictions: The exponential smoothing case. *Biometrika*, **80**: 623–641.

Tsay, R. S. (2010). *Analysis of Financial Time Series*, 2nd Edition. John Wiley & Sons, Hoboken, NJ.

Tsay, R. S. (2014). *Multivariate Time Series Analysis: With R and financial applications*. John Wiley & Sons, Hoboken, NJ.

Tsay, R. S. (2020). Testing serial correlations in high-dimensional time series via extreme value theory. *Journal of Econometrics*, **216**: 106–117.

Tsay, R. S. and Chen, R. (2018). Nonlinear Time Series Analysis. John Wiley & Sons, Hoboken, NJ.

West, M. and Harrison, J. (2006). Bayesian Forecasting and Dynamic Models. Springer Science and Business Media, New York, NY.

Young, P. (1984). Recursive Estimation and Time-Series Analysis. Springer-Verlag, Berlin, Germany.

CHAPTER 3

ANALYSIS OF MULTIVARIATE TIME SERIES

In this chapter, we consider jointly multiple time series with the goal of finding the linear dynamic relationships between them and exploring ways to improve the accuracy in forecasting. Our focus is on low-dimensional time series. The extension to high-dimensional case will be discussed later. See Chapters 6 and 7. In the literature, the analysis of multiple time series is also called the multivariate time series analysis and has been extensively studied. See, for instance, Hannan and Deistler (1988), Tiao and Box (1981), Reinsel (1993), Lütkepohl (2005), Tsay (2014), and the references therein. Our discussion here is brief and emphasizes on applications.

Let $\boldsymbol{z}_t = (z_{1t}, \ldots, z_{kt})'$ be a k-dimensional time series. A general class of linear models for $\boldsymbol{z}_t$ is the vector autoregressive moving-average (VARMA) model

$$\boldsymbol{\phi}(B)\boldsymbol{z}_t = \boldsymbol{c}_0 + \theta(B)\boldsymbol{a}_t, \tag{3.1}$$

where $\boldsymbol{\phi}(B) = \boldsymbol{I} - \sum_{i=1}^{p} \boldsymbol{\phi}_i B^i$ and $\theta(B) = \boldsymbol{I} - \sum_{i=1}^{q} \theta_i B^i$ are matrix polynomials of degrees p and q, respectively, B is the back-shift operator such that $B\boldsymbol{z}_t = \boldsymbol{z}_{t-1}$, $\boldsymbol{c}_0$ is a k-dimensional constant, and $\{\boldsymbol{a}_t\}$ is a sequence of independent and identically distributed random vectors with mean zero and positive-definite covariance matrix $\boldsymbol{\Sigma}_a$. We assume that the absolute values of the solutions of $|\boldsymbol{\phi}(B)| = 0$ are greater than or equal to 1 and those of $|\theta(B)| = 0$ are greater than 1. In other words, we focus on the case that $\boldsymbol{z}_t$ is a causal time series. Furthermore, for identifiability, we assume that (a) the rank of $[\boldsymbol{\phi}_p, \theta_q]$ is k and (b) $\boldsymbol{\phi}(B)$ and $\theta(B)$ are left coprime, i.e. if $\boldsymbol{\phi}(B) = \boldsymbol{U}(B)\boldsymbol{\phi}_*(B)$ and $\theta(B) = \boldsymbol{U}(B)\theta_*(B)$, then $|\boldsymbol{U}(B)|$ is a non-zero constant. Assumption (b) is easy to understand as there should be no non-trivial left common factor between $\boldsymbol{\phi}(B)$ and $\theta(B)$. Assumption (a) is more involved and has been

Statistical Learning for Big Dependent Data, First Edition. Daniel Peña and Ruey S. Tsay.
© 2021 John Wiley & Sons, Inc. Published 2021 by John Wiley & Sons, Inc.

discussed in Dunsmuir and Hannan (1976), Hannan and Deistler (1998, section 2.7), and Tsay (2014). These two assumptions are referred to as the *block identifiability* conditions.

If the absolute values of the solutions of $|\boldsymbol{\phi}(B)| = 0$ are all greater than 1, then $\boldsymbol{z}_t$ is a weakly stationary time series; otherwise, $\boldsymbol{z}_t$ is a unit-root nonstationary series. Analysis of unit-root nonstationary multivariate time series is harder in the sense that there exists the possibility of co-integration. By co-integration, we mean that if some linear combinations of unit-root nonstationary series become stationary, then the nonstationary series are co-integrated. See Box and Tiao (1977) and Engle and Granger (1987). We discuss analysis of co-integrated series later. In what follows, we consider some commonly used multivariate time series models and discuss their properties and applications.

3.1 TRANSFER FUNCTION MODELS

Transfer function models (TFMs), or distributed-lag models in the econometric literature, are useful in control engineering and in forecasting. See Box and Jenkins (1976). They are special cases of the VARMA models of Eq. (3.1) by dividing $\boldsymbol{z}_t = (\boldsymbol{x}_t', \boldsymbol{y}_t')'$ with $\boldsymbol{x}_t$ being the input (or exogenous) variable and $\boldsymbol{y}_t$ the output (or endogenous) variable. The input variable does not depend dynamically on the output variable, but the output variable depends on the input variable. This type of relationships is called a unidirectional relationship. In the econometric literature, the relationship is related to Granger causality, see Section 3.2.2, because $\boldsymbol{x}_t$ causes $\boldsymbol{y}_t$, but $\boldsymbol{y}_t$ does not cause $\boldsymbol{x}_t$.

3.1.1 Single Input and Single Output

For simplicity, we start with the case of a single input x_t and a single output y_t. The TFM then becomes

$$\phi_x(B)x_t = c_x + \theta_x(B)b_t, \tag{3.2}$$

$$y_t = c_y + \frac{\omega(B)}{\delta(B)}x_t + n_t \tag{3.3}$$

where c_x and c_y are constants, $\phi_x(B) = 1 - \sum_{i=1}^{p_x} \phi_{x,i}B^i$ and $\theta_x(B) = 1 - \sum_{i=1}^{q_x} \theta_{x,i}B^i$ are autoregressive (AR) and moving-average (MA) polynomials of x_t with degrees p_x and q_x, which are nonnegative integers, b_t is a sequence of independent and identically distributed random variables with mean zero and variance $\sigma_b^2 > 0$, $\omega(B) = B^b(\omega_0 + \cdots + \omega_s B^s)$ and $\delta(B) = 1 - \delta_1 B - \cdots - \delta_r B^r$, b is a nonnegative integer and the absolute values of the solutions of $\delta(B) = 0$ are greater than or equal to 1, and n_t is a disturbance term which follows a scalar ARMA model and is independent of x_t. In Eq. (3.2), $\phi_x(B)$ and $\theta_x(B)$ have no common factors. Similarly, in Eq. (3.3), $\omega(B)$ and $\delta(B)$ have no common factors. Since the scale of $\omega(B)$ is not fixed, we can start $\delta(B)$ with 1 without causing any problem. The parameter b of operator $\omega(B)$ in Eq. (3.3) is called the *dead time* or *delay* of the model and it plays an important role in the application of TF models. It denotes the time periods needed for the effect of the input x_t to appear in the output y_t.

Let $v(B) = \frac{\omega(B)}{\delta(B)} = v_0 + v_1 B + v_2 B^2 + \cdots$. The coefficients v_i are called the *impulse response function* of the TFM. They are the main subject of interest in transfer function (TF) modeling. Specifically, v_i is the effect on y_{t+i} when there is a unit change in x_t and $\sum_{j=0}^{i} v_j$ is the cumulative effect on $y_t, \ldots, y_{t+i}$ when there is a unit change in x_t. In particular, $g = \sum_{j=0}^{\infty} v_j = v(1)$ is the *steady-state gain* of the TFM provided that it exists. Under the assumptions of the model in Eq. (3.3), g is finite if no solutions of $\delta(B) = 0$ lie on the unit circle.

The impulse response function v_i can be obtained by equating the coefficients of B^i in the following equation

$$(1 - \delta_1 B - \cdots - \delta_r B^r)(v_0 + v_1 B + v_2 B^2 + \cdots) = \omega_0 B^b + \omega_1 B^{b+1} + \cdots + \omega_s B^{b+s}.$$

Specifically, we have

$$v_i = \begin{cases} 0, & i < b, \\ \omega_0, & i = b, \\ \delta_1 v_{i-1} + \delta_2 v_{i-2} + \cdots + \delta_r v_{i-r} + \omega_{i-b}, & i = b+1, \ldots, b+s, \\ \delta_1 v_{i-1} + \delta_2 v_{i-2} + \cdots + \delta_r v_{i-r}, & i > b+s. \end{cases} \quad (3.4)$$

In general, the coefficients v_i consist of

1. b zero values $v_0, v_1, \ldots, v_{b-1}$ provided that $b > 0$.
2. A further $s - r + 1$ values $v_b, v_{b+1}, \ldots, v_{b+s-r}$ following no fixed pattern provided that $s \geq r$.
3. Values v_i with $i \geq b + s - r + 1$ satisfying the rth order difference equation $\delta(B)v_i = 0$ with starting values $v_{b+s}, b_{b+s-1}, \ldots, v_{b+s-r+1}$.

It pays to study the relationships between v_i, ω_j, and δ_ℓ as they are useful in specifying b, r, and s in practice.

The polynomial ratio $\frac{B^b \omega(B)}{\delta(B)}$ of Eq. (3.3) is sufficiently flexible for handling the impulse response function v_i in most applications, even with lower orders r and s. For instance, Figure 3.1 shows the impulse response functions and cumulative impulse response functions of (a) $v(B) = \frac{(0.4+0.4B)B^3}{1-0.5B}$ (left panel) and (b) $v(B) = \frac{0.5B^2}{1-1.3B+0.4B^2}$ (right panel). The steady-state gain of Example (a) is $g = 1.6$, whereas that for Example (b) is $g = 5$.

Given the observations $\{x_t, y_t | t = 1, \ldots, T\}$, TF modeling is concerned with identifying a suitable model in Eqs. (3.2) and (3.3). The model for the input variable x_t can be obtained by applying the methods of scalar series discussed in Chapter 2. Thus, we focus on identifying the model in Eq. (3.3). For simplicity, assume $c_y = c_x = 0$ and rewrite the model as

$$y_t = v(B)x_t + n_t = v_0 x_t + v_1 x_{t-1} + v_2 x_{t-2} + \cdots + n_t. \quad (3.5)$$

Since n_t and x_t are independent, one can multiply Eq. (3.5) by x_{t-j} ($j \geq 0$) and take expectation to obtain

$$\gamma_{yx}(j) = v_0 \gamma_{xx}(j) + v_1 \gamma_{xx}(j-1) + v_2 \gamma_{xx}(j-2) + \cdots,$$

where $\gamma_{yx}(j) = \text{cov}(y_t, x_{t-j})$ and $\gamma_{xx}(j) = \text{cov}(x_t, x_{t-j})$. The prior equation, however, is not particularly useful because x_t is typically serially dependent with many non-zero

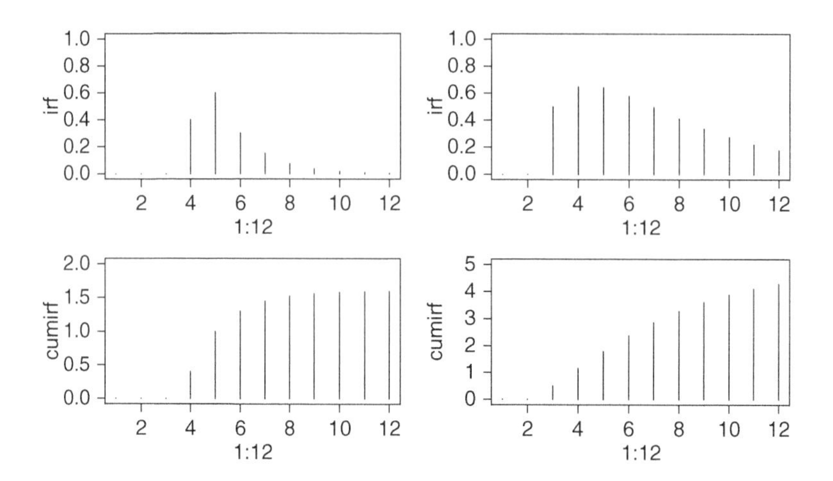

Figure 3.1 Impulse response and cumulative impulse response functions of two examples of TFs. Left panel: $v(B) = \frac{(0.4+0.4B)B^3}{1-0.5B}$ and right panel: $v(B) = \frac{0.5B^2}{1-1.3B+0.4B^2}$.

$\gamma_{xx}(j)$. A solution to mitigate the impact of serial dependence in x_t is the method of *pre-whiting*, which we present next. More details of pre-whitening can be found in Box et al. (2015).

Pre-whiting: Multiplying Eq. (3.5) by $\frac{\phi_x(B)}{\theta_x(B)}$, we have

$$\frac{\phi_x(B)}{\theta_x(B)}y_t = v(B)\frac{\phi_x(B)}{\theta_x(B)}x_t + \frac{\phi_x(B)}{\theta_x(B)}n_t.$$

Let $\eta_t = \frac{\phi_x(B)}{\theta_x(B)}y_t$ and $\tilde{n}_t = \frac{\phi_x(B)}{\theta_x(B)}n_t$. The prior equation then becomes

$$\eta_t = v(B)b_t + \tilde{n}_t = v_0 b_t + v_1 b_{t-1} + v_2 b_{t-2} + \cdots + \tilde{n}_t, \tag{3.6}$$

where b_t is a white noise series and independent of $\tilde{n}_t$, and η_t is called the pre-whitened series of y_t. Multiplying Eq. (3.6) by b_{t-j} ($j \geq 0$) and taking expectation, we have

$$\gamma_{\eta b}(j) = v_j \gamma_{bb}(0), \quad j \geq 0, \tag{3.7}$$

where, again, $\gamma_{\eta b}(j) = \text{cov}(\eta_t, b_{t-j})$ and $\gamma_{bb}(j) = \text{cov}(b_t, b_{t-j})$. Eq. (3.7) provides a consistent estimator of the impulse response weight v_j. Specifically, we have

$$\rho_{\eta b}(j) = v_j \sqrt{\frac{\gamma_{bb}(0)}{\gamma_{\eta\eta}(0)}}, \quad j = 0, 1, 2, \ldots, \tag{3.8}$$

where $\rho_{\eta b}(j)$ is the correlation between η_t and b_{t-j}. In other words,

$$v_j = c \times \rho_{\eta b}(j), \quad j \geq 0, \tag{3.9}$$

where c denotes the ratio of the standard error of η_t to that of b_t. In practice, b_t is estimated by the residual series of x_t and η_t is the pre-whitened series of the output variable y_t via the model of x_t.

To specify the orders r, s, and b, one can make use of the patterns of v_i shown in Eq. (3.4), especially the difference equation $v_i - \sum_{j=1}^{r} \delta_j v_{i-j} = 0$ for $i > b + s$ and $v_i = 0$ for $i < b$. This can be done via the *corner method*, which is given below.

Corner Method. The corner method is designed to show the patterns of the impulse response weights v_j of a TFM. A two-way table is built where the rows are numbered $0, 1, 2, \ldots$ and the columns $1, 2, 3, \ldots$. Each element of this table is the determinant of a matrix of standardized impulse function values $u(B) = v(B)/v_{\max}$, where $v_{\max} = \max_j \{|v_j|\}$. The (i, j)th element of the two-way table is the determinant of the $j \times j$ matrix

$$M(i,j) = \begin{bmatrix} u_i & u_{i-1} & \cdots & u_{i-j+1} \\ u_{i+1} & u_i & \cdots & u_{i+j+2} \\ \vdots & \vdots & \ddots & \vdots \\ u_{i+j-1} & u_{i+j-2} & \cdots & u_i \end{bmatrix},$$

where $u_i = 0$ if $i < 0$. From the patterns of v_i in Eq. (3.4), we see that (a) $|M(i,j)| = 0$ for $i \leq b - 1$, because all elements of the first row $M(i,j)$ are zero for $i \leq b - 1$, (b) $|M(i,j)| = 0$ for $j > r$ and $i > b + s$, because the rows of $M(i,j)$ are linearly dependence introduced by the difference equation of v_i, and (c) $|M(i,j)| \neq 0$ for $i = b, \ldots, b + s$, because of the presence of w_{i-b}. Consequently, the table of corner method should exhibit the pattern shown in Figure 3.1 from which the values of (r, s, b) can be identified.

Specifically, from Table 3.1, we can identity b, r, s as follows:

1. b denotes the first row of the table with some non-zero entries.
2. There exists a rectangle of zeros with upper-left corner located at $(s + b + 1, r + 1)$, from which r and s can be identified.

To demonstrate, Table 3.2 shows the corner method for the TF $v(B) = \frac{(0.4+0.4B)B^3}{1-0.5B}$ of the left panel of Figure 3.1. From the table, it is easily seen that $b = 3$, $r + 1 = 2$, and $s + b + 1 = 5$. Therefore, as expected, $(r, s, b) = (1, 1, 3)$.

TABLE 3.1 Theoretical Pattern of Impulse Response Weights of a TFM

(i,j)	1	2	$\cdots$	$r-1$	r	$r+1$	$r+2$	$\cdots$
0	0	0	$\cdots$	0	0	0	0	$\cdots$
1	0	0	$\cdots$	0	0	0	0	$\cdots$
$\vdots$	$\vdots$	$\vdots$		$\vdots$	$\vdots$	$\vdots$	$\vdots$	$\cdots$
$b-1$	0	0	$\cdots$	0	0	0	0	$\cdots$
b	X	X	$\cdots$	X	X	X	X	$\cdots$
$\vdots$	$\vdots$	$\vdots$		$\vdots$	$\vdots$	$\vdots$	$\vdots$	$\cdots$
$s+b$	X	X	$\cdots$	X	X	X	X	$\cdots$
$s+b+1$	*	*	$\cdots$	*	X	0	0	$\cdots$
$s+b+2$	*	*	$\cdots$	*	X	0	0	$\cdots$
$\vdots$	$\vdots$	$\vdots$		$\vdots$	$\vdots$	$\vdots$	$\vdots$	$\cdots$

* Denotes an arbitrary real number.

TABLE 3.2 The Corner Method Table for Transfer Function Model $v(B) = (0.4 + 0.4B)B^3/(1 - 0.5B)$ of Figure 3.1

	r						
	1	2	3	4	5	6	7
0	0.000 00	0.000 00	0.000 00	0.000 00	0.000 00	0.000 00	0.000 00
1	0.000 00	0.000 00	0.000 00	0.000 00	0.000 00	0.000 00	0.000 00
2	0.000 00	0.000 00	0.000 00	0.000 00	0.000 00	0.000 00	0.000 00
3	0.666 67	0.444 44	0.296 30	0.197 53	0.131 69	0.087 79	0.058 53
4	1.000 00	0.666 67	0.444 44	0.296 30	0.197 53	0.131 69	0.087 79
5	0.500 00	0.000 00	0.000 00	0.000 00	0.000 00	0.000 00	0.000 00
6	0.250 00	0.000 00	0.000 00	0.000 00	0.000 00	0.000 00	0.000 00
7	0.125 00	0.000 00	0.000 00	0.000 00	0.000 00	0.000 00	0.000 00

Information criteria: An alternative approach to specify the order of a TFM is to use information criteria. The delay b is relatively easy to identify. For a given b, the order (r, s) can be selected by any information criterion such as Akaike's information criterion (AIC) or Bayesian information criterion (BIC) with some pre-specified upper bounds $(r_{\max}, s_{\max})$, see Section 2.8.4.

Example 3.1

To demonstrate TF modeling, we consider a toy example. The data consist of 36 monthly observations of sales and expenditures in advertising of a company; see Abraham and Ledolter (1983) and Blattberg and Jeuland (1981). The data are in the file `advertising-and-sales-data.csv`. Figure 3.2 shows the time plots of the data from which one can see that the sales roughly follow the advertising expenditures.

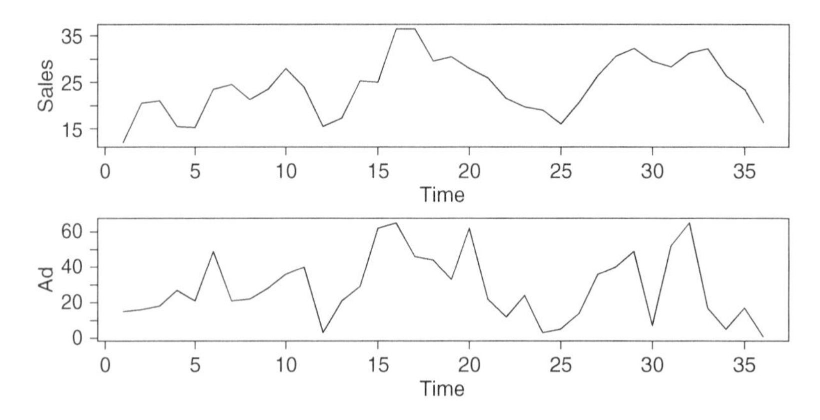

Figure 3.2 Time plots of sales and advertising expenditures, in thousands, of a company for 36 months.

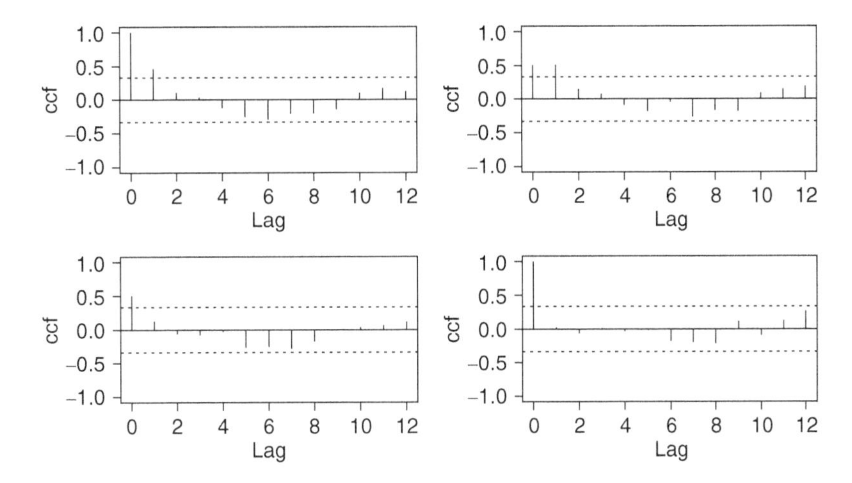

Figure 3.3 Sample cross-correlations of the pre-whitened bivariate series of sales and advertising.

Simple inspection of the sample partial autocorrelation function indicates that an AR(1) model is adequate for the advertising expenditures. The fitted model is

$$(1 - 0.3417B)(x_t - 27.95) = b_t, \quad \hat{\sigma}_b^2 = 303.7, \tag{3.10}$$

where x_t denotes the monthly expenditure of advertising. Following the modeling procedure of transfer function model (TFM), we compute the pre-whitened sales series

$$\tilde{y}_t \equiv= (1 - 0.3417B)y_t, \tag{3.11}$$

where y_t is the monthly sales. Let $z_t = (\tilde{y}_t, b_t)'$ be the pre-whitened bivariate series of the data. Figure 3.3 shows the plots of cross-correlation matrices of z_t. Of particular interest is the upper-right and lower-left plots of Figure 3.3. The upper-right plot shows the dependence of $\tilde{y}_t$ on the lagged values of b_t, whereas the lower-left plot shows the dependence of b_t on the lagged values of $\tilde{y}_t$. From the plots, it is reassuring to see that sales depends on the lagged values of advertising expenditures, but advertising expenditures do not depend on the lagged values of sales. This confirms the TF relationship between the two time series. Using the properties of pre-whitening, the upper-right plot also shows that the first two impulse response weights are statistically significant, but all higher-order weights are not. Consequently, the plot suggests that $b = 0$, $r = 0$ and $s = 1$.

Table 3.3 provides the corner method of the pre-whitened series. Using $2/\sqrt{36} \approx 0.33$ as a rough threshold, the table indicates that all entries of row 3 and beyond are not significant, confirming the choice of $(b, r, s) = (0, 0, 1)$. The fitted TFM is

$$y_t = 16.91 + 0.12x_t + 0.14x_{t-1} + \frac{1}{1 - 0.497B} e_t, \tag{3.12}$$

where all estimates are significant at the 5% level and e_t is the residual series. This model fits the data well as there exist no serial correlations in the residuals with $Q(12) = 5.57$ for the Ljung-Box statistics. ■

TABLE 3.3 Table of Corner Method for the Pre-whitened Series of Sales and Advertising Expenditures of Example 3.1

	r						
	1	2	3	4	5	6	7
0	0.983 12	0.966 52	0.950 20	0.934 16	0.918 38	0.902 88	0.887 63
1	1.000 00	0.715 45	0.570 78	0.673 98	0.455 87	0.217 40	−0.335 50
2	0.289 44	−0.060 95	−0.164 61	0.207 73	0.066 74	0.221 74	0.254 38
3	0.144 73	0.071 79	0.069 66	0.080 33	−0.091 27	0.148 08	−0.200 47
4	−0.175 65	0.080 85	0.005 55	0.061 67	−0.053 40	0.016 38	0.150 58
5	−0.345 42	0.104 64	−0.071 13	0.051 03	−0.020 18	0.056 11	−0.102 78
6	−0.083 55	−0.168 47	−0.050 55	0.018 96	0.046 00	0.065 61	0.049 85

The approximated 2-standard error limit is 0.36

3.1.2 Multiple Inputs and Multiple Outputs

In some applications of TFM, the input variable or the output variable or both can be multivariate. In such cases, the pre-whitening process becomes tedious. A more direct and simple approach is to apply multivariate time series analysis. In fact, TFMs can be regarded as special cases of the VARMA models. Details are given below.

3.2 VECTOR AR MODELS

The vector AR (VAR) models are, arguably, the most widely used multivariate time series models in practice. A VAR(p) model for a k-dimensional time series $\boldsymbol{z}_t$ is

$$\boldsymbol{z}_t = \boldsymbol{\phi}_0 + \sum_{i=1}^{p} \boldsymbol{\phi}_i \boldsymbol{z}_{t-i} + \boldsymbol{a}_t, \tag{3.13}$$

where $\boldsymbol{\phi}_0$ is a k-dimensional vector, $\boldsymbol{\phi}_i$ are $k \times k$ real matrices, and $\boldsymbol{a}_t$ is a white noise series with mean zero and positive-definite covariance matrix $\boldsymbol{\Sigma}_a > 0$. Clearly, we assume that the order p is nonnegative and $\boldsymbol{\phi}_p \neq \boldsymbol{0}$ if $p > 0$. Let $\phi(B) = \boldsymbol{I} - \sum_{i=1}^{p} \boldsymbol{\phi}_i B^i$ be the AR matrix polynomial. The series $\boldsymbol{z}_t$ is weakly stationary if all zeros of the determinant equation $|\boldsymbol{\phi}(B)| = 0$ are greater than 1 in absolute value. Consider a stationary series. Taking expectation of Eq. (3.13), we have

$$E(\boldsymbol{z}_t) \equiv \boldsymbol{\mu} = [\boldsymbol{\phi}(1)]^{-1}\boldsymbol{\phi}_0,$$

where $\boldsymbol{\phi}(1) = \boldsymbol{I} - \sum_{i=1}^{p} \boldsymbol{\phi}_i$, which is nonsingular. Using the prior equation, a stationary VAR(p) model can be rewritten as

$$\boldsymbol{\phi}(B)(\boldsymbol{z}_t - \boldsymbol{\mu}) = \boldsymbol{a}_t. \tag{3.14}$$

This model is often used in finance because $\boldsymbol{z}_t - \boldsymbol{\mu}$ is the deviation from the mean and a stationary series $\boldsymbol{z}_t$ is mean reverting. Post-multiplying Eq. (3.14) by $(\boldsymbol{z}_{t-j} - \boldsymbol{\mu})'$ and taking expectation, we have

$$\Gamma_z(j) = \begin{cases} \sum_{i=1}^{p} \phi_j \Gamma_z(j-i) + \Sigma_a & \text{if } j = 0, \\ \sum_{i=1}^{p} \phi_i \Gamma_z(j-i) & \text{if } j > 0, \end{cases} \tag{3.15}$$

where, as before, $\Gamma_z(j)$ is the lag-j autocovariance matrix of z_t. Eq. (3.15) is the *moment equation* of a stationary VAR(p) model. In particular, considering the equations jointly for $j = 1, \ldots, p$, we have

$$[\Gamma_z(1), \ldots, \Gamma_z(p)] = [\phi_1, \ldots, \phi_p] \begin{bmatrix} \Gamma_z(0) & \Gamma_z(1) & \cdots & \Gamma_z(p-1) \\ \Gamma_z(-1) & \Gamma_z(0) & \cdots & \Gamma_z(p-2) \\ \vdots & \vdots & \ddots & \vdots \\ \Gamma_z(1-p) & \Gamma_z(2-p) & \cdots & \Gamma_z(0) \end{bmatrix}, \tag{3.16}$$

which is referred to as the multivariate Yule-Walker Equation and can be used to obtain the AR-coefficient matrices ϕ_j given the autocovariance matrices $\Gamma_z(j)$. For instance, for an AR(1) we obtain

$$\phi_1 = \Gamma_z(1)[\Gamma_z(0)]^{-1}. \tag{3.17}$$

On the other hand, given a stationary VAR(p) model, we can compute the autocovariance matrices $\Gamma_z(j)$ recursively via the moment equation in (3.15). The initial values of the recursion can be obtained by solving a system of linear equations. To illustrate, consider a stationary (zero-mean) VAR(1) model, $z_t = \phi_1 z_{t-1} + a_t$. Taking the covariance on both sides of the model and noting that a_t is uncorrelated with z_{t-1}, we have

$$\Gamma_z(0) = \phi_1 \Gamma_z(0) \phi_1' + \Sigma_a.$$

Consequently,

$$\text{vec}[\Gamma_z(0)] = (\phi_1 \otimes \phi_1)\text{vec}[\Gamma_z(0)] + \text{vec}(\Sigma_a),$$

and

$$\text{vec}[\Gamma_z(0)] = [I - \phi_1 \otimes \phi_1]^{-1} \text{vec}(\Sigma_a),$$

where I is a $k^2 \times k^2$ identity matrix and $\otimes$ denotes the Kronecker product. This is a generalization of the scalar case for which $\text{var}(z_t) = \sigma_a^2/(1 - \phi_1^2)$. With $\Gamma_z(0)$ given, one can compute $\Gamma_z(j)$ recursively via the moment equation in (3.15) for a VAR(1) model.

Since a k-dimensional VAR(p) model can always be rewritten as a kp-dimensional VAR(1) model, the prior discussion applies to all stationary VAR(p) models. Specifically, let $Z_t = (z_t', z_{t-1}', \ldots, z_{t-p+1}')'$ be a kp-dimensional stagged time series. Then, it is easy to see that

$$Z_t = \begin{bmatrix} \phi_1 & \phi_2 & \cdots & \phi_{p-1} & \phi_p \\ I & 0 & \cdots & 0 & 0 \\ 0 & I & \cdots & 0 & 0 \\ \vdots & \vdots & \ddots & \vdots & \vdots \\ 0 & 0 & \cdots & I & 0 \end{bmatrix} Z_{t-1} + \begin{bmatrix} a_t \\ 0 \\ 0 \\ \vdots \\ 0 \end{bmatrix}, \tag{3.18}$$

which is in the form of a kp-dimensional VAR(1) model. The coefficient matrix of $\boldsymbol{Z}_{t-1}$ in Eq. (3.18) is called the companion matrix of the AR matrix polynomial $\boldsymbol{\phi}(B)$. It can be shown that eigenvalues of the companion matrix are the inverses of the solutions of the equation $|\boldsymbol{\phi}(B)| = 0$. This provides an alternative way to describe the condition of weak stationarity of a VAR(p) model, namely all eigenvalues of the companion matrix of $\boldsymbol{\phi}(B)$ are less than 1 in absolute value.

3.2.1 Impulse Response Function

A stationary VAR(p) model has an MA representation,

$$\boldsymbol{z}_t = \boldsymbol{\mu} + \sum_{i=0}^{\infty} \boldsymbol{\psi}_i \boldsymbol{a}_{t-i} \equiv \boldsymbol{\mu} + \boldsymbol{\psi}(B)\boldsymbol{a}_t, \tag{3.19}$$

where $\boldsymbol{\psi}(B) = \sum_{i=0}^{\infty} \boldsymbol{\psi}_i B^i$ such that $\boldsymbol{\phi}(B)\boldsymbol{\psi}(B) = \boldsymbol{I}$. The coefficient matrix $\boldsymbol{\psi}_i$ is the effect of $\boldsymbol{a}_t$ on $\boldsymbol{z}_{t+i}$ and can be regarded as an impulse response matrix. For a given VAR model, the $\boldsymbol{\psi}_i$ matrix can be obtained by equating the coefficient matrices of B^i in $\boldsymbol{\phi}(B)\boldsymbol{\psi}(B) = \boldsymbol{I}$. Also, by Eq. (3.19), we have

$$\boldsymbol{\Gamma}_z(j) = \sum_{i=0}^{\infty} \boldsymbol{\psi}_{i+j} \boldsymbol{\Sigma}_a \boldsymbol{\psi}_i', \quad j \geq 0.$$

The impulse response matrices $\boldsymbol{\psi}_i$ of Eq. (3.19) might be hard to interpret because the innovations a_{it} are typically correlated. To mitigate the difficulty, one can perform transformation, say

$$\boldsymbol{b}_t = \boldsymbol{P}'\boldsymbol{a}_t,$$

such that var($\boldsymbol{b}_t$) is a diagonal matrix. There are many ways to achieve this, one possibility is to consider the eigenvalue-eigenvector analysis of $\boldsymbol{\Sigma}_a$, i.e.

$$\boldsymbol{\Sigma}_a \boldsymbol{P} = \boldsymbol{P}\boldsymbol{\Lambda},$$

where $\boldsymbol{\Lambda} = \text{diag}\{\lambda_1, \ldots, \lambda_k\}$ with λ_i being the ith ordered eigenvalues of $\boldsymbol{\Sigma}_a$ and $\boldsymbol{P}$ is the matrix of corresponding eigenvectors. Since $\boldsymbol{\Sigma}_a$ is positive-definite, $\boldsymbol{P}$ is an orthonormal matrix, e.g. $\boldsymbol{P}\boldsymbol{P}' = \boldsymbol{P}'\boldsymbol{P} = \boldsymbol{I}$. Letting $\boldsymbol{\psi}_i^* = \boldsymbol{P}\boldsymbol{\psi}$ for $i \geq 0$, then we have

$$\boldsymbol{z}_t = \boldsymbol{\mu} + \sum_{i=0}^{\infty} \boldsymbol{\psi}_i^* \boldsymbol{b}_t. \tag{3.20}$$

The coefficient matrices $\boldsymbol{\psi}_i^*$ are called the impulse response matrices of orthogonal innovations $\boldsymbol{b}_t$. This is so because it is easy to see that $\partial z_{ut}/\partial b_{v,t-i} = \psi_{i,uv}^*$, where $\psi_{i,uv}^*$ denotes the (u, v)th element of the matrix $\boldsymbol{\psi}_i^*$, implying that one unit change in $b_{v,t-i}$ would affect z_{ut} by the amount $\psi_{i,uv}^*$. The partial sum matrices

$$\boldsymbol{\Psi}_v = \sum_{i=0}^{v} \boldsymbol{\psi}_i^*, \tag{3.21}$$

are the cumulative impulse response matrices. Note that the (i, j)th element of $\boldsymbol{\psi}_v$ and $\boldsymbol{\Psi}_v$ are the impact and cumulative impact of $b_{j,t-v}$ on z_{it}. The matrix $\boldsymbol{\Psi}_\infty$ denotes the *total multipliers* of the VAR model. In many applications, researchers are interested in the impulse response functions of a fitted VAR model.

3.2.2 Some Special Cases

The structure of the coefficient matrices $\boldsymbol{\phi}_i$ of a VAR(p) model may convey useful information about the underlying time series. We consider some special cases in this section.

Case 1: TFM. In this case, without loss of generality, we assume that $\boldsymbol{z}_t$ can be partitioned into $\boldsymbol{z}_{1t}$ and $\boldsymbol{z}_{2t}$ such that the model becomes

$$\begin{bmatrix} \boldsymbol{\phi}_{11}(B) & \boldsymbol{0} \\ \boldsymbol{\phi}_{21}(B) & \boldsymbol{\phi}_{22}(B) \end{bmatrix} \begin{bmatrix} \boldsymbol{z}_{1t} \\ \boldsymbol{z}_{2t} \end{bmatrix} = \begin{bmatrix} \boldsymbol{a}_{1t} \\ \boldsymbol{a}_{2t} \end{bmatrix}. \tag{3.22}$$

Here $\boldsymbol{z}_{1t}$ is the input vector and follows the VAR model $\boldsymbol{\phi}_{11}(B)\boldsymbol{z}_{1t} = \boldsymbol{a}_{1t}$. The output vector is $\boldsymbol{z}_{2t}$ and satisfies the equation

$$\boldsymbol{\phi}_{22}(B)\boldsymbol{z}_{2t} = -\boldsymbol{\phi}_{21}(B)\boldsymbol{z}_{1t} + \boldsymbol{a}_{2t} = -\boldsymbol{\phi}_{21}(B)[\boldsymbol{\phi}_{11}(B)]^{-1}\boldsymbol{a}_{1t} + \boldsymbol{a}_{2t}. \tag{3.23}$$

However, Eq. (3.23) is not in a TF formulation, because $\boldsymbol{a}_{2t}$ and $\boldsymbol{a}_{1t}$ might be correlated. To put the model in a TF form, we need to perform some orthogonalization. Specifically, consider the multivariate multiple linear regression

$$\boldsymbol{a}_{2t} = \boldsymbol{\beta}\boldsymbol{a}_{1t} + \boldsymbol{e}_t, \tag{3.24}$$

and obtain the least squares estimate of $\boldsymbol{\beta}$, denoted by $\hat{\boldsymbol{\beta}}$. The residual series $\hat{\boldsymbol{e}}_t = \boldsymbol{a}_{2t} - \hat{\boldsymbol{\beta}}\boldsymbol{a}_{1t}$ is orthogonal to $\boldsymbol{a}_{1t}$. Consequently, we have a TFM

$$\boldsymbol{z}_{2t} = [\boldsymbol{\phi}_{22}(B)]^{-1}[-\boldsymbol{\phi}_{21}(B) + \hat{\boldsymbol{\beta}}\boldsymbol{\phi}_{11}(B)]\boldsymbol{z}_{1t} + [\boldsymbol{\phi}_{22}(B)]^{-1}\hat{\boldsymbol{e}}_t. \tag{3.25}$$

In practice, one might be interested in testing the null hypothesis $H_0 : \boldsymbol{\phi}_{12}(B) = \boldsymbol{0}$ versus the alternative hypothesis $H_a : \boldsymbol{\phi}_{12}(B) \neq \boldsymbol{0}$. If the distribution of $\boldsymbol{a}_t$ is known, e.g. multivariate Gaussian, then the likelihood ratio test can be used. This is referred to as the *Granger causality* test in the literature. For the test, the full model is the k-dimensional VAR(p) model and the reduced model is in Eq. (3.22). The resulting likelihood ratio statistic would asymptotically follow a χ^2_d, where the degrees of freedom $d = pk_1k_2$ with k_i being the dimension of $\boldsymbol{z}_{it}$. See, for instance, Tsay (2014) for further information.

There are multiple advantages in using VAR models directly in handling TFMs. First, both $\boldsymbol{z}_{1t}$ and $\boldsymbol{z}_{2t}$ are random vectors so the approach can treat multiple inputs and multiple outputs without the requirement of pre-whitening. Second, there is no need to identify which variable is input and which one is output in data analysis. Third, one can perform the Granger causality test to confirm the unidirectional relationship from $\boldsymbol{z}_{1t}$ to $\boldsymbol{z}_{2t}$. On the other hand, the model in Eq. (3.25) might not be parsimonious for the true underlying TFM.

Case 2: Block structure: If $\boldsymbol{\phi}_{21}(B)$ of Eq. (3.22) is also zero, then $\boldsymbol{z}_{1t}$ and $\boldsymbol{z}_{2t}$ follow two separate VAR models, even though the two vector series might be contemporaneously correlated. By contemporaneous correlations, we mean the correlations among elements of $\boldsymbol{a}_t$. In the special case that $\boldsymbol{\phi}(B)$ is a diagonal matrix, then $\boldsymbol{z}_t$ consists of k scalar AR time series that are not dynamically dependent.

Case 3: Hierarchical TF: If $z_t = (z'_{1t}, z'_{2t}, z'_{3t})'$ and satisfies the model

$$\begin{bmatrix} \phi_{11}(B) & \mathbf{0} & \mathbf{0} \\ \phi_{21}(B) & \phi_{22}(B) & \mathbf{0} \\ \phi_{31}(B) & \phi_{32}(B) & \phi_{33}(B) \end{bmatrix} \begin{bmatrix} z_{1t} \\ z_{2t} \\ z_{3t} \end{bmatrix} = \begin{bmatrix} a_{1t} \\ a_{2t} \\ a_{3t} \end{bmatrix}. \quad (3.26)$$

Then, z_{1t} is an input vector to z_{2t} and both z_{1t} and z_{2t} are input vectors to z_{3t}. The implied TFMs can be obtained by the same technique mentioned in Case 1.

Case 4: Suppose that the VAR model can be decomposed as:

$$z_t = \mathbf{\Phi}_*(B) z_{t-1} + \mathbf{a}_t = \boldsymbol{A}(B)\boldsymbol{C}(B) z_{t-1} + \mathbf{a}_t \quad (3.27)$$

where $\mathbf{\Phi}_*(B) = \sum_{i=1}^{p} \mathbf{\Phi}_i B^{i-1}$, and $\boldsymbol{A}(B) = \sum_{i=0}^{p_1} \boldsymbol{A}_i B^i$ and $\boldsymbol{C}(B) = \sum_{i=0}^{p_2} \boldsymbol{C}_i B^i$ are matrix polynomial operators of dimensions $k \times r$ and $r \times k$ with degrees p_1 and p_2, respectively, and $p = p_1 + p_2 + 1$. Letting $\boldsymbol{y}_t = \boldsymbol{C}(B) z_t$ be a r-dimensional time series, we can write the model for z_t as

$$z_t = \boldsymbol{A}(B)\boldsymbol{y}_{t-1} + \mathbf{a}_t.$$

This is the reduced-rank AR model proposed by Velu et al. (1986) and is useful when r is much smaller than k. See also Ahn and Reinsel (1988). It is related to the factor models studied in Chapter 6.

3.2.3 Estimation

Given the realizations $\{z_1, \ldots, z_T\}$, one can estimate the VAR(p) model by the ordinary least squares (OLS) method, the generalized least squares (GLS) method, or the maximum likelihood (ML) method. It turns out that the OLS estimators are the same as the GLS estimators for VAR models and, under the normality assumption, OLS estimators are asymptotically equivalent to the ML estimators. See, for instance, Tsay (2014, chapter 2) and the references therein. Therefore, we shall only consider the OLS estimation here.

The VAR(p) model in Eq. (3.13) can be written as

$$z'_t = \boldsymbol{x}'_t \boldsymbol{\beta} + \boldsymbol{a}'_t, \quad (3.28)$$

where $\boldsymbol{x}_t = (1, z'_{t-1}, \ldots, z'_{t-p})'$ and $\boldsymbol{\beta}' = [\boldsymbol{\phi}_0, \boldsymbol{\phi}_1, \ldots, \boldsymbol{\phi}_p]$ is a $(kp+1) \times p$ coefficient matrix. Using Eq. (3.28) for $t = p+1, \ldots, T$, we have the following matrix equation of data

$$\boldsymbol{Z} = \boldsymbol{X}\boldsymbol{\beta} + \boldsymbol{A}, \quad (3.29)$$

where the ith rows of $\boldsymbol{Z}, \boldsymbol{X}$, and $\boldsymbol{A}$ are $z'_{p+i}, \boldsymbol{x}'_{p+i}$, and $\boldsymbol{a}'_{p+i}$, respectively. The dimensions of $\boldsymbol{Z}$ and $\boldsymbol{A}$ are $(T-p) \times k$, whereas that of $\boldsymbol{X}$ is $(T-p) \times (kp+1)$. The objective function of the OLS estimation of the model is then

$$S(\boldsymbol{\beta}) = \text{tr}[(\boldsymbol{Z} - \boldsymbol{X}\boldsymbol{\beta})(\boldsymbol{Z} - \boldsymbol{X}\boldsymbol{\beta})'],$$

where tr($\boldsymbol{C}$) is the trace of the square-matrix $\boldsymbol{C}$. Taking partial derivatives of $S(\boldsymbol{\beta})$ with respective to $\boldsymbol{\beta}$, we can obtain a linear system of equations of the first-order conditions from which the OLS estimate is

$$\hat{\boldsymbol{\beta}} = (\boldsymbol{X}'\boldsymbol{X})^{-1}(\boldsymbol{X}'\boldsymbol{Z}) = \left(\sum_{t=p+1}^{T} \boldsymbol{x}_t \boldsymbol{x}'_t \right)^{-1} \left(\sum_{t=p+1}^{T} \boldsymbol{x}_t z'_t \right). \quad (3.30)$$

The OLS estimate of $\boldsymbol{\Sigma}_a$ is

$$\tilde{\boldsymbol{\Sigma}}_a = \frac{1}{T-(k+1)p-1}\sum_{t=p+1}^{T} \hat{\boldsymbol{a}}_t \hat{\boldsymbol{a}}_t' \tag{3.31}$$

where $\hat{\boldsymbol{a}}_t = \boldsymbol{z}_t - \hat{\boldsymbol{\phi}}_0 - \sum_{i=1}^{p} \hat{\boldsymbol{\phi}}_i \boldsymbol{z}_{t-i}$ is the residual series. If ML estimation is used, then the ML estimate of $\boldsymbol{\Sigma}_a$ is

$$\hat{\boldsymbol{\Sigma}}_a = \frac{1}{T-p}\sum_{t=p+1}^{T} \hat{\boldsymbol{a}}_t \hat{\boldsymbol{a}}_t'. \tag{3.32}$$

Theorem 3.1 Assume that $\boldsymbol{a}_t$ of the VAR(p) model is a sequence of independent and identically distributed random vectors with mean zero and positive-definite covariance matrix $\boldsymbol{\Sigma}_a$. Then, the OLS estimation gives (a) $E(\hat{\boldsymbol{\beta}}) = \boldsymbol{\beta}$, (b) $E(\tilde{\boldsymbol{\Sigma}}_a) = \boldsymbol{\Sigma}_a$, (c) the residual $\hat{\boldsymbol{A}}$ is uncorrelated with $\hat{\boldsymbol{\beta}}$, and (d) the covariance matrix of the parameter estimates is

$$\text{Cov}[\text{vec}(\hat{\boldsymbol{\beta}})] = \tilde{\boldsymbol{\Sigma}}_a \otimes (\boldsymbol{X}'\boldsymbol{X})^{-1}.$$

To establish the limiting normal distribution of the estimate $\hat{\boldsymbol{\beta}}$, we further assume that

$$E|a_{it}a_{jt}a_{ut}a_{vt}| < \infty, \quad \text{for all } 1 \le i,j,u,v \le k \text{ and } t, \tag{3.33}$$

which implies that all fourth moments of $\boldsymbol{a}_t$ are finite.

Theorem 3.2 Assume that the condition of Theorem 3.1 and Eq. (3.33) hold. Then, as $T \to \infty$,

(i) $\hat{\boldsymbol{\beta}} \to_p \boldsymbol{\beta}$,

(ii) $\sqrt{T-p}[\text{vec}(\hat{\boldsymbol{\beta}}) - \text{vec}(\boldsymbol{\beta})] \to_d N(\boldsymbol{0}, \boldsymbol{\Sigma}_a \otimes \boldsymbol{G}^{-1})$,

where $\to_p$ and $\to_d$ denote convergence in probability and in distribution, respectively, and

$$\boldsymbol{G} = \begin{bmatrix} 1 & \boldsymbol{0}' \\ \boldsymbol{0} & \boldsymbol{\Gamma}_0^* \end{bmatrix} + \begin{bmatrix} 0 \\ \boldsymbol{u} \end{bmatrix} [0, \boldsymbol{u}']$$

where $\boldsymbol{0}$ is a kp-dimensional zero vector, $\boldsymbol{u} = \boldsymbol{1}_p \otimes \boldsymbol{\mu}$ with $\boldsymbol{1}_p$ being the p-dimensional vector of ones, and

$$\boldsymbol{\Gamma}_0^* = \begin{bmatrix} \boldsymbol{\Gamma}_z(0) & \boldsymbol{\Gamma}_z(1) & \cdots & \boldsymbol{\Gamma}_z(p-1) \\ \boldsymbol{\Gamma}_z(-1) & \boldsymbol{\Gamma}_z(0) & \cdots & \boldsymbol{\Gamma}_z(p-2) \\ \vdots & \vdots & \ddots & \vdots \\ \boldsymbol{\Gamma}_z(1-p) & \boldsymbol{\Gamma}_z(2-p) & \cdots & \boldsymbol{\Gamma}_z(0) \end{bmatrix}.$$

3.2.4 Model Building

Order selection: The first step in building a VAR model for a time series $\boldsymbol{z}_t$ is to select the order p. There are several methods available in the literature. See, for instance, Tsay (2014, chapter 2). The most commonly used method is to employ one of the following criterion functions:

$$\text{AIC}(p) = \ln|\hat{\boldsymbol{\Sigma}}_a| + \frac{2pk^2}{T},$$

$$\text{BIC}(p) = \ln|\hat{\boldsymbol{\Sigma}}_a| + \frac{\ln(T)pk^2}{T},$$

$$\text{Hannan and Quinn (HQ)}(p) = \ln|\hat{\boldsymbol{\Sigma}}_a| + \frac{2\ln(\ln(T))pk^2}{T},$$

where $\hat{\boldsymbol{\Sigma}}_a$ is the ML estimate of $\boldsymbol{\Sigma}$ in Eq. (3.32), T is the sample size, and k is the dimension. These criteria are the multivariate generalizations of those studied in Section 2.8.4. See Akaike (1973), Schwarz (1978), and Hannan and Quinn (1979). In practice, with a prespecified maximum order P, one selects the order that gives rise to the minimum value of a criterion function. For instance, the order selected by AIC is

$$\hat{p} = \arg\min_{0\le p\le P} \text{AIC}(p),$$

where $p = 0$ corresponding to fitting a model with constant term only. Under the assumption that z_t follows a finite-order VAR model, both BIC and HQ criteria are shown to be consistent, but AIC has some small probability to specify an order greater than the true order p. However, AIC is often used because there exists no true VAR model in any real application. For further discussion of properties of AIC, see Shibata (1980).

Model checking: Let $\boldsymbol{R}_j$ be the lag-j cross-correlation matrix of the innovation series $\boldsymbol{a}_t$ of the VAR(p) model in Eq. (3.13). For model checking, the null hypothesis of interest is $H_0 : \boldsymbol{R}_1 = \cdots = \boldsymbol{R}_m = \boldsymbol{0}$ versus the alternative hypothesis $H_a : \boldsymbol{R}_i \neq \boldsymbol{0}$ for some $1 \le i \le m$, where m is a prespecified positive integer. Let $\hat{\boldsymbol{a}}_t$ be the residual series of the model, e.g. the least squares residuals. A natural estimate of $\boldsymbol{R}_j$ is

$$\hat{\boldsymbol{R}}_j = \hat{\boldsymbol{D}}^{-1/2}\hat{\boldsymbol{C}}_j\hat{\boldsymbol{D}}^{-1/2},$$

where $\hat{\boldsymbol{C}}_v$ is the lag-v residual covariance matrix given by

$$\hat{\boldsymbol{C}}_v = \frac{1}{T-p}\sum_{t=p+v+1}^{T} \hat{\boldsymbol{a}}_t\hat{\boldsymbol{a}}_{t-v}, \quad v = 0, 1, \ldots,$$

and $\hat{\boldsymbol{D}} = \text{diag}\{\hat{\boldsymbol{C}}_{0,11}, \ldots, \hat{\boldsymbol{C}}_{0,kk}\}$, consisting of the diagonal elements of $\hat{\boldsymbol{C}}_0$. A commonly used test statistic is the multivariate portmanteau statistic defined as

$$\begin{aligned}
Q_k(m) &= T^2\sum_{\ell=1}^{m}\frac{1}{T-\ell}\,\text{tr}(\hat{\boldsymbol{R}}_\ell'\hat{\boldsymbol{R}}_0^{-1}\hat{\boldsymbol{R}}_\ell\hat{\boldsymbol{R}}_0^{-1}) \\
&= T^2\sum_{\ell=1}^{m}\frac{1}{T-\ell}\,\text{tr}(\hat{\boldsymbol{R}}_\ell'\hat{\boldsymbol{R}}_0^{-1}\hat{\boldsymbol{R}}_\ell\hat{\boldsymbol{R}}_0^{-1}\hat{\boldsymbol{D}}^{-1}\hat{\boldsymbol{D}}) \\
&= T^2\sum_{\ell=1}^{m}\frac{1}{T-\ell}\,\text{tr}(\hat{\boldsymbol{D}}\hat{\boldsymbol{R}}_\ell'\hat{\boldsymbol{D}}\hat{\boldsymbol{D}}^{-1}\hat{\boldsymbol{R}}_0^{-1}\hat{\boldsymbol{D}}^{-1}\hat{\boldsymbol{D}}\hat{\boldsymbol{R}}_\ell\hat{\boldsymbol{D}}\hat{\boldsymbol{D}}^{-1}\hat{\boldsymbol{R}}_0^{-1}\hat{\boldsymbol{D}}^{-1}) \\
&= T^2\sum_{\ell=1}^{m}\frac{1}{T-\ell}\,\text{tr}(\hat{\boldsymbol{C}}_\ell'\hat{\boldsymbol{C}}_0^{-1}\hat{\boldsymbol{C}}_\ell\hat{\boldsymbol{C}}_0^{-1}).
\end{aligned} \tag{3.34}$$

Under the assumptions of Theorem 3.2, one can show that the test statistic $Q_k(m)$ is asymptotically distributed as χ^2_d distribution with degrees of freedom $d = (m - p)k^2$ provided that $m > p$. Readers are referred to Li and McLeod (1981) and Hosking (1981) for further information.

Model simplification: One of the difficulties in multivariate time series modeling is that a fitted model often contains many parameters that are highly correlated, but marginally insignificant at the usual 5% level. Model simplification becomes a necessity. However, it would be too tedious to remove insignificant parameters one by one. There are several ways to overcome this difficulty. One approach is to use regularization such as the ℓ_1 penalized estimation. We shall discuss this approach in Chapter 7 in analysis of high-dimensional time series. The second approach is *thresholding* by simultaneously setting all parameters with t-ratios less than a given threshold to zero. This approach is used by Tsay (2014) and available in the **MTS** package of R. To implement thresholding, care must be exercised in choosing a proper threshold. A large threshold can easily invalidate the adequacy of a fitted model, whereas a small threshold often results in insufficient simplification. Following Tsay (2014), we examine the resulting information criteria after re-fitting the model for a given threshold. If the information criterion, e.g. AIC, increases, then the threshold is too high. On the other hand, if the selected information criterion drops, then one can increase the threshold. Limited experience shows that this thresholding approach works well in many applications.

3.2.5 Prediction

Consider the VAR(p) model in Eq. (3.13). Suppose that one is interested in predicting z_{n+h} at the forecast origin $t = n$. That is, given the information available at time $t = n$, denoted by F_t, one wants to learn $P(z_{t+h}|F_t)$, where h is a positive integer referring to as the forecast horizon in the literature. For simplicity, one often focuses on $E(z_{t+h}|F_t)$ and $\text{Var}(z_{t+h}|F_t)$ instead of the conditional distribution $P(z_{t+h}|F_t)$. To this end, we use the criterion of minimizing the mean squares of forecast errors (MSFE) and consider two cases. Other forecasting criteria can be used with some complication of no closed-form solution to produce forecasts.

Case 1: Assume that the model is known, i.e. estimated parameters of the VAR(p) model are treated as the true ones. From the model, we have

$$z_{n+h} = \boldsymbol{\phi}_0 + \sum_{i=1}^{p} \boldsymbol{\phi}_i z_{n+h-i} + \boldsymbol{a}_{n+h}. \tag{3.35}$$

For $h = 1$, Eq. (3.35) becomes

$$z_{n+1} = \boldsymbol{\phi}_0 + \boldsymbol{\phi}_1 z_n + \cdots + \boldsymbol{\phi}_p z_{n+1-p} + \boldsymbol{a}_{n+1}.$$

Under the MSFE criterion, the 1-step ahead point forecast is simply

$$z_n(1) = E(z_{n+1}|F_t) = \boldsymbol{\phi}_0 + \boldsymbol{\phi}_1 z_n + \cdots + \boldsymbol{\phi}_p z_{n+1-p}, \tag{3.36}$$

and the associated forecast error and its variance are

$$\boldsymbol{e}_n(1) = z_{n+1} - z_n(1) = \boldsymbol{a}_{n+1}, \quad \text{V}[\boldsymbol{e}_n(1)] = \boldsymbol{\Sigma}_a. \tag{3.37}$$

By Eqs. (3.36) and (3.37), we have

$$\boldsymbol{z}_{n+1} = \boldsymbol{z}_n(1) + \boldsymbol{a}_{n+1}. \tag{3.38}$$

For $h = 2$, using Eqs. (3.35) and (3.38), we have

$$\begin{aligned}
\boldsymbol{z}_{n+2} &= \boldsymbol{\phi}_0 + \boldsymbol{\phi}_1 \boldsymbol{z}_{t+1} + \boldsymbol{\phi}_2 \boldsymbol{z}_n + \cdots + \boldsymbol{\phi}_p \boldsymbol{z}_{n+2-p} + \boldsymbol{a}_{n+2} \\
&= \boldsymbol{\phi}_0 + \boldsymbol{\phi}_1 [\boldsymbol{z}_n(1) + \boldsymbol{a}_{n+1}] + \boldsymbol{\phi}_2 \boldsymbol{z}_n + \cdots + \boldsymbol{\phi}_p \boldsymbol{z}_{n+2-p} + \boldsymbol{a}_{n+2} \\
&= \boldsymbol{\phi}_0 + \boldsymbol{\phi}_1 \boldsymbol{z}_n(1) + \boldsymbol{\phi}_2 \boldsymbol{z}_n + \cdots + \boldsymbol{\phi}_p \boldsymbol{z}_{n+2-p} + \boldsymbol{a}_{n+2} + \boldsymbol{\phi}_1 \boldsymbol{a}_{n+1}.
\end{aligned}$$

It is then clear that the 2-step ahead prediction and its associated error are

$$\boldsymbol{z}_n(2) = E(\boldsymbol{z}_{n+2}|F_n) = \boldsymbol{\phi}_0 + \boldsymbol{\phi}_1 \boldsymbol{z}_n(1) + \boldsymbol{\phi}_2 \boldsymbol{z}_n + \cdots + \boldsymbol{\phi}_p \boldsymbol{z}_{n+2-p} \tag{3.39}$$

$$\boldsymbol{e}_n(2) = \boldsymbol{z}_{n+2} - \boldsymbol{z}_n(2) = \boldsymbol{a}_{n+2} + \boldsymbol{\phi}_1 \boldsymbol{a}_{n+1} \tag{3.40}$$

$$\mathrm{V}[\boldsymbol{e}_n(2)] = \boldsymbol{\Sigma}_a + \boldsymbol{\phi}_1 \boldsymbol{\Sigma}_a \boldsymbol{\phi}_1'.$$

In general, one can repeat the process to obtain that, for the h-step ahead prediction,

$$\boldsymbol{z}_n(h) = \boldsymbol{\phi}_0 + \sum_{i=1}^{p} \boldsymbol{\phi}_i \boldsymbol{z}_n(h-i) \tag{3.41}$$

$$\boldsymbol{e}_n(h) = \boldsymbol{a}_{n+h} + \sum_{j=1}^{h-1} \boldsymbol{\psi}_j \boldsymbol{a}_{n+h-j} \tag{3.42}$$

$$\mathrm{V}[\boldsymbol{e}_n(h)] = \boldsymbol{\Sigma}_a + \sum_{j=1}^{h-1} \boldsymbol{\psi}_j \boldsymbol{\Sigma}_a \boldsymbol{\psi}_j', \tag{3.43}$$

where $\boldsymbol{\psi}_i$ is the ith ψ-weight coefficient matrix of $\boldsymbol{z}_t$; see Eq. (3.19). In Eq. (3.41), it is understood that $\boldsymbol{z}_n(\ell) = \boldsymbol{z}_{t+\ell}$ for $\ell \leq 0$, i.e. $\boldsymbol{z}_n(0) = \boldsymbol{z}_n$ and $\boldsymbol{z}_n(-1) = \boldsymbol{z}_{n-1}$. Eqs. (3.41) and (3.43) imply that the forecasts and their associated covariance matrices of a VAR(p) model can be calculated recursively.

Using $\boldsymbol{\phi}_0 = (\boldsymbol{I} - \sum_{i=1}^{p} \boldsymbol{\phi}_i)\boldsymbol{\mu}$, Eq. (3.41) can be rewritten as

$$\boldsymbol{z}_n(h) - \boldsymbol{\mu} = \sum_{i=1}^{p} [\boldsymbol{z}_n(h-i) - \boldsymbol{\mu}].$$

Then, by using the kp-dimensional model representation in Eq. (3.18), one can easily derive that, if $\boldsymbol{z}_t$ is a stationary time series, we have $\boldsymbol{z}_n(h) - \boldsymbol{\mu} \to \boldsymbol{0}$ as $h \to \infty$. That is, $\boldsymbol{z}_n(h) \to \boldsymbol{\mu}$ as $h \to \infty$. This is referred to as the *mean reverting* of a stationary vector AR process. Furthermore, from Eq. (3.43), we have $\mathrm{V}[\boldsymbol{e}_n(h)] \to \mathrm{V}(\boldsymbol{z}_t)$ as $h \to \infty$.

Case 2: Parameter uncertainty. In practice, the coefficient matrices $\boldsymbol{\phi}_i$ are unknown and one may consider parameter uncertainty in forecasting. For VAR models, the result is relatively simple. Note that the estimates $\hat{\boldsymbol{\phi}}_i$ and $\hat{\boldsymbol{\Sigma}}_a$ are available given F_n so that the model can be written as

$$\boldsymbol{z}_{n+h} = \hat{\boldsymbol{\phi}}_0 + \sum_{i=1}^{p} \hat{\boldsymbol{\phi}}_i \boldsymbol{z}_{n+h-i} + [\boldsymbol{\phi}_0 - \hat{\boldsymbol{\phi}}_0] + \sum_{i=1}^{p} [\boldsymbol{\phi}_i - \hat{\boldsymbol{\phi}}_i] \boldsymbol{z}_{n+h-i} + \boldsymbol{a}_{n+h}. \tag{3.44}$$

For $j = 1, \ldots, h$, define recursively

$$\hat{z}_n(j) = \hat{\phi}_0 + \sum_{i=1}^{p} \hat{\phi}_i \hat{z}_n(j-i), \tag{3.45}$$

where it is understood that $\hat{z}_n(\ell) = z_{n+\ell}$ for $\ell \leq 0$. Also, let

$$\hat{e}_n(j) = z_{n+j} - \hat{z}_n(j), \quad j = 1, \ldots, h \tag{3.46}$$

be the j-step ahead forecast error under parameter uncertainty. By Eq. (3.46), we have

$$\hat{e}_n(j) = z_{n+j} - z_n(j) + z_n(j) - \hat{z}_n(j) = e_n(j) + [z_n(j) - \hat{z}_n(j)]. \tag{3.47}$$

Since $e_n(j) = a_{n+j} + \sum_{v=1}^{j-1} \psi_v a_{n+v}$ and both $z_n(j)$ and $\hat{z}_n(j)$ are functions of F_n, the two terms on the right side of Eq. (3.47) are uncorrelated for $j \geq 1$ and we have

$$\begin{aligned} \mathrm{V}[\hat{e}_n(j)] &= \mathrm{V}[e_n(j)] + E\{[z_n(j) - \hat{z}_n(j)][z_n(j) - \hat{z}_n(j)]'\} \\ &\equiv \mathrm{V}[e_n(j)] + \mathrm{MSE}[z_n(j) - \hat{z}_n(j)], \end{aligned} \tag{3.48}$$

where $\equiv$ denotes equivalence. It remains to derive the MSE term. For the VAR models, one can derive a closed-form solution for the MSE. See, for instance, Tsay (2014, section 2.9.2) and the references therein. For a VAR(p) model and 1-step ahead prediction, we have $\mathrm{MSE}[z_n(1) - \hat{z}_n(1)] = (kp+1)\Sigma_a/(T-p)$. Therefore, we obtain

$$\mathrm{V}[\hat{e}_n(1)] = \frac{(T-p)+kp+1}{T-p}\Sigma. \tag{3.49}$$

Since $kp+1$ is the number of parameters for the component z_{it} of a fitted VAR(p) model, Eq. (3.49) implies that for each estimated parameter of an VAR(p) model, it increases the uncertainty in 1-step ahead prediction by a factor of $1/(T-p)$, where $T-p$ is the effective number of observations used in fitting the VAR(p) model. In practice, the result says that it pays to use an adequate, parsimonious VAR model in prediction.

3.2.6 Forecast Error Variance Decomposition

Consider the h-step ahead forecast error in Eq. (3.42). In addition, let $\eta_t = \Sigma_a^{-1/2} a_t$, where $\Sigma_a^{1/2}$ denotes the square-root matrix of Σ_a. Then, we have $\mathrm{V}(\eta_t) = I_k$, the $k \times k$ identity matrix. The forecast error can then be written as

$$e_n(h) = \Psi_0 \eta_{n+h} + \sum_{j=1}^{h-1} \Psi_j \eta_{n+h-j}, \tag{3.50}$$

where $\Psi_0 = \Sigma_a^{1/2}$ and $\Psi_j = \psi_j \Sigma_a^{1/2}$. By Eq. (3.50), we have

$$\mathrm{V}[e_n(h)] = \sum_{j=0}^{h-1} \Psi_j \Psi_j'. \tag{3.51}$$

Consequently, for the ith component $e_{n,i}(h)$, we have

$$\text{Var}[e_{n,i}(h)] = \sum_{j=0}^{h-1} \sum_{v=1}^{k} \Psi_{j,iv}^2 = \sum_{v=1}^{k} \left[\sum_{j=0}^{h-1} \Psi_{j,iv}^2 \right], \tag{3.52}$$

where $\Psi_{j,iv}$ is the (i, v)th element of $\boldsymbol{\Psi}_j$. Letting $w_{iv}(h) = \sum_{j=0}^{h-1} \Psi_{j,iv}^2$, we have

$$\text{Var}[e_{n,i}(h)] = \sum_{v=1}^{k} w_{iv}(h), \tag{3.53}$$

which is referred to as the forecast error variance decomposition, because $w_{iv}(h)$ is the impact of the vth component on the forecast error variance of $z_{i,n+h}$. In the literature, the ratio $w_{iv}(h)/\text{Var}[e_{n,i}(h)]$ is the proportion of uncertainty in forecasting $z_{i,n+h}$ attributed to the vth component.

Example 3.2

To illustrate the analysis of multivariate time series using VAR models, we consider the quarterly gross domestic product (GDP) by expenditures in constant prices of six countries from the first quarter of 1980 to the first quarter of 2018. The original data are available from the FRED, website of the Federal Reserve Bank of St. Louis. The countries considered are (i) United States (US), (ii) United Kingdom (UK), (iii) France (FR), (iv) Australia (AU), (v) Germany (GE), and (vi) Canada (CA). Let $\boldsymbol{z}_t = (z_{1t}, \ldots, z_{6t})'$ be the series of growth rates from the previous quarter, i.e. $z_{it} = 100(Z_{it} - Z_{i,t-1})/Z_{i,t-1}$, where Z_{it} is the GDP index with 2010 being 100 and seasonally adjusted. Figure 3.4 shows the time plots of $\boldsymbol{z}_t$ with the left panel consisting of US, UK, and FR (top to bottom) and the right panel consisting of AU, GE, and CA. From the plot, the drops in GDP caused by the 2008 financial crisis are clearly seen (the last quarter of 2008). The only exception is AU.

In what follows, we use the **MTS** package of R to conduct the analysis. Details of the analysis are given in the attached R output. To begin, we use information criteria to select the VAR order for $\boldsymbol{z}_t$. In this particular instance, the orders selected by AIC, BIC, and HQ are 13, 1, and 1, respectively. For simplicity, we entertain a VAR(1) model. Using the least squares method, we obtain the fitted model

$$\boldsymbol{z}_t = \begin{bmatrix} 0.32 \\ 0.28 \\ 0.13 \\ 0.54 \\ 0.12 \\ 0.07 \end{bmatrix} + \begin{bmatrix} 0.16 & 0.44 & -0.18 & -0.02 & -0.02 & 0.17 \\ 0.15 & 0.50 & -0.00 & -0.14 & -0.05 & 0.06 \\ 0.08 & 0.16 & 0.43 & -0.02 & -0.05 & 0.04 \\ 0.04 & 0.19 & -0.34 & 0.10 & 0.03 & 0.29 \\ 0.21 & 0.08 & 0.64 & -0.16 & -0.13 & 0.03 \\ 0.31 & 0.30 & -0.24 & 0.10 & 0.09 & 0.24 \end{bmatrix} \boldsymbol{z}_{t-1} + \boldsymbol{a}_t,$$

where some of the estimates are not significant at the 5% level. The information criteria of the fitted model are −6.87, −6.16, and −6.59, respectively, for AIC, BIC, and HQ. Model checking indicates that the residuals of this fitted model do not

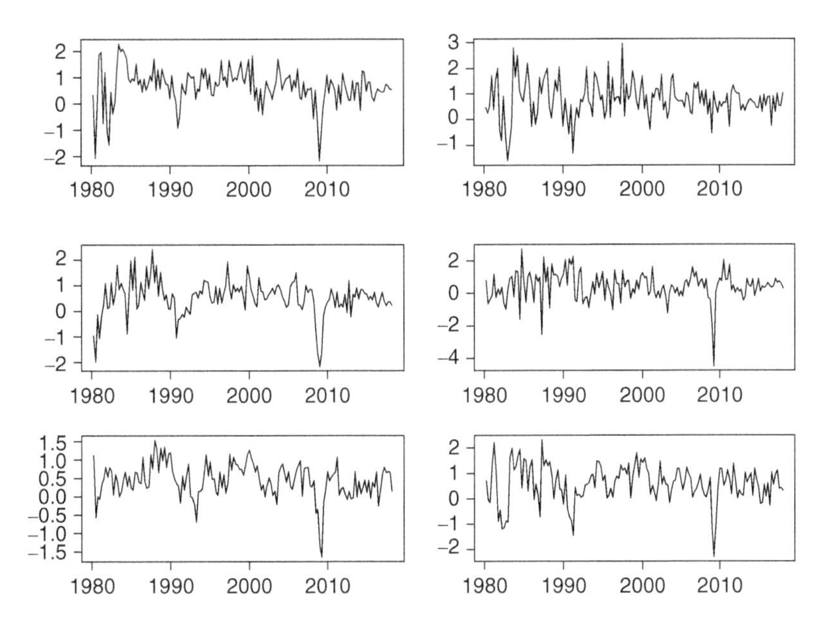

Figure 3.4 Time plots of the quarterly growth rates of the GDP by expenditures in constant prices of United States, United Kingdom, and France (left panel, top to bottom) and Australia, Germany, and Canada (right panel, top to bottom) from 1980.I to 2018.I. The data were seasonally adjusted.

have significant serial correlations. However, the model contains several insignificant parameter estimates. Following the suggested method of model simplification, we use thresholding to refine the fitted VAR(1) model.

We start with a threshold 1 for the t-ratio and the resulting information criteria all decrease. See the R output for details. We increase the threshold to 1.3 and the associated model is

$$\boldsymbol{z}_t = \begin{bmatrix} 0.31 \\ 0.27 \\ 0.13 \\ 0.54 \\ - \\ - \end{bmatrix} + \begin{bmatrix} 0.15 & 0.43 & -0.20 & - & - & 0.17 \\ 0.17 & 0.49 & - & -0.13 & - & - \\ 0.08 & 0.16 & 0.39 & - & - & - \\ - & 0.20 & -0.31 & 0.11 & - & 0.31 \\ 0.19 & - & 0.61 & - & - & - \\ 0.32 & 0.31 & -0.21 & 0.13 & 0.09 & 0.25 \end{bmatrix} \boldsymbol{z}_{t-1} + \boldsymbol{a}_t. \qquad (3.54)$$

The information criteria of the refined model are −7.00, −6.57, and −6.83, respectively, for AIC, BIC, and HQ. The thresholding removes 16 insignificant parameters from the VAR(1) model. Figure 3.5 shows the p-values of the Ljung-Box statistics of the residuals of the model in Eq. (3.54). These p-values show that the model provides a reasonable fit. Figure 3.6 shows the time plots of the residuals of the VAR(1) model in Eq. (3.54).

To understand the relationships between the GDP growth rates of the six countries, we can study the impulse response functions of the fitted model in Eq. (3.54).

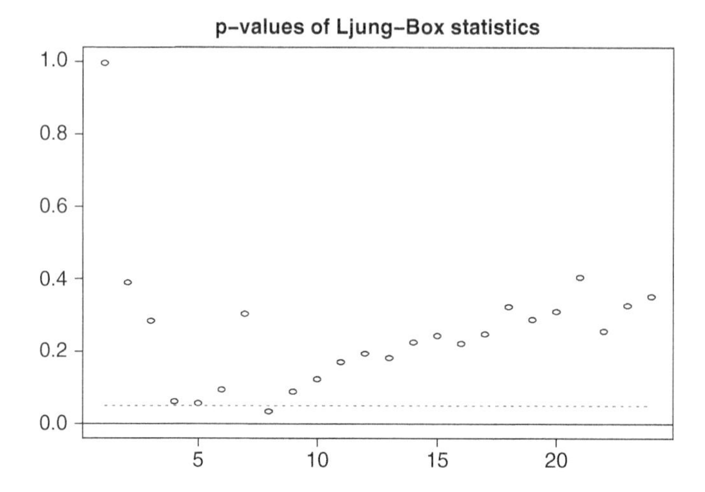

Figure 3.5 p-Values of the Ljung-Box statistics for the residuals of the model in Eq. (3.54). The horizontal axis denotes the lag.

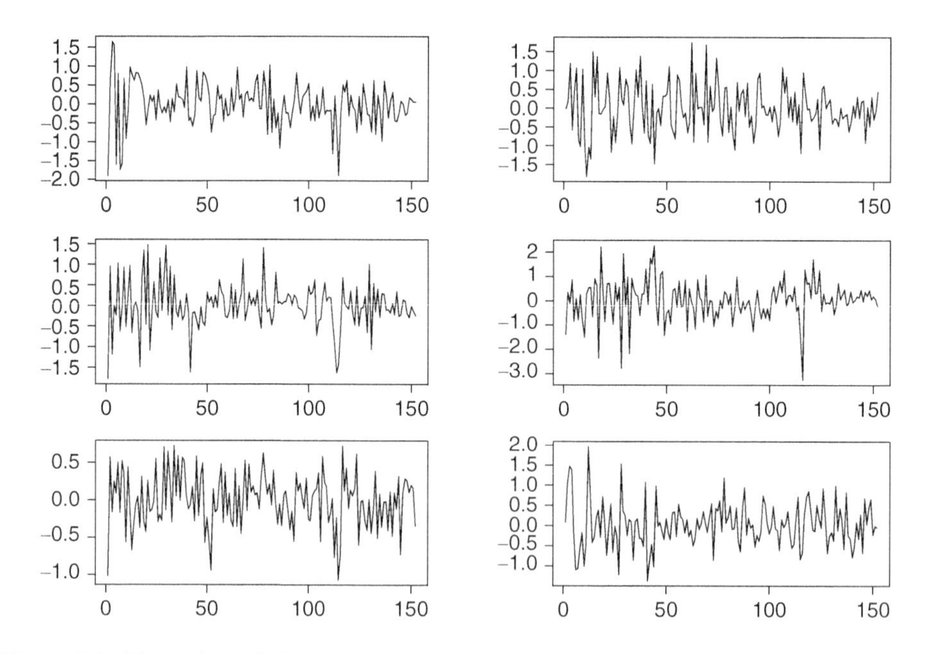

Figure 3.6 Time plots of the residuals of Model (3.54). The horizontal axis denotes time sequence.

Figures 3.7 and 3.8 show the impulse response functions and the cumulative impulse response functions of the model, respectively. These functions are computed using the orthogonal innovations. From the plots, we see that the growth rate of US quarterly GDP hardly depends on those of Australia and Germany, but is affected by those of United Kingdom, Canada, and France. Also, as expected, the growth rates of GDP depend heavily on its own past values. Finally, one can use the fitted model

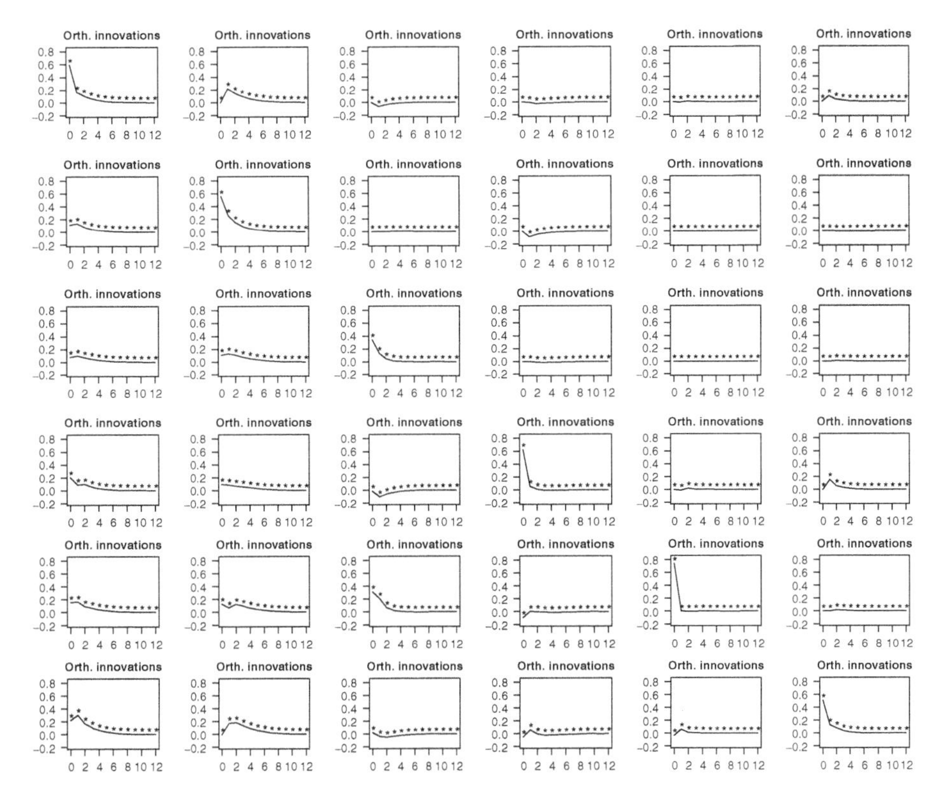

Figure 3.7 Impulse response functions of the model in Eq. (3.54). Orthogonal innovations are used.

to produce forecasts. Details of 1-step to 4-step ahead forecasts and their associated errors are shown in the R output. The forecast variance decomposition is also given.

R output for GDP Data: Output edited to save space.

```
> gdp <- read.table("gdpsimple6c8018.txt",header=T)
> dim(gdp)
[1] 153   6
> tdx <- c(1:153)/4 + 1980 ## Define calendar time (quarters)
> MTSplot(gdp,tdx)   ### Time series plots
> VARorder(gdp) ## Order selection
selected order: aic =  13
selected order: bic =  1
selected order: hq =  1
Summary table:
       p     AIC     BIC      HQ      M(p) p-value
 [1,]  0 -6.8887 -6.8887 -6.8887   0.0000  0.0000
 [2,]  1 -7.7959 -7.0829 -7.5063 182.5630  0.0000
 [3,]  2 -7.7156 -6.2895 -7.1362  49.3614  0.0681
 [4,]  3 -7.5512 -5.4121 -6.6823  36.9038  0.4269
 [5,]  4 -7.5679 -4.7157 -6.4093  55.7927  0.0187
 [6,]  5 -7.4977 -3.9324 -6.0494  43.4377  0.1841
 [7,]  6 -7.3380 -3.0598 -5.6001  31.8750  0.6652
 [8,]  7 -7.1339 -2.1425 -5.1063  25.7069  0.8983
 [9,]  8 -7.3742 -1.6699 -5.0570  64.3401  0.0025
```

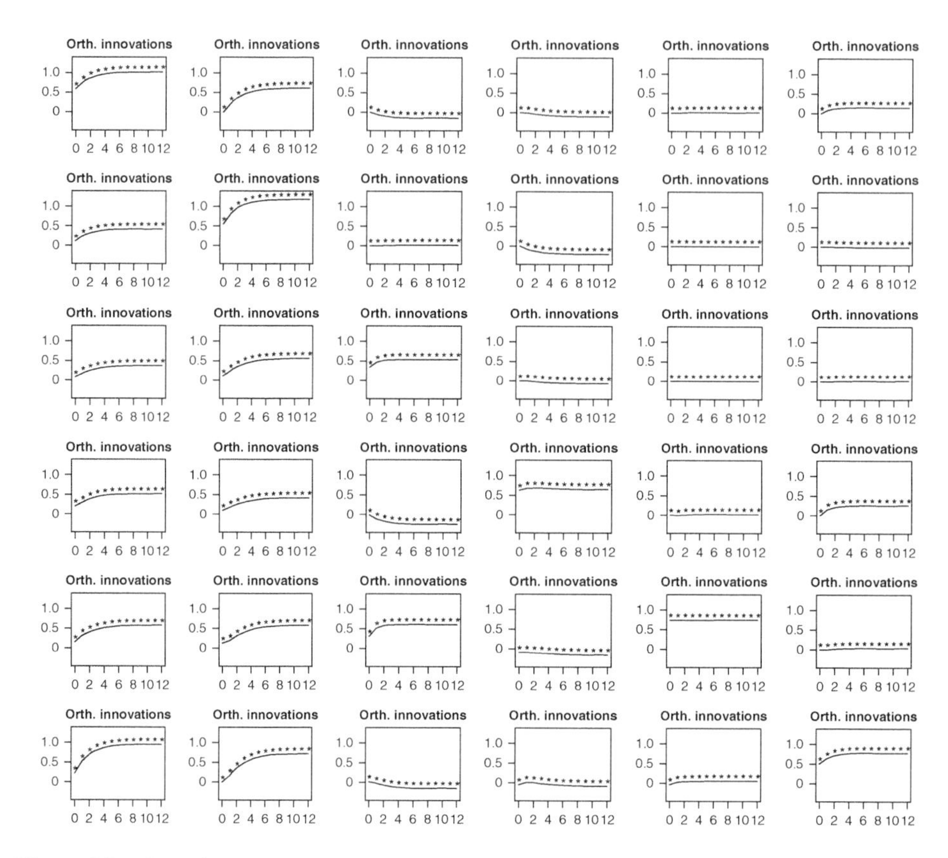

Figure 3.8 Cumulative impulse response functions of the model in Eq. (3.54). Orthogonal innovations are used.

```
[10,]  9 -7.5061 -1.0887 -4.8993  50.9105  0.0509
[11,] 10 -7.4301 -0.2997 -4.5336  30.9780  0.7062
[12,] 11 -7.6115  0.2320 -4.4253  47.2625  0.0991
[13,] 12 -7.8454  0.7112 -4.3696  46.8496  0.1064
[14,] 13 -8.1084  1.1612 -4.3429  44.3816  0.1593
> m1 <- VAR(gdp,1)  ## Estimation
Constant term:
Estimates:  0.3163104 0.2756356 0.1320467 0.5399103 0.1174305 0.0735754
Std. Error:  0.0830115 0.0790354 0.0511402 0.0931643 0.1156071 0.0781213
AR coefficient matrix
AR( 1)-matrix
       [,1]  [,2]     [,3]    [,4]    [,5]   [,6]
[1,] 0.1641 0.437 -0.17598 -0.0212 -0.0245 0.1708
[2,] 0.1517 0.497 -0.00266 -0.1428 -0.0455 0.0564
[3,] 0.0764 0.160  0.42812 -0.0230 -0.0513 0.0384
[4,] 0.0369 0.189 -0.34466  0.1027  0.0278 0.2910
[5,] 0.2095 0.083  0.63957 -0.1587 -0.1284 0.0287
[6,] 0.3096 0.303 -0.24379  0.1031  0.0936 0.2437
standard error
       [,1]   [,2]   [,3]   [,4]   [,5]   [,6]
[1,] 0.0956 0.0852 0.1391 0.0762 0.0660 0.0828
[2,] 0.0910 0.0812 0.1324 0.0725 0.0628 0.0788
[3,] 0.0589 0.0525 0.0857 0.0469 0.0407 0.0510
[4,] 0.1073 0.0957 0.1561 0.0855 0.0741 0.0929
```

```
[5,] 0.1332 0.1187 0.1937 0.1061 0.0919 0.1153
[6,] 0.0900 0.0802 0.1309 0.0717 0.0621 0.0779

Residuals cov-mtx:
           [,1]       [,2]       [,3]        [,4]        [,5]       [,6]
[1,] 0.34833145 0.06573605 0.04571734  0.12044573  0.08966860 0.13046158
[2,] 0.06573605 0.31576212 0.06765773  0.07606172  0.08271335 0.02379309
[3,] 0.04571734 0.06765773 0.13220287  0.02157057  0.12784988 0.02467035
[4,] 0.12044573 0.07606172 0.02157057  0.43874844 -0.01641509 0.01504688
[5,] 0.08966860 0.08271335 0.12784988 -0.01641509  0.67559359 0.01948580
[6,] 0.13046158 0.02379309 0.02467035  0.01504688  0.01948580 0.30850029

det(SSE) =  0.0006456201
AIC =  -6.874711
BIC =  -6.161667
HQ  =  -6.585061
>
> m1a <- refVAR(m1,thr=1) ## Model refinement
Constant term:
Estimates:  0.3081411 0.2735526 0.1258255 0.5426269 0.124787 0
Std. Error:  0.07492052 0.07314373 0.04599156 0.09241243 0.1142533 0
AR coefficient matrix
AR( 1)-matrix
       [,1]  [,2]   [,3]   [,4]    [,5]  [,6]
[1,] 0.1528 0.433 -0.198  0.000  0.0000 0.173
[2,] 0.1698 0.490  0.000 -0.133  0.0000 0.000
[3,] 0.0883 0.158  0.438  0.000 -0.0508 0.000
[4,] 0.0000 0.197 -0.310  0.108  0.0000 0.306
[5,] 0.2441 0.000  0.688 -0.147 -0.1275 0.000
[6,] 0.3152 0.306 -0.208  0.131  0.0940 0.251
standard error
       [,1]   [,2]   [,3]   [,4]   [,5]   [,6]
[1,] 0.0898 0.0837 0.1223 0.0000 0.0000 0.0820
[2,] 0.0771 0.0752 0.0000 0.0713 0.0000 0.0000
[3,] 0.0483 0.0516 0.0842 0.0000 0.0400 0.0000
[4,] 0.0000 0.0934 0.1363 0.0803 0.0000 0.0815
[5,] 0.1138 0.0000 0.1815 0.1043 0.0912 0.0000
[6,] 0.0898 0.0801 0.1252 0.0652 0.0621 0.0775

det(SSE) =  0.0006646203
AIC =  -6.989497
BIC =  -6.494328
HQ  =  -6.788351
>
> m1b <- refVAR(m1a,thr=1.3)  ## Model refinement
Constant term:
Estimates:  0.3081411 0.2735526 0.1296568 0.5426269 0 0
Std. Error:  0.07492052 0.07314373 0.04598706 0.09241243 0 0
AR coefficient matrix
AR( 1)-matrix
      [,1]  [,2]   [,3]   [,4]  [,5]  [,6]
[1,] 0.153 0.433 -0.198  0.000 0.000 0.173
[2,] 0.170 0.490  0.000 -0.133 0.000 0.000
[3,] 0.084 0.157  0.389  0.000 0.000 0.000
[4,] 0.000 0.197 -0.310  0.108 0.000 0.306
[5,] 0.193 0.000  0.606  0.000 0.000 0.000
[6,] 0.315 0.306 -0.208  0.131 0.094 0.251
standard error
       [,1]   [,2]   [,3]   [,4]   [,5]   [,6]
[1,] 0.0898 0.0837 0.1223 0.0000 0.0000 0.0820
[2,] 0.0771 0.0752 0.0000 0.0713 0.0000 0.0000
[3,] 0.0482 0.0517 0.0748 0.0000 0.0000 0.0000
[4,] 0.0000 0.0934 0.1363 0.0803 0.0000 0.0815
```

```
[5,] 0.0953 0.0000 0.1420 0.0000 0.0000 0.0000
[6,] 0.0898 0.0801 0.1252 0.0652 0.0621 0.0775

Residuals cov-mtx:
           [,1]       [,2]       [,3]        [,4]        [,5]       [,6]
[1,] 0.34878812 0.06636693 0.04651763  0.12014493  0.09241242 0.13067112
[2,] 0.06636693 0.31837874 0.06986894  0.07505788  0.08724666 0.02384652
[3,] 0.04651763 0.06986894 0.13438894  0.02058539  0.13290600 0.02473165
[4,] 0.12014493 0.07505788 0.02058539  0.43965478 -0.01848134 0.01497720
[5,] 0.09241242 0.08724666 0.13290600 -0.01848134  0.69600443 0.02249785
[6,] 0.13067112 0.02384652 0.02473165  0.01497720  0.02249785 0.31038747

det(SSE) =  0.0006813695
AIC =  -7.003824
BIC =  -6.568075
HQ  =  -6.826815
>
> MTSdiag(m1b)  ### Model checking
[1] "Covariance matrix:"
       US     UK     FR      AU      GE     CA
US 0.3511 0.0668 0.0468  0.1209  0.0930 0.1315
UK 0.0668 0.3205 0.0703  0.0756  0.0878 0.0240
FR 0.0468 0.0703 0.1353  0.0207  0.1338 0.0249
AU 0.1209 0.0756 0.0207  0.4426 -0.0186 0.0151
GE 0.0930 0.0878 0.1338 -0.0186  0.6998 0.0219
CA 0.1315 0.0240 0.0249  0.0151  0.0219 0.3118
CCM at lag:  0
      [,1]   [,2]   [,3]    [,4]    [,5]   [,6]
[1,] 1.000 0.1992 0.2149  0.3068  0.1877 0.3976
[2,] 0.199 1.0000 0.3378  0.2006  0.1854 0.0759
[3,] 0.215 0.3378 1.0000  0.0847  0.4348 0.1212
[4,] 0.307 0.2006 0.0847  1.0000 -0.0334 0.0406
[5,] 0.188 0.1854 0.4348 -0.0334  1.0000 0.0470
[6,] 0.398 0.0759 0.1212  0.0406  0.0470 1.0000

Hit Enter for more plots:
Hit Enter for more plots:
Hit Enter for p-value plot of individual ccm:
Hit Enter to compute MQ statistics:

Ljung-Box Statistics:
        m       Q(m)      df    p-value
 [1,]   1.0      17.2    36.0     1.00
 [2,]   2.0      74.7    72.0     0.39
 [3,]   3.0     115.9   108.0     0.28
 [4,]   4.0     171.0   144.0     0.06
 .....
[23,]  23.0     845.6   828.0     0.33
[24,]  24.0     879.2   864.0     0.35
Hit Enter to obtain residual plots:

> Phi <- m1b$Phi
> Sig <- m1b$Sigma

> VARirf(Phi,Sig)
Press return to continue

> VARpred(m1b,4)
orig  153
Forecasts at origin:  153
         US     UK     FR     AU     GE     CA
[1,] 0.5209 0.3444 0.2724 0.7550 0.1993 0.4602
[2,] 0.5622 0.4299 0.3332 0.7486 0.2655 0.4464
```

```
[3,] 0.5911 0.4796 0.3738 0.7417 0.3103 0.4748
[4,] 0.6139 0.5098 0.3997 0.7468 0.3404 0.5012
Standard Errors of predictions:
       [,1]   [,2]   [,3]   [,4]   [,5]   [,6]
[1,] 0.5906 0.5643 0.3666 0.6631 0.8343 0.5571
[2,] 0.6586 0.6390 0.4222 0.7011 0.8770 0.6748
[3,] 0.6873 0.6634 0.4446 0.7166 0.8926 0.7258
[4,] 0.6984 0.6716 0.4548 0.7227 0.9004 0.7487
Root mean square errors of predictions:
       [,1]   [,2]   [,3]   [,4]   [,5]   [,6]
[1,] 0.6039 0.5770 0.3749 0.6781 0.8531 0.5697
[2,] 0.6864 0.6692 0.4444 0.7172 0.8951 0.7204
[3,] 0.6995 0.6738 0.4541 0.7233 0.8993 0.7470
[4,] 0.7032 0.6752 0.4592 0.7254 0.9037 0.7585
>
> fev1 <- FEVdec(Phi,Theta = NULL,Sig)  ## Decomposition
Dimension:  6
Order of the ARMA mdoel:
[1] 1 0
Standard deviation of forecast error:
          [,1]      [,2]      [,3]      [,4]      [,5]      [,6]      [,7]      [,8]
[1,] 0.5905829 0.6586473 0.6872954 0.6983882 0.7025126 0.7039905 0.7045060 0.7046824
[2,] 0.5642506 0.6390180 0.6633696 0.6716175 0.6744252 0.6753770 0.6756982 0.6758062
[3,] 0.3665910 0.4222175 0.4446271 0.4547874 0.4591659 0.4609381 0.4616191 0.4618706
[4,] 0.6630647 0.7011405 0.7166383 0.7227332 0.7250744 0.7259287 0.7262292 0.7263321
[5,] 0.8342688 0.8770128 0.8925819 0.9003501 0.9039646 0.9054998 0.9061076 0.9063364
[6,] 0.5571243 0.6747937 0.7257590 0.7486805 0.7580397 0.7616258 0.7629370 0.7634007
          [,9]     [,10]     [,11]     [,12]     [,13]
[1,] 0.7047419 0.7047618 0.7047684 0.7047706 0.7047714
[2,] 0.6758424 0.6758545 0.6758586 0.6758600 0.6758604
[3,] 0.4619608 0.4619924 0.4620034 0.4620071 0.4620084
[4,] 0.7263667 0.7263782 0.7263820 0.7263832 0.7263836
[5,] 0.9064194 0.9064487 0.9064589 0.9064624 0.9064636
[6,] 0.7635609 0.7636154 0.7636337 0.7636399 0.7636419
Forecast-Error-Variance Decomposition
Forecast horizon:  1
           [,1]         [,2]         [,3]        [,4]        [,5]      [,6]
[1,] 1.00000000 0.000000e+00 0.0000000000 0.000000000 0.000000000 0.0000000
[2,] 0.03966412 9.603359e-01 0.0000000000 0.000000000 0.000000000 0.0000000
[3,] 0.04616469 9.061078e-02 0.8632245291 0.000000000 0.000000000 0.0000000
[4,] 0.09413203 2.026795e-02 0.0006721359 0.884927881 0.000000000 0.0000000
[5,] 0.03517928 2.280445e-02 0.1409470776 0.011923650 0.789145544 0.0000000
[6,] 0.15772239 1.090734e-05 0.0015651704 0.007197495 0.003130395 0.8303736
Forecast horizon:  2
           [,1]       [,2]         [,3]         [,4]         [,5]       [,6]
[1,] 0.86408805 0.10861660 9.364737e-03 0.0001535092 6.676552e-05 0.01771033
[2,] 0.07113831 0.91190366 1.287049e-05 0.0169451645 0.000000e+00 0.00000000
[3,] 0.08847041 0.16246642 7.490632e-01 0.0000000000 0.000000e+00 0.00000000
[4,] 0.09973589 0.03263652 2.125301e-02 0.7971758411 1.847759e-04 0.04901396
[5,] 0.06583187 0.02644026 1.828411e-01 0.0107896993 7.140970e-01 0.00000000
[6,] 0.30672433 0.06351395 4.263206e-03 0.0131768793 1.052668e-02 0.60179496
```

■

3.3 VECTOR MOVING-AVERAGE MODELS

When $p = 0$, the model in Eq. (3.1) reduces to

$$z_t = c_0 + \theta(B)a_t, \tag{3.55}$$

which is a vector moving-average (VMA) model of order q. The VMA models can occur in many situations. For example, temporal smoothing of independent data introduces short-term serial correlations. Consider a white noise process $\boldsymbol{a}_t$. Averaging two consequent observations, say $\boldsymbol{z}_t = (\boldsymbol{a}_t + \boldsymbol{a}_{t-1})/2$, introduces lag-1 serial correlations, resulting in a VMA(1) model. Another example is a random walk with measurement errors, commonly seen in modeling asset prices. Suppose that $z_t = z_{t-1} + a_t$, where a_t is a white noise series, but we observe $y_t = z_t + \epsilon_t$, where ϵ_t denotes an independent measurement error. Then, $w_t \equiv y_t - y_{t-1} = a_t + \epsilon_t - \epsilon_{t-1}$, which follows an MA(1) model.

3.3.1 Properties of VMA Models

Properties of VMA(q) model of Eq. (3.55) are easy to derive and understand. First, the process $\boldsymbol{z}_t$ is stationary and $E(\boldsymbol{z}_t) = \boldsymbol{c}_0$. Second, the autocovariance matrices of $\boldsymbol{z}_t$ are

$$\boldsymbol{\Gamma}_z(0) = \boldsymbol{\Sigma}_a + \sum_{i=1}^{q} \boldsymbol{\theta}_i \boldsymbol{\Sigma}_a \boldsymbol{\theta}_i'$$

$$\boldsymbol{\Gamma}_z(\ell) = \sum_{i=\ell}^{q} \boldsymbol{\theta}_i \boldsymbol{\Sigma}_a \boldsymbol{\theta}_{i-\ell}', \quad \text{where} \quad \boldsymbol{\theta}_0 = -\boldsymbol{I}, \quad \ell = 1, \ldots, q, \tag{3.56}$$

$$\boldsymbol{\Gamma}_z(\ell) = \boldsymbol{0}, \quad \ell > q. \tag{3.57}$$

From Eqs. (3.56) and (3.57), a VMA(q) model has, at most, q non-zero autocorrelation matrices, and the model is referred to as a finite memory model in the literature.

If all zeros of the determinant equation $|\boldsymbol{\theta}(B)| = 0$ are greater than 1 in absolute value, then $\boldsymbol{z}_t$ has a VAR representation

$$\boldsymbol{z}_t = \boldsymbol{\phi}_0 + \boldsymbol{a}_t + \sum_{i=1}^{\infty} \boldsymbol{\pi}_i \boldsymbol{z}_{t-i}, \tag{3.58}$$

where $\boldsymbol{\phi}_0$ is a constant vector given by $[\boldsymbol{\theta}(1)]^{-1}\boldsymbol{c}_0$ and the coefficient matrices $\boldsymbol{\pi}_i$ are determined by $\boldsymbol{\pi}(B) = \boldsymbol{I} - \sum_{i=1}^{\infty} \boldsymbol{\pi}_i B^i$ such that $[\boldsymbol{\pi}(B)]^{-1} = \boldsymbol{\theta}(B)$. More specifically,

$$\boldsymbol{\pi}_i = \sum_{j=1}^{\min(i,q)} \boldsymbol{\theta}_i \boldsymbol{\pi}_{i-j}, \quad i > 0.$$

It is easy to see that $\boldsymbol{\pi}_i \to \boldsymbol{0}$ as $i \to \infty$. Any VMA model that has a meaningful VAR representation in Eq. (3.58) is said to be *invertible*.

Finally, for a k-dimensional VMA(q) model and any full-rank matrix $\boldsymbol{C}_{g\times k}$, $\boldsymbol{w}_t = \boldsymbol{C}\boldsymbol{z}_t$ is a g-dimensional VMA(q) process. In particular, any component z_{it} of $\boldsymbol{z}_t$ is a scalar MA(q) process.

3.3.2 VMA Modeling

For a given realization $\{\boldsymbol{z}_t | t = 1, \ldots, T\}$ of a VMA process, the order q can be determined by examining the sample autocorrelation matrices with the help of some Portmanteau test of serial correlation. Furthermore, one often employs the ML method to estimate a specified VMA model. The estimation is, in general, time-consuming because it involves nonlinear optimization.

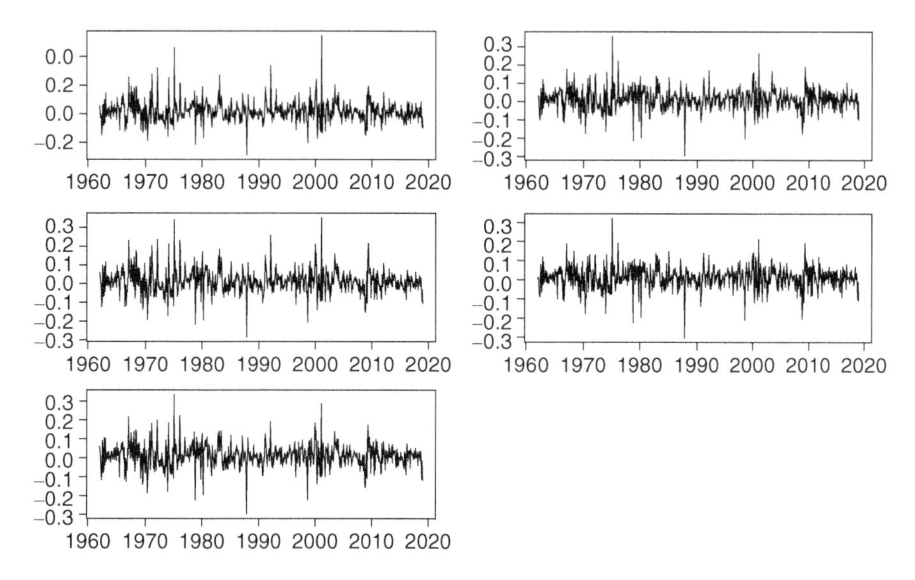

Figure 3.9 Time plots of simple returns of five market indexes from January 1962 to December 2018.

To demonstrate VMA model, we consider the monthly simple returns of some US market indexes from January 1962 to December 2018 for 684 observations. The market indexes considered are the market cap 1 to 5 portfolios of NYSE/AMEX/NASDAQ and the data are obtained by the Center for Research in Security Prices (CRSP). Figure 3.9 shows the time plots of the returns of five market indexes used. As expected, except for some outlying observations, the series are weakly stationary. In what follows, we use the **MTS** package of R to perform the analysis.

Let $\boldsymbol{z}_t = (z_{1t}, \ldots, z_{5t})'$ be the monthly return series. The order selection selects a VMA(12) model, but the sample cross-correlation matrices of $\boldsymbol{z}_t$ show strong serial correlations at lags 1 and 12. This is evident from the sample autocorrelations of the five return series of Figure 3.10. Therefore, we entertain the following VMA model

$$\boldsymbol{z}_t = \boldsymbol{\mu} + \boldsymbol{a}_t + \boldsymbol{\theta}_1 \boldsymbol{a}_{t-1} + \boldsymbol{\theta}_{12} \boldsymbol{a}_{t-12}, \tag{3.59}$$

where we change the sign of $\boldsymbol{\theta}_i$ to save space below. The parameter estimates of Model (3.59) are given in the attached R output. Since both estimates of $\boldsymbol{\theta}_1$ and $\boldsymbol{\theta}_{12}$ contain some insignificant parameters, we perform model refinement to simplify the fitted model. The estimates are

$$\hat{\boldsymbol{\mu}} = \begin{bmatrix} 0.02 \\ 0.01 \\ 0.01 \\ 0.01 \\ 0.01 \end{bmatrix}, \hat{\boldsymbol{\theta}}_1 = \begin{bmatrix} 0 & 0 & 0 & 0 & 0.41 \\ 0 & 0.05 & 0 & 0 & 0.30 \\ 0 & 0 & 0 & 0 & 0.29 \\ 0 & 0 & 0 & 0 & 0.25 \\ 0 & 0 & 0 & 0 & 0.21 \end{bmatrix}, \hat{\boldsymbol{\theta}}_{12} = \begin{bmatrix} 0.56 & 0 & -0.52 & 0 & 0 \\ 0.39 & 0 & -0.35 & 0 & 0 \\ 0.36 & 0 & -0.46 & 0 & 0.14 \\ 0.29 & 0 & -0.45 & 0 & 0.19 \\ 0.26 & 0 & -0.47 & 0 & 0.23 \end{bmatrix}$$

and $\hat{\boldsymbol{\Sigma}}_a$ is given in the R output. Model checking, not shown, of the fitted model indicates that the model fits the data reasonably well.

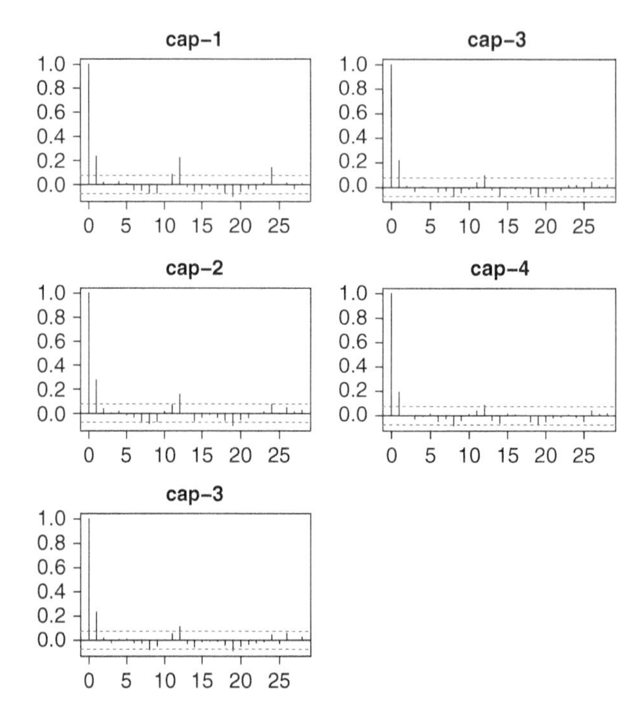

Figure 3.10 Sample autocorrelation functions of the simple returns of five US market indexes.

Remark. Some remarks are in order. Readers who are familiar with seasonal time series models may wish to entertain a multiplicative seasonal model for the five return series. We did not consider any multiplicative model because there exist no significant serial correlations at lag 13. Also, in the R output, we refined the estimation three times using the same threshold 1.5. This is because the model changes after each refinement, so do the t-ratios of coefficient estimates.

R Demonstration: VMA modeling.

```
> VMAorder(rtn)
Q(j,m) Statistics:
         j     Q(j,m)    p-value
 [1,]   1.0     780.8      0.00
 [2,]   2.0     642.8      0.00
 [3,]   3.0     609.0      0.00
 [4,]   4.0     580.4      0.00
 [5,]   5.0     565.4      0.00
 [6,]   6.0     539.3      0.00
 [7,]   7.0     510.6      0.00
 [8,]   8.0     466.3      0.00
 [9,]   9.0     446.6      0.00
[10,]  10.0     420.8      0.00
[11,]  11.0     375.9      0.00
[12,]  12.0     340.0      0.00
[13,]  13.0     213.6      0.24
```

```
[14,]  14.0     183.5     0.31
[15,]  15.0     161.7     0.24
[16,]  16.0     135.6     0.24
[17,]  17.0     106.4     0.31
[18,]  18.0      90.2     0.11
[19,]  19.0      40.4     0.83
[20,]  20.0      14.5     0.95
> par(mfcol=c(3,2))
> acf(rtn[,1],main="cap-1")
> acf(rtn[,2],main="cap-2")
> acf(rtn[,3],main="cap-3")
> acf(rtn[,4],main="cap-4")
> acf(rtn[,5],main="cap-5")
### Estimation
> m3 <- VMAs(rtn,malags=c(1,12))
> m3$secoef[is.na(m3$secoef)] <- 1
> m3a <- refVMAs(m3,thr=1.5) ## We use threshold 1.5 three times,
> m3b <- refVMAs(m3a,thr=1.5) ## because the t-ratios change after
> m3c <- refVMAs(m3b,thr=1.5) ## each refinement.
Coefficient(s):
         Estimate  Std. Error  t value Pr(>|t|)
CAP1RET  0.015306    0.003795    4.033 5.50e-05 ***
CAP2RET  0.011547    0.003274    3.527 0.000421 ***
CAP3RET  0.010202    0.003001    3.400 0.000674 ***
CAP4RET  0.010423    0.002858    3.647 0.000265 ***
CAP5RET  0.009976    0.002733    3.650 0.000262 ***
        -0.408258    0.041311   -9.883  < 2e-16 ***
        -0.557096    0.091418   -6.094 1.10e-09 ***
         0.517168    0.111725    4.629 3.68e-06 ***
        -0.049634    0.021062   -2.357 0.018443 *
        -0.302076    0.042213   -7.156 8.30e-13 ***
        -0.386759    0.083062   -4.656 3.22e-06 ***
         0.347225    0.099952    3.474 0.000513 ***
        -0.292933    0.035620   -8.224 2.22e-16 ***
        -0.356795    0.080206   -4.448 8.65e-06 ***
         0.457800    0.101440    4.513 6.39e-06 ***
        -0.137197    0.037085   -3.700 0.000216 ***
        -0.246584    0.035070   -7.031 2.05e-12 ***
        -0.294384    0.079091   -3.722 0.000198 ***
         0.452605    0.104004    4.352 1.35e-05 ***
        -0.188642    0.045144   -4.179 2.93e-05 ***
        -0.206023    0.035654   -5.778 7.54e-09 ***
        -0.258952    0.078335   -3.306 0.000947 ***
         0.466077    0.108915    4.279 1.88e-05 ***
        -0.228562    0.054749   -4.175 2.98e-05 ***
---
Signif. codes:  0 '***' 0.001 '**' 0.01 '*' 0.05 '.' 0.1 ' ' 1
---
Estimates in matrix form:
Constant term:
Estimates:  0.01530554 0.01154678 0.01020212 0.0104232 0.009975902
MA coefficient matrix
MA( 1)-matrix
     [,1]    [,2] [,3] [,4]   [,5]
[1,]    0  0.0000    0    0 -0.408
[2,]    0 -0.0496    0    0 -0.302
[3,]    0  0.0000    0    0 -0.293
```

```
[4,]      0  0.0000     0      0 -0.247
[5,]      0  0.0000     0      0 -0.206
MA( 12)-matrix
        [,1] [,2]   [,3] [,4]    [,5]
[1,] -0.557     0 0.517     0   0.000
[2,] -0.387     0 0.347     0   0.000
[3,] -0.357     0 0.458     0  -0.137
[4,] -0.294     0 0.453     0  -0.189
[5,] -0.259     0 0.466     0  -0.229

Residuals cov-matrix:
            [,1]        [,2]        [,3]        [,4]        [,5]
[1,] 0.004724746 0.003946767 0.003694343 0.003546549 0.003372278
[2,] 0.003946767 0.003700418 0.003438245 0.003334797 0.003221117
[3,] 0.003694343 0.003438245 0.003408408 0.003289334 0.003213296
[4,] 0.003546549 0.003334797 0.003289334 0.003327253 0.003235831
[5,] 0.003372278 0.003221117 0.003213296 0.003235831 0.003276711
---
aic =  -39.42939
bic =  -39.27051
```

3.4 STATIONARY VARMA MODELS

3.4.1 Are VAR Models Sufficient?

The VAR model is widely used in multivariate time series analysis for various reasons. First, it is relatively simple. The model can be estimated by the OLS method and, hence, does not depend critically on the normality assumption. Second, it avoids the difficulty of identifiability encountered by the VARMA model of Eq. (3.1). On the other hand, the VAR model also has some weaknesses. First, as seen in the scalar case of Section 2.2.4, the class of AR models is not close in the sense that a linear combination of two AR processes is in general not an AR process. The same result applies to the vector case. Second, VAR models may not result in a parsimonious parameterization. This is particularly so when the dimension k is large. To illustrate, consider a two-dimensional VMA process

$$\boldsymbol{z}_t = (\boldsymbol{I} - \boldsymbol{\theta}B)\boldsymbol{a}_t, \quad \boldsymbol{\theta} = \begin{bmatrix} 0.2 & 0.3 \\ -0.6 & 1.1 \end{bmatrix}, \quad \boldsymbol{\Sigma}_a = \boldsymbol{I}_2.$$

It is easy to see that the two eigenvalues of $\boldsymbol{\theta}$ are 0.8 and 0.5 so that $\boldsymbol{z}_t$ is an invertible time series. The VAR presentation of $\boldsymbol{z}_t$ is

$$\boldsymbol{z}_t = \boldsymbol{a}_t + \boldsymbol{\theta}\boldsymbol{z}_{t-1} + \boldsymbol{\theta}^2\boldsymbol{z}_{t-2} + \boldsymbol{\theta}^3\boldsymbol{z}_{t-3} + \cdots .$$

It is also easy to calculate that

$$\boldsymbol{\theta}^6 \approx \begin{bmatrix} -0.23 & 0.25 \\ -0.49 & 0.51 \end{bmatrix}, \quad \boldsymbol{\theta}^7 \approx \begin{bmatrix} -0.19 & 0.20 \\ -0.40 & 0.41 \end{bmatrix}.$$

Consequently, one would need a VAR(p) model with $p \geq 7$ to provide a good approximation in modeling $\boldsymbol{z}_t$. The number of AR coefficients used would then be more than 28, whereas the number of coefficients of the data generating process is only 4.

In practice, the choice between using VAR or VARMA models depends on the dimension and sample size of the problem under study. If the dimension k is not large, say $k \leq 10$, then VARMA models can be used. For other cases, VAR models are often preferred for their simplicity both in understanding the dynamic structure of the data and in estimation. We discuss VARMA models for high-dimensional time series later.

3.4.2 Properties of VARMA Models

We summarize the key properties of VARMA models of Eq. (3.1) below:

- Stationarity condition: All zeros of the determinant equation $|\boldsymbol{\phi}(B)| = 0$ are greater than 1 in absolute value.
  The VMA representation of a stationary VARMA model is

$$\boldsymbol{z}_t = \boldsymbol{\mu} + \boldsymbol{a}_t + \sum_{i=1}^{\infty} \boldsymbol{\psi}_i \boldsymbol{a}_{t-i}, \tag{3.60}$$

  where $\boldsymbol{\mu} = E(\boldsymbol{z}_t)$ and the ψ-weight coefficient $\boldsymbol{\psi}_i$ is given by $[\boldsymbol{\phi}(B)]^{-1}\boldsymbol{\theta}(B) = \boldsymbol{\psi}(B) = \boldsymbol{I} + \sum_{i=1}^{\infty} \boldsymbol{\psi}_i B^i$.
- Invertibility condition: All zeros of the determinant equation $|\boldsymbol{\theta}(B)| = 0$ are greater than 1 in absolute value.
  The VAR representation of an invertible VARMA model is

$$\boldsymbol{z}_t = \boldsymbol{c} + \sum_{i=1}^{\infty} \boldsymbol{\pi}_i \boldsymbol{z}_{t-i} + \boldsymbol{a}_t, \tag{3.61}$$

  where $\boldsymbol{c} = [\boldsymbol{\theta}(1)]^{-1}\boldsymbol{c}_0$ and the π-weight coefficient matrix $\boldsymbol{\pi}_i$ is given by $\boldsymbol{\pi}(B) = \boldsymbol{I} - \sum_{i=1}^{\infty} \boldsymbol{\pi}_i B^i = [\boldsymbol{\theta}(B)]^{-1}\boldsymbol{\phi}(B)$.
- Moment condition: For a stationary VARMA process $\boldsymbol{z}_t$ of Eq. (3.1), we have
  1. $E(\boldsymbol{z}_t) = \boldsymbol{\mu}$ such that $\boldsymbol{\phi}(1)\boldsymbol{\mu} = \boldsymbol{c}_0$. That is, $\boldsymbol{\mu} = [\boldsymbol{\phi}(1)]^{-1}\boldsymbol{c}_0$.
  2. $\boldsymbol{\Gamma}_z(\ell) = \sum_{i=1}^{p} \boldsymbol{\phi}_i \boldsymbol{\Gamma}_z(\ell - i) - \sum_{j=\ell}^{q} \boldsymbol{\theta}_j \boldsymbol{\Sigma}_a \boldsymbol{\psi}'_{\ell-j}$, for $\ell = 1, \ldots, q$, where $\boldsymbol{\psi}_0 = \boldsymbol{I}$.
  3. $\boldsymbol{\Gamma}_z(\ell) = \sum_{i=1}^{p} \boldsymbol{\phi}_i \boldsymbol{\Gamma}(\ell - i)$ for $\ell > q$.

  If we consider the third moment condition jointly for $\ell = q+1, \ldots, q+p$, then we have a system of p moment conditions, which is referred to as the *generalized Yule-Walker equations* of a VARMA(p, q) model.

3.4.3 Modeling VARMA Process

There are several approaches available in the literature for modeling VARMA processes. One approach, proposed by Tiao and Tsay (1983), is to fit a large VAR that provides consistent estimates of the residuals and incorporate these estimated residuals to find the MA order. They proposed to use a table of extended cross-correlation matrices to obtain the order of the ARMA process. The second approach is based on the structure of covariance matrix of two long vectors of time series, consisting of the past and the future of the process under study, referred to the Hankel matrix, and uses properties of the Hankel matrix and its associated Kronecker indexes to determine the model. See Akaike (1976), Hannan and Deistler (1988), and Tsay (1991). A third approach is the scalar component models (SCMs) proposed by Tiao and Tsay (1989) that are linear combinations of the time series with specific structure. For instance, an important concept of the SCMs is a series that is white noise. Let $\boldsymbol{v}$ be a k-dimensional non-zero real vector. The linear combination $w_t = \boldsymbol{v}'\boldsymbol{z}_t$ is called a SCM of order (0,0), denoted by SCM(0,0), if w_t is uncorrelated with all lagged values of $\boldsymbol{z}_t$ and, therefore, is a white noise process. In other words, $\text{corr}(w_t, \boldsymbol{z}_{t-j}) = \boldsymbol{0}$ for $j \geq 1$. Consequently, the existence of a SCM(0,0) w_t means that a linear combination of $\boldsymbol{z}_t$ is uncorrelated with F_{t-1}, the information available at time $t-1$. The SCM(0,0) w_t can be used to simplify the model specification. It turns out that SCM(0,0) is also useful in factor models of time series data, see the discussions in Chapter 6. A useful tool for VARMA modeling is the canonical correlation analysis. Tiao and Tsay (1989) use this tool to search for the number of SCM(0,0) and their associated linear vectors. See Tsay (2014) for a presentation of these methods.

3.4.4 Use of VARMA Models

Once a VARMA model is built for a vector time series, one can use the model for statistical inference. If the relationships between the component series are of interest, then the ψ-weight coefficient matrices in the VMA representation can be used to obtain the *impulse response function*, similar to that discussed in the VAR model. In the econometric literature, such a study is referred to as the *multiplier study*. It plays an important role in policy decision and simulation.

Another application of VARMA models is prediction. The idea and computation involved are similar to those of the univariate time series discussed in Chapter 2. Suppose that the forecast origin is $t = n$. Let F_n denote the information available at $t = n$ and h be the forecast origin, then

$$\hat{\boldsymbol{z}}_n(h) = E(\boldsymbol{z}_{n+h}|F_n) = \boldsymbol{c}_0 + \sum_{i=1}^{p} \boldsymbol{\phi}_i \hat{\boldsymbol{z}}_n(h-i) - \sum_{j=1}^{q} \boldsymbol{\theta}_j \hat{\boldsymbol{a}}_n(h-j), \tag{3.62}$$

where

$$\hat{\boldsymbol{z}}_n(\ell) = \begin{cases} \hat{\boldsymbol{z}}_n(\ell) & \text{if } \ell > 0, \\ \boldsymbol{z}_{n-\ell} & \text{if } \ell \leq 0, \end{cases} \quad \hat{\boldsymbol{a}}_n(\ell) = \begin{cases} \boldsymbol{0} & \text{if } \ell > 0, \\ \boldsymbol{a}_{n-\ell} & \text{if } \ell \leq 0. \end{cases}$$

The associated forecast error is

$$\boldsymbol{e}_n(h) = \boldsymbol{z}_{n+h} - \hat{\boldsymbol{z}}_n(h) = \boldsymbol{a}_{n+h} + \sum_{i=1}^{h-1} \boldsymbol{\psi}_i \boldsymbol{a}_{n+h-i},$$

and, hence, the covariance matrix of forecast error is

$$\text{Cov}[\boldsymbol{e}_n(h)] = \boldsymbol{\Sigma}_a + \sum_{i=1}^{h-1} \boldsymbol{\psi}_i \boldsymbol{\Sigma}_a \boldsymbol{\psi}_i',$$

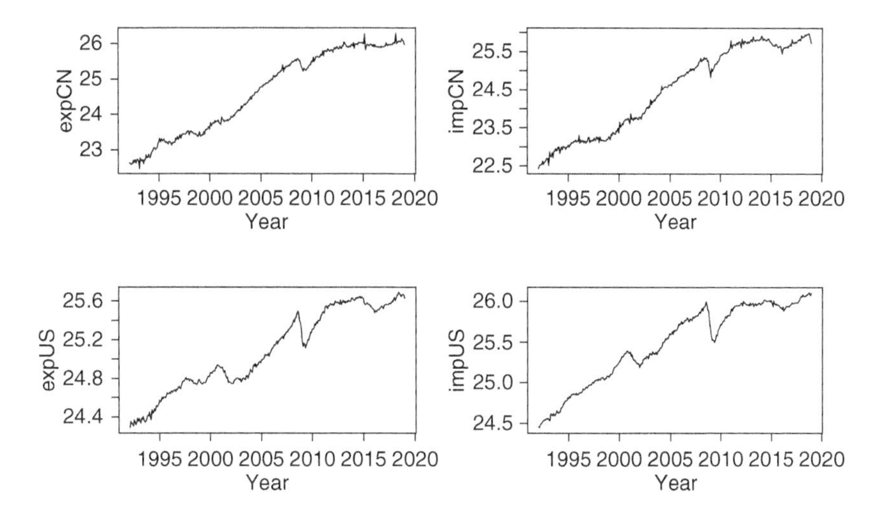

Figure 3.11 Time plots of log series of monthly exports and imports of China and United States from January 1992 to December 2018. The original data are from FRED and in US dollars. The left panel shows the export series and the upper plots are for China. Source: Data from Federal Reserve Economic Data (FRED).

where it is understood that we treat the model as given and ψ_i is the ψ-weight coefficient matrix of Eq. (3.60). In practice, the forecasts and the covariance matrices of forecast errors can be calculated recursively.

Example 3.3

Consider the monthly exports and imports of China and United States from January 1992 to December 2018. The data, downloaded from the FRED, are in US dollars and seasonally adjusted. We take the natural log transformation and the transformed data are in the file `m-expimpcnus.csv`. Figure 3.11 shows the time plots of the four log time series, where the left panel shows the exports and the upper figures are for China. As expected, these plots confirm the increasing trend in international trade of the two countries. Therefore, we take the first difference. Let $z_t = (\text{exp-CN}_t, \text{exp-US}_t, \text{imp-CN}_t, \text{imp-US}_t)'$ be the growth series of the monthly international trade. The z_t series are shown in Figure 3.12. From the plots, there exist some outlying observations, but, for simplicity, we do not consider any treatment for those aberrant values.

Following the modeling approach in Tsay (2014, chapter 3), we use the extended cross-correlation matrices of Tiao and Tsay (1983) to specify the VARMA order. Details of the output are in the attached R output. For this particular instance, a VARMA(1,2) model is specified. The maximum likelihood estimation (MLE) shows that there are many insignificant estimates in the fitted VARMA(1,2) model so we refine the model by removing simultaneously all estimates with t-ratio less than 0.8

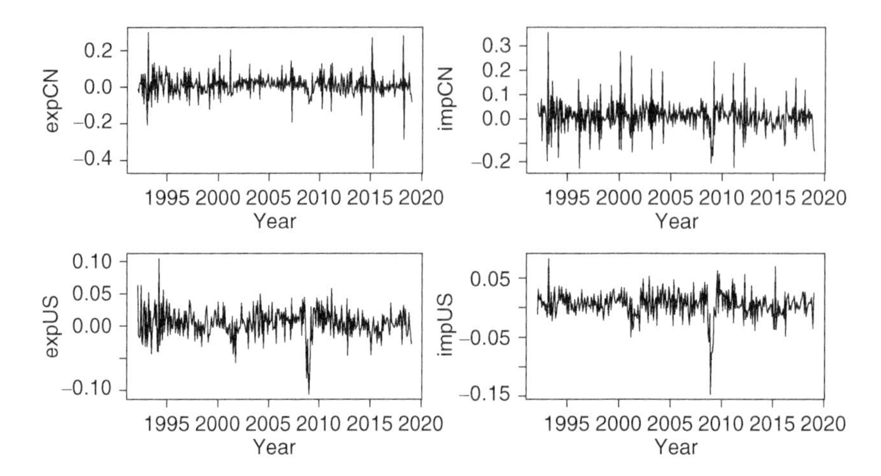

Figure 3.12 Growth series of monthly exports and imports of China and United States from January 1992 to December 2018. The original data are from FRED and in US dollars. The left panel shows the export series and the upper plots are for China. Source: Data from Federal Reserve Economic Data (FRED).

in absolute value. The resulting model is

$$
\begin{aligned}
\boldsymbol{z}_t &= \begin{bmatrix} 0.01 \\ 0 \\ 0.01 \\ 0 \end{bmatrix} + \begin{bmatrix} -0.18 & 0 & 0 & 0.45 \\ 0.28 & -0.27 & 0 & 0.28 \\ 0 & 0.27 & -0.53 & 0 \\ 0.14 & 0.07 & 0.29 & 0 \end{bmatrix} \boldsymbol{z}_{t-1} + \boldsymbol{a}_t \\
&- \begin{bmatrix} 0.49 & 0 & -0.02 & 0 \\ 0.27 & 0 & 0 & 0 \\ 0 & 0 & 0 & -0.37 \\ 0.13 & 0 & 0.23 & 0 \end{bmatrix} \boldsymbol{a}_{t-1} + \begin{bmatrix} 0 & 0.22 & -0.06 & 0.63 \\ 0.20 & -0.22 & 0 & 0.32 \\ 0 & 0.36 & -0.28 & 0.57 \\ 0.06 & 0 & 0.18 & 0 \end{bmatrix} \boldsymbol{a}_{t-2}
\end{aligned}
$$

and the covariance matrix of $\boldsymbol{a}_t$ is

$$
\hat{\boldsymbol{\Sigma}}_a = 10^{-3} \begin{bmatrix} 2.987 & 0.182 & 1.325 & 0.108 \\ 0.182 & 0.447 & 0.346 & 0.205 \\ 1.325 & 0.346 & 3.448 & 0.253 \\ 0.108 & 0.205 & 0.253 & 0.424 \end{bmatrix}.
$$

Model checking indicates that the residual series have some cross-correlations at the seasonal lag 12, indicating that some seasonality remains. However, there are no significant serial correlations at the first 10 lags.

From the fitted model, the four series of international trade are inter-related. This is confirmed by the impulse response functions of the model; see Figure 3.13. However, the impulse response functions show that the dependence of China imports on other series is rather weak. The model also shows that the export and import series of China have significant constant terms, supporting the common belief that the strong

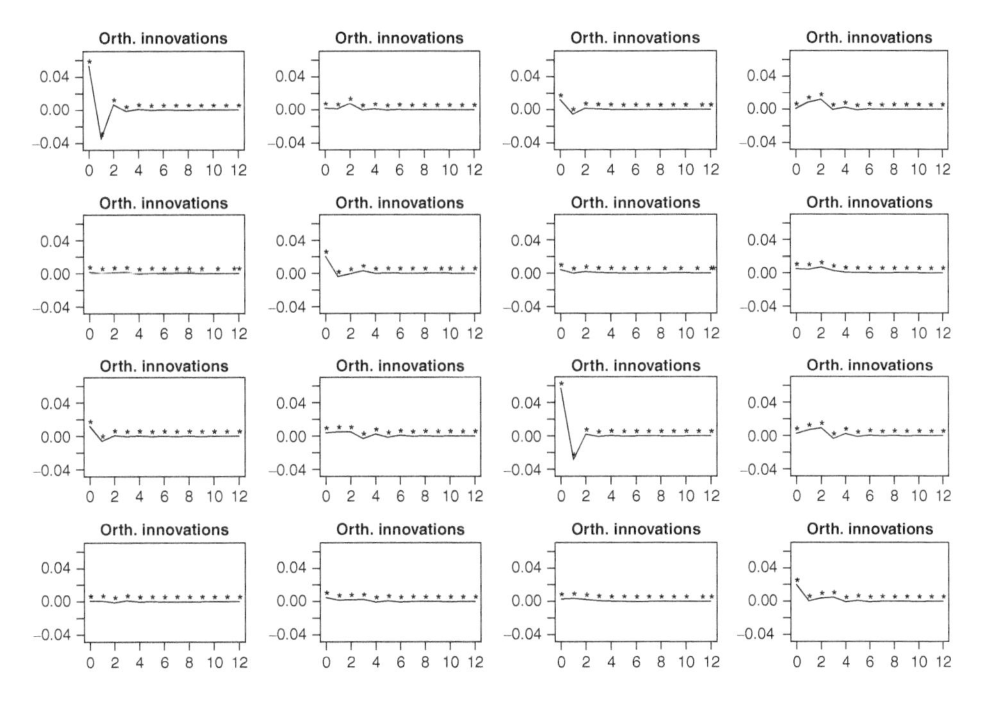

Figure 3.13 Impulse response functions of the fitted VARMA(1,2) model for the monthly growth rates of exports and imports of China and United States.

growth of China economy over the past two decades dramatically increased its international trade. The variabilities of the two Chinese series are also greater than those of the United States.

Finally, the fitted model can be used to produce forecasts. In this particular instance, the 1-step ahead predictions of the four series are 0.063(0.055), −0.001(0.021), 0.091(0.059), and −0.021(0.021), respectively, where the numbers in parentheses denote forecast standard errors. While the approximate 95% prediction intervals contain zero, the predictions suggest that the exports and imports of China would continue to grow despite trade war with the United States. In fact, CNBC reported on 13 February 2019, that the China exports increased by 9.1% from a year ago in January, whereas imports declined by 1.5%. For this particular instance, the fitted model seems to provide reasonable forecasts.

R Demonstration: International trade of China and the United States

```
> da <- read.csv("m-expimpcnus.csv'')   ## logged data
> da <- da[,-1] ## remove observation sequence
> require(MTS)
> xt <- diffM(da) ## May use diff(da) provided that "da''
is a matrix.
> tdx <- c(2:325)/12+1992 ## Create calendar time
> MTSplot(da,tdx)

> Eccm(xt)
```

```
p-values table of Extended Cross-correlation Matrices:
Column: MA order
Row: AR order
       0      1      2      3      4      5      6
0 0.0000 0.0000 0.0007 0.6472 0.5537 0.8951 0.6524
1 0.0000 0.0000 0.026  0.8999 0.7330 0.9054 0.9040
2 0.1933 0.9199 0.5442 0.9776 0.8144 1.0000 0.9995
3 1.0000 0.9994 1.0000 1.0000 0.9995 0.9999 0.9999
4 1.0000 1.0000 1.0000 1.0000 1.0000 1.0000 0.8856
5 1.0000 1.0000 1.0000 1.0000 1.0000 0.9998 0.9929

> m1 <- VARMA(xt,p=1,q=2) ## Output omitted.
> m1a <- refVARMA(m1,thr=0.8)
Estimates in matrix form:
Constant term:
Estimates:  0.009814692 0 0.0135986 0
AR coefficient matrix
AR( 1)-matrix
       [,1]    [,2]   [,3]  [,4]
[1,] -0.177  0.0000  0.000 0.449
[2,]  0.277 -0.2673  0.000 0.280
[3,]  0.000  0.2675 -0.532 0.000
[4,]  0.136  0.0724  0.288 0.000
MA coefficient matrix
MA( 1)-matrix
      [,1] [,2]    [,3]   [,4]
[1,] 0.486    0 -0.0224  0.000
[2,] 0.272    0  0.0000  0.000
[3,] 0.000    0  0.0000 -0.374
[4,] 0.130    0  0.2332  0.000
MA( 2)-matrix
        [,1]   [,2]    [,3]   [,4]
[1,]  0.0000 -0.224  0.0648 -0.632
[2,] -0.1951  0.219  0.0000 -0.316
[3,]  0.0000 -0.360  0.2812 -0.570
[4,] -0.0621  0.000 -0.1811  0.000

Residuals cov-matrix:
             [,1]         [,2]         [,3]         [,4]
[1,] 0.0029774472 0.0001818597 0.0013214291 0.0001074424
[2,] 0.0001818597 0.0004460318 0.0003451413 0.0002049293
[3,] 0.0013214291 0.0003451413 0.0034373700 0.0002521699
[4,] 0.0001074424 0.0002049293 0.0002521699 0.0004224317
----
aic=  -27.32588
bic=  -26.98671

> MTSdiag(m1a)
```

```
Ljung-Box Statistics:
          m       Q(m)      df     p-value
 [1,]    1.0      15.7     16.0      0.47
 [2,]    2.0      20.2     32.0      0.95
 [3,]    3.0      44.1     48.0      0.63
 [4,]    4.0      64.6     64.0      0.46
 [5,]    5.0      83.5     80.0      0.37
 [6,]    6.0      94.6     96.0      0.52
 [7,]    7.0     108.6    112.0      0.57
 [8,]    8.0     121.6    128.0      0.64
 [9,]    9.0     156.7    144.0      0.22
[10,]   10.0     185.1    160.0      0.09
[11,]   11.0     219.8    176.0      0.01
[12,]   12.0     289.1    192.0      0.00

> VARMApred(m1a,4)
Predictions at origin  323
         [,1]       [,2]      [,3]       [,4]
[1,] 0.063025 -0.001014 0.091208 -0.020800
[2,] 0.011863  0.004853 0.021111 -0.004543
[3,] 0.005677  0.000716 0.003671  0.008047
[4,] 0.012422  0.003634 0.011838  0.001880
Standard errors of predictions
        [,1]    [,2]    [,3]    [,4]
[1,] 0.05457 0.02112 0.05863 0.02055
[2,] 0.06537 0.02191 0.06590 0.02095
[3,] 0.06731 0.02302 0.06681 0.02159
[4,] 0.06733 0.02342 0.06696 0.02232
```

■

3.5 UNIT ROOTS AND CO-INTEGRATION

As discussed in Chapter 2, a scalar time series z_t is a unit-root process of order 1 if $w_t = (1 - B)z_t$ is a stationary and invertible process. In the economic literature, z_t is called an I(1) process, meaning that it is an integrated process of order 1. In general, z_t is an I(d) process if $w_t = (1 - B)^d z_t$ is stationary and invertible. The I(1) processes play an important role in analyzing business and economic time series. For instance, the quarterly GDP of an economy often forms an I(1) process. Similarly, most prices (or log prices) of stocks are also I(1) processes. In this section, we consider multivariate integrated processes with emphasis on the issues of spurious relation and co-integration.

A multivariate time series $\boldsymbol{z}_t$ of Eq. (3.1) is an integrated process if $|\boldsymbol{\phi}(1)| = 0$, but $|\boldsymbol{\theta}(1)| \neq 0$. In other words, "1" is a solution of the determinant equation $|\boldsymbol{\phi}(B)| = 0$, but not $|\boldsymbol{\theta}(B)| = 0$. Integrated processes exist in many real applications, especially the I(1) processes, and their properties have been widely studied in the literature. See, for instance, Phillips (1987), Chan and Wei (1988), and Tsay and Tiao (1990), among many others.

3.5.1 Spurious Regression

Suppose that z_{1t} and z_{2t} are two scalar I(1) processes, and we are interested in their linear relationship. To this end, the simple linear regression is often used, i.e.

$$z_{2t} = \alpha + \beta z_{1t} + \epsilon_t, \tag{3.63}$$

where ϵ_t denotes the error term. Intuitively, if $\beta \neq 0$, then the two processes are linearly correlated. Consequently, the hypothesis testing of $H_0 : \beta = 0$ versus $H_a : \beta \neq 0$ is often used to draw inference. However, because both z_{1t} and z_{2t} are I(1) processes, the conventional reference distribution for the t-ratio of the least squares estimate of β is not applicable. Use of the conventional reference distribution often leads to false discovery, implying that one rejects H_0 too often when it is true. To demonstrate, we generate two independent random walk series as below:

$$z_{1t} = z_{1,t-1} + a_{1t}, \quad z_{10} = 0,$$

$$z_{2t} = z_{2,t-1} + a_{2t}, \quad z_{20} = 0, \quad t = 1, \ldots, 100,$$

where $\{a_{1t}\}$ and $\{a_{2t}\}$ are two independent standard Gaussian processes. Then, we fit the linear regression of Eq. (3.63). Using the conventional critical value of 1.96, we tabulate the frequency of rejecting the null hypothesis of $\beta = 0$. With 1000 replications, the empirical frequency of rejecting $\beta = 0$ is approximately 77%, which is much higher than the 5% Type-I error. This simple simulation demonstrates clearly that care must be exercised when one employs linear regression that involves integrated processes. The fact that the conventional test can easily misidentify the linear relationship between I(1) processes is referred to *spurious linear regression*. To avoid spurious regression, one should always check the residuals of the linear regression of Eq. (3.63). If the residual series itself is integrated, then the linear relationship is spurious. See, for instance, Tsay (2014, chapter 5).

3.5.2 Linear Combinations of a Vector Process

Linear combinations of a vector time series in the form $y_t = \boldsymbol{b}'\boldsymbol{z}_t$ may lead to some interesting scalar processes with important applications. We have seen in Chapter 1 that, by choosing $\boldsymbol{b}$ as one of the eigenvectors of the covariance matrix of $\boldsymbol{z}_t$, we obtain a principal component. Principal component analysis (PCA) decomposes the k-dimensional $\boldsymbol{z}_t$ into a set of contemporaneously uncorrelated series ordered by the magnitudes of their variances. We can also decompose $\boldsymbol{z}_t$ into a set of linear combinations based on their predictabilities. This is the approach proposed by Box and Tiao (1977). Suppose we want to find linear combinations of $\boldsymbol{z}_t$ based on their 1-step ahead predictabilities. As discussed before, we can write the vector time series as

$$\boldsymbol{z}_t = \boldsymbol{z}_{t-1}(1) + \boldsymbol{a}_t, \tag{3.64}$$

where $\boldsymbol{z}_{t-1}(1)$ is the 1-step ahead prediction of $\boldsymbol{z}_t$ at the forecast origin $t-1$. Taking variance of Eq. (3.64), we have $\boldsymbol{\Gamma}_z(0 = \boldsymbol{\Gamma}_p(1) + \boldsymbol{\Sigma}_a$, where $\boldsymbol{\Gamma}_p(1)$ is the covariance matrix of $\boldsymbol{z}_{t-1}(1)$ and $\boldsymbol{z}_{t-1}(1)$ and $\boldsymbol{a}_t$ are uncorrelated. Pre-multiplying Eq. (3.64) by $\boldsymbol{b}'$, we have

$$y_t = y_{t-1}(1) + u_t \tag{3.65}$$

where $y_{t-1}(1)$ is the 1-step ahead forecast of y_t at the forecast origin $t-1$ and $u_t = \boldsymbol{b}'\boldsymbol{a}_t$. From Eq. (3.65), we have $\sigma_y^2 = p_y^2 + \sigma_u^2$, where $\sigma_y^2 = \boldsymbol{b}'\boldsymbol{\Gamma}_z(0)\boldsymbol{b}$ is the variance of the scalar series y_t, σ_u^2 is the variance of its innovation process, and $p_y^2 = \boldsymbol{b}'\boldsymbol{\Gamma}_p(1)\boldsymbol{b}$ is the explained variance. The predictability of a scalar series $y_t = \boldsymbol{b}'\boldsymbol{z}_t$ is defined in Chapter 2 by

$$\lambda = \frac{p_y^2}{\sigma_y^2} = \frac{\boldsymbol{b}'\boldsymbol{\Gamma}_p(1)\boldsymbol{b}}{\boldsymbol{b}'\boldsymbol{\Gamma}_z(0)\boldsymbol{b}}$$

and the linear combination of maximum predictability can be obtained by solving

$$\frac{\partial\lambda}{\partial\boldsymbol{b}} = \frac{2\boldsymbol{\Gamma}_p(1)\boldsymbol{b}\sigma_y^2 - 2\boldsymbol{\Gamma}_z(0)\boldsymbol{b}p_y^2}{(\boldsymbol{b}'\boldsymbol{\Gamma}_z(0)\boldsymbol{b})^2} = \mathbf{0},$$

which implies that

$$\boldsymbol{\Gamma}_z^{-1}(0)\boldsymbol{\Gamma}_p(1)\boldsymbol{b} = \lambda\boldsymbol{b}$$

and the vector $\boldsymbol{b}$ must be an eigenvector of the matrix $\mathbf{M} = \boldsymbol{\Gamma}_z^{-1}(\mathbf{0})\boldsymbol{\Gamma}_1$ (1) and, as the eigenvalue measures the predictability, we must choose the eigenvector corresponding to the largest eigenvalue. The least predictable linear combination is the one formed by the eigenvector associated with the smallest eigenvalue. This analysis can be applied for any forecast horizon h. For instance, for a VAR(1) model $\hat{z}_{t-1}(1) = \phi z_{t-1}$, $\boldsymbol{\Gamma}_p(1) = \boldsymbol{\phi}\boldsymbol{\Gamma}_z(0)\boldsymbol{\phi}'$, and $\boldsymbol{b}$ is the largest eigenvalue of the matrix

$$\boldsymbol{M} = \boldsymbol{\Gamma}_z^{-1}(0)\boldsymbol{\phi}\boldsymbol{\phi}\boldsymbol{\Gamma}_z(0)\boldsymbol{\phi}'.$$

Because $\boldsymbol{\phi} = \boldsymbol{\Gamma}_z(1)\boldsymbol{\Gamma}_z^{-1}(0)$, the prior matrix can be written as

$$\boldsymbol{M} = \boldsymbol{\Gamma}_z^{-1}(0)\boldsymbol{\Gamma}_z(1)\boldsymbol{\Gamma}_z^{-1}(0)\boldsymbol{\Gamma}_z(1)',$$

which is the canonical correlation matrix between $\boldsymbol{z}_t$ and $\boldsymbol{z}_{t-1}$. Thus, the linear combination of maximum predictability is equivalent to the canonical variate corresponding to the largest canonical correlation between $\boldsymbol{z}_t$ and its past. See Velu et al. (1987) for a general proof of this result. Canonical correlation analysis and linear combinations with interesting properties are key ideas behind some useful methods for building VARMA models, as the SCMs. An interesting case, illustrated by Box and Tiao (1977), occurs when the vector series is integrated but some linear combinations of it are stationary. This is the idea of co-integration discussed in the next section.

3.5.3 Co-integration

An important issue of integrated processes is co-integration. Consider two univariate I(1) processes z_{1t} and z_{2t}. If there exists a non-zero linear combination $y_t = \beta_1 z_{1t} + \beta_2 z_{2t}$, which is stationary, then z_{1t} and z_{2t} are co-integrated. The linear combination $\boldsymbol{\beta} = (\beta_1, \beta_2)'$ is referred to as the co-integrating vector. In general, consider a k-dimensional integrated process $\boldsymbol{z}_t$. If there exists a non-zero k-dimensional vector $\boldsymbol{\beta}$ such that $y_t = \boldsymbol{\beta}'\boldsymbol{z}_t$ is stationary, then $\boldsymbol{z}_t$ is co-integrated with co-integrating vector $\boldsymbol{\beta}$. If there exists a $k \times m$ matrix $\boldsymbol{\beta}$ of full rank m such that the m-dimensional process $\boldsymbol{y}_t = \boldsymbol{\beta}'\boldsymbol{z}_t$ is stationary, then $\boldsymbol{z}_t$ is co-integrated with co-integrating rank m and $\boldsymbol{\beta}$ is the co-integrating matrix. In other words, m is the number of co-integrating vectors. Obviously, $m \le k$.

Co-integration is of particular interest in economics and finance because they are related to the concept of long-term equilibrium in economics and to the idea of pairs trading in finance. Roughly speaking, integrated processes are not predictable but stationary ones are. Co-integration means that while individual components of $\boldsymbol{z}_t$ are not predictable, but their linear combinations $\boldsymbol{y}_t = \boldsymbol{\beta}'\boldsymbol{z}_t$ are predictable. Long-term equilibrium and pairs-trading make use of such predictability of integrated processes. See Tsay (2010) for a demonstration of pairs trading.

3.5.4 Over-Differencing

The conventional approach for handling unit-root nonstationarity in time series analysis is to take difference. By definition, if the scalar series z_t is an I(1) process, then $w_t = (1 - B)z_t$ is stationary and invertible. Therefore, one can analyze w_t and produce the 1-step ahead forecast $\hat{z}_T(1) = E(z_{T+1}|F_T) = z_T + \hat{w}_T(1)$. For the multivariate case, the situation is more complex. If $\boldsymbol{z}_t$ is a co-integrated process of co-integrating rank m with $m < k$, then there are only $k - m$ unit roots in the k-dimensional process $\boldsymbol{z}_t$. If one considers $\boldsymbol{w}_t = (1 - B)\boldsymbol{z}_t$, i.e. $w_{it} = z_{it} - z_{i,t-1}$, then $\boldsymbol{w}_t$ would be non-invertible. To demonstrate, consider the bivariate ARMA((1,1) model

$$\boldsymbol{z}_t - \begin{bmatrix} 1.05 & -0.05 \\ 0.45 & 0.55 \end{bmatrix} \boldsymbol{z}_{t-1} = \boldsymbol{a}_t - \begin{bmatrix} -0.05 & 0.45 \\ 0.45 & -0.05 \end{bmatrix} \boldsymbol{a}_{t-1} \tag{3.66}$$

where $\boldsymbol{a}_t$ is a two-dimensional white noise series with mean zero and positive-definite covariance matrix $\boldsymbol{\Sigma}_a$. It is easy to verify that, for the model in Eq. (3.66),

$$|\boldsymbol{\phi}(B)| = (1 - B)(1 - 0.6B), \quad |\boldsymbol{\theta}(B)| = (1 - 0.4B)(1 + 0.5B),$$

so that $|\boldsymbol{\phi}(1)| = 0$, but $|\boldsymbol{\theta}(1)| \neq 0$. Consequently, $\boldsymbol{z}_t$ is unit-root nonstationary, but invertible. Furthermore, it is also easy to verify, from Eq. (3.66), that

$$z_{1t} - z_{2t} = 0.6(z_{1,t-1} - z_{2,t-1}) + (a_{1t} - a_{2t}) + 0.5(a_{1,t-1} - a_{2,t-1}).$$

Letting $x_t = z_{1t} - z_{2t}$ and $b_t = a_{1t} - a_{2t}$, we have

$$x_t = 0.6x_{t-1} + b_t + 0.5b_{t-1},$$

which is a stationary and invertible time series. Therefore, z_{1t} and z_{2t} are co-integrated with co-integrating vector $(1, -1)'$.

Next, let $\boldsymbol{w}_t = (1 - B)\boldsymbol{z}_t$, i.e. taking the first difference of each series. Pre-multiplying Eq. (3.66) by

$$\begin{bmatrix} 1 - 0.55B & -0.05B \\ 0.45B & 1 - 1.05B \end{bmatrix},$$

we obtain

$$(1 - 0.6B)\boldsymbol{w}_t = \begin{bmatrix} 1 - 0.5B - 0.005B^2 & -0.5B + 0.245B^2 \\ 0.495B^2 & 1 - B - 0.255B^2 \end{bmatrix} \boldsymbol{a}_t. \tag{3.67}$$

Thus, $\boldsymbol{w}_t$ follows a stationary VARMA(1,2) model. However, from Eq. (3.67) and denoting the MA matrix polynomial of $\boldsymbol{w}_t$ by $\boldsymbol{\theta}^*(B)$, we can see that

$$|\boldsymbol{\theta}^*(1)| = \begin{vmatrix} 1 - 0.5 - 0.005 & -0.5 + 0.245 \\ 0.495 & 1 - 1 - 0.255 \end{vmatrix} = \begin{vmatrix} 0.495 & -0.255 \\ 0.495 & -0.255 \end{vmatrix} = 0.$$

This implies that one is a solution of the determinant equation $|\boldsymbol{\theta}^*(B)| = 0$. Consequently, the VARMA(1,2) model for $\boldsymbol{w}_t$ is non-invertible.

The fact that differencing individual components of a co-integrated vector process with co-integrating rank less than the dimension results in a non-invertible VARMA model is referred to as *over-differencing* in the time series literature. Over-differencing can lead to complications in modeling co-integrated processes. For instance, since $\boldsymbol{w}_t$ is non-invertible, it does not have a finite-order VAR approximation. Therefore, one often needs a high-order VAR model to achieve good approximation in modeling $\boldsymbol{w}_t$ in practice. Another complication is that the solutions of the ML function of a non-invertible model are on the boundary of the parameter space and, hence, the conventional asymptotic theory for MLE would not apply. Interested readers are referred to Tanaka (2017, chapter 7) and the references therein.

To avoid over-differencing, the idea of error-correction was proposed in Engle and Granger (1987). Details are given in the next section.

3.6 ERROR-CORRECTION MODELS

Consider the VARMA model in Eq. (3.1), but assume that 1 might be a solution of the determinant equation $|\boldsymbol{\phi}(B)| = 0$. In other words, we assume $\boldsymbol{z}_t$ can be an integrated process of order 1. Let $\nabla \boldsymbol{z}_t = \boldsymbol{z}_t - \boldsymbol{z}_{t-1} = (1 - B)\boldsymbol{z}_t$ be the first-differenced series of $\boldsymbol{z}_t$. By straightforward algebraic calculation, e.g. Tsay (2014, chapter 5), we have

$$\nabla \boldsymbol{z}_t = \boldsymbol{\Pi} \boldsymbol{z}_{t-1} + \sum_{i=1}^{p-1} \boldsymbol{\phi}_i^* \nabla \boldsymbol{z}_{t-i} + \boldsymbol{c}_0 + \boldsymbol{\theta}(B)\boldsymbol{a}_t, \tag{3.68}$$

where $\boldsymbol{\Pi} = \sum_{i=1}^{p} \boldsymbol{\phi}_i - \boldsymbol{I} = -\boldsymbol{\phi}(1)$ and $\boldsymbol{\phi}_i^* = -(\boldsymbol{\phi}_{i+1} + \cdots + \boldsymbol{\phi}_p)$ for $i = 1, \ldots, p-1$. Eq. (3.68) is called an *error-correction model* (ECM) for the integrated process $\boldsymbol{z}_t$. Note that this model is invertible because we do not alter the MA matrix polynomial. However, if $|\boldsymbol{\phi}(1)| = 0$, then the coefficient matrix $\boldsymbol{\Pi}$ is singular. In general, we have the following cases:

1. If rank($\boldsymbol{\Pi}$) $= 0$, then $\boldsymbol{\Pi} = \boldsymbol{0}$ and $\boldsymbol{z}_t$ has k unit roots. It is not co-integrated.
2. If rank($\boldsymbol{\Pi}$) $= k$, then $\boldsymbol{\Pi}$ is nonsingular and $\boldsymbol{z}_t$ is a stationary process.
3. If rank($\boldsymbol{\Pi}$) $= m$ with $1 \leq m < k$, then $\boldsymbol{z}_t$ has $v - k - m$ unit roots and there exist m co-integrating vectors. In this case, one can write $\boldsymbol{\Pi} = \boldsymbol{\alpha}\boldsymbol{\beta}'$, where both $\boldsymbol{\alpha}$ and $\boldsymbol{\beta}$ are $k \times m$ real-valued matrices of rank m. Let $\boldsymbol{w}_t = \boldsymbol{\beta}'\boldsymbol{z}_t$, which is an m-dimensional stationary process. In other words, columns of $\boldsymbol{\beta}$ are the co-integrating vectors of $\boldsymbol{z}_t$. Note that $\boldsymbol{\alpha}\boldsymbol{\beta}' = (\boldsymbol{\alpha}\boldsymbol{P})(\boldsymbol{P}\boldsymbol{\beta})'$ for any $m \times m$ orthogonal matrix $\boldsymbol{P}$. Therefore, co-integrating vectors are not uniquely defined if $m > 1$. The column space of $\boldsymbol{\beta}$ is unique, however.

The stationarity of $\boldsymbol{w}_t$ of a co-integrated process $\boldsymbol{z}_t$ can easily be seen by re-writing the model as

$$\nabla \boldsymbol{z}_t = \boldsymbol{\alpha} \boldsymbol{w}_{t-1} + \sum_{i=1}^{p-1} \boldsymbol{\phi}_i^* \nabla \boldsymbol{z}_{t-i} + \boldsymbol{c}_0 + \boldsymbol{\theta}(B)\boldsymbol{a}_t. \tag{3.69}$$

Here $\boldsymbol{\alpha}$ is of full rank and $\nabla \boldsymbol{z}_t$ is stationary, $\boldsymbol{w}_t$ must be stationary, because a stationary process cannot depend on any unit-root process.

The error-correction model of Eq. (3.68) is referred to as ECM in the *transitory form* because the coefficient matrices $\boldsymbol{\phi}_i^*$ show the transitory effects. An alternative form of ECM is given by

$$\nabla \boldsymbol{z}_t = \boldsymbol{\Pi} \boldsymbol{z}_{t-p} + \sum_{i=1}^{p-1} \boldsymbol{\phi}_i^* \nabla \boldsymbol{z}_{t-i} + \boldsymbol{c}_0 + \boldsymbol{\theta}(B)\boldsymbol{a}_t, \tag{3.70}$$

where $\boldsymbol{\Pi}$ remains unchanged, i.e. $\boldsymbol{\Pi} = -\boldsymbol{\phi}(1)$, but $\boldsymbol{\phi}_i^* = -(\boldsymbol{I} - \boldsymbol{\phi}_1 - \cdots - \boldsymbol{\phi}_i)$ for $i = 1, \ldots, p-1$. These new $\boldsymbol{\phi}_i^*$ matrices contain the long-run cumulative effects, and the ECM in Eq. (3.70) is referred to as ECM in the *long-run* form.

3.6.1 Co-integration Test

Detecting the existence (and the number) of co-integrating vectors of a multivariate process is referred to as *co-integration* test in the econometric literature. There are several approaches available in the literature to co-integration test. For VAR models, the most commonly used co-integration test is the Johansen test (Johansen, 1991; Johansen and Juselius, 1990). The test can be carried out using either ECM of (3.68) or (3.70). We use the transitory form in our discussion and expand the ECM of Eq. (3.68) to include possible time trend. That is,

$$\nabla \boldsymbol{z}_t = \boldsymbol{\Pi} \boldsymbol{z}_{t-1} + \sum_{i=1}^{p-1} \boldsymbol{\phi}_i^* \nabla \boldsymbol{z}_{t-i} + \boldsymbol{d}_t + \boldsymbol{a}_t, \tag{3.71}$$

where $\boldsymbol{d}_t = \boldsymbol{c}_0 + \boldsymbol{c}_1 t$ with $\boldsymbol{c}_0$ and $\boldsymbol{c}_1$ are k-dimensional real vectors and other coefficients as the same as those of Eq. (3.68). The co-integration test is then concerned with the rank of the matrix $\boldsymbol{\Pi}$, which is related to the correlation matrix between $\boldsymbol{z}_{t-1}$ and $\nabla \boldsymbol{z}_t$. Assuming multivariate normality, Johansen (1991, 1995) uses the ML test to detect the rank of $\boldsymbol{\Pi}$. This is essentially based on canonical correlation analysis. Specifically, to mitigate the effect of the stationary part on co-integration test, one considers two preliminary multivariate linear regressions

$$\nabla \boldsymbol{z}_t = \sum_{i=1}^{p-1} \boldsymbol{\gamma}_i \nabla \boldsymbol{z}_{t-i} + \boldsymbol{d}_t + \boldsymbol{u}_t \tag{3.72}$$

$$\boldsymbol{z}_{t-1} = \sum_{i=1}^{p-1} \boldsymbol{\gamma}_i^* \nabla \boldsymbol{z}_{t-i} + \boldsymbol{d}_t^* + \boldsymbol{v}_t \tag{3.73}$$

where a superscript "*" is used for the coefficient matrix of Eq. (3.73), the deterministic trend is pre-specified, and $\boldsymbol{u}_t$ and $\boldsymbol{v}_t$ denote the error term, respectively. Let $\hat{\boldsymbol{u}}_t$

and $\hat{\boldsymbol{v}}_t$ be the residual of the OLS estimation of Eqs. (3.72) and (3.73), respectively. Then, one considers the multivariate linear regression

$$\hat{\boldsymbol{u}}_t = \boldsymbol{\Pi}\hat{\boldsymbol{v}}_t + \boldsymbol{e}_t, \tag{3.74}$$

where $\boldsymbol{e}_t$ is the error term. The least squares estimates of $\boldsymbol{\Pi}$ are equivalent for Eqs. (3.71) and (3.74). Therefore, under multivariate normality, one can check the rank of $\boldsymbol{\Pi}$ by testing the canonical correlation coefficients between $\hat{\boldsymbol{u}}_t$ and $\hat{\boldsymbol{v}}_t$. Let $\lambda_1^2 \geq \lambda_2 \geq \cdots \geq \lambda_k^2$ be the ordered squared canonical correlations between $\hat{\boldsymbol{u}}_t$ and $\hat{\boldsymbol{v}}_t$. Johansen (1991) considered two test statistics for identifying rank($\boldsymbol{\Pi}$).

For $m = 0, \ldots, k-1$, consider the null hypothesis $H(m)$: Rank($\boldsymbol{\Pi}$) $= m$ versus the alternative hypothesis H_a : rank($\boldsymbol{\Pi}$) $> m$. The test statistic is

$$L_{\text{tr}}(m) = -(T - kp) \sum_{i=m+1}^{k} \ln(1 - \lambda_i^2), \tag{3.75}$$

where T is the sample size, p is the VAR order, k is the dimension. The idea of the test is that, under the null hypothesis $H(m)$, rank($\boldsymbol{\Pi}$) $= m$ so that the largest m canonical correlations between $\hat{\boldsymbol{u}}_t$ and $\hat{\boldsymbol{v}}_t$ are positive, but the remaining $k - m$ canonical correlations are zero. This test statistic is referred to as the trace test. Unlike the stationary case, the limiting distribution of $L_{\text{tr}}(m)$ under $H(m)$ is not chi-square. It is a function of standard Brownian motion. For details, see Johansen (1991) and Tsay (2014, chapter 5) and the references therein. In practice, the asymptotic critical values of $L_{\text{tr}}(m)$ are obtained by simulation.

The second test statistic considered by Johansen (1991) is concerned with the test

$$H_0 : \text{Rank}(\boldsymbol{\Pi}) = m \quad \text{vs} \quad \text{Rank}(\boldsymbol{\Pi}) = m + 1.$$

In this case, the test statistic is

$$L_{\max}(m) = -(T - kp)\ln(1 - \lambda_{m+1}^2). \tag{3.76}$$

This test statistic is referred to as the maximum test. Similarly to $L_{\text{tr}}(m)$, the limiting distribution of $L_{\max}(m)$ is also not chi-square, but a function of the standard Brownian motion.

Once the number of co-integrating vectors m is determined, the eigenvectors associated with the largest m eigenvalues $\{\lambda_i^2 | i = 1, \ldots, m\}$ of the canonical correlation analysis between $\hat{\boldsymbol{u}}_t$ and $\hat{\boldsymbol{v}}_t$ can be used to estimate the co-integrating vectors.

Example 3.4

Consider the monthly Moody's seasoned Aaa and Baa bond yields from May 1952 to June 2019. The data are available from FRED of the Federal Reserve Bank of St. Louis and the shown in Figure 3.14. From the plots, the two bond yields move in unison with some occasional deviations.

To demonstrate co-integration test, we start with scalar unit-root testing. For the Aaa bond yields, an AR(3) model is selected and the augmented Dickey-Fuller test is −0.45. Compared with the critical values of 1 or 5%, the unit-root null hypothesis cannot be rejected. For the Baa bond yields, an AR(2) model is selected and the augmented Dickey-Fuller test statistic is −0.40, which, again, fails to reject the unit-root

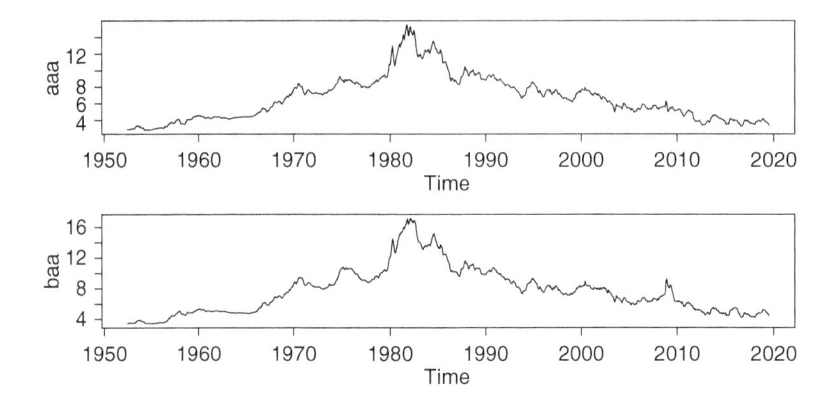

Figure 3.14 Time plots of monthly Moody's Aaa and Baa bond yields from May 1952 to June 2019. The upper plot is for Aaa bond.

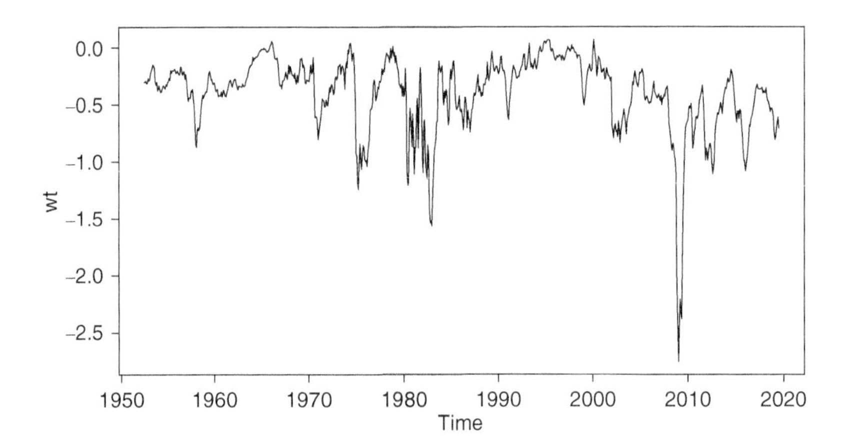

Figure 3.15 Time plot of the co-integrated series between the monthly Moody's Aaa and Baa bond yields.

null hypothesis. Therefore, each of the two bond yield series appears to be unit-root non-stationary.

Consider the two series jointly. The AIC selects a VAR(8) model for the two series. The Johansen test statistics show that $L_{\max}(0) = 27.19$ and $L_{\max}(1) = 2.01$. Compared with their critical values, the test statistics show that $\text{Rank}(\mathbf{\Pi}) = 1$ implying that the two bond yield series are co-integrated. This is not surprising as any significant and prolonged deviation between the two bond yields would lead to arbitrage opportunities. If the trace statistics are used, the test statistics are $L_{\text{tr}}(0) = 29.20$ and $L_{\text{tr}}(1) = 2.01$, which, again, lead to the same conclusion of co-integration. The co-integrating vector is $(1, -0.926)'$. Figure 3.15 shows the time plot of the co-integrated series of the two bond yields. The stationarity of the series is evident except for the big drop during the 2008 financial crisis.

R output: Co-integration analysis. Output edited to save space. R packages **urca** and **MTS** are used in the analysis.

```
> da <- read.csv("m-usbnd.csv")
> bond <- da[,2:3]
> ar(bond[,1])
Call: ar(x = bond[, 1])
Coefficients:
      1        2        3
 1.1940   -0.2906   0.0913
Order selected 3  sigma^2 estimated as  0.07097
> ar(bond[,2])
Call: ar(x = bond[, 2])
Coefficients:
      1        2
 1.2100   -0.2151
Order selected 2  sigma^2 estimated as  0.07096

> require(urca)
> nn1 <- ur.df(bond[,1],lags=3,type="none")
> summary(nn1)
###############################################
# Augmented Dickey-Fuller Test Unit Root Test #
###############################################
Test regression none
Call: lm(formula = z.diff ~ z.lag.1 - 1 + z.diff.lag)
Coefficients:
             Estimate Std. Error t value Pr(>|t|)
z.lag.1     -0.0004245  0.0009386  -0.452    0.651
z.diff.lag1  0.4208651  0.0353942  11.891  < 2e-16 ***
z.diff.lag2 -0.2535622  0.0373338  -6.792 2.16e-11 ***
z.diff.lag3  0.0503625  0.0354006   1.423    0.155
---
Value of test-statistic is: -0.4523

Critical values for test statistics:
      1pct  5pct 10pct
tau1 -2.58 -1.95 -1.62

> nn2 <- ur.df(bond[,2],lags=2,type="none")
> summary(nn2)
###############################################
# Augmented Dickey-Fuller Test Unit Root Test #
###############################################
Test regression none
Call:lm(formula = z.diff ~ z.lag.1 - 1 + z.diff.lag)
Coefficients:
             Estimate Std. Error t value Pr(>|t|)
z.lag.1     -0.0003144  0.0007815  -0.402    0.688
z.diff.lag1  0.4718882  0.0349516  13.501  < 2e-16 ***
z.diff.lag2 -0.1522756  0.0349617  -4.355  1.5e-05 ***
---
Value of test-statistic is: -0.4022

Critical values for test statistics:
```

```
      1pct  5pct 10pct
tau1 -2.58 -1.95 -1.62

> VARorder(bond)
selected order: aic =  8
selected order: bic =  3
selected order: hq =  3
Summary table:
       p     AIC     BIC      HQ      M(p) p-value
 [1,]  0 -0.0011 -0.0011 -0.0011    0.0000  0.0000
 [2,]  1 -7.7485 -7.7252 -7.7396 6124.4128  0.0000
 [3,]  2 -8.0868 -8.0402 -8.0689  274.1698  0.0000
 [4,]  3 -8.1345 -8.0647 -8.1077   45.3339  0.0000
 .........
[14,] 13 -8.1381 -7.8354 -8.0219    1.9572  0.7436

> n1 <- ca.jo(bond,ecdet="const",K=8)
> summary(n1)
######################
# Johansen-Procedure #
######################
Test type: maximal eigenvalue statistic (lambda max), without
linear trend and constant in cointegration

Eigenvalues (lambda):
[1] 3.349441e-02 2.516686e-03 5.160725e-19

Values of teststatistic and critical values of test:
          test 10pct  5pct  1pct
r <= 1 |  2.01  7.52  9.24 12.97
r = 0  | 27.19 13.75 15.67 20.20

Eigenvectors, normalized to the first column:
(These are the cointegration relations)

             AAA.l8     BAA.l8    constant
AAA.l8    1.0000000  1.0000000   1.0000000
BAA.l8   -0.9259745 -0.3516517  -0.9459504
constant  0.4098622 -4.0114581 -13.0158264

Weights W:
(This is the loading matrix)

          AAA.l8       BAA.l8      constant
AAA.d 0.03582889 -0.005378767 -3.068444e-18
BAA.d 0.08131706 -0.003636711 -8.823027e-18

>n2 <- ca.jo(bond,type="trace",ecdet=c("const"),K=8,spec=c("transitory"))
> summary(n2)
######################
# Johansen-Procedure #
######################
Test type: trace statistic, without linear trend and constant in
cointegration

Eigenvalues (lambda):
```

```
[1] 3.349441e-02 2.516686e-03 1.360635e-19

Values of teststatistic and critical values of test:
          test 10pct  5pct  1pct
r <= 1 |  2.01  7.52  9.24 12.97
r = 0  | 29.20 17.85 19.96 24.60
```

Finally, once the number of co-integrating series is identified, one can estimate a specified ECM by the ML method or the least squares method. Details can be found, for instance, in Tsay (2014, chapter 5). ■

EXERCISES

1. Consider the log series of the US monthly exports and imports data in the file `m-expimpcnus.csv`. (a) Are the two log series unit-root non-stationary? Perform unit-root tests to draw conclusions. (b) Are the two series co-integrated? Perform co-integration test to verify the answer.
2. Consider, again, the US monthly export and import series of Problem 1. (a) Build a bivariate time series model for the two series. Perform model simplification and model checking to justify the model used. (b) Obtain the impulse response function of the model built using orthogonal innovations. (c) Obtain 1-step to 4-step ahead forecasts using the model built at the forecast origin $t = 324$, which is the last data point (or December 2018).
3. Consider the Taiwan AirBox data in `TaiwanAirBox032017.csv`. The file contains 514 series with 744 observations. Focus on Series 2, 3, and 4. Build a multivariate time series for the three-dimensional time series. Perform model checking and obtain 1-step to 3-step ahead forecasts for the series at the forecast origin T=744.
4. Consider the November temperatures of Europe, North America, and South America discussed in Chapter 1. See the file `temperatures.txt`. (a) Is there a unit root in the individual time series? Why? (b) Are the three series co-integrated? Why?
5. Again consider the three temperature series of Europe, North America, and South America. Build a vector ARMA model (AR or MA model is allowed) for the three series. Perform model checking. Obtain impulse response functions, with orthogonal innovations, based on the model obtained.

REFERENCES

Abraham, B. and Ledolter, J. (1983). *Statistical Methods for Forecasting*. Wiley, New York, NY.

Ahn, S. K. and Reinsel, G. C. (1988). Nested reduced rank autoregressive models for multiple time series. *Journal of the American Statistical Association*, **83**: 849–856.

Ahn, S. K. and Reinsel, G. C. (1990). Estimation for partially nonstationary multivariate autoregressive models. *Journal of the American Statistical Association*, **85**: 849–856.

Akaike, H. (1973). *Information theory and an extension of the maximum likelihood principle.* In B. N. Petrov and F. Csaki (eds.). *2nd International Symposium on Information Theory*, pp. 267–281. Akademia Kiado, Budapest.

Akaike, H. (1976). *Canonical correlation analysis of time series and the use of an information criterion.* In R. K. Methra and D. G. Lainiotis (eds.). *Systems Identification: Advances and Case Studies*, pp. 27–96. Academic Press, New York, NY.

Blattberg, R. C. and Jeuland, A. P. (1981). A micromodeling approach to investigate the advertising-sales relationship. *Management Science*, **27**: 988–1005.

Box, G. E. P. and Jenkins, G. M. (1976). *Time Series Analysis: Forecasting and Control*, revised ed., Holden-Day, San Francisco, CA.

Box, G. E. P., Jenkins, G. M., Reinsel, G. C., and Ljung, G. M. (2015). *Time Series Analysis: Forecasting and Control*, 5th Edition. Wiley, Hoboken, NJ.

Box, G. E. P. and Tiao, G. C. (1977). A canonical analysis of multiple time series. *Biometrika*, **64**: 355–366.

Chan, N. H. and Wei, C. Z. (1988). Limiting distributions of least squares estimates of unstable autoregressive processes. *Annals of Statistics*, **16**: 367–401.

Dunsmuir, W. and Hannan, E. J. (1976). Vector linear time series models. *Advanced Applied Probability*, **8**: 339–364.

Engle, R. F. and Granger, C. W. J. (1987). Cointegration and error correction: representations, estimation and testing. *Econometrica*, **55**: 251–276.

Hannan, E. J. and Deistler, M. (1988). *The Statistical Theory of Linear Systems.* Wiley, New York, NY.

Hannan, E. J. and Quinn, B. G. (1979). The determination of the order of an autoregression. *Journal of the Royal Statistical Society, Series B*, **41**: 190–195.

Hosking, J. R. M. (1981). Lagrange-multiplier tests of multivariate time series model. *Journal of the Royal Statistical Society, Series B*, **43**: 219–230.

Johansen, S. (1991). Estimation and hypothesis testing of cointegration vectors in Gaussian vector autoregressive models. *Econometrica*, **59**: 1551–1580.

Johansen, S. (1995). *Likelihood Based Inference in Cointegrated Vector Error Correction Models.* Oxford University Press, Oxford, UK.

Johansen, S. and Juselius, K. (1990). Maximum likelihood estimation and inference on cointegration-with applications to the demand for money. *Oxford Bulletin of Economics and Statistics*, **52**: 169–210.

Li, W. K. and McLeod, A. I. (1981). Distribution of the residual autocorrelations in multivariate time series models. *Journal of the Royal Statistical Society, Series B*, **43**: 231–239.

Lütkepohl, H. (2005). *New Introduction to Multiple Time Series Analysis.* Springer, New York, NY.

Phillips, P. C. B. (1987). Time series regression with a unit root. *Econometrica*, **55**: 277–301.

Reinsel, G. (1993). *Elements of Multivariate Time Series Analysis.* Springer-Verlag, New York, NY.

Schwarz, G. (1978). Estimating the dimension of a model. *Annals of Statistics*, **6**: 461–464.

Shibata, R. (1980). Asymptotically efficient selection of the order of the model for estimating parameters of a linear process. *Annals of Statistics*, **8**: 147–164.

Tanaka, K. (2017). *Time Series Analysis: Nonstationary and Noninvertible Distribution Theory*, 2nd Edition. Wiley, Hoboken, NJ.

Tiao, G. C. and Box, G. E. P. (1981). Modeling multivariate time series with applications. *Journal of American Statistical Association*, **72**: 802–816.

Tiao, G. C. and Tsay, R. S. (1983). Multiple time series modeling and extended sample cross-correlations. *Journal of Business & Economic Statistics*, **1**: 43–56.

Tiao, G. C. and Tsay, R. S. (1989). Model specification in multivariate time series (with discussion). *Journal of the Royal Statistical Society, Series B*, **51**: 157–213.

Tsay, R. S. (1991). Two canonical forms for vector ARMA processes. *Statistica Sinica*, **1**: 247–269.

Tsay, R. S. (2010). *Analysis of Financial Time Series*, 3rd Edition. Wiley, Hoboken, NJ.

Tsay, R. S. (2014). *Multivariate Time Series Analysis with R and Financial Applications*. Wiley, Hoboken, NJ.

Tsay, R. S. and Tiao, G. C. (1990). Asymptotic properties of multivariate nonstationary processes with application to autoregressions. *Annals of Statistics*, **18**: 220–250.

Velu, R. P., Reinsel, G. C., and Wichern, D. W. (1986). Reduced rank models for multiple time series. *Biometrika*, **73**: 105–118.

Velu, R. P., Wichern, D. W., and Reinsel, G. C. (1987). A note on non?stationarity and canonical analysis of multiple time series models. *Journal of Time Series Analysis*, **8**: 479–487.

CHAPTER 4

HANDLING HETEROGENEITY IN MANY TIME SERIES

Empirical time series are often affected by multiple occasional events that produce different causes of heterogeneity. The impacts of these events should be considered in the modeling process because they may lead to biased parameter estimates and/or model misspecification, resulting in poor forecasting performance. The first class of events we consider consists of a known cause such as strikes on a set of production series, a leap year on monthly sales series, a tariff change on import and export series, or an extreme climatological event on air pollution indexes. The second class of events deals with missing values in a time series caused by instrument failure or human errors. Here we assume that the missing values occur at random and, for a vector series, the whole data point or some elements of the vector may be missing. For these two types of events, we know their time of occurrence and want to remove its impact from the set of time series under study. How to carry out such adjustments is the focus of the first part of this chapter.

There are other situations in which we do not know if any atypical event has occurred, nor the time of its occurrence, if any. However, from the aberrant observations of the data, we can infer that some unexpected events did occur during the sampling period. Those aberrant data points are commonly referred to as *outliers* and were first studied in time series by Fox (1972). They happen often in real time series. For instance, Maravall et al. (2016) analyzed more than 15,000 monthly economic and social time series from many countries and found that more than 60% of the series have at least one clear outlier and the expected number of outliers in a time series is around 2% of the data points. Outliers may appear isolated or in groups forming patches of outliers. Some events can produce a permanent change in the level of the series and they are called *level shifts* or structural breaks. On the

Statistical Learning for Big Dependent Data, First Edition. Daniel Peña and Ruey S. Tsay.
© 2021 John Wiley & Sons, Inc. Published 2021 by John Wiley & Sons, Inc.

other hand, others only cause a temporary change to the underlying time series. The second part of this chapter focuses on outlier detection in a vector time series. Following the literature, we classify outliers into level shift, transitory change, additive outlier, innovation outlier, and variance change, and discuss methods for detecting these outliers and adjusting their impacts. The methods discussed include the traditional ones available in the time series literature and some recent developments in high-dimensional data analysis. For a set of independent time series, we can clean the series one by one. However, for a set of related time series the information about an event that may affect all of them is distributed among the series. It is then better off to develop detection procedures that make use of information in the series jointly. In this chapter, we first define univariate and multivariate outliers and present a procedure to adjust their effects in each series or in a vector of time series modeled by a vector autoregressive moving-average (VARMA) model. Then, we present a method based on projections to clean a large set of time series without the need of estimating a multivariate model.

An alternative approach to deal with outliers or structural changes is to use robust estimation methods that are not sensitive to atypical events. The outliers can be detected by examining the residuals computed using some robust parameter estimates. There is substantial literature on robust estimation of univariate time series, but only a few articles deal with multivariate series. A brief summary of these results is presented in this chapter. Heterogeneity may also appear because some parameters of the model are changing over time, or they change at a particular time point. Again, this is an important and extensively studied topic, and some available results for multivariate time series are presented in the last section of this chapter. Finally, another type of heterogeneity is introduced by certain nonlinear effects in the time series, but this topic is not considered here and we refer the readers to the book by Tsay and Chen (2018).

4.1 INTERVENTION ANALYSIS

Fixed effects of an external event to a time series can be considered as impacts of some explanatory variables in a model. Box and Tiao (1975) called such an approach *intervention analysis* in which various dummy or indicator variables serve as exogenous variables to a time series model. The parameters of the exogenous variables quantify the impact of the intervention. In the economic literature, intervention analysis is referred to as *event study*. To illustrate, suppose that we have a monthly production series z_t that was subject to the impact of strike on a working day in month $t = h$. Since the strike occurred only on a working day, we expect that its impact on the series, if any, would only happen in month $t = h$. It is then reasonable to introduce a dummy (or indicator) variable

$$I_t^{(h)} = \begin{cases} 1, & \text{if } t = h, \\ 0, & \text{otherwise,} \end{cases} \tag{4.1}$$

and postulate the observed data z_t as $z_t = y_t + \omega I_t^{(h)}$, where ω denotes the impact of the strike and y_t is the underlying outlier-free time series. Clearly, for $t \neq h$, we observe $z_t = y_t$, but for $t = h$, we have $z_h = y_h + \omega$, indicating that what we observed is contaminated by the strike at time $t = h$. As a second example, consider the monthly exports z_t of an economy. Suppose that a major trading partner of the economy imposed a permanent heavy tariff increase starting at month $t = h$ and we are interested in assessing the impact of the tariff on the export series z_t. Here we can define a step function as

$$S_t^{(h)} = \begin{cases} 1, & \text{if } t \geq h, \\ 0, & \text{otherwise,} \end{cases} \tag{4.2}$$

and postulate the observed monthly export data as $z_t = y_t + \omega S_t^{(h)}$, where ω signifies the impact of the new tariff. Clearly, for $t < h$, we have $z_t = y_t$, but, for $t \geq h$, we have $z_t = y_t + \omega$. Here the impact of the new heavy tariff is expected to last once it occurred. Readers can immediately see that this second example can also be described as $z_t = y_t + \omega \frac{1}{(1-B)} I_t^{(h)}$, where B is the back-shift operator. This is so because $(1 - B)S_t^{(h)} = I_t^{(h)}$. The function $\frac{1}{1-B}$ represents a ratio of two polynomials in B and is widely used in intervention analysis or outlier detection.

By combining the indicator and step functions together and using ratios of polynomials to describe the impact of an intervention, we obtain a flexible approach to intervention analysis. Details are given in next section and in Box and Tiao (1975). In fact, the exogenous variables in intervention analysis are not necessarily dummy or step functions. Other deterministic variables can also be used. For example, in many monthly economic or production time series the number of working days in each month is an important factor to consider. We could take such working-day effects into account by dividing the observations by the number of working days in each month, resulting in an average daily production per month series for analysis. A better solution is to introduce days of the week in each month as explanatory variables and incorporate them into the model. This is so because days of a week are likely to have different levels of activity, for example, supermarket sales are not the same on Mondays as they are on Fridays. We can measure the daily effect by including seven explanatory variables showing the number of Mondays, Tuesdays, etc. in each month. Furthermore, many monthly series are affected by holidays that fall irregularly from one year to another, such as the Easter in Europe and the lunar new year in China. We can measure such effects by introducing an exogenous variable that takes a value of one in the month in which the holiday falls and zero for other months of the year.

To better understand ways to handle heterogeneity in a time series, it is useful to consider that the observed time series z_t is contaminated by some exogenous effects. As such, we say that corresponding to the observed z_t, there exists an uncontaminated (or outlier-free) series y_t. In practice, we only observed z_t, but the goal of this chapter is to recover y_t as accurately as possible. Similar notation applies to the vector case.

4.1.1 Intervention with Indicator Variables

We start with an observed scalar time series z_t and assume that the corresponding uncontaminated series y_t follows an ARIMA(p, d, q) model, say $\phi(B)(1 - B)^d y_t = \theta(B)a_t$. If $d > 0$, then we assume, for simplicity, that the series starts with $t = 1$ with some zero initial values. Therefore, we can use the autoregressive

(AR) representation of the series $\pi(B)y_t = a_t$ or the moving-average (MA) representation of the series $y_t = \psi(B)a_t$, where it is understood that $\pi(B) = \phi(B)(1 - B)^d/\theta(B)$ and $\psi(B) = \theta(B)/[\phi(B)(1 - B)^d]$ and $\pi(B)\psi(B) = 1$. In practice, both π_i and ψ_j can be obtained by long division of two polynomials in B.

We assume that an unusual event of size w_0 happened at time $t = h$. In this case, by using the indicator variable in Eq. (4.1), we can write the observed time series as

$$z_t = w_0 I_t^{(h)} + y_t = w_0 I_t^{(h)} + \psi(B)a_t. \tag{4.3}$$

In other words, we have

$$z_t = \begin{cases} y_t, & \text{if } t \neq h, \\ y_h + w_0, & \text{if } t = h. \end{cases}$$

Representation (4.3) is a special case of the *transfer functions* presented in Chapter 3 with input variable $I_t^{(h)}$. It is also possible that an unusual event at time $t = h$ can affect the time series under study over several time periods. For instance, a big winter snow storm in Chicago is likely to interrupt the air traffic at the O'Hare International Airport for several days. In this case, we can write the observed time series as

$$z_t = \sum_{i=0}^{m} w_i I_t^{(h+i)} + \psi(B)a_t, \tag{4.4}$$

where m is a finite positive integer and w_i is the impact of the event on time period $t = h + i, i = 0, \dots, m$. Here we use a parameter w_i to denote the impact of the event on the series at $t = h + i$. In practice, it might be hard to know *a priori* the value of m. Also, it is likely that the impacts will decay smoothly as time elapses. It is then logical to postulate the following model for z_t:

$$z_t = \frac{w_0}{1 - \delta B} I_t^{(h)} + \psi(B)a_t = \frac{w_0}{1 - \delta B} I_t^{(h)} + y_t, \tag{4.5}$$

where w_0 is the initial impact of the unusual event and $0 < \delta < 1$ denotes the discounting rate of the impact. This is so because $w_0/(1 - \delta B) = w_0(1 + \delta B + \delta^2 B^2 + \cdots)$. More specifically, Eq. (4.5) shows that

$$z_t = \begin{cases} y_t, & \text{if } t < h, \\ y_h + w_0, & \text{if } t = h, \\ y_{h+1} + w_0\delta, & \text{if } t = h + 1, \\ y_{h+2} + w_0\delta^2, & \text{if } t = h + 2, \\ \vdots & \vdots \end{cases}$$

where y_t is the unobserved uncontaminated time series.

In general, a flexible model to describe the impact of an unusual event on a time series at time index $t = h$ is

$$z_t = \frac{w(B)}{\delta(B)} I_t^{(h)} + y_t, \tag{4.6}$$

where $w(B) = w_0 + w_1 B + \cdots + w_m B^m$ and $\delta(B) = 1 - \delta_1 B - \cdots - \delta_r B^r$ are two polynomials in B with finite orders m and r, and $y_t = \psi(B)a_t$ is the uncontaminated time series. The flexibility of $w(B)/\delta(B)$ can be found in Box and Tiao (1975) and Chapter 3. Clearly, we require that $w(B)$ and $\delta(B)$ have no common factors in Eq. (4.6).

In real applications, a time series may be subject to the impacts of multiple unusual events at time $t = h_i$, for $i = 1, \dots, g$, with g being a nonnegative integer. In this case, we postulate the following model for z_t:

$$z_t = \sum_{i=1}^{g} \frac{w_i(B)}{\delta_i(B)} I_t^{(h_i)} + \psi(B)a_t, \tag{4.7}$$

where $w_i(B) = w_{i,0} + w_{i,1}B + \cdots + w_{i,m_i}B^{m_i}$ and $\delta_i(B) = 1 - \delta_{i,1}B - \cdots - \delta_{i,r_i}B^{r_i}$ with m_i and r_i are nonnegative integers.

The prior discussion of modeling impacts of unusual events on a scalar time series can easily be generalized to the vector case. For instance, for an observed vector series the model in Eq. (4.3) would become,

$$\boldsymbol{z}_t = \boldsymbol{w}_0 I_t^{(h)} + \boldsymbol{y}_t = \boldsymbol{w}_0 I_t^{(h)} + \boldsymbol{\psi}(B)\boldsymbol{a}_t,$$

where $\boldsymbol{w}_0 = (w_{01}, \dots, w_{0k})'$ is the vector of effects of the event with w_{0i} being the impact on the ith component z_{it}, $\boldsymbol{y}_t$ is the corresponding uncontaminated vector time series, and $\boldsymbol{\psi}(B) = [\boldsymbol{\Phi}(B)]^{-1}\boldsymbol{\theta}(B)$ denotes the ψ-weight matrices of $\boldsymbol{y}_t$, which follows the VARMA model $\boldsymbol{\Phi}(B)\boldsymbol{y}_t = \boldsymbol{\theta}(B)\boldsymbol{a}_t$. See Chapter 3 for further discussion on the ψ-weight matrices. Similarly, the model in Eq. (4.6) can be generalized as

$$\boldsymbol{z}_t = [\boldsymbol{\delta}(B)]^{-1}\boldsymbol{w}(B)I_t^{(h)} + \boldsymbol{y}_t, \tag{4.8}$$

where $\boldsymbol{w}(B) = \boldsymbol{w}_0 + \boldsymbol{w}_1 B + \cdots + \boldsymbol{w}_m B^m$ and $\boldsymbol{\delta}(B) = \boldsymbol{I} - \boldsymbol{\delta}_1 B - \cdots - \boldsymbol{\delta}_r B^r$ are two matrix polynomials of finite orders m and r, respectively, and $\boldsymbol{w}_i$ and $\boldsymbol{\delta}_j$ are $k \times k$ real matrices with k being the dimension of $\boldsymbol{z}_t$. Here we require that $\boldsymbol{w}(B)$ and $\boldsymbol{\delta}(B)$ are left co-prime. A special case of Eq. (4.8) that is useful in application is

$$\boldsymbol{z}_t = (1 - \delta B)^{-1}\boldsymbol{w}_0 I_t^{(h)} + \boldsymbol{y}_t,$$

in which the impacts of the unusual event on each component of $\boldsymbol{z}_t$ decay at the same rate.

4.1.2 Intervention with Step Functions

In some applications, an unusual event can permanently affect the time series under study. For instance, consider the market share of Brand A of a product, and a big sale promotion by its manufacturer at time index $t = h$. If the promotion was announced prior to the effective time of sale promotion, consumers might delay any purchase in an anticipation of a price drop. Then, they purchased a large quantity of Brand A during the promotion period, resulting in a drop in sales immediately after the promotion was over. In theory, such a big promotion may permanently change the market share of Brand A. To describe such a complex market responses, it would become harder to use the indicator variable $I_t^{(h)}$. Instead, the step function of Eq. (4.2) could be used to simplify the model. A candidate model for such an intervention is

$$z_t = (w_{-1}F + w_0 + w_1 B)S_t^{(h)} + y_t,$$

where $F = B^{-1}$ is the forward-shift operator, y_t is the underlying market share without the promotion and w_i are parameters used to quantify the effect of sale promotion. In this particular case, one would expect $w_{-1} < 0$, $w_0 > 0$ and $w_1 < 0$, and the permanent effect of the promotion is given by $w_{-1} + w_0 + w_1$.

Turn to the case of vector time series. The effect of an unusual event that begins at $t = h$ can be modeled as

$$z_t = \boldsymbol{w}_0 S_t^{(h)} + \boldsymbol{\Psi}(B)\boldsymbol{a}_t,$$

where, as before, $\boldsymbol{w}_0 = (w_{0,1}, \ldots, w_{0,k})'$ with k being the dimension of z_t. Here, it is easy to see that we have $\boldsymbol{z}_t = \boldsymbol{y}_t$ for $t < h$ and $\boldsymbol{z}_t = \boldsymbol{w}_0 + \boldsymbol{y}_t$ for $t \geq h$, where $\boldsymbol{y}_t$ is the uncontaminated vector time series. Obviously, we can generalize the model to include cases with multiple interventions, each with complicated impacts on the underlying time series.

Since

$$I_t^{(h)} = S_t^{(h)} - S_{t-1}^{h} = \nabla S_t^{(h)}, \tag{4.9}$$

the choice of using an indicator variable or a step function for a given intervention may not be critical in real applications. One can let the data choose a proper variable to use. For instance, if we estimate the model

$$\boldsymbol{z}_t = (\boldsymbol{w}_0 - \boldsymbol{w}_1 B)S_t^{(h)} + \boldsymbol{\Psi}(B)\boldsymbol{a}_t,$$

and find that $\hat{\boldsymbol{w}}_0 \simeq \hat{\boldsymbol{w}}_1$, then we can simplify the model as

$$z_t = \hat{\boldsymbol{w}}_0 \nabla S_t^{(h)} + \boldsymbol{\Psi}(B)\boldsymbol{a} = \hat{\boldsymbol{w}}_0 I_t^{(h)} + \boldsymbol{\Psi}(B)\boldsymbol{a}_t,$$

indicating that the intervention effect only occurs at $t = h$. If the choice between using indicator variables and step functions is in doubt, one can always use model selection methods such as information criteria to select a proper model.

4.1.3 Intervention with General Exogenous Variables

Intervention analysis is not limited to using either indicator variables or step functions. In fact, one can use any exogenous variable for intervention analysis so long as the variable is known at time t. For instance, let x_t be the monthly expenditure of advertisement of a company on its products and z_t be the total monthly sales of the company. Then, the effect of advertisement on sales can be modeled as

$$z_t = v(B)x_t + y_t,$$

where y_t is the monthly sales without any advertisement and $v(B) = v_0 + v_1 B + v_2 B^2 + \cdots$ is the transfer function. In practice, $v(B)$ can often be well approximated by ratio of two polynomials such as $v(B) = w(B)B^b/\delta(B)$, where $w(B) = w_0 + w_1 B + \cdots + w_m B^m$ and $\delta(B) = 1 - \delta_1 B - \cdots - \delta_r B^r$ with m and r are nonnegative integers, and b is an integer commonly referred to as the *delay* of the intervention. This belongs to the transfer function models discussed in Chapter 3.

4.1.4 Building an Intervention Model

Again, we start with the scalar case with a single intervention at time $t = h$ and consider the model

$$z_t = v(B)I_t^{(h)} + \psi(B)a_t. \tag{4.10}$$

If h is sufficiently large, one can build a model for the uncontaminated series y_t using data $\{y_1, \ldots, y_{h-1}\}$, equivalently $\{z_1, \ldots, z_{h-1}\}$. In other words, we use the data prior to the intervention to infer $\psi(B)$. The model is then used to predict $\hat{z}_{h-1}(\ell)$ for $\ell = 1, \ldots, T-h+1$, where T is the sample size. The forecasting errors $e_{h-1}(\ell) = z_{h-1+\ell} - \hat{z}_{h-1}(\ell)$ show the impacts of the intervention and provide information concerning the functional form of $v(b)$ which, in turn, can be used to specify $w(B)$ and $\delta(B)$. The specified model in Eq. (4.10) is then estimated by the conditional maximum likelihood method, including the parameters of $\psi(B)$.

On the other hand, if h is not sufficiently large to be useful in specifying $\psi(B)$, then an iterative procedure involving the following three steps is often used.

Modeling procedure:

1. Start with an initial estimation of the transfer function, $\hat{v}^{(0)}(B)$, to obtain an estimation of y_t, by $\hat{y}_t = z_t - \hat{v}^{(0)}(B)I_t^{(h)}$.
2. Estimate the parameters of the ARIMA model, $\hat{\psi}(B)$, using the series $\hat{y}_t$, and calculate the residuals
$$\hat{e}_t = \hat{y}_t - \sum_{i=1}^{t-1} \hat{\psi}_i \hat{e}_{t-i} = \hat{\psi}^{-1}(B)\hat{y}_t = \hat{\pi}(B)\hat{y}_t.$$
3. Estimate the parameters of the transfer function of the intervention using the relation between these residuals and the parameters of the transfer function, via (4.10) and assuming that $\hat{\psi}(B) \approx \psi(B)$. Specifically, use the regression equation:
$$\hat{e}_t = v(B)\hat{\pi}(B)I_t^{(h)} + a_t,$$
with parameters v_i and regressor $x_t = \hat{\pi}(B)I_t^{(h)}$. With the new estimates $\hat{v}_i$, go to step 1.

This three-step procedure is repeated until convergence. In practice, we may begin with $\hat{v}^{(0)}(B) = 0$ and $\hat{y}_t = z_t$.

The modeling procedure discussed above also applies to the multivariate time series with an intervention. However, instead of using VARMA models, one often employ a VAR model for the uncontaminated series $\boldsymbol{y}_t$ to simplify the estimation, especially when the dimension k of $\boldsymbol{z}_t$ is large. In addition, one also uses $\boldsymbol{v}(B) \approx \boldsymbol{w}(B) = \boldsymbol{w}_0 + \boldsymbol{w}_1 B + \cdots + \boldsymbol{w}_m B^m$ to simplify further the estimation. Under these simplifying assumptions, the estimation of a multivariate intervention model can be carried out via an iterative ordinary least squares method.

Example 4.1

We illustrate intervention analysis using the command `VARX` of the **MTS** package in R. The data used are the series of stock indexes in 99 most important stock markets in the world, and we consider two external events represented by two step functions as interventions. The first intervention is on 21 November 2007 (with $h = 2058$), marking the very beginning of the 2008 financial crisis when the US Treasury created a superfund to deal with the subprime mortgage market. The second is on 16

September 2008 (with $h = 2273$), marking the first trading day after the collapse of the investment bank Lehman Brothers on 15 September 2008.

Figures 4.1 and 4.2 show the time plots of the log stock indexes and returns, respectively. As mentioned in Chapter 2, the data contain some outlying returns. See also Figure 4.2. However, for simplicity, we do not deal with outliers here. Since both the auto- and cross-correlations of the daily index returns are small, we assume that the return series are white noise. Note that a step function for the index series corresponds to an indicator variable for the return series. Therefore, we employ the

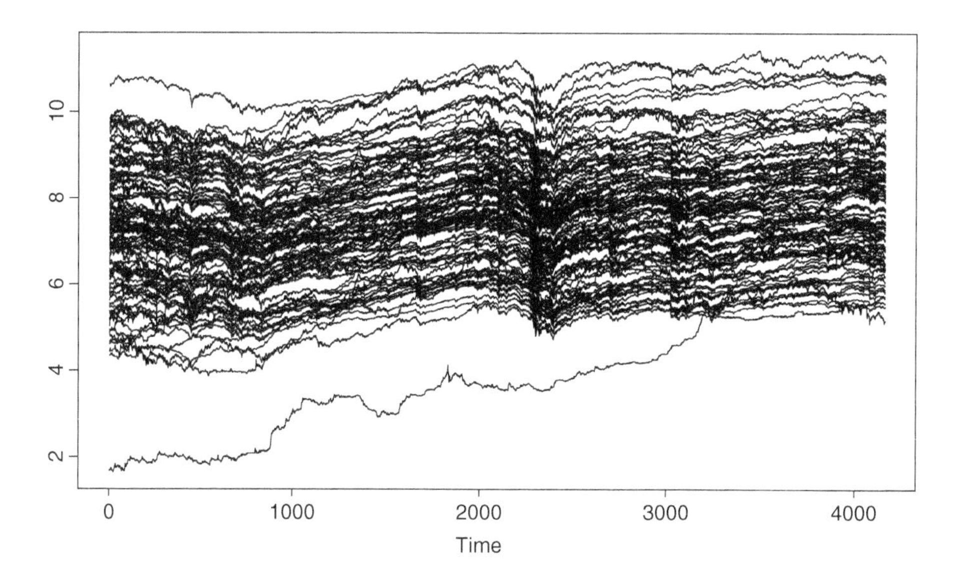

Figure 4.1 Time plots of the log series of world stock indexes in 99 markets.

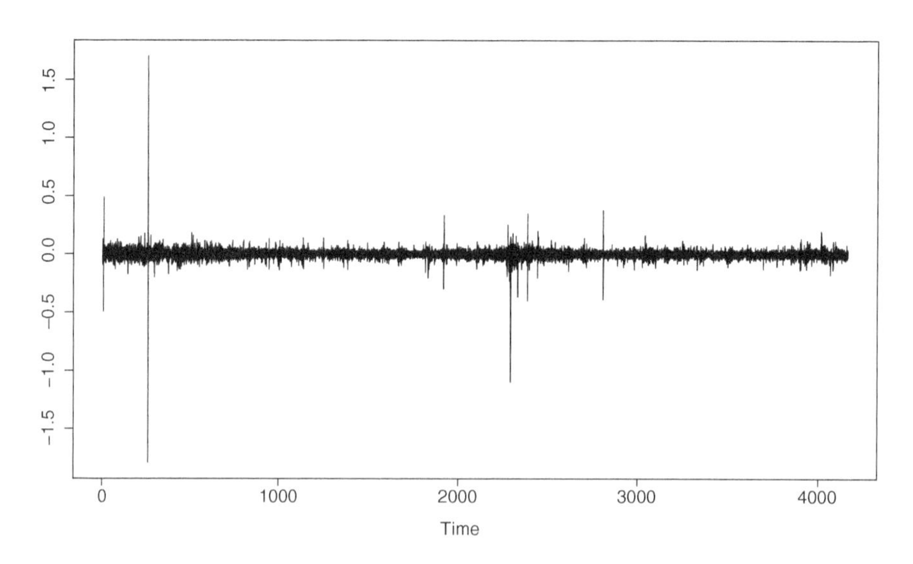

Figure 4.2 Time plots of the return series of world stock indexes in 99 markets.

following intervention model for the 99 return series z_t:

$$z_t = w_1 I_t^{(2057)} + w_2 I_t^{(2272)} + \mu + a_t.$$

Figures 4.3 and 4.4 plot the t-ratios of the estimates $\hat{w}_1$ and $\hat{w}_2$, respectively. From the plots, we see that, as expected, the two interventions have negative impacts on

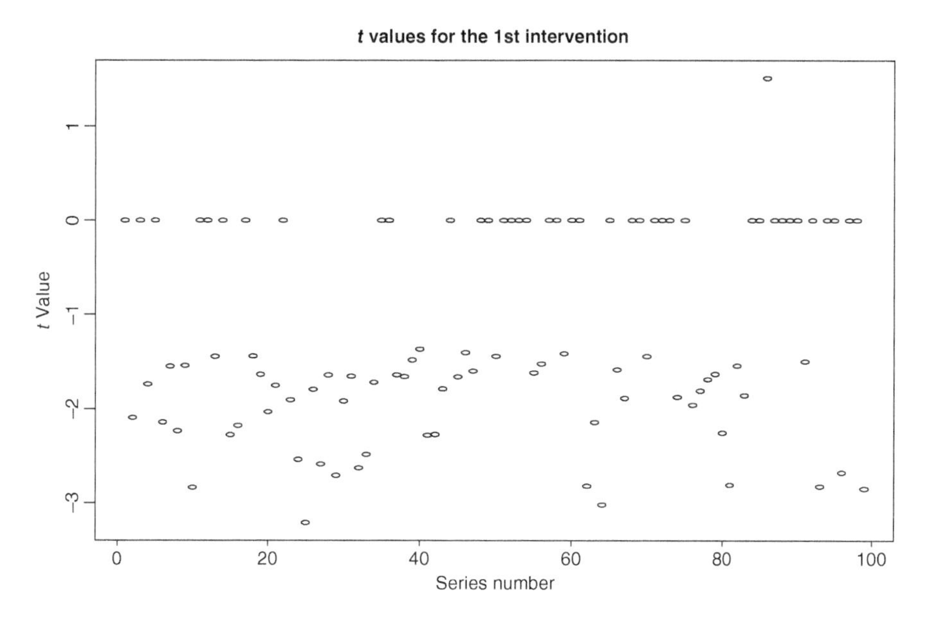

Figure 4.3 t-ratios for the coefficient estimates of the intervention variable on 21 November 2007 for the 99 log return series of the world stock indexes.

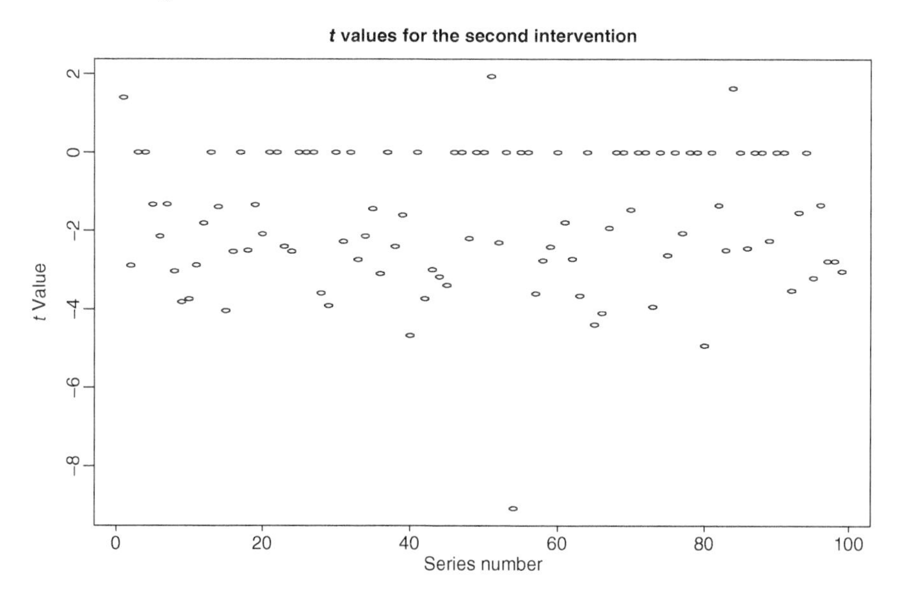

Figure 4.4 t-ratios of the coefficient estimates of the intervention variable on 16 September 2008 for the 99 log return series of the world stock indexes.

the return series. This is particularly so for the second intervention (the collapse of Lehman Brothers) as Figure 4.4 contains many t-ratios less than −2.0.

R commands used for Intervention Analysis:

```
> require(MTS)
>da=read.csv2("Stock_indexes_99world.csv", header=FALSE)
>x=da[,-1] # remove time index
>lx=as.matrix(log(x))
>T=dim(lx)[1]
>k=dim(lx)[2]
>y=diff(lx) # log returns
>sx=matrix(0,T,1)
>sx1=sx
>sx2=sx
>sx1[2058:T,1]=1 ## first step function
>sx2[2273:T,1]=1 ## 2nd step function
>iy1=diff(sx1) ## Indicator
>iy2=diff(sx2)
>iy=cbind(iy1,iy2)
#Estimation of the intervention model for y, the return series,
#with 2 indicators.
#The model is refined to eliminate coefficients with t-ratio
#smaller than 1.3
>r=VARX(y,0,iy,0)
>r1=refVARX(r, thres=1.3)
#The variable coef of Object r1 (r1$coef) contains the coefficient
#estimate. It is a 99-by-2 matrix.
#The variable se.coef of Object r1 (r1$se.coef) contains the standard
# errors of all parameter estimates, including the constant in row 1.
# It is a 3-by-99 matrix.
>s=r1$se.coef[-1,] # remove standard errors of the constant terms.
>M=r1$coef/t(s)
#>tt1=which(abs(M)>=3, arr.ind=T)
>plot(M[,1], main="t ratios for the 1st intervention ", ylab="t-value",
         xlab="series number")
>plot(M[,2], main="t ratios for the 2nd intervention ", ylab="t-value",
         xlab="series number")
```

■

4.2 ESTIMATION OF MISSING VALUES

4.2.1 Univariate Interpolation

Missing values of a time series can occur at a single time point or during an interval. Failure of instruments used to collect the data is an example. In this chapter, we assume that missing values occur randomly in the sample. That is, there is no systematic pattern governing the occurrences nor the magnitudes of missing values. Under such an assumption, we discuss ways to estimate the missing values so that traditional statistical methods developed for complete data set can be applied. In some cases, one may treat the missing values as additional parameters of a statistical model and estimates the missing values along with other model parameters.

We begin with a single missing value in a zero-mean stationary time series z_t. Suppose that z_h is missing. Obviously, h cannot be the first or the last time point.

Otherwise, the sample size can be reduced to avoid the missing value. We use the notation $\boldsymbol{z}_{(-h)} = \{z_1, \ldots, z_{h-1}, z_{h+1}, \ldots, z_T\}$ to denote the available data. The subscript $(-h)$ signifies that z_h is missing. To estimate z_h, one can minimize its square estimation error given the data and the model. Let M denote the model for z_t, the estimate of z_h, call $\hat{z}_h$, is obtained by minimizing $E[(z_h - z^*)^2|M, \boldsymbol{z}_{(-h)}]$ over z^*. It turns out that the solution is simply the conditional expectation of z_h given the model and the available data. That is, $\hat{z}_h = E(z_h|M, \boldsymbol{z}_{(-h)})$. This conditional expectation is readily available if the model M is known. For instance, for an ARMA model

$$E(z_h|\boldsymbol{z}_{(-h)}) = -\sum_{j=1}^{\infty} \rho_j^i (z_{h+j} + z_{h-j}), \tag{4.11}$$

where the coefficients ρ_j^i are the inverse autocorrelation coefficients, that are the ACF of the inverse ARMA model. That is, given the ARMA(p, q) model, $\phi(B)z_t = \theta(B)a_t$, the inverse autocorrelation function is the conventional autocorrelation function of the dual or inverse process $\theta(B)z_t = \phi(B)a_t$. These coefficients measure the correlation between observations of the series separated by j periods when we eliminate the effects of all the remaining observations. See Abraham and Ledolter (1984). Eq. (4.11) represents a symmetric filter.

If the parameters of the model M are unknown, or the series is nonstationary, we can estimate the missing value and the model parameters jointly via the intervention analysis discussed in Section 4.1. The estimation algorithm is as follows:

Scalar filter algorithm:

1. Insert an arbitrary value z_h^* for z_h, e.g. $z_h^* = 0$, to have a complete data set, $\boldsymbol{z}^* = (z_1, \ldots, z_{h-1}, z_h^*, z_{h+1}, \ldots, z_T)$.
2. Estimate an intervention model for the completed series, $\boldsymbol{z}^*$, including the indicator variable $I_t^{(h)}$ of Eq. (4.1). Let $\hat{w}_0$ be the estimate of the intervention effect.
3. Estimate the missing value by

$$\hat{z}_h = E(z_h|\boldsymbol{z}_{(-h)}) = z_h^* - \hat{w}_0. \tag{4.12}$$

In the prior algorithm, we entertain the model

$$z_t^* = w_0 I_t^{(h)} + \psi(B)a_t = w_0 I_t^{(h)} + z_t. \tag{4.13}$$

Taking expectations at time $t = h$ conditional on the values $\boldsymbol{z}_{(-h)}$, we have

$$z_h^* = w_0 + E(z_h|\boldsymbol{z}_{(-h)}),$$

which is Eq. (4.12).

It is interesting to see that the estimation of the missing value does not depend on the arbitrary initial value z_h^*. To illustrate this, suppose the z_t follows a random walk. The intervention model is

$$z_t^* = w_0 I_t^{(h)} + y_t,$$

where z_t^* is used to denote the complete data set and $(1 - B)y_t = a_t$ under the random walk model. Taking the first difference of the prior equation, we have

$$\nabla z_t^* = w_0 x_t + a_t,$$

where $x_t = (1 - B)I_t^{(h)}$, so that $x_h = 1$, $x_{h+1} = -1$, and $x_t = 0$, otherwise. The least squares estimate of w_0 in this regression equation is given by

$$\begin{aligned}\hat{w}_0 &= \frac{\sum_t x_t \nabla z_t^*}{\sum (x_t)^2} = \frac{\nabla z_h^* - \nabla z_{h+1}^*}{2} \\ &= \frac{z_h^* - z_{h-1} - z_{h+1} + z_h^*}{2} = z_h^* - \frac{z_{h-1} + z_{h+1}}{2},\end{aligned}$$

and the estimate of the missing value, via Eq. (4.12), is

$$\hat{z}_h = \frac{z_{h-1} + z_{h+1}}{2},$$

which does not depend on z_h^*. Note that this equation is a particular case of (4.11) because the inverse process of a random walk is an MA(1) process with unit parameter and, therefore, its first-order autocorrelation coefficient is well defined and equal to −0.5. Thus, the missing value is simply estimated by the average of the observations z_{h-1} and z_{h+1}. The fact that $\hat{z}_h$ does not depend on z_h^* continues to hold when the model parameters are estimated simultaneously using the aforementioned scalar filter algorithm.

This filtering approach can easily be generalized if a block of g observations is missing, say from h and $h + g - 1$. Let $\boldsymbol{z}_{[-h:(h+g-1)]} = (z_1, \ldots, z_{h-1}, z_{h+g}, \ldots, z_T)'$ be the available data. Putting zeros in these g missing values to construct the complete data set $\boldsymbol{z}^*$, and defining g indicator variables $I_t^{(h+i-1)}$ for $i = 1, \ldots, g$, we estimate the intervention model

$$z_t^* = \sum_{i=1}^{g} w_i I_t^{(h+i-1)} + z_t, \tag{4.14}$$

and the missing values are obtained by

$$\hat{z}_{h+i} = z_{h+i}^* - \hat{w}_i, \quad i = 0, \ldots, h + g - 1.$$

This idea can be generalized to any combination of missing values. Of course, the efficacy of the algorithm depends on the random patterns and numbers of missing values. Note that if the series is stationary with a non-zero mean, this average value should be subtracted from all the observations and the interpolation obtained, as explained before, will be added to the sample mean to obtain the original value. With integrated time series the intervention analysis approach can always be used.

4.2.2 Multivariate Interpolation

Given a set of related time series $\boldsymbol{z}_t$, missing values may occur at $\boldsymbol{z}_h$ for the whole vector or for some of its components. One can apply the scalar filter algorithm of the previous section to individual time series separately, but such an approach fails to make use of the interdependence between the series. A better procedure is to use all information among the series in estimating the missing values jointly. Similar to the scalar case, one can replace all missing values with zeros to obtain a complete data set, which is denoted by z^*. Equipped with this complete data, one can apply intervention analysis to estimate the model parameters and the missing values. Here the intervention model would involve some indicator variables constructed based on the locations of missing values.

Consider the case that z_h is completely missing. Then z_h^* is a zero vector and the intervention model is

$$z_t^* = \boldsymbol{w} I_t^{(h)} + z_t,$$

where z_t follows a vector ARMA model. If the model of z_t is known, then we can easily estimate $\boldsymbol{w}$. The estimate of the missing value is similar to that of Eq. (4.12) and is given by

$$\hat{z}_h = z_h^* - \hat{\boldsymbol{w}} = -\hat{\boldsymbol{w}}.$$

Next, consider the case that z_h is only partially missing. For simplicity, we assume that z_t follows a VAR(1) model and the first k_0 components of z_h are missing. Partition the vector $z_t = (z_{1t}', z_{2t}')'$, where z_{1t} is $k_0 \times 1$ and z_{2t} is $(k - k_0) \times 1$, and apply the same to the innovation $\boldsymbol{a}_t = (\boldsymbol{a}_{1t}', \boldsymbol{a}_{2t}')'$. Write the AR(1) coefficient matrix as $\boldsymbol{\phi}_1 = [\phi_{ij}]$. Then, we have

$$z_{1h} = \boldsymbol{\phi}_{11} z_{1,h-1} + \boldsymbol{\phi}_{12} z_{2,h-1} + \boldsymbol{a}_{1h},$$

$$z_{1,h+1} = \boldsymbol{\phi}_{11} z_{1h} + \boldsymbol{\phi}_{12} z_{2h} + \boldsymbol{a}_{1,h+1}.$$

Therefore, the missing values are related to the observed data via

$$\begin{bmatrix} \boldsymbol{\phi}_{11} z_{1,h-1} + \boldsymbol{\phi}_{12} z_{2,h-1} \\ z_{1,h+1} - \boldsymbol{\phi}_{12} z_{2h} \end{bmatrix} = \begin{bmatrix} z_{1h} \\ -\boldsymbol{\phi}_{11} z_{1h} \end{bmatrix} + \begin{bmatrix} -\boldsymbol{a}_{1h} \\ \boldsymbol{a}_{1,h+1} \end{bmatrix}. \quad (4.15)$$

Equation (4.15) assumes the form of a multivariate multiple linear regression in which the response variable is known, the explanatory variable is the matrix $[\boldsymbol{I}, -\boldsymbol{\phi}_{11}']'$, and the unknown parameters are in z_{1h}. Note that $-\boldsymbol{a}_{1h}$ has the same distribution as $\boldsymbol{a}_{1h}$. Consequently, z_{1h} can be estimated by the least squares method provided that $\boldsymbol{\phi}_1$ is known. Specifically, if we rewrite Eq. (4.15) as

$$\boldsymbol{Y} = \boldsymbol{X} z_{1h} + \boldsymbol{e},$$

then we have

$$\hat{z}_{1h} = (\boldsymbol{X}'\boldsymbol{X})^{-1}(\boldsymbol{X}'\boldsymbol{Y}).$$

In practice, $\boldsymbol{\phi}_1$ is unknown so that one can use an iterative algorithm to obtain estimates of z_{1h} and $\boldsymbol{\phi}_1$ jointly. Thus, the interpolation approach to missing values takes into account all the relationships among the series.

In the literature, the problem of missing values has been widely studied. Time series analysis is no exception. In general, there are two approaches to handle missing values. One approach is the expectation-maximization (EM) algorithm of Dempster et al. (1977) and the other is Markov chain Monte Carlo (MCMC) methods. The interpolation methods discussed in the section belong to EM algorithm. Readers are referred to Tsay (2014, chapter 6) for additional discussion of missing values in vector time series.

Example 4.2

We illustrate the estimation of missing values using the series of imports and exports of the United States and China; see Chapter 3. We focus on the first differenced series $\boldsymbol{z}_t$ and assume that, for illustrative purpose, $\boldsymbol{z}_{125}$ is missing. The observed value is $\boldsymbol{z}_{125} = (0.04149, 0.0101, 0.01314, 0.01496)'$. First, we consider the first component z_{1t} and apply the scalar filter algorithm to estimate the missing value using an ARMA(3,1) model. In this scalar case, the estimate of missing value is 0.00495.

Next, we use the four series jointly and apply the multivariate filter algorithm with a VAR(3) model to estimate $\boldsymbol{z}_{125}$. The estimated missing value for $z_{1,125}$ is 0.0384, which is much closer to the observed value. This simple illustration shows that it pays to use multivariate interpolation when the time series are related.

R analysis for missing values:

```
> require(MTS)
> da <- read.csv("m-expimpcnus.csv") # data in natural log
##Mising value estimation example 4.2
> da=da[,-1] # remove the time index
> da=as.ts(da)
> xt=diff(da)
> T=dim(xt)[1]
#missing value is estimated at position tm for the first series
# using an arma(3,1) model.
> tm=125
> iv=matrix(0,T,1)
> iv[tm]=1
> xmis=xt
> xmis[tm,1]=0
> ou1=arima(xmis[,1],order=c(3,0,1),xreg=iv)
> ou1$coef
       ar1        ar2        ar3        ma1   intercept        iv
 0.2569381  0.2881239  0.2053035 -0.807763   0.010267 -0.00495
> xt[tm,1]
     expcn
0.04148874
#The true value, 0.04148874, is estimated as 0.004949534.
#A multivariate estimation with a VAR(3) model is used.
> xmis[tm,]=c(0,0,0,0)
> rr=VARX(xmis,3,iv,0,output=F) # output not shown
> rr$beta
            [,1]
expcn -0.03837700
expus -0.01457980
impcn -0.02445329
impus -0.01323786
 > xt[tm,]
     expcn      expus      impcn      impus
0.04148874 0.01008952 0.01317370 0.01496092
```

■

4.3 OUTLIERS IN VECTOR TIME SERIES

Aberrant observations or outliers are common in time series applications. They occurred for various reasons including unanticipated external events such as failure in instruments, strikes, and outbreaks of virus. These aberrant data points appear randomly at unknown time points, and if overlooked, they may lead to model misspecification, biased parameter estimates, and suboptimal forecasts. It is, therefore, important in time series analysis to detect and handle the heterogeneity caused by such aberrant observations.

Detecting outliers in time series analysis serves multiple purposes. First, the existence of an outlier at time $t = h$ may lead to the discovery of unusual event that was relevant, but overlooked before. Second, by studying the impact of a detected outlier, one can improve model estimation and forecasting. However, outlier detection is complicated because the number, the locations, and the types of outliers are unknown *a priori* in addition to the fact that the model is also unknown.

In this section, we study how to identify the occurrence of an unusual event by examining the trace it left behind in the data. We focus mainly on two types of outliers, namely additive outliers (AO) and level shift. Additive outliers represent aberrant observations caused by an external event such as a measurement or recording error. Level shifts are caused by events that have a permanent impact on the time series under study.

We also consider other types of outliers including innovative outliers, transitory changes, and ramp shifts. The impacts of these three types of outliers on time series analysis tend to be relatively minor compared with those of AOs and level shifts. Transitory changes denote outliers whose impact on the time series only lasts for a few time periods, whereas innovative outliers represent unusual innovations, i.e. outliers in the a_t series. Outliers can occur in stationary or nonstationary series, but the ramp shifts are only useful in modeling integrated series.

4.3.1 Multivariate Additive Outliers

We say that at time $t = h$ a vector time series has a multivariate additive outlier (MAO), if the value of the series at that time point is contaminated by some external event. As before, let $\boldsymbol{z}_t$ be the observed vector time series and $\boldsymbol{y}_t$ the uncontaminated one. Then, an MAO at time $t = h$ can be defined as

$$\boldsymbol{z}_t = \begin{cases} \boldsymbol{y}_t, & t \neq h, \\ \boldsymbol{y}_t + \boldsymbol{\omega}_A, & t = h, \end{cases}$$

where $\boldsymbol{\omega}_A = (\omega_{1A}, \dots, \omega_{kA})'$ denotes the size of the outlier. If $k = 1$, we have a scalar time series, the outlier is simply an AO. Assume, for simplicity, that $E(\boldsymbol{y}_t) = \boldsymbol{0}$ and $\boldsymbol{y}_t$ follows the VAR(p) model

$$\boldsymbol{\Phi}(B)\boldsymbol{y}_t = \boldsymbol{a}_t, \tag{4.16}$$

with MA representation $\boldsymbol{y}_t = \boldsymbol{\Psi}(B)\boldsymbol{a}_t$. Then, using the indicator variable $I_t^{(h)}$, we have the following model

$$\boldsymbol{z}_t = \boldsymbol{\omega}_A I_t^{(h)} + \boldsymbol{\Psi}(B)\boldsymbol{a}_t = \boldsymbol{\omega}_A I_t^{(h)} + \boldsymbol{y}_t \tag{4.17}$$

This equation is in the form of an intervention model but here the time index h is unknown. Applying the AR operator to Eq. (4.17), we have

$$\boldsymbol{\Phi}(B)\boldsymbol{z}_t = \boldsymbol{\Phi}(B)\boldsymbol{\omega}_A I_t^{(h)} + \boldsymbol{a}_t. \tag{4.18}$$

Equations (4.17) and (4.18) are interchangeable: either of them can be used to define the MAO.

4.3.1.1 Effects on Residuals and Estimation An MAO leaves certain trace in the residual series of z_t that can be used to detect the outlier. For simplicity in discussion, suppose that the true parameters of the VAR model are known. Then, from Eq. (4.18), we have

$$\boldsymbol{e}_t = \boldsymbol{X}_t^A \boldsymbol{\omega}_A + \boldsymbol{a}_t, \tag{4.19}$$

where $\boldsymbol{e}_t = \boldsymbol{\Phi}(B)\boldsymbol{z}_t$ are the residuals of the observed time series and $\boldsymbol{X}_t^A$ is a square matrix given by $\boldsymbol{X}_t^A = \boldsymbol{\Phi}(B)I_t^{(h)}$. Using $B^j\, I_t^{(h)} = \boldsymbol{I}_{t-j}^{(h)} = I_t^{(h+j)}$, it is easy to see that

$$\boldsymbol{X}_t^A = \begin{cases} \boldsymbol{0}, & \text{if } t < h, \\ \boldsymbol{I}_k, & \text{if } t = h, \\ -\boldsymbol{\Phi}_j, & \text{if } t = h + j, \text{ with } j = 1, \ldots, p, \\ \boldsymbol{0}, & \text{if } t > h + p, \end{cases}$$

where $\boldsymbol{I}_k$ is the $k \times k$ identity matrix. Consequently, Eq. (4.19) implies that an MAO at $t = h$ affects $p + 1$ residuals of z_t starting at $t = h$. Specifically,

$$\boldsymbol{e}_{h+j} = -\boldsymbol{\Phi}_j \boldsymbol{\omega}_A + \boldsymbol{a}_{h+j}, \quad j = 0, \ldots, p \tag{4.20}$$

with $\boldsymbol{\Phi}_0 = -\boldsymbol{I}_k$. Equation (4.20) shows that an AO at $t = h$ has a complicated impact on the residuals $\boldsymbol{e}_t$ of the series $\boldsymbol{z}_t$. In theory, the MAO affects $\boldsymbol{e}_t$ for $t = h, \ldots, h + p$. However, it is possible that $\boldsymbol{\Phi}_j \boldsymbol{\omega}_A = \boldsymbol{0}$ for some j. In this case, the MAO does not affect $\boldsymbol{e}_{h+j}$. Also, if $\boldsymbol{\omega}_A$ is the ith unit vector, i.e. $\omega_{jA} = 1$ if $j = i$ and $\omega_{jA} = 0$, otherwise, then the product $-\boldsymbol{\Phi}_j \boldsymbol{\omega}_A$ is the ith column of $-\boldsymbol{\Phi}_j$, indicating that an MAO that only happens in a single series can affect all the other time series.

These results can easily be generalized to a VARMA process via its AR representation. For $\boldsymbol{\pi}(B)\boldsymbol{y}_t = \boldsymbol{a}_t$, provided that $\boldsymbol{\pi}(B)$ is known, the observed residuals are $\boldsymbol{e}_t = \boldsymbol{\pi}(B)\boldsymbol{z}_t$ and Eq. (4.20) becomes

$$\boldsymbol{e}_{h+j} = -\boldsymbol{\pi}_j \boldsymbol{\omega}_A + \boldsymbol{a}_{h+j}, \quad j = 0, 1, \ldots,$$

where $\boldsymbol{\pi}_0 = -\boldsymbol{I}_k$. Consequently, in theory, an MAO at $t = h$ for a vector time series $\boldsymbol{z}_t$ may affect all of its residuals $\boldsymbol{e}_t$ for $t \geq h$. In the presence of multiple MAOs, effect on the residuals of the observed vector time series, $\boldsymbol{z}_t$, can become complicated.

The global effect of an AO or an MAO in a time series is to bias its sample ACFs toward zero and, depending on the components of $\boldsymbol{\omega}_A$, also bias the cross-correlations toward zero. Therefore, parameter estimates of a model for the series are also biased. We illustrate the effect using the simple case of a scalar zero-mean AR(1) process with an AO of size ω_A at time h. The least squares estimate of the AR parameter is $\hat{\phi} = \sum(z_t - \bar{z})(z_{t-1} - \bar{z}) / \sum(z_t - \bar{z})^2$, where $\bar{z}$ denotes the sample mean. As $z_h = y_h + \omega_A$, and $z_t = y_t$, $t \neq h$, then $\bar{z} = \bar{y} + \frac{1}{T}\omega_A$, and assuming T large, so that $\omega_A/T \simeq 0$ and $\bar{z} \simeq 0$, we have

$$\hat{\phi} = \frac{\sum y_t y_{t-1} + \omega_A(y_{h-1} + y_{h+1})}{\sum y_t^2 + \omega_A^2 + 2\omega_A y_h}.$$

Therefore, the size of the outlier appears linearly in the numerator, but quadratically in the denominator, suggesting that when ω_A is very large the value of $\hat{\phi}$ will be small. This result can be generalized to the vector case depending on the values $\boldsymbol{\phi}_j \boldsymbol{\omega}_A$. In general, the effect of the AO decreases with the sample size and the magnitude of the outlier.

4.3.2 Multivariate Level Shift or Structural Break

Level shift (LS), also known as structural break, is another type of disturbance that can significantly affect a time series. We say that a vector time series has undergone a level shift at time $t = h$ if it follows the model:

$$\boldsymbol{z}_t = \boldsymbol{\omega}_L S_t^{(h)} + \boldsymbol{\psi}(B)\boldsymbol{a}_t = \boldsymbol{\omega}_L S_t^{(h)} + \boldsymbol{y}_t, \tag{4.21}$$

where $S_t^{(h)}$ is a step-function defined in Eq. (4.2) and $\boldsymbol{\omega}_L = (\omega_{1L}, \ldots, \omega_{kL})'$ is the size of the shift. The values of the observed vector series are related to the uncontaminated ones via

$$\boldsymbol{z}_t = \begin{cases} \boldsymbol{y}_t, & t < h, \\ \boldsymbol{y}_t + \boldsymbol{\omega}_L, & t \geq h. \end{cases}$$

If the vector process is stationary with mean $\boldsymbol{\mu}$, a level shift makes the series non-stationary, since the expectation of each observation is $\boldsymbol{\mu}$ for $t < h$ and $\boldsymbol{\mu} + \boldsymbol{\omega}_L$ for $t \geq h$.

4.3.2.1 *Effects on Residuals and Estimation*

A LS at time h of a time series $\boldsymbol{z}_t$ can be written as

$$\boldsymbol{\pi}(B)\boldsymbol{z}_t = \boldsymbol{\pi}(B)\boldsymbol{\omega}_L S_t^{(h)} + \boldsymbol{a}_t.$$

Therefore, the residuals $\boldsymbol{e}_t = \boldsymbol{\pi}(B)\boldsymbol{z}_t$ become

$$\boldsymbol{e}_t = \boldsymbol{X}_t^{LS}\boldsymbol{\omega}_L + \boldsymbol{a}_t, \tag{4.22}$$

where the square matrix $\boldsymbol{X}_t^{LS}$ is now given by

$$\boldsymbol{X}_t^{LS} = \begin{cases} \boldsymbol{0}, & t < h, \\ \boldsymbol{\pi}(B)S_t^{(h)} = [\boldsymbol{I}_k - \sum_{i=1}^{j} \boldsymbol{\pi}_i], & t = h + j, \quad j \geq 0 \end{cases} \tag{4.23}$$

This expression shows that all the residuals after a LS are affected, but the amount depends on the model and the distance between the time indexes h and T, the sample size. As a matter of fact, if $h = T$, there is no difference between a LS and an AO.

A notable impact of a LS to a scalar time series is that it moves the series toward a unit-root process. To illustrate, we consider a LS at h of a stationary scalar time series z_t, which follows an AR(1) model with mean zero. It can be shown that the estimate of the AR parameter can be written as

$$\widehat{\phi} = \frac{\widehat{\phi}_0 + T^{-1}S_3 + T^{-1}\tilde{\omega}_L^2(T - h)}{1 + T^{-1}2S_4 + T^{-1}\tilde{\omega}_L^2(T - h + 1)},$$

where $\tilde{\omega}_L = \omega_L/s_y$ and $\tilde{y}_t = y_t/s_y$ with s_y being the sample standard error of y_t, $\widehat{\phi}_0 = \sum y_t y_{t-1} / \sum y_t^2$ is the estimate obtained from the uncontaminated series y_t, $S_3 = \tilde{\omega}_L(\tilde{y}_{h-1} + \tilde{y}_h + 2\tilde{y}_{h+1} + \cdots + 2\tilde{y}_T)$, and $S_4 = \tilde{\omega}_L \sum_{j=0} \tilde{y}_{h+j}$. If $T - h$ is not very small, we have

$$\widehat{\phi} \to 1 \quad \text{as} \quad \tilde{\omega}_L \to \infty.$$

This result implies that so long as the LS occurs not too close to the end of the sample, the lag-1 sample ACF of the observed series z_t will approach 1 if the size of the shift is large. This result continues to hold for other sample ACFs. Consequently, a LS that occurs sufficiently far away from the end of the series tends to bias the sample ACFs toward those of a unit-root process.

For the vector time series, a large multivariate level shift (MLS) sufficiently far away from the end of the series would move the series not only to unit-root stationarity, but also toward a cointegrated system. To see this, let $\boldsymbol{u}$ be a vector orthogonal to $\boldsymbol{\omega}_L$ such that $\boldsymbol{u}'\boldsymbol{\omega}_L = \boldsymbol{0}$. Then, by Eq. (4.21), we have

$$w_t \equiv \boldsymbol{u}'\boldsymbol{z}_t = \boldsymbol{u}'\boldsymbol{y}_t,$$

which is a stationary time series. In general, for a given $\boldsymbol{\omega}_L$, we have $k-1$ such orthogonal vectors, implying that a large level shift occurred far away from end of the sample would lead $\boldsymbol{z}_t$ to behave as a cointegrated system with $k-1$ cointegrating vectors. In summary, an MLS can have profound impacts on a stationary vector time series.

4.3.3 Other Types of Outliers

Next, we consider three other types of outliers. They are multivariate innovative outlier, multivariate transitory change (MTC), and multivariate ramp change.

4.3.3.1 Multivariate Innovative Outliers A multivariate innovative outlier (MIO) in a vector series denotes that the innovation $\boldsymbol{a}_t$ is affected by certain unknown event at time $t = h$. Since the innovations represent one-step ahead prediction errors of a time series, we can say that an MIO is an unusual increment in the prediction errors of the series at time index h. The model for an MIO of size $\boldsymbol{\omega}_I$ at time h is

$$\boldsymbol{z}_t = \boldsymbol{\Psi}(B)[\boldsymbol{\omega}_I I_t^{(h)} + \boldsymbol{a}_t], \tag{4.24}$$

or in the VAR(∞) representation

$$\boldsymbol{\pi}(B)\boldsymbol{z}_t = \boldsymbol{\omega}_I I_t^{(h)} + \boldsymbol{a}_t.$$

As before, letting $\boldsymbol{e}_t = \boldsymbol{\pi}(B)\boldsymbol{z}_t$ be the residuals of the observed data, we have

$$\boldsymbol{e}_t = \boldsymbol{\omega}_I I_t^{(h)} + \boldsymbol{a}_t = \boldsymbol{X}_t^I \boldsymbol{\omega}_I + \boldsymbol{a}_t, \tag{4.25}$$

where

$$\boldsymbol{X}_t^I = \boldsymbol{I}_k I_t^{(h)}.$$

Therefore, $\boldsymbol{e}_h = \boldsymbol{\omega}_I + \boldsymbol{a}_h$ and $\boldsymbol{e}_t = \boldsymbol{a}_t$ if $t \neq h$. This implies that an MIO only affects a single residual at the time of occurrence. Note that if $\boldsymbol{z}_t$ is a white noise series, then an MIO is equivalent to an MAO.

Using $\boldsymbol{\Psi}(B)\boldsymbol{y}_t = \boldsymbol{a}_t$ for the uncontaminated series $\boldsymbol{y}_t$, we can rewrite Eq. (4.24) as

$$\boldsymbol{z}_t = \boldsymbol{y}_t + \boldsymbol{\Psi}(B)\boldsymbol{\omega}_I I_t^{(h)}. \tag{4.26}$$

Therefore, the relationship between $\boldsymbol{z}_t$ and $\boldsymbol{y}_t$ is

$$\boldsymbol{z}_t = \begin{cases} \boldsymbol{y}_t, & t < h, \\ \boldsymbol{y}_t + \boldsymbol{\Psi}_j \boldsymbol{\omega}_I, & t = h + j, \quad j \geq 0, \end{cases} \tag{4.27}$$

where $\boldsymbol{\Psi}_0 = \boldsymbol{I}_k$. Consequently, an MIO affects the observed series $\boldsymbol{z}_t$ for all $t \geq h$. Specifically, we observed that: (1) for $t = h$, the observation is contaminated by the size of the outlier; (2) for stationary vector series, the effects of an MIO on the observations would decrease as t moves far past h, because $\boldsymbol{\Psi}_i \to \boldsymbol{0}$ as $i \to \infty$; (3) for a unit-root nonstationary series, the effects of an MIO can be complex depending on the order of cointegration of the series. An MIO can introduce different configurations for the marginal models of individual components. The implications depend on the vector model, the outlier size and the interaction between both (see Tsay et al., 2000). In some cases, it leads to a patch of outliers in the marginal component models with patch length determined by various factors. This result can help explain the empirical finding that univariate outlier detection often identifies consecutive outliers. These results demonstrate the advantages of studying outliers in a multivariate framework.

For illustration, consider the effect of an MIO on a vector MA(1) model $\boldsymbol{y}_t = (\boldsymbol{I} - \boldsymbol{\Theta}_1 B)\boldsymbol{a}_t$ with $|\boldsymbol{\Theta}_1| = 0$. This vector MA(1) model can occur in practice, especially when the dimension k is large and many of the elements in $\boldsymbol{\Theta}_1$ are zeros. If an MIO occurs at time h with size $\boldsymbol{\omega}_I$, which belongs to the right null space of $\boldsymbol{\Theta}_1$, then as $\boldsymbol{\Theta}_1\boldsymbol{\omega} = \boldsymbol{0}$, the zero vector, the outlier only affects a single observation at time h and, hence, it is equivalent to an MAO. For higher order models, the differences between multivariate and univariate cases can be more substantial.

4.3.3.2 Transitory Change Another interesting type of outlier in vector time series analysis is the MTC defined as

$$\boldsymbol{z}_t = \frac{\boldsymbol{\omega}_{TC}}{1 - \delta B} I_t^{(h)} + \boldsymbol{\Psi}(B)\boldsymbol{a}_t, \tag{4.28}$$

where $0 \leq \delta \leq 1$ represents the discounting rate of the outlier impact. If $\delta = 1$, we have $\nabla^{-1} I_t^{(h)} = S_t^{(h)}$ so that the outlier becomes an MLS and if $\delta = 0$, we have an MAO. In practice, this type of outlier is used by fixing the value of δ (often $\delta = 0.7$) to avoid confusion with an MAO or an MLS. Although this outlier is useful in univariate time series analysis, it is less so in multivariate case. This is because the common value of δ may not be suitable for all components of $\boldsymbol{z}_t$.

4.3.3.3 Ramp Shift An outlier that is of interest in nonstationary time series analysis, especially in I(2) series, is the ramp shift defined as

$$\boldsymbol{z}_t = \boldsymbol{\omega}_R R_t^{(h)} + \boldsymbol{\Psi}(B)\boldsymbol{a}_t,$$

where the ramp variable $R_t^{(h)}$ is given by

$$R_t^{(h)} = \begin{cases} 0, & t < h, \\ t + 1 - h, & t \geq h. \end{cases}$$

The ramp variable introduces a deterministic trend with slopes $\boldsymbol{\omega}_R$ into the vector time series from time h on. The variable is related to the step function and indicator variable as $\nabla R_t^{(h)} = S_t^{(h)}$ and $\nabla^2 R_t^{(h)} = I_t^{(h)}$. A ramp effect is only expected in nonstationary series, and can easily be confused with an MIO in series that require a second-order difference.

4.3.4 Masking and Swamping

Masking appears when an outlier, or group of outliers, is not identified due to the bias in the parameter estimation introduced by the presence of other outliers. Swamping is the opposite effect, where the bias generated by outliers makes some good observations identified as outliers. These effects are especially serious with groups of outliers, but it may happen also with isolated outliers. For instance, consider a scalar AR(1) process with zero mean and unit variance and parameter ϕ. Suppose we have an AO of size w_h at time $t = h$ and let $\hat{\phi}$ be the estimated parameter with the contaminated data, z_t. The estimated residual at this point is

$$\hat{e}_h = z_h - \hat{\phi} z_{h-1} = a_h + w_h + (\phi - \hat{\phi}) y_{h-1}.$$

Because of the outlier, we typically have $\phi > \hat{\phi}$. If $y_{h-1} < 0$, then the effect of the outlier, w_h, can be masked (reduced) by the bias in the parameter estimation. For the next observation, we have

$$\hat{e}_{h+1} = z_{h+1} - \hat{\phi} z_h = a_{h+1} - \hat{\phi} w_h + (\phi - \hat{\phi}) y_h.$$

Therefore, even if the observation z_{h+1} is uncontaminated, its residual could be large if the outlier size $\hat{\phi} w_h$ is substantial. In this situation, the good observation z_{h+1} would be swamped by the outlier.

These effects become more pronounced with a group of consecutive outliers. Suppose that we have outliers at both $t = h$ and $t = h + 1$, with sizes w_h and w_{h+1}, respectively. Then

$$\hat{e}_{h+1} = z_{h+1} - \hat{\phi} z_h = a_{h+1} + w_{h+1} - \hat{\phi} w_h + (\phi - \hat{\phi}) y_h,$$

and

$$\hat{e}_{h+2} = z_{h+2} - \hat{\phi} z_{h+1} = a_{h+2} - \hat{\phi} w_{h+1} + (\phi - \hat{\phi}) y_{h+1}.$$

If $w_{h+1} \simeq \hat{\phi} w_h$, the effect of the outlier at $t = h + 1$ is completely masked. If $-\hat{\phi} w_{h+1}$ is large, observation z_{h+2} is swamped. The same phenomenon can occur in the multivariate case, where the possibilities of these two effects increase for various configurations of parameter matrices and vectors of outlier effects.

4.4 UNIVARIATE OUTLIER DETECTION

Many procedures for univariate outlier detection have been proposed. They are also useful to search for multivariate outliers through projections, to be discussed later. The most commonly used procedure was proposed by Chang et al. (1988), Tsay (1986) and Chen and Liu (1993). It was further improved to make it more robust and able to deal with multiple outliers avoiding masking and swamping by Bianco et al. (2001), Sánchez and Peña (2003), and Galeano et al. (2006), among others. The procedure presented here incorporates these improvements.

Given an observed univariate time series, the detection of possible outliers is carried out in four steps: (1) obtain a robust estimate of the parameters and compute the residuals; (2) identify the locations of the level shifts (LS), if any, and clean the series from the detected effects; (3) identify the locations and sizes of the other outliers; and (4) estimate jointly the model parameters of the series and all detected outlier effects. For simplicity, we only consider LS, AO, and IO in the detecting procedure. Other types of outlier can be easily incorporated if needed.

The procedure is based on estimating the outlier size using the residuals. We have shown (see Eqs. (4.19), (4.25) and (4.22)) that the residuals of a univariate time series affected by an outlier can be written as

$$e_t = x_t^C \omega_C + a_t, \tag{4.29}$$

where C denotes the type of outlier, $C \in (LS, AO, IO)$, ω_C is the outlier size and, as we have shown, for LS, $x_t^{LS} = \pi(B)S_t^{(h)}$, for AO, $x_t^A = \pi(B)I_t^{(h)}$, and for IO, $x_t^I = I_t^{(h)}$. Therefore, the estimated outlier size is $\widehat{\omega}_C = \sum \widehat{e}_t x_t^C / \sum (x_t^C)^2$ with standard deviation $s(\widehat{\omega}) = \widehat{\sigma}_e / \sqrt{\sum (x_t^C)^2}$, and the likelihood ratio (LR) test statistic for $H_0 : \omega = 0$ is $\lambda_h = \widehat{\omega}_C / s(\widehat{\omega}_C)$, where the subscript h is used to signify that the test statistic is for time index h. Details of the detection procedure are given below.

1. *Initial robust estimation of model parameters and residuals*: An initial model for the series is identified and its parameters estimated, $\widehat{\pi}^{(0)}(B)z_t = \widehat{e}_t^{(0)}$. The residual variance of this first model is estimated by the normalized *mad*, $\widehat{\sigma}_e^{(0)} =$ 1.4815∗Median($\widehat{e}_t^{(0)}$), and all residuals with $|\widehat{e}_t^{(0)}| > 2.5\widehat{\sigma}_e$ are identified as potential outliers. Let Ω denote the set of time indexes of these suspicious residuals. The following initial robust model is then estimated

$$z_t = \sum_{t \in \Omega} w_i I_t^{(h_i)} + \pi(B)^{-1} a_t. \tag{4.30}$$

   Let $\widehat{\pi}_R(B)$ be the AR representation with the parameter estimates of Eq. (4.30). The residuals are recalculated via $\widehat{e}_t = \widehat{\pi}_R(B)z_t$, and their mad, $\widehat{\sigma}_e$, is updated.
2. *Search for Level Shifts*. The search for outliers starts with level shifts because they affect all residuals after time index h. The possible effect of a level shift is estimated at every time index h using Eq. (4.29). Let h_1 be the time index at which $|\lambda_h|$ attains its maximum. If $|\lambda_{h_1}| > c_1$, where c_1 is a given critical value, then a potential level shift at h_1 is detected. The size of the LS at h_1 is estimated jointly with model parameters via

$$z_t = w_L S_t^{(h_1)} + \pi(B)^{-1} a_t. \tag{4.31}$$

   One can then test again the significance of w_L. If the estimate $\widehat{w}_L$ is indeed statistically significant, then a LS at h_1 is confirmed, the series is adjusted via $z_t^c = z_t - \widehat{w}_L S_t^{(h_1)}$, and the residuals are recomputed using the new parameter estimates in Eq. (4.31). The search for LS is repeated with the newly adjusted series $z_t^{(c)}$. If the size of a new level shift is significant, a new adjusted series is obtained via $z_t^c = z_t - \sum \widehat{w}_i S_t^{(h_i)}$ and the residuals recalculated accordingly. The detection is repeated again until no significant LS is detected.
3. *Search for AO or IO*. At each time index h, the sizes of these two possible outliers, A and I, are estimated using Eq. (4.29). The corresponding estimates of the effects are $\widehat{\omega}_A = (\widehat{e}_h - \pi_1 \widehat{e}_{h+1} - \cdots - \pi_{T-h}\widehat{e}_T)/\rho_A^2$ for AO, where

$\rho_A^2 = (1 + \pi_1^2 + \cdots + \pi_{T-h}^2)$, and $\hat{\omega}_I = \hat{e}_h$ for IO. At each time h, the likelihood ratio test (LRT) statistics $\lambda_h^C = \hat{\omega}_i / s(\hat{\omega}_i)$ for $C = AO, IO$, where the standard error is computed following the linear regression model. Let g be the time point such that $|\lambda_g^A| = \max_h |\lambda_h^A|$ and m be the time point such that $|\lambda_m^I| = \max_h |\lambda_h^I|$. Let $\lambda_{\text{Max}} = \max\{|\lambda_g^{AO}|, |\lambda_m^{IO}|\}$. If $\lambda_{\text{Max}} > c_2$, where c_2 is a prespecified critical value, we conclude that an outlier is detected for the series. Otherwise, no outliers exist in the series. The effect of a detected outlier is corrected in the series, for AO $\hat{z}_g = z_g - \hat{\omega}_A$ and for IO the residuals are corrected by $\hat{e}_n = \hat{e}_n - \hat{\omega}_I$. The search for AO or IO is continued until no more outliers are detected.

4. *Final joint estimation of outlier effects and parameters*. When the detection procedure ends, all the detected outlier effects and model parameters are estimated jointly. Assume that level shifts have been detected at times $t_1, t_2, \ldots, t_u$, and outliers at times $t_1^*, t_2^*, \ldots, t_v^*$. The final model becomes

$$z_t = \sum_{j=1}^{u} \omega_j S_t^{(t_i)} + \sum_{j=1}^{v} \omega_j v_{t_i^*}(B) I_t^{(t_i^*)} + \psi(B) a_t,$$

where $v_{t_i^*}(B) = 1$ if the outlier is an AO and $v_{t_i^*}(B) = \psi(B)$ if it is an IO. The effects that are not statistically significant are eliminated one by one and the model is re-estimated until all the effects and estimated coefficients are significant.

The critical values c_1 and c_2 are usually taken to be in the interval [3, 3.5] and were obtained by simulation in Chang et al. (1988). These cutoffs must depend on the sample size because the test statistics used are the maximum value over each time index. If the statistics λ_h were independent, then

$$\Pr(\lambda_1 \leq a, \ldots, \lambda_T \leq a) = \Pr(\lambda_1 \leq a)^T = 1 - \alpha,$$

where α is the type-I error. For instance, if we want the probability of no false outlier detection to be 0.95 or, $\alpha = 0.05$, we must choose the critical value $a = z_{0.95^{1/T}}$, where z_v is the $100v$ quantile of the standard Gaussian distribution. However, the test statistics λ_i used are dependent, and the previous value a becomes a conservative choice. The dependence among the test statistics makes it hard, if not impossible, to determine proper critical values for c_1 and c_2. In theory, one can choose them using the concept of false discovery rate (FDR); see, for instance, chapter 15 of Efron and Hastie (2016). However, Monte Carlos simulations have shown that such an approach does not provide better results than using the distribution of the maximum.

When applying this detection procedure, it is important to carry out the corrections of level shifts one by one and to estimate their effects jointly. For instance, suppose a series with two LS of sizes 5 and 10. If the second shift is detected first, its size will not be well estimated, due to the presence of the first shift. However, when the two level shifts are identified and their effects are estimated jointly the bias problem does not occur. A similar bias problem can appear with the remaining outliers, although it is normally not as serious as that of level shifts. To avoid the bias and false detections, it is advisable to correct the outliers one by one and always include the detected outliers in a joint estimation. Note that if the series are nonstationary the IO may be confused with other outlying effects such as with level shifts in an I(1) series and with ramp effects in an I(2) process. Before accepting an IO in an integrated time series, one should always check to ensure that it is not masking with another type of outlier.

4.4.1 Other Procedures for Univariate Outlier Detection

There exist some alternative approaches to outlier detection. We briefly mention some of them below. To begin, there is substantial evidence that using the same critical value for all the test statistics can misidentify a level shift as an innovative outlier; see Balke (1993) and Sánchez and Peña (2003). The latter authors showed that the critical values for the test statistic for detecting level shift are different from those for testing additive or innovative outliers. Therefore, it is important to identify level shifts in a series before checking for other types of outlier. Instead of testing for level shift with an intervention model, Bai (1994) proposed a cusum statistic that has several advantages over the LRT intervention model. First, it is not necessary to specify the order of the ARMA model, which can be difficult in the presence of level shifts. Second, the statistic seems to be robust to the presence of other outliers whereas the LRT statistic is not. Galeano et al. (2006) showed in a Monte Carlo study that this approach works better than the LRT. The method is explained in the Appendix.

Next, if a series has several outliers, the parameter estimates can encounter marked biases that, in turn, affect the power of test statistics. Some alternatives to the initial cleaning procedure of the previous section have been studied by Luceño (1998) and Justel et al. (2000). Hendry et al. (2008) proposed a fast way to detect AOs and level shifts using highly saturated intervention models. To simplify the presentation, suppose we are interested in detecting AOs in an AR(p) process. These authors propose to estimate the model via

$$z_t = \sum_{i=1}^{p} \phi_i z_{t-i} + \sum_{j=h_1}^{h_2} w_j I_t^{(j)} + u_t, \quad i = 1, \dots, T, \tag{4.32}$$

where $I_t^{(j)}$ is the indicator variable for time index j and h_1 and h_2 are two positive integers. The procedure works as follows: (1) introduce indicator variables in the first half of the observations and estimate Model (4.32) with $h_1 = 1$ and $h_2 = [n/2]$, which is the integer part of $n/2$. Identify observations with significant $\widehat{w}_j$ estimates and treat those data points as a set of potential outliers; (2) estimate Model (4.32) with $h_1 = [n/2] + 1$ and $h_2 = T$ in the second half of the sample. Augment the set of potential outliers by including data points with significant estimates $\widehat{w}_j$; (3) estimate a model with indicator variables for observations in the set of potential outliers and check the new estimates $\widehat{w}_i$ for significance. In order to control for the large number of tests made in this procedure, the significance level of the t tests for $H_0 : w = 0$ is chosen as $\alpha = 1/T$, so that we expect to find $T\alpha = 1$ false outlier in the sample. An advantage of the method is that it simultaneously takes care of robust parameter estimation and outlier detection, because the impacts of suspicious observations are taken care of. A drawback, however, is that it may fail to identify correctly the number of outliers. Suppose an AO occurs at $t = h$. The information about its size is contained in the p residuals $\widehat{e}_h, \dots, \widehat{e}_{h+p}$ and, therefore, we may find several outliers at time points $h + j$ that are caused by the single outlier at $t = h$. Nevertheless, the procedure provides a fast way to carry out robust estimation to AOs with many time series. For LS, the efficiency is smaller because all the step variables are highly correlated and it becomes more difficult to separate their effects to identify the level shifts.

Example 4.3

We use the package **tsoutliers** to detect outliers in univariate time series. The package is based on the detection procedure of Chen and Liu (1993), which is similar to the one described in Section 4.4. We employ the first two series of hourly AirBox measurements of $PM_{2.5}$ in March 2017, Taiwan. The model selected by the program is an ARIMA(1,0,1) for the first series and an ARIMA(2,0,0) for the second one.

```
> require(tsoutliers)
>x= read.csv("TaiwanAirBox032017.csv")
>x=as.matrix(x)
>x=x[,-1]  ## Remove the row number
>y1=as.ts(x[,1])
>y2=as.ts(x[,2])
>M1=tso(y1)
>M1$outliers
type ind time coefhat tstat
1 TC 14 14 -16.71296 -4.207933
2 LS 86 86 17.30870 6.124887
3 LS 464 464 -10.35496 -4.818682
4 TC 608 608 21.41543 5.225062
5 AO 609 609 20.87007 7.108145
>y1ad=M1$yadj
>ts.plot(y1,main="original and adjusted series", ylab="PM2.5",
      xlab="First series Taiwan AirBox data",pch=0, lty=1)
>lines(y1ad,col="red")
>M2=tso(y2)
>M2$outliers
Eight outliers were detected
```

Thus, five outliers are found in the first series of AirBox data and eight in the second series. The detected outliers of the first series include two level shifts, two temporary level changes, and one AO. Figure 4.5 presents the time plots of the original series and the outlier-adjusted series. ■

4.4.2 New Approaches to Outlier Detection

With advances in high-dimensional statistical analysis, one can use penalized linear regression methods to detect outliers in a more direct way, making the new methods more computationally efficient for handling a collection of many time series. The method used in our study is the Lasso regression, which is briefly mentioned in Chapter 1 and discussed in detail in Chapter 7. Here we use Lasso with step functions to detect level-shifts and, with indicator variables, to detect AOs. The nice feature of Lasso linear regression is that the number of predictors can be larger than the sample size. Consequently, we can consider simultaneously all possible level-shifts or all AOs in a time series.

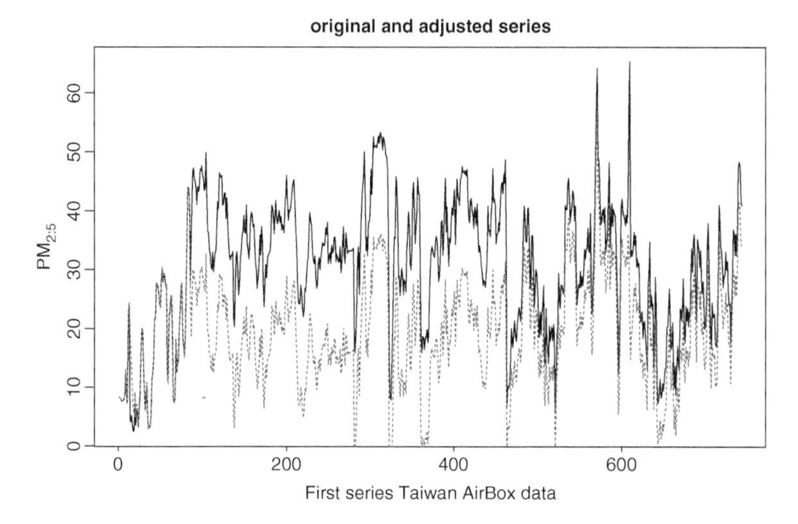

Figure 4.5 Time plots of the first series of AirBox data set. The original series is in black, and the adjusted series after removing the impacts of two level shifts and three other outliers is in dashed red line.

In what follows, we introduce the basic concepts of Lasso regression to detect level-shifts and AOs. For computational simplicity, we assume an AR model for the series under study and use cross-validation to select the Lasso penalty parameter. As shown in Equation (4.23), the residuals of an AR(p) process with a level shift at time $t = h$ can be written as

$$\begin{bmatrix} e_1 \\ \vdots \\ e_{h-1} \\ e_h \\ e_{h+1} \\ \vdots \\ e_T \end{bmatrix} = \omega_h \begin{bmatrix} 0 \\ \vdots \\ 0 \\ 1 \\ 1-\phi_1 \\ \vdots \\ 1-\sum_{j=1}^{T-h}\phi_j \end{bmatrix} + \begin{bmatrix} a_1 \\ \vdots \\ a_{h-1} \\ a_h \\ a_{h+1} \\ \vdots \\ a_T \end{bmatrix}.$$

or, in vector forms, we have

$$\boldsymbol{e} = \omega_h \boldsymbol{x}_h + \boldsymbol{a}.$$

Consequently, as we have seen, we create the predictor $\boldsymbol{x}_h$ and estimate ω_h by LS. Since h is unknown in practice, we can create many predictors $\{\boldsymbol{x}_h | h = j_o, \ldots, T - j_o\}$ where j_o is a small positive integer. The reason for using j_o is obvious. For instance, we cannot have a level shift at $t = 1$ or $t = T$. In the literature, one can perform such estimation at each h. However, with penalized estimation such as Lasso, one can estimate all ω_h simultaneously with the linear regression model

$$\boldsymbol{e} = \sum_{h=j_o}^{T-j_o} \omega_h \boldsymbol{x}_h + \boldsymbol{a}, \tag{4.33}$$

and perform penalized estimation with penalty $\lambda \sum_{h=j_o}^{T-j_o} |\omega_h|$, where λ is the penalty parameter with its value selected by cross-validation.[1] As the predictors are highly correlated and we encounter strong multicollinearity, we propose a two-step procedure. We assume, as in the usual Lasso regression, that the predictors are standardized and the two-step procedure is as follows:

Step 1. Let $\widehat{\omega}_h$ be the Lasso estimates of the parameters in model (4.33). Select those $|\widehat{\omega}_{h_i}|$ which are greater than the 90% quantile of all $|\widehat{\omega}_h|$ as possible candidates for a level shift. Denote the selected time points by $\{h_1, \ldots, h_m\}$, where $m \approx [T/10]$ with T being the sample size.

Step 2. Perform the linear regression

$$\boldsymbol{e} = \sum_{i=1}^{m} \omega_{h_i} \boldsymbol{x}_{h_i} + \boldsymbol{a}. \tag{4.34}$$

Let t_{h_i} be the t-ratio of the least squares estimate of ω_{h_i} in Eq. (4.34). Then, z_t has a level shift at $t = h_i$ with a magnitude $\widehat{\omega}_{h_i}$ if and only if $|t_{h_i}| > c_1$, where c_1 is a critical value. We use $c_1 = 3.5$ based on critical values used in the literature of outlier detection in time series analysis. The series z_t has no level shift if all t_{h_i} are less than c_1 in magnitude.

Finally, a level-shift-adjusted innovation series can be obtained by

$$\tilde{\boldsymbol{e}} = \boldsymbol{e} - \sum_{i \in S} \widehat{\omega}_{h_i} \boldsymbol{x}_{h_i},$$

where $S = \{h_i | \, |t_{h_i}| > c_1\}$ and $\widehat{\omega}_{h_i}$ is the least squares estimates in Eq. (4.34). A level-shift- adjusted time series $\tilde{z}_t$ can then be obtained by filtering $\tilde{\boldsymbol{e}}$ with the AR polynomial $\phi(B)$. In practice, $\phi(B)$ is unknown, but we use the estimates of fitting an AR(p) model to z_t. If needed, one can use a robust estimate of an AR(p) model to mitigate the impact of level shifts and other outliers in z_t or use an iterative method based on the detected level shifts.

Note that one can also use the penalized linear regression method to detect AOs or temporary changes or innovation outliers in a time series. For instance, consider the AO at time point $t = h$, then the $\boldsymbol{x}_h$ would become $(0, \ldots, 0, 1, -\phi_1, \ldots, -\phi_p, 0, \ldots, 0)'$, where 1 is located at $t = h$. In this case the multicollinearity problem is less severe and we can apply the aforementioned two-step procedure for detecting AOs.

Example 4.4

To demonstrate the detection of level shifts and AOs with Lasso we employ the Taiwan AirBox $PM_{2.5}$ series. Due to the special series shown in Chapter 2, we remove series 29 and 70 from this study, because these series only have a few observed $PM_{2.5}$ measurements.

[1]Instead of using cross-validation, one might want to use out-of-sample prediction to select λ in a time series setting.

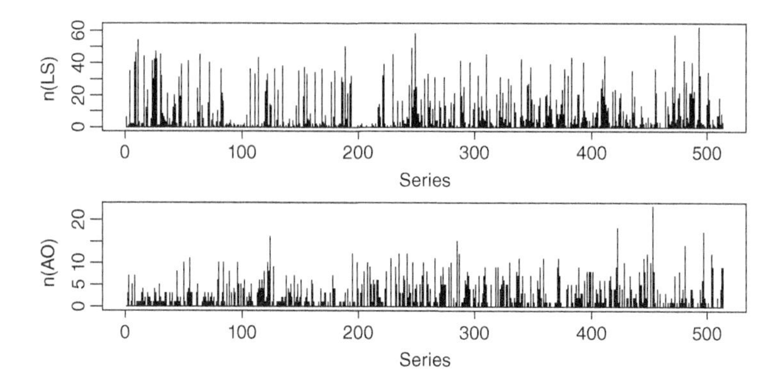

Figure 4.6 AirBox data set: outlier detection via Lasso. The upper plot is the number of level shifts and the lower plot is the number of AOs.

We apply the proposed level shift procedure first to the first two series, as in Example 4.3, and then to all the series. The AR order is $p = 2$ for the first two series in order to use a similar model to compare the results. The number of outliers detected remains at 5. In the first series all the outliers are identified as LS whereas in Example 4.3 two are LS, two TC and an AO. Since TC has not been considered here, the two TC are identified as LS. In the second series, only three AOs are found.

To detect outliers for all time series, we set $p = 5$ to cover different dynamic behaviors of the data. If a series contains detected level shifts, we remove the impact of level shifts based on the proposed procedure. We then use the level-shift-adjusted series to detect AOs. The results are summarized in Figure 4.6. The upper plot shows the number of level shifts of each series, whereas the lower plot shows the number of additive outliers. For the 514 time series considered in our study, there are 5014 detected level shifts and 1524 AOs. The percentages for level shifts and AOs are 1.31% and 0.40%, respectively. From the plot, the maximum number of level shifts detected occurs in Series 495 with 62 level shifts. This phenomenon might be affected by the fact that some aberrant observations are classified as level shifts as we focus on level shifts and AOs only. Overall, the percentage of level shifts or AOs is about 1.71%, which seems reasonable. Finally, the **R** commands used are given below. The command `outlierLasso` is available from the package **SLBDD** associated with the book.

Lasso approach to outlier detection:

```
>air <- read.csv("TaiwanAirBox032017.csv")
>air <- as.matrix(air)
>air <- air[,-1]
>air <- air[, -c(29,70)] # remove series 29 and 70.
>set.seed(11)
>m1 <- outlierLasso(air[,1:2],p=2,crit=3.5)
>m1$nLS
 [1] 5 0
>m1$nAO
[1] 0 3
```

```
>m2 <- outlierLasso(air, p=5, crit=3.5)
>sum(m2$nLS)
 [1] 4600
>sum(m2$nAO)
[1] 1676
> 1676/(514*744)
[1] 0.004382662
> 4600/(514*744)
[1] 0.01202879
```

■

Example 4.5

Next, we compare the iterative detection procedure in **tsoutliers** with the Lasso method using the series of imports and exports of the United States and China; see Chapter 3 for details of the data. We focus on the first differenced series. In this instance, the iterative procedure finds a total of 12 outliers, many in the time indexes from 11 to 15 and from 200 to 205. The models used range from a simple AR(1) to a complex ARIMA(1,0,2) model. The Lasso approach with an AR(5) model detects one level shift and one additive outliers for the first series and a level shift for the third series of the data. The commands used are given below.

```
>da<- read.csv("m-expimpcnus.csv")
>da=da[,-1]
>require(MTS)
>require(tsoutliers)
>require(glmnet)
>da=as.ts(da)
>xt=diff(da)
>r1=tso(xt[,1])
>r2=tso(xt[,2])
>r3=tso(xt[,3])
>r4=tso(xt[,4])
>r1$outliers
type ind time coefhat tstat
1 TC 11  12  -0.1420 -4.308
2 AO 13  14   0.2963  5.255
3 AO 278 279 -0.3341 -6.720
4 AO 314 315 -0.1906 -3.852
>r2$outliers
type ind time coefhat tstat
1 AO 26  27  0.08224  4.103
2 TC 113 114 -0.04902 -3.930
3 TC 200 201 -0.08349 -6.371
4 AO 203 204 -0.10017 -4.742
>r3$outliers
type ind time coefhat tstat
1 TC 201 202 -0.1963 -8.194
2 TC 205 206  0.1019  4.265
```

```
>r4$outliers
type ind time coefhat tstat
1 AO 14  15    0.07118   4.043
2 TC 202 203 -0.12295 -8.917
>set.seed(100)
>m3 <- outlierLasso(xt,p=5,crit=3.5)
> names(m3)
[1] "nAO" "nLS"
> m3$nAO
[1] 1 0 0 0
> m3$nLS
[1] 0 0 0 0
```

■

4.5 MULTIVARIATE OUTLIERS DETECTION

4.5.1 VARMA Outlier Detection

Multivariate outliers can be detected by an iterative procedure similar to that of the univariate case proposed by Tsay et al (2000). First, an initial VARIMA model for the series under study assuming no outliers is built. The parameters of this initial model are estimated by using series adjusted by an initial cleaning procedure of the univariate time series in order to have robust parameter estimates. Let $\hat{\boldsymbol{e}}_t$ be the estimated residuals and $\hat{\boldsymbol{\Pi}}_i$ the estimated coefficients of the AR representation. Then, at each time point the effect of each type of outlier can be estimated as follows. For an MIO at time index h, all information about the outlier is contained in $\hat{\boldsymbol{e}}_h$, and we estimate the outlier size by using $\hat{\boldsymbol{\omega}}_{I,h} = \hat{\boldsymbol{e}}_h$, where the subscript I indicates MIO. For the other types of outliers, the same estimation idea applies, and we shall give details for the MAO case only. In this case, from (4.20), we have

$$\hat{\boldsymbol{e}}_t = \left[\boldsymbol{I} - \sum_{i=1}^{\infty} \hat{\boldsymbol{\Pi}}_i B^i\right] I_t^{(h)} \boldsymbol{\omega}_{A,h} + \boldsymbol{a}_t = \left[I_t^{(h)} - \sum_{i=1}^{\infty} \hat{\boldsymbol{\Pi}}_i I_{t-i}^{(h)}\right] \boldsymbol{\omega}_{A,h} + \boldsymbol{a}_t.$$

Since $\boldsymbol{a}_t$ has in general a full covariance matrix Σ_a, we use generalized least squares (GLS) to estimate $\boldsymbol{\omega}_{A,h}$:

$$\hat{\boldsymbol{\omega}}_{A,h} = \left[\sum_{i=0}^{T-h} \hat{\boldsymbol{\Pi}}_i \Sigma_a^{-1} \hat{\boldsymbol{\Pi}}_i\right]^{-1} \sum_{i=0}^{T-h} \hat{\boldsymbol{\Pi}}_i \Sigma_a^{-1} \hat{\boldsymbol{e}}_{h+i}, \qquad (\hat{\boldsymbol{\Pi}}_0 = -\boldsymbol{I}),$$

and the covariance matrix of the estimator is $\Sigma_{A,h} = \left[\sum_{i=0}^{T-h} \hat{\boldsymbol{\Pi}}_i \Sigma_a^{-1} \hat{\boldsymbol{\Pi}}_i\right]^{-1}$.

To test the significance of a multivariate outlier at time index h, we consider the null hypothesis H_o: $\boldsymbol{\omega}_h = \boldsymbol{0}$ versus the alternative hypothesis H_a: $\boldsymbol{\omega}_h \neq \boldsymbol{0}$. Two test statistics are used. The first test statistic is

$$J_{i,h} = \hat{\boldsymbol{\omega}}_{i,h}' \Sigma_{i,h}^{-1} \hat{\boldsymbol{\omega}}_{i,h},$$

where $i = AO, IO, LS$ depending on the type of outlier. This statistic treats all the components jointly. For a fixed h and assuming that the model is known, $J_{i,h}$ is distributed as a chi-square random variable with k degrees of freedom under the null hypothesis. The second test statistic is the maximum in absolute value of components of $\hat{\boldsymbol{\omega}}_{i,h}$ when $\Sigma_{i,h}$ is known. That is,

$$C_{i,h} = \max_{1 \le j \le k} |\hat{\omega}_{j,i,h}| / \sqrt{\sigma_{j,i,h}},$$

where $\hat{\omega}_{j,i,h}$ and $\sigma_{j,i,h}$ are the jth element of $\hat{\boldsymbol{\omega}}_{i,h}$ and the (j,j)th element of $\Sigma_{i,h}$ respectively.

We define the overall test statistics as

$$J_{\max}(i, h_i) = \max_h J_{i,h}, \quad C_{\max}(i, h_i^*) = \max_h C_{i,h}, \quad i = AO, IO, LS, \tag{4.35}$$

where it is understood that h_i denotes the time index when the maximum of test statistic $J_{i,h}$ occurs and h_i^* denotes the time index when the maximum of $C_{i,h}$ occurs. Under the null hypothesis of no outlier in the sample and assuming that the model of z_t is known, $J_{\max}(I, h_I)$ is the maximum of a random sample of size T from a chi-square distribution with k degrees of freedom and its asymptotic distribution can be obtained using the extreme value distribution. Each of the other three joint test statistics in (4.35) is the maximum of a dependent sample from a chi-square distribution with k degrees of freedom and their asymptotic distributions are, therefore, more complicated. In practice, simulation is required to generate finite-sample critical values of the two test statistics.

As in the univariate case, if a single joint statistic $J_{\max}(i, h_o)$ is significant at time index h_o, we identify a multivariate outlier of type i at h_o, where $i = AO, IO, LS$. In the case of multiple significant joint test statistics, we identify the outlier type based on the test statistic that has the smallest empirical p-value. For example, if $J_{\max}(A, h_A)$ has the smallest p-value at time index h_o and the p-value is smaller than 0.05, then we identify an AO at time index h_o at the 5% significance level. When all of the four joint statistics are insignificant at a given level, we use the component statistics $C_{\max}(i, h_i^*)$ to check for additional outliers. This step ensures that no component outliers are overlooked.

Once an outlier is identified, its impact on the underlying time series is removed. The adjusted series is treated as a new data set and the detecting procedure is iterated. We terminate the detection procedure when no significant outliers are detected. Finally, we recommend a joint estimation of the model parameters and all detected outliers. If some outlier parameters are found to be insignificant in this joint estimation, they are deleted. The joint estimation is repeated until all detected outliers are significant at the given significance level.

4.5.2 Outlier Detection by Projections

For a large set of time series, we need a fast way to clean the data before any modeling exercise. To this end, we classify any series with 30% or more observations missing as an outlying series and remove it from the data set for further consideration. Similarly, we also remove any series from the data set if it contains 30% or more observations with the same value. Series with high proportion of missing values or of data assuming the same value can occur when the series were collected automatically as some instruments may breakdown. Once outlying series have been removed, we use simple linear interpolations of individual series to fill in any missing values remained in the data set. Consider a scalar time series z_t with a patch of missing values, say $z_{h+1}, \ldots, z_{h+g}$, where g is the number of missing values. Here two nearby non-missing values are z_h and z_{h+g+1}, and we fill in the missing values with $z_{h+j} = z_h + j(z_{h+g+1} - z_h)/(g+1)$, for $j = 1, \ldots, g$.

Next, we consider a procedure to outlier detection that does not require any specification of multivariate model and, hence, can be applied to a large collection of time series. This procedure makes use of univariate outlier detection procedures and projections of multivariate time series. The basic idea of the procedure was proposed by Galeano et al. (2006), which showed that a multivariate outlier produces at least a univariate outlier in almost every projected series, and by detecting the univariate outliers we can identify the multivariate ones. Also, it is better in some situations to find multivariate outliers by applying univariate test statistics to optimal projections than by using multivariate statistics on the original series. The first advantage of using projection to search multivariate outliers is simplicity. By using projected univariate series we do not need to prespecify a multivariate model for the underlying series. Second, it can be shown that a convenient projection direction will lead to test statistics that are more powerful than the multivariate ones.

To begin, we analyze properties of univariate series obtained by the projection of a VARMA series, $\boldsymbol{\Phi}(B)\boldsymbol{y}_t = \boldsymbol{c} + \boldsymbol{\Theta}(B)\boldsymbol{a}_t$. It is well known that a non-zero linear combination of the components of a VARMA model follows a univariate ARMA model; see, for instance, Lütkepohl (1993) and Tsay (2014). Consider a nontrivial k-dimensional vector $\boldsymbol{v}$. Let

$$y_t = \boldsymbol{v}'\boldsymbol{y}_t.$$

If $\boldsymbol{y}_t$ follows a VARMA(p, q) process, then y_t follows an ARMA(p^*, q^*) model with $p^* \leqslant kp$ and $q^* \leqslant (k-1)p + q$. In particular, if $\boldsymbol{y}_t$ is a VMA(q) series, then y_t is an MA(q^*) with $q^* \leqslant q$, and if it is a VAR(p) process, y_t is ARMA(p^*, q^*) with $p^* \leqslant kp$ and $q^* \leqslant (k-1)p$. The general form of the model for y_t is $\phi(B)y_t = c + \theta(B)e_t$, where $\phi(B) = |\boldsymbol{\Phi}(B)|$, $c = \boldsymbol{v}'\boldsymbol{\Phi}(1)^*\boldsymbol{c}$, $\theta(B)e_t = \boldsymbol{v}'\boldsymbol{\Phi}(B)^*\boldsymbol{\Theta}(B)\boldsymbol{a}_t$, where $\boldsymbol{\Phi}(B)^*$ is the adjoint matrix of $\boldsymbol{\Phi}(B)$ and e_t is a scalar white noise process with mean zero and variance $\sigma_e^2 > 0$.

When the observed series $\boldsymbol{z}_t$ is affected by an outlier of size $\boldsymbol{w}$ we have seen, from Eqs. (4.17) and (4.26), that the model can be written as

$$\boldsymbol{z}_t = \boldsymbol{y}_t + \boldsymbol{\alpha}(B)\boldsymbol{w}I_t^{(h)},$$

where for MAO, $\boldsymbol{\alpha}(B) = \boldsymbol{I}$ and for MLS, $\boldsymbol{\alpha}(B) = \boldsymbol{I}/(1-B)$. The projected observed series is $z_t = \boldsymbol{v}'\boldsymbol{z}_t$, which satisfies $z_t = y_t + \boldsymbol{v}'\boldsymbol{\alpha}(B)\boldsymbol{w}I_t^{(h)}$. If $\boldsymbol{z}_t$ has an MAO, the projected observed series is $z_t = y_t + \beta I_t^{(h)}$ with $\beta = \boldsymbol{v}'\boldsymbol{w}$. If $\boldsymbol{z}_t$ has an MLS, the projected series z_t has a level shift with size $\beta = \boldsymbol{v}'\boldsymbol{w}$ at time $t = h$. Thus, the two basic types of multivariate outliers, MAO and MLS, will appear as univariate outliers of the same type, AO and LS, in the projected series.

An MIO produces a more complicated effect in projection. It leads to a patch of consecutive outliers starting at time index $t = h$ with sizes $\boldsymbol{v}'\boldsymbol{w}, \boldsymbol{v}'\boldsymbol{\Psi}_1\boldsymbol{w}, \ldots, \boldsymbol{v}'\boldsymbol{\Psi}_{T-h}\boldsymbol{w}$. Assuming that h is not close to T and because $\boldsymbol{\Psi}_j \to 0$ as $j \to \infty$, the sizes of the outliers in the patch approach zero. In the particular case that $\boldsymbol{v}'\boldsymbol{\Psi}_i\boldsymbol{w} = \psi_i\boldsymbol{v}'\boldsymbol{w}$, for

$i = 1, \ldots, T - h$, then z_t has an innovational outlier at $t = h$ with size $\beta = \boldsymbol{v}'\boldsymbol{w}$. However, if $\boldsymbol{v}'\boldsymbol{\Psi}_i\boldsymbol{w} = 0$ for $i = 1, \ldots, T - h$, then z_t has an AO at $t = h$ with size $\beta = \boldsymbol{v}'\boldsymbol{w}$, and if $\boldsymbol{v}'\boldsymbol{\Psi}_i\boldsymbol{w} = \boldsymbol{v}'\boldsymbol{w}$ for $i = 0, \ldots, T - h$, then z_t has a level shift at $t = h$ with size $\beta = \boldsymbol{v}'\boldsymbol{w}$. Therefore, the univariate series z_t obtained by projection can be affected by an AO, a patch of outliers, or a level shift. For this reason, we do not consider MIO in this section.

Peña and Prieto (2001) proposed a procedure for multivariate outlier detection based on projections that maximize or minimize the kurtosis coefficient of the projected data. This result can be generalized to multivariate time series by defining the maximum discrimination direction as the direction that maximizes the size of the univariate outlier, $\boldsymbol{v}'\boldsymbol{w}$, with respect to the variance of the projected series. It was shown by Galeano et al. (2006) that for MAO and MLS the direction of maximum discrimination is the one defined by the outlier size $\boldsymbol{w}$ and that this direction can be obtained by finding the extremes of the kurtosis coefficient of the projected series.

Multiple outliers can produce, as we have seen in section 4.3.4, masking and swamping effects. For instance, a projection might effectively reveal one outlier, but mask the effects of others. To overcome such a difficulty, we iterate the search process after an outlier is found and its effect adjusted from the vector series. Also, Peña and Prieto (2007) proved that the search for multivariate outliers can be improved if in addition to the extreme kurtosis directions we include some random directions computed in a slightly modified StahelDonoho method. Thus, after removing the effects of detected outliers we propose to iterate the search procedure using new directions until no more outliers are found. Therefore, if a set of outliers are masked in one direction, they may be revealed in a later iteration after removing the detected outliers. Swamping implies that one outlier affects the series in such a way that other "good" data points appear like outliers. The procedure proposed in the next section includes iteration to avoid swamping effects, which can appear in the univariate searches using the projection statistics. The idea is to delete insignificant outliers after a joint estimation of the parameters and detected outliers. This will be clarified further in the next section.

4.5.3 A Projection Algorithm for Outliers Detection

The procedure proposed here is based on those of Galeano et al. (2006, 2020) to identify multivariate and univariate LS and AO without requiring a multivariate model for the vector time series. As explained in the previous section, MIOs are not included because their effects depend on the specified multivariate model. The algorithm consists of two parts. In the first one an initial cleaning is applied to the set of time series without assuming any multivariate model. In the second one, the specified model, either a VARMA or a dynamic factor model (DFM, see Chapter 6), is estimated with the clean series obtained in the first step and used to detect outliers. Of course, when the assumed model is correct, using it in the outlier search will lead to a more powerful procedure. We describe next the algorithm to clean the series from outliers without a model.

1. Given $\boldsymbol{z}_1, \ldots, \boldsymbol{z}_T$, the two directions that maximize or minimize the kurtosis coefficient of the projected time series are found. Then, project the observed vector series into these directions and search for outliers in the projected time series. To this end, we make use of the algorithm for univariate time series analysis mentioned before.

2. If no outliers are found in step 1, go to step 4. Otherwise, let $T_o = (t_1, \dots, t_o)'$ be the set of locations of the outliers found in step 1(a) and $C_o = (c_1, \dots, c_o)'$ be the labels of the type of detected outliers (LS or AO). Assume that each of the k time series has an outlier at time index $t_i \in T_o$ of the type $c_i \in C_o$, and remove their effects on each of the k time series. Then, repeat steps 1 and 2 with the cleaned vector time series until no more outliers in the projected time series are found.
3. Let d_o be the number of directions obtained in steps 1 and 2 to which the projected time series contains detected outliers. Compute d_o random directions as explained in Peña and Prieto (2007), project the vector time series obtained after step 2, and search for outliers in the projected time series as in step 1. Remove outlier effects, if any, in each of the k times series at the locations of the detected outliers as in step 2.
4. Search for univariate outliers in the k univariate time series that have been already cleaned from multivariate outliers after step 2

Some comments on the proposed procedure are in order. First, when computing the directions of maximum kurtosis and random directions, respectively, in steps 1 and 3, it is assumed that $T > k$. Therefore, if $k \geq T$, these directions are computed by using as data the main principal components of the time series. Second, for large number of time series instead of using 2 and 4 to perform corrections of detected outlier effects, one can apply a fast interpolation to all k time series. A fast way to do so for an MAO is to substitute the vector observations at the time indexes of occurrence of the MAO found in the projections by their interpolated values using exponentially weighted moving average (EWMA) smoothing with moving average window of width 4. This approach reduces the heavy computational burden of cleaning when the dimension of the time series is high. Third, in step 1 the interpolation of MAO outliers found for all series may lead to some small loss of efficiency as we may drop good points. However, when the number of multivariate outliers is a small fraction of the sample size, this effect can be neglected. Fourth, the proposed algorithm depends on the choice of the cutoff to determine whether a time series observation is labeled as outlier or not, that should be taken as the $(1 - \alpha/2)^{1/(Tm)}$ quantile of the standard normal distribution, denoted by $z_{(1-\alpha/2)^{1/(Tm)}}$.

When we want to fit a model for the vector of time series, for instance a DFM (see Chapter 6), using the set of outlier positions found in the initial cleaning, the model is fitted to the original vector time series estimating jointly the outlier effects and the model parameters. The fitted model is refined if necessary, e.g. removing insignificant outliers, if any. For moderate dimension k, a VAR(p) models can also be used. If some effect is found not significant at a given level, we remove the least significant one and repeat the joint estimation until all the effects are significant.

4.5.4 The Nonstationary Case

Assume $\boldsymbol{y}_t \sim I(d_1, \dots, d_k)$, where d_is are nonnegative integers denoting the orders of differencing of the components. Let $d = \max(d_1, \dots, d_k)$, and consider first the case of $d = 1$, which we denote simply by $\boldsymbol{y}_t \sim I(1)$. For such a series, in addition to the two basic outliers, MAO and MLS, we also entertain the multivariate ramp shift (MRS), that implies a slope change in the multivariate series and it may occur in an $I(1)$ series.

Consequently, for an MRS, we assume that it only applies to the components of $\boldsymbol{y}_t$ with $d_j = 1$, that is, the size of the outlier $\boldsymbol{w} = (w_1, \ldots, w_k)'$ satisfies $w_j = 0$ if $d_j = 0$.

The observed series $\boldsymbol{z}_t$ can be transformed into stationarity by taking the first difference. This differencing affects the existing outliers as follows. In the MIO case, $\nabla \boldsymbol{z}_t = \nabla \boldsymbol{y}_t + \tilde{\boldsymbol{\Psi}}(B)\boldsymbol{w}I_t^{(h)}$, where $\tilde{\boldsymbol{\Psi}}(B) = \nabla\boldsymbol{\Psi}(B)$, so that an MIO produces an MIO in the differenced series. In the MAO case, $\nabla \boldsymbol{z}_t = \nabla \boldsymbol{y}_t + \boldsymbol{w}(I_t^{(h)} - I_{t-1}^{(h)})$, producing two consecutive MAOs with the same size but opposite signs. In the MLS case, $\nabla \boldsymbol{z}_t = \nabla \boldsymbol{y}_t + \boldsymbol{w}I_t^{(h)}$, resulting in an MAO of the same size. In the MTC case, $\nabla \boldsymbol{z}_t = \nabla \boldsymbol{y}_t + \zeta(B)\boldsymbol{w}I_t^{(h)}$, where $\boldsymbol{\zeta}(B) = 1 + \zeta_1 B + \zeta_2 B^2 + \cdots$ such that $\zeta_j = \delta^{j-1}(1-\delta)$. Thus, an MTC produces an MTC with decreasing coefficients ζ_j. In the MRS case, $\nabla \boldsymbol{z}_t = \nabla \boldsymbol{y}_t + \boldsymbol{w}S_t^{(h)}$, which produces an MLS with size equal to that of the MRS.

Therefore, in the $I(1)$ case, we propose a procedure similar to step 1 for the stationary case but now applied to the first differenced series. The procedure takes the first difference of $\boldsymbol{z}_t$ and check for MLS as in the stationary case. However, now the MLS detected in the differenced series are incorporated as ramp shifts in the original series and are estimated jointly with the model parameters. If any of the ramp shifts is not significant, it is removed from the model and the detecting process is repeated until all the ramp shifts are significant. Finally, we obtain a series $\boldsymbol{z}_t^* = \boldsymbol{z}_t - \sum_{i=1}^{r_R} \boldsymbol{w}_i R_t^{(h)}$ which is free of ramp shifts. Then we take the first difference of $\boldsymbol{z}_t^*$ and proceed to detect isolated MAOs, that correspond to LS, and two consecutive MAOs, that correspond to one MAO in the original series. All the outliers detected in the differenced series are incorporated by the corresponding effects in the original series and are estimated jointly with the model parameters. If any of the outliers becomes insignificant, it is removed from the model. We repeat the process until all the outliers are statistically significant.

The procedure can also be applied to cointegrated series. In this case, $\nabla \boldsymbol{z}_t$ is over-differenced, implying that its moving average component contains unit roots. Nevertheless, this is not a problem for the proposed procedure, because the directions of the outliers will be in general different from the directions of cointegration. In other words, if $\boldsymbol{v}$ is a vector obtained by maximizing or minimizing the kurtosis coefficient, then it is unlikely to be a cointegration vector, and $\boldsymbol{v}'\nabla \boldsymbol{z}_t = \nabla(\boldsymbol{v}'\boldsymbol{z}_t)$ is stationary and invertible because $\boldsymbol{v}'\boldsymbol{z}_t$ is a nonstationary series. However, if the series are cointegrated, then the final estimation should be carried out using the error correction model as explained in Chapter 3. Note that if $\boldsymbol{v}$ is the cointegration vector, then $\boldsymbol{v}'\boldsymbol{z}_t$ is stationary and $\nabla \boldsymbol{v}'\boldsymbol{z}_t$ is over-differenced.

Example 4.6

To illustrate, we search for multivariate outliers in the series of imports and exports of the United States and China. See Chapter 3 for further information of the data. We use the program `outliers.hdts` that detects multivariate AOs without a model in the first step of the proposed projection procedure. The second step of the program will be explained in Chapter 6.

In this particular instance, the program detected seven multivariate outliers and six univariate ones. These seven MAO and six AO generate more outlier effects than the six AO and six TC found in the univariate case of Example 4.5. Note that the time indexes of the outliers coincide broadly with those found in Example 4.5.

In summary, the univariate outliers are again detected, but some detected multivariate ones, such as $h = 109$, were not identified in the univariate search.

R commands for multivariate outlier detection:

```
>da<- read.csv("m-expimpcnus.csv")
>da=da[,-1]
>da=as.ts(da)
>xt=diff(da)
# The inputs of  outliers.hdts are (Yt,r.maxn,ind).
# Yt is the T-by-k data matrix,
# r.max is the number of factors and ind is an indicator.
# Here r.max = 1. We set ind=1 for taking no difference.
# If ind=2, the first difference is taken.
>out=outliers.hdts(xt,1,1)
[1] "First step - Running the first substep:
                          projections of maximum kurtosis"
Maximum kurtosis:  Direction number =  1 Number of outliers = 4
[1] "First step - Running the second substep: random projections"
Random projections:  Number =  1 Number of outliers = 3
[1] "First step - Running the third substep: univariate cleaning"
Univariate cleaning:  Number =  1 Number of outliers =  0
Univariate cleaning:  Number =  2 Number of outliers =  3
Univariate cleaning:  Number =  3 Number of outliers =  1
Univariate cleaning:  Number =  4 Number of outliers =  2
>out$times.kurt
[1] 277 278 313 314
out$times.rand
[1]  12  13 109
out$times.uni
[1] 26 200 203 203  14 202
#Note that this number may be different in another run because
# the random directions used are different.
```

■

Example 4.7

As a second illustration, we use the consumer price index (CPI) data of European countries. See Chapter 1. We search for multivariate and univariate outliers using the program `outliers.hdts` that search for MAOs using projections, followed by search for AO in the individual series. We concentrate on the initial cleaning of outliers without using any multivariate model. The series analyzed are the differenced data $\nabla\nabla_{12} \log z_t$. The commands used are given below:

```
>x=read.csv2("CPIEurope2000-15.csv")
>x=as.ts(x)
>lx=log(x)
>x1=diff(lx,lag=12)
>x12=diff(x1)
>out=outliers.hdts(x12,1,1)

The output obtained first is

[1] "First step - Running the first substep: projections of maximum
       kurtosis"
```

```
Maximum kurtosis: Direction number = 1 Number of outliers = 3
________
[1] "First step - Running the second substep: random projections"
Random projections: Number = 1 Number of outliers = 0
________
Random projections: Number = 3 Number of outliers = 0

#The number of MAO identified by kurtosis and random projection
   is obtained

> out$times.kurt
[1]  24  48 120   1   3  14  15 108
> out$times.rand
[1] 114
# and for the univariate outliers
 out$times.uni
 [1]  96 110 144   6 126 173  13  69  72  94
```

The output of the univariate search indicates 18 outliers. The largest number appear in Series 23 (Romania), then Series 25 (Slovakia), and Series 3 (Czech. Rep.). ■

4.6 ROBUST ESTIMATION

An alternative approach to handling outliers is to use estimation methods that are insensitive to aberrant data points. The approach is called *robust estimation* and is a complementary way to outlier detection for handling data heterogeneity. Outlier detection emphasizes on searching and estimating effects for possible outliers whereas robust estimation investigates methods that mitigate their impacts in statistical inference. Thus, the two approaches complement each other.

It is well known that the least squares (LS) estimation is optimal under normality but it may become inefficient when the observations are from a heavy-tailed distribution, which tends to produce outlying data points. The *breakdown point* of an estimator is defined as the minimum fraction of outliers in the sample that makes the estimate to assume any arbitrary value. For instance, the LS estimate of an AR(1) process has a breakdown point of $1/T$ because a single outlier may change ϕ to any arbitrary value. If instead of the LS method we use a robust estimate with high breakdown point, b/T and we have a fraction of outlier smaller than this breakdown point, b/T, the outliers can be identified as large residuals of the robust model. Usually robust estimates minimize a function of the residuals that grows slower than the square or, even better, that is bounded, so that the maximum effect of an outlier is limited. Consider a scalar AR(p) model and let $\boldsymbol{\phi}_p$ be the parameter vector. A general class of robust estimates, introduced by Huber (1981), are M-estimators that minimize

$$M = \sum_{t=1}^{T} \gamma \left[\frac{e_t(\boldsymbol{\phi}_p)}{\hat{\sigma}} \right], \tag{4.36}$$

where γ is a nondecreasing bounded function with $\gamma(0) = 0$ and $\gamma(\infty) = 1$, so that the effect of each residual, $e_t(\boldsymbol{\phi}_p) = z_t - \boldsymbol{\phi}_p' \boldsymbol{z}_{t-1,p}$, where $\boldsymbol{z}_{t-1,p} = (z_{t-1}, \dots, z_{t-p})'$, is bounded, and $\hat{\sigma}$ is an initial estimate of σ that makes the M-estimate independent of the scale of the observations. Note that in Eq. (4.36) the LS estimate is obtained for $\gamma(x) = x^2$, and the L_1 estimate for $\gamma(x) = |x|$, and these two estimates do not require an initial value of $\widehat{\sigma}$ in contrast to M-estimates. This initial value of the variability can be obtained by, for instance, interpolating in a simple way all suspicious observations and computing an initial estimating $\widehat{\sigma}$ by a robust measure of scale as the mad (median of absolute deviation) of these residuals:

$$\widehat{\sigma} = \frac{1}{0.675}\,\text{Med}(|e_i|, e_i \neq 0).$$

The function (4.36) is minimized by an iterative algorithm as follows. Taking the derivative of M in (4.36) with respect $\boldsymbol{\phi}_p$, and dropping constant terms, the estimates must verify

$$\sum_{t=p+1}^{T} \gamma'\left(\frac{z_t - \boldsymbol{\phi}'_p \mathbf{z}_{t-1,p}}{\widehat{\sigma}}\right)\mathbf{z}_{t-1,p} = 0, \tag{4.37}$$

where $\gamma'(x) = d\gamma(x)/dx$. Next, define the weight function

$$\omega(x) = \gamma'(x)/x \qquad \text{if } x \neq 0; \qquad \omega(0) = 0,$$

and the solution of Eq. (4.37) can be written as

$$\sum_{i=1}^{n} \omega_i \mathbf{z}_{t-1,p}(z_t - \mathbf{z}'_{t-1,p}\boldsymbol{\phi}_p) = 0,$$

where $\omega_i = \omega(e_i(\boldsymbol{\phi}_p)/\widehat{\sigma})$. The estimator is

$$\widehat{\boldsymbol{\phi}}_p = \left[\sum_{i=1}^{T} \omega_i \mathbf{z}_{t-1,p}\mathbf{z}'_{t-1,p}\right]^{-1}\left(\sum_{i=1}^{T} \omega_i \mathbf{z}_{t-1,p} z_t\right), \tag{4.38}$$

which is a weighted LS estimate with weights depending on the standardized residuals $e_i(\boldsymbol{\phi}_p)/\widehat{\sigma}$ so that observations with large residuals will have smaller weight in the estimation. The function (4.36) is minimized by using some initial estimate $\boldsymbol{\phi}_p^{(0)}$ to obtain the residuals and then computing a new value of the estimate by (4.38). This estimate will produce new residuals and new weights, and the procedure is iterated until convergence.

The M-estimates are useful when the sample only have low leverage outliers. By low leverage, we mean that the outlying observation does not markedly affect the parameter estimate. A better alternative in the general situation is to use MM-estimates, in which the estimation is carried out in two steps: (1) an initial consistent estimate $\widehat{\boldsymbol{\phi}}_p^{(0)}$ is computed that has a high breakdown point, although it may not be efficient, and a robust scale is obtained using this estimate; (2) Eq. (4.36) is minimized by the previous iterative procedure starting with $\widehat{\boldsymbol{\phi}}_p^{(0)}$.

A key problem in robust estimation is to avoid the propagation of outliers when computing the residuals. For instance, as seen before, an AO in an AR(p) at $t = h$ contaminates all p consecutive residuals starting at $t = h$, because $e_t(\boldsymbol{\phi}_p) = z_t - \boldsymbol{\phi}'_p \mathbf{z}_{t-1,p}$. Thus, robust filters to avoid this effect have been proposed, see, for instance, Muler et al. (2009) in the univariate case and Muler (2013) for the vector case, and the references therein. A good presentation of robust methods for time series analysis can be found in Maronna et al. (2019).

Example 4.8

Here, we search for outliers using the robust estimation R package **robustarima**. Its command `arima.rob` can estimate a specified time series model with outlier detection. We also compare the results with those of `tsoutliers` in Example 4.5. Again, the data used are the series of imports and exports of the United States and China. The commands used are:

```
> da <- read.csv("m-expimpcnus.csv")
> da=da[,-1]
> da=as.ts(da)
> xt=diff(da)
> T=dim(xt)[1]
> k=dim(xt)[2]
>require(robustarima)ith
>Otable <- matrix(0,k,4)
>for (i in 1:k){
> z <-as.ts(xt[,i])
> mm <- arima.rob(z~1,p=3)
> mm1 <- mm$outliers
>Otable[i,1] <- i
>for (j in 1:3){
> jdx <- c(1:mm1$nout)[mm1$outlier.type==j]
>Otable[i,(j+1)]=length(jdx)} }
>colnames(Otable) <- c("Series","IO","AO","LS")
> print(Otable)
     Series IO AO LS
[1,]      1  0  9  0
[2,]      2  0  4  0
[3,]      3  0  6  0
[4,]      4  0  5  0
```

Comparing the results with those of Example 4.5, we see that the two approaches produce similar results. The main difference is that the robust method identified several consecutive AO outliers whereas `tsoutliers` treats these AOs as transitory changes. For instance, see the results of the fourth series below:

```
> z <-as.ts(xt[,4])
> mm <- arima.rob(z~1,p=3)
>  mm$outliers
 Number of outliers detected:  5
Outlier index
[1]  14 200 202 203 204
Outlier type
```

```
[1] "AO" "AO" "AO" "AO" "AO"
Outlier impact
[1]  0.0718 -0.0699 -0.1195 -0.0579 -0.0910
Outlier t-statistics
[1] 4.3659 4.2716 7.1753 3.5323 5.4793
```

The result shows five AOs with three consecutive ones from $t = 202$ to 204. On the other hand, the program `tsoutliers` in Example 4.5 detected the AO at $t = 14$, but a TC at $t = 202$. ■

4.7 HETEROGENEITY FOR PARAMETER CHANGES

Another form of heterogeneity in time series analysis is caused by changes on the model parameters. Their effects can be observed gradually throughout the sampling period. There is a large literature on monitoring parameter changes in univariate time series models. First, mean and variance shifts in AR process have been studied from the Bayesian point of view by McCulloch and Tsay (1993). Changes in AR parameters have been considered by Lee et al. (2003), Gombay and Serban (2009), and Chan et al. (2014), among many others. Second, changes in the marginal variance were studied by Inclán and Tiao (1994), who proposed an iterative procedure based in a cumulative sum of squares statistics and by Bai (1994) and Bai and Perron (1998). Much less work has been carried out for multivariate time series analysis. Galeano and Peña (2007) have studied the detection of step changes in the variance and in the correlation structure of the components of a VARMA model using both the LR and cusum approaches. In what follows, we present procedures for detecting parameter changes in a univariate times series and covariance changes in a multivariate time series.

4.7.1 Parameter Changes in Univariate Time Series

Consider a univariate AR(p) times series that follows the model $z_t = \boldsymbol{\phi}'\boldsymbol{Z}_{t-1} + a_t$ with $1 \leq t \leq T$, where $\boldsymbol{\phi} = (\phi_0, \phi_1, \dots, \phi_p)'$ and $\boldsymbol{Z}_{t-1} = (1, z_{t-1}, \dots, z_{t-p})'$. Suppose that the series suffers m parameter changes at time indexes $\tau_1 < \tau_2 < \cdots < \tau_m$ with $\tau_1 > 1$ and $\tau_m < T$. The changes can be both in the AR coefficients and in the noise variance. For simplicity, we consider $\text{Var}(a_t) = 1$ to avoid any confusion in variance parameterization. Thus, when $1 \leq t < \tau_1$ the series follows the AR(p) model $z_t = \boldsymbol{\phi}_1'\boldsymbol{Z}_{t-1} + \sigma_1 a_t$, but when $t = \tau_1$ the model changes to $z_t = \boldsymbol{\phi}_2'\boldsymbol{Z}_{t-1} + \sigma_2 a_t$. This second model applies to data in the interval $\tau_1 \leq t < \tau_2$. Again, at time $t = \tau_2$, the AR parameter shift to $\boldsymbol{\phi}_3'$ and the variance of the noise to σ_3, and so on.

Defining $\tau_0 = 1$ and $\tau_{m+1} = T$, this model can be written as

$$z_t = \sum_{i=1}^{m+1} (\boldsymbol{\phi}_i'\boldsymbol{Z}_{t-1} + \sigma_i a_t) I(\tau_{j-1} \leq t < \tau_j), \tag{4.39}$$

where $I(A) = 1$ if A is true and 0 otherwise. Chan et al. (2014) proposed to write this model in a more compact form using as new parameters the vectors of parameter changes. Define at time t the vector $\boldsymbol{\beta}_t$ of parameter changes and take $\boldsymbol{\beta}_1 = \boldsymbol{\phi}_1$.

Then, for $t < \tau_1$, the parameters are unchanged so that $\boldsymbol{\beta}_t = \mathbf{0}$. However, at $t = \tau_1$, the parameters become $\boldsymbol{\phi}_2$, so that $\boldsymbol{\beta}_{\tau_1} = \boldsymbol{\phi}_2 - \boldsymbol{\phi}_1$. In general, $\boldsymbol{\beta}_t = \mathbf{0}$ if t does not belong to the set of change points $(\tau_1, \tau_2, \ldots, \tau_m)$, and $\boldsymbol{\beta}_t = \boldsymbol{\phi}_{i+1} - \boldsymbol{\phi}_i$ when $t = \tau_i$. Thus, we can write Eq. (4.39) as

$$\boldsymbol{Y} = \boldsymbol{X}\boldsymbol{B} + \boldsymbol{U}, \tag{4.40}$$

where $\boldsymbol{Y} = (z_1, \ldots, z_T)'$, $\boldsymbol{X}$ is the $T \times T(p+1)$ matrix of lag values of the variable defined as

$$\boldsymbol{X} = \begin{bmatrix} \boldsymbol{Y}_0' & \mathbf{0}' & \mathbf{0}' & \cdots & \mathbf{0}' \\ \boldsymbol{Y}_1' & \boldsymbol{Y}_1' & \mathbf{0}' & \cdots & \mathbf{0}' \\ \boldsymbol{Y}_2' & \boldsymbol{Y}_2' & \boldsymbol{Y}_2' & \cdots & \mathbf{0}' \\ \vdots & \vdots & \vdots & \ddots & \vdots \\ \boldsymbol{Y}_{T-1}' & \boldsymbol{Y}_{T-1}' & \boldsymbol{Y}_{T-1}' & \cdots & \boldsymbol{Y}_{T-1}' \end{bmatrix},$$

with $\boldsymbol{Y}_i = (1, z_i, z_{i-1}, \ldots, z_{i-p+1})'$ and $\mathbf{0}$ is a $p+1$ vector of zeros, $\boldsymbol{B} = (\boldsymbol{\beta}_1', \ldots, \boldsymbol{\beta}_T')$ is a $T(p+1)$ vector of parameter changes and $\boldsymbol{U} = (\sigma_1 a_t, \ldots, \sigma_t a_t)$.

Chan et al. (2014) proposed to estimate this model that has only $m+1$ coefficients $\boldsymbol{\beta}_i'$ different from zero by the group Lasso estimation, that will be explained in more detail in Chapter 7. In this case, we have group structure in the model parameters, that is, we are interested in the T vectors $\boldsymbol{\beta}_t$ and not in the individual values of the $T(p+1)$ parameters β_{tj}, where $\boldsymbol{\beta}_t = (\beta_{t0}, \beta_{t1}, \ldots, \beta_{tp})$. The standard Lasso minimization attempts to make many β_{tj} equal to zero, whereas here we are interested in making all the coefficients in the group $\boldsymbol{\beta}_t$ equal to zero. To this end, Yuan and Lin (2006) proposed the Group Lasso in which the estimation equation is

$$\widehat{\boldsymbol{B}} = \arg\min \frac{1}{T} \|\boldsymbol{Y} - \boldsymbol{X}\boldsymbol{B}\|^2 + \lambda_T \sum_{t=1}^{T} \|\boldsymbol{\beta}_t\|,$$

where λ_T is a penalty parameter and $\|.\|$ is the Euclidean or ℓ_2 norm. Note that if we use in the penalty the ℓ_1 norm we will be back to the standard Lasso. However, by using the ℓ_2 norm all the coefficients at time t are considered together. Thus, if $\widehat{\boldsymbol{\beta}}_t \neq \mathbf{0}$ a parameter change is detected at time index t, and the estimated number of changes, $\widehat{m}$, is equal to the number of non-zero elements $\widehat{\boldsymbol{\beta}}_t$ minus one. Let $(\widehat{\tau}_1, \widehat{\tau}_2, \ldots, \widehat{\tau}_m)$ be the time indexes of these changes. The AR parameters are estimated by

$$\widehat{\boldsymbol{\phi}}_1 = \widehat{\boldsymbol{\beta}}_1 \qquad \text{and} \qquad \widehat{\boldsymbol{\phi}}_j = \sum_{t=1}^{\widehat{\tau}_j} \widehat{\boldsymbol{\beta}}_t, \quad j = 2, \ldots, \widehat{m}+1.$$

Chan et al. (2014) proved that the procedure is consistent in terms of prediction error, but overestimates the number of change points. They also propose to apply a model selection criterion to select the best set of change points among the $\widehat{m}$ found by the Group Lasso estimation. Finally, there are some works focusing on nonparametric approach to detecting change points of multivariate data. See, for instance, Matteson and James (2014).

4.7.2 Covariance Changes in Multivariate Time Series

Suppose that instead of observing a time series $\boldsymbol{y}_t = (y_{1t}, \ldots, y_{kt})'$ that follows a VARMA model $\boldsymbol{\Phi}(B)\boldsymbol{y}_t = \boldsymbol{\Theta}(B)\boldsymbol{a}_t$, with $\boldsymbol{a}_t \sim N_k(\mathbf{0}, \Sigma)$, we observe $\boldsymbol{z}_t$, given by

$$\boldsymbol{\Phi}(B)\boldsymbol{z}_t = \boldsymbol{\Theta}(B)\boldsymbol{e}_t,$$

with

$$\boldsymbol{e}_t = \boldsymbol{a}_t + \boldsymbol{W} S_t^{(h)} \boldsymbol{a}_t,$$

where $S_t^{(h)}$ is a step function and $\boldsymbol{W}$ a lower triangular matrix of size $k \times k$ denoting the impact of the covariance change. This equation implies that the covariance of the innovations $\boldsymbol{e}_t$ of the observed time series changes at the time point $t = h$ from Σ to $\boldsymbol{\Omega} = (\boldsymbol{I} + \boldsymbol{W})\Sigma(\boldsymbol{I} + \boldsymbol{W})'$. The relation between the observed series, $\boldsymbol{z}_t$, and the unobserved vector ARMA time series, $\boldsymbol{y}_t$, is given by

$$\boldsymbol{z}_t = \boldsymbol{y}_t + \boldsymbol{\Psi}(B) \boldsymbol{W} S_t^{(h)} \boldsymbol{a}_t, \tag{4.41}$$

with $\boldsymbol{\Psi}(B) = \boldsymbol{\Phi}^{-1}(B)\boldsymbol{\Theta}(B)$.

Without loss of generality, we assume that $(\boldsymbol{I} + \boldsymbol{W})$ is a positive-definite matrix, so that the matrix $\boldsymbol{W}$ is well identified. Let $\Sigma = \boldsymbol{L}_\Sigma \boldsymbol{L}_\Sigma'$ and $\boldsymbol{\Omega} = \boldsymbol{L}_\Omega \boldsymbol{L}_\Omega'$ be the Cholesky decompositions of the two covariance matrices. Then

$$\boldsymbol{\Omega} = \boldsymbol{L}_\Omega \boldsymbol{L}_\Omega' = (\boldsymbol{I} + \boldsymbol{W})\boldsymbol{L}_\Sigma \boldsymbol{L}_\Sigma'(\boldsymbol{I} + \boldsymbol{W})',$$

and

$$\boldsymbol{L}_\Omega = (\boldsymbol{I} + \boldsymbol{W})\boldsymbol{L}_\Sigma,$$

which implies

$$\boldsymbol{W} = \boldsymbol{L}_\Omega \boldsymbol{L}_\Sigma^{-1} - \boldsymbol{I}. \tag{4.42}$$

Thus, estimating $\boldsymbol{W}$ by (4.42) results in a well-defined matrix.

To test the significance of a covariance change at $t = h$, suppose that the parameters of the vector ARMA model are known and we compute the residuals via

$$\boldsymbol{e}_t = \boldsymbol{z}_t - \sum_{i=1}^{p} \boldsymbol{\Phi}_i \boldsymbol{z}_{t-i} + \sum_{j=1}^{q} \boldsymbol{\Theta}_j \boldsymbol{e}_{t-j}. \tag{4.43}$$

We want to test the null hypothesis $H_0 : \boldsymbol{W} = \boldsymbol{0}$, which implies that these residuals are iid homoscedastic, versus the alternative hypothesis $H_1 : \boldsymbol{W} \neq \boldsymbol{0}$, implying that they are heteroskedastic. We propose two statistics: the LR test, which is asymptotically the most powerful test, and the cusum test. Let us define three matrices $\boldsymbol{S} = \sum_{t=1}^{T}(\boldsymbol{e}_t\boldsymbol{e}_t')/T$, $\boldsymbol{S}_1 = \sum_{t=1}^{h-1}(\boldsymbol{e}_t\boldsymbol{e}_t')/(h-1)$ and $\boldsymbol{S}_2 = \sum_{t=h}^{T}(\boldsymbol{e}_t\boldsymbol{e}_t')/(T-h+1)$. The LR statistic for a variance change at and after the time point $t = h$ is given by

$$\mathrm{LR}_h = \log \frac{|\boldsymbol{S}|^T}{|\boldsymbol{S}_1|^{h-1}|\boldsymbol{S}_2|^{T-h+1}}, \tag{4.44}$$

and, under the null hypothesis of no covariance change and that the model is known, the LR_h statistic has an asymptotic chi-square distribution with $\frac{1}{2}k(k+1)$ degrees of freedom.

An alternative cusum test statistic can be built as follows. Let $A_h = \sum_{t=1}^{h} \boldsymbol{e}_t'\Sigma^{-1}\boldsymbol{e}_t$ be the multivariate cumulative sum of squares of the residuals, where h is any given value $1 \leq h \leq T$. Let,

$$C_h = \frac{h}{\sqrt{2kT}}\left(\frac{A_h}{h} - \frac{A_T}{T}\right), \tag{4.45}$$

be the centered and normalized cumulative sum of squares of the sequence $\boldsymbol{e}_t$. The asymptotic distribution of this statistic under the hypothesis of no change in the covariance matrix is a Brownian Bridge at $\upsilon = h/T$, which is normal with zero mean and variance $\upsilon(1-\upsilon)$. Thus, to test for the presence of a covariance change at the time point h the value of the statistic C_h is compared with percentiles of a zero mean normal distribution with variance $\upsilon(1-\upsilon)$. The impact of the covariance change is estimated by computing the Cholesky decomposition of the matrices $\boldsymbol{S}_1$ and $\boldsymbol{S}_2$ and using Eq. (4.42)

$$\widehat{\boldsymbol{W}} = \boldsymbol{L}_{S_2}\boldsymbol{L}_{S_1}^{-1} - \boldsymbol{I}. \tag{4.46}$$

4.7.2.1 Detecting Multiple Covariance Changes

Multiple covariance changes may occur in a vector time series. With r changes the observed time series is given by

$$\boldsymbol{z}_t = \boldsymbol{y}_t + \boldsymbol{\Psi}(B)(I + \boldsymbol{W}_r S_t^{(h_r)}) \cdots (I + \boldsymbol{W}_1 S_t^{(h_1)})\boldsymbol{a}_t,$$

where $\{h_1, \dots, h_r\}$ are the time indexes of the r change points and $\boldsymbol{W}_1, \dots, \boldsymbol{W}_r$ are $k \times k$ lower triangular matrices denoting the impact of the r changes. Assuming that the parameters are known, the filtered series of residuals is given by

$$\boldsymbol{e}_t = (I + \boldsymbol{W}_r S_t^{(h_r)}) \cdots (I + \boldsymbol{W}_1 S_t^{(h_1)})\boldsymbol{a}_t,$$

and the residual covariance matrix of $\boldsymbol{e}_t$ changes from Σ to $(\boldsymbol{I} + \boldsymbol{W}_1)\Sigma(\boldsymbol{I} + \boldsymbol{W}_1)'$ at $t = h_1$, and to $(\boldsymbol{I} + \boldsymbol{W}_2)(\boldsymbol{I} + \boldsymbol{W}_1)\Sigma(\boldsymbol{I} + \boldsymbol{W}_1)'(\boldsymbol{I} + \boldsymbol{W}_2)'$ at $t = h_2$, and so on. In practice, the parameters of the VARMA model, the number, locations and the sizes of the variance changes are unknown. Let $\widehat{\mathrm{LR}}_t$ and $\widehat{C}_t$ be the statistics (4.44) and (4.45), respectively, computed using the estimated residuals which are obtained by (4.43). We define the maximum of these statistics in the series as,

$$\Lambda_{\max}(h_{\max}^{\mathrm{LR}}) = \max\{|\widehat{\mathrm{LR}}_t|, 1 \le t \le T\},\ \Gamma_{\max}(h_{\max}^{C}) = \max\{|\widehat{C}_t|, 1 \le t \le T\}, \tag{4.47}$$

where $h_{\max}^{\mathrm{LR}}$ and $h_{\max}^{C} + 1$ are the estimates of the times of change using the LR test and the cusum statistic, respectively. The distribution of $\Lambda_{\max}$ in Eq. (4.47) is intractable and critical values are obtained by simulation. The distribution of $\Gamma_{\max}$ in Eq. (4.47) is asymptotically the distribution of $\sup\{|M_r^0| : 0 \le r \le 1\}$, which is given by (see Billingsley, 2013),

$$P\{\sup|M_r^0| \le a : 0 \le r \le 1\} = 1 + 2\sum_{i=1}^{\infty}(-1)^i \exp(-2i^2a^2),$$

and critical values can be obtained from this distribution. If several changes have occurred in the series, we propose two iterative procedures to detect them and estimate their impacts based on the statistics $\Lambda_{\max}$ and $\Gamma_{\max}$.

4.7.2.2 LR Test

The following procedure is a generalization to the one proposed by Tsay (1988). The algorithm is based on cleaning the series after finding a variance change and proceeds as follows:

1. Assuming no variance changes, a vector ARMA model is specified for the observed series $\boldsymbol{z}_t$. The maximum likelihood estimates of the model are obtained as well as the filtered series of residuals denoted by $\widehat{\boldsymbol{e}}_t$. Define $\boldsymbol{z}_t^* = \boldsymbol{z}_t$.

2. Compute the statistics LR_h, $h = d + 1, \ldots, T - d$, for a given positive integer d, using the residuals obtained in step 1. The number d denotes the minimum number of residuals needed to estimate the covariance matrix and we have taken d as the number of parameters to estimate plus one,

$$d = k + k(p+q) + \frac{k(k+1)}{2} + 1.$$

With the statistics LR_h, the statistic $\Lambda_{\max}(h_{\max}^{\text{LR}})$ in Eq. (4.47) is obtained.

3. Compare $\Lambda_{\max}(h_{\max}^{\text{LR}})$ with a specified critical value C for a given critical level. If $\Lambda_{\max}(h_{\max}^{\text{LR}}) < C$, it is concluded that there is no covariance change and the procedure ends. If $\Lambda_{\max}(h_{\max}^{\text{LR}}) \geq C$, it is assumed that there is a significant covariance change at time $t = h_{\max}^{\text{LR}}$.

4. The matrix $\widehat{\boldsymbol{W}}$ is estimated with Eq. (4.46) and a modified residual series is computed as follows:

$$\boldsymbol{e}_t^* = \begin{cases} \widehat{\boldsymbol{e}}_t, & t < h_{\max}^{\text{LR}}, \\ (\boldsymbol{I} + \widehat{\boldsymbol{W}})^{-1}\widehat{\boldsymbol{e}}_t, & t \geq h_{\max}^{\text{LR}}, \end{cases}$$

and, with this residual series, a corrected time series is defined by

$$\boldsymbol{z}_t^* = \begin{cases} \boldsymbol{z}_t, & t < h_{\max}^{\text{LR}} \\ \widehat{\boldsymbol{\Phi}}_1 \boldsymbol{z}_{t-1} + \cdots + \widehat{\boldsymbol{\Phi}}_p \boldsymbol{z}_{t-p} + \boldsymbol{e}_t^* - \widehat{\boldsymbol{\Theta}}_1 \boldsymbol{e}_{t-1}^* - \cdots - \widehat{\boldsymbol{\Theta}}_q \boldsymbol{e}_{t-q}^*, & t \geq h_{\max}^{\text{LR}} \end{cases}$$

where the polynomial matrices $\widehat{\boldsymbol{\Phi}}(B)$ and $\widehat{\boldsymbol{\Theta}}(B)$ are the maximum likelihood estimates of the parameters. Then, go back to step 1 considering $\boldsymbol{z}_t^*$ as the observed process.

5. When no more covariance changes are found, the model parameters and all the covariance changes detected in the previous steps are estimated jointly, using the model

$$\boldsymbol{\Phi}(B)\boldsymbol{z}_t = \boldsymbol{\Theta}(B)(\boldsymbol{I} + \boldsymbol{W}_r S_t^{(h_r)}) \cdots (\boldsymbol{I} + \boldsymbol{W}_1 S_t^{(h_1)})\boldsymbol{a}_t. \tag{4.48}$$

This joint estimation is carried out in two steps. First, estimate the parameters assuming no variance changes and then estimate the matrices $\boldsymbol{W}_i$. After that, correct the series, and repeat these two steps until convergence.

The cusum procedure is a generalization to the one proposed by Inclán and Tiao (1994). The algorithm is based on successive divisions of the series into two pieces when a change is detected and is explained in the Appendix. Some comments on these algorithms are in order. First, the critical values for LR statistic have to be obtained by simulation, while those of the cusum procedure are the asymptotic critical values of the maximum of the absolute value of a Brownian Bridge. Second, in both algorithms we require a minimum distance between variance changes larger than d, so that the covariance matrix can be estimated. If several changes were found in an interval smaller than d, these changes will be considered as multivariate outliers and estimated by the procedures discussed before. Third, the last step in the LR procedure is needed for avoiding bias in the size of the estimated covariance changes. Note that in step 4, the size of the covariance change is estimated after detecting it. Thus, if there are two covariance changes the impact of the first detected change is estimated without taking into account the second one. Therefore, a joint estimation is needed to reduce biases.

APPENDIX 4.A: CUSUM ALGORITHMS

4.A.1 Detecting Univariate LS

Bai (1994) proposed to identify LS by the cusum statistic,

$$C_t = \frac{t}{\sqrt{T}\psi(1)\sigma_e}\left(\frac{1}{t}\sum_{i=1}^{t} z_i - \bar{z}\right), \tag{4.49}$$

to test for a level shift at $t = h + 1$ in a linear process, and showed that the statistic converges weakly to a standard Brownian Bridge on $[0, 1]$. In practice, the quantity $\psi(1)\sigma_e$ is replaced by a consistent estimator,

$$\widehat{\psi(1)\sigma_e} = \left[\widehat{\gamma(0)} + 2\sum_{i=1}^{K}\left(1 - \frac{|i|}{K}\right)\widehat{\gamma(i)}\right]^{\frac{1}{2}},$$

where $\widehat{\gamma(i)} = \text{Cov}(y_t, y_{t-i})$ and K is a quantity such that $K \to \infty$ and $K/T \to 0$ as $T \to \infty$. Under the assumption of no level shifts in the sample, the statistic $\max_{1\leq t\leq T}|C_t|$ is asymptotically distributed as the supremum of the absolute value of a Brownian Bridge with cumulative distribution function, $F(x) = 1 + 2\sum_{i=1}^{\infty}(-1)^i e^{-2i^2x^2}$, for $x > 0$. The main advantages of the cusum statistic (4.49) over the LRT statistic for detecting level shifts include that (1) the ARMA model is not needed, (2) the cusum statistic has been shown to be more powerful than the LRT in several Monte Carlo studies, and (3) the statistic (4.49) seems to be robust to the presence of other outliers whereas the LRT statistic is not.

4.A.2 Detecting Multivariate Level Shift

Galeano et al. (2006) proposed an iterative procedure to identify multivariate level shifts based on the algorithm proposed in Inclán and Tiao (1994) for detecting several variance changes. For a vector of k series compute projected univariate series, $z_{it} = \boldsymbol{v}_i'\boldsymbol{z}_t$, for $i = 1, \ldots, N$, where $N \geq k$, and let H be the prespecified minimum distance between two level shifts. The proposed algorithm divides the series into pieces after detecting a level shift and proceeds as follows:

1. Let $t_1 = 1$ and $t_2 = T$. Obtain

   $$D_L = \max_{1\leq i\leq N} \max_{t_1\leq t\leq t_2} |C_t^i|, \tag{4.50}$$

   where C_t^i is the cusum statistic (4.49) applied to the ith projected series for $i = 1, \ldots, k$. Let

   $$(i_{\max}, t_{\max}) = \arg\max_{1\leq i\leq N} \arg\max_{t_1\leq t\leq t_2}|C_t^i|. \tag{4.51}$$

   If $D_L > D_{L,\alpha}$, where $D_{L,\alpha}$ is the critical value for the significance level α, then there is a possible level shift at $t = t_{\max} + 1$, and we go to 2a. If $D_L < D_{L,\alpha}$, then there is no level shift in the series and the algorithm stops.

2a. Define $t_2 = t_{\max}$ of step 1, and obtain new values of D_L and $(i_{\max}, t_{\max})$ of (4.50) and (4.51), respectively. If $D_L > D_{L,\alpha}$, and $t_2 - t_{\max} > H$, then we redefine $t_2 = t_{\max}$ and repeat step 2a until $D_L < D_{L,\alpha}$ or $t_2 - t_{\max} \le H$. Define $t_{\text{first}} = t_2$ where t_2 is the last time index that attains the maximum of the cusum statistic that is larger than $D_{L,\alpha}$ and satisfies $t_2 - t_{\max} > H$. The point $t_{\text{first}} + 1$ is the first time point with a possible level shift.

2b. Define $t_1 = t_{\max}$ of step 1 and $t_2 = T$, and obtain new values of D_L and $(i_{\max}, t_{\max})$ of (4.50) and (4.51). If $D_L > D_{L,\alpha}$, and $t_{\max} - t_1 > H$, then we redefine $t_1 = t_{\max}$ and repeat step 2b until $D_L < D_{L,\alpha}$ or $t_{\max} - t_1 \le H$. Define $t_{\text{last}} = t_1$ where t_1 is the last time index that attains the maximum of the cusum statistics which is larger than $D_{L,\alpha}$ and satisfies $t_{\max} - t_1 > H$. The point $t_{\text{last}} + 1$ is the last time point with a possible level shift.

2c. If $t_{\text{last}} - t_{\text{first}} < H$, there is just a level shift and the algorithm stops. If not, keep both values as possible change points and repeat steps 2a and 2b for $t_1 = t_{\text{first}}$ and $t_2 = t_{\text{last}}$ until no more possible change points are detected. Then, go to step 3.

3. Define a vector $h^L = (h_0^L, \ldots, h_{r_L+1}^L)$, where $h_0^L = 1$, $h_{r_L+1}^L = T$ and $h_1^L < \cdots < h_{r_L}^L$ are the change points detected in step 2. Obtain the statistic D_L in each sub-intervals (h_i^L, h_{i+2}^L) and check its statistical significance. If it is not significant, eliminate the corresponding possible change point. Repeat step 3 until the number of possible change points remains unchanged and the time indexes of change points are the same between iterations. Removing $h_0^L = 1$ and $h_{r_{L+1}}^L = T$ from the final vector of time indexes, we obtain r_L level shifts in the series at time indexes $h_i^L + 1$ for $i = 1, \ldots, L$.

4. Let $\{h_1^L, \ldots, h_{r_L}^L\}$ be the time indexes of r_L detected level shifts. To remove the impacts of level shifts, fit the model:

$$(\boldsymbol{I} - \boldsymbol{\Pi}_1 B - \cdots - \boldsymbol{\Pi}_{\hat{p}} B^{\hat{p}})\boldsymbol{z}_t^* = \boldsymbol{a}_t^*, \tag{4.52}$$

where $\boldsymbol{z}_t^* = \boldsymbol{z}_t - \sum_{i=1}^{r_L} \boldsymbol{w}_i S_t^{(h_i^L)}$, and the order $\hat{p}$ is chosen such that

$$\hat{p} = \arg\min_{0 \le p \le p_{\max}} AIC(p) = \arg\min_{0 \le p \le p_{\max}} \left\{ \log|\hat{\Sigma}_p| + 2\frac{k^2 p}{T} \right\},$$

where $\hat{\Sigma}_p = \frac{1}{T-2p-1} \sum_{t=p+1}^{T} \boldsymbol{a}_t^* \boldsymbol{a}_t^{*\prime}$ and $p_{\max}$ is a prespecified upper bound. If some of the effects of level shifts are not significant, the least significant one is removed from the model in (4.52) and the effects of the remaining $r_L - 1$ level shifts are re-estimated. This process is repeated until all the level shifts are significant.

Critical values of the statistic D_L, which is the maximum of dependent random variables, can be obtained by simulation (see Galeano et al., 2006). Note that the test statistics (4.49) are highly correlated for close observations and consecutive large values of C_t might be caused by a single level shift. To avoid over detection two level shifts should not be closer than the number H in steps 2 and 3; s $H = 10$ seems to works well in practice.

4.A.3 Detecting Multiple Covariance Changes

The algorithm is a generalization of Inclán and Tiao (1994) and proceeds as follows:

1. Assuming no covariance changes, a vector ARMA model is specified for the observed series $\boldsymbol{z}_t$. The maximum likelihood estimates of the model are obtained as well as the residual series that we denote by $\hat{\boldsymbol{e}}_t$. Let $t_1 = 1$.
2. Obtain $\Gamma_{\max}(h^C_{\max})$ in (4.47) for $t = 1, \ldots, T$. If $\Gamma_{\max}(h^C_{\max}) > C$, where C is a specified critical value for a given critical level, go to step 3. If $\Gamma_{\max}(h^C_{\max}) < C$, it is assumed that there is no covariance change in the series and the procedure ends.
3. Step 3 contains three substeps:
   a. Obtain $\Gamma_{\max}(h^C_{\max})$ for $t = 1, \ldots, t_2$, where $t_2 = h^C_{\max}$. If $\Gamma_{\max}(h^C_{\max}) > C$, redefine $t_2 = h^C_{\max}$ and repeat step 3(a) until $\Gamma_{\max}(h^C_{\max}) < C$. When this happens, define $h_{\text{first}} = t_2$, where t_2 is the last value such that $\Gamma_{\max}(h^C_{\max}) > C$.
   b. Repeat a similar search in the interval $t_2 \le t \le T$, where t_2 is the point $h^C_{\max}$ obtained in step 2. For that, define $t_1 = h^C_{\max} + 1$, where $h^C_{\max} = \arg\max\{C_t : t = t_1, \ldots, T\}$ and repeat it until $\Gamma_{\max}(h^C_{\max}) < C$. Define $h_{\text{last}} = t_1 - 1$, where t_1 is the last value such that $\Gamma_{\max}(h^C_{\max}) > C$.
   c. If $|h_{\text{last}} - h_{\text{first}}| < d$, there is just one change point and the algorithm ends. Otherwise, keep both values as possible change points and repeat steps 2 and 3 for $t_1 = h_{\text{first}}$ and $T = h_{\text{last}}$, until no more possible change points are detected. Then, go to step 4.
4. Define a vector $\boldsymbol{\ell} = (\ell_1, \ldots, \ell_s)$, where $\ell_1 = 1$, $\ell_s = T$ and $\ell_2, \ldots, \ell_{s-1}$ are the points detected in steps 2 and 3 in increasing order. Obtain the statistic C_t in each one of the intervals (ℓ_i, ℓ_{i+2}) and check if its maximum is still significant. If it is not, eliminate the corresponding point. Repeat step 4 until the number of possible change points does not change, and the points found in previous iterations do not differ from those in the last one. The vector $(\ell_2 + 1, \ldots, \ell_{s-1} + 1)$ are the points of covariance change.
5. Finally, estimate the parameters of the model and the covariance changes detected in the previous steps jointly by using (4.48).

EXERCISES

1. Compute the optimal interpolation for the univariate ARMA process $(1 - 0.6B - 0.3B^2)z_t = 5 + a_t$ at time h as a function of the observations before and after $t = h$. How many values are used? How are they weighted?
2. Prove that the optimal interpolation of the vector process $(\boldsymbol{I} - \boldsymbol{\Phi}B)\boldsymbol{z}_t = \boldsymbol{a}_t$ at time $t = h$ is given by $\hat{\boldsymbol{z}}_h = (\boldsymbol{I} + \boldsymbol{\Phi}'\boldsymbol{\Phi})^{-1}\boldsymbol{\Phi}(\boldsymbol{z}_{h-1} + \boldsymbol{z}_{h+1})$.
3. Use the package **tsoutliers** to detect outliers in the 9th and 10th series of the data set `TaiwanAirBox032017.csv`. Then, use the Lasso approach to detect level shifts and AOs in the same two time series. Compare the detection results.
4. Compare the results of Exercise 3 with those obtained via the program **arima.rob.**
5. With the three series of world temperature (`Temperatures.csv`) find outliers using the programs **tso, arima.rob** and **outlierLasso.**

6. Simulate, as in Example 2.2, 100 values of the three series that follow an ARMA model. For instance, an AR(2) or ARMA(1,1). Introduce in the three series an outlier of size 3 and compare the results of univariate detection using the programs **tso, arima.rob** and **outlierLasso** with multivariate detection using the program **outliers.hdts.** Verify that the the multivariate detection is more powerful than the univariate detection.
7. In a white noise series a_t of variance σ^2, an outlier of size ω is identified by the ratio ω/σ. In a vector white noise $\boldsymbol{a}_t$ of k time series with the same variance σ^2 and a multivariate outlier $\boldsymbol{\omega}$, the univariate time series obtained by projecting $\boldsymbol{a}_t$ in the direction $\boldsymbol{\omega}$, $y_t = \boldsymbol{\omega}'\boldsymbol{a}_t$ has an outlier of size $\boldsymbol{\omega}'\boldsymbol{\omega}$ and variance $\boldsymbol{\omega}'\boldsymbol{\omega}\,\sigma^2$, and the ratio to identify the outlier is $\sqrt{\boldsymbol{\omega}'\boldsymbol{\omega}}/\sigma = \sqrt{k}\bar{\omega}/\sigma$, where $\bar{\omega} = \sqrt{\boldsymbol{\omega}'\boldsymbol{\omega}/k}$. Explain why these results prove that the multivariate outlier detection by projections can be more powerful than univariate detection if the direction of the outlier is well identified.
8. Suppose that k is large and that the series follows a VAR(1) with a sparse parameter matrix with many coefficients equal to zero and rank $r << k$. If the vector of series is affected by two MAO at time h and $h + 1$. Justify that only a small part of random projections of the vector of series will be able to show the presence of MAOs.

REFERENCES

Abraham, B. and Ledolter, J. (1984). A note on inverse autocorrelations. *Biometrika*, **71**: 609–614.

Bai, J. (1994). Least squares estimation of a shift in linear processes. *Journal of Time Series Analysis*, **15**: 453–472.

Bai, J. and Perron, P. (1998). Estimating and testing linear models with multiple structural changes. *Econometrica*, **66**: 47–78.

Balke, N. S. (1993). Detecting level shifts in time series. *Journal of Business and Economic Statistics*, **11**: 81–92.

Bianco, A. M., Garcia Ben, M., Martínez, E. J., and Yohai, V. J. (2001). Outlier detection in regression models with ARIMA errors using robust estimates. *Journal of Forecasting*, **20**: 565–579.

Billingsley, P. (2013). *Convergence of Probability Measures*. John Wiley & Sons, Hoboken, NJ.

Box, G. E. P. and Tiao, G. C. (1975). Intervention analysis with applications to economic and environmental problems. *Journal of the American Statistical Association*, **70**: 70–79.

Chan, N. H., Yau, C. Y., and Zhang, R. M. (2014). Group LASSO for structural break time series. *Journal of the American Statistical Association*, **109**: 590–599.

Chang, I., Tiao, G. C., and Chen, C. (1988). Estimation of time series parameters in the presence of outliers. *Technometrics*, **30**: 193–204.

Chen, C. and Liu, L. M. (1993). Joint estimation of model parameters and outlier effects in time series. *Journal of the American Statistical Association*, **88**: 284–297.

Dempster, A. P., Laird, N. M., and Rubin, D. B. (1977). Maximum likelihood from incomplete data via the EM algorithm (with discussion). *Journal of the Royal Statistical Society*, Series B, **39**: 1–38.

Galeano, P. and Peña, D. (2007). Covariance changes detection in multivariate time series. *Journal of Statistical Planning and Inference*, **137**: 194–211.

Galeano, P., Peña, D., and Tsay, R. S. (2006). Outlier detection in multivariate time series by projection pursuit. *Journal of the American Statistical Association*, **101**: 654–669.

Galeano, P., Peña, D., and Tsay, R. S. (2020). Outlier detection in High Dimensional time series. Working Paper. Universidad Carlos III de Madrid.

Gombay, E. and Serban, D. (2009). Monitoring parameter change in AR (p) time series models. *Journal of Multivariate Analysis*, **100**: 715–725.

Efron, B. and Hastie, T. (2016). *Computer Age Statistical Inference*. Cambridge University Press, Cambridge, UK.

Fox, A. J. (1972). Outliers in time series. *Journal of the Royal Statistical Society B*, **34**: 350–363.

Hendry, D. F., Johansen, S., and Santos, C. (2008). Automatic selection of indicators in a fully saturated regression. *Computational Statistics*, **33**: 317–335, Erratum, 337–339.

Huber, P. J. (1981). *Robust Statistics*. John Wiley & Sons, New York, NY.

Inclán, C. and Tiao, G. C. (1994). Use of cumulative sums of squares for retrospective detection of changes of variance. *Journal of the American Statistical Association*, **89**: 913–923.

Justel, A., Peña, D., and Tsay, R. S. (2000). Detection of outlier patches in autoregressive time series. *Statistica Sinica*, **11**: 651–673.

Lee, S., Ha, J., Na, O., and Na, S. (2003). The cusum test for parameter change in time series models. *Scandinavian Journal of Statistics*, **30**: 781–796.

Luceño, A. (1998). Detecting possibly non-consecutive outliers in industrial time series. *Journal of the Royal Statistical Society, Series B*, **60**: 295–310.

Lütkepohl, H. (1993) *Introduction to Multiple Time Series Analysis*, 2nd Edition. Springer-Verlag, New York, NY.

Maravall, A., Pavón, R. L., and Cañete, D. P. (2016). Reg-ARIMA model identification: empirical evidence. *Statistica Sinica*, **26**: 1365–1388.

Maronna, R. A., Martin, R. D., Yohai, V. J., and Salibián-Barrera, M. (2019). *Robust Statistics: Theory and Methods (with R)*. John Wiley & Sons, Hoboken, NJ.

Matteson, D. S. and James, N. A. (2014). A nonparametric approach for multiple change point analysis of multivariate data. *Journal of the American Statistical Association*, **109**: 334–345.

McCulloch, R. E. and Tsay, R. S. (1993). Bayesian inference and prediction for mean and variance shifts in autoregressive time series. *Journal of the American Statistical Association*, **88**: 968–978.

Muler, N. (2013). Robust estimation for vector autoregressive models. *Computational Statistics & Data Analysis*, **65**: 68–79.

Muler, N., Peña, D., and Yohai, V. J. (2009). Robust estimation for ARMA models. *The Annals of Statistics*, **37**: 816–840.

Peña, D. and Prieto, F. J. (2007). Combining random and specific directions for outlier detection and robust estimation in high-dimensional multivariate data. *Journal of Computational and Graphical Statistics*, **16**(1): 228–254.

Peña, D. and Prieto, F. J. (2001). Multivariate outlier detection and robust covariance matrix estimation (with discussion). *Technometrics*, **43**: 286–310.

Sánchez, M. J. and Peña, D. (2003). The identification of multiple outliers in ARIMA models. *Communications in Statistics: Theory and Methods*, **32**: 1265–1287.

Tsay, R. S. (1986). Time series model specification in the presence of outliers. *Journal of the American Statistical Association*, **81**: 132–141.

Tsay, R. S. (1988). Outliers, level shifts and variance changes in time series. *Journal of Forecasting*, **7**: 1–20.

Tsay, R. S. (2014). *Multivariate Time Series Analysis with R and Financial Applications*. John Wiley & Sons, Hoboken, NJ.

Tsay, R. S. and Chen, R. (2018). *Nonlinear Time Series Analysis*. John Wiley & Sons, Hoboken, NJ.

Tsay, R. S., Peña, D., and Pankratz, A. E. (2000). Outliers in multivariate time series. *Biometrika*, **87**: 789–804.

Yuan, M. and Lin, Y. (2006). Model selection and estimation in regression with grouped variables. *Journal of the Royal Statistical Society, Series B*, **68**: 49–67.

CHAPTER 5

CLUSTERING AND CLASSIFICATION OF TIME SERIES

In this chapter, we study how to divide a set of time series into homogeneous groups of series with similar properties and how to classify a time series into one cluster among several possible clusters. The procedures to achieve these objectives are called *classification methods*. When the classification is made without a reference set, in which similar series are organized in a known number of groups, the procedures are called *clustering methods*. Such a classification is also called *unsupervised classification* or *unsupervised pattern recognition*. When we have series that are classified into groups with labels, and the objective is to classify new observed series, the procedures are called *discrimination methods*. The classification is then called *supervised classification* or *supervised pattern recognition*.

In time series clustering, we have a set of k time series $(z_{i1}, \dots, z_{iT})$ for $i = 1, \dots, k$, and would like to partition these k series into groups (or clusters) such that (1) all series are classified, (2) each series belongs to one and only one group, and (3) each group is internally as homogeneous as possible. Clustering is an important method in data analysis and has applications in many scientific fields. The procedures for clustering (i) select a measure of proximity (distance or dissimilarity) between two time series and use these proximities to form groups or (ii) define a set of features or variables $\boldsymbol{x}_i$, where $\boldsymbol{x}_i \in \boldsymbol{R}^p$, for the ith series that summarize its properties and use these k vectors of variables to perform grouping. Then, clusters are built by (a) agglomerating the data using certain proximities between series as with hierarchical methods; (b) partitioning the data by similarity in groups as in k-means; (c) estimating a mixture of several data generating models, including the mixing probabilities, as in model-based clustering; and (d) projecting the data in a smaller space to find the clusters.

Statistical Learning for Big Dependent Data, First Edition. Daniel Peña and Ruey S. Tsay.
© 2021 John Wiley & Sons, Inc. Published 2021 by John Wiley & Sons, Inc.

Classification or discrimination appears as follows. Assume that a system under study belongs to one of c possible states, say $S = (s_1, \ldots, s_c)$, and each observed time series is generated from one and only one of the states. Some characteristics of the observed time series are available for each state so that we can infer the expected output of the system for a given state. Next, a new time series is observed, but its system state is unknown. The objective then is to make a decision on the system state that generated the observed time series, or to classify the time series into one of the c possible states. Some classical applications of classification with two states, $c = 2$, include the discrimination between nuclear explosions and earthquakes using seismic data, individuals being affected or unaffected by a neurological disease using electrocardiogram (ECG), expansion or contraction period of an economy using a set of economic time series, and success or bankruptcy of a company using a set of time series quantifying its credit worthiness.

The procedures used for classification also depend on the information available about the output from each state. There are three possible cases. The first case is when we know the time series model used by each state to generate data. The problem then is to classify the output into the state with the highest probability of generating the observed series. In the second case we do not know the models, but the distribution of a set of features (variables) of the time series given the state can be obtained. Then, we can compute the features of the observed time series and used their distributions to do classification. This case is similar to the standard discriminant analysis in multivariate analysis. In the third case, we have a sample of series generated by each state that are summarized in features with an unknown joint distribution and some nonparametric methods are applied to carry out the classification.

5.1 DISTANCES AND DISSIMILARITIES

Finding series that are similar to each other is a common problem with many applications. Indeed, similarity between series is a key concept in all classification problems. In this section, we discuss some measures of distance or dissimilarities between time series. With a given distance measure, one can form groups of series by putting together those series that are close in distance. The choice of the proper distance measure to use depends on the objective of the classification. Also, all distances or dissimilarities between pairs of series are usually summarized in a proximity matrix and analyzing properties of this matrix is important in characterizing the set of time series.

5.1.1 Distance Between Univariate Time Series

Suppose that we want to define a measure of distance between two stationary time series x_t and y_t. We assume that, without loss of generality, the series are standardized so that they have zero mean and unit variance. A distance between two vectors $\boldsymbol{x}_t$ and $\boldsymbol{y}_t$ is a function $d(\boldsymbol{x}_t, \boldsymbol{y}_t)$ with the following three properties: (1) the function is nonnegative, $d(\boldsymbol{x}_t, \boldsymbol{y}_t) \geq 0$ and assumes the value 0 if, and only if, $\boldsymbol{x}_t = \boldsymbol{y}_t$; (2) the function is symmetric, $d(\boldsymbol{x}_t, \boldsymbol{y}_t) = d(\boldsymbol{y}_t, \boldsymbol{x}_t)$; and (3) the function verifies the triangle inequality, $d(\boldsymbol{x}_t, \boldsymbol{y}_t) \leq d(\boldsymbol{x}_t, \boldsymbol{z}_t) + d(\boldsymbol{z}_t, \boldsymbol{y}_t)$. A simple measure of distance between two time series can be built by comparing directly its observed values. Of course, this is only useful if the series are standardized, as assumed here, so that their values do not depend on the scale. Then, given $\boldsymbol{x}_t = (x_1, \ldots, x_T)'$ and $\boldsymbol{y}_t = (y_1, \ldots, y_T)'$, the Euclidean distance, or Euclidean metric, between the two series is

$$d_E(x_t, y_t) = \sqrt{(\boldsymbol{x}_t - \boldsymbol{y}_t)'(\boldsymbol{x}_t - \boldsymbol{y}_t)} = \sqrt{2T(1 - \hat{\rho}_{xy}(0))}. \tag{5.1}$$

This distance only depends on the lag-zero autocorrelation among the series, $\hat{\rho}_{xy}(0)$, and, therefore, it does not take into account any possible lag relationship among the series.

A useful procedure to define distance is to summarize a time series $\boldsymbol{x}_t$ in a vector of characteristics $\boldsymbol{\theta}_x = (\theta_{x1}, \ldots, \theta_{xp})'$ and define a distance using the resulting vectors. A common way to build a distance between two vectors of standardized data is by using the de Minkowski family

$$d_m(\boldsymbol{\theta}_x, \boldsymbol{\theta}_y) = \left(\sum_{i=1}^{p} |\theta_{xi} - \theta_{yi}|^m \right)^{1/m}, \tag{5.2}$$

where $m > 0$. For instance, for $m = 1$, we have the L_1 or "city block" distance. For $m = 2$, the Euclidean or L_2 distance, and for $m \to \infty$, $d_\infty(\boldsymbol{\theta}_x, \boldsymbol{\theta}_y) \to \sup\{|\theta_{xi} - \theta_{yi}||, 1 \leq i \leq p\}$. It can be shown that $d_a(\boldsymbol{\theta}_x, \boldsymbol{\theta}_y) \leq d_b(\boldsymbol{\theta}_x, \boldsymbol{\theta}_y)$ if $a \geq b$ and $b \geq 1$.

The linear dynamic dependency of a stationary univariate time series is characterized by its autocorrelation structure. Thus, we can build a measure of distance between two series by comparing their theoretical or estimated autocorrelations. We usually compare the autocorrelation coefficient functions (ACFs), but can also use the partial or inverse autocorrelations or a combination of these coefficients as the periodogram. Calling $\hat{\boldsymbol{\rho}}_x = (\hat{\rho}_x(1), \ldots, \hat{\rho}_x(h))'$ the vector of the first h lags of sample ACF, a distance between two series x_t and y_t can be defined by

$$d(x_t, y_t, h) = \sqrt{\sum_{j=1}^{h} [\hat{\rho}_x(j) - \hat{\rho}_y(j)]^2}, \tag{5.3}$$

which is the Euclidean distance between the vectors $\hat{\boldsymbol{\rho}}_x$ and $\hat{\boldsymbol{\rho}}_y$ and is referred to as the *ACF distance*. In Eq. (5.3), all ACFs have the same weight. This can be modified by introducing a matrix of weights which is a decreasing function of the order of ACF. For instance, the weights may decrease linearly. For sample ACFs, the weights may depend on their asymptotic variances so that high-order coefficients have lower weights, because they tend to have higher variabilities. A third and more complex alternative is to take into account that the sample ACFs are correlated and employ a Mahalanobis distance between two vectors of sample ACFs. Thus, a general measure of distance between the ACFs is

$$d_{\boldsymbol{W}}(x_t, y_t, h) = \sqrt{(\hat{\boldsymbol{\rho}}_x - \hat{\boldsymbol{\rho}}_y)' \boldsymbol{W} (\hat{\boldsymbol{\rho}}_x - \hat{\boldsymbol{\rho}}_y)},$$

where the matrix $\boldsymbol{W}$ is the identity matrix if all the coefficients have the same weight, as in Eq. (5.3), is a diagonal matrix if the weights decrease with the lag, and is a full-rank matrix if it is the inverse covariance matrix of the sample ACFs.

An alternative approach to summarize a time series is to fit an AR(h) model to the series and employ the vector of estimated autoregressive (AR) coefficients $\hat{\boldsymbol{\phi}}_x = [\hat{\phi}_x(1), \ldots, \hat{\phi}_x(h)]'$. Then, compute the Euclidean distance between the AR coefficient vectors. This measure was proposed by Piccolo (1990). See also Corduas and Piccolo (2008). It assumes the form

$$d_{\phi}(x_t, y_t, h) = \sqrt{\sum_{j=1}^{h} [\hat{\phi}_x(j) - \hat{\phi}_y(j)]^2}. \tag{5.4}$$

The distance based on ACF or AR coefficient vectors is nonparametric, because the ACFs or an AR fitting do not implicitly assume a parametric model for the underlying time series. In practice, the lag or order h used must be sufficiently large to describe properly the linear dynamic dependence of the data.

A parametric approach is to compare the vectors of parameters of a given model specified for the two time series. Care must be exercised though. For instance, it is not a good idea to put AR and MA parameters together to form a parameter vector of an ARIMA model, because similar ARIMA models can assume rather different parameter values. For instance, an MA(1) model with coefficient 0.4 is close to an AR(3) model with coefficients (−0.4, −0.16, −0.064). Peña (1990) proposed to define the distance between two ARIMA models using their AR representations, namely,

$$d_{\pi}(x_t, y_t, h) = \sqrt{(\boldsymbol{\pi}_x - \boldsymbol{\pi}_y)' \boldsymbol{V}^{-1} (\boldsymbol{\pi}_x - \boldsymbol{\pi}_y)}, \tag{5.5}$$

where $\boldsymbol{V}^{-1}$ is a measure of the average precision matrix of the two π-weight vectors.

A different way to summarize the dynamic dependence of a time series is to use its normalized periodogram, defined for stationary process in Chapter 2 as

$$\mathrm{NI}(w_j) = 2 \left[1 + 2 \sum_{h=1}^{T-1} \hat{\rho}_z(h) \cos(w_j h) \right], \quad 0 \le w_j \le \pi.$$

A distance between two periodograms was proposed by Caiado et al. (2006) using the logarithm of the normalized periodogram as

$$d_{\mathrm{NP}}(x_t, y_t) = \sqrt{\sum_{j=1}^{[T/2]} [\log \mathrm{NI}_x(2\pi j/T) - \log \mathrm{NI}_y(2\pi j/T)]^2}. \tag{5.6}$$

It is straightforward to show that all previous distance measures satisfy the three properties of a distance.

For nonstationary time series we cannot use their ACFs. To cluster integrated time series, we can either compare the parameters with Eq. (5.5) or take the proper difference and compare the ACFs of the resulting stationary series. In other cases, one may be interested in a vector of features, as trend or seasonal coefficients, and use the distance between these features to cluster time series. One can also use the forecast of a series to perform clustering, such as comparing the forecasting densities (Alonso et al., 2006) or a vector of out-of-sample forecasts. Also, as the data may contain outliers, it is important to clean the series or use some robust methods to compute the ACFs or AR coefficients. For ACFs, Spearman's rank correlations can be used in which one uses the corresponding rank series to compute ACFs; see, for instance, Tsay (2020). All the distances can be summarized in a symmetric matrix of distances that has zeros on the diagonal and positive values in the off-diagonal elements.

The previous distances are based on linear properties of time series. See Maharaj et al. (2019) for further analysis. Some authors have proposed distance measures based on nonlinear characteristics (see Montero and Vilar, 2014), or with the objective to compare the shape of two time series with possible different numbers of data points. For instance, in speech recognition we want to identify words that have been recorded with different speaking speeds. In this case, we search for an optimal alignments into the two series and this is the objective of the *dynamic time warping* measure, which minimizes some distance between values of the series taken in sequence.

5.1.2 Dissimilarities Between Univariate Series

A problem with distances is that they depend on the scale of the variables and are not invariant under monotonic transformations. For instance, the square of the Euclidean metric d_2^2 does not verify the triangle inequality and it is not a metric. To verify this fact, consider the three points $A = (0, 0)$, $B = (5, 5)$, and $C = (2, 0)$. With the Euclidean distance $d_2(AB) = 5\sqrt{2} \leq d_2(AC) + d_2(CB) = 2 + \sqrt{34}$, but with squares $d_2^2(AB) = 50 > d_2^2(AC) + d_2^2(CB) = 4 + 34$, and it is not a metric. For this reason, measures of proximity based on the ordering of the observations, that is not affected by monotonic transformations, are often used.

A function of two time series is a *dissimilarity* if it verifies: (1) $D(x_t, y_t) \geq 0$ and $D(x_t, y_t) = 0$ if and only if $x_t = ay_t + c$, for some constants a and c; (2) $D(x_t, y_t) = D(y_t, x_t)$. Note that $D(x_t, y_t)$ does not need to verify the triangle inequality. Also, for standardized time series $D(x_t, y_t) = 0$ implies $x_t = y_t$. Given a dissimilarity we can build a similarity measure as follows: let M be the maximum dissimilarity between two time series in the set, the function

$$S(x_t, y_t) = M - D(x_t, y_t),$$

is always positive and symmetric, and it is called a similarity measure. Often the dissimilarities verify $0 \leq D(x_t, y_t) \leq 1$, where 0 means identity and 1 independence, so that the similarities are also between 0 and 1.

Distances based on vectors of serial correlations (ordinary, partial, or inverse) compare the coefficients lag by lag and two series with the same predictability from its past, but distributed in different lags, may appear as very different. If we are interested in the global dependency of a series from its past, for standardized Gaussian variables the likelihood function only depends on the determinant of the correlation matrix. This suggests summarizing all ACFs in the correlation matrix of the observations given by

$$\boldsymbol{R}_T = \begin{bmatrix} 1 & \rho(1) & \cdots & \rho(T-2) & \rho(T-1) \\ \rho(1) & 1 & \cdots & \rho(T-3) & \rho(T-2) \\ \vdots & \vdots & \ddots & \vdots & \vdots \\ \rho(T-2) & \rho(T-3) & \cdots & 1 & \rho(1) \\ \rho(T-1) & \rho(T-2) & \cdots & \rho(1) & 1 \end{bmatrix}$$

$$\equiv \begin{bmatrix} 1 & \boldsymbol{\rho}'_{(1:T-1)} \\ \boldsymbol{\rho}_{(1:T-1)} & \boldsymbol{R}_{T-1} \end{bmatrix}. \tag{5.7}$$

This Toeplitz matrix is also denoted as $\boldsymbol{R}_{z,T}$ if we want to signify the underlying time series. By properties of the determinant of a partitioned matrix, we have

$$|\boldsymbol{R}_T| = |\boldsymbol{R}_{T-1}|(1 - R^2_{T-1}), \tag{5.8}$$

where $R^2_{T-1} = \boldsymbol{\rho}'_{(1:T-1)}\boldsymbol{R}^{-1}_{T-1}\,\boldsymbol{\rho}_{(1:T-1)}$ is the square of the multiple correlation coefficient of an AR$(T-1)$ fit of the series, z_t, by all possible lags, $z_t = \sum_{j=1}^{T-1} b_j z_{t-j} + a_t$. By recursive use of this expression, we have

$$|\boldsymbol{R}_T| = \left[\prod_{i=1}^{T-1}(1 - R^2_i)\right]. \tag{5.9}$$

Suppose all time series can be well approximated by an AR(h) model with h much smaller than T. Then, $R^2_{T-1} = R^2_{T-2} = \cdots = R^2_h$ and

$$|\boldsymbol{R}_T| = (1 - R^2_h)^{T-h+1}\left[\prod_{i=1}^{h-1}(1 - R^2_i)\right].$$

For instance, suppose that $h = 1$. Then $\rho(j) = \rho_1^j$ and it is easy to see, by induction, that $|\boldsymbol{R}_T| = (1 - \rho_1^2)^{T-1}$. Therefore, we can compute the determinant of the large $T \times T$ matrix $\boldsymbol{R}_T$ of all the observations if we know the determinant of the smaller $(h+1) \times (h+1)$ correlation matrix of the vector $\boldsymbol{Z}_{t,h} = (z_t, z_{t-1}, \ldots, z_{t-h})'$. This matrix, $\boldsymbol{R}_h$ is given in Eq. (5.7) using h instead of T. The determinant of this smaller matrix has an interesting interpretation as a global measure of all autocorrelations and, as in Eq. (5.8), we have

$$|\boldsymbol{R}_h| = |\boldsymbol{R}_{h-1}|(1 - R^2_h),$$

where R^2_h is the square of the multiple correlation coefficient of an AR(h) fit of z_t and also

$$|\boldsymbol{R}_h|^{1/h} = \left[\prod_{i=1}^{h}(1 - R^2_i)\right]^{1/h}. \tag{5.10}$$

Note that $1 - R^2_i$ is a measure of the linear dependence of the series on its past from z_{t-1} to z_{t-i}, and $|\boldsymbol{R}_h|^{1/h}$ is the geometric average of these dependence measures. Thus, $1 - |\boldsymbol{R}_h|^{1/h}$ can be interpreted as an average squared correlation coefficient, obtained when fitting AR models with increasing orders to the series. We have $0 \leq |\boldsymbol{R}_h| \leq 1$, with equality to one holding if $\boldsymbol{R}_h$ is diagonal and all lagged ACFs are zero and equality to zero holding if there exists a linear combination, $\boldsymbol{a}'\boldsymbol{Z}_{t,h} = c$ (a constant), so that the series is deterministic and all future values are determined by the past ones.

Based on these properties, Peña and Rodriguez (2002) define the *total correlation* of a stationary time series x_t by

$$\text{TC}_{x,h} = 1 - |\boldsymbol{R}_h|^{1/(h+1)}, \quad (5.11)$$

and it is a measure of the distance of the series from a univariate white noise process, that has TC = 0. We can build a dissimilarity measure between two series by the absolute value of the difference between their total correlations, that is,

$$D_U(x_t, y_t, h) = |\text{TC}_{x,h} - \text{TC}_{y,h}| = \left| |\boldsymbol{R}_{x,h}|^{1/(h+1)} - |\boldsymbol{R}_{y,h}|^{1/(h+1)} \right|, \quad (5.12)$$

and use the proximity matrix of these dissimilarities to perform clustering.

The proximity matrix is built as follows. Let $d(i,j)$ be the measure chosen to define the proximity between Series i and j. Then build a squared and symmetric matrix with zeros in the diagonal and off-diagonal element of the ith row and jth column equal to $d(i,j)$.

Example 5.1

To illustrate the computation of distances and dissimilarities, we use the Taiwan AirBox data in the file `TaiwanAirBox032017.csv` that contains hourly $PM_{2.5}$ measurements from Air-Box devices for March 2017. There are 744 observations and 516 series in different locations. The data are loaded in a matrix that does not contain the first column. The time plots of the series are shown in Figure 5.1 with three empirical dynamic quantiles (EDQs). See Chapter 2 for details of the plot. As most of

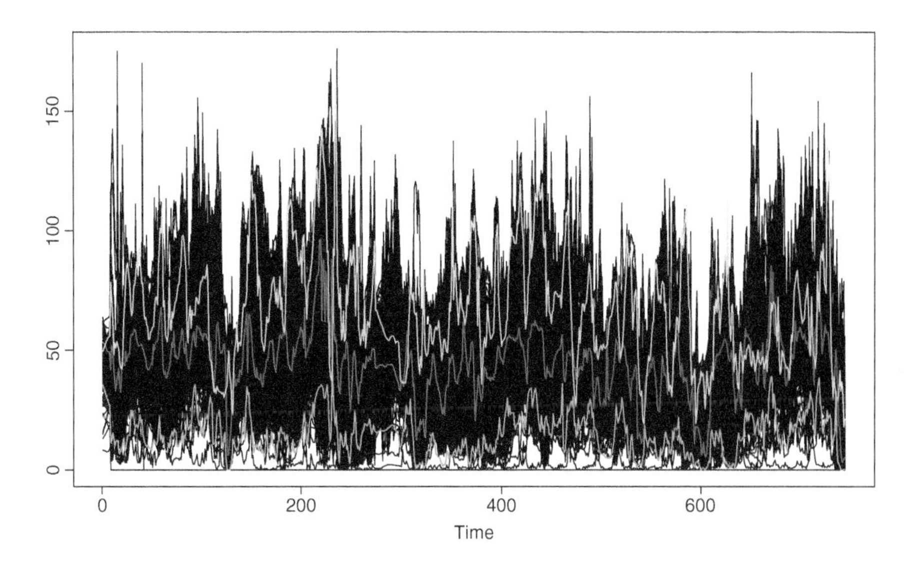

Figure 5.1 The AirBox Taiwan Data and the empirical dynamic quantiles with probabilities 0.05, 0.5, and 0.95.

the series are nonstationary, the first difference is applied and the series $z_t = \nabla x_t$ are employed to compute the distances. Four distances are computed. The ACF and partial autocorrelation coefficient (PACF) distances with $h = 5$, Piccolo distance (PIC) with default order, and the periodogram distance (PER). This is carried out using the R package **TSclust** with command `diss`. Details are given below:

R commands for time series distance:

```
>x=as.matrix(TaiwanAirBox032017)
>x=x[,-1]
>ts.plot(x)
>T=dim(x)[1]
>k=dim(x)[2]
>z=diff(x)
> h=5
> aa=acf(z,lag.max=h,plot=FALSE)
#a matrix with columns the acf of each series is formed
>H <- c(2:6)
>nl<-h
>ACM=matrix(0,nl,k)
>for (i in 1:k){
    ACM[,i]=aa$acf[H,i,i]
}
###All distances can be computed via the package TSclust
> library(TSclust)
>Macf=diss(t(z),METHOD ="ACF", lag.max=h)
>Macf=as.matrix(Macf)
##Similarity matrices based on PACF, Piccolo measure.
## and normalized periodogram are computed
## with the package TSclust
>Mpacf=diss(t(z),METHOD ="PACF", lag.max=h)
>Mpacf=as.matrix(Mpacf)
>Mpic=diss(t(z), METHOD ="AR.PIC")
>Mpic=as.matrix(Mpic)
>MPer=diss(t(z),METHOD ="PER", logarithm=TRUE, normal-
ize=TRUE)
>MPer=as.matrix(MPer)
###Histogram of the distances dropping the zeros
## for each of the four methods and find
## the maximum distance between two series
> par(mfcol=c(2,2))
> hist(Macf[Macf>0])
> max(Macf)
[1] 0.9031959
> arrayInd(which.max(Macf),dim(Macf))
     [,1] [,2]
[1,]  510   29
> hist(Mpacf[Mpacf>0])
```

```
> max(Macf)
[1] 0.9031959
> arrayInd(which.max(Mpacf),dim(Mpacf))
     [,1] [,2]
[1,]  378   29
> hist(Mpic[Mpic>0])
> max(Mpic)
[1] 1.440145
> arrayInd(which.max(Mpic),dim(Mpic))
     [,1] [,2]
[1,]  232   29
> hist(MPer[MPer>0])
> max(MPer)
[1] 0.3543029
> arrayInd(which.max(MPer),dim(MPer))
     [,1] [,2]
[1,]  140   70
> ###The average distances of each method are computed
> md1=apply(Macf,1,mean)
> md2=apply(Mpacf,1,mean)
> md3=apply(Mpic,1,mean)
> md4=apply(MPer,1,mean)
> plot(md1, main="Average acf distances")
> plot(md2, main="Average pacf distances")
> plot(md3, main="Average pic distances")
> plot(md4, main="Average per distances")
> ##identification of extreme series
> which(md1>.5)
V29
 29
> which(md2>.5)
V29
 29
> which(md3>.8)
V29
 29
> which(md4>.12)
V29 V70
 29  70
```

Figure 5.2 shows the histograms of the four distances. The ACF, PACF, and PIC distances give similar results, with most distances following an asymmetric distribution that seems to skew to the right (long distances), indicating the presence of multiple groups. The histogram of the PER, on the other hand, shows a concentration of distances with a small number of large distances, indicating the existence of a large group and a small group. The maximum distance produced by ACF, PACF, and PIC occurs with Series 29. Note that Series 29 and 70 were found as outlying time series in Chapter 2 because most of their values are zero. We have also computed

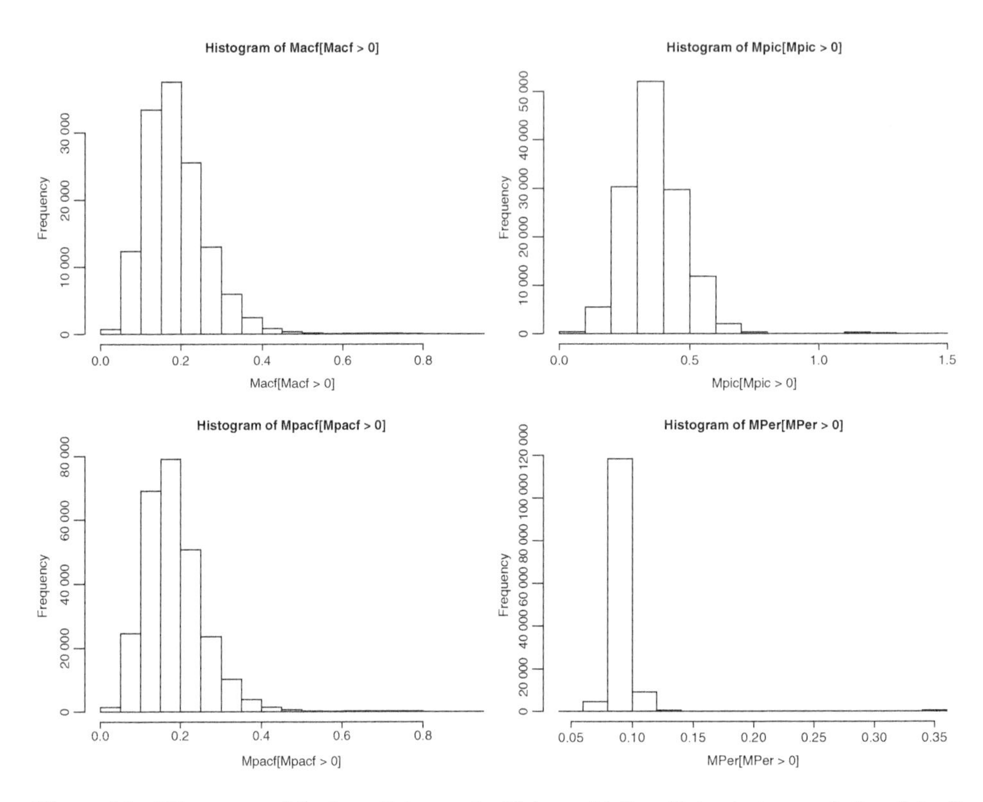

Figure 5.2 Histogram of the four distances for Taiwan AirBox Data. Autocorrelation (Macf) and partial autocorrelation (Mpacf) coefficients, Piccolo distance (Mpic) and periodograms distance (MPer).

the average distance of each time series to all the others and the values are given in Figure 5.3. In all four methods the largest values appear to Series 29. The distances for Series 70 are also large, although not so much. But the PER is the only measure that clearly separates these two outlying series from the others.

We also compute the dissimilarities between the correlation determinants using Eq. (5.12) with $h = 5$. See Figures 5.4 and 5.5. The conclusions are similar to those found via the distances between autocorrelation vectors. The R commands used are given below:

R commands for dissimilarity:

```
## Continue with Example 5.1.
# We compute the autocorrelation matrices for each series
# to compute the difference between the determinants
>nR=h+1
>R=array(0,dim=c(k,nR,nR))
>for (i in 1:k){
> for (j in 1:nR){
>  for (s in 1:nR){
> d=abs(j-s)
```

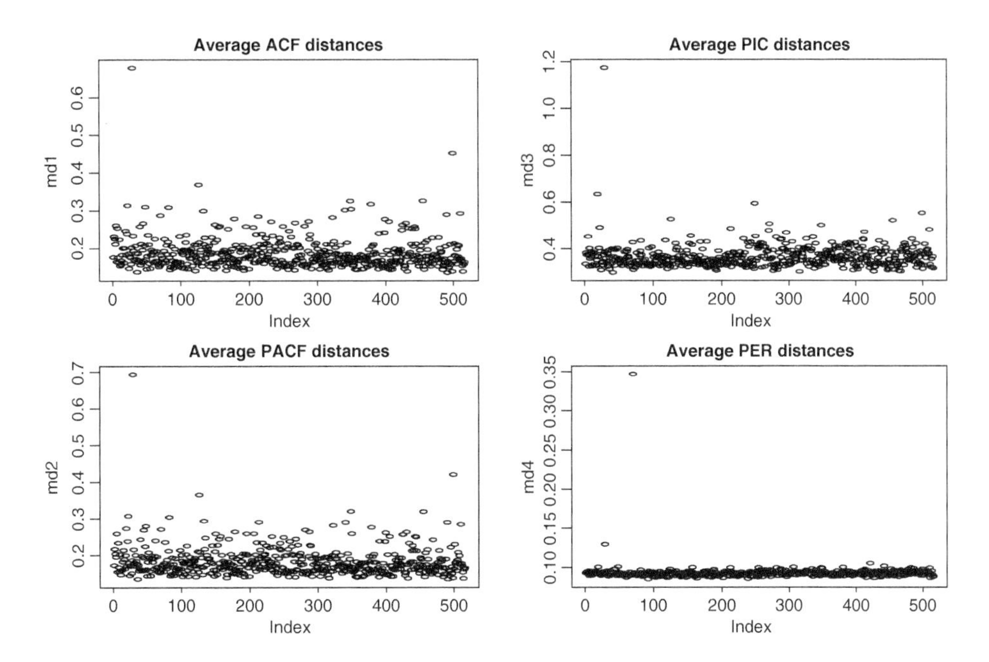

Figure 5.3 Average distances for the Taiwan AirBox Data.

```
> if(d==0) R[i,j,s]=1 else R[i,j,s]<-ACM[d,i]    }  }  }
>DetR=matrix(0,k,1)
>for (i in 1:k){
>  DetR[i]=det(R[i,,])}
>Mdis=matrix(0,k,k)
>for (i in 1:k){
>  for (j in 1:k) {
>    Mdis[i,j]=abs(DetR[i]^(1/nl)-DetR[j]^(1/nl))   }}
>MdM=Mdis[Mdis>0]
>hist(MdM)
>max(MdM)
>arrayInd(which.max(Mdis),dim(Mdis))
>mD=apply(Mdis,1,mean)
>plot(mD, main="Average distances determinant")
```

As a second example, we use the electricity price data in the file `Pelectricity1344.csv`. The data are the weekly series of electricity price each hour of each day during 678 weeks in eight regions of New England (NE), the United States, and we have 1344 series. Note that here $k > T$. The series correspond to each of the seven days of the week, at each of the 24 hours of the day, and for one of the eight regions in NE. That is, $7 \times 24 \times 8 = 1344$. Using the same R commands, Figures 5.6 and 5.7 are obtained. The plots are similar between the two data sets, but their scales indicate that the electricity prices appear to be more homogeneous than the $PM_{2.5}$ measurements. ■

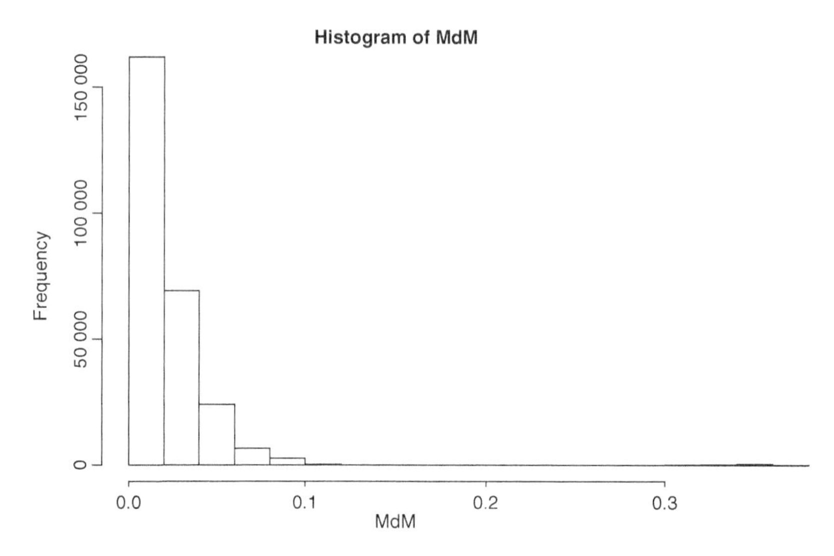

Figure 5.4 Histogram of the distances between correlation determinants for the Taiwan AirBox Data.

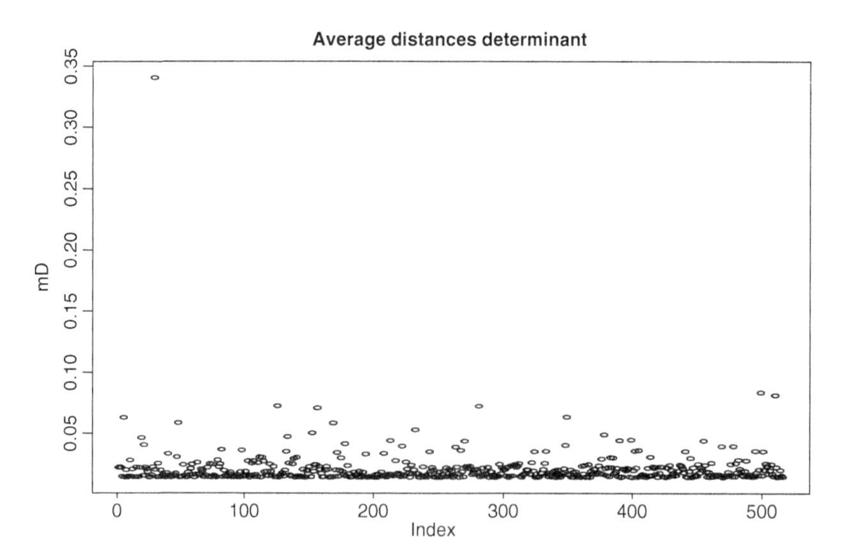

Figure 5.5 Average distances between correlation determinants in the Taiwan AirBox Data.

5.1.3 Dissimilarities Based on Cross-Linear Dependency

The previous measures compare some features that characterize the serial dependence of individual time series, but they do not consider the dependency between any two series. We have seen that the Euclidean distance between the values of two time series only depends on the lag-zero autocorrelation. An alternative way to compare the linear dependency of two series is to use dissimilarities, $D(x_t, y_t)$, which are

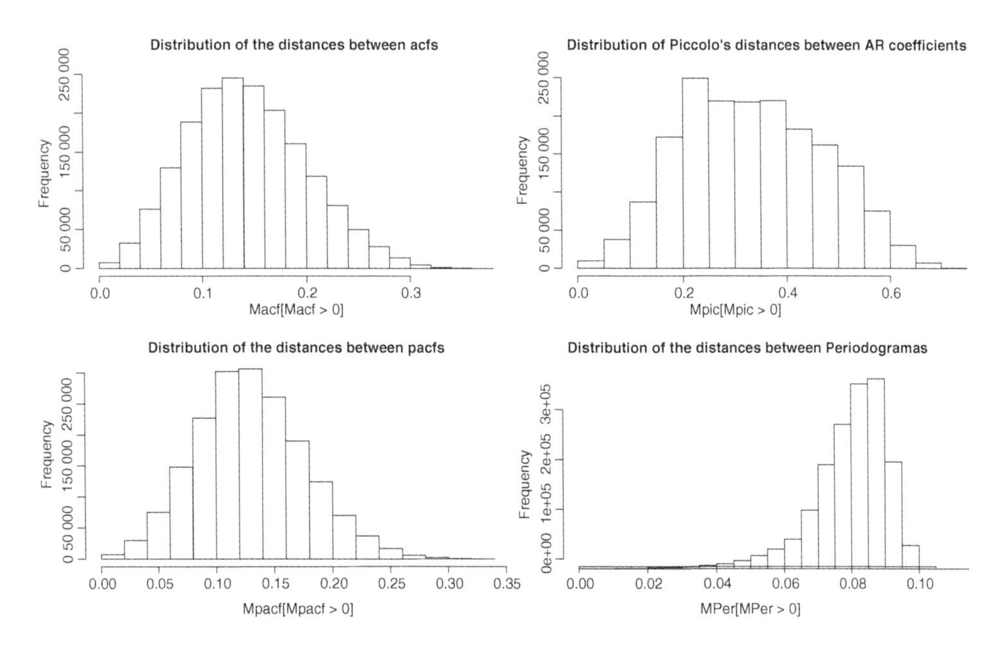

Figure 5.6 Histograms of the four distance measures between univariate series for the electricity price data.

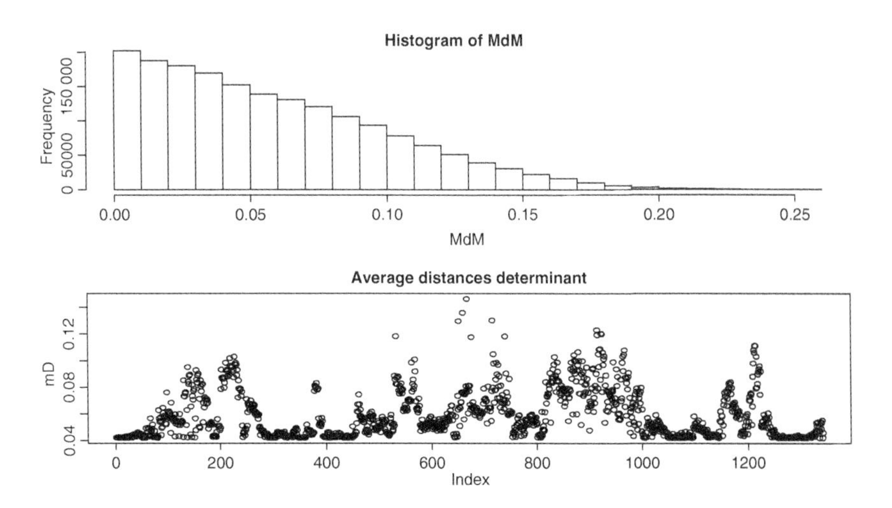

Figure 5.7 Histogram and the distances between determinants for the electricity price data.

positive and symmetric. In addition, we want $0 \leq D(x_t, y_t) \leq 1$ with value 0 if the two series are identical and 1 if and only if all the cross-correlation coefficients between the two time series are zero. Thus, the extreme values of $D(x_t, y_t)$ imply complete linear dependency or complete lack of linear dependence. Finally, we are often interested in the intensity, not the sign, of the dependence between two series. These considerations lead to the following measures.

Consider the cross-correlation vector

$$\boldsymbol{\rho}_{xy} = [\rho_{xy}(-h), \ldots, \rho_{xy}(-1), \rho_{xy}(0), \rho_{xy}(1), \ldots, \rho_{xy}(h)]', \tag{5.13}$$

where h is a positive integer and $\rho_{xy}(-h) = \rho_{yx}(h)$. As a measure of the linear dependency between the two series, we can use the Euclidean norm

$$\|\boldsymbol{\rho}_{xy}\| = \sqrt{\boldsymbol{\rho}'_{xy}\boldsymbol{\rho}_{xy}} = \sqrt{\sum_{j=-h}^{h} \rho^2_{xy}(j)}.$$

But $\|\boldsymbol{\rho}_{xy}\|$ is not a distance between x_t and y_t, because it does not verify the triangle inequality, $\|\boldsymbol{\rho}_{xy}\| \leq \|\boldsymbol{\rho}_{xz}\| + \|\boldsymbol{\rho}_{yz}\|$. For instance, suppose that x_t and y_t are correlated so that $\|\boldsymbol{\rho}_{xy}\| > 0$, but z_t is uncorrelated with both series so that $\|\boldsymbol{\rho}_{xz}\| = \|\boldsymbol{\rho}_{yz}\| = 0$. On the other hand, we have $\|\boldsymbol{\rho}_{xy}\| = 0$ if x_t and y_t are independent. Also, if $x_t = ay_t + b$, where a and b are real numbers with $a \neq 0$, then $\|\boldsymbol{\rho}_{xy}\| = \|\boldsymbol{\rho}_{xx}\| = \|\boldsymbol{\rho}_{yy}\|$, where, according to (5.13)

$$\boldsymbol{\rho}_{yy} = [\rho_y(-h), \ldots, \rho_y(-1), 1, \rho_y(1), \ldots, \rho_y(h)]'.$$

Therefore, we can build a dissimilarity between two series based on their linear dependency by

$$D_{CC2}(x_t, y_t) = 1 - \frac{\|\boldsymbol{\rho}_{xy}\|^2}{\|\boldsymbol{\rho}_{xx}\|\|\boldsymbol{\rho}_{xy}\|}. \tag{5.14}$$

Clearly, if the two series are independent, then $D_{CC2}(x_t, y_t) = 1$, and if the two series are identical, then $D_{CC2}(x_t, x_t) = 0$. However, a drawback of summarizing the dependency with the norm of the vector of cross-correlations is that this measure is invariant to a permutation of the lags. Thus, two series with the same global dependency may have very different distribution of the coefficients among the lags. Also, two identical time series shifted on time may appear as different. See, for instance, $y_t = x_{t-1}$.

A better way to summarize the global linear dependency is with the matrix $\boldsymbol{R}_{xy,h}$ that corresponds to the correlation matrix of the stationary vector process given by $\boldsymbol{Z}_t = (y_t, y_{t-1}, \ldots, y_{t-h}, x_t, x_{t-1}, \ldots, x_{t-h})'$, namely

$$\boldsymbol{R}_{yx,h} = \begin{pmatrix} \boldsymbol{R}_{y,h} & \boldsymbol{C}'_{xy,h} \\ \boldsymbol{C}_{xy,h} & \boldsymbol{R}_{x,h} \end{pmatrix}, \tag{5.15}$$

where $\boldsymbol{R}_{x,h}$ is the $(h+1) \times (h+1)$ non-negative definite correlation matrix of x_t defined before, $\boldsymbol{R}_{y,h}$ corresponds to the correlation matrix of $\boldsymbol{Y}_{t,h} = (y_t, y_{t-1}, \ldots, y_{t-h})'$ and $\boldsymbol{C}_{xy,h}$ includes the cross-correlations between the two vector series. Note that $|\boldsymbol{R}_{yx,h}| = |\boldsymbol{R}_{xy,h}|$, and the determinant verifies that (1) $0 \leq |\boldsymbol{R}_{yx,h}| \leq 1$, with equality to one holding when $\boldsymbol{R}_{yx,h}$ is diagonal and the two series are both serially uncorrelated and linearly unrelated; and (2) $|\boldsymbol{R}_{yx,h}| = 0$ when there exists a non-trivial linear combination $\boldsymbol{a}'\boldsymbol{Z}_t = 0$ so that the series are exactly linearly related. A possible measure of similarity is the *total correlation* between the two time series, as in (5.11), given by

$$\text{TC}_h = 1 - |\boldsymbol{R}_{yx,h}|^{1/2(h+1)}, \tag{5.16}$$

but this measure depends on both the cross-correlations and the autocorrelations of the two series. As

$$|\boldsymbol{R}_{yx,h}| = |\boldsymbol{R}_{x,h}| \left|\boldsymbol{R}_{y,h} - \boldsymbol{C}_{xy,h}\boldsymbol{R}_{x,h}^{-1}\boldsymbol{C}'_{xy,h}\right|, \tag{5.17}$$

if the series are unrelated, and $\boldsymbol{C}_{xy,h} = 0$, then $|\boldsymbol{R}_{yx,h}| = |\boldsymbol{R}_{x,h}||\boldsymbol{R}_{y,h}|$ and this value can be far away from one. In fact, this value will be very small if both series have strong autocorrelations. For instance, $|\boldsymbol{R}_{x,1}| = 1 - \rho_x^2$ will be very small if the first autocorrelation coefficient is close to 1. Second, if the series are identical then $|\boldsymbol{R}_{yx,h}| = 0$ and TC = 1, but a small value of the determinant, or a large value of TC, does not imply a strong relationship between the series. For instance, if $|\boldsymbol{R}_{x,h}|$ is very small, because there are strong autocorrelations in series x_t, then, by (5.17), $|\boldsymbol{R}_{yx,h}|$ will also be small.

A measure of the *average linear dependency* between two series x_t and y_t is the *generalized cross-correlation measure*, GCC(x_t, y_t), defined by Alonso and Peña (2019) as

$$\begin{aligned}
\text{GCC}(x_t, y_t) &= 1 - \left(\frac{|\boldsymbol{R}_{yx,h}|}{|\boldsymbol{R}_{x,h}||\boldsymbol{R}_{y,h}|}\right)^{1/(h+1)} \\
&= 1 - \frac{|\boldsymbol{R}_{y,h} - \boldsymbol{C}_{xy,h}\boldsymbol{R}_{x,h}^{-1}\boldsymbol{C}'_{xy,h}|^{1/(h+1)}}{|\boldsymbol{R}_{y,h}|^{1/(h+1)}}.
\end{aligned} \tag{5.18}$$

The GCC verifies: (1) $\text{GCC}(x_t, y_t) = \text{GCC}(y_t, x_t)$ so the measure is symmetric; (2) $0 \leq \text{GCC}(x_t, y_t) \leq 1$ for Fischer's inequality (see Lütkepohl, 1996); (3) $\text{GCC}(x_t, y_t) = 1$ if and only if there is a perfect linear dependency among the series; and (4) $\text{GCC}(x_t, y_t) = 0$ if and only if all the cross-correlation coefficients are zero.

To prove (3), as $\text{GCC}(x_t, y_t) = 1$ implies $|\boldsymbol{R}_{yx,h}| = 0$, because the denominator is bounded, and then there exists at least a row (column) which is a linear combination of the others rows (columns). To prove (4), write $\text{GCC}(x_t, y_t)$ as

$$\begin{aligned}
\text{GCC}(x_t, y_t) &= 1 - \left|\begin{pmatrix} \boldsymbol{R}_{x,h}^{-1} & \boldsymbol{0}_{xy,h} \\ \boldsymbol{0}_{yx,h} & \boldsymbol{I}_{y,h} \end{pmatrix}\begin{pmatrix} \boldsymbol{R}_{x,h} & \boldsymbol{C}_{xy,h} \\ \boldsymbol{C}'_{xy,h} & \boldsymbol{R}_{y,h} \end{pmatrix}\begin{pmatrix} \boldsymbol{I}_{x,h} & \boldsymbol{0}_{xy,h} \\ \boldsymbol{0}_{yx,h} & \boldsymbol{R}_{y,h}^{-1} \end{pmatrix}\right|^{1/(h+1)} \\
&= 1 - \left|\begin{pmatrix} \boldsymbol{I}_{x,h} & \boldsymbol{R}_{x,h}^{-1}\boldsymbol{C}_{xy,h}\boldsymbol{R}_{y,h}^{-1} \\ \boldsymbol{C}_{xy,h} & \boldsymbol{I}_{y,h} \end{pmatrix}\right|^{1/(h+1)}.
\end{aligned}$$

By the Hadamard's inequality, the right-hand side determinant is smaller or equal to 1 and equality is achieved if and only if the matrix is diagonal, that is, if and only if $\boldsymbol{C}_{xy,h} = \boldsymbol{0}_{xy,h}$.

Notice that for $h = 0$ the $\text{GCC}(x_t, y_t)$ is just the squared correlation coefficient between the two variables. Also, for any h, when both series are white noise and $\rho_{xy}(h) \neq 0$ for some $h \neq 0$ and $\rho_{xy}(j) = 0$ for all $j \neq h$, then $\text{GCC}(x_t, y_t) = \rho_{xy}^2(h)$. In general, for $h > 0$, it can be shown (see Alonso and Peña, 2019) that the $\text{GCC}(x_t, y_t)$ represents the increase in accuracy in prediction of the bivariate model with respect to the univariate models and it can be interpreted as *an average squared correlation coefficient* when we explain the residuals of an AR fitting of one variable by the values of the other.

In summary, a useful measure of dissimilarity between two stationary time series is

$$D_{\mathrm{GCC}} = \left(\frac{|\boldsymbol{R}_{yx,h}|}{|\boldsymbol{R}_{x,h}||\boldsymbol{R}_{y,h}|} \right)^{1/(h+1)}, \tag{5.19}$$

which takes the value 0 when the two series are linearly related (or identical if standardized) and the value 1 if all their cross-correlation coefficients are 0. For two white noise processes with only instantaneous correlation $\rho_{xy}(0)$, $D_{\mathrm{GCC}} = 1 - \rho_{xy}^2(0) = D_{CC2}(x_t, y_t)$.

Example 5.2

To illustrate the computation of dissimilarities based on linear dependency, we use, again, the Taiwan AirBox data and the New England electricity price data of Example 5.1. Below are the R commands used to compute the dissimilarity measures in Eqs. (5.14) and (5.19). The command `GCCmatrix` is part of the book package.

```
#############Compute the GCC measure
## Continue from Example 5.1.
> MCC=matrix(0,k,k)
> for (i in 1:k){
> for (j in 1:k){
> ab=ccf(x[,i],x[,j],lag.max=h,plot=FALSE)
> aa=acf(x[,i],lag.max=h, plot=FALSE)
> bb=acf(x[,j],lag.max=h, plot=FALSE)
> aa2=c(aa$acf,aa$acf)
> aa2=aa2[-1]
> bb2=c(bb$acf,bb$acf)
> bb2=bb2[-1]
> ma=sqrt(sum(aa2*aa2))
> mb=sqrt(sum(bb2*bb2))
> mab=sum(ab$acf*ab$acf)
> MCC[i,j]=1-mab/(ma*mb)          }     }
> par(mfrow=c(1,2))
> hist(MCC[MCC>.01],main="distribution of D(CC)")
##Compute the GCC matrix
>res1=GCCmatrix(z, 4)
>DM=res1$DM
>DMm=DM[DM<1]
>hist(DMm, main="distribution of D(GCC)")
```

Figure 5.8 gives the histogram of the GCC dissimilarities for the Taiwan AirBox data. Most of the series have large dissimilarities indicating that they are uncorrelated, but there is a small group of highly related series. For instance, the average squared correlation between the most related series, occurred between Series 312 and 315, is GCC $= 1 - 0.202 \simeq 0.8$. The situation is rather different for the electricity price data, where most of the series are cross-correlated and the dissimilarities are smaller as shown in Figure 5.9. ■

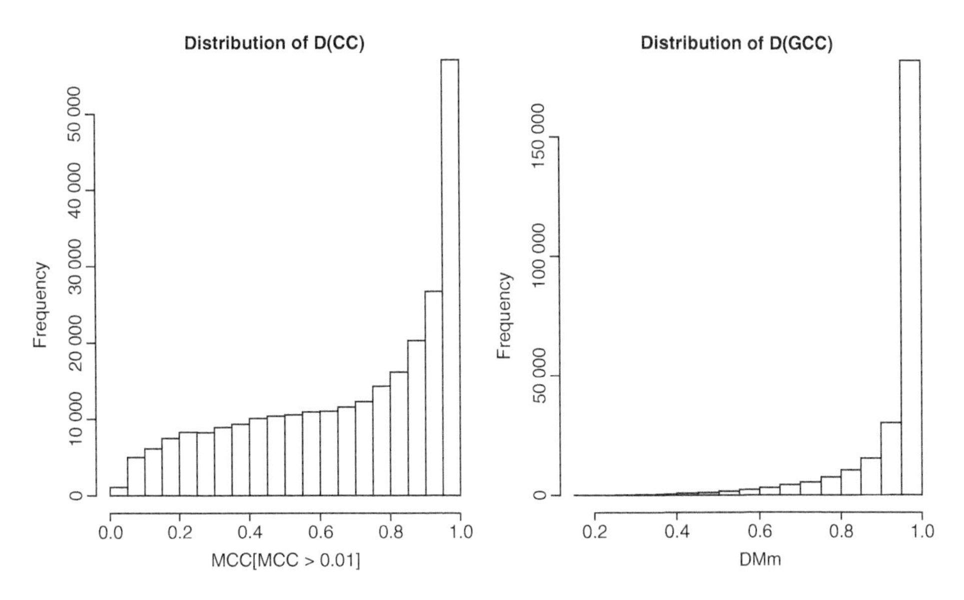

Figure 5.8 Histograms of the dissimilarities with CC and GCC between series in Taiwan Air-Box Data. See Eqs. (5.14) and (5.19).

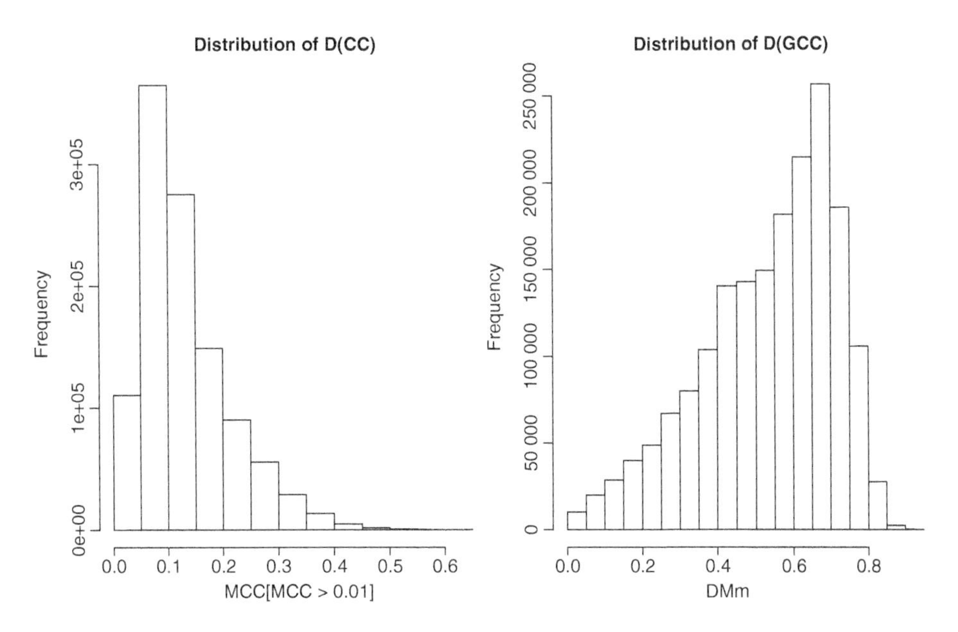

Figure 5.9 Histograms of the dissimilarities with CC and GCC between series in the Electricity Price Data. See Eqs. (5.14) and (5.19).

5.2 HIERARCHICAL CLUSTERING OF TIME SERIES

Hierarchical clustering methods start from a matrix of proximities (distances or dissimilarities), $\boldsymbol{D}$, between the time series, then build a hierarchy based on the matrix. The algorithms used are of the following two types. The first type is agglomerative method that starts with as many groups as individual series and in each subsequent step two groups are merged using proximity. The process continues until all the series belong to a single group. The second type is divisive method that starts with all the series in a single group followed by successive divisions until individual series are reached. In both types of methods, a criterion is used to select the number of groups suitable for the observed data. There is no provision, however, for reallocation of any series which has been incorrectly grouped in earlier steps. In other words, once two groups are merged or divided, they cannot be changed in any subsequent step. For large sets of time series agglomerative algorithms require less computing time and, hence, are most frequently used in practice. From now on, hierarchical methods will refer to them. A hierarchical agglomerative algorithm typically consists of the following three steps:

1. Define as elements to be clustered the k original time series and build a proximity matrix $\boldsymbol{D}$ between elements u and v, where $u, v = 1, \ldots, k$. Put $i = 1$ and $\boldsymbol{D}_i = \boldsymbol{D}$. Denote the (u, v)th element of $\boldsymbol{D}_i$ as $d_{i(uv)}$.
2. Find the minimum distance in $\boldsymbol{D}_i$, say $d^{(i)} = \min_{u,v}\{d_{i(uv)}\}$. Choose a pair of elements such that their distance is equal to the minimum $d^{(i)}$. Group the pair to form a new element. Compute the distances between the new element and other elements. (The methods to compute these distances will be discussed in the next section). The total number of elements is reduced by 1. Advance i by 1, and denote the updated distance matrix also by $\boldsymbol{D}_i$.
3. Go back to Step 2 with the updated $\boldsymbol{D}_i$ matrix and repeat the merging process until the number of elements becomes 1.

This clustering process takes k iterations to complete and the sequence of minimum distances can be denoted by $d^{(1)} \leq d^{(2)} \leq \cdots \leq d^{(k)}$. We use these minimum distances later to help identify the number of clusters.

5.2.1 Criteria for Defining Distances Between Groups

Suppose we have an element A, which is a group with $n_a \geq 1$ time series, and another element B, with $n_b \geq 1$ series. They merge to form a new element, (AB), which has $n_{ab} = n_a + n_b$ time series. The distance between this new element, (AB), to another element C, which has n_c time series, is usually calculated using one of the following three methods:

1. *Simple linkage or nearest neighbor*: The distance between C and (AB) is the minimum of the distances between the time series in C and those in A or B. Mathematically, we have

$$d(C; AB) = \min\{d_{CA}, d_{CB}\}.$$

Since the criterion only takes the minimum distance into account, it may happen that a series in C and a series in (AB) are far apart, even though $d(C; AB)$ is small. In practice, this criterion tends to produce long drawn out clusters, which may contain quite dissimilar elements at the opposite ends.

2. *Complete linkage or furthest neighbor*: The distance between C and the new element (AB) is the largest distance between a series in C and a series in A or B. That is,

$$d(C; AB) = \max\{d_{CA}, d_{CB}\}.$$

As this criterion finds the maximum distance, it is possible that an element C is closer to elements in A than to elements in B, although most time series in C are closer to the time series in B than those to A. In practice, this criterion tends to produce spherical clusters.

3. *Average linkage*: The distance between C and the new group (AB) is average of the distances between C and elements in (AB). That is

$$d(C; AB) = \frac{\sum_{c \in C} \sum_{g \in AB} d_{cg}}{n_c n_{ab}},$$

where n_{ab} is the number of time series in (AB). This criterion is a compromise between the previous two criteria. The main drawback is that the resulting distance depends on the number of elements in C and in AB whereas the other two criteria do not dependent on the number of elements in a group.

5.2.2 The Dendrogram

The dendrogram, or hierarchical tree diagram, is a tree-type graphical display of clustering results with a hierarchical algorithm. This display is useful if the series follow a hierarchical structure, but can be misleading otherwise. Also, it is hard to comprehend with a large data set. The dendrogram is constructed in the following way:

1. The k initial series are *displayed* in the bottom part of the graph.
2. The fusion between the series are represented by three straight lines: two vertical ones connecting the series and a horizontal line that indicates the (distance) level at which the fusion occurs.
3. The process is repeated until all series are connected by straight lines.

If we cut the dendrogram at any given distance (height or deep), we obtain a classification of time series into groups at that distance level. Members of each group are also available. In the next section, we discuss some procedures to determine the number of groups.

5.2.3 Selecting the Number of Groups

5.2.3.1 *The Height and Step Plots* The output of a hierarchical clustering includes the sequence of dissimilarities or distances $d^{(1)} \leq d^{(2)} \leq \cdots \leq d^{(k)}$ at which two elements are merged. These distances correspond to the *heights* in the dendrogram. It is not hard to imagine that if $d^{(i)}$ and $d^{(i+1)}$ are close then the difference between elements formed with height $d^{(i)}$ and those formed with height $d^{(i+1)}$ is relatively small. Therefore, the heights in a dendrogram and their increments

$\Delta_i = d^{(i+1)} - d^{(i)}$ provide certain information about the number of clusters. To make use of this information in determining the number of clusters, we propose a *Height plot* and a *Step plot* for a set of time series $\{z_1, \ldots, z_T\}$, where z_t is a k-dimensional time series:

1. Select one of the distance measures discussed before, such as the ACF distance, to build a dendrogram for z_t.
2. Let y_t be a randomly selected series from z_t.
3. Build a time series model for y_t. For simplicity, we focus on the AR model with order selected by an information criterion, such as Akaike's information criterion (AIC) or Bayesian information criterion (BIC).
4. Perform the following simulation M iterations:
   a. For the jth iteration, use the AR model for the randomly selected series y_t to generate k time series of sample size T. Denoted the simulated data for the jth iteration as $\{z_t^{(j)}\}$.
   b. Apply the same distance measure to $\{z_t^{(j)}\}$ to obtain a dendrogram and call its set of minimum distances $\{d_j^{(1)} \leq d_j^{(2)} \leq \cdots \leq d_j^{(k)}\}$.
5. Let $\lambda_{0.95}^{(i)}$ be the 95%th quantile of the set $\{d_j^{(i)} | j = 1, \ldots, M\}$.
6. Consider a scatterplot of $\{d^{(i)} | i = 1, \ldots, k\}$ of the observed data $\{z_t\}$ versus i. Add to the scatterplot the line $\{\lambda_{0.95}^{(i)}\}$ versus i. The number of clusters is then $g + 1$, where g is the number of points at the right end of the scatterplot that are above the $\lambda_{0.95}^{(i)}$ line.

We referred to this plot the *Height plot* of a dendrogram. Since the simulated time series are all from a single model, every set of the series $\{z_t^{(j)}\}$ should have a single cluster. Therefore, the heights $\{d_j^{(i)}\}$ have no particular meaning except for the scale of the heights. It is then reasonable to use these heights to provide a reference point for the behavior of the heights $\{d^{(i)}\}$ of the observed data. If the observed data came from a single cluster, the $d^{(i)}$ should behave in a similar manner as the simulated reference. Therefore, a deviation from the reference line, given by the 95%th quantiles $\{\lambda_{0.95}^{(i)} | i = 1, \ldots, k\}$, signifies the existence of clusters.

Instead of using the heights $d^{(i)}$ directly, one can obtain a scatterplot of the increments in heights $\{D_i | i = 1, \ldots, k\}$, or steps, versus the index i. This is called the *Step plot*. The reference line is then the increments in the 95%th quantiles of the heights.

To illustrate, we consider the Taiwan AirBox data set that consists of hourly series of $PM_{2.5}$ data. After removing the outlying series (29 and 70), we have $k = 514$ and $T = 744$. Figure 5.10 shows the resulting height plot. From the plot, there are two points that appear to stand out with respective to the reference line obtained from the suggested simulation procedure with $M = 100$ iterations. Therefore, the height plot suggests that there are three clusters in the data set. Figure 5.11 shows the resulting step plot. In this particular case, the step plot suggests two clusters for the AirBox $PM_{2.5}$ data set, but also lends some support for three clusters. The plots are obtained by the command `stepp` of the package for the book.

5.2.3.2 *Silhouette Statistic* The Silhouette statistic, proposed by Rousseeuw (1987), computes a measure of the quality of the final clustering result and selects the

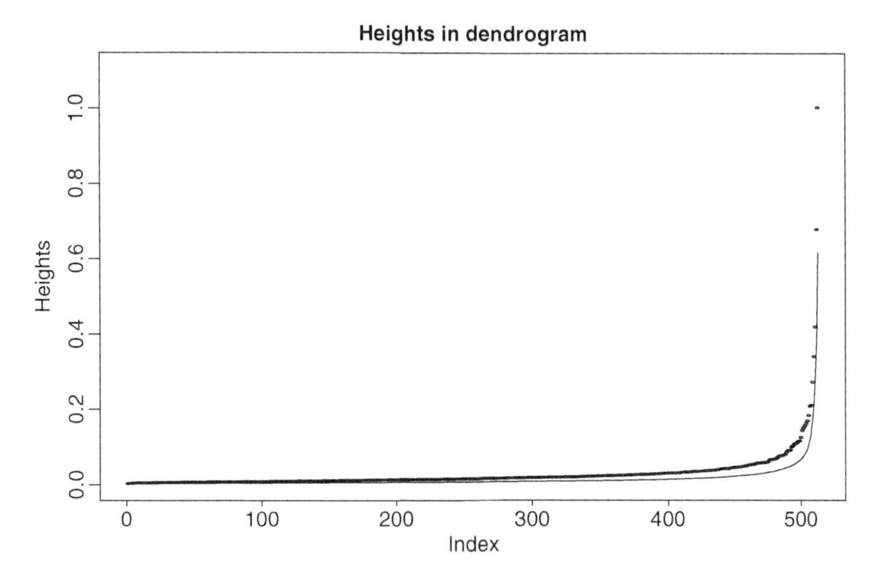

Figure 5.10 The scatterplot of heights of dendrogram for the Taiwan AirBox $PM_{2.5}$ data set. The solid line is a reference line obtained by simulation using a single model for a randomly selected series of the data. ACF is used in distance measure.

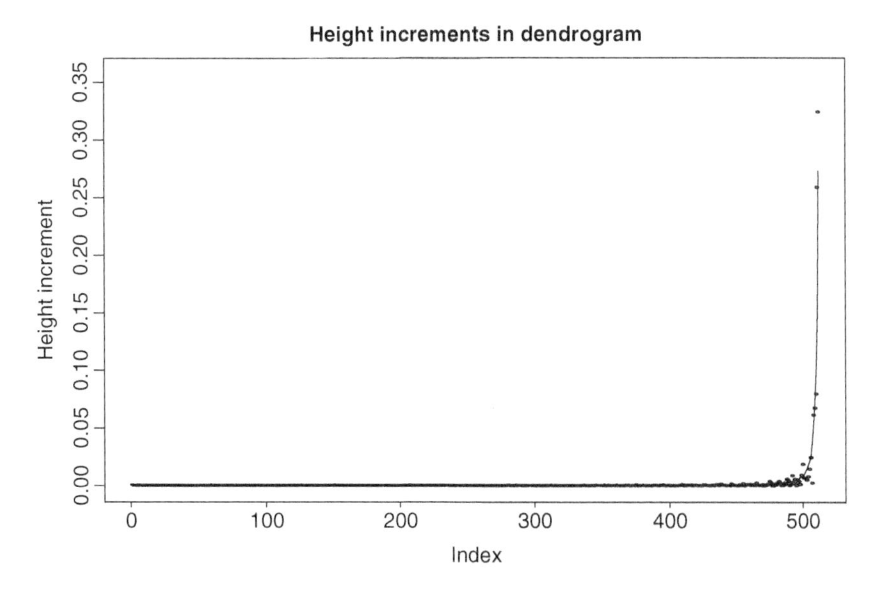

Figure 5.11 The scatterplot of increments in heights of dendrogram for the Taiwan AirBox $PM_{2.5}$ data set. The solid line is a reference line obtained by simulation using a single model for a randomly selected series of the data. ACF is used in distance measure.

number of clusters to maximize the quality measure. The quality of the classification of one element can be defined as the difference between its average distance to members of its group and its average distance to members of the group closest to the one it belongs. Thus, given a distance measure between elements and a clustering result, the average distance of an element to all members of its group is compared with the average distance to members in its closest group. The closest group of an element is defined as a group, different from the one the element is included, that have members with the smallest average distance from the element. If the element is well classified, the average distance to elements in its group should be smaller than that to elements in its closest group, and the larger the difference the better the quality of the classification of this element. This idea can be extended to all the elements in the set and define the quality of the given classification as the average quality of the classification of all the elements. It is computed as follows.

Suppose we have a classification of the observations in g groups, say $G_1, \ldots, G_g$ with $g > 1$. Let $d(i, i')$ be the distance or dissimilarity used to build the groups. Let $a(i)$ be the average distance of the ith element to other elements in its group. For instance, if the ith observation belongs to group G_j with n_j members and i' is any of the other members of this group, $a(i)$ is given by

$$a(i) = \frac{\sum_{i' \in G_j} d(i, i')}{n_j - 1}.$$

To judge the value $a(i)$, whether the element i is well classified, we compare $a(i)$ to the average distance of the ith element to members in other groups G_h, $h = (1, \ldots, g, \mid h \neq j)$. The average distance to group G_h, with n_h elements is given by

$$b_{G_h}(i) = \frac{\sum_{j \in G_h} d(i, j)}{n_h}.$$

Define the closest group to the ith observation as the group with minimum average distance. Note that the concept of closest group is defined for a particular element, and that different members of the same group may have, in general, different closest groups. Let $c(i)$ be the average distance of the ith element in the G_j group to its closest group computed as

$$c(i) = \min_{h \neq j} [b_{G_h}(i)].$$

The quality of the classification of the ith observation is then defined

$$s(i) = \frac{c(i) - a(i)}{\max(a(i), c(i))}.$$

This measure is called the *silhouette measure* of element i. If $c(i) > a(i)$, the element is well classified and $s(i) > 0$. The maximum value of the silhouette measure is $s(i) = 1$, which occurs when $a(i) = 0$, because then the observation is identical to all other members of its group, showing a perfect fit. In general, for observations well classified, $a(i) < c(i)$ and $0 < s(i) < 1$. On the other hand, when the observation is not well classified $s(i) < 0$, which means that the average distance to its closest group is smaller than that to the group it belongs. Then, we should reclassify the ith observation in its closest group. With this reallocation the new value of the silhouette statistic moves from $s(i)$ to $-s(i)$ and would be positive. The minimum value of the statistic $s(i)$ is -1, which occurs when $c(i) = 0$, that is, the ith element is identical to members of some other group, implying a complete lack of similarity with the group it has been classified.

We can look at the distribution of the values of $s(i)$ to decide about the number of clusters. An obvious global measure of the quality of the classification is the average value of this measure over all the observations. This is the Silhouette statistic

$$S = \frac{\sum_{i=1}^{n} s(i)}{n}, \tag{5.20}$$

where $n = \sum n_j$ is the total number of elements to classify. The larger this number, the better the global classification and the number of groups is selected with this criterion as the value that maximizes S in (5.20).

In practice, this criterion is applied in hierarchical clustering by fixing the maximum number of groups, $K_{\max}$, cutting the dendrogram at different heights until reaching the maximum number of groups, comparing the resulting Silhouette statistics, and choosing the clustering with maximum Silhouette statistic. Also, we can use the whole distribution of $s(i)$ to make a choice. Note that the Silhouette statistic is not appropriate to decide if we should form clusters or not, because it is not defined when the number of clusters is 1.

The Silhouette criterion works well when the sizes of the groups are similar but may fail to find groups with a small number of elements. The reason is that a few elements with low negative values in (5.20) would have a small effect in the average. Also, the presence of outliers may affect the statistic. A *robust Silhouette statistic* (Rsilh from now) can be built by first cleaning the set from outlying series with respect to the group. The outlying series can be found as follows: (1) series that join the groups at a distance larger than a given threshold of the distribution of the distances and (2) series that form a group of size smaller than a given minimum size. After identifying any outlying series, the groups are obtained in two steps. First, the silhouette statistic is applied to the set of time series that are not isolated and, second, the isolated series are candidates to be assigned to its closest group according to some chosen distance, that could be the simple linkage (closest element in the group), or other distance. If the series is considered an outlier, it is included in a group 0 of outlying series.

5.2.3.3 *The Gap Statistic* An alternative method to select the number of groups is the Gap statistics introduced by Tibshirani et al. (2001). This statistic compares a measure of the average distance inside the clusters with the expected value of this measure when the data have been generated by a uniform distribution over the space covered by the data. Given a classification of the observations in $g > 1$ groups, $G_1, \ldots, G_g$, an indicator of the homogeneity of the classification in the jth group is the average value of the squared distances between the elements of the group

$$D_j = \frac{1}{2n_j} \sum_{i,i' \in G_j} d^2(i, i'),$$

where n_j the number of elements in G_j and each distance is included twice and for that reason the sum is divided by $2n_j$. A quality measure of the clustering is the sum of average squared distances:

$$W_g = \sum_{j=1}^{g} D_j. \tag{5.21}$$

For instance, suppose that each time series is described by the first p autocorrelation coefficients and call θ_{jr}, the $p \times 1$ vector of ACFs of the series $r = 1, \ldots, n_j$ in group $j = 1, \ldots, g$. The Euclidean distance between pairs of elements in the jth group is

$$D_j = \sum_{r=1}^{n_j} \sum_{s=1}^{n_j} (\theta_{jr} - \theta_{js})'(\theta_{jr} - \theta_{js}) = 2n_j \sum_{r=1}^{n_j} (\theta_{jr} - \bar{\theta}_j)'(\theta_{jr} - \bar{\theta}_j)$$

where $\bar{\theta}_j$ is the mean vector of the jth group. Thus

$$W_g = \sum_{j=1}^{g} \sum_{r=1}^{n_j} (\theta_{jr} - \bar{\theta}_j)'(\theta_{jr} - \bar{\theta}_j) = \sum_{j=1}^{g} n_j s_j^2 \tag{5.22}$$

where $s_j^2 = \sum_{r=1}^{n_j} (\theta_{jr} - \bar{\theta}_j)'(\theta_{jr} - \bar{\theta}_j)/n_j$ is the variance in the jth group. Thus, W_g is the pooled within-groups sum of squares around the group means, which is a standard measure of the quality of the grouping. The gap statistic considers the difference

$$\text{Gap}(g) = E(\log W_g) - \log W_g, \tag{5.23}$$

where the expectation is computed assuming that all the data comes from the same reference distribution. The larger the difference between the expected value under this hypothesis and the observed sum of squares, the better the classification. The reference distribution is usually taken as uniform, which is proved to be the most likely one to produce spurious clusters. The procedure is as follows:

1. Cluster the k time series in $g = 1, 2, \ldots, M$ groups and compute the values $W_1, \ldots, W_M$.
2. Generate C samples assuming that each variable has an uniform distribution over the range of values in the sample and apply the clustering method to each simulated sample to compute the values W_{gb} and $\log(W_{gb})$ for $1 \leq g \leq M$ and $1 \leq b \leq C$. Compute the mean for the gth group

$$\widehat{\log(W_g)} = (1/C) \sum_{b=1}^{C} \log(W_{gb})$$

and the standard deviation

$$\text{sd}_g = \sqrt{(1/C) \sum_{b=1}^{C} [\log(W_{gb}) - \widehat{\log(W_g)}]^2}$$

and the Gap statistic

$$\text{Gap}(g) = \widehat{\log(W_g)} - \log(W_g)$$

3. Choose the number of clusters by the smallest value of g such that

$$\text{Gap}(g) \geq \text{Gap}(g+1) - \text{sd}_g\sqrt{1+(1/C)}$$

Example 5.3

Consider, again, the Taiwan AirBox and the New England electricity price data in the files `TaiwanAirBox032017.csv` and `Pelectricity1344.csv`, respectively, but use the first differenced series, that is, $z_t = \nabla \boldsymbol{x}_t$, where $\boldsymbol{x}_t$ is the observed time series. As before, we removed Series 29 and 70 from the AirBox data set. Figure 5.12 shows the dendrograms of the hierarchical clustering results for the AirBox data based on four distance measures, which are the first 5 lags of ACF and PACF, the PIC distance with AR order selected automatically, and the PER. From the plots, the PACF and PIC distances seem to provide similar clustering results. This is not surprising as PACF is closely related to the AR coefficients. The PER distance, on the other hand, seems to provide slightly different results. We then apply the silhouette and gap statistics with the ACF distance to the data set. The silhouette statistic suggests two clusters whereas the gap statistic indicates four clusters. Furthermore, based on the ACF distance, if we select three clusters, then the cluster sizes are 390, 107, and 17, respectively. The R commands and packages used

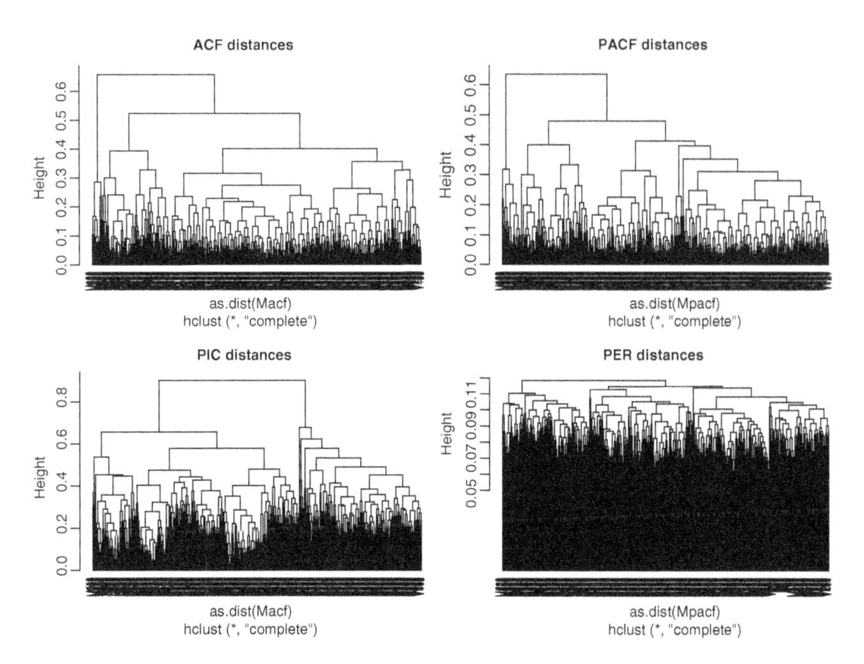

Figure 5.12 Dendrograms of clustering the first differenced series of Taiwan AirBox Data with four distance measures.

and some outputs are given below, where the command `hclust`, for hierarchical clustering, is part of the R package **stats**.

Distance measures and hierarchical clustering: Taiwan AirBox data set

```
>x=as.matrix(TaiwanAirBox032017)
>x=x[,-1]
>x=x[,-c(29,70)]
>z=diff(x)
##All the distances are computed with the package TSclust
>require(TSclust)
>Macf=diss(t(z),METHOD ="ACF", lag.max=5)
>Macf=as.matrix(Macf)
>Mpacf=diss(t(z),METHOD ="PACF", lag.max=5)
>Mpacf=as.matrix(Mpacf)
>Mpic=diss(t(z), METHOD ="AR.PIC")
>Mpic=as.matrix(Mpic)
>MPer=diss(t(z),METHOD ="PER", logarithm=TRUE, normal-
ize=TRUE)
>MPer=as.matrix(MPer)
##A Hierarchical clustering with complete distance is applied
> sc1=hclust(as.dist(Macf),method="complete")
>.........
> sc4=hclust(as.dist(MPer),method="complete")
> par(mfrow=c(2,2)
> plot(sc1,main="ACF distances",hang=-1)
> .........
> plot(sc4,main="PER distances",hang=-1)
##Make the Step plot for each method
> wd1=diff(sc1$height)
> plot(wd1,main="ACF distances")
>.........
> wd4=diff(sc4$height)
> plot(wd4,main="PER distances")
#Apply the Silhouette and gap criteria in the acf cluster
> silh.clus(8,as.dist(Macf), method="complete")
$nClus
[1] 2
$coef
nClus silIndex
1      1 0.0000000
2      2 0.3819963
3      3 0.2006368
4      4 0.1826210
5      5 0.1419417
6      6 0.1313848
7      7 0.1278480
8      8 0.1194526
```

```
###The maximum value of the Silhouette is for 2 groups
> memb=cutree(sc1,1:10)
> gap.clus(Macf,memb, 100)
$optim.k
[1] 9
$gap.values
gap
1 -0.037521773
2 -0.097965198
3 -0.087936795
4 -0.005051419
5 -0.018106280
6 -0.006000857
7 -0.004245895
8 -0.034548137
9  0.001659488
###The maximum Gap is first 4 then  9.
##compute the size of the 3 groups
> memb=cutree(sc1,3)
> g1u=which(memb==1)
> g2u=which(memb==2)
> g3u=which(memb==3)
> ng1=length(which(memb==1))
> ng1
[1] 390
> ng2=length(which(memb==2))
> ng2
[1] 107
> ng3=length(which(memb==3))
> ng3
[1] 17
> ##compares the groups with location data
> ##The location are loaded and analyzed
> locations032017 <- read.csv("XXX/data/locations032017.csv")
> lo=as.matrix(locations032017)
> lo=lo[,-1]
> par(mfrow=c(2,2))
> g1lo=lo[g1u,]
> g2lo=lo[g2u,]
> g3lo=lo[g3u,]
> plot(lo, main="All locations")
> plot(g1lo, main="Group 1 locations")
> plot(g2lo, main="Group 2 locations")
> plot(g3lo,main="Group 3 locations")
```

To understand better the clustering results of the Taiwan AirBox data, we examine carefully the locations of Air Boxes. The longitude and latitude of each Air Box are in the file `locations032017.csv`. The left plot of Figure 5.13 shows the locations

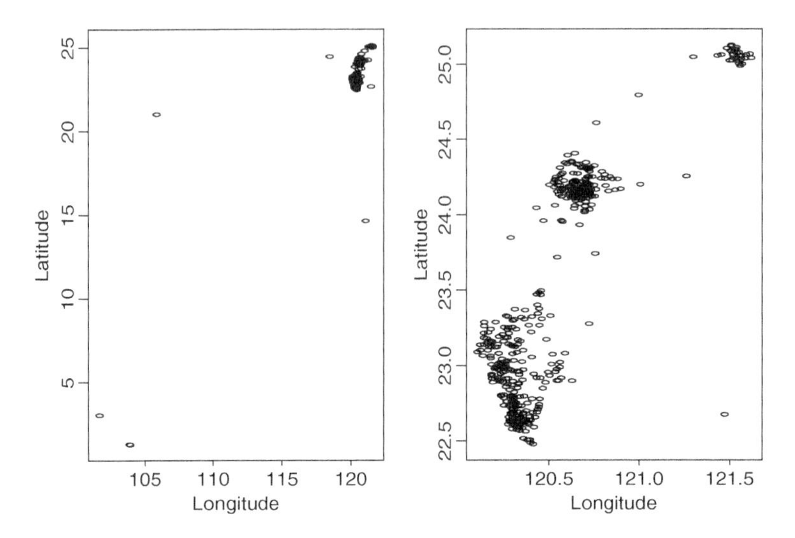

Figure 5.13 Longitude and latitude of Air Boxes for the Taiwan AirBox data. The left plot shows the locations of all 516 series whereas the right plot shows the locations of 508 series with six outlier series removed.

of the Air Boxes, from which we see that there exist outlying locations in the location file because the island is located roughly between 120–122 in longitude and 22–25 in latitude. Therefore, we remove eight series from the data set in our further analysis. These are Series 1, 29, 35, 46, 70, 118, 155, and 157, where the removal of Series 29 and 70 were explained before. The right plot of Figure 5.13 shows the locations of the remaining 508 $PM_{2.5}$ time series around Taiwan.

Using the 508 differenced series of $PM_{2.5}$ series, we apply the four distance measures and the proposed height and step plots of Section 5.2.3.1 to identity the number of clusters for the dataset. The four distance measures are the distances based on the first 5 lags of ACF and PACF, the PIC distance with order selected automatically, and the PER. Figure 5.14 shows the results of height plots. The height plot of the ACF distance is in part (a), from which we see a clear indication of three clusters. The height plot of PACF distance is in part (b), which indicates about five clusters. The height plot of PIC distance is in part (c), which suggests two clusters and the result for PER in part (d) also suggests five clusters.

Figure 5.15 shows the step plots of the 508 differenced Taiwan AirBox series. Again, the same four distance measures are used in the analysis. In this particular instance, the step plots of PACF and PER distance measures, i.e. parts (b) and (d), suggest three and five clusters, respectively. The result of PIC distance is in part (c) and it suggests two clusters. The step plot of ACF distance, on the other hand, indicates there is only a single cluster. We then apply PACF distance measure with three clusters to the data. The sizes of the three clusters are 132, 346, and 40, respectively. Figure 5.16 shows the locations (longitude and latitude) of members of the three clusters. In this particular instance, the membership locations scatter across the Island, even though there appear to have certain concentration. For example, most of the right-upper locations belong to Cluster 1. Overall, for the first-differenced $PM_{2.5}$ series, the PACF distance does not provide clusters based purely on the locations of Air Boxes.

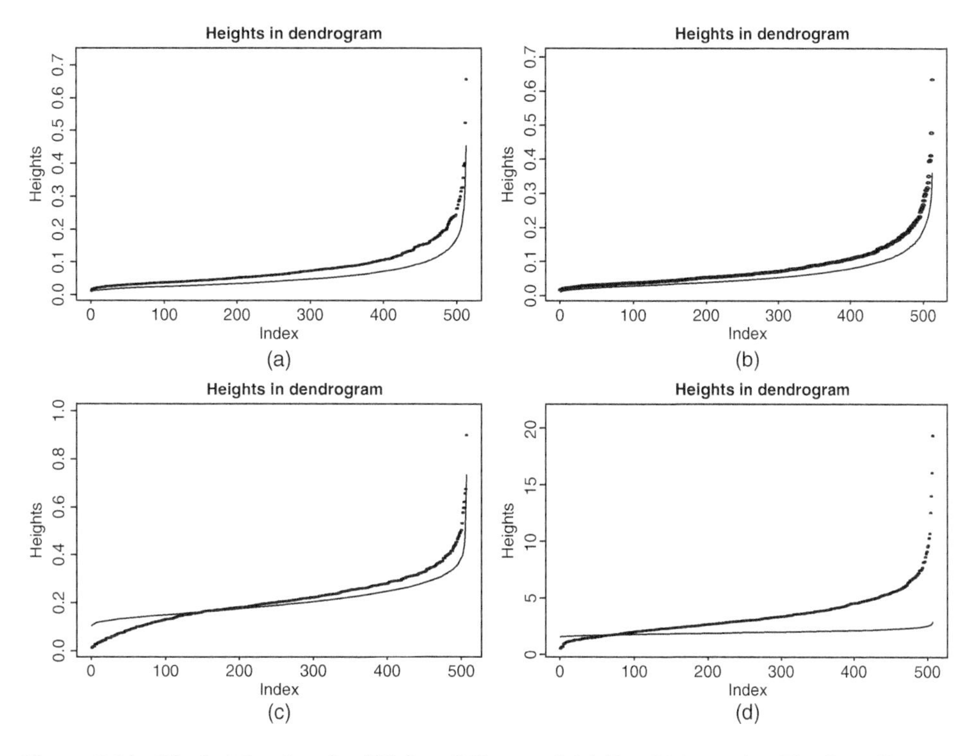

Figure 5.14 The height plots for 508 first-differenced AirBox $PM_{2.5}$ series. The four distance measures used are the first 5 lags of ACF in part (a), and PACF in part (b), the PIC with order selected automatically in part (c), and the PER in part (d).

As discussed in Section 5.1.3, an alternative dissimilarity measure between time series is to use cross-correlations. To demonstrate, we also apply the GCC measure of Eq. (5.18) to the differenced AirBox $PM_{2.5}$ data. The command used is `GCCclus`. The number of clusters selected by the Silhouette statistics is equal to three. The membership locations of the three clusters are show in Figure 5.17, where the upper-left plot shows the locations of all 508 series. From the plots, we see that the three clusters now coincide well with the geographical locations of the Air Boxes. Therefore, in this particular instance, clustering by cross-linear dependency seems to be much more informative than that based on the PACF of individual series.

R commands for locations of Taiwan AirBox data

```
##The locations data are cleaned, 508 series remain
> loc <- read.csv("locations032017.csv")
> loc <- loc[,-1]
> loc1 <- loc[-c(1,29,35,46,70,118,155,157),]
> dim(loc1)
> par(mfcol=c(1,2))
> plot(loc[,2],loc[,1],xlab="longitude",ylab="latitude")
> plot(loc1[,2],loc1[,1],xlab="longitude",ylab="latitude")
> require(TSclust)
> da <- read.csv("TaiwanAirBox032017.csv'')
```

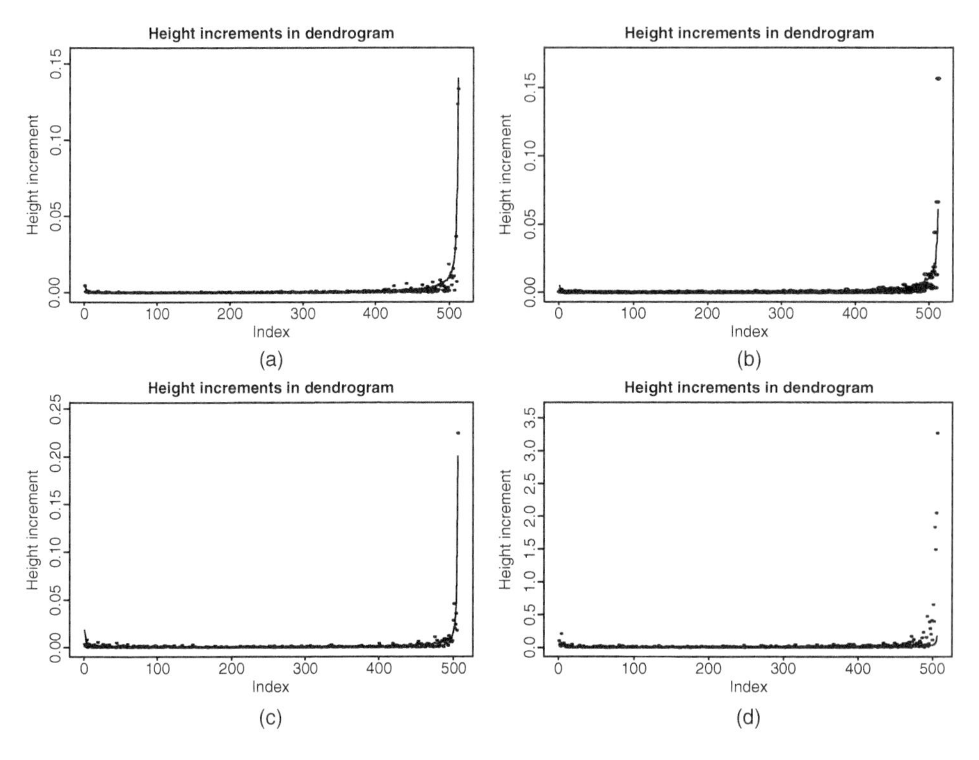

Figure 5.15 The step lots for 508 first-differenced AirBox $PM_{2.5}$ series. The four distance measures used are the first 5 lags of ACF in part (a), and PACF in part (b), the PIC with order selected automatically in part (c), and the PER in part(d).

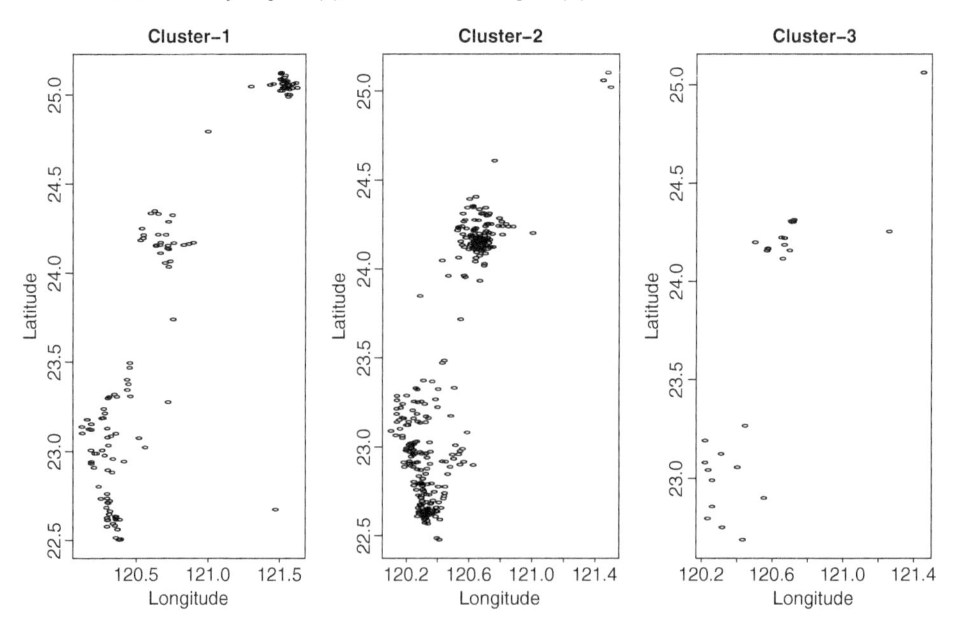

Figure 5.16 Membership locations of the differenced Taiwan AirBox $PM_{2.5}$ data that have 508 series with 743 observations. There are three clusters based on the PACF distance similarity measure.

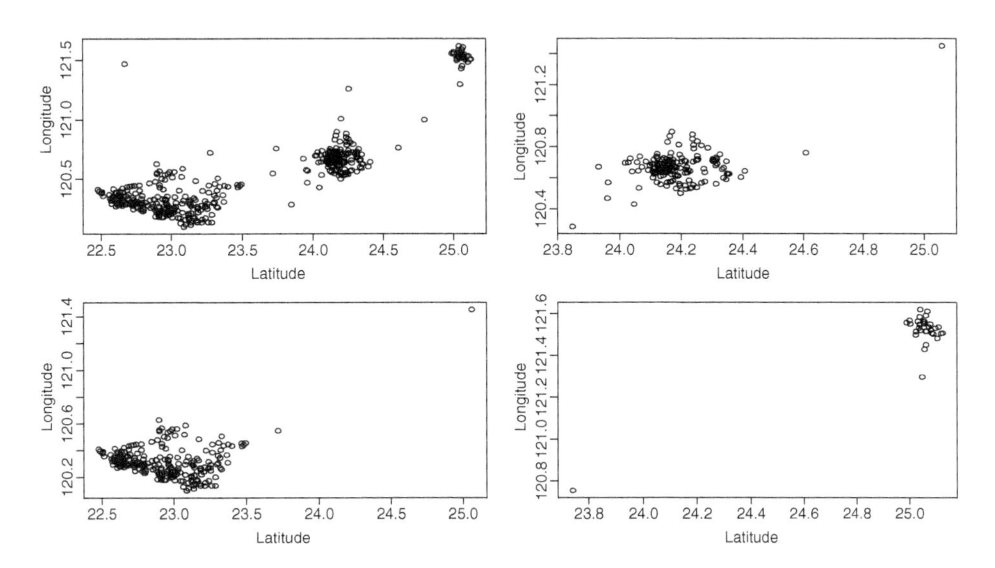

Figure 5.17 Distribution of the cleaned locations of the Taiwan Air Boxes for all 508 series in upper-left, and for three groups selected by cross dependency with the GCC for the increments of the series

```
> da <- as.matrix(da[,-1])
> da <- da[,-c(1,29,35,46,70,118,155,157)]
> dda <- diffM(da)
> mPACF <- diss(t(dda),METHOD="PACF",lag.max=5)
> cPACF <- hclust(as.dist(mPACF),method="complete")
> names(cPACF)
[1] "merge"       "height"      "order"      "labels"      "method"      "call"
[7] "dist.method"
> memb <- cutree(cPACF,3)
> locg1 <- which(memb==1)
> locg2 <- which(memb==2)
> locg3 <- which(memb==3)
> length(locg1)
[1] 132
> length(locg2)
[1] 346
> length(locg3)
[1] 30
> par(mfcol=c(1,3))
> plot(loc1[locg1,2],loc1[locg1,1],xlab="Longitude",ylab="Latitude",
main="Cluster-1")
> plot(loc1[locg2,2],loc1[locg2,1],xlab="Longitude",ylab="Latitude",
main="Cluster-2")
> plot(loc1[locg3,2],loc1[locg3,1],xlab="Longitude",ylab="Latitude",
main="Cluster-3")
>
##A similar analysis with GCC is applied.
##(Not all commands are shown)
##GCC dissimilarity cluster is applied to the cleaned 508 series
> sc1=GCCclus(dda,lag=2,silh=1)
> Number of clusters by using Silhouette statistic 14
```

```
labels abs.freq    rel.freq
1       1       34 0.066929134
2       2      184 0.362204724
3       3        1 0.001968504
4       4      279 0.549212598
5       5        1 0.001968504
6       6        1 0.001968504
7       7        1 0.001968504
8       8        1 0.001968504
9       9        1 0.001968504
10     10        1 0.001968504
11     11        1 0.001968504
12     12        1 0.001968504
13     13        1 0.001968504
14     14        1 0.001968504
##The result is three clusters and several outlier series.
##The three clusters are plotted with their location
>g1=which(sc1$labels==1)
>g2=which(sc1$labels==2)
>g3=which(sc1$labels==4)
>g1lo=loc1[g1,]
>g2lo=loc1[g2,]
>g3lo=loc1[g3,]
>par(mfrow=c(2,2))
>plot(loc1)
>plot(g2lo)
>plot(g3lo)
>plot(g1lo)
```

Turn to the first-differenced series of electricity price of New England regions. Figure 5.18 shows the step plots of Section 5.2.3.1. Again, the same four distance measures are used. They are based on 5 lags of ACF and PACF, the PIC distance with AR order elected automatically, and the PER. The ACF, PACF, and PER distances selects 2, 2, and 3 clusters, respectively. On the other hand, the PIC distance shows a single cluster. We also apply the GCC similarity measure. Using the program `GCCclus`, the number of lags selected is 10 and the number of clusters with the Silhouette statistic is 14. The Dendrogram is shown in Figure 5.19 from Alonso and Peña (2019). Four out of the 14 clusters have 192 observations and correspond to all the series of Tuesday, Wednesday, Thursday, and Friday. Monday series are split into a main group of 191 series and an isolated series (Monday, 10 a.m., Maine), that contains several outlier points. The weekends series, of Saturday and Sunday, are also split into two clusters with (i) sleeping hours from 01 to 06th (or 01 to 07th) hours and (ii) the awake hours from 07 to 24th (08 to 24th) hours.

R commands used for Electricity price data

```
> sc2=GCCclus(z,toPlot=TRUE) #z: denotes the differenced data
> k used for GCC: 10
> Number of clusters by using Silhouette statistic 14
labels abs.freq     rel.freq
1       1      192 0.1428571429
2       2      192 0.1428571429
3       3       56 0.0416666667
```

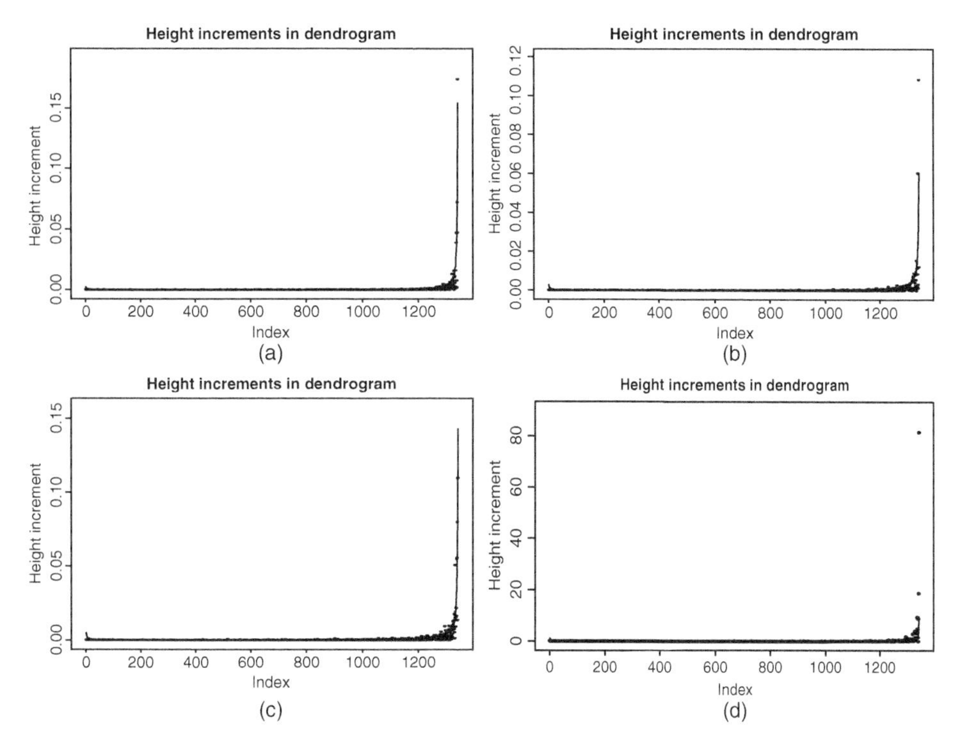

Figure 5.18 The step plots for 1344 first-differenced New England electricity price series. The four distance measures used are the first 5 lags of ACF in part (a), and PACF in part (b), the PIC with order selected automatically in part (c), and the PER in part(d).

```
4         4       135 0.1004464286
5         5         1 0.0007440476
6         6       191 0.1421130952
7         7         1 0.0007440476
8         8        48 0.0357142857
9         9       133 0.0989583333
10       10         1 0.0007440476
11       11         1 0.0007440476
12       12         9 0.0066964286
13       13       192 0.1428571429
14       14       192 0.1428571429
```

■

5.3 CLUSTERING BY VARIABLES

Instead of using a distance matrix $\boldsymbol{D}$ to cluster time series, we can represent each time series by a vector of selected characteristics or features and use these features as variables for classification. For instance, we can represent a time series via the vector of its AR coefficients. Also, when the number of series is large, the distance matrix $\boldsymbol{D}$ is a large squared matrix and we can apply multidimensional scaling techniques (see, for instance, Seber, 1984) to obtain a new $k \times p$ matrix, where $p \ll k$ is the number of

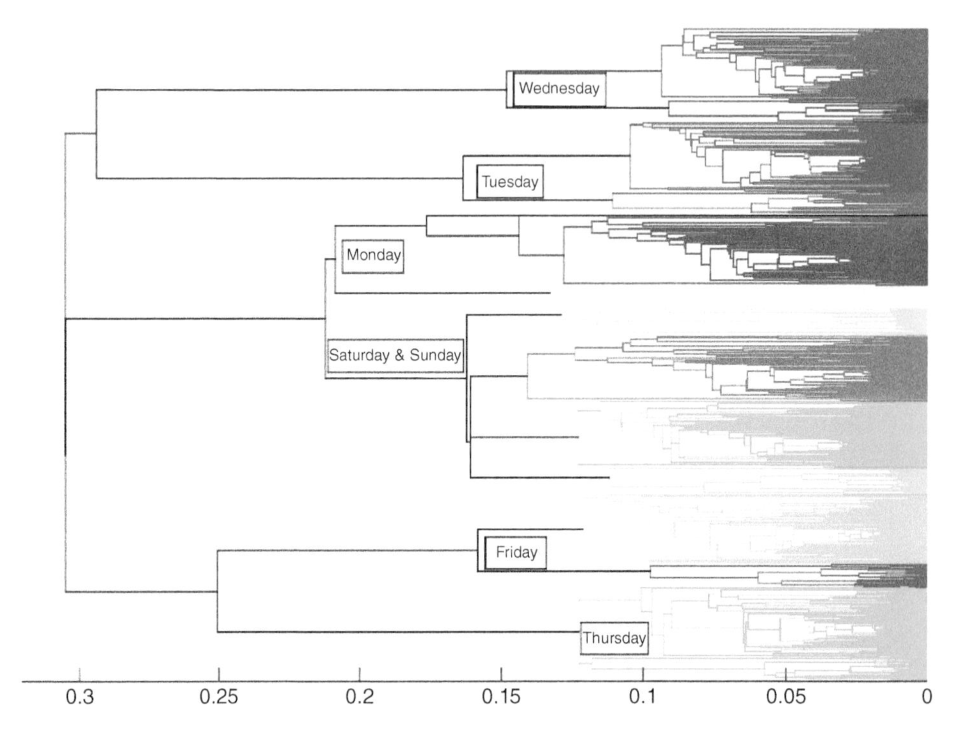

Figure 5.19 Dendrogram for the differenced electricity price data with the GCC dissimilarity of Eq. (5.18).

principal coordinates selected as variables, and use these new variables to perform clustering analysis. In this section, we present four procedures for clustering analysis based on variables. These methods are k-means, k-medoids that can also be used with dissimilarities, projection methods, and mixtures of distributions.

5.3.1 The k-means Algorithm

Given a set of k objects (or k time series in our case) characterized by p variables with observations $(\boldsymbol{x}_1, \dots, \boldsymbol{x}_k)$, we want to partition these objects into a given number of groups, say G, so that each group is as homogeneous as possible. The vectors $\boldsymbol{x}_i$, where $i = 1, \dots, k$, have components x_{ij}, representing the value of the jth variable, $j = 1, \dots, p$, of the ith object. The groups must be mutually exclusive, and each object belongs to one and only one group. The basic idea of k-means is that the grouping is carried out so that the final groups are as homogeneous as possible. This is measured by the within-group variation or *sum of squares within groups* (SSW) for all variables, which is the quantity to be minimized:

$$\text{SSW} = \sum_{g=1}^{G} \sum_{j=1}^{p} \sum_{i=1}^{n_g} (x_{ijg} - \bar{x}_{jg})^2, \tag{5.24}$$

where x_{ijg} is the value of the variable j of subject i in group g and $\bar{x}_{jg}$ is the mean of the jth variable in the gth group that has n_g objects (or time series). The criterion can also be written as

$$\min \text{SSW} = \min \sum_{g=1}^{G} \sum_{j=1}^{p} n_g s_{jg}^2, \tag{5.25}$$

where s_{jg}^2 is the variance of variable j in the gth group. The variances of the variables in the groups are clearly a measure of heterogeneity in the classification, and by minimizing them, we obtain homogeneous groups. An alternative criterion would be to minimize the squared distances between objects and their group centroids. If we measure the distances using the Euclidean distance, this criterion is written as

$$\min \sum_{g=1}^{G} \sum_{i=1}^{n_g} (\boldsymbol{x}_{ig} - \bar{\boldsymbol{x}}_g)'(\boldsymbol{x}_{ig} - \bar{\boldsymbol{x}}_g) = \min \sum_{g=1}^{G} \sum_{i=1}^{n_g} d^2(i, g),$$

where $d^2(i, g)$ is the squared Euclidean distance between object i of group g and the mean of the group. It is easy to prove that the two criteria are identical. Since a scalar is equal to its trace, we can write the latest criterion as

$$\min \sum_{g=1}^{G} \sum_{i=1}^{n_g} \text{tr}[d^2(i, g)] = \min \text{tr} \left[\sum_{g=1}^{G} \sum_{i=1}^{n_g} (\boldsymbol{x}_{ig} - \bar{\boldsymbol{x}}_g)(\boldsymbol{x}_{ig} - \bar{\boldsymbol{x}}_g)' \right].$$

Letting $\boldsymbol{W}$ be the matrix of the sum of squares between groups,

$$\boldsymbol{W} = \sum_{g=1}^{G} \sum_{i=1}^{n_g} (\boldsymbol{x}_{ig} - \bar{\boldsymbol{x}}_g)(\boldsymbol{x}_{ig} - \bar{\boldsymbol{x}}_g)', \tag{5.26}$$

we get

$$\min \text{tr}(\boldsymbol{W}) = \min \text{SSW}$$

and the two criteria coincide. This criterion is also called the *trace criterion* and was proposed by Ward in 1963.

Minimization of the criterion would entail calculating it for all possible partitions in G groups, a clearly impossible task, except with very small number of objects. The k-means algorithm looks for an optimal partition with the constraint that in each iteration you can only move one item from one group to another. Note also that this criterion makes sense if G is fixed, as we can always reduce SSW by increasing the number of groups. In fact, we can always make SSW = 0 by choosing each object to form its own group.

The k-means algorithm is as follows:

1. Choose G points as centroids of the initial groups. This can be done by:
   (a) randomly assigning objects to the G groups and taking the centroids of the groups;
   (b) taking as the centroids the G points farthest apart from each other;
   (c) constructing initial groups using some prior information and calculating their centroids, or selecting their centroids *a priori*.
2. Calculate the Euclidean distances of each object to the G centroids, and assign an object to the group with the shortest distance. Once an object is reassigned, the centroids of affected groups are recalculated.

3. Check whether reassigning some of the objects improves the criterion SSW. This would happen if an object in a group is closer to the centroid of another group. In this case, it is moved to the new group and the centroids of the two groups affected by the change are recomputed.
4. When it is no longer possible to improve the criterion, stop the algorithm.

Consequently, the result of the algorithm can depend on the initial allocation and order of the items. The algorithm should always be repeated from different starting values and permuting the items in the sample. The effect of the order is usually small, but it is advisable to make sure it has no influence in each case.

Both the SSW and trace criteria used by k-means have two important properties. First, they are not invariant to scaling changes. Second, minimizing the Euclidean distance produces approximately spherical groups. With respect to the scale, if the variables are in different units, it is better to standardize them so that the result of the k-means does not depend on irrelevant changes in the scale of the measurements. When they are in the same units, it is usually better not to standardize, as larger variance in one variable may be due precisely to the fact that there are several groups of items with different values of this variable, and such information could be hidden if the data are standardized.

The expected spherical groups produced by k-means can be problematic when the groups are expected to have very different shapes. In this case, we can apply variable transformations so that their joint distribution inside the groups is expected to be approximately normal with diagonal covariance matrix.

5.3.1.1 *Number of Groups* The Silhouette and Gap statistics of Section 5.2 can be applied to determine the number of groups in k-means. In these cases, the distances are the Euclidean distances between the coordinates of the features that represent the series. There are some additional criteria that are specific for variables. Hartigan (1975) proposed to carry out an approximate F-test of variability reduction, comparing SSW_g to SSW_{g+1} and calculating the relative reduction in variability with the increase of an additional group. The test is

$$F = \frac{\text{SSW}_g - \text{SSW}_{g+1}}{\text{SSW}_{g+1}/(n-g-1)}, \tag{5.27}$$

which measures the decrease in variability by increasing one group with the average variability in the clustering. The value of F is compared to an F distribution with p and $p(n-g-1)$ degrees of freedom. However, such a comparison may not be justified in applications. As a general rule, Hartigan (1975) suggested to introduce an additional group if this quotient is greater than 10.

Calinski and Harabasz (1974) proposed to maximize the ratio of the between-groups sum of squares and the within-groups sum of squares. Note that the covariance matrix of the data, $\boldsymbol{V}$, can be written as

$$\mathbf{V} = \sum_{i=1}^{k}(\boldsymbol{x}_i - \bar{\boldsymbol{x}})(\boldsymbol{x}_i - \bar{\boldsymbol{x}})' = \sum_{g=1}^{G}\sum_{i=1}^{n_g}(\boldsymbol{x}_{ig} - \bar{\boldsymbol{x}}_g)(\boldsymbol{x}_{ig} - \bar{\boldsymbol{x}}_g)' + \sum_{g=1}^{G} n_g(\bar{\boldsymbol{x}}_g - \bar{\boldsymbol{x}})(\bar{\boldsymbol{x}}_g - \bar{\boldsymbol{x}})',$$

and applying the trace operator we have

$$\text{tr}(\boldsymbol{V}) = \text{SSW} + \text{SSB}.$$

The CH criterion selects the number of groups by maximizing

$$\mathrm{CH}(g) = \frac{\mathrm{SSB}_g/(g-1)}{\mathrm{SSW}_g/(n-g)}.$$

A useful graphical display is to plot this ratio as a function of the number of groups.

Example 5.4

We apply k-means to the `TaiwanAirBox` data. We use the set of 508 time series after removing eight series with six of them associated with outlying locations and two series with only a few non-zero data points. See Example 5.3. The R command `kmeans` is used, which is in the **stats** package. Three types of variables of different dimensions are used. The first type consists of the first 5 lags of ACF of each series so $p = 5$; the second type is the average value of the series in each of the 31 days in the sample so that $p = 31$; and the third type uses every observation of the time series as a variable so that we have $p = T = 743$. The number of clusters is selected using the Calinski and Harabasz criterion (CHC), that is computed below. A maximum of 20 clusters are allowed and, for each number of groups, k-means is applied with 10 random starting values and the value of the criterion is saved in the vector CHC.

R commands for k-means:

```
##- means with the first five autocorrelation coefficients.
## The outlier-cleaned series of Example 5.3 is used.
> x=as.matrix(TaiwanAirBox032017)[,-1]
> lo=as.matrix(locations032017)[,-1]
> remv = c(1,29,35,46,70,118,155,157) # outlying locations
> x =x[,-remv]
> lo = lo[-remv,]
> z=diff(x)
#the acf of the stationary series z are computed
> h=6
> aa=acf(z,lag.max=h,plot=FALSE)
> H=c(2:6)
> nl=length(H)
> k = dim(z)[2]
#a matrix with columns the acf of each series is formed
> ACM=matrix(0,nl,k)
> for (i in 1:k){
> ACM[,i]=aa$acf[H,i,i]    }
##the k-means is applied
> Kmax=20
> bbb=rep(0,Kmax)
> Wk=as.vector(bbb)
> Bk=Wk
> HC=Bk[-1]
> CHC=HC
> n=dim(ACM)[2]
> for (i in 1:Kmax){
+   center=i
```

```
+    AA=kmeans(t(ACM),centers=i,nstart=10)
+    Wk[i]=AA$tot.withinss
+    Bk[i]=AA$betweens }
> for (j in 1:(Kmax-1)){
>HC[j]=(Wk[j]-Wk[j+1])/(Wk[j+1]/(n-j))
>CHC[j]=(Bk[j]/(j))/(Wk[j]/(n-j+1))      }
> par(mfrow=c(1,1))
> plot(CHC)
###Three groups are identified
> AA=kmeans(t(ACM),centers=3,nstart=10)
> AA

k-means clustering with 3 clusters of sizes 148, 191, 169

Cluster means:
[,1]        [,2]        [,3]        [,4]        [,5]
1  0.13979848 -0.02746675 -0.04727573 -0.06181010 -0.061825575
2  0.08036432 -0.12370731 -0.09944578 -0.03522905 -0.004116231
3 -0.03754526 -0.05904016 -0.03590138 -0.02665427 -0.024233302
(output truncated)
```

Figure 5.20 shows the plot of the CHC criterion, from which we see that three clusters seems reasonable. The sizes of the clusters are 148, 191, and 169, respectively. These groups split the series based on the strength of the first five ACFs. Based on the center of each group, which is the average ACFs, we see that Group 3 seems to contain white noise series, that is, an ARIMA(0,1,0) model for the observed series. However, the groups are not related to the locations of Air Boxes, as shown in Figure 5.21.

Next, we construct variables for each series by taking the daily average of $PM_{2.5}$ measurements. Thus, we have 31 variables for each series and each variable is an average of 24 hourly $PM_{2.5}$ measurements. These variables perform better in showing the locations of the Air Boxes than does the ACF. The group sizes are 196, 171, and 141, respectively. Figure 5.22 shows the locations of the three groups.

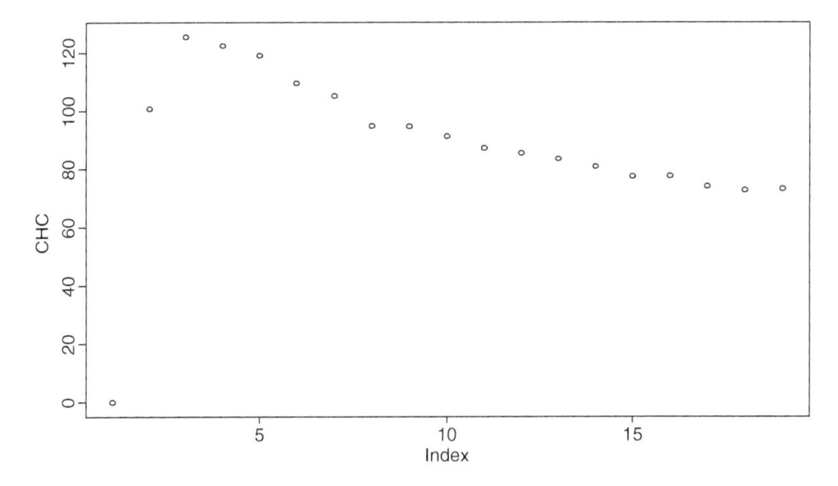

Figure 5.20 The CHC criterion for the Taiwan AirBox Data. The series is differenced.

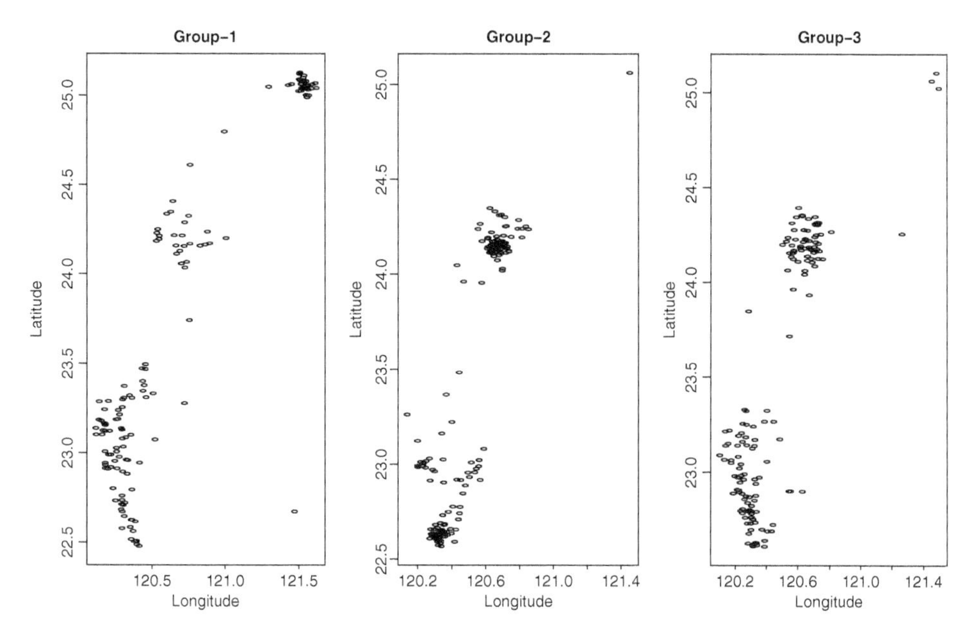

Figure 5.21 Locations (longitude and latitude) of Air-Boxes for the three groups obtained by k-means with the first 5 lags of ACFs.

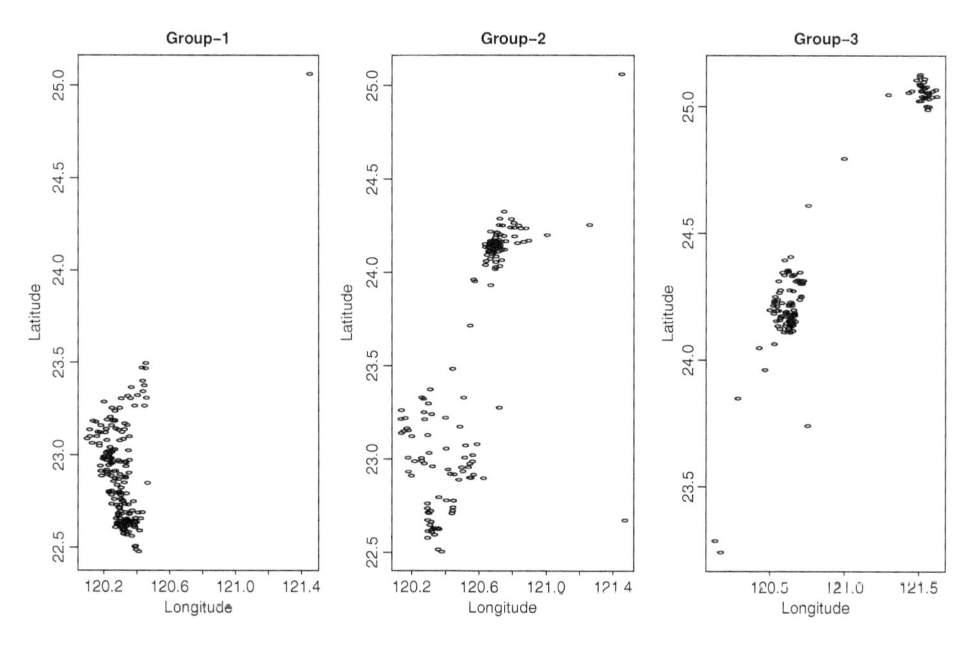

Figure 5.22 Locations (longitude and latitude) of Air Boxes for the three groups obtained by daily average of $PM_{2.5}$ measurements.

R commands for clustering:

```
##A set of variables is built as the average value of series each day
> ke=dim(x)[2]
> Md=matrix(0,31,ke)
> for (i in 1:31){
+   for (j in 1:ke){
+     s1=1+12*(i-1)
+     s2=s1+11
+     Md[i,j]=sum(x[s1:s2,j])/12
+   }
+ }
#For comparison purpose, we also use 3 groups.
> Data=t(Md)
> AD=kmeans(Data,centers=3,nstart=10)
> AD$size
[1] 196 171 141
> lc1 <- which(AD$cluster==1)
> lc2 <- which(AD$cluster==2)
> lc3 <- which(AD$cluster==3)
> par(mfcol=c(1,3))
> plot(loc1[lc1,2],loc1[lc1,1],xlab="longitude",ylab="latitude",
main="Group-1")
# Similar commands apply to Groups 2 and 3.
```

Finally, we take every observation as a variable and apply k-means directly to the observed time series. In this case we have more variables than observations, $p = T = 743 > k = 508$. Again, the CHC criterion selects three groups with sizes 152, 103, and 253. Figure 5.23 shows the locations of the three groups of Air Boxes. It is interesting to see that the locations of Air Boxes in Group 1 and Group 2 are more concentrated, as one would expect. This result is in agreement with Example 5.3, where we showed that taking into account the cross-correlations between the series led to a more meaningful classification in terms of locations of the Air Boxes. The Euclidean distance between two series, by (5.1), would be small if they have a strong positive cross correlation and would increase if the cross-correlation is weak or negative. Thus, the SSW criterion puts in the same group series with strong positive cross-correlations. This is shown in Figure 5.24, that presents the histograms of the correlation between the time series in each group: two of the groups are formed by series strongly correlated, and the third one incorporates series with low or negative correlation. As we found before, geographical neighbors are expected to have high cross-correlations. ■

5.3.2 *k*-Medoids

The k-means algorithm may be affected by outliers and several alternatives have been proposed to address the issue. Kaufman and Rousseeuw (1987) proposed a partition algorithm that is related to k-means but with several differences. First, the center of a group is its medoid, instead of the sample mean. The medoid of a group is its member that minimizes the average dissimilarity (or average distance) to all members of the group. For a scalar variable, the medoid of a group is the median of the L_1 distance. However, for multivariate variables, the median of a group can be

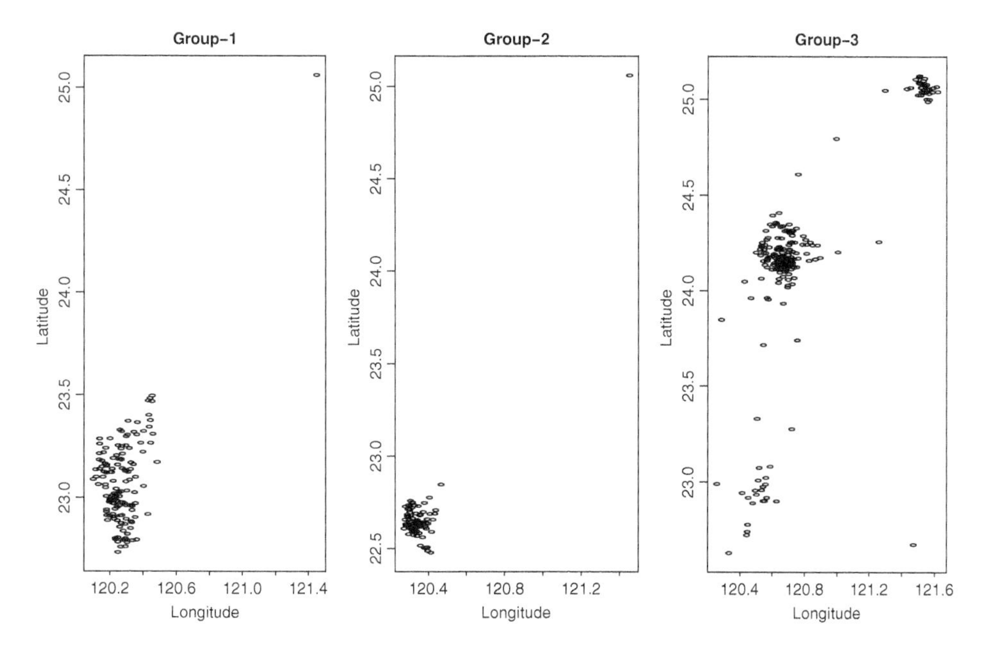

Figure 5.23 Locations (longitude and latitude) of Air Boxes for the three groups obtained by using directly the hourly $PM_{2.5}$ measurements.

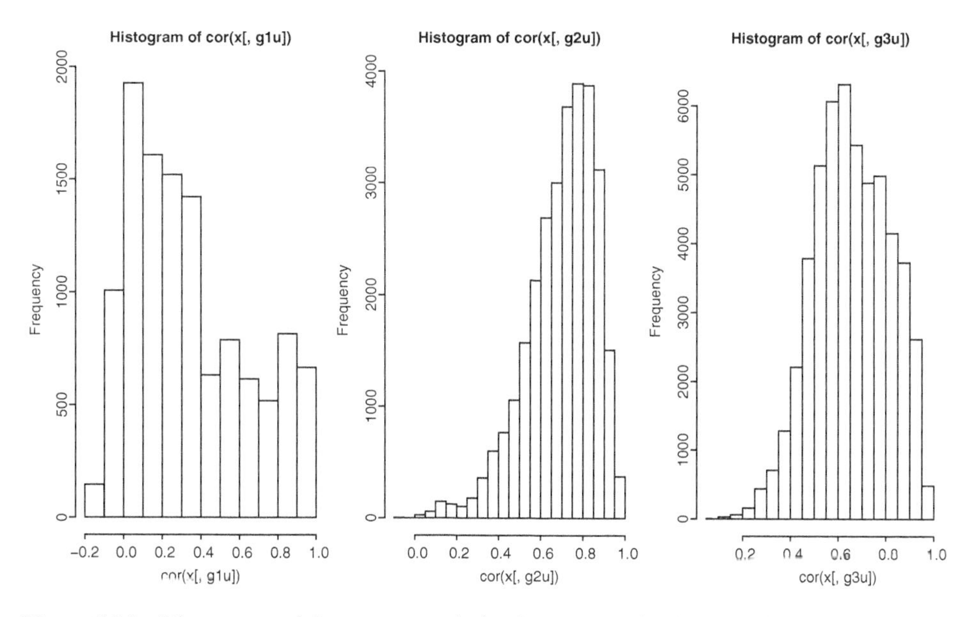

Figure 5.24 Histograms of the cross-correlation between series in each of the groups for the Air Boxes clustered by Euclidean distance between the series.

defined in several ways and the medoid is not unique. Second, the way to swap objects among the groups is different than that of the k-means. Third, it can be applied with any dissimilarity or distance measure between points, $D(i,j)$, whereas the k-means is built with the Euclidean distance. Once the measure $D(i,j)$ is selected, the criterion is to minimize

$$\text{SM} = \sum_{j=1}^{G} \sum_{i \in G_j} D(i, M(j)), \tag{5.28}$$

where $M(j)$ is the medoid of the jth group to which ith object belongs, G is the number of groups and G_j denotes the jth group. The k-medoids algorithm, or partitioning around medoids (PAM) algorithm, works as follows (see Kaufman and Rousseeuw, 2009, for more details):

1. Build: Select G of the k objects (or time series in our case) as leaders (medoids) of the initial groups and associate each object to its closest medoid. This selection is done as follows: choose as the first leader the object that produces the largest decrease in the objective function (5.28). Each object that is not a leader is used to compute the decrease in the objective function and the object with the smallest value of (5.28) is selected as a new leader. Continue in the same way until G leaders are found.
2. Swap: check whether swapping the leader and a member of its group reduces (5.28). Check this for all objects and groups.
3. If it is possible to reduce the objective function SM in Eq. (5.28) by swapping two objects then do so. When it is not possible to reduce the objective function, stop the process.

For a large number of objects, the algorithm is slow, and some approximate algorithms such as the Clustering Large Applications (CLARA) in R and others have been developed. **CLARA** applies PAM to multiple subsamples and give the best result, others only explore a subset of the possible swaps between objects and leaders by sampling.

5.3.3 Model-Based Clustering by Variables

For model-based methods we assume that the k time series were generated by G different models and the objective is to classify the series into these G groups such that every member of a group follows the same model. Typically, a model is characterized by a vector of p parameters. They can be the coefficients of the AR representation of a model or by a general vector of features, that usually are assumed to follows a multivariate normal distribution. A natural approach to carry out the subdivision of the series into groups is to assume that the parameter vectors $\boldsymbol{\theta}_i$ have been generated by a mixture of G multivariate normal distributions. Then, the objective is to estimate jointly the parameters of the distributions in the mixture and the posterior probabilities that each series belongs to one and only one component of the mixture. This approach, called model-based clustering, was proposed, among others, by Banfield and Raftery (1993). These authors also developed the algorithm `MCLUST`, based on mixtures of normal distributions for multivariate data.

For time series analysis, the approach has been studied for ARMA mixtures by Xiong and Yeung (2004) and from a Bayesian point of view by Fruhwirth-Schnatter and Kaufmann (2008). These procedures start with some initial estimation of the allocation of the series to the groups. Then, compute the parameters of the mixture and with those parameters make a new allocation of the series to the groups, and iterate these two steps until convergence. The estimation is carried out group by group and the best solution is selected by the BIC criterion or the posterior probability.

5.3.3.1 *Maximum Likelihood (ML) Estimation of the AR Mixture Model* Suppose that each series is generated by a Gaussian AR(p) model, where $p < k$, of one of G groups with different AR parameters and residual variances. The series are independent and each one $\mathbf{z}_i = (z_{i1}, \ldots, z_{iT})'$, $i = 1, \ldots, k$, can be considered as generated by a mixture of distributions:

$$f(\mathbf{z}_i) = \sum_{g=1}^{G} \pi_g f_g(\mathbf{z}_i | \boldsymbol{\phi}_g, \sigma_g^2), \tag{5.29}$$

where π_g is the unknown probability of the g group, $\boldsymbol{\phi}_g = (\phi_{1g}, \ldots, \phi_{pg})'$ is the vector of AR parameters and σ_g^2 is the variance of the noise in group g, and f_g is the joint density of the univariate observations generated by a Gaussian AR(p), see section 2.8.3. Calling $\boldsymbol{\psi}_g = (\boldsymbol{\phi}_g', \sigma_g^2)$, and $\boldsymbol{\beta} = (\pi_1, \ldots, \pi_G, \boldsymbol{\phi}_1', \ldots, \boldsymbol{\phi}_G', \sigma_1^2, \ldots, \sigma_G^2)'$ to all the parameter vector of the mixture model in Eq. (5.29), the joint conditional log-likelihood for $\widetilde{\mathbf{z}}_i = (z_{i,p+1}, \ldots, z_{i,T})'$, for $i = 1, \ldots, k$, is

$$L_C(\boldsymbol{\beta}) = \sum_{i=1}^{k} \log\left(\sum_{g=1}^{G} \pi_g f_g(\widetilde{\mathbf{z}}_i | \boldsymbol{\psi}_g)\right), \tag{5.30}$$

where $f_g(\widetilde{\mathbf{z}}_i | \boldsymbol{\psi}_g)$ is the joint density function of $\widetilde{\mathbf{z}}_i$ conditioned on the first p observations. This function has many maximums, linked to solutions where each density is determined precisely by an observation. In order to avoid these singularities, we assume that there are at least $p + 1$ observations from each group and try to find a local maximum of this function. An additional problem with this likelihood function is that the normal distributions are not identified, since the ordering $1, \ldots, G$ is arbitrary. To solve this problem, we can assume that the distributions $1, \ldots, G$ correspond to $\pi_1 \geq \pi_2 \geq \cdots \geq \pi_G$.

To maximize the function in Eq. (5.30) with respect to the probabilities π_g, the restriction $\sum_{g=1}^{G} \pi_g = 1$ is introduced with a Lagrange multiplier and the function to be maximized is

$$L_C(\boldsymbol{\beta}) - \sum_{i=1}^{k} \log \sum_{g=1}^{G} \pi_g f_g(\widetilde{\mathbf{z}}_i | \boldsymbol{\psi}_g) - \lambda\left(\sum_{g=1}^{G} \pi_g - 1\right). \tag{5.31}$$

Taking the derivatives of the function with respect to the group probabilities π_g:

$$\frac{\partial L_C(\boldsymbol{\beta})}{\partial \pi_g} = \sum_{i=1}^{k} \frac{f_g(\widetilde{\mathbf{z}}_i | \boldsymbol{\psi}_g)}{\sum_{g=1}^{G} \pi_g f_g(\widetilde{\mathbf{z}}_i | \boldsymbol{\psi}_g)} - \lambda = 0$$

and multiplying by π_g (assuming that $\pi_g \neq 0$, as otherwise the model g is redundant), we write

$$\lambda \widehat{\pi}_g = \sum_{i=1}^{k} \widehat{\pi}_{ig}, \tag{5.32}$$

where $\widehat{\pi}_{ig}$ is the posterior probability, given the data $\widetilde{z}_i$, that series i was generated by the gth distribution. These probabilities are calculated by Bayes' theorem

$$\widehat{\pi}_{ig} = \frac{\pi_g f_g(\widetilde{z}_i|\boldsymbol{\psi}_g)}{\sum_{g=1}^{G} \pi_g f_g(\widetilde{z}_i|\boldsymbol{\psi}_g)}, \tag{5.33}$$

and $\sum_{g=1}^{G} \widehat{\pi}_{ig} = 1$. In order to determine the value of λ, adding up (5.32) for all the groups, we find $\lambda = k$, and, substituting this value in (5.32), the prior group probabilities are estimated by

$$\widehat{\pi}_g = \frac{1}{k} \sum_{i=1}^{k} \widehat{\pi}_{ig}. \tag{5.34}$$

To estimate the AR parameters, we take the derivative of Eq. (5.31) with respect to them that can be written as

$$\frac{\partial L_C(\boldsymbol{\beta})}{\partial \boldsymbol{\phi}_g} = \sum_{i=1}^{k} \frac{\pi_g f_g(\widetilde{z}_i|\boldsymbol{\psi}_g)}{\sum_{g=1}^{G} \pi_g f_g(\widetilde{z}_i|\boldsymbol{\psi}_g)} \frac{\partial \log f_g(\widetilde{z}_i|\boldsymbol{\psi}_g)}{\partial \boldsymbol{\phi}_g} = 0.$$

Using Eq. (5.33) and the results of section 2.8.1, we have

$$\frac{\partial L_C(\boldsymbol{\beta})}{\partial \boldsymbol{\phi}_g} = 0 = \sum_{i=1}^{k} \widehat{\pi}_{ig}(\widetilde{\mathbf{Z}}_i' \widetilde{z}_i - \widetilde{\mathbf{Z}}_i' \widetilde{\mathbf{Z}}_i \boldsymbol{\phi}_g),$$

where $\widetilde{\mathbf{Z}}_i$ is the $(T-p) \times p$ matrix of lagged values. Therefore,

$$\widehat{\boldsymbol{\phi}}_g = \left(\sum_{i=1}^{k} \widehat{\pi}_{ig} \widetilde{\mathbf{Z}}_i' \widetilde{\mathbf{Z}}_i \right)^{-1} \left(\sum_{i=1}^{k} \widehat{\pi}_{ig} \widetilde{\mathbf{Z}}_i' \widetilde{z}_i \right), \tag{5.35}$$

which is the standard equation to estimate the parameters using all the series but each series is weighted by the estimated posterior probability of the series belonging to the groups. A similar equation is obtained for the innovation variances. However, to compute $\widehat{\boldsymbol{\phi}}_g$ we need the $\widehat{\pi}_{ig}$, that are computed by (5.33) and requires the model parameters. Intuitively, we could iterate between both steps, which is the solution provided by the expectation-maximization (EM) algorithm.

5.3.3.2 *The EM Algorithm* The EM algorithm is a powerful iterative method to find ML estimates of missing values and also missing variables (latent variables). It was introduced by Dempster et al. (1977). With missing variables, as in our case, the algorithm iterates between an expectation (E) step, finding the expectation of the log-likelihood with respect to the missing variables using the current estimates of the parameters, and a maximization (M) step, maximizing the log-likelihood with respect to the parameters, given the values of the missing variables.

In order to apply the EM algorithm we transform the problem by introducing a set of unobserved or missing vector variables $(\boldsymbol{s}_1, \ldots, \boldsymbol{s}_k)$, where $\boldsymbol{s}_i$ is a G-dimensional vector of multinomial variables with only one component equal to one, that corresponds to the group to which $\widetilde{\boldsymbol{z}}_i$ belongs, and all the rest equal zero. For example, if $\widetilde{\boldsymbol{z}}_i$ comes from the first model $s_{i1} = 1$ and $s_{i2} = \cdots = s_{iG} = 0$. Then $\sum_{g=1}^{G} s_{ig} = 1$ and $\sum_{i=1}^{k} \sum_{g=1}^{G} s_{ig} = k$. With these new variables, the density function of $\widetilde{\boldsymbol{z}}_i$ conditional on s_i can be written

$$f(\widetilde{\boldsymbol{z}}_i|\boldsymbol{s}_i) = \prod_{g=1}^{G} f_g(\widetilde{\boldsymbol{z}}_i|\boldsymbol{\psi}_g)^{s_{ig}}. \tag{5.36}$$

Analogously, the probability function of the missing variables $\boldsymbol{s}_i$ will be

$$p(\boldsymbol{s}_i) = \prod_{g=1}^{G} \pi_g{}^{s_{ig}}. \tag{5.37}$$

On the other hand, the joint density function of the observed variables $\widetilde{\boldsymbol{z}}_i$ and the missing variables $\boldsymbol{s}_i$ is

$$f(\widetilde{\boldsymbol{z}}_i, \boldsymbol{s}_i) = f(\widetilde{\boldsymbol{z}}_i|\boldsymbol{s}_i)p(\boldsymbol{s}_i),$$

that can be written, by Eqs. (5.36) and (5.37), as

$$f(\widetilde{\boldsymbol{z}}_i, \boldsymbol{s}_i) = \prod_{g=1}^{G} (\pi_g f_g(\widetilde{\boldsymbol{z}}_i|\boldsymbol{\psi}_g))^{s_{ig}},$$

and the joint log-likelihood function is

$$L(\boldsymbol{\beta}) = \sum_{i=1}^{k} \log f(\widetilde{\boldsymbol{z}}_i, \boldsymbol{s}_i) = \sum_{i=1}^{k} \sum_{g=1}^{G} s_{ig} \log \pi_g + \sum_{i=1}^{k} \sum_{g=1}^{G} s_{ig} \log f_g(\widetilde{\boldsymbol{z}}_i|\boldsymbol{\psi}_g). \tag{5.38}$$

The EM algorithm begins with an initial estimate $\widehat{\boldsymbol{\beta}}^{(0)}$ of the vector of parameters, usually computed with Hierarchical clustering or k-means. In step E, we calculate the expectations of the missing values in the complete likelihood (5.38) conditional on the initial parameters and on the observed data. Since the likelihood is linear in s_{ig}, this is equivalent to substituting the missing variables by its expectations. The missing variables, s_{ig}, are binomial with values 0,1, and

$$E(s_{ig}|\widehat{\boldsymbol{\beta}}^{(0)}, \boldsymbol{Z}) = p(s_{ig} = 1|\widehat{\boldsymbol{\beta}}^{(0)}, \boldsymbol{Z}) = \widehat{\pi}_{ig}^{(0)}$$

where $\widehat{\pi}_{ig}^{(0)}$ is the probability that object $\boldsymbol{z}_i$ comes from model g when $\boldsymbol{z}_i$ has already been observed and the parameters of the models are those given by $\widehat{\boldsymbol{\beta}}^{(0)}$. These are the a posteriori probabilities that can be computed by (5.33) using $\widehat{\boldsymbol{\beta}}^{(0)}$ as the parameter values. By substituting the missing variables with their expectations, we obtain

$$L_C^*(\boldsymbol{\beta}) = \sum_{i=1}^{k}\sum_{g=1}^{G} \widehat{\pi}_{ig}^{(0)} \log \pi_g + \sum_{i=1}^{k}\sum_{g=1}^{G} \pi_{ig}^{(0)} \log f_g(\widetilde{z}_i|\boldsymbol{\psi}_g)$$

In step M, this function is maximized for the parameters $\boldsymbol{\beta}$. We see that the parameters π_g appear only in the first term and those of the normal distributions only in the second and, therefore, they can be obtained independently. Starting with the π_g, those parameters are subject to their sum being one, and they are computed by (5.34). Then, we compute the new estimates of the AR parameters and the residual variances with the values $\widehat{\pi}_{ig}^{(0)}$. In this way a new vector of parameters, $\widehat{\boldsymbol{\beta}}^{(1)}$, is obtained and the algorithm is iterated until convergence.

5.3.3.3 Estimation of Mixture of Multivariate Normals

The EM algorithm can also be applied if we use a vector of features $\boldsymbol{x}_i$ to summarize each time series. Assuming that this vector follows a mixture of multivariate normal distribution, with mean vector $\boldsymbol{\mu}_g$, covariance matrix $\boldsymbol{V}_g$, and probabilities π_g, for $g = 1, \ldots, G$, and calling $\boldsymbol{\psi} = (\boldsymbol{\pi}, \boldsymbol{\mu}_1, \ldots, \boldsymbol{\mu}_G, \boldsymbol{V}_1, \ldots, \boldsymbol{V}_G)$ to the vector of parameters, the estimation algorithm is as follows:

1. Start with an initial value $\widehat{\boldsymbol{\psi}}^{(0)}$. Let $s = 1$ and calculate $\widehat{\pi}_{ig}^{(s)}$ with

$$\pi_{ig}^{(s)} = \frac{\pi_g^{(s-1)} f_g^N(\boldsymbol{x}_i|\widehat{\boldsymbol{\mu}}_g^{(s-1)}, \widehat{\boldsymbol{V}}_g^{(s-1)})}{\sum_{g=1}^{G} \pi_g^{(s-1)} f_g^N(\boldsymbol{x}_i|\widehat{\boldsymbol{\mu}}_g^{(s-1)}, \widehat{\boldsymbol{V}}_g^{(s-1)})} \tag{5.39}$$

   and

$$\widehat{\pi}_g^{(s)} = \frac{1}{k}\sum_{i=1}^{k} \pi_{ig}^{(s)}. \tag{5.40}$$

2. Estimate the parameters of the multivariate normals using

$$\widehat{\boldsymbol{\mu}}_g^{(s)} = \sum_{i=1}^{k} \frac{\pi_{ig}^{(s)}}{\sum_{i=1}^{n} \pi_{ig}^{(s)}} \boldsymbol{x}_i, \tag{5.41}$$

$$\widehat{\boldsymbol{V}}_g = \sum_{i=1}^{k} \frac{\pi_{ig}^{(s)}}{\sum_{i=1}^{k} \pi_{ig}^{(s)}} (\boldsymbol{x}_i - \widehat{\boldsymbol{\mu}}_g)(\boldsymbol{x}_i - \widehat{\boldsymbol{\mu}}_g)'. \tag{5.42}$$

3. Go to 1 with $s \leftarrow s + 1$ and iterate the two steps until convergence.

The application of the prior method for a large number of groups requires some simplification to avoid the growth in the number of parameters that require estimation and to ensure the positive-definiteness of covariance matrices. The program **MCLUST**, Fraley and Raftery (1999) and Fraley et al. (2012), uses the EM algorithm by parameterizing the covariance matrices by their spectral decomposition

$$\boldsymbol{V}_g = \lambda_g \boldsymbol{C}_g \boldsymbol{A}_g \boldsymbol{C}_g' \tag{5.43}$$

where $\boldsymbol{C}_g$ is an orthogonal matrix with the eigenvectors of $\boldsymbol{V}_g$ and $\lambda_g \boldsymbol{A}_g$ is the matrix of its eigenvalues, with the scalar λ_g being the largest value of the matrix. The scalar λ_g indicates the volume of the ellipsoid, the eigenvectors of the matrix in $\boldsymbol{C}_g$ indicate the directions and the eigenvalues in $\boldsymbol{A}_g$ indicate the shape of the ellipsoid. In this way, we can let the directions of some groups, or the shape of others, be different. The program (1) chooses a value M for the maximum number of groups, (2) estimates the parameters of the mixture via the EM algorithm for $G = 1, \dots, M$, and for different assumptions on the decomposition (5.43), fixing the initial conditions by hierarchical clustering, (3) uses the BIC criterion to select the most convenient number of groups and conditions for the covariance matrices. The number of parameters used in BIC calculation depends on G and the number of parameters for the model in each group.

5.3.3.4 Bayesian Estimation Model-based clustering can also be carried out from a Bayesian point of view using Markov chain Monte Carlo (MCMC) methods in estimation, as proposed by Fruhwirth-Schnatter and Kaufmann (2008). Suppose, as before, we have G groups each with a unique AR(p) model for its member series. We need to estimate the model parameters in each group, where $\boldsymbol{\psi}_g = (\boldsymbol{\phi}'_g, \sigma_g^2)$, $g = 1, \dots, G$, and we call $\boldsymbol{\psi} = (\boldsymbol{\psi}_1, \dots, \boldsymbol{\psi}_G)$ the parameter vector for the mixture. We also need the classification variables, $\boldsymbol{S} = (s_1, \dots, s_k)$, that are assumed now to be independent variables each follows a multinomial distribution with probabilities $p(s_i = g) = n_g/k = \pi_i$, where n_g is the number of series in the gth group, which is unknown and to be estimated, and call $\boldsymbol{\pi} = (\pi_1, \dots, \pi_G)'$. The distribution of series z_i, $i = 1, \dots, k$, given the parameters and the classification variables, $p(z_i|\boldsymbol{\psi}_g, \boldsymbol{S})$, is assumed to be normal (this can be generalized). We employ conjugate priors for the parameter vector $\boldsymbol{\psi}_g$ and a Dirichlet prior for $\boldsymbol{S}$. Now, a natural decomposition of the parameters is as follows: (1) given the classification variables, estimate the AR parameters in each group as well as the residual variance and (2) given the model parameters estimate the classification variables. Then, the application of a MCMC procedure (or Gibbs sampler) is as follows:

1. Initialization: Draw the initial estimates from the prior distributions or apply the *k-means* method with G clusters to the k time series. In the latter case, obtain the initial estimates of the AR(p) model by the pooling series in each group.
2. For iteration $m = 1, \dots, M$, where M is a pre-specified positive integer, (a) update the classification variables by drawing random samples from the posterior $p(s_i|\, z_i, \boldsymbol{\psi}, \boldsymbol{\pi})$, using the current parameter values, and (b) update the AR coefficients and residual variance by drawing from their conditional posterior distributions, $p(\boldsymbol{\psi}|\, \boldsymbol{Z}, \boldsymbol{S})$, and update $\boldsymbol{\pi}$ from the posterior $p(\boldsymbol{\pi}|\, \boldsymbol{Z}, \boldsymbol{S})$
3. Discard the first N iterations of the random draws and use the remaining draws of $M - N$ iterations to make inference.

In practice, one should use a sufficiently large M and check for the convergence of the MCMC algorithm.

The number of groups is selected as follows. Let M_G be the model specification with G groups, where $G = 1, \dots, G_{max}$, where G_{max} is a prespecified positive integer, and use a uniform priori $p(M_G) = 1/G_{max}$. Compute the marginal likelihood

$$p(\mathbf{Z}|M_G) = \int p(\mathbf{Z}|\boldsymbol{\psi}, \boldsymbol{\pi}, \boldsymbol{S}, G)p(\boldsymbol{\psi}, \boldsymbol{\pi}, \boldsymbol{S})d\boldsymbol{\psi}d\boldsymbol{\pi}d\boldsymbol{S},$$

and combine it with the prior $p(M_G)$ with Bayes' theorem to compute the posterior $p(M_G|\mathbf{Z})$ to determine the number of groups. See Fruhwirth-Schnatter and Kaufmann (2008) and Tsay (2014, chapter 6) for details of the method used. As in all MCMC procedures convergence of the iterations must be checked and the readers are referred to Gamerman and Lopes (2006) and the references therein. This approach has been used to improve forecasting of many time series by Wang et al. (2013).

5.3.3.5 *Clustering with Structural Breaks* Wang and Tsay (2019) proposed a Bayesian approach for clustering time series with structural breaks generalizing the work of Fruhwirth-Schnatter and Kaufmann (2008). The number of breaks is treated as a random variable, with group membership and group-specific parameters allowed to change on these breaks. Once breaks and group memberships are found, it is possible to pool information across all series in the same group for estimation and prediction. When observations of a new period are available, the estimation results for regime sequence and parameters can be used as initial values for updating without rerunning the estimation.

Example 5.5

We used the **mclust** package for model-based clustering of the Taiwan AirBox data. The series are summarized by the first five autocorrelation coefficients. The R commands used are given below. We do not repeat the computation of the ACM matrix as it is the same as that of Example 5.4. The main command is `mclustBIC`, that provides a summary of the values of the BIC as approximations to the posterior probabilities of the model under different scenarios, given by the structure of the covariance matrices in the mixture of normals. Once the number of clusters is chosen, the command `Mclust` gives the estimation result of the mixture model.

R commands for model-based clustering with package **mclust**:

```
##application of mclust to ACM of the AirBox data
> require(mclust)
> Dat=t(ACM)
> st=mclustBIC(Dat, G=1:12)
fitting ...
|============================
> plot(st)
plot(st)
> st
Top 3 models based on the BIC criterion:
VVE,3     EEV,2     EVE,3
7362.747 7353.859 7342.181
##Given that the maximum values of the BIC for most of the models
##is for G=3, three clusters are selected.
```

```
> s1=Mclust(Dat, G=3)
fittin
====================
| 100
> summary(s1)
----------------------------------------------------
Gaussian finite mixture model fitted by EM algorithm
----------------------------------------------------
## output edited to save space
Clustering table:
1   2   3
35 196 277
> s1$BIC
Bayesian Information Criterion (BIC):
EII      VII      EEI      VEI      EVI     VVI      EEE
3 7209.426 7224.402 7322.735 7341.534 7306.48 7317.736 7298.159
EVE      VEE      VVE      EEV      VEV      EVV      VVV
3 7342.181 7303.551 7362.747 7297.843 7299.565 7266.905 7255.187

Top 3 models based on the BIC criterion:
VVE,3    EVE,3    VEI,3
7362.747 7342.181 7341.534
> pp=s1$parameters
> pp$mean
[,1]        [,2]        [,3]
[1,]  0.19604984  0.10948809  0.007026086
[2,] -0.01327302 -0.07975918 -0.078649900
[3,] -0.05420812 -0.08667588 -0.049024126
[4,] -0.08677588 -0.04970037 -0.027687299
[5,] -0.05664926 -0.02840346 -0.023241266
```

Figure 5.25 shows the output of the program `mclustBIC` that includes the BIC values (proportional to the posterior probabilities) for different numbers of groups and covariance matrices structures. The largest posterior probabilities or BIC values appear with three groups and the covariance matrices VVE (ellipsoidal with eigenvalues and eigenvectors depending on the group), EVE (ellipsoidal with eigenvalues that depend on the group but the same eigenvectors), and VEI (diagonal with variances that depend on the group). As expected, they separate the series by the strength of the autocorrelations. The model with the highest probability, VVE,3, has three groups and general covariance matrices for the autocorrelation coefficients. ■

5.3.4 Clustering by Projections

When the dimension p of the feature vector $\boldsymbol{x}_i$ and the number of series, k, are high, with $k > p$, it would be beneficial to project the feature vector into a lower dimensional subspace on which clustering can easily be carried out. The idea of finding a low-dimensional projection space that is useful to summarize the high-dimensional data is referred to as the *projection pursuit* by Friedman and Tukey (1974).

Peña and Prieto (2001) proposed a one-dimensional projection pursuit algorithm based on directions obtained by either maximizing or minimizing the kurtosis

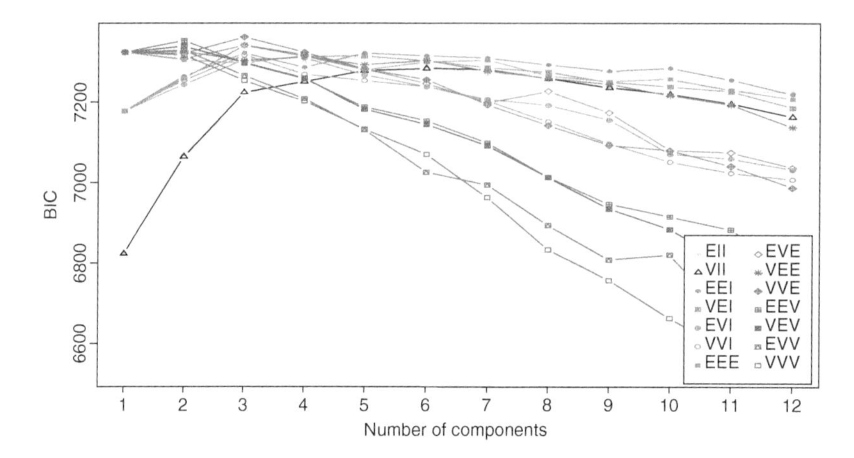

Figure 5.25 BIC values for different numbers of groups and covariance structures for the Taiwan AirBox data.

coefficient of the projected data. Minimizing the kurtosis coefficient implies maximizing the bimodality of the projection, whereas maximizing the kurtosis coefficient implies detecting groups of outliers in the projection well separated from the bulk of the data. Searching for bimodality leads to breaking the sample into two large clusters that are further analyzed. Searching for groups of outliers with respect to a central distribution leads to the identification of small clusters that are clearly separated from the rest of the data along some specific projection. They proved that, if the data were generated by a mixture of two normal distribution with the same covariance matrix, the direction that minimizes the kurtosis is the Fisher best linear discrimination direction that we will discuss in Section 5.5. They also showed other optimality properties of these directions when the data comes from a mixture of several elliptical distributions. Based on these results, they proposed an iterative procedure in which the data are projected onto these extreme kurtosis directions and a unidimensional search for clusters is carried out along them. We describe briefly their algorithm below, the details can be found in Peña and Prieto (2001), as the discussion of the cut-offs and the optimization methods to maximize or minimize the kurtosis of the projections.

The projection directions $\boldsymbol{d}_s$ are obtained through the following steps. Start with $s = 1$, let $\boldsymbol{y}_i^{(1)} = \boldsymbol{x}_i$ and define

$$\bar{\boldsymbol{y}}^{(s)} = \frac{1}{k} \sum_{i=1}^{k} \boldsymbol{y}_i^{(s)},$$

$$\boldsymbol{S}_s = \frac{1}{(k-1)} \sum_{i=1}^{k} [\boldsymbol{y}_i^{(s)} - \bar{\boldsymbol{y}}^{(s)}][\boldsymbol{y}_i^{(s)} - \bar{\boldsymbol{y}}^{(s)}]'.$$

1. Find a direction $\boldsymbol{d}_s$ that solves the problem

$$\begin{aligned} \max \quad & s(\boldsymbol{d}_s) = \frac{1}{k} \sum_{i=1}^{k} (\boldsymbol{d}_s' \boldsymbol{y}_i^{(s)} - \boldsymbol{d}_s' \bar{\boldsymbol{y}}_s)^4 \\ \text{s.t.} \quad & \boldsymbol{d}_s' \boldsymbol{S}_s \boldsymbol{d}_s = 1, \end{aligned} \tag{5.44}$$

that is, a direction that maximizes the kurtosis coefficient of the projected data.

2. Project the observations onto a subspace that is $\boldsymbol{S}_s$-orthogonal to the directions $\boldsymbol{d}_1, \ldots, \boldsymbol{d}_s$. If $s < p$, define

$$\boldsymbol{y}_i^{(s+1)} = (\boldsymbol{I} - \boldsymbol{d}_s \boldsymbol{d}_s' \boldsymbol{S}_s)\boldsymbol{y}_i^{(s)},$$

let $s \leftarrow s + 1$ and compute a new direction by repeating step 1. Otherwise, stop.
3. Compute another set of p directions $\boldsymbol{d}_{p+1}, \ldots, \boldsymbol{d}_{2p}$ by repeating steps 1 and 2, except that now the objective function in Eq. (5.44) is minimized instead of maximized.

Note that in each step s for $s > 1$ the direction found is $\boldsymbol{S}_s$-orthogonal to the previous one (rather than just orthogonal). This choice is made to ensure that each step of the algorithm is affine equivalent and, therefore, the complete procedure (see Peña and Prieto, 2001, for details).

The search for groups in the projected data is made with the first-order gaps or spacings, that are the differences between the order statistics. We consider that a set of time series can be split into two clusters when we find a sufficiently large first-order gap in one projection. Before testing for a significant value in the gaps, we first standardize the projected data and transform these observations using the inverse of the standard univariate normal distribution function Φ. In this manner, if the projected data would follow a normal distribution, the transformed data would be uniformly distributed. We can then use the fact that for uniform data the spacings are identically distributed. The resulting algorithm to identify significant gaps has been implemented as follows:

1. For each one of the directions $\boldsymbol{d}_s$, $s = 1, \ldots, 2p$, compute the univariate projections of the original observations $u_{si} = \boldsymbol{x}_i' \boldsymbol{d}_s$.
2. Standardize these observations, $p_{si} = (u_{si} - m_s)/s_s$, where $m_s = \sum_i u_{si}/k$ and $s_s = \sum_i (u_{si} - m_s)^2/(k-1)$.
3. Sort the projections p_{si} for each value of s, to obtain the order statistics $p_{s(i)}$ and transform then using the inverse of the standard normal distribution function $p_{si} = \Phi^{-1}(p_{s(i)})$
4. Compute the gaps between consecutive values, $w_{si} = p_{s,i+1} - p_{si}$.
5. Search for the presence of significant gaps in w_{si}. These large gaps will be indications of the presence of more than one cluster. In particular, we introduce a threshold $\kappa = \nu(c)$, where $\nu(c) = 1 - (1-c)^{1/k}$ denotes the cth percentile of the distribution of the spacings, define $i_{0s} = 0$ and

$$r = \inf_j \{k > j > i_{0s} : w_{sj} > \kappa\}.$$

If $r \neq 0$ and one or more of the gaps w_{sj} is greater than κ, then the presence of several possible clusters has been detected in this direction and go to step 6. Otherwise, go to the next projection direction.
6. Label all observations l with $p_{sl} \leq p_{sr}$ as belonging to clusters different to those having $p_{sl} > p_{sr}$. Let $i_{0s} = r$ and repeat step 5.

Note that in each direction if there are nn gaps with values greater than the threshold $nn + 1$ groups are detected. This is applied to all the $2p$ directions. After completing the analysis of the gaps, the algorithm carries out a final step to reassign observations within the clusters identified in the data by using the Mahalanobis distance. The clusters are sorted by size, so that $n_1 \geq n_2 \geq \cdots \geq n_G$, and the observations are labeled so that the first n_1 belong to the first group and the last n_G to the G group. Then, starting with the largest cluster and for $g = 1, \ldots, G$, the mean $\boldsymbol{m}_g$ and covariance matrix $\boldsymbol{S}_g$ of the observations in the cluster and the Mahalanobis distances $\delta_j = (\boldsymbol{x}_j - \boldsymbol{m}_j)'\boldsymbol{S}_g^{-1}(\boldsymbol{x}_j - \boldsymbol{m}_j)$ to all observations not assigned to this cluster g are computed. If any observation not in this cluster satisfy $\delta_j \leq \chi^2_{p,0.99}$, then this observation is added to the cluster. This procedure seems to have better performance than k-means and MCLUST in the Monte Carlo results presented in Peña and Prieto (2001) when $p < k$. However, when p is large as the algorithm computes the covariance matrix of the variables its performance may decrease unless k/p is moderate, let us say larger than 10.

Example 5.6

We analyze, again, the Taiwan AirBox data using as variables the average value of each day, so that $p = 31$. As in Example 5.3, $k = 508$ series were used. The clustering by projections is applied with the commands `ClusKur` and `ClusMain` included in the **SLBDD** book package. The data used are in matrix `Data`, computed as in Example 5.4. The output of the command `ClusKur` is the number of groups found and the labels of the observations. In this particular instance, four groups are found and several outliers, identified with the label '-1', do not correspond to any of the previous groups. Figure 5.26 shows the locations of the groups and Figure 5.27 three of the directions used in finding clusters.

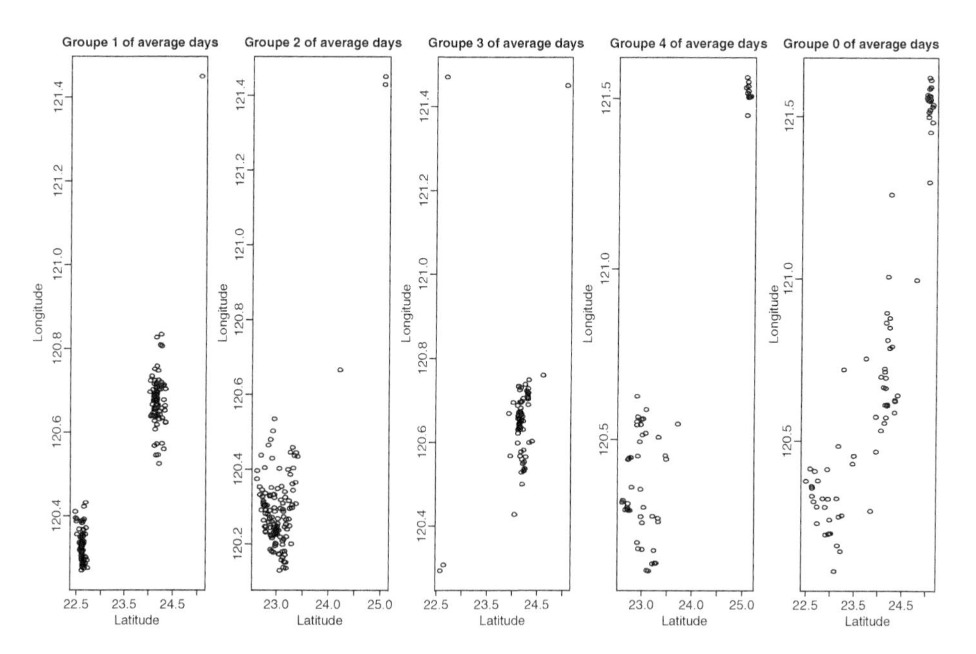

Figure 5.26 Locations of the five groups found by projections for the Taiwan AirBox data with the daily average of $PM_{2.5}$ measurements as variables.

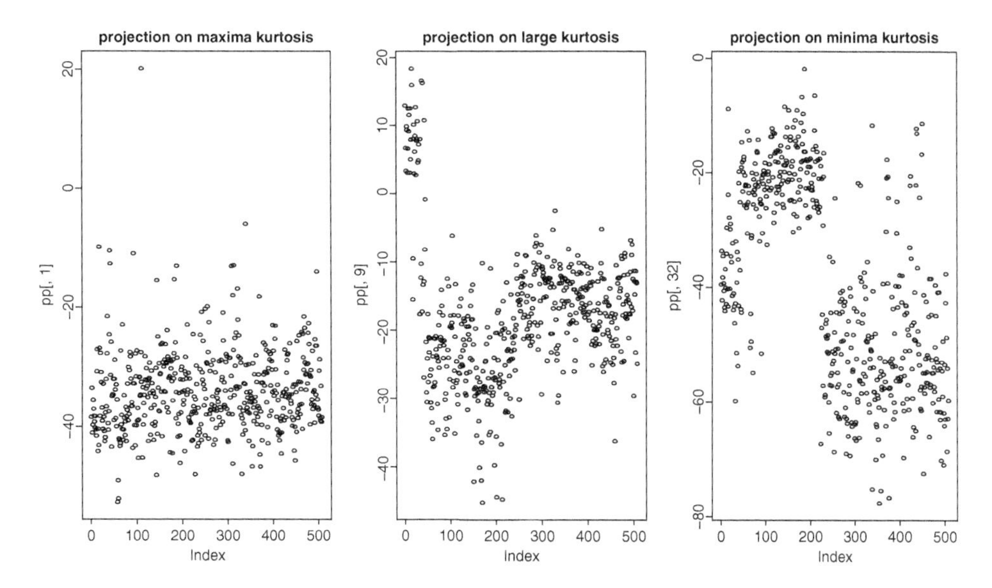

Figure 5.27 Three of the projection directions computed to find clusters for the Taiwan AirBox data with the average daily measurements as variables.

R commands for clustering via projection.

```
> par(mfrow=c(1,1))
> sal1=ClusKur(Data) # Data is the data matrix from Example 5.4.
> plot(sal1$lbl,main="AirBoxes groups,-1 are outliers of these 4 groups")
> bb=ClusMain(Data)
##The columns of matrix pp are the projections of the data in the 31
##directions of largest and the 31 of smallest kurtosis
> pp=Data\%*\%bb$V
> par(mfrow=c(1,3))
##Some projections are shown
> plot(pp[,1], main="projection on maxima kurtosis")
> plot(pp[,9], main="projection on large kurtosis")
> plot(pp[,32], main="projection on minima kurtosis")
> memb=sal1$lbl
> g1u=which(memb==1)
> g2u=which(memb==2)
> g3u=which(memb==3)
> g4u=which(memb==4)
> g0u=which(memb==-1)
## The number of series in each group are shown
> ng1=length(g1u)
> ng1
[1] 154
> ng2=length(g2u)
> ng2
[1] 130
> ng3=length(g3u)
> ng3
[1] 76
> ng4=length(g4u)
```

```
> ng4
[1] 63
> ng0=length(g0u)
> ng0
[1] 85
> g1lo=lo[g1u,]
> g2lo=lo[g2u,]
> g3lo=lo[g3u,]
> g4lo=lo[g4u,]
> g0lo=lo[g0u,]
> par(mfrow=c(1,5))
##The relation between groups and location is displayed
> plot(g1lo, main="Groupe 1 of average days")
> plot(g2lo, main="Groupe 2 of average days")
> plot(g3lo, main="Groupe 3 of average days")
> plot(g4lo, main="Groupe 4 of average days")
> plot(g0lo, main="Groupe 0 of average days")
```

■

5.4 CLASSIFICATION WITH TIME SERIES

Discrimination, or classification, appears in many areas of human activity: from medical diagnosis to credit scoring and from fraud detection to recognizing art forgeries. The first application of discriminant analysis, invented by R. A. Fisher in 1935 (see Fisher Box, 1978), consists of classifying the remains of a skull found in an excavation as human or not, by utilizing the distribution of physical cranial measurements for human skulls and those of other anthropoids.

In time series analysis, assume that we have a system that is in one of c possible states, say $S = (s_1, \ldots, s_c)$, and it generates a vector of series with properties that depend on the state. The characteristics of these vectors of time series are known given the state, or can be estimated by a sample of series generated in each state, and our goal is to forecast the expected output of the system, given the state. Now, a new vector of time series is observed but the state is unknown and we want to estimate the state of the system. In what follows, we use states and clusters interchangeably.

The methods used depend on the available information of the clusters. For the first type of methods, we know the time series model for each and every cluster. Therefore, we can compute the probability that the observed time series is generated by the model of a given cluster and classify the series into the cluster with the highest probability. In the second type of methods, we know the distribution of a vector of features (or variables) of the time series in each cluster and use these features to perform classification. This case is similar to the standard discriminant problem in multivariate analysis. In the third type of methods, we know some properties of the vector of features in a given cluster, but not their joint distribution, and employ nonparametric methods to carry out the classification.

5.4.1 Classification Among a Set of Models

Suppose first that the output of the system is a univariate time series and that the model generating the series is known given the state. Thus, we observe a series, $z = (z_1, \ldots, z_T)'$, that has been generated by one of c processes, with vector of marginal means $\boldsymbol{\mu}_j = (\mu_{j1}, \ldots, \mu_{jT})'$ for $j = 1, \ldots, c$ and covariance matrices that depend on the lag and are denoted by $\boldsymbol{\Sigma}_j = \{\gamma_j(\ell) : 0 \leq \ell \leq T-1\}$, where $\gamma_j(\ell) = E[(z_t - \mu_{j,t})(z_{t-\ell} - \mu_{j,t-\ell})]$. Suppose that we have prior probabilities $p(s_i)$ of the system being in the ith state, where $\sum_{i=1}^{c} p(s_i) = 1$. The posterior probability of the system being in state s_i is, by Bayes' theorem

$$p(s_i|\boldsymbol{z}) = \frac{p(\boldsymbol{z}|s_i)p(s_i)}{\sum_{j=1}^{c} p(\boldsymbol{z}|s_j)p(s_j)}. \tag{5.45}$$

To compute $p(\boldsymbol{z}|s_i)$, one typically assumes that the process is Gaussian so that the probability is proportional to the likelihood, and we have

$$p(\boldsymbol{z}|s_j) \propto |\boldsymbol{\Sigma}_j|^{-\frac{1}{2}} \exp\left[-\frac{1}{2}(\boldsymbol{z} - \boldsymbol{\mu}_j)'\boldsymbol{\Sigma}_j^{-1}(\boldsymbol{z} - \boldsymbol{\mu}_j)\right], \tag{5.46}$$

and the state is estimated as the one with the highest posterior probability

$$\widehat{s}_i = \arg\max_{s_j \in S}[p(\boldsymbol{z}|s_j)p(s_j)].$$

When the prior probabilities are equal, this is equivalent to fitting the series $\boldsymbol{z}$ with the c models and selecting the state that leads to the best fit. As we saw in Chapter 2 (section 2.8.3), by the prediction error decomposition,

$$p(\boldsymbol{z}|s_i) = p(z_1|s_i)p(z_2|z_1, s_i)....p(z_T|z_1, \ldots, z_{T-1}, s_i).$$

Suppose that the time series model in each state is an ARMA (p_j, q_j) model, for $j = 1, \ldots, c$. Calling $e_{jt} = z_t - \mu_j$, the model for e_{jt} can be written as $\pi_j(B)e_{jt} = \epsilon_{jt}$, where $\pi_j(B) = 1 - \pi_{j1}B - \pi_{j2}B^2 - \cdots$. Thus,

$$\begin{bmatrix} 1 & 0 & 0 & \cdots & 0 \\ -\pi_{j1} & 1 & 0 & \cdots & 0 \\ -\pi_{j2} & -\pi_{j1} & 1 & \cdots & 0 \\ \vdots & \vdots & \vdots & \ddots & \vdots \\ -\pi_{jp} & -\pi_{j,p-1} & \cdots & \cdots & 0 \\ \vdots & \vdots & \cdots & \cdots & \vdots \\ -\pi_{j,T-1} & \pi_{j,T-2} & -\pi_{jp} & \cdots & 1 \end{bmatrix} \begin{bmatrix} e_{j1} \\ e_{j2} \\ e_{j3} \\ \vdots \\ e_{jp} \\ \vdots \\ e_{jT} \end{bmatrix} = \begin{bmatrix} \epsilon_{j1} \\ \epsilon_{j2} \\ \epsilon_{j3} \\ \vdots \\ \epsilon_{jp} \\ \vdots \\ \epsilon_{jT} \end{bmatrix}.$$

or

$$\boldsymbol{\Pi}_j \boldsymbol{e}_j = \boldsymbol{\epsilon}_j$$

where $\boldsymbol{\Pi}_j$ is a Toeplitz triangular matrix with the jth ARMA model parameters. Letting $\sigma_j^2 \boldsymbol{I}$ be the covariance matrix of the white noise process $\{\epsilon_{jt}\}$ and calling $\boldsymbol{\Sigma}_j$ the covariance matrix of $\boldsymbol{e}_j$, we have,

$$\boldsymbol{\Pi}_j \boldsymbol{\Sigma}_j \boldsymbol{\Pi}_j' = \sigma_j^2 \boldsymbol{I},$$

and, therefore,

$$\boldsymbol{\Sigma}_j^{-1} = \frac{1}{\sigma_j^2}\boldsymbol{\Pi}_j'\boldsymbol{\Pi}_j.$$

As $|\mathbf{\Pi}_j| = 1$, for being a triangular matrix with all diagonal elements being one, then $|\mathbf{\Sigma}_j^{-1}| = \sigma_j^{-2}$ and in the likelihood Eq. (5.46) we substitute $|\mathbf{\Sigma}_j|^{-\frac{1}{2}}$ by σ_j^{-1}. The exponent is

$$\boldsymbol{e}_j'\mathbf{\Sigma}_j^{-1}\boldsymbol{e}_j = \boldsymbol{e}_j'\frac{1}{\sigma_j^2}(\mathbf{\Pi}_j'\mathbf{\Pi}_j)\boldsymbol{e}_j = \frac{1}{\sigma_j^2}\sum_{t=1}^{T} e_{jt}^2.$$

As σ_j^2 is estimated by $\hat{\sigma}_j^2 = \sum_{t=1}^{T} e_{jt}^2/T$, the probability in Eq. (5.46) only depends on $\hat{\sigma}_j^2$, and the state with highest posterior probability is the one with smallest $\hat{\sigma}_j^2$. In this case, discriminant analysis assigns the observed time series to the state (model) that provides the best fit of the time series, or the smallest one step ahead squared forecast error. It is easy to see that by this procedure we also minimize the classification error. Suppose, for simplicity, $c = 2$, the classification error when the series has been generated by state s_1 but we classify it in state s_2 is

$$p(z \in s_2|s_1) \propto p(s_1|z \in s_2)p(z \in s_2)$$

and $p(z \in s_2|s_1)$ would be minimum if $p(s_1|z \in s_2)$ is minimum which implies that the probability of s_1 given z must be minimum, or the probability of s_2 given z is maximum. This explains the criterion used.

The extension to multivariate time series is straightforward. Suppose that the output given the state is a vector of k time series and the data are in $\boldsymbol{Z}$, that is a $T \times k$ matrix. We want to select the state that maximize $p(\boldsymbol{Z}|s_i)$. Assuming that the prior probabilities are the same for all the states, then we like to maximize the likelihood of the data given each of the possible models. Suppose, for simplicity, that the models generated are $VAR(p_j)$. Then, using the notation in Section 3.2.3 and the results in Chapter 3, we have

$$\log p(\boldsymbol{Z}|s_j) \propto -\frac{T-p}{2}\log|\mathbf{\Sigma}_j^a| - \frac{1}{2}S(\boldsymbol{\beta}_j),$$

where $\mathbf{\Sigma}_j^a$ is the covariance matrix of the residuals estimated under model VAR(p_j) and $\boldsymbol{\beta}_j$ the parameters of this model. Therefore, the estimated state is the one with the best fit, as measured by $|\mathbf{\Sigma}_j^a|$.

When the models for the states are unknown, the classification can be carried out by using the correlation matrices of the series in the states or the spectrum. Kakizawa et al. (1998) propose to use the Kullblack–Leibler (KL) divergence from stationary vectors of time series to discriminate in seismology earthquakes from mining or nuclear explosions. Assuming joint normality and series standardized to zero mean and unit variance, the KL divergence between a new observed series $(\boldsymbol{x}_{01}, \ldots, \boldsymbol{x}_{0T})$ with correlation matrix $\boldsymbol{R}_0$ and the $\boldsymbol{R}_j$ matrix of state s_j is

$$D_{\mathrm{KL}}(0;\, s_j) = \frac{1}{2}\left[\mathrm{tr}(\boldsymbol{R}_j\boldsymbol{R}_0^{-1}) - \log\frac{|\boldsymbol{R}_j|}{|\boldsymbol{R}_0|} - kT\right], \tag{5.47}$$

and the series is classified to the state that minimizes Eq. (5.47). An alternative is to substitute the large kT squared covariance matrices for kh squared covariance matrix with a small value of h, as explained before.

5.4.2 Checking the Classification Rule

When the classification rule is estimated, it is useful to check its performance out of sample using some form of cross-validation (CV): the sample is divided into two parts, one is used for estimating the classification models and the other for assessing the classification error. With independent data, this is usually done with h-fold CV, or leave-h-out CV, where h observations are selected at random from the sample as a test subsample, the model is estimated using the remaining data of size $T - h$, that is called the training sample, and the classification error is computed with the estimated model in the test subsample of size h. In order to improve the estimation of the classification error, we can repeat the process several times and obtain an average out-of-sample classification error. The optimal way to split the sample into training and test sets depends on the model and a general rule that seems to work well in practice is to use 60–70% of the sample for estimation and 30–40% to estimate the classification error.

With time series, if we delete observations at random, we face the problem of missing values. Then, the model is estimated as discussed in Chapter 4, and it provides interpolations for the deleted observations. Finally, using the true values the interpolation errors can be computed. However, with correlated data when we delete x_{t_0} the next values, $x_{t_0+1}, \dots, x_{t_0+k}$ are correlated with x_{t_0} and, therefore, we do not delete all the information in x_{t_0}. A consequence of this is that the interpolation error will be smaller than the out-of-sample forecast error. A better way to do CV with time series is to split the sample into two consecutive periods, $(1, \dots, t_0)$ and $(t_0 + 1, \dots, T)$, estimate the models in the first part and compute the forecast errors in the second part. To avoid the dependency of the origin of the forecasts this process is repeated for all origins with $t_{\min} \leq t_0 \leq T - 1$, where $t_{\min}$ is taken between $T/2$ and $2T/3$ of the sample. This procedure is called rolling forecast.

When we use features of the series for the classification to be discussed next, then the standard CV can be applied by splitting the sample into training and testing samples.

5.5 CLASSIFICATION WITH FEATURES

Suppose that the vector time series can be summarized in a vector of features $\boldsymbol{x}$ of dimension p. For instance, if the output is a stationary univariate time series, we can summarize it by its autocorrelations or the periodogram of the coefficients by fitting wavelets, among others. For stationary vector time series, we can summarize them by the cross-correlation matrices or the coefficients of fitting some base functions. For nonstationary time series we can select as features the trend in selected intervals, the seasonal coefficients, and so on. In general, the selection of the features depends on the specific application. For seismic time series Shumway and Stoffer (2000) use the peak to peak amplitudes as features for discrimination, and Huang et al. (2004) use local spectral properties for the same type of data. Maharaj and Alonso (2014) use the vector of features of the variance and correlations of the wavelets fitted to ECG time series to discriminate between ill persons and healthy ones.

We assume first that the vector of features has a known distribution that depends on the state. Assuming normality, we have that $\boldsymbol{x} \in N_p(\boldsymbol{\mu}_j, \boldsymbol{V}_j)$, for $j = 1, 2, \dots, c$. The probability of each state given the information of the observed features, $\boldsymbol{x}_0$, is $p(s_i|\boldsymbol{x}_0)$, and the posterior probabilities becomes $p(\boldsymbol{x}_0|s_j)$. These probabilities replacing $\boldsymbol{z}$ for $\boldsymbol{x}_0$ are given in Eqs. (5.45) and (5.46). The solution then is to classify using the maximum posterior probability or, if the prior probabilities are assumed equal, the ML.

If the distributions of the features are unknown, we assume that a sample of n vector time series generated by each state, of the form $(s_i, \boldsymbol{z}_1^T(i))$, where s_i indicates the state and $\boldsymbol{z}_1^T(i)$ a vector of time series that is transformed into the sample $(s_i, \boldsymbol{x}_i)$, where $\boldsymbol{x}_i$ is the vector of features. With this information we estimate the joint distribution of the features given the state, $f(\boldsymbol{x}|s_j)$, that is often assumed to be multivariate normal. In order to simplify the exposition in the next sections, we consider the parameters $\boldsymbol{\mu}_j$ and $\boldsymbol{V}_j$ as known, although often they are estimated in the usual way.

5.5.1 Linear Discriminant Function

Suppose that the prior probabilities and the covariance matrices are the same in all states, that is, $\boldsymbol{V}_j = \boldsymbol{V}$, and consider first the case $c = 2$. Assuming normality, the vector of features $\boldsymbol{x}_0$ is classified in the first state if

$$(\boldsymbol{x}_0 - \boldsymbol{\mu}_1)'\boldsymbol{V}^{-1}(\boldsymbol{x}_0 - \boldsymbol{\mu}_1) < (\boldsymbol{x}_0 - \boldsymbol{\mu}_2)'\boldsymbol{V}^{-1}(\boldsymbol{x}_0 - \boldsymbol{\mu}_2).$$

Let

$$D_i = (\boldsymbol{x}_0 - \boldsymbol{\mu}_i)'\boldsymbol{V}^{-1}(\boldsymbol{x}_0 - \boldsymbol{\mu}_i), \quad i = 1, 2, \tag{5.48}$$

be the Mahalanobis distance between $\boldsymbol{x}_0$ and the vector of means, the output series will be classified in the first population if $D_2 > D_1$. A well-known alternative interpretation of this rule can be obtained by writing the equation as

$$(\boldsymbol{\mu}_1 - \boldsymbol{\mu}_2)'\boldsymbol{V}^{-1}\boldsymbol{x}_0 > \frac{1}{2}(\boldsymbol{\mu}_1 - \boldsymbol{\mu}_2)'\boldsymbol{V}^{-1}(\boldsymbol{\mu}_1 + \boldsymbol{\mu}_2),$$

implying that the features are combined to form a linear indicator, $v = \boldsymbol{w}'\boldsymbol{x}_0$, where $\boldsymbol{w} = \boldsymbol{V}^{-1}(\boldsymbol{\mu}_1 - \boldsymbol{\mu}_2)$, and if this indicator is larger than v_0, given by

$$v_0 = \frac{1}{2}(\boldsymbol{\mu}_1 - \boldsymbol{\mu}_2)'\boldsymbol{V}^{-1}(\boldsymbol{\mu}_1 + \boldsymbol{\mu}_2),$$

the first state is selected. This is the well know Fisher's linear discriminant function (LDF). It is easy to see that this rule minimizes the sum of the classification errors $p(\boldsymbol{z} \in s_2|s_1) + p(\boldsymbol{z} \in s_1|s_2)$, where the notation $\boldsymbol{z} \in s_2$ means the series is classified in state s_2.

In the general case of c states, assuming that the covariance matrix is the same, the extension of the previous procedure is to compute the Mahalanobis distances (5.48) for $i = 1, \ldots, c$ and classify the series in the state whose mean vector is closer to the observed vector of features. This rule can also be expressed as finding $c - 1$ separating hyperplanes into c regions.

When the parameters are unknown and estimated from a training sample of size n with n_i elements from the ith state and $n = \sum_{i=1}^{c} n_i$, the mean vectors are estimated by the sample means and the covariance matrix by the pooled covariance matrix

$$\widehat{\boldsymbol{V}} = \sum_{i=1}^{c}\sum_{j=1}^{n_i}(\boldsymbol{x}_{ij} - \bar{\boldsymbol{x}}_i)(\boldsymbol{x}_{ij} - \bar{\boldsymbol{x}}_i)'/(n - c). \tag{5.49}$$

5.5.2 Quadratic Classification and Admissible Functions

In time series classification, often the covariance matrices of the states are different. For instance, we have seen that two stationary time series models are generated by two Gaussian populations with different covariance matrices. Here $c = 2$. Then, we classify the series x_{0t} in state 1 if it is more likely that the observation $\boldsymbol{x}_0$ is generated by this state, which implies

$$\log|\boldsymbol{V}_1| + (\boldsymbol{x}_0 - \boldsymbol{\mu}_1)'\boldsymbol{V}_1^{-1}(\boldsymbol{x}_0 - \boldsymbol{\mu}_1) > \log|\boldsymbol{V}_2| + (\boldsymbol{x}_0 - \boldsymbol{\mu}_2)'\boldsymbol{V}_2^{-1}(\boldsymbol{x}_0 - \boldsymbol{\mu}_2),$$

and, therefore, if

$$\boldsymbol{x}_0'(\boldsymbol{V}_1^{-1} - \boldsymbol{V}_2^{-1})\boldsymbol{x}_0 - 2\boldsymbol{x}_0'(\boldsymbol{V}_1^{-1}\boldsymbol{\mu}_1 - \boldsymbol{V}_2^{-1}\boldsymbol{\mu}_2) > c, \tag{5.50}$$

where $c = \log(|\boldsymbol{V}_2|/|\boldsymbol{V}_1|) + \boldsymbol{\mu}_2'\boldsymbol{V}_2^{-1}\boldsymbol{\mu}_2 - \boldsymbol{\mu}_1'\boldsymbol{V}_1^{-1}\boldsymbol{\mu}_1$. This is a quadratic discriminant rule. Specifically, define

$$\boldsymbol{V}_d^{-1} = (\boldsymbol{V}_1^{-1} - \boldsymbol{V}_2^{-1}),$$

and the new variable

$$\boldsymbol{y}_0 = \boldsymbol{V}_d^{-1/2}\boldsymbol{x}_0.$$

Calling $\boldsymbol{y}_0 = (y_{01}, \ldots, y_{0p})'$ and defining $\boldsymbol{m} = (m_1, \ldots, m_p)'$ as

$$\boldsymbol{y} = \boldsymbol{V}_d^{1/2}(\boldsymbol{V}_1^{-1}\boldsymbol{\mu}_1 - \boldsymbol{V}_2^{-1}\boldsymbol{\mu}_2),$$

Eq. (5.50) can be written as $\boldsymbol{y}_0'\boldsymbol{y}_0 - 2\boldsymbol{y}_0'\boldsymbol{m} > c$, or

$$\sum_{i=1}^{p} y_{0i}^2 - 2\sum_{i=1}^{p} y_{0i}m_i > c,$$

which is a quadratic equation if the new variables y_{0i}.

In the general case of c states, an observation $\boldsymbol{x}_0$ is classified in the state in which

$$D_i(\boldsymbol{x}_0) = \log|\boldsymbol{V}_i| + (\boldsymbol{x}_0 - \boldsymbol{\mu}_i)'\boldsymbol{V}_i^{-1}(\boldsymbol{x}_0 - \boldsymbol{\mu}_i) \tag{5.51}$$

is the largest.

When the parameters must be estimated, in the linear case we have to estimate cp mean parameters and $p(p+1)/2$ parameters for the covariance matrix. In the quadratic case this number grows to $c(p + p(p+1)/2)$ and, if the dimension of the vector of features is large, this estimation requires a large sample. Friedman (1989) proposed to use a shrinkage estimator of the covariance matrices by

$$\widehat{\boldsymbol{V}}_i(\alpha) = \alpha\widehat{\boldsymbol{V}}_i + (1-\alpha)\widehat{\boldsymbol{V}},$$

where $\widehat{\boldsymbol{V}}_i$ is the estimation of the covariance matrix in the ith group, and $\widehat{\boldsymbol{V}}$ the pooled covariance matrix in Eq. (5.49).

An alternative approach in these situations is to obtain a good linear discriminant rule according to some criteria. This is the idea of admissible linear procedures by Anderson and Bahadur (1962). The set of admissible linear procedures are in the form:

$$\boldsymbol{w} = (t_1\boldsymbol{V}_1 + t_2\boldsymbol{V}_2)^{-1}(\boldsymbol{\mu}_2 - \boldsymbol{\mu}_1), \tag{5.52}$$

where t_1 and t_2 are positive numbers such that the matrix $(t_1 \boldsymbol{V}_1 + t_2 \boldsymbol{V}_2)$ is positive definite. Then the cutoff point τ for the classification is $\tau = \boldsymbol{w}'\boldsymbol{\mu}_1 + t_1 \boldsymbol{w}'\boldsymbol{V}_1\boldsymbol{w} = \boldsymbol{w}'\boldsymbol{\mu}_2 - t_2\boldsymbol{w}'\boldsymbol{V}_2\boldsymbol{w}$. In order to find t_1 and t_2, we can fix $t_1 = 1$ and found t_2 by computing the classification errors and making the sum as small as possible. Admissible linear procedures can be obtained by information measures. An alternative of these linear admissible procedures is to find linear procedures in a larger space that includes nonlinear transformations of the variables, which is discussed via the Support Vector Machines.

5.5.3 Logistic Regression

Often the parameters of the feature distribution are unknown and must be estimated from previous time series data. When this distribution is multivariate normal, the Fisher LDF with the estimated Mahalanobis distances is optimal with equal covariance matrices and usually works well in large samples. However, when the distribution of the features used for classification is not normal, we have no guarantee that LDF will be optimal.

One possibility is to construct a model to predict the values of the classification variable using the features as explanatory variables. For example, if we wish to discriminate between those loans which will be repaid and those that will be difficult to collect, a new variable y, can be added to the database which takes the value zero when the loan is repaid and the value of one, otherwise. The problem of classification becomes one of predicting the value of the dummy variable, y, in a new element from which we know the vector of variables $\boldsymbol{x}$. If the predicted variable is closer to zero than to one, we classify the element in the first state; otherwise, we place it in the second.

Suppose that the sample to build the classifier consists of elements $(y_i, \boldsymbol{x}_i)$, where y_i is the value of the binary classification variable in this element and $\boldsymbol{x}_i$ is the vector of p features used as explanatory variables. The simplest approach to predict the value of the classification variable in a new element whose variables $\boldsymbol{x}$ are known is to formulate a regression model:

$$y = \beta_0 + \boldsymbol{\beta}_1'\boldsymbol{x} + \boldsymbol{u}, \tag{5.53}$$

and it is easy to see that a classification with this equation estimated by least squares is equivalent to Fisher's LDF. Note that given a fix value of $\boldsymbol{x} = \boldsymbol{x}_0$ the variable y is binomial with possible values one and zero, probabilities $p_0(\boldsymbol{x}_0) = P(y = 0|\boldsymbol{x}_0)$ and $p_1(\boldsymbol{x}_0) = P(y = 1|\boldsymbol{x}_0)$ that must add to one and expected value $p_1(\boldsymbol{x}_0)$. Taking expectations in (5.53) we obtain

$$p_1(\boldsymbol{x}_0) = \beta_0 + \boldsymbol{\beta}_1'\boldsymbol{x}_0. \tag{5.54}$$

However, we cannot guarantee that the value $\widehat{\beta}_0 + \widehat{\boldsymbol{\beta}}_1'\boldsymbol{x}_0$ will be between zero and one to be interpreted as a probability. To avoid this problem, we transform (5.54) to

$$p_1(\boldsymbol{x}) = F(\beta_0 + \boldsymbol{\beta}_1'\boldsymbol{x}), \tag{5.55}$$

where taking as F any distribution function $p_1(\boldsymbol{x})$ will be between zero and one. The logistic distribution function is often selected because then

$$p_1(\boldsymbol{x}) = \frac{e^{\beta_0 + \boldsymbol{\beta}_1'\boldsymbol{x}}}{1 + e^{\beta_0 + \boldsymbol{\beta}_1'\boldsymbol{x}}}. \tag{5.56}$$

and $p_1(\boldsymbol{x}_0) = 1/(1 + e^{\beta_0 + \boldsymbol{\beta}_1' \boldsymbol{x}})$. We can estimate a linear equation by the *Logit* variable, g_i defined by:

$$g_i = \log \frac{p_1(\boldsymbol{x}_i)}{p_0(\boldsymbol{x}_i)} = \beta_0 + \boldsymbol{\beta}_1' \boldsymbol{x}_i \tag{5.57}$$

that represents the difference between the probabilities of belonging to both populations in a log scale. It is easy to see that if the features follow a normal distribution, they follow the logistic model but, in this case, estimating this model is less efficient than estimating the Fisher LDF. The logistic model defined by probabilities (5.57) can be estimated by ML; see Hosmer and Lemeshow (1989). Logistic models are specific cases of the generalized linear models; see McCullagh (2018).

The model is easily extended for G classes, taking one as a reference, for instance the last one, and assuming that the probabilities of each class are given by, generalizing (5.57),

$$p_g(\mathbf{x}) = \frac{e^{\beta_{0g} + \boldsymbol{\beta}_{1g}' \boldsymbol{x}}}{1 + \sum_{j=1}^{G-1} e^{\beta_{0j} + \boldsymbol{\beta}_{1j}' \boldsymbol{x}}}, \qquad g = 1, \ldots, (G-1). \tag{5.58}$$

These probabilities imply $G - 1$ logit models as (5.57), where now g_i is the log ratio of the probabilities of the ith class divided by the probability of the reference class. Thus, the multinomial model defined by (5.58) has $(G - 1)(p + 1)$ parameters, the coefficients of the explanatory variables, and can be estimated by ML. Once the parameters that define the probabilities (5.58) are estimated new data is classified in the most likely class. With many variables and classes it is recommended, as in regression with many correlated variables, to use a shrinkage estimate, the Lasso method, discussed in Chapter 7, which is a good alternative. See McLachlan (2004) for the ML estimation of the logistic model and references about its applications.

Example 5.7

We consider the series of electroencephalograms (EEG) studied by Andrzejak et al. (2001) and Maharaj and Alonso (2007). An EEG is a recording of the electrical activity of the brain. We have 200 time series of individuals with $T = 4096$ observations of EEG. The first 100 series, type A, correspond to healthy volunteers whereas the second 100 series, type E, consist of EEG recordings during seizure activity of patients. Figures 5.28 and 5.29 show the time series plots along with three EDQ (prob = 0.01,0.5,0.95) of Chapter 2. From the plots, we see that the EEG series for seizure patients, type E, have much more extreme values than those of healthy individuals, type A. This feature can be useful for discrimination. We fit ARIMA models to the series in both groups using the command `SummaryModel` and found that series of type A mostly follow (76%) an ARMA(5,3) or ARMA(5,2) model, whereas series in type E mostly follow (55%) other ARMA models. As an illustration, in this example we use the first 5 autocorrelation coefficients to classify the data. Figure 5.30 show that the histograms of the lag-1 ACF for both groups are different, but the differences are small. We apply the LDF and logistic regression (LG) to the EEG data. These two methods are first applied to all the data, then they are estimated in a training sample that includes 65% of the data and evaluated in the test sample with the remaining 35% of the data. Figure 5.31 shows the projection

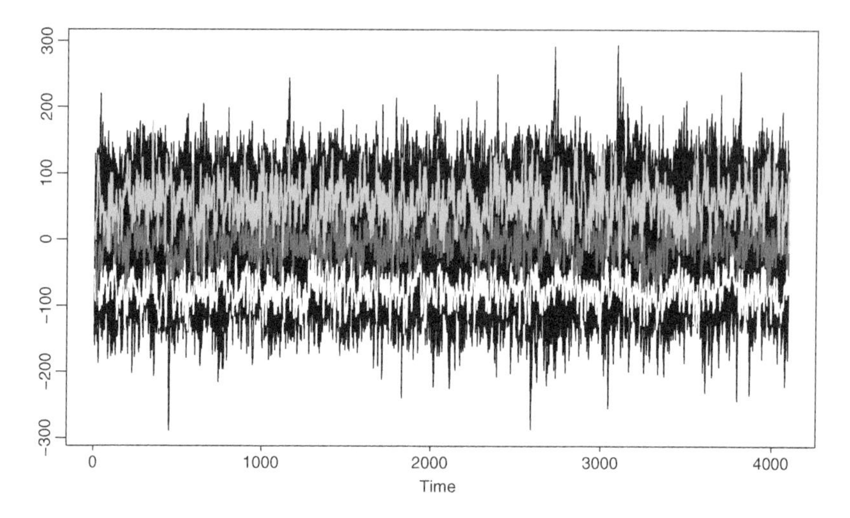

Figure 5.28 Empirical Dynamic Quantiles with probabilities 0.05, 0.5,and 0.95 for the EEG data of healthy individuals.

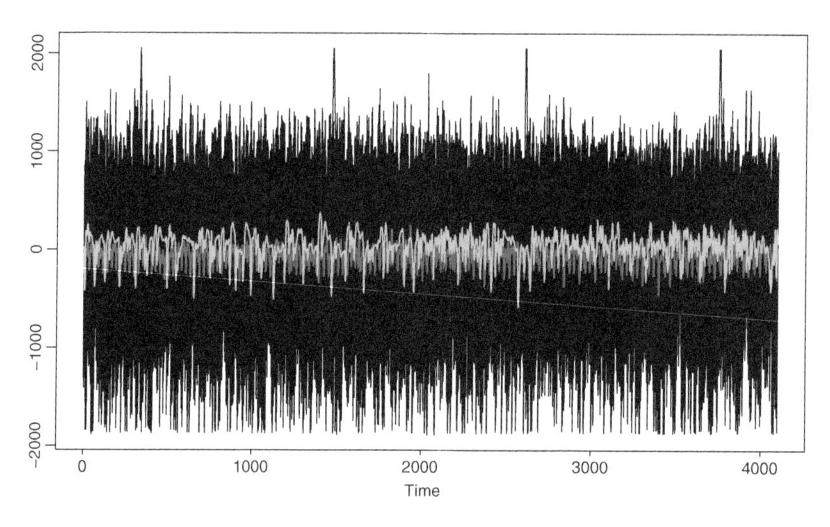

Figure 5.29 Empirical Dynamic Quantiles with probabilities 0.05, 0.5, 0.95 for the EEG data of seizure patients.

of the data in the direction of maximum discrimination computed with LDF. The estimated model for LR has all coefficients statistically significant and is shown in the output. With all the data, the discrimination errors are 9 for LDF and 2.5% for LR. However, when the split sample LDF has 8% error in the training sample and 16% in the test sample, whereas LR has 3% in the training sample and 33% in the test sample. These results suggest some overfitting in the LR. To compare the two methods with just one sample may be misleading and it is useful to replicate the analysis with several splits in training and test data. With 50 splits and using the program

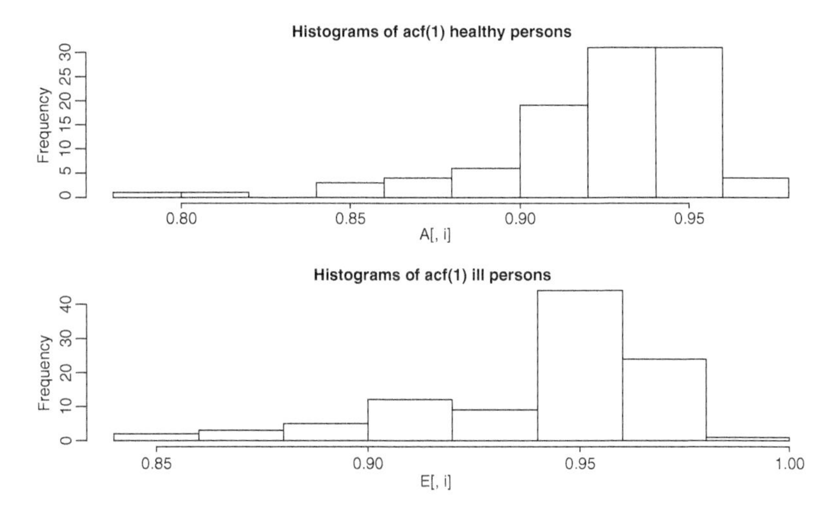

Figure 5.30 Histograms of the lag-1 autocorrelation coefficients of the EEG data of healthy and seizure people.

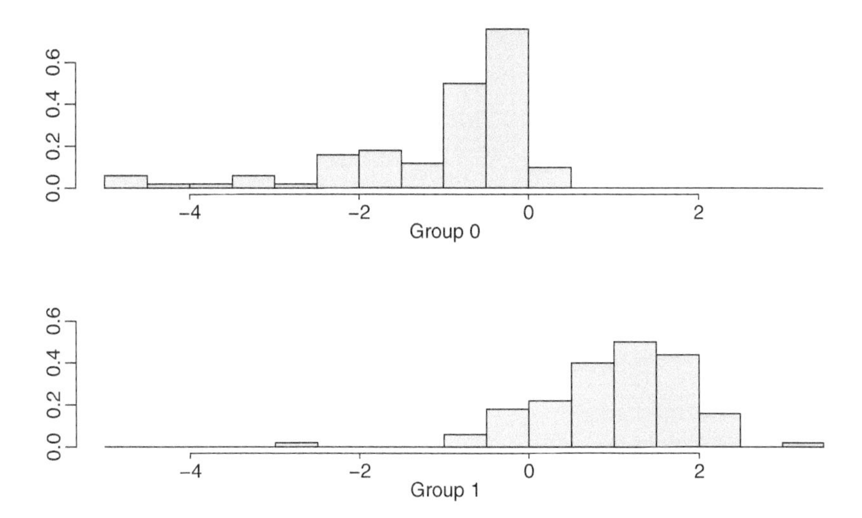

Figure 5.31 Projection of the EEG data on the direction of maximum discrimination.

`discrim` that carries out all the analysis, the average results for LDF is 8% in sample and 13% in the out of sample and for LR 3% in sample and 13% out of sample. Thus, both methods perform similarly in this example.

R commands and analysis of EEG data

```
> # Example of discrimination with the EEG data
> library(caTools)
> library(MASS)
> library(e1071)
> XA <- read.csv("EEGsetA.csv", header=FALSE)
```

```
> XE <- read.csv("EEGsetE.csv", header=FALSE)
> XA=as.matrix(XA)
> XE=as.matrix(XE)
> k=dim(XA)[2]
> T=dim(XA)[1]
> #The EDQ(.05,.5,.95) shows differences between the two groups
> edqplot(XA)
> edqplot(XE)
> edqplot(XA)
> edqplot(XE)
#There are also differences among models in both groups
>SummaryModel(XA)
Number of stationary series:  100
q=0 q=1 q=2 q=3
p=0   0   0   0   0
p=1   0   0   0   0
p=2   0   0   2   0
p=3   0   0   4   3
p=4   0   0   6   9
p=5   0   0  31  45
>SummaryModel(XE)
Number of stationary series:  100
q=0 q=1 q=2 q=3
p=0   0   0   0   0
p=1   0   0   0   0
p=2   0   0   0   4
p=3   0   0   3   9
p=4   0   0  15  23
p=5   0   1  22  23
> h=5
> #Compute acf and ccf of the stationary series XA.
> aa=acf(XA,lag.max=h,plot=FALSE)
> #a matrix with columns the acf of each series is formed
> MA=matrix(0,h,k)
> nl=h+1
> for(i in 1:k)
+   { MA[,i]=aa$acf[c(2:nl),i,i] }
> k=dim(XE)[2]
> T=dim(XE)[1]
> #Compute the acf and ccf of the stationary series XE
> aa=acf(XE,lag.max=h,plot=FALSE)
> #a matrix with columns the acf of each series is formed
> ME=matrix(0,h,k)
> nl=h+1
> for(i in 1:k)
+ { ME[,i]=aa$acf[c(2:nl),i,i] }
> A=t(MA)
> E=t(ME)
> ACM=rbind(A,E)
> par(mfrow=c(2,1))
> #The histogram of the acf(1) for healthy and ill patients
> i=1
> plot(hist(A[,i],plot=FALSE),main="histograms of acf(1) healthy persons")
> plot(hist(E[,i],plot=FALSE),main="histograms of acf(1) ill persons")
> y1=matrix(0,100,1)
> y2=matrix(1,100,1)
> y=rbind(y1,y2)
```

```
> yf=as.factor(y)
> data=data.frame(yf,ACM)
> names(data)
[1] "yf" "X1" "X2" "X3" "X4" "X5"
> ##Discrimination with  LDF
> ##First with all the data
> slda1=lda(x=ACM,grouping = yf, cv=TRUE)
> plot(slda1)
> plot(slda1)
> c1=predict(slda1,ACM)$class
> slda1.data <- cbind(data, predict(slda1)$x)
> tf0 = table(yf, c1)
> tf0
c1
yf   0  1
0 95  5
1 13 87
> ###logistic discrimination with all the data
> lg1=glm(yf~., data=data,family=binomial(link="logit"))
> summary(lg1)
Call:
glm(formula = yf ~., family = binomial(link = "logit"), data = data)
Deviance Residuals:
Min        1Q   Median       3Q      Max
-1.1993  -0.1308   0.0161   0.1821   5.6475

Coefficients:
Estimate Std. Error z value Pr(>|z|)
(Intercept) -1246.9      207.0   -6.024 1.70e-09 ***
X1            439.1      130.8    3.356 0.000792 ***
X2           2932.6      511.0    5.739 9.54e-09 ***
X3          -3595.0      627.6   -5.728 1.01e-08 ***
X4           1874.0      350.7    5.344 9.07e-08 ***
X5           -402.7       86.5   -4.656 3.23e-06 ***
- - -
Signif. codes:  0 '***' 0.001 '**' 0.01 '*' 0.05 '.' 0.1 ' ' 1
(Dispersion parameter for binomial family taken to be 1)

Null deviance: 277.259  on 199  degrees of freedom
Residual deviance:  70.907  on 194  degrees of freedom
AIC: 82.907

Number of Fisher Scoring iterations: 8
> #Compute probabilities in the form of P(y=1|X).
> #Our decision boundary will be 0.5. If P(y=1|X) > 0.5
> #then y = 1 otherwise y=0.
> y_lg1 <- predict(lg1,newdata=data,type='response')
> y_lg1r <- ifelse(y_lg1 > 0.5,1,0)
> misClasificError <- mean(y_lg1r != yf)
> print(paste('Accuracy',1-misClasificError))
[1] "Accuracy 0.975"
> tl0 = table(yf, y_lg1r)
> tl0
y_lg1r
yf   0  1
0 99  1
1  4 96
```

```
> # Discrimination via splitting the sample
> #Splitting the data set into a training set and a test set
> set.seed(123)
> split = sample.split(yf, SplitRatio = 0.65)
> training_set = subset(data, split == TRUE)
> test_set = subset(data, split == FALSE)
> training_set[,-1] = scale(training_set[,-1])
> test_set[,-1] = scale(test_set[,-1])
> ## LDF with the splitted sample
> fld1=lda(x=training_set[,-1],grouping = training_set[,1])
> plot(fld1)
> y_fld1 = predict(fld1, newdata = training_set[,-1])
> misClasificError <- mean(y_fld1$class != training_set[,1])
> print(paste('Accuracy of LDF in sample',1-misClasificError))
[1] "Accuracy of LDF in sample 0.923076923076923"
> y_fld2 = predict(fld1, newdata = test_set[,-1])
> misClasificError <- mean(y_fld2$class != test_set[,1])
> print(paste('Accuracy of LDF out of sample',1-misClasificError))
[1] "Accuracy of LDF out of sample 0.842857142857143"
> tf1 = table(training_set[, 1], y_fld1$class)
> tf1
0  1
0 61  4
1  6 59
> tf2 = table(test_set[, 1], y_fld2$class)
> tf2
0  1
0 28  7
1  4 31
> #Logistic discrimination splitting the sample
> lg2=glm(yf~.,data=training_set,family=binomial(link="logit"),
control = list(maxit = 50))
> summary(lg2)
Call:
glm(formula=yf ~.,family=binomial(link="logit"),data=training_set,
control = list(maxit = 50))

Deviance Residuals:
Min        1Q      Median        3Q        Max
-1.62439  -0.00012   0.00000   0.00313   1.90223

Coefficients:
Estimate Std. Error z value Pr(>|z|)
(Intercept)     -8.138      4.281  -1.901    0.0573.
X1              84.759     45.209   1.875    0.0608.
X2             710.381    410.286   1.731    0.0834.
X3           -1509.568    822.208  -1.836    0.0664.
X4             910.752    491.394   1.853    0.0638.
X5            -179.072     96.971  -1.847    0.0648.
- - -
Signif. codes:  0 '***' 0.001 '**' 0.01 '*' 0.05 '.' 0.1 ' ' 1
(Dispersion parameter for binomial family taken to be 1)
Null deviance: 180.218  on 129  degrees of freedom
Residual deviance:  10.136  on 124  degrees of freedom
AIC: 22.136
```

```
Number of Fisher Scoring iterations: 12

> y_lg2 = predict(lg2, newdata = training_set[,-1])
> y_lg2r <- ifelse(y_lg2 > 0.5,1,0)
> misClasificError <- mean(y_lg2r != training_set[,1])
> print(paste('Accuracy of LR in sample',1-misClasificError))
[1] "Accuracy of LR in sample 0.976923076923077"
> tl1 = table(training_set[, 1], y_lg2r)
> tl1
y_lg2r
0  1
0 64  1
1  2 63
> y_lg3 = predict(lg2, newdata = test_set[,-1])
> y_lg3r <- ifelse(y_lg3 > 0.5,1,0)
> misClasificError <- mean(y_lg3r != test_set[,1])
> print(paste('Accuracy of LR out of sample',1-misClasificError))
[1] "Accuracy of LR out of sample 0.671428571428571"
> tl2 = table(test_set[, 1], y_lg3r)
> tl2
y_lg3r
0  1
0 31  4
1 19 16
[1] "Accuracy LDF in sample 0.903538461538462"
[1] "Accuracy LDF out of sample 0.871142857142857"
[1] "Accuracy LR in sample 0.968615384615385"
[1] "Accuracy LR out of sample 0.876571428571429"
```

■

5.6 NONPARAMETRIC CLASSIFICATION

When the distribution of the features is unknown, several nonparametric methods can be used for discrimination. The first one is nearest neighbors (NN), that is very simple to apply and often used as the benchmark in comparison between methods (see, for instance, Dau et al. 2018). The second method is support vector machines (SVM), proposed by Vapnik (2013) and related to the theory of construction of classification machines in machine learning as presented in Cherkassky and Mulier (1998), and Vapnik (2013). Third, there are some classification methods that can be applied to a broader set of prediction problems, including time series forecasting, and these procedures are explained in Chapters 7 and 8. They include the classification and regression tree (CART), random forest, boosting, neural networks, and deep learning. Finally, for a large data set we can estimate the joint probability distribution of the features and use this empirical distribution to classify new observations. The efficiency of these procedures in different settings of time series classification problems is still under intensive research.

5.6.1 Nearest Neighbors

A simple and general classification procedure which has provided good results in many cases is as follows:

1. Define a measure of distance between two points.
2. Calculate the distances from the point to be classified, $\boldsymbol{x}_0$, to all the points in the sample.
3. Choose the m sample points closest to $\boldsymbol{x}_0$. Calculate the proportions of these m points which belong to each of the populations. Classify the point $\boldsymbol{x}_0$ to the population with the greatest frequency among the m points.

This method is known as the m- NN. In the specific case of $m = 1$ NN classifies a new element in the population to which its closest element belongs. A key problem is the choice of m. A possibility is to try different values, obtain the classification error as a function of m, and choose the value of m which leads to the smallest observed classification error.

One of the advantages of this simple method is that it only looks at the distribution of points around the new data to be classified. Thus, if some of the features are not relevant and the information useful for classification is contained in a space of smaller dimension, the local approximation makes that the distance computed could be a good representation of the larger space. For time series analysis, any of the distances and dissimilarity measures presented in Section 5.1 can be used.

NN does not work well when the dimension of the feature space, p, is large. Then, the distance of a point to all the others is usually relatively high and this deteriorates the performance of the method. For instance, suppose data defined by a uniform distribution of points in a hypercube of side one and dimension p. Consider an internal hypercube of side $L < 1$. It will include a proportion of points of the population (the initial one) equal to $pr = L^p$. For large p, for instance $p = 500$, a large hypercube with $L = 0.99$ will include points of the population with probability $p = 0.0065$. Thus, if we take a sample of the population with $n = 1000$, the expected number of points in this hypercube is 6.5. The high-dimensional space is very wide in relative terms and local distances may not work well in high dimension. Some adaptive nearest neighbors have been proposed for this situation; see Hastie et al. (2001). Another possibility is to reduce the dimension of the space by projecting the data in some useful directions, as the extreme kurtosis directions; see Peña and Prieto (2001) and Peña et al. (2010). The efficiency and usefulness of these methods requires further research.

5.6.2 Support Vector Machines

A popular method for discrimination is the SVM; see Vapnik and Chapelle (2000), Cherkassky and Mulier (1998), Scholkopf and Smola (2001), and the references therein. We present the method here first for the case of two states. This procedure approaches the classification problem slightly different from the previous ones: (1) instead of looking for a global solution that depends on all the points in the sample, it searches for a local solution that depends mostly on the points that define the border between the two states; (2) instead of looking for a reduction of the space of variables and solving the problem in the smaller space, it looks for a larger space, including nonlinear transformations of the original variables, where the points may be separated linearly.

A set of n data points, $\{\boldsymbol{x}_i \in R^p | i = 1, \ldots, n\}$, that belong to two classes is linearly separable if it is possible to find an hyperplane that separates the observations perfectly. A hyperplane in R^p is a subspace of dimension $p - 1$ and is defined by a linear equation

$$\boldsymbol{w}'\boldsymbol{x} = b, \tag{5.59}$$

where $\boldsymbol{w} = (w_1, \ldots, w_p) \neq \mathbf{0}$, $\boldsymbol{x} = (x_1, \ldots, x_p)'$ and b is a real number. If the hyperplane contains the origin, its equation becomes $\boldsymbol{w}'\boldsymbol{x} = 0$: all vectors going from the origin to a point in the hyperplane, $\boldsymbol{x}$, are orthogonal to $\boldsymbol{w}$, which is the vector defining the hyperplane. If the hyperplane does not contain the origin, $b \neq 0$. Let $\boldsymbol{x}_0$ be the projection of the origin onto the hyperplane, then $\boldsymbol{w}'\boldsymbol{x}_0 = b$. Consequently, we can rewrite the Eq. (5.59) as

$$\boldsymbol{w}'\boldsymbol{x} = \boldsymbol{w}'\boldsymbol{x}_0 \quad \text{or} \quad \boldsymbol{w}'\boldsymbol{x} + w_0 = 0,$$

where $w_0 = -\boldsymbol{w}'\boldsymbol{x}_0$. In general, a hyperplane in R^p partitions R^p into two separated subspace, namely $\boldsymbol{w}'\boldsymbol{x} + w_0 > 0$ and $\boldsymbol{w}'\boldsymbol{x} + w_0 < 0$.

5.6.2.1 *Linearly Separable Problems* Suppose that we have a training set with n elements, $\{(y_i, \boldsymbol{x}_i)\}$, where y_i is the binary classification variable, which, to simplify things here, takes the possible values −1 and +1. If the set is linearly separable all the observations of a group, for example, those of $y_i = -1$, are found on one side of a hyperplane and all the points of the other group, with $y_i = 1$, are on the other side. Therefore, points in one side verify that $\boldsymbol{w}'\boldsymbol{x}_i + w_0 < 0$ and those at the other that $\boldsymbol{w}'\boldsymbol{x}_i + w_0 > 0$. Also, for some value m, that depends on the data, we have the equivalent relations $\boldsymbol{w}'\boldsymbol{x}_i + w_0 \leq -m$, and $\boldsymbol{w}'\boldsymbol{x}_i + w_0 \geq m$. These two inequalities can be written jointly as

$$y_i(\boldsymbol{w}'\boldsymbol{x}_i + w_0) \geq m, \qquad \text{for } i = 1, \ldots, n, \tag{5.60}$$

and we want to find the hyperplane that makes m, the margin or separation between the groups as large as possible. Figure 5.32 illustrates the problem we want to solve. In the maximization problem we must take into account that the hyperplane that

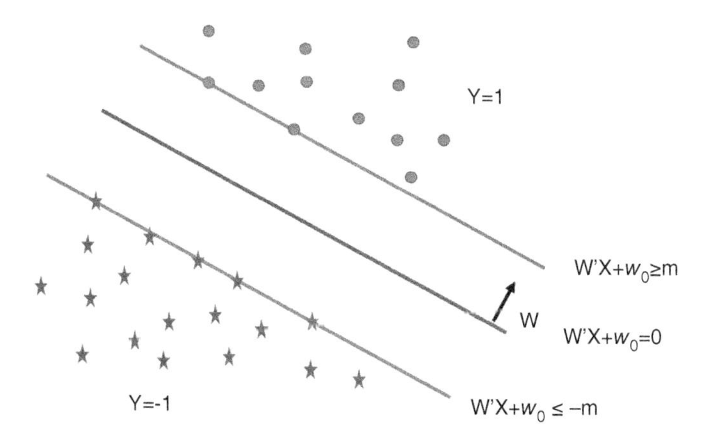

Figure 5.32 Two linearly separable groups in R^2, the separating plane and the margin. The blue circles group has two support vectors and the red star one has five.

maximizes m with the condition in Eq. (5.60) is not unique, as any other hyperplane, $c(\boldsymbol{w}'\boldsymbol{x}_i + w_0) = 0$, for any $c \neq 0$ is also a solution. To identify the hyperplane, we define $\mathbf{u} = \boldsymbol{w}/\|\boldsymbol{w}\|$, and the problem to be solved is

$$\begin{gathered} \max(m) \\ \text{such that } \|\boldsymbol{u}\| = 1 \quad \text{and} \quad y_i(\boldsymbol{u}'\boldsymbol{x}_i + w_0) \geq m. \end{gathered} \tag{5.61}$$

We write this optimization problem in a slightly different way, which is easier to solve. First, note that the distance of a point $\boldsymbol{x}_i$ to a general hyperplane, $\boldsymbol{w}'\boldsymbol{x} + w_0 = 0$ is the distance between this point and its projection on the hyperplane $\boldsymbol{x}_i(p)$. The vector $\boldsymbol{x}_i - \boldsymbol{x}_i(p)$ is orthogonal to the hyperplane, so that we can write

$$\boldsymbol{x}_i - \boldsymbol{x}_i(p) = d_i\boldsymbol{u},$$

where $\boldsymbol{u}$ is a vector of unit length orthogonal to the hyperplane and d_i is the distance to the hyperplane. If we multiply this equation by $\boldsymbol{u}'$ and use that $\boldsymbol{x}_i(p)$ is in the hyperplane, so that $\boldsymbol{w}'\boldsymbol{x}_i(p) + w_0 = 0$, then

$$\boldsymbol{u}'(\boldsymbol{x}_i - \boldsymbol{x}_i(p)) = \frac{\boldsymbol{w}'}{\|\boldsymbol{w}\|}[\boldsymbol{x}_i - \boldsymbol{x}_i(p)] = \frac{\boldsymbol{w}'\boldsymbol{x}_i + w_0}{\|\boldsymbol{w}\|} = d_i.$$

Therefore, the equivalent condition to Eq. (5.60) is, for some M value

$$y_i\frac{(\boldsymbol{w}'\boldsymbol{x}_i + w_0)}{\|\boldsymbol{w}\|} \geq M \qquad \text{for } i = 1, \ldots, n$$

or

$$y_i(\boldsymbol{w}'\boldsymbol{x}_i + w_0) \geq \|\boldsymbol{w}\|M, \qquad \text{for } i = 1, \ldots, n.$$

The norm of the vector $\boldsymbol{w}$ is arbitrary, but has to be defined to identify the hyperplane. In fact, comparing with Eq. (5.61) if $\|\boldsymbol{w}\| = 1$, then $M = m$. we need to fix the scale with any of these variables and a simple solution is to take

$$\|\boldsymbol{w}\| = 1/M,$$

and maximizing the distance M is the same as minimizing the norm $\|\boldsymbol{w}\|$. This approach also has the advantage that we do not need the restriction $\|\boldsymbol{w}\| = 1$, because the hyperplane is perfectly defined by the solution. The optimization problem is then

$$\begin{gathered} \min \frac{1}{2}\|\boldsymbol{w}\|^2, \\ \text{subject to} \quad y_i(\boldsymbol{w}'\boldsymbol{x}_i + w_0) \geq 1, \quad i = 1, \ldots, n. \end{gathered} \tag{5.62}$$

This is a quadratic optimization problem with $p + 1$ variables, the components of the vector $\boldsymbol{w}$ plus w_0. The function to be minimized can be written using the Lagrange multipliers as

$$F_L = \frac{1}{2}\boldsymbol{w}'\boldsymbol{w} - \sum_{i=1}^{n} \lambda_i[y_i(\boldsymbol{w}'\boldsymbol{x}_i + w_0) - 1], \tag{5.63}$$

where the $\lambda_i \geq 0$ so that when the constraints hold for all points $F_L \leq \boldsymbol{w}'\boldsymbol{w}/2$, and if the constraint $y_i(\boldsymbol{w}'\boldsymbol{x}_i + w_0) \geq 1$ is not satisfied for the variables $(y_j, \boldsymbol{x}_j)$, it increases F_L. Taking the derivatives with respect to $\boldsymbol{w}$ and w_0

$$\frac{\partial F_L}{\partial \boldsymbol{w}} = 0 = \boldsymbol{w} - \sum_{i=1}^{n} \lambda_i y_i \boldsymbol{x}_i, \tag{5.64}$$

$$\frac{\partial F_L}{\partial w_0} = 0 = \sum_{i=1}^{n} \lambda_i y_i, \tag{5.65}$$

and inserting in Eq. (5.63) these equations, we obtain the dual problem

$$F_D = \frac{1}{2}\boldsymbol{w}'\boldsymbol{w} - \boldsymbol{w}'\boldsymbol{w} + \sum_{i=1}^{n} \lambda_i = \sum_{i=1}^{n} \lambda_i - \frac{1}{2}\sum_{i=1}^{n}\sum_{j=1}^{n} \lambda_i \lambda_j y_i y_j \boldsymbol{x}_i' \boldsymbol{x}_j. \tag{5.66}$$

This is a problem in the n variables λ_i that verify the restrictions $\lambda_i \geq 0$ and $\sum_{i=1}^{n} \lambda_i y_i = 0$. It is called the dual problem and minimizing F_L to find the $p+1$ variables with their restrictions that define the hyperplane is equivalent to maximizing F_D with respect the n variables λ_i with their restrictions. The second problem is easier to solve and also the objective function only depends on the data through the scalar products of the feature vectors. It is a convex optimization problem with a quadratic criterion and linear constraints that can be solved by standard optimization methods. See, for instance, Hastie et al. (2001).

Note also that the hyperplane is defined by the scalar products between the sample points, because by Eq. (5.64) it can be written as

$$f(\boldsymbol{x}) = \boldsymbol{w}'\boldsymbol{x} + w_0 = \sum_{i=1}^{n} \lambda_i y_i \boldsymbol{x}_i' \boldsymbol{x} + w_0. \tag{5.67}$$

The hyperplane is defined by the points with $\lambda_i > 0$ and those with $\lambda_i = 0$ have no role in the equation of the hyperplane. Points with $\lambda_i > 0$ are called support vectors and verify $y_i(\boldsymbol{w}'\boldsymbol{x}_i + w_0) = 1$. See Figure 5.32. The classification is made with the sign of $\boldsymbol{w}'\boldsymbol{x}_i + w_0$. Note that the restrictions define a space bounded by the limiting hyperplanes $\boldsymbol{w}'\boldsymbol{x}_i + w_0 = 1$ and $\boldsymbol{w}'\boldsymbol{x}_i + w_0 = -1$ and inside this space there exists no data points. The data contained in these limiting hyperplanes are the support vectors, that play a key role because if they are modified the classification will change, a property that apply only to the support vectors. This is an important difference with Fisher's LDF or logistic regression. In these approaches a small perturbation of any point, from $\boldsymbol{x}_i$ to $\boldsymbol{x}_i + \epsilon \boldsymbol{u}$ for small ϵ, may produce a change in the classification. In SVM, this would happen for points close to the limiting hyperplanes but has no effect for points far away from both hyperplanes.

For normal separable data with the same covariance matrix, the solution provided by Fisher LDF is optimal, but when the covariance matrices are different and the data is not normal, SVM provides a useful alternative. Usually in these cases, the logistic regression would provide a solution similar to that of SVM.

5.6.2.2 Nonlinearly Separable Problems If the data are not linearly separable, we can modify the problem and find the hyperplane that maximizes the separation of the population for most of the data, but allowing a small number of points to be incorrectly classified. With this objective, we introduce a variable for each point $\alpha_i \geq 0$ that will be zero if the point is well classified and positive otherwise. The optimization problem of Eq. (5.62) is modified as

$$\min \left(\frac{1}{2}\|\boldsymbol{w}\|^2 + C\sum_{i=1}^{n} \alpha_i \right), \tag{5.68}$$

$$\text{subject to } y_i(\boldsymbol{w}'\boldsymbol{x}_i + w_0) \geq 1 - \alpha_i, \quad i = 1, \ldots, n,$$
$$\alpha_i \geq 0, \tag{5.69}$$

where the parameter C needs to be given. Usually different values are tried and the one with the smallest classification error is chosen. It can also be obtained by CV, as explained in Section 5.4.2. The Lagrange function is

$$F_L = \frac{1}{2}\boldsymbol{w}'\boldsymbol{w}w + C\sum_{i=1}^{n} \alpha_i - \sum_{i=1}^{n} \lambda_i[y_i(\boldsymbol{w}'\boldsymbol{x}_i + w_0) - (1 - \alpha_i)] - \sum_{i=1}^{n} \mu_i\alpha_i \tag{5.70}$$

with $\lambda_i \geq 0$. Taking the derivatives with respect to $\boldsymbol{w}$, w_0, and α_i, we have

$$\frac{\partial F_L}{\partial \boldsymbol{w}} = 0 = \boldsymbol{w} - \sum_{i=1}^{n} \lambda_i y_i \boldsymbol{x}_i,$$

$$\frac{\partial F_L}{\partial w_0} = 0 = \sum_{i=1}^{n} \lambda_i y_i,$$

$$\frac{\partial F_L}{\partial \alpha_i} = 0 = C - \lambda_i - \mu_i$$

and, inserting in Eq. (5.70) these equations, we have the dual function

$$F_D = \sum_{i=1}^{n} \lambda_i - \frac{1}{2}\sum_{i=1}^{n}\sum_{j=1}^{n} \lambda_i\lambda_j y_i y_j \boldsymbol{x}_i'\boldsymbol{x}_j, \tag{5.71}$$

that has to be maximized with the restrictions $\mu_i \geq 0$, $0 \leq \lambda_i \leq C$ and $\sum_{i=1}^{n} \lambda_i y_i = 0$. The problem is similar to Eq. (5.66) and, again, the function F_D, as well as the hyperplane, depends on the data by the scalar products $\boldsymbol{x}_i'\boldsymbol{x}$, as shown in Eq. (5.67).

In addition to allowing for some points that are not linearly separable, the SVM approach can deal with nonlinearity by increasing the dimension of the space. This is made by introducing new features that are nonlinear functions of the previous ones and searching for an optimal linear separation in this larger space. This approach is the opposite to the one followed by classical procedures that project the data to a smaller space and use nonlinear functions to separate the states. For example, in Figure 5.33 the two groups cannot be separated in dimension one but they are linearly separable in a space of dimension two. When we introduce new variables, it is important to realize that, as shown in Eq. (5.66), the objective function only depends

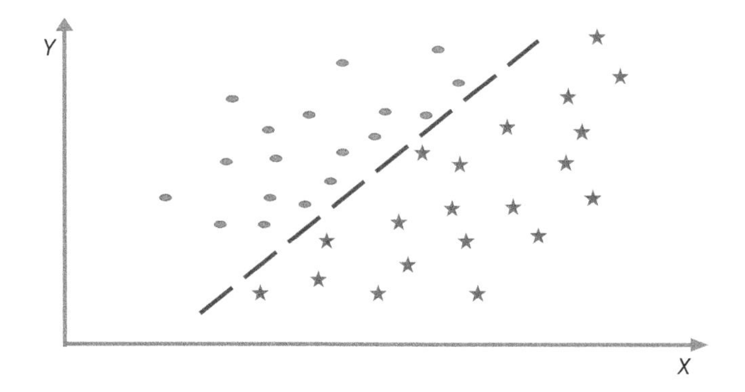

Figure 5.33 Two variables that can be linearly separated in two dimensions and not in one.

on the scalar products between the observations. Therefore, if we introduce nonlinear transformations of the features and go from the variables $\boldsymbol{x} \in R^p$ to a new set of variables $\boldsymbol{h}(\boldsymbol{x}) \in R^q$, with $q > p$, we have to compute the scalar products $h(\boldsymbol{x})'h(\boldsymbol{x})$. A fast way to do this is using an inner product kernel function that gives the scalar products directly. For instance, the polynomial inner product kernel function is

$$K(\boldsymbol{x}_i, \boldsymbol{x}_j) = (1 + \boldsymbol{x}_i'\boldsymbol{x}_j)^d = \boldsymbol{h}(\boldsymbol{x}_i)'\boldsymbol{h}(\boldsymbol{x}_j),$$

and provides the scalar product of the two vectors $\boldsymbol{h}(\boldsymbol{x}_i)'\boldsymbol{h}(\boldsymbol{x}_j)$ that includes the polynomial of order d in each of the p features of $\boldsymbol{x}$. For instance, suppose $d = 2$ and $p = 2$. Then, $\boldsymbol{x}_i = (x_{1i}, x_{2i})'$ and $\boldsymbol{x}_i'\boldsymbol{x}_i = x_{1i}^2 + x_{2i}^2$, and

$$K(\boldsymbol{x}_i, \boldsymbol{x}_i) = (1 + x_{1i}^2 + x_{2i}^2)^2 = (1 + 2x_{1i}^2 + 2x_{2i}^2 + 2x_{1i}^2x_{2i}^2 + x_{1i}^4 + x_{2i}^4).$$

This inner product includes all the terms of the polynomial of degree $d = 2$ in the original variables. Note that $K(\boldsymbol{x}_i, \boldsymbol{x}_i)$ is the squared norm of the vector $\boldsymbol{h}(\boldsymbol{x}_i) = (1, \sqrt{2}x_{1i}, \sqrt{2}x_{2i}, \sqrt{2}x_{1i}x_{2i}, x_{1i}^2, x_{2i}^2)$, $\boldsymbol{h}(\boldsymbol{x}_i) \in \mathbf{R}^6$. In general, $K(\boldsymbol{x}_i, \boldsymbol{x}_j)$ gives the scalar product of the vectors $h(\boldsymbol{x}_i)$ and $h(\boldsymbol{x}_j)$, including all the terms until order d. Therefore, if we use this kernel to increase the dimension of the feature space, the hyperplane is

$$f(\boldsymbol{h}(\boldsymbol{x})) = \boldsymbol{w}'\mathbf{h}(\mathbf{x}) + w_0 = \sum_{i=1}^{n} \lambda_i y_i K(\boldsymbol{x}_i, \boldsymbol{x}_j) + w_0,$$

where now $\boldsymbol{x} \in \boldsymbol{R}^q$ and $\boldsymbol{h}(\boldsymbol{x}) \in \boldsymbol{R}^q$, and the optimization problem to be solved is Eq. (5.71) with the new variables $\boldsymbol{h}(\boldsymbol{x}_i)$, has the form

$$\max F_D = \sum_{i=1}^{n} \lambda_i - \frac{1}{2} \sum_{i=1}^{n} \sum_{j=1}^{n} \lambda_i \lambda_j y_i y_j K(\boldsymbol{x}_i, \boldsymbol{x}_j),$$

$$C \geq \lambda_i \geq 0, \qquad \mu_i \geq 0, \qquad \sum_{i=1}^{n} \lambda_i y_i = 0.$$

Other inner product kernel functions available are those based on Radial basis, Splines, Fourier coefficients, Sigmoids, and others (Cherkassky and Mulier, 1998).

For the case of more than two states, the classification is made by steps. In step 1, the classification is made between the first state against the rest. If the element is classified in the first state, the procedure stops. Otherwise, we move to step 2 and classify between the second state against the rest. The procedure continues until the element is classified.

5.6.3 Density Estimation

If the density of the features is not normal and the number of features considered, p, is not too large, we can use the training set of previous outputs from the system to estimate the distribution of the features directly. Suppose we have n_i time series generated by state s_i and a total of $n = n_1 + \cdots + n_c$ observations. To classify a point, $\boldsymbol{x}_0$, of dimension p we do not need to estimate all the density but rather just the density at this point. Then, the classification is carried out maximizing the probability a posteriori, that is, by maximizing

$$\max_{i \in S} \pi_i f(\boldsymbol{x}_0 | s_i).$$

One way of estimating the density for a given state at point $\boldsymbol{x}_0$ is to construct a hypercube in p dimensions with $\boldsymbol{x}_0$ as its center and h as its side, count the points originating from the state included in it, $n_i(\boldsymbol{x}_0)$, and estimate the density by the relative frequency of points in the volume of the hypercube, as

$$\hat{f}(\boldsymbol{x}_0 | s_i) = \frac{n_i(\boldsymbol{x}_0)}{n_i h^p}.$$

This procedure is similar to that of NN, but instead of setting the m number of nearest points we establish a setting h and count how many points from each distribution are in this interval. The above rule can be written as

$$\hat{f}(\boldsymbol{x}_0 | s_i) = \frac{1}{n_i h^p} \sum_{i=1}^{n} \prod_{j=1}^{p} k\left(\frac{x_{ij} - x_{0j}}{h}\right),$$

where the function k is called the kernel and verifies

$$k\left(\frac{x_{ij} - x_{0j}}{h}\right) = \begin{cases} 1, & \text{if } |x_{ij} - x_{0j}| \leq h, \\ 0, & \text{otherwise.} \end{cases}$$

This density estimator is rather irregular since the points either enter the hypercube and contribute with a value of one to the density of the point, or they contribute nothing whatsoever to determining the density. A better estimator is to allow the points to contribute to the estimation of the density function at a given point according to their distance to it, which can be done by using a kernel which averages the information of the points according to their distance. A commonly used kernel is the normal or Gaussian one, given by

$$k_G\left(\frac{x_{ij} - x_{0j}}{h}\right) = \frac{1}{\sqrt{2\pi}} \exp\left[-\frac{1}{2}\left(\frac{x_{ij} - x_{0j}}{h}\right)^2\right].$$

and then, assuming that the features are independent, we obtain the so called *Naive Bayes classifier* given by

$$\hat{f}^{NB}(\mathbf{x}_0|s_i) = \frac{1}{n_i h^p} \sum_{i=1}^{n_i} \prod_{j=1}^{p} k_G\left(\frac{x_{ij} - x_{0j}}{h}\right).$$

The procedure can be improved taking into account the dependency among the variables and using a normal multivariate kernel, leading to:

$$\hat{f}(\mathbf{x}_0|s_i) = \frac{1}{n_i h^p |\mathbf{S}_i|^{1/2}} \sum_{i=1}^{n_i} \exp\left\{-\frac{1}{2h^2}(\mathbf{x}_i - \mathbf{x}_0)'\mathbf{S}_i^{-1}(\mathbf{x}_i - \mathbf{x}_0)\right\},$$

where $\mathbf{S}_i$ is an estimate of the covariance matrix in the state s_i. In general, the same $\mathbf{S}_i$ matrix is frequently taken in all the states using weighted average of the matrices estimated in each state.

The problem with all the kernels methods is that the density estimation is critically dependent on the choice of the h parameter, which is unknown, but can be estimated by CV.

Example 5.8

We use, again, the Taiwan AirBox Data and their locations. The plot of the locations indicates that some of the Air Boxes were in Taipei, in the North of the Island (those with latitude ≤ 22.7 and longitude ≤ 129.5) and some others in the Condado de Yilan, in the south-east of Taipei (latitude ≥ 24.5 and longitude ≥ 121.4). Using the set of series in both locations we develop a discriminant procedure to classify a new time series as being from the North (Taipei) or from the South (Yilan). The variables available for discrimination are the first five ACFs computed in Example 5.1. The **R** commands are as follows, but we have not repeated those used in Examples 5.1 and 5.6.

The results show that the ACFs have low power for discrimination. The classification errors are around 30–35% in SVM and slightly higher in LDF and LR. We have tried other nonlinear kernels in SVM but the results are similar and, in this example, the nonlinear features of SVM do not provide a clear advantage.

R commands for discriminant analysis:

```
> #The two groups of data, Taipei and Yilan, are obtained.
> library(caTools)
> library(MASS)
> library(e1071)
> lo=as.matrix(locations032017)
> x=as.matrix(TaiwanAirBox032017)
> #First, we build the matrix of clean data
> x=x[,-c(1,29,70)]
> lo=lo[,-1]
> lo=lo[-c(29,70),]
> #Data frame with classification variable for Taipei and Yilan.
> i1=which(lo[,1]<22.7)
> i2=which(lo[,2]<120.5)
> i3=sort(intersect(i1,i2))
```

```
> x1=x[,i3]
> i4=length(i3)
> y1=matrix(0,i4,1)
> j1=which(lo[,1]>24.5)
> j2=which(lo[,2]>121.4)
> j3=sort(intersect(j1,j2))
> j4=length(j3)
> intersect(i3,j3)
> x2=x[,j3]
> y2=matrix(1,j4,1)
> y=c(y1,y2)
> xc=cbind(x1,x2)
##The computation of ACF is similar to Examples 5.1 and 5.6.
## The data, prepared as in Example 5.6 is in data.frame=data
##Splitting the data set into the Training set and Test set
> set.seed(123)
> split = sample.split(yf, SplitRatio = 0.65)
> training_set = subset(data, split == TRUE)
> test_set = subset(data, split == FALSE)
> training_set[,-1] = scale(training_set[,-1])
> test_set[,-1] = scale(test_set[,-1])
## Discrimination with Support Vector Machine
> svo = svm(y=training_set[,1], x=training_set[,-1],
type = 'C-classification', kernel = 'sigmoid')
> y_sv1 = predict(svo, newdata = training_set[,-1])
> misClasificError <- mean(y_sv1 != training_set[,1])
> print(paste('Accuracy of SVM in sample',1-misClasificError))
[1] "Accuracy of SVM in sample 0.708333333333333"
> y_sv2 = predict(svo, newdata = test_set[,-1])
> misClasificError <- mean(y_sv2 != test_set[,1])
> print(paste('Accuracy of SVM out of sample',1-misClasificError))
[1] "Accuracy of SVM out of sample 0.666666666666667"
## As a reference we present the results with LDF and LR for this data
## LDF and LR are computes as in Example 5.6
> print(paste('Accuracy of LDF in sample',1-misClasificError))
[1] "Accuracy of LDF in sample 0.75"
>print(paste('Accuracyof LDF out of sample',1-misClasificError))
[1] "Accuracyof LDF out of sample 0.705882352941176"
> print(paste('Accuracy of LR in sample',1-misClasificError))
[1] "Accuracy of LR in sample 0.729166666666667"
> print(paste('Accuracy of LR out of sample',1-misClasificError))
[1] "Accuracy of LR out of sample 0.705882352941176"
```
■

5.7 OTHER CLASSIFICATION PROBLEMS AND METHODS

Often, we have situations in which only some parts of the available data have labels. For instance, we have temperature time series of hospital patients and we know that some of them have been confirmed by a test to be infected by virus COVID-19, whereas others have not received a virus test. In other situations, we have the evolution of a vector of time series but the labels of the possible classes or groups are only known for a small part of the sample. The problem in these cases has been called of *partially unsupervised classification* and appears specially when obtaining the labels for the classification of the objects is expensive. In these situations, we have to use cluster and discrimination methods together to estimate the probabilities of the classes, applying cluster analysis to some part of the data and a discrimination model to the other, see for instance, Liu et al. (2002).

In other situations, we have data and labels for all the objects but most of them belong to the same class and only a few are from the other uncommon classes. For instance, in problems of fraud detection we may have many regular operations but only a few that describe a fraud. In this situation, a classifier may minimize the error of classification by allocating all objects to the larger class. This is called the imbalanced classification problem and several solutions have been proposed. See Ganganwar (2012) for a review of the topic.

The first and simplest solution to this problem is to use a Bayesian approach and compute the posterior probability of an object belonging to each class, as in Eq. (5.19), taking the prior probabilities into account. We can, also, incorporate the consequences of the errors of classification and introduce them as in a decision problem. The second solution applies to very large data sets where we can sample the data to provide a more balanced class distribution, either by random over-sampling with random replication of examples of the minority class, or by random under-sampling, removing cases from the overrepresented class. The third one is to approach the problem as one of outlier detection, where we try to identify atypical events in the data set, or series that are atypical from the rest, as we have discussed in Chapter 4.

Time series classification has also been approached from the point of view of functional data analysis, often using the concept of depth. See Lopez-Pintado and Romo (2006) and Tupper et al. (2015) for a band depth approach for clustering nonstationary time series. A common distance with functional data is the *dynamic time warping* that considers distances between values of the series taken in sequence, but for different time points. See Jeong et al. (2011).

EXERCISES

1. Consider the World Stock Indexes. Apply a hierarchical clustering using as dissimilarity measure the cross linear dependency. Compare the results obtained with different number of lags in the dissimilarities. Discuss the different criteria used to select the number of clusters.
2. Apply k-means and k-medoids to cluster the World Stock Indexes using the Euclidean distance among the standardized series. Comment on the differences between the clusters found and those obtained in Exercise 1.
3. Consider quarterly economic series of European Union from 2000 to 2019 in the file `UMEdata2000_2018.csv`. Compare the results of a hierarchical clustering using dissimilarities between univariate properties of the series, say the first five autocorrelation coefficients, and using cross-correlations.
4. Follow the analysis of Example 5.7 to discriminate between the five files with EEG data. Apply a similar analysis to discriminate between the data in the files `EEGsetB.csv`, `EEGsetC.csv` and `EEGsetE.csv`. Select the best way to classify the five files between seizure and healthy individuals and compare different discrimination methods using the five files, A to E.
5. Apply SVM to discriminate between the data in the files `EEGsetA.csv` and `EEGsetB.csv`, `EEGsetC.csv`,`EEGsetD.csv` and `EEGsetE.csv` in order to classify the results from healthy individuals or seizure patients.

6. Using the Taiwan AirBox Data in Example 5.5, compare the clustering results obtained using the ACF and using the coefficients of an AR fitting to the series.

REFERENCES

Alonso, A. M., Berrendero, J. R., Hernández, A., and Justel, A. (2006). Time series clustering based on forecast densities. *Computational Statistics and Data Analysis*, **51**: 762–766.

Alonso, A. M. and Peña, D. (2019). Clustering time series by linear dependency. *Statistics and Computing*, **29**: 655–676.

Anderson, T. W. and Bahadur, R. R. (1962). Classification into two multivariate normal distributions with different covariance matrices. *The Annals of Mathematical Statistics*, **33**: 420–431.

Andrzejak, R. G., Lehnertz, K., Rieke, C., Mormann, F., David, P., and Elger, C. E. (2001). Indications of nonlinear deterministic and finite dimensional structures in time series of brain electrical activity: dependence on recording region and brain state. *Physical Review E* **64**, 061907.

Banfield, J. D. and Raftery, A. E. (1993). Model-based Gaussian and non-Gaussian clustering. *Biometrics*, **49**: 803–821.

Caiado, J., Crato, N., and Peña, D. (2006). A periodogram-based metric for time series classification. *Computational Statistics and Data Analysis*, **50**: 2668–2684.

Calinski, T. and Harabasz, J. (1974). A dendrite method for cluster analysis. *Communications in Statistics-Theory and Methods*, **3**: 1–27.

Cherkassky, V. and Mulier, F. (1998). *Learning From Data: Concepts, Theory, and Methods*. John Wiley & Sons, New York, NY.

Corduas, M. and Piccolo, D. (2008). Time series clustering and classification by the autoregressive metric. *Computational Statistics and Data Analysis*, **52**: 1860–1872

Dau, H. A., Bagnall, A., Kamgar, K., Yeh, C. C. M., Zhu, Y., Gharghabi, S., Ratanamahatana, C. A., and Keogh, E. (2018). The UCR time series archive. arXiv preprint arXiv:1810.07758.

Dempster, A. P., Laird, N. M., and Rubin, D. B. (1977). Maximum likelihood from incomplete data via the EM algorithm. *Journal of the Royal Statistical Society, Series B*, **39**: 1–38.

Fraley, C. and Raftery, A. E. (1999). MCLUST: Software for model-based cluster analysis. *Journal of Classification*, **16**: 297–306.

Fraley, C., Raftery, A. E., Murphy, T. B., and Scrucca, L. (2012). mclust version 4 for R: Normal mixture modeling for model-based clustering, classification, and density estimation. *Technical report No. 597*. Department of Statistics, University of Washington.

Friedman, J. (1989). Regularized discriminant analysis. *Journal of the American Statistical Association*, **84**: 165–175.

Fisher Box, J. (1978). *R. A. Fisher, the Life of a Scientist*. John Wiley & Sons, New York, NY.

Friedman, J. H. and Tukey, J. W. (1974). A projection pursuit algorithm for exploratory analysis. *IEEE Transactions on Computers*, **23**: 881–889.

Fruhwirth-Schnatter, S. and Kaufmann, S. (2008). Model-based clustering of multiple time series. *Journal of Business and Economic Statistics*, **26**: 78–89.

Ganganwar, V. (2012). An overview of classification algorithms for imbalanced datasets. *International Journal of Emerging Technology and Advanced Engineering*, **2**: 42–47.

Gamerman, D. and Lopes, H. F. (2006). *Markov Chain Monte Carlo: Stochastic Simulation for Bayesian Inference*. Chapman and Hall/CRC, London, UK.

Hartigan, J. A. (1975). *Clustering Algorithms*. John Wiley & Sons, New York, NY.

Hastie, T., Tibshirani, R., and Friedman, J. (2001). *The Elements of Statistical Learning*. New York: Springer Series in Statistics.

Hosmer, D. W. and Lemeshow, S. S. (1989). *Applied Logistic Regression*. John Wiley & Sons, Hoboken, NJ.

Huang, H. Y., Ombao, H., and Stoffer, D. S. (2004). Discrimination and classification of non-stationary time series using the SLEX model. *Journal of the American Statistical Association*, **99**: 763–774.

Jeong, Y. S., Jeong, M. K., and Omitaomu, O. A. (2011). Weighted dynamic time warping for time series classification. *Pattern Recognition*, **44**: 2231–2240.

Kakizawa, Y., Shumway, R. H., and Taniguchi, M. (1998). Discrimination and clustering for multivariate time series. *Journal of the American Statistical Association*, **93**: 328–340.

Kaufman, L. and Rousseeuw, P. J. (1987). Clustering by means of Medoids. In edited by Y. Dodge (eds.). *Statistical Data Analysis Based on the L_1 Norm and Related Methods*. North-Holland.

Kaufman, L. and Rousseeuw, P. J. (2009). *Finding Groups in Data: An Introduction to Cluster Analysis*. John Wiley & Sons, Hoboken, NJ.

Lopez-Pintado, S. and Romo, J. (2006). Depth-Based Classification for Functional Data. *DIMACS Series in Discrete Mathematics and Theoretical Computer Science*, **72**, 103.

Liu, B., Lee, W. S., Yu, P. S., and Li, X. (2002). Partially supervised classification of text documents. *Proceedings 19th International Conference in Machine Learning, Sydney, Australia*, 387–394.

Lütkepohl, H. (1996) *Handbook of Matrices*. John Wiley & Sons, Hoboken, NJ.

McLachlan, G. J. (2004). *Discriminant Analysis and Statistical Pattern Recognition*. John Wiley & Sons, New York, NY.

Maharaj, E. A. and Alonso, A. M. (2007). Discrimination of locally stationary time series using wavelets. *Computational Statistics and Data Analysis*, **52**: 879–895.

Maharaj, E. A. and Alonso, A. M. (2014). Discriminant analysis of multivariate time series: Application to diagnosis based on ECG signals. *Computational Statistics and Data Analysis*, **70**: 67–87.

Maharaj, E. A., D'Urso, P., and Caiado, J. (2019). *Time Series Clustering and Classification*. Chapman and Hall/CRC, London, UK.

McCullagh, P. (2018). *Generalized Linear Models*. Routledge, Abingdon, UK.

Montero, P. and Vilar, J. A. (2014). TSclust: An R package for time series clustering. *Journal of Statistical Software*, **62**: 1–43.

Peña, D. (1990). Influential observations in time series. *Journal of Business and Economic Statistics*, **8**: 235–241.

Peña, D. and Prieto, F. J. (2001). Cluster identification using projections. *Journal of the American Statistical Association*, **96**: 1433–1445.

Peña, D. and Rodriguez, J. (2002). A powerful portmanteau test of lack of fit for time series. *Journal of the American Statistical Association*, **97**: 601–610.

Peña, D., Prieto, F. J., and Viladomat, J. (2010). Eigenvectors of a kurtosis matrix as interesting directions to reveal cluster structure. *Journal of Multivariate Analysis*, **109**: 1995–2007.

Piccolo, D. (1990). A distance measure for classifying ARMA models. *Jounal of Time Series Analysis*, **2**: 153–163.

Rousseeuw, P. J. (1987). Silhouettes: A graphical aid to the interpretation and validation of cluster analysis. *Journal of Computational and Applied Mathematics*, **20**: 53–65.

Scholkopf, B. and Smola, A. J. (2001). *Learning with Kernels: Support Vector Machines, Regularization,Optimization, and Beyond*. MIT Press, Cambridge, MA.

Seber, G. A. F. (1984). *Multivariate Observations*. John Wiley & Sons, Hoboken, NJ.

Shumway, R. H. and Stoffer, D. S. (2000). *Time Series Analysis and Its Applications*. Springer.

Tibshirani, R., Walther, G., and Hastie, T. (2001). Estimating the number of clusters in a data set via the gap statistic. *Journal of the Royal Statistical Society, Series B*, **63**: 411–423.

Tsay, R. S. (2014). *Multivariate Time Series Analysis: With R and Financial Applications*. John Wiley & Sons, Hoboken, NJ.

Tsay, R. S. (2020). Testing serial correlations in high-dimensional time series via extreme value theory. *Journal of Econometrics*, **216**: 106–117

Tupper, L. L., Matteson, D. S., and Anderson, C. L. (2015). Band depth clustering for nonstationary time series and wind speed behavior. arXiv: 1509.00051.

Vapnik, V. (2013). *The Nature of Statistical Learning Theory*. Springer Science & Business Media, New York, NY.

Vapnik, V. and Chapelle, O. (2000). Bounds on error expectation for support vector machines. *Neural Computation*, **12**: 2013–2036.

Xiong, Y. and Yeung, D. (2004). Time series clustering with ARMA mixtures. *Pattern Recognition*, **37**: 1675–1689.

Wang, Y. and Tsay, R. S. (2019). Clustering multiple time series with structural breaks. *Journal of Time Series Analysis*, **40**: 182–202.

Wang, Y., Tsay, R. S., Ledolter, J., and Shrestha, K. M. (2013). Forecasting simultaneously high-dimensional time series: A robust model-based clustering approach. *Journal of Forecasting*, **32**: 673–684.

CHAPTER 6

DYNAMIC FACTOR MODELS

Factor models (FM) were introduced by Charles Spearman, a British psychologist, in the first quarter of the twentieth century to explain the concept of intelligence. He found that the results of different people in a set of tests of mental capacity could be explained by a general factor, which he called factor G for intelligence, and a specific component, that depends on the class of test. FMs were mainly studied first in psychology, then in economics for studying the dimension of economic development, and in sociology for understanding social and economic concepts. See, Harman (1976) for further information.

Early factor analysis for time series includes Anderson (1963), that recognized the need of developing specific results for time series data such as lag factor effects, and Brillinger (1964), who introduced dynamic principal components (DPCs) for time series. Geweke (1977) proposed a dynamic factor model (DFM) for k stationary time series assuming that each variable is the sum of two independent components: a common component, generated by r factors, plus a specific component, or noise. The r factors are assumed to follow independent linear processes and can affect the variables with lags. He also proposed a frequency-domain procedure to estimate the model by adopting the classic Jöreskog's maximum likelihood (ML) method for FMs (Jöreskog, 1967). Sargent and Sims (1977) called the model an index model and discussed their applications in economics by using Geweke's estimation approach. See also, Geweke and Singleton (1981). Engle and Watson (1981) considered a one-FM as a special case of the state space representation of a linear system and showed how to estimate the model using the Kalman filter. Chamberlain (1983) and Chamberlain and Rothschild (1983) introduced the approximate dynamic factor models (ADFMs), where the noises have dynamics, but with bounded eigenvalues

Statistical Learning for Big Dependent Data, First Edition. Daniel Peña and Ruey S. Tsay.
© 2021 John Wiley & Sons, Inc. Published 2021 by John Wiley & Sons, Inc.

and applied them to arbitrage pricing models, emphasizing the difference between bounded and unbounded eigenvalues. Connor and Korajczyk (1986) discussed these models, also for the arbitrage problem, and showed that a consistent estimate of the factors can be obtained by principal component analysis (PCA).

Peña and Box (1987) proposed a FM in which all the dynamics is driven by the factors and the noises are serially uncorrelated. This model is usually called the exact DFM (EDFM). They proposed an eigenvalue analysis of the lagged covariance matrices for estimating the factors. The authors also related, for the first time, properties of DFMs to vector autoregressive moving-average (VARMA) processes showing the lack of identification of a VARMA model when the data are generated by a DFM. Stock and Watson (1988) presented a test statistic for common trends that opens the way to the nonstationary analysis of DFM and Molenaar et al. (1992) studied the nonstationary DFM, where the factors follow a linear trend. They used Jöreskog's ML method to estimate the model. Forni et al. (2000) proposed a general approach allowing infinite dynamics in the factors and nonorthogonal idiosyncratic components and estimated the model by using the DPCs of Brillinger (1964). They called this model the generalized dynamic factor model (GDFM) and developed its properties for structural analysis and forecasting (Forni et al. 2005, 2009, 2015). For some discussions of various FMs, see, Tsay (2014, chapter 6).

In recent years, DFM has become an active area of research in statistics and econometrics with many important contributions, and the field is still under rapid development. Some recent references include Creal and Tsay (2015) that considers high-dimensional stochastic copula models with time-varying loadings, Bai and Li (2016) that studies ML inference in approximate FMs, Chan et al. (2017) that considers inference and model selection in DFM, Gao and Tsay (2019) that presents a structural-factor approach for high-dimensional time series, Zhang et al. (2018) that proposes identifying cointegration via DFM, McAlinn et al. (2018) that considers sparse DFM from a Bayesian point of view, Wang et al. (2019), Chen et al. (2020) that studies FMs for matrix-valued high-dimensional time series, and Gao and Tsay (2020) that investigates structural-FMs for high-dimensional unit-root time series.

The advantage of DFMs for modeling multivariate time series is that they can represent the dynamic evolution with a small number of parameters. This is a clear advantage over the VARMA models where the number of parameters grows quadratically with the number of series. In this chapter, we present DFMs that have been proposed for both stationary and nonstationary time series. We study first the EDFM for stationary data, where all the dynamics is driven by the factors whereas the noises are white noise. Under some general conditions this model is well identified and consistent estimates can be obtained. The dynamics of the noises can be generalized to allow for some weak autocorrelation or cross-correlations as in the ADFM. This model is not identified in finite samples, but with some strong assumptions on weak noise dependency it can be asymptotically identified. Another extension is to allow that the factors can influence the series with a more complex dynamic, even including an infinite number of lags, and this model is called the GDFM. This framework usually includes the noise assumptions of the ADFM. The model can then be estimated by generalized dynamic principal components (GDPCs).

The DFM has also been extended to nonstationary data, mainly for integrated processes. However, the merits of different procedures that have been proposed are still under research. We also discuss other FM formulations and extensions that are available in the literature. Finally, we address the case in which the sample size is smaller than the number of time series under study.

6.1 THE DFM FOR STATIONARY SERIES

Consider realizations of a k-dimensional stationary time series $\boldsymbol{z}_1, \ldots, \boldsymbol{z}_T$, where $\boldsymbol{z}_t = (z_{1t}, \ldots, z_{kt})'$. We assume that each time series has been centered by subtracting its sample mean so that $\sum_{t=1}^{T} z_{jt} = 0$ for all $j = 1, \ldots, k$, and let $\boldsymbol{Z}$ be the data matrix of dimension $T \times k$, where the tth row is $\boldsymbol{z}_t'$. We assume that the dynamic evolution of the series can be explained as the sum of two orthogonal components. The first one is common among all series and, hence, is responsible for the autocorrelations and cross-correlations among the series. The second component is the noise, also called specific or idiosyncratic component, and may take into account some smaller dynamics that are specific to each series. The DFM is defined as follows:

$$\boldsymbol{z}_t = \boldsymbol{P}\boldsymbol{f}_t + \boldsymbol{n}_t, \tag{6.1}$$

where $\boldsymbol{f}_t = (f_{1t}, \ldots, f_{rt})'$ is the r-dimensional vector of common factors, $\boldsymbol{P} = [p_{ij}]$ is a $k \times r$ factor loading matrix, and $\boldsymbol{n}_t = (n_{1t}, \ldots, n_{kt})'$ is a k-dimensional idiosyncratic series. The two components $\boldsymbol{f}_t$ and $\boldsymbol{n}_t$ are independent. Letting $\boldsymbol{p}_{i.}$ be the ith row of the $\boldsymbol{P}$ matrix, which is a $1 \times r$ row vector, we have

$$z_{it} = \boldsymbol{p}_{i.}\boldsymbol{f}_t + n_{it}. \tag{6.2}$$

Let $\boldsymbol{F}$ be the $T \times r$ matrix of factor values where the tth row is $\boldsymbol{f}_t'$, and $\boldsymbol{N}$ be the $T \times k$ matrix of noises. In matrix notation, the FM becomes

$$\boldsymbol{Z} = \boldsymbol{F}\boldsymbol{P}' + \boldsymbol{N}. \tag{6.3}$$

The $k \times 1$ vector $\boldsymbol{c}_t = \boldsymbol{P}\boldsymbol{f}_t = (c_{1t}, \ldots, c_{kt})'$ represents the common component of each series and is generated by the r unobserved common factors in $\boldsymbol{f}_t$, where, for practice purpose, $r \ll k$. In the particular case that both $\boldsymbol{f}_t$ and $\boldsymbol{n}_t$ are white noise processes with diagonal covariance matrices, the model reduces to the classic factor model in multivariate analysis.

Similarly to the classic FM, we need conditions on the loadings and the factors to identify the FM in Eq. (6.1). For any invertible $r \times r$ matrix $\boldsymbol{Q}$, we have

$$\boldsymbol{c}_t = \boldsymbol{P}\boldsymbol{f}_t = \boldsymbol{P}\boldsymbol{Q}^{-1}\boldsymbol{Q}\boldsymbol{f}_t \equiv \boldsymbol{P}^*\boldsymbol{f}_t^*,$$

where $\boldsymbol{P}^* = \boldsymbol{P}\boldsymbol{Q}^{-1}$ and $\boldsymbol{f}_t^* = \boldsymbol{Q}\boldsymbol{f}_t$. Thus, we need to fix the scales of $\boldsymbol{P}$ and $\boldsymbol{f}_t$ and we can choose between two approaches. The first one is to require that $\boldsymbol{\Gamma}_f(0) = E(\boldsymbol{f}_t\boldsymbol{f}_t') = \boldsymbol{I}$, the $r \times r$ identity matrix, as in the classic FM. This requirement implies that $\boldsymbol{\Gamma}_{f^*}(0)$ must also be the $r \times r$ identity matrix. Consequently,

$$\boldsymbol{I} = \boldsymbol{\Gamma}_{f^*}(0) = E[\boldsymbol{f}_t^*(\boldsymbol{f}_t^*)'] = \boldsymbol{Q}E[\boldsymbol{f}_t\boldsymbol{f}_t']\boldsymbol{Q}' = \boldsymbol{Q}\boldsymbol{Q}',$$

implying that $\boldsymbol{Q}$ is an orthonormal matrix. The second approach is to require that $\boldsymbol{P}'\boldsymbol{P} = \boldsymbol{I}$. In this case, we also need $(\boldsymbol{P}^*)'\boldsymbol{P}^* = \boldsymbol{I}$, and consequently, we have

$$\boldsymbol{I} = (\boldsymbol{P}^*)'\boldsymbol{P}^* = (\boldsymbol{Q}^{-1})'\boldsymbol{P}'\boldsymbol{P}\boldsymbol{Q}^{-1} = (\boldsymbol{Q}^{-1})'\boldsymbol{Q}^{-1},$$

which implies that $(\boldsymbol{QQ}')^{-1} = \boldsymbol{I}$ and, hence, $\boldsymbol{QQ}' = \boldsymbol{I}$. Again, $\boldsymbol{Q}$ is an orthonormal matrix. The two approaches lead to the same requirement on $\boldsymbol{Q}$. The prior discussion states that the FM in Eq. (6.1) is not yet uniquely identified. In fact, for any $r \times r$ orthonormal matrix $\boldsymbol{Q}$, we have

$$z_t = \boldsymbol{P}\boldsymbol{f}_t + \boldsymbol{n}_t = \boldsymbol{P}^*\boldsymbol{f}_t^* + \boldsymbol{n}_t,$$

where $\boldsymbol{P}^* = \boldsymbol{PQ}^{-1}$ and $\boldsymbol{f}_t^* = \boldsymbol{Q}\boldsymbol{f}_t$. Note that a $r \times r$ orthonormal matrix $\boldsymbol{Q}$ has $r(r+1)/2$ independent elements and it represents a rotation in the r-dimensional space.

In this chapter, we adopt the condition that $\boldsymbol{P}'\boldsymbol{P} = \boldsymbol{I}$. Since the common factor $\boldsymbol{f}_t$ is not observable (also known as latent), we further assume, without loss of generality, that $\boldsymbol{\Gamma}_f(0)$ is a diagonal matrix. In other words, we can perform orthogonalization on the latent common factor $\boldsymbol{f}_t$ if needed. With the aforementioned assumptions, the DFM is identified under rotations. Note that a rotation of the uncorrelated factors with an arbitrary orthogonal matrix $\boldsymbol{f}_t^* = \boldsymbol{A}\boldsymbol{f}_t$, where $\boldsymbol{AA}' = \boldsymbol{I}$ is orthogonal, may lead to a covariance matrix of the new factors, $E(\boldsymbol{f}_t^*\boldsymbol{f}_t^{*\prime}) = \boldsymbol{A}\boldsymbol{\Gamma}_f(0)\boldsymbol{A}'$ that need not be diagonal. However, the rotation would keep the orthonormal condition of the new loading matrix, because $(\boldsymbol{P}^*)'\boldsymbol{P}^* = (\boldsymbol{A}^{-1})'\boldsymbol{P}'\boldsymbol{P}\boldsymbol{A}^{-1} = (\boldsymbol{AA}')^{-1} = \boldsymbol{I}$. Of course, once the identified model is fitted to the data we can always explore if a rotation of the factors leads to a model that is easier to understand. Also, as mentioned before, one can always perform orthogonalization on $\boldsymbol{f}_t$ so that the resulting common factors have a diagonal covariance matrix.

In addition to the structural Eq. (6.1), we also need to specify the model for the common factors and the noises. The factors are assumed to be linear stationary series that can be well approximated by a VARMA model,

$$\boldsymbol{\Phi}(B)\boldsymbol{f}_t = \boldsymbol{\Theta}(B)\boldsymbol{a}_t, \tag{6.4}$$

where $\boldsymbol{\Phi}(B) = \boldsymbol{I} - \boldsymbol{\Phi}_1 B - \cdots - \boldsymbol{\Phi}_p B^p$ and $\boldsymbol{\Theta}(B) = \boldsymbol{I} - \boldsymbol{\Theta}_1 B - \cdots - \boldsymbol{\Theta}_q B^q$ and, for stationarity and invertibility, the roots of the determinant equations $|\boldsymbol{\Phi}(B)| = 0$ and $|\boldsymbol{\Theta}(B)| = 0$ are outside the unit circle. The factors must have a diagonal covariance matrix and, for simplicity, all the polynomial matrices of the VARMA representation (6.4) are diagonal $r \times r$ matrices and $\boldsymbol{a}_t \sim N_r(\boldsymbol{0}, \boldsymbol{\Sigma}_a)$, where $\boldsymbol{\Sigma}_a$ is the diagonal variance-covariance matrix and $E(\boldsymbol{a}_t\boldsymbol{a}'_{t-h}) = \boldsymbol{0}$, for $h \neq 0$. Also, we assume that the factors are uncorrelated with the idiosyncratic noise, that is, $E(\boldsymbol{a}_t\boldsymbol{n}'_{t-h}) = \boldsymbol{0}$, for all $h = 0, \pm 1, \pm 2, \ldots$.

There are two types of assumptions that are commonly imposed on the idiosyncratic or specific component $\boldsymbol{n}_t$. In the first case the idiosyncratic component is simply a white noise process, that is, $\boldsymbol{n}_t = \boldsymbol{e}_t$, where $\boldsymbol{e}_t$ is a sequence of independent and identically distributed random vectors with mean zero and covariance matrix $\boldsymbol{\Sigma}_e$. In addition, as stated before, $\{\boldsymbol{e}_t\}$ is uncorrelated with $\boldsymbol{f}_t$ or equivalently $\{\boldsymbol{e}_t\}$ is uncorrelated with $\{\boldsymbol{a}_t\}$, i.e. $E(\boldsymbol{a}_t\boldsymbol{e}'_{t-h}) = \boldsymbol{0}$, for all $h = 0, \pm 1, \pm 2, \ldots$. With this assumption for $\boldsymbol{n}_t$, the model is perfectly identified (except for rotations) and it is called the exact DFM or EDFM. This model was proposed by Peña and Box (1987). For large k the number of parameters in the covariance matrix is $k(k+1)/2$ that can be larger than T, and in this case one often further assumes that $\boldsymbol{\Sigma}_e$ is diagonal to improve the efficacy in estimation. In real applications, the sample covariance matrix of the estimated noise term $\hat{\boldsymbol{n}}_t$ is singular regardless of the sample size T.

The second type of assumptions is for the high-dimensional case in which k can approach infinity. Then, $\boldsymbol{n}_t$ may have some weak serial and cross-sectional correlations, so that the largest eigenvalue of the covariance matrix of $\boldsymbol{n}_t$ is relatively small compared with the smallest variance of the common factors. Under such assumptions, the DFM is identifiable when both T and k go to infinity. The resulting model is called approximate DFM or ADFM. The ADFM even allows for some weak cross-correlations between the common factors and the noises if those correlations converge to zero asymptotically. In general, to identify an ADFM, one requires that all serial or cross-dependencies disappear for large k and T, so that the model would approach an EDFM in the limit. We provide some further details of the model in the next section. The ADFM was proposed by Geweke (1977), Chamberlain (1983), and Chamberlain and Rothschild (1983), and studied by Forni et al (2000), Stock and Watson (2002a, 2002b) and Bai and Ng (2002).

6.1.1 Properties of the Covariance Matrices

For a stationary time series, the available information to estimate a DFM is contained in the covariance matrices of the series. Then, from Eq. (6.1), calling $\boldsymbol{\Gamma}_f(h)$ and $\boldsymbol{\Gamma}_n(h)$ to the lag h covariances of the factors and the noises, we have

$$\boldsymbol{\Gamma}_z(h) = \boldsymbol{P}\boldsymbol{\Gamma}_f(h)\boldsymbol{P}' + \boldsymbol{\Gamma}_n(h), \quad h \geq 0. \tag{6.5}$$

This moment equation plays an important role in studying DFM. In practice, care must be exercised because autocovariance matrices depend on the scale. If one uses standardized process of $\boldsymbol{z}_t$, then the results are likely to be different from those based on the original scale. As in classic factor analysis, if the series are not in the same measurement units standardizing them makes the result free from scaling effects. When all series are in the same scale, those with larger variances have heavier effects on the results. Thus, depending on the data and the objective of the analysis we can decide whether one should or should not standardize the observed process $\boldsymbol{z}_t$. We recommend to try both and compare the results.

6.1.1.1 *The Exact DFM*

In this case, k is finite and $\boldsymbol{n}_t = \boldsymbol{e}_t$ is white noise so that Eq. (6.5) gives

$$\boldsymbol{\Gamma}_z(0) = \boldsymbol{P}\boldsymbol{\Gamma}_f(0)\boldsymbol{P}' + \boldsymbol{\Sigma}_e, \tag{6.6}$$

$$\boldsymbol{\Gamma}_z(h) = \boldsymbol{P}\boldsymbol{\Gamma}_f(h)\boldsymbol{P}', \quad h > 0. \tag{6.7}$$

From Eq. (6.7), the rank of $\boldsymbol{\Gamma}_z(h)$ is the same as that of $\boldsymbol{\Gamma}_f(h)$ because $\boldsymbol{P}$ is full rank. Consequently, the number of common factors $r = \max_h\{\text{rank}[\boldsymbol{\Gamma}_z(h)]|h > 0\}$. Therefore, checking ranks of the sample autocovariance matrices of $\boldsymbol{z}_t$ provides information about the number of common factors. More details are given in the next section.

Next, since all dynamic dependence of the model is driven by the common factor $\boldsymbol{f}_t$, it seems natural to further assume that no linear combination of the common factors is a white noise series. Furthermore, if the latent common factors f_{it} of $\boldsymbol{f}_t$ are independent of each other, then $\boldsymbol{\Gamma}_f(h)$ is a diagonal matrix for all h. Consequently, from Eqs. (6.6) and (6.7), $\boldsymbol{\Gamma}_z(h)$ is symmetric for all $h \geq 0$. These features of the autocovariance matrices of $\boldsymbol{z}_t$ can be used to verify the existence of an EDFM with independent common factors.

With further assumptions on the covariance matrix $\boldsymbol{\Sigma}_e$ of the specific component $\boldsymbol{n}_t$ in Eq. (6.6), we have the following two interesting cases:

I. The covariance matrix $\boldsymbol{\Sigma}_e = \sigma^2\boldsymbol{I}$: Post-multiplying Eq. (6.6) by the jth column $\boldsymbol{p}_{.j}$ of $\boldsymbol{P}$, and using $\boldsymbol{P}'\boldsymbol{P} = \boldsymbol{I}$ and some algebra, we have

$$\boldsymbol{\Gamma}_z(0)\boldsymbol{p}_{.j} = (\gamma^2_{f,j} + \sigma^2)\boldsymbol{p}_{.j},$$

where $\gamma^2_{f,j}$ is the variance of the jth common factor f_{jt} and $1 \leq j \leq r$. Therefore, in this particular instance, the jth column $\boldsymbol{p}_{.j}$ is an eigenvector of $\boldsymbol{\Gamma}_z(0)$ associated with eigenvalue $\gamma^2_{f,j} + \sigma^2$. Moreover, let $\boldsymbol{P}_\perp$ be the $k \times (k-r)$ orthogonal complement of $\boldsymbol{P}$ in R^k such that $(\boldsymbol{P}_\perp)'\boldsymbol{P} = \boldsymbol{0}$, a $(k-r) \times r$ zero matrix. Then, post-multiplying Eq. (6.6) by the jth column $\boldsymbol{p}^*_{.j}$ of $\boldsymbol{P}_\perp$, we have

$$\boldsymbol{\Gamma}_z(0)\boldsymbol{p}^*_{.j} = \sigma^2\boldsymbol{p}^*_{.j},$$

where $1 \leq j \leq k-r$. Thus, $\boldsymbol{p}^*_{.j}$ is an eigenvector of $\boldsymbol{\Gamma}_z(0)$ associated with eigenvalue σ^2. Based on the prior discussion, we see that, in this particular case, the r eigenvectors of $\boldsymbol{\Gamma}_z(0)$ associated with the r largest eigenvalues form the loading matrix of the EDFM.

II. The covariance matrix $\boldsymbol{\Sigma}_e = \text{diag}\{\sigma^2_1, \ldots, \sigma^2_k\}$. Let $\sigma^2 = \max_{1\leq j\leq k}\{\sigma^2_j\}$ be the maximum variance of the noises. We have

$$\boldsymbol{\Sigma}_e = \sigma^2\boldsymbol{I} + \boldsymbol{D}_0,$$

where $\boldsymbol{D}_0 = \text{diag}\{\sigma^2_1 - \sigma^2, \ldots, \sigma^2_k - \sigma^2\}$. Without loss of generality, assume that $\gamma^2_{f,r}$ be the smallest variance of the common factors f_{jt} for $1 \leq j \leq r$. Then, using the same method as that of Case I, we obtain

$$\boldsymbol{\Gamma}_z(0)\boldsymbol{p}_{.r} = (\gamma^2_{f,r} + \sigma^2)(\boldsymbol{p}_{.r} + \boldsymbol{\delta}_r), \quad (6.8)$$

where $\boldsymbol{\delta}_r = \boldsymbol{D}_0\boldsymbol{p}_{.r}/(\gamma^2_{f,r} + \sigma^2)$. From Eq. (6.8), we see that $\boldsymbol{p}_{.r}$ is no longer an eigenvector of $\boldsymbol{\Gamma}_z(0)$ because $\boldsymbol{\delta}_r$ is not a zero vector in general. However, if $\gamma^2_{f,r}$ is much larger than σ^2, then $\boldsymbol{\delta}_r$ would be small or even ignorable, and $\boldsymbol{p}_{.r}$ would approximately be an eigenvector of $\boldsymbol{\Gamma}_z(0)$ with eigenvalue $\gamma^2_{f,r} + \sigma^2$.

To summarize, let $\sigma^2/\gamma^2_{f,r}$ be the noise-to-signal ratio of the EDFM under study, where $\gamma^2_{f,r}$ is the smallest variance of the common factor and σ^2 is the largest variance of the noises. Then, the results of Case I would continue to hold in Case II approximately provided that the noise-to-signal ratio $\sigma^2/\gamma_{f,r}$ is very small. In summary, the model would be more useful and easier to identify if the noise-to-signal ratio is small (or equivalently, the signal-to-noise ratio is large).

6.1.1.2 The Approximate DFM In this case, the model is not identifiable in finite samples, but it can be identified when both k and T go to infinity under certain conditions. When $\boldsymbol{n}_t$ has dynamic dependence, Eq. (6.7) becomes

$$\boldsymbol{\Gamma}_z(h) = \boldsymbol{P}\boldsymbol{\Gamma}_f(h)\boldsymbol{P}' + \boldsymbol{\Gamma}_n(h), \quad h > 0, \tag{6.9}$$

provided that $\boldsymbol{f}_t$ is uncorrelated with $\boldsymbol{n}_{t-h}$. Then, the rank of $\boldsymbol{\Gamma}_z(h)$ no longer provides information concerning the number of common factors r. However, under similar conditions as those of Case II for the EDFM, we still can identify the loading matrix. This can be achieved if we require that

$$\lim_{k\to\infty} \frac{1}{k^2} \sum_{i=1}^{k} \sum_{j=1}^{k} |\gamma_{n,ij}| \to 0, \tag{6.10}$$

where $\gamma_{n,ij}$ is the covariance of n_{it} and n_{jt}. This condition is rather strong. It implies that the serial and cross-dependence of the noise component $\boldsymbol{n}_t$ should be weak and most of them are zero when k increases. Interested readers are referred to Forni et al. (2000) for a careful investigation on the identification of ADFM.

Example 6.1

To demonstrate how autocovariance matrices show features of DFMs, we consider the growth rates of standardized gross domestic product of 19 Euro countries. The original data are in the file `UMEdata20002018.csv`, which has 57 variables and 76 observations. The data span is from the first quarter of 2000 to the fourth quarter of 2018. For each country, there are three economic variables, which are (a) the gross domestic product at market prices (GDP), (b) final consumption expenditure (CON), and (c) gross fixed capital formation (INV).

For stationarity, we consider the growth rates of the GDP series. Thus, in this example, the 19-dimensional GDP growth rate series is $\boldsymbol{z}_t = \nabla \log(\text{GDP}_t)$ Figure 6.1 shows the time plots of this series $\boldsymbol{z}_t$. From the plots, it is evident that, as expected, all 19 economies were affected by the 2008 financial crisis. The plot also indicates some countries experienced negative growth around 2013. We perform eigenvalue-eigenvector analysis of $\widehat{\boldsymbol{\Gamma}}_z(h)$ for $h = 0, \ldots, 4$. Except for the covariance matrix, the lagged autocovariance matrices are not symmetric so that their eigenvalues (and eigenvectors) may contain complex numbers. Our goal here is to show the consistency of the patterns exhibited by the sample autocovariance matrices and these patterns provide some justification for using DFMs for the 19 GDP growth rate series.

Table 6.1 provides the absolute values of the four largest eigenvalues of $\widehat{\boldsymbol{\Gamma}}_z(h)$ for $h = 0, \ldots, 4$. From the table, it is clear that the first eigenvalue dominates the others for all h considered and its value decreases as h increases. More interestingly, the first eigenvalue is real-valued. The dominating nature of the first eigenvalue for various lags, h, suggests that there is, at least, a single common factor in the system, i.e. $r = 1$.

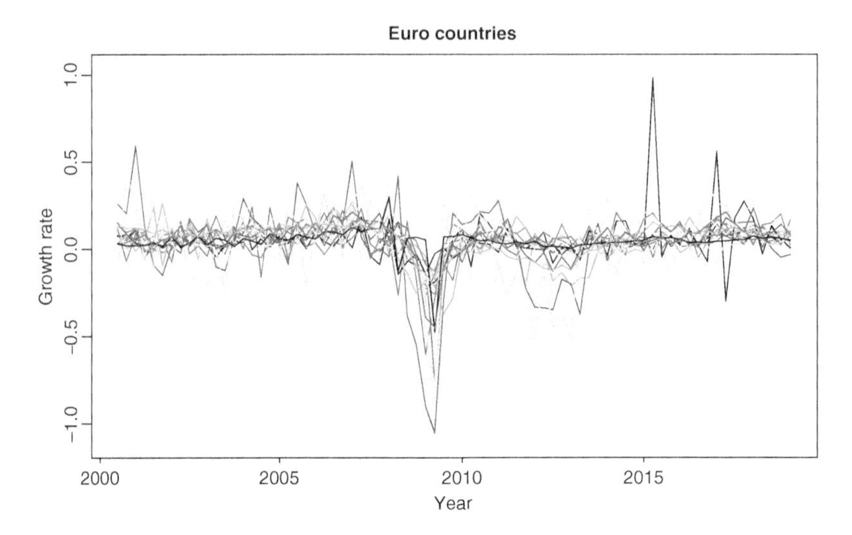

Figure 6.1 Time plots of the growth rates of the GDP of 19 Euro countries from 2000.II to 2018.IV.

TABLE 6.1 Absolute Values of the First Four Eigenvalues of Sample Autocovariance Matrices for the Growth Rates of the GDP Series of 19 Euro Countries from 2000.II to 2018.IV, Where * Indicates Complex Eigenvalues

lag(h)	0	1	2	3	4
1st	0.178	0.125	0.075	0.044	0.020
2nd	0.034	0.014	0.010*	0.009*	0.010*
3rd	0.024	0.009	0.010*	0.009*	0.010*
4th	0.018	0.007	0.004*	0.006*	0.007

Figure 6.2 plots the loadings of the first eigenvalue of $\hat{\Gamma}_z(h)$ for $h = 0, 1, 2$ and 3. Since the sign of an eigenvector cannot be determined, we adjust the sign so that the first element of each loading vector is positive. From the plot, the loadings are rather consistent. This result indicates the first column of the loading matrix $\boldsymbol{P}$ is stable for $h = 0$, 1, 2, and 3. Consequently, the eigenvalue-eigenvector analysis of the sample autocovariance matrices seems to suggest that, for this particular instance, a DFM with a single factor might be appropriate for the 19-dimensional GDP growth rates. Furthermore, the loadings indicate that the common factor is a weighted average of the GDP growth rates.

R commands used in Example 6.1: R output is edited to save space.

```
> UMEdata = read.csv("UMEdata20002018.csv", header=FALSE)
>x=as.matrix(UMEdata)
>n=nrow(x)
>G=matrix(0,n,19)
>for (i in 1:19){G[,i]=x[,1+3*(i-1)]}
```

```
>G0=log(G)
>sG0=scale(G0)
>G1=diff(sG0)
> y1 <- range(G1)
> tdx <- c(2:76)/4+2000
>plot(tdx,G1[,1],xlab="year",ylab="Growth rate", main="Euro countries",
type="l", ylim=y1)
>for (i in 2:19){lines(tdx,G1[,i],col=i)}
>out=acf(G1,lag.max=4,type="covariance",plot=FALSE)
>S0=eigen(out$acf[1,,])
>E0=S0$values[1:4]
>EVf0=S0$vectors[,1]
>EVs0=S0$vectors[,2]
>S1=eigen(out$acf[2,,])
>S1$values[1:4] # May have complex eigenvalues
>Mod(S1$values[1:4])
>E1=S1$values[1:4]
>EVf1=S1$vectors[,1]
>EVs1=S1$vectors[,2]
.....
>EVf=cbind(EVf0,EVf1,EVf2,EVf3)
>EVs=cbind(EVs0,EVs1,EVs2,EVs3) # Not shown
>ts.plot(EVf)
>EVf=cbind(-EVf0,EVf1,EVf2,-EVf3) # Change some signs
>ts.plot(EVf) # re-plot.
```

■

6.1.2 Dynamic Factor and VARMA Models

The EDFM of Eq. (6.1) with $\boldsymbol{n}_t$ being white noise implies that $k-r$ linear combinations of the observed series are white noise and r linear combinations contain all the dynamics. Let $\boldsymbol{P}_\perp$ be the $k \times (k-r)$ matrix that verifies $(\boldsymbol{P}_\perp)'\boldsymbol{P} = \boldsymbol{0}$. Then, the $k-r$ linear combinations $\boldsymbol{y}_{1t} = (\boldsymbol{P}_\perp)'\boldsymbol{z}_t$ are white noise and the r linear combinations

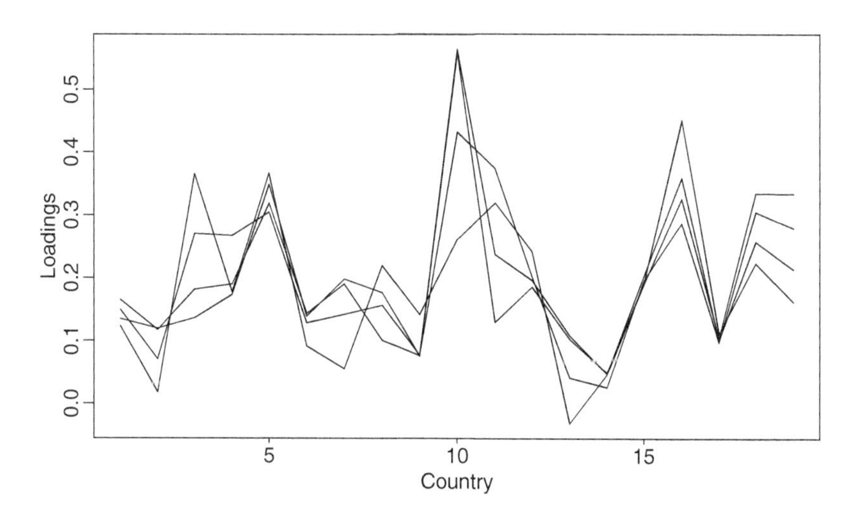

Figure 6.2 Loadings of the first eigenvalues of the autocovariance matrices, lag 0 to lag 3, of the GDP growth rates of 19 Euro countries.

$y_{2t} = P' z_t$ contain the common factors. Hu (2005) proved that the opposite is also true. That is, consider a VAR(p) model for z_t

$$z_t = \sum_{i=1}^{p} \Phi_i z_{t-i} + a_t,$$

if we are able to find a $k \times k$ matrix G such that $y_t = G z_t$ satisfying the model

$$y_t = \sum_{i=1}^{p} \Phi_i^* y_{t-i} + a_t^*,$$

where $\Phi_i^* = G \Phi_i G^{-1}$ and $\Sigma_a^* = G \Phi_i G'$, and this new series y_t can be partitioned as $y_t = (y'_{1t}, y'_{2t})'$, where y_{1t} is a $(k-r)$-dimensional white noise process and y_{2t} follows a VAR(p) model, then z_t follows the DFM in Eq. (6.1).

Next, Box and Tiao (1977) and Tiao and Tsay (1989) studied ways to simplify the dynamics of a vector of time series that follows a VARMA model. As seen in Chapter 3, Box and Tiao (1977) proposed searching for linear combinations $v_t = C' z_t$ with elements arranged according to their predictabilities. From

$$z_t = \hat{z}_{t-1}(1) + \epsilon_t, \tag{6.11}$$

and calling $\Gamma_p(1) = E[\hat{z}_{t-1}(1)\hat{z}'_{t-1}(1)]$, we have that $\Gamma_z(0) = \Gamma_p(1) + \Sigma_e$, and it was shown in Section 3.5.2 that the linear combination of maximum predictability is given by the largest eigenvalue of the matrix

$$M = \Gamma_z^{-1}(0)\Gamma_p(1).$$

Assuming an EDFM and from Eq. (6.6), we have $\Gamma_p(1) = P\Gamma_f(0)P'$. Consequently, the linear combination of maximum predictability is given by the eigenvector corresponding to the largest eigenvalue of the matrix

$$M = \Gamma_z^{-1}(0) P \Gamma_f(0) P',$$

which is of rank r. Then,

$$MP = \Gamma_z^{-1}(0) P \Gamma_f(0).$$

Assuming for simplicity that $\Sigma_\epsilon = \sigma^2 I$ and using $\Gamma_f(0) = \text{diag}\{\gamma_{f,1}^2, \ldots, \gamma_{f,r}^2\}$ and the matrix inversion formula

$$(A + BDB')^{-1} = A^{-1} - A^{-1}B(B'A^{-1}B + D^{-1})^{-1}B'A^{-1},$$

with some algebra, we can obtain

$$\Gamma_z^{-1}(0)P = [\sigma^2 I + P\Gamma_f(0)P']^{-1}P = P[\sigma^2 I + \Gamma_f(0)]^{-1} = PV,$$

where V is a diagonal matrix with jth diagonal element being $(\gamma_{f,j}^2 + \sigma^2)^{-1}$, where $\gamma_{f,j}^2$ is the variance of f_{jt}. Therefore, as $MP = PV\Gamma_f(0) = PD$, where D is a diagonal matrix with elements $\gamma_{f,j}^2/(\gamma_{f,j}^2 + \sigma^2)$, the direction of maximum predictability is given by the column of the P matrix corresponding to the factor with the largest variance.

This provides a direct link between the EDFM and the work of Box and Tiao (1977).

As explained in Chapter 3, Tiao and Tsay (1989) introduced simplification of VARMA models via the scalar component models that are linear combinations of the observed time series with specific dynamics. In particular, an SCM(0,0) is a white noise process and under the EDFM, the number of linear combinations that are white noise, or number of SCM(0,0), is $k - r$. Thus, the SCM(0,0) model is also linked to the EDFM.

Peña and Box (1987) proved that if the series z_t follows a DFM in Eq. (6.1) with factors given by a VARMA(p_1, q_1) process, then the series also follows a VARMA(p_2, q_2) process with $p_1 = p_2$ and $q_2 = \max(p_1, q_1)$. However, the parameter matrices of the VARMA(p_2, q_2) representation of the series z_t are not identified, which can make the estimation of the VARMA model difficult and unstable.

6.2 FITTING A STATIONARY DFM TO DATA

Given the data $\{z_1, \ldots, z_T\}$, where $z_t = (z_{t1}, \ldots, z_{tk})'$, we assume, for simplicity, that each series z_{it} is mean-adjusted. Let Z be the $T \times k$ data matrix, where $Z' = [z_1, \ldots, z_T]$. Then $\hat{\Gamma}_z(h) = \sum_{t=h+1}^{T} z_t z_{t-h}'/T$ and, in particular, $\hat{\Gamma}_z(0) = Z'Z/T$. In this section, we discuss different procedures for estimating the $k \times r$ loading matrix P, the $T \times r$ matrix F of common factors, and the parameters of the models for the factors and the noises. We start with the case that the number of common factors r is known, then discuss methods to select r.

In this section, we assume that $T > k$. The case of $T \leq k$ will be discussed in Section 6.10, where we adopt a different formulation and use projected PCA to improve the efficiency in estimating the common factors.

6.2.1 Principal Components (PC) Estimation

Under certain conditions, consistent estimation for both, the EDFM and ADFM, can be obtained via the eigenvalue-eigenvector analysis of the sample covariance matrix $\hat{\Gamma}_z(0)$. With r given, the loading matrix P is estimated by the normalized eigenvectors of $\hat{\Gamma}_z(0)$ corresponding to the largest r eigenvalues. Denote the estimate by $\hat{P}$ and since $\hat{\Gamma}_z(0)$ is symmetric and positive definite for $T > k$, we have $\hat{P}'\hat{P} = I$. The latent factors are then estimated by $\hat{f}_t = \hat{P}' z_t$, which are the first r PCs of the data matrix Z. Note that pre-multiplying in (6.1), by $\hat{P}'$, we have that

$$\hat{P}' z_t = \hat{P}' P f_t + \hat{P}' n_t,$$

and, for fixed k, $\hat{P}$ would converge to P as $T \to \infty$. Therefore, we have, approximately, $\hat{P}' P = I$. For the EDFM, the noises n_t are white noise and the linear combinations $\hat{P}' n_t$ are zero mean variables and their sample mean would converge to zero as T increases via the law of large numbers. For the ADFM, the proper conditions on the structure of the noises and f_t are imposed so that the eigenvectors of $\hat{\Gamma}_z(0)$ continue to provide consistent estimates of the loadings and the factors $\hat{f}_t = \hat{P}' z_t$. See, Bai and Ng (2002, 2013).

Formally, we want to find a $k \times r$ matrix $\widehat{\boldsymbol{P}}$ and a $T \times r$ matrix of factors $\widehat{\boldsymbol{F}}$, with $\widehat{\boldsymbol{F}}' = [\widehat{\boldsymbol{f}}_1, \ldots, \widehat{\boldsymbol{f}}_T]$ that minimize

$$\text{MSE}(\boldsymbol{P}, \boldsymbol{F}) = \frac{1}{kT} \sum_{t=1}^{T} (\boldsymbol{z}_t - \boldsymbol{P}\boldsymbol{f}_t)'(\boldsymbol{z}_t - \boldsymbol{P}\boldsymbol{f}_t), \tag{6.12}$$

with the restrictions $\widehat{\boldsymbol{P}}'\widehat{\boldsymbol{P}} = \boldsymbol{I}$ and $\widehat{\boldsymbol{F}}'\widehat{\boldsymbol{F}} = \boldsymbol{D}$, where $\boldsymbol{D}$ is a positive definite diagonal matrix. Then

$$\frac{\partial \text{MSE}(\boldsymbol{P}, \boldsymbol{F})}{\partial \boldsymbol{f}_t} = -2\widehat{\boldsymbol{P}}'\boldsymbol{z}_t + 2\widehat{\boldsymbol{P}}'\widehat{\boldsymbol{P}}\widehat{\boldsymbol{f}}_t = \boldsymbol{0},$$

and imposing the restriction $\widehat{\boldsymbol{P}}'\widehat{\boldsymbol{P}} = \boldsymbol{I}$, we have

$$\widehat{\boldsymbol{f}}_t = \widehat{\boldsymbol{P}}'\boldsymbol{z}_t. \tag{6.13}$$

Also

$$\frac{\partial \text{MSE}(\boldsymbol{P}, \boldsymbol{F})}{\partial \boldsymbol{P}'} = -2\sum_{t=1}^{T} \boldsymbol{z}_t\widehat{\boldsymbol{f}}_t' + 2\widehat{\boldsymbol{P}}\sum_{t=1}^{T} \widehat{\boldsymbol{f}}_t\widehat{\boldsymbol{f}}_t' = \boldsymbol{0},$$

and with the restriction $\widehat{\boldsymbol{F}}'\widehat{\boldsymbol{F}} = \sum_{t=1}^{T} \widehat{\boldsymbol{f}}_t\widehat{\boldsymbol{f}}_t' = \boldsymbol{D}$, and using Eq. (6.13), we have that $\boldsymbol{Z}'\boldsymbol{Z}\widehat{\boldsymbol{P}} = \widehat{\boldsymbol{P}}\boldsymbol{D}$, and the columns of $\widehat{\mathbf{P}}$, that we denote by $\widehat{\boldsymbol{p}}_{\cdot j}$, must verify

$$\widehat{\boldsymbol{\Gamma}}_z(0)\widehat{\boldsymbol{p}}_{\cdot j} = \lambda_j\widehat{\boldsymbol{p}}_{\cdot j}. \tag{6.14}$$

Thus, the columns of $\widehat{\mathbf{P}}$ are eigenvectors corresponding to the r largest eigenvalues of $\widehat{\boldsymbol{\Gamma}}_z(0)$. It is easy to see that

$$\text{MSE}(\widehat{\boldsymbol{P}}, \widehat{\boldsymbol{F}}) = \sum_{j=1}^{r} \lambda_j,$$

and, therefore, we choose the r largest eigenvalues of the matrix $\widehat{\boldsymbol{\Gamma}}_z(0)$. A similar analysis was presented for PCs in Chapter 1.

Stock and Watson (2002a, 2002b) and Bai and Ng (2002) proved that PC estimation is consistent when both k and T go to infinity under weak autocorrelations and cross-correlations of the noises. However, as we see later, consistency is not obtained when $c = \lim_{T,k\to\infty} k/T$ is greater than one, which means that $k > T$. The conditions about the structure of the idiosyncratic component can be made slightly more general by allowing that the autocorrelations and cross-correlations depend on t and then taking the supremum overall t. Also, the distribution of the noises must verify some broad conditions to prove consistency.

Once the factors are estimated, the idiosyncratic components can be computed as $\widehat{\boldsymbol{n}}_t = (\boldsymbol{I} - \widehat{\boldsymbol{P}}\widehat{\boldsymbol{P}}')\boldsymbol{z}_t$ and we can estimate scalar autoregressive moving-average (ARMA) models for each component n_{it}. Note that the estimated noises are $k \times 1$ vectors in a linear space of dimension $k - r$ because there are r linear combinations of them that must be equal to zero, as $\widehat{\boldsymbol{P}}'\widehat{\boldsymbol{n}}_t = (\widehat{\boldsymbol{P}}' - (\widehat{\boldsymbol{P}}'\widehat{\boldsymbol{P}})\widehat{\boldsymbol{P}}')\boldsymbol{z}_t = 0$. Calling $\widehat{\boldsymbol{N}}' = [\widehat{\boldsymbol{n}}_1, \ldots, \widehat{\boldsymbol{n}}_T]$ the matrix of estimated noises, the estimated noise covariance matrix $\widehat{\boldsymbol{\Gamma}}_n(0) = \boldsymbol{N}'\boldsymbol{N}/T$

is singular. However, its diagonal elements, the variances of the noises, can be estimated by $\sum_{t=1}^{T} \hat{n}_{it}^2/T$. Note that, from Eq. (6.6),

$$\sum_{i=1}^{k} \gamma_{z_i}(0) = \sum_{j=1}^{r} \gamma_{f_j}(0) + \sum_{i=1}^{k} \gamma_{n_i}(0),$$

the sum of variances of the noises is equal to the sum of the variances of the series minus the sum of the variances of the factors.

The models for the factors are also obtained by the univariate analysis of each factor time series $\hat{f}_{it}$.

6.2.2 Pooled PC Estimator

For the EDFM, a more efficient estimation than PCA, suggested by Peña and Box (1987), can be obtained by pooling the information from the lag autocovariance matrices. The properties of these matrices show that we can estimate $\boldsymbol{P}$ by pooling the r eigenvectors linked to the largest eigenvalues of the estimated lag autocovariance matrices of the data. This could be carried out by forming the pooled matrix

$$\boldsymbol{C} = \sum_{i=1}^{h_0} \hat{\boldsymbol{\Gamma}}_z(h),$$

where h_0 is a prespecified positive integer, and computing the eigenvectors of $\boldsymbol{C}$ corresponding to the r largest eigenvalues to form the matrix $\hat{\boldsymbol{P}}$. However, as the estimated matrices $\hat{\boldsymbol{\Gamma}}_z(h)$ need not be symmetric they may have complex eigenvalues. A better alternative, proposed by Lam et al. (2011), is to use a symmetric matrix that has the same eigenvectors as those of the $\hat{\boldsymbol{\Gamma}}_z(h)$. Define

$$\boldsymbol{L} = \sum_{i=1}^{h_0} \hat{\boldsymbol{\Gamma}}_z(h)\hat{\boldsymbol{\Gamma}}_z(h)'. \tag{6.15}$$

and use the eigenvectors of $\boldsymbol{L}$ corresponding to the largest r eigenvalues to estimate the loading and the factors. For the properties of the covariance matrices shown in Section 6.2.1 pooling information can be useful not only when the noises are white but also when they are heteroscedastic, or have similar autocorrelation structure. See, Lam et al. (2011), Chan et al. (2017), Gao and Tsay (2019), and Caro and Peña (2020) for the advantages of using the matrix (6.15) to estimate the factors.

6.2.3 Generalized PC Estimator

In the heterogeneous case, the variances of the noise components are different and one can use the generalized least squares method to improve the efficiency in estimation. The objective function then becomes

$$\text{MSE}(\boldsymbol{P}, \boldsymbol{F}) = \frac{1}{kT} \sum_{t=1}^{T} (\boldsymbol{z}_t - \boldsymbol{P}\boldsymbol{f}_t)' \boldsymbol{\Sigma}_n^{-1} (\boldsymbol{z}_t - \boldsymbol{P}\boldsymbol{f}_t). \tag{6.16}$$

As $\boldsymbol{\Sigma}_n^{-1}$ is unknown, the minimization of Eq. (6.16) is usually carried out in two steps. First, the estimators in Eqs. (6.13) and (6.14) are used to compute $\hat{\boldsymbol{n}}_t = (\boldsymbol{I} - \hat{\boldsymbol{P}}\hat{\boldsymbol{P}}')\boldsymbol{z}_t$. As mentioned before, estimating the noise covariance matrix with $\hat{\boldsymbol{n}}_t$ results in a singular matrix, because the idiosyncratic space is of dimension $k - r$. However, assuming that $\boldsymbol{\Sigma}_n$ is diagonal, the variances, $\{\sigma_i^2\}$, can be estimated using each component n_{it}. By transforming the variables in Eq. (6.16) by $\boldsymbol{z}_t^* = \hat{\boldsymbol{\Sigma}}_n^{-1/2}\boldsymbol{z}_t = \{z_{it}/\hat{\sigma}_i, i = 1, \ldots, k\}$ and letting $\boldsymbol{P}^* = \hat{\boldsymbol{\Sigma}}_n^{-1/2}\boldsymbol{P}$, we have the standard least squares problem with homoscedastic variables. We can then apply PCA to the new variables $\boldsymbol{z}_t^*$ to obtain a new loading matrix $\hat{\boldsymbol{P}}^*$ that verifies $\boldsymbol{P}^* = \hat{\boldsymbol{\Sigma}}_n^{-1/2}\boldsymbol{P}$. To obtain the original loading matrix for the observed variables, we can transform with $\hat{\boldsymbol{P}} = \hat{\boldsymbol{\Sigma}}_n^{1/2}\hat{\boldsymbol{P}}^*$. Unless there exist large differences among the variances of the noises, the advantage of using generalized least squares is usually small. A second simple alternative is to estimate the covariance matrix associated with common factors by $\hat{\boldsymbol{\Gamma}}_c(0) = \hat{\boldsymbol{\Gamma}}_z(0) - \hat{\boldsymbol{\Sigma}}_n$, and then compute the PCs of $\hat{\boldsymbol{\Gamma}}_c(0)$. A more efficient and complex two step procedure has been proposed by Breitung and Tenhofen (2011).

6.2.4 ML Estimation

The ML estimation requires some assumptions on the distribution of the observed time series. Often, $\boldsymbol{z}_t$ is supposed to be Gaussian and two ML estimation approaches have been proposed. The first one is an extension of the classical FM and assumes that $\boldsymbol{f}_t$ is a random variable with zero mean and covariance matrix $\boldsymbol{\Gamma}_f(0)$, but has no dynamics. That is, $\boldsymbol{f}_t$ is uncorrelated with its past values. Then, $\boldsymbol{z}_t$ is $N_k(\boldsymbol{0}, \boldsymbol{\Gamma}_z(\boldsymbol{0}))$ with $\boldsymbol{\Gamma}_z(0) = \boldsymbol{P}\boldsymbol{\Gamma}_f(\boldsymbol{0})\boldsymbol{P}' + \boldsymbol{\Sigma}_e$. In this particular instance, the parameters include (a) the $k \times r$ elements of the loading matrix $\boldsymbol{P}$, (b) the r diagonal elements of $\boldsymbol{\Gamma}_f(\boldsymbol{0})$, and (c) the k diagonal elements of $\boldsymbol{\Sigma}_e$. The series $\boldsymbol{z}_t$ is independent of its lagged values, and the log-likelihood function is

$$L(\boldsymbol{P}, \boldsymbol{\Gamma}_f(\boldsymbol{0}), \boldsymbol{\Sigma}_n) = -\frac{T}{2}\log|\boldsymbol{\Gamma}_z(0)| - \frac{1}{2}\sum_{t=1}^{T}\boldsymbol{z}_t'[\boldsymbol{\Gamma}_z(0)]^{-1}\boldsymbol{z}_t,$$

that can be written as

$$L(\boldsymbol{P}, \boldsymbol{\Gamma}_f(\boldsymbol{0}), \boldsymbol{\Sigma}_n) = -\frac{T}{2}\log|\boldsymbol{\Gamma}_z(0)| - \frac{T}{2}\,\mathrm{tr}[\hat{\boldsymbol{\Gamma}}_z(0)\boldsymbol{\Gamma}_z^{-1}(0)], \tag{6.17}$$

where $\mathrm{tr}(\boldsymbol{A})$ denotes the trace of the matrix $\boldsymbol{A}$. The maximization of Eq. (6.17) can be carried out by taking the partial derivatives with respect to the parameters. The situation is similar to that of the standard factor analysis when $T > k$, see Seber (1984), and has been studied in the time series literature by Bai and Li (2012). For $k \geq T$, these authors proposed a quasi-likelihood function in which instead of the covariance matrix of the observations, which is singular, an approximate full rank matrix is used.

The second approach, see, Choi (2012), Doz et al. (2012), and Bai and Li (2016), is to take $\boldsymbol{f}_t$ as parameters, or random variables, with dynamical dependence, and to use the conditional distribution of $\boldsymbol{z}_t$ as $N_k(\boldsymbol{P}\boldsymbol{f}_t, \boldsymbol{\Sigma}_e)$. The log-likelihood then becomes

$$L(\boldsymbol{P}, \boldsymbol{F}, \boldsymbol{\Sigma}_e) = -\frac{T}{2}\log|\boldsymbol{\Sigma}_e| - \frac{1}{2}\sum_{t=1}^{T}(\boldsymbol{z}_t - \boldsymbol{P}\boldsymbol{f}_t)'\boldsymbol{\Sigma}_e^{-1}(\boldsymbol{z}_t - \boldsymbol{P}\boldsymbol{f}_t), \tag{6.18}$$

where the parameters include (a) the $k \times r$ elements of $\boldsymbol{P}$, (b) the $T \times r$ elements of $\boldsymbol{F}$, and (c) the k diagonal elements of $\boldsymbol{\Sigma}_e$. Assuming $\boldsymbol{\Sigma}_e = \sigma^2 \boldsymbol{I}$, maximizing Eq. (6.18) is equivalent to PCA, and assuming $\boldsymbol{\Sigma}_e$ a positive definite matrix to generalized PCs. The minimization of Eq. (6.18) can be carried out by putting the model in a state space form and computing the likelihood function using the Kalman filter. The observation equation for the state space representation is simply Eq. (6.1). The transition equation is obtained by assuming a VAR(p_1) and a VAR(p_2) model for the factors and noises, respectively. See, Chapter 2 for state-space model and Kalman filter. This approach is useful for a small k and has the advantage of allowing the series to contain different periodicities, but it is less useful when k is large. The EM algorithm has also been used for estimating FMs but, again, is not practical for large k.

The following iterative procedure, proposed by Bai and Li (2016), can be carried out easily with some initial estimates of $\boldsymbol{P}$ and $\boldsymbol{F}$ to perform the approximate ML estimation:

1. Given the present estimates $\widehat{\boldsymbol{P}}, \widehat{\boldsymbol{F}}$, the residuals are computed by $\widehat{\boldsymbol{n}}_t = (\boldsymbol{I} - \widehat{\boldsymbol{P}}\widehat{\boldsymbol{P}}')\boldsymbol{z}_t$, and the noise variances, $\text{var}(n_{it}) = \widehat{\sigma}_i^2$, which are used to form a diagonal matrix $\widehat{\boldsymbol{\Sigma}}_e$.
2. Autoregressive (AR) models are fitted to each residual series, $\widehat{\phi}_i(B)\widehat{n}_{it} = a_{it}$, $i = 1, \ldots, k$.
3. The vector of loadings for each series, $(\boldsymbol{p}_{i.})'$, that is, the ith row of the matrix $\widehat{\boldsymbol{P}}$, is updated with the following regression

$$\widehat{\phi}_i(B)z_{it} = (\boldsymbol{p}_{i.})'\widehat{\phi}_i(B)\widehat{\boldsymbol{f}}_t + \epsilon_{it}.$$

4. The factors are updated by using the current vector of loadings and running the regression

$$\widehat{z}_{it}/\widehat{\sigma}_i = (\boldsymbol{p}_{i.})'\widehat{\boldsymbol{f}}_t/\widehat{\sigma}_i + \epsilon_{it}^*.$$

5. With the new factors and loadings, go to step 1 and iterate until convergence.

6.2.5 Selecting the Number of Factors

Several tests have been proposed in the literature to select the number of factors and we will present here the most often used ones. The first class of tests is based on checking the rank of the autocovariance matrices. Note that for the EDFM, the maximum rank must be r and this maximum is achieved also by the sum of some autocovariance matrices with $h > 0$. These tests are less useful if there is strong autocorrelation in the noises (e.g. the ADFM), because then the lag covariance matrices can be full rank. Also, these types of tests work well for $T >> k$, that is, with a small number of long time series, but are less powerful when k is relatively large compared with T. A second class of tests make use of the idea of separating the large eigenvalues of the covariance matrices from the small ones and selecting r as the number of large eigenvalues. A third way to select the number of factors is to use a model selection criterion. We describe these three methods next. The selection of r for the case of $k \geq T$ can be carried out by testing for high-dimensional white noises discussed in Section 6.10.

6.2.5.1 Rank Testing via Canonical Correlation Tiao and Tsay (1989) proposed to check the rank of some moment matrices for both the stationary and nonstationary vector time series using canonical correlation analysis and a chi-square statistic. A similar test for EDFM was proposed by Peña and Poncela (2006) as follows. For all $h > 0$, there exists a $k \times (k-r)$ matrix $\boldsymbol{P}_\perp$ such that $\boldsymbol{\Gamma}_z(h)\boldsymbol{P}_\perp = \boldsymbol{0}$. Thus, the $k-r$ independent linear combinations of the observed series given by $\boldsymbol{P}'_\perp \boldsymbol{z}_t$ are cross-sectionally and serially uncorrelated for all lags, and also uncorrelated with $\boldsymbol{P}'_\perp \boldsymbol{z}_{t-h}$. Consider now the $k \times k$ canonical correlation matrix

$$\boldsymbol{M}(h) = [E(\boldsymbol{z}_t\boldsymbol{z}_t')]^{-1}E(\boldsymbol{z}_t\boldsymbol{z}_{t-h}')[E(\boldsymbol{z}_{t-h}\boldsymbol{z}_{t-h}')]^{-1}E(\boldsymbol{z}_{t-h}\boldsymbol{z}_t), \tag{6.19}$$

where, for simplicity, we assume $E(\boldsymbol{z}_t) = \boldsymbol{0}$. Assuming an EDFM satisfying $\boldsymbol{\Gamma}_z(h) = \boldsymbol{\Gamma}_z(-h)$, we have that

$$\boldsymbol{M}(h) = [\boldsymbol{\Gamma}_z^{-1}(0)\boldsymbol{\Gamma}_z(h)]^2.$$

Suppose that $\text{rank}[\boldsymbol{\Gamma}_f(h)] = r$, which is also the rank of $\boldsymbol{\Gamma}_z(h)$, then, we have $\text{rank}[\boldsymbol{M}(h)] = r$. The number of zero canonical correlations between $\boldsymbol{z}_{t-h}$ and $\boldsymbol{z}_t$ is given by the number of zero eigenvalues of the $\boldsymbol{M}(h)$ matrix, that is $k-r$. Thus, the number of common factors r is equivalent to the number of non-zero canonical correlations between $\boldsymbol{z}_{t-h}$ and $\boldsymbol{z}_t$.

The test for the number of factors is based on the aforementioned result. Let $\hat{\lambda}_1 \geq \hat{\lambda}_2 \geq \cdots \geq \hat{\lambda}_k$ be the ordered eigenvalues of the matrix $\widehat{\boldsymbol{M}}(h)$. If we have r factors, the eigenvalues $\hat{\lambda}_{r+1}, \ldots, \hat{\lambda}_k$ are estimates of squared correlations equal to zero and have asymptotic variance $1/(T-h)$. Therefore, the statistics $-(T-h)\log(1-\hat{\lambda}_j) \simeq (T-h)\hat{\lambda}_j$ follow a chi-square distribution. It can be shown (see Peña and Poncela, 2006) that the statistic

$$S_{h-r} = -(T-h)\sum_{j=r+1}^{k}\log(1-\hat{\lambda}_j), \tag{6.20}$$

is asymptotically a $\chi^2_{(k-r)^2}$. The test is applied sequentially. It starts with $r = 0$ and if the hypothesis is rejected, the value $r = 1$ is tried and so on. The testing procedure stops when the hypothesis of r factor, or r non-zero canonical correlation cannot be rejected.

This test works well when T/k is large, but deteriorates when this ratio is small. Also, when the lag covariance matrices of the factors are not full rank, the test may fail. For instance, suppose two factors that follow moving-average (MA) processes. The first is MA(1) and the second is MA(3), but with coefficients for lags one and two equal to zero. Then, the lag-1 covariance matrix of the two factors is of rank one, the lag-2 matrix is rank zero, and the lag-3 has rank one. Thus, the test would indicate a single factor, instead of two. However, the test can be modified to overcome such difficulties by either checking the equivalence of identified common factors for different lags or applying canonical correlation analysis simultaneously to multiple lag autocovariance matrices. See, Bolivar et al. (2020).

6.2.5.2 Testing a Jump in Eigenvalues The tests here are based on the ratios of consecutive eigenvalues (in a proper ordering) of the covariance matrix or the matrix sum of lag covariance matrices. The idea originates from the elbow plot for selecting the number of PC and the random matrix theory and the tests try to find a jump in the ratio of eigenvalues which would occur when the true number of factors is reached. See, Onatski (2009, 2010), Lam and Yao (2012), and Ahn and Horenstein (2013).

Lam and Yao (2012) proposed to compute the ordered eigenvalues $\lambda_1 \geq \lambda_2 \geq \cdots \geq \lambda_k$ of the pooled covariance matrices

$$\mathbf{M} = \sum_{h=1}^{h_0} \widehat{\boldsymbol{\Gamma}}_z(h)\widehat{\boldsymbol{\Gamma}}_z(h)', \tag{6.21}$$

where h_0 is a prespecified positive integer, and select r as

$$\widehat{r} = \arg \min_{1\leq i\leq r^*} \frac{\lambda_{i+1}}{\lambda_i},$$

for some $r^* = \alpha k$, where k is the number of series and $0 < \alpha < 1$ such as $\alpha = 0.2$. Suppose that the first r eigenvalues are large and the remaining eigenvalues are small. Then, the ratios $\lambda_{i+1}/\lambda_i \leq 1$ would have a big decrease for $i = r$.

A similar test has been proposed by Ahn and Horenstein (2013) by using the ordered eigenvalues $\nu_1 \geq \nu_2 \geq \cdots \geq \nu_k$ of the covariance matrix $\widehat{\boldsymbol{\Gamma}}_z(0)$. The criterion is

$$\widehat{r} = \arg \max_{1\leq i\leq r^*} \frac{\nu_i}{\nu_{i+1}}.$$

The Lam and Yao criterion cannot be applied if some factors are white noise because then there is no information about those factors in the lag covariance matrices. However, the Ahn and Horenstein test continues to work. Also, none of the two tests can be used if there exist no common factors. Limited experience indicates that the ratios tend to be dominated by a dominating factor, which often occurs when there exist strong serial correlations in z_t. As a matter of fact, the test tends to find $r = 1$ in application. The test can be iterated and applied to the residuals once the first dominant factor is eliminated.

6.2.5.3 Using Information Criteria Like many other statistical problems, information criteria can be used to select the number of common factors. The Bayesian information criterion (BIC) criterion explained in Chapter 3 can be used when $T > k$, as

$$\text{BIC}(r) = \log \widehat{\sigma}_r^2 + r\frac{\log T}{T}$$

but this criterion does not always provide a consistent estimation of the number of factors when k is large or close to T. A consistent model selection criterion that seems to work well in practice was proposed by Bai and Ng (2002), and it can be written as

$$\text{BNG}(r) = \log \widehat{\sigma}_r^2 + r\frac{\log(\min(T, k))}{\min(T, k)}.$$

where

$$\widehat{\sigma}_r^2 = \text{MSE}(\widehat{\boldsymbol{P}}_r, \widehat{\boldsymbol{F}}_r) = \frac{1}{kT}\sum_{t=1}^{T}(\boldsymbol{z}_t - \widehat{\boldsymbol{P}}_r \boldsymbol{f}_t^r)'(\boldsymbol{z}_t - \widehat{\boldsymbol{P}}_r \boldsymbol{f}_t^r)$$

is the mean squared error (MSE) when r factors are fitted, where we use $\boldsymbol{f}_t^r$ to signify that the dimension of $\boldsymbol{f}_t$ is r. The number of factors is chosen so as to minimize BNG(r).

6.2.6 Forecasting with DFM

Forecasting with DFM models is carried out by fitting proper scalar ARMA models to each factor and to each noise series, and generating forecasts from the fitted models, say $\widehat{\boldsymbol{f}}_t(h)$ and $\widehat{\boldsymbol{n}}_t(h)$, where h is the forecast horizon and t is the forecast origin. Then

$$\widehat{\boldsymbol{z}}_t(h) = \widehat{\boldsymbol{P}}\widehat{\boldsymbol{f}}_t(h) + \widehat{\boldsymbol{n}}_t(h). \tag{6.22}$$

When forecasting an individual time series with an EDFM the factors summarize all information about the dynamics and

$$E(z_{i,t+h}|\boldsymbol{Z},\boldsymbol{F}) = E[(\boldsymbol{p}_{i.})'\boldsymbol{f}_{t+h} + n_{i,t+h}|\boldsymbol{Z},\boldsymbol{F}] = (\boldsymbol{p}_{i.})'E(\boldsymbol{f}_{t+h}|\boldsymbol{F}) + E(n_{i,t+h}|\boldsymbol{Z},\boldsymbol{F}).$$

Therefore, to forecast each individual time series we only need the forecasts of all factors plus the forecast of the specific noise.

The forecasts generated by DFM can be written (see Peña and Poncela, 2004) as incorporating a pooling term similar to the one derived from hierarchical Bayesian models. For instance, suppose a one factor DFM where the factor follows an AR(1) process with parameter ϕ. According to (6.22) the forecast of each series is $\widehat{z}_{i,t+h|t} = \phi^h p_i f_t$ and this equation can be also written as

$$\widehat{z}_{i,t+h|t} = \phi^h \left[\alpha_t p_i f_{t|t-1} + (1-\alpha_t)\sum_{i=1}^{k}\delta_i z_{it}\right],$$

where $\delta_i = (p_i^2/\sigma^2)/\sum_{i=1}^{k}(p_i^2/\sigma^2)$ is the relative signal-to-noise ratio in the ith series. Thus, the forecast can be seen as a weighted combination of two terms: the first is the forecast of the factor at time t with information until $t-1$, the second is a weighted average of the values of the series at time t, where the weights of each series depend on its relative signal-to-noise ratio. The weight of each of the two terms in the forecast depend on its relative precision, that is $\alpha_t = \text{var}^{-1}(f_{t|t-1})/[\text{var}^{-1}(f_{t|t-1}) + \sum_{i=1}^{k}(p_i^2/\sigma^2)]$; see, Peña and Poncela (2004) for a deeper analysis. These authors derive the expected gains, in the one-factor case, of the FM forecasts with respect to univariate and shrinkage models and show that the advantage of the FM increases with the dimension of the time series vector and with the strength of the dynamic relationship among the components.

Boivin and Ng (2006) have studied the advantages for forecasting of using a larger number of series to extract the factors. This advantage could be very small and even a shortcoming when the idiosyncratic errors are cross-correlated. In Chapter 7, we will see how DFM can be combined with other approaches to forecast large sets of time series.

DFMs can also be combined with other approaches to produce non-linear forecast. Fan et al. (2017) have proposed sufficient forecasting, an interesting forecasting method, which in addition to the linear DFM prediction incorporates a set of sufficient predictive indices, inferred from high-dimensional predictors. The method extends the sufficient dimension reduction based on sliced inverse regression (Li, 1991) and the central subspace analysis by Cook (2018) to high-dimensional regimes by summarizing the cross-correlation through FMs.

Example 6.2

To illustrate, we consider, again, the growth rates of the quarterly GDP of 19 Euro countries used in Example 6.1. Applying the Lam and Yao's test to a pooled covariance matrix in Eq. (6.21), we found $\hat{r} = 1$. The estimated common factor is shown in Figure 6.3. Broadly speaking, the estimated common factor is similar to the one obtained from the sample covariance matrix. The factor explains about 51% of the variability of the series. The analysis is carried out by the command `dfmpc` in the SLBDD R package associated with the book. The program also gives the loading vector, see Figure 6.4, and the models for the factor and the idiosyncratic noises.

R commands for DFM analysis: Continue from Example 6.1.

```
>out=dfmpc(G1)
>out$r # output is not shown.
>out$F
>out$L
>plot(tdx,-out$F[,1],xlab='year',ylab='1st factor',type='l')
>plot(1:19,-out$L,xlab='country',ylab='weight',type='l')
>out$MarmaF
>out$MarmaE
```

■

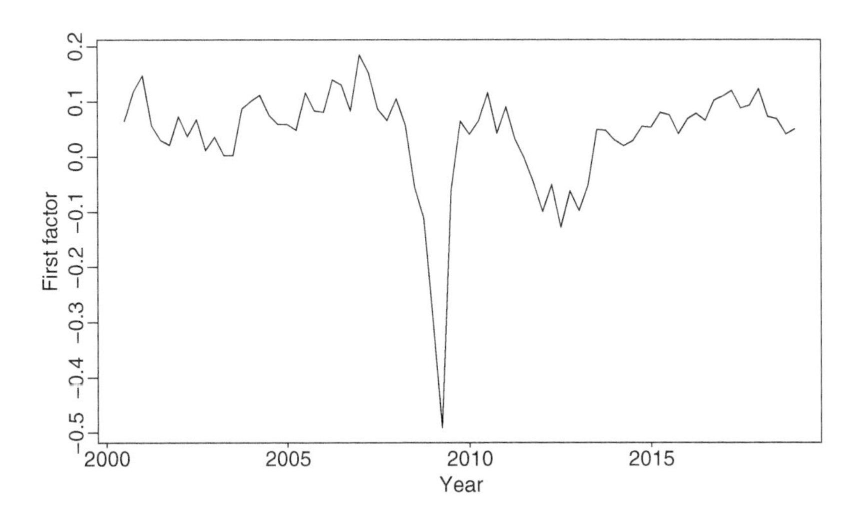

Figure 6.3 Time plot of the first common factor for the growth rates of the GDP series of 19 Euro countries.

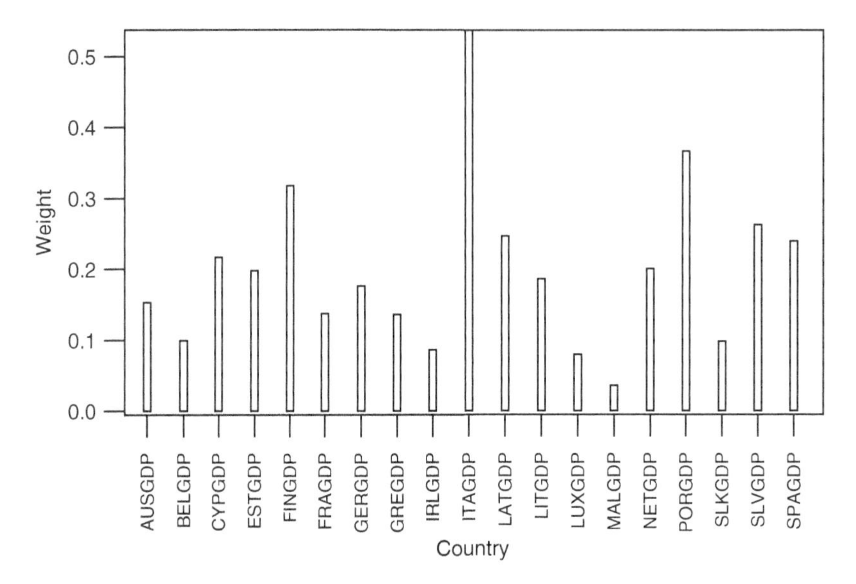

Figure 6.4 Loadings for the first common factor for the growth rates of the GDP series of 19 Euro countries.

Example 6.3

We continue the study of the quarterly series of GDP at market prices, final consumption expenditure (CON), and gross fixed capital formation (INV) for the 19 countries that formed the European Monetary Union (EMU) with a total of 57 time series. The data span is from Q1 2000 to Q4 2018 for a total of 76 quarterly observations. See, Examples 6.1. and 6.2. Figure 6.5 shows the time plots of the growth rates $\nabla \log \boldsymbol{x}_t$. Applying the DFM analysis, we found a single common factor is selected, which is shown in Figure 6.6. The corresponding loadings are shown in Figure 6.7, from

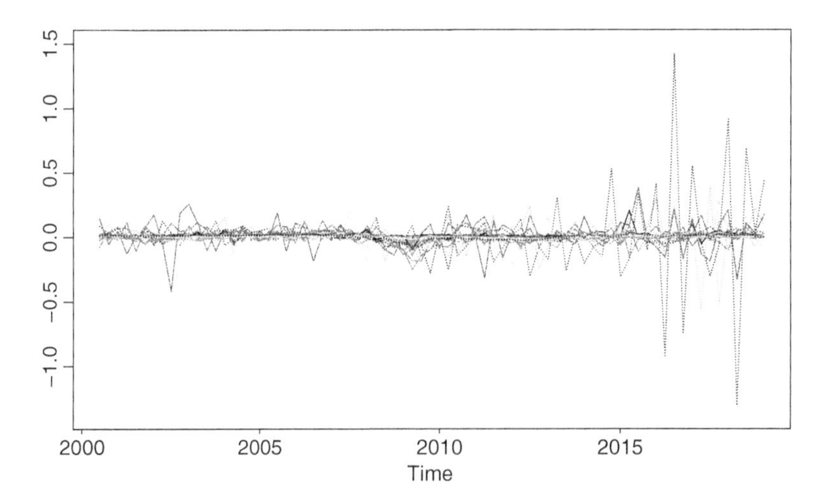

Figure 6.5 Time plots of the growth rates of 57 economic variables of the 19 countries in European Monetary Union. Quarterly data from 2000 to 2018.

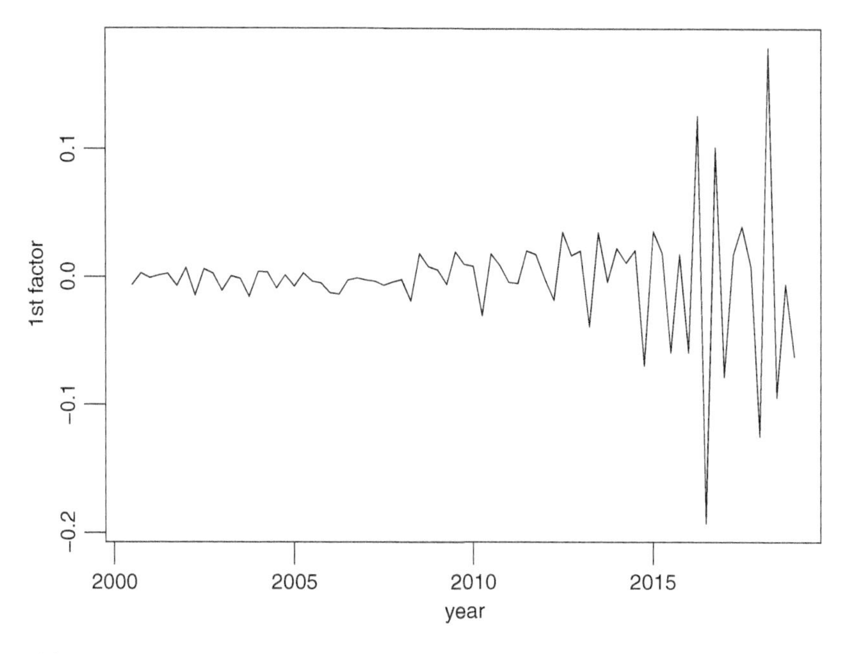

Figure 6.6 Time plot of the first common factor of the 57 economic growth rates of the EMU countries.

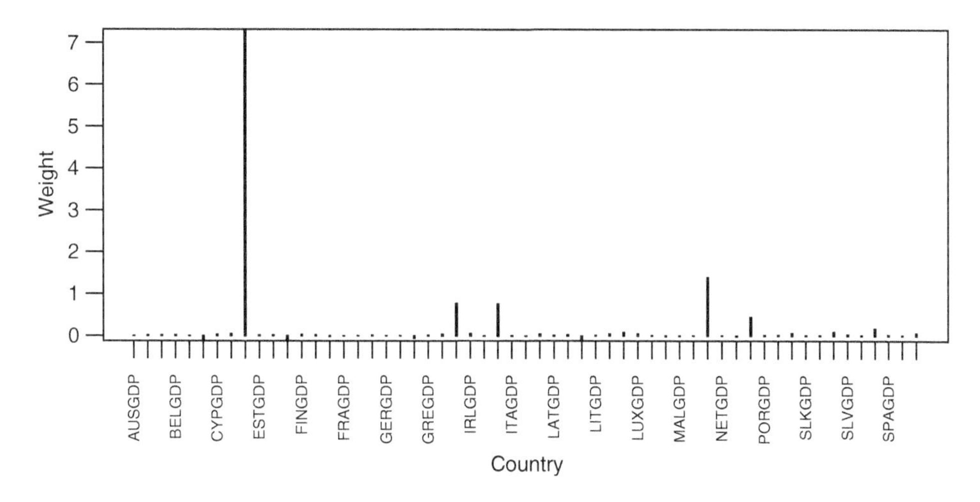

Figure 6.7 Loadings of the first common factor of the 57 economic growth rates of the EMU countries.

which we see that the loadings place a heavy weight on the ninth series, which is the investment series of Cyprus. This is unusual and we found that the sample correlation between the ninth series and the common factor is 0.996. Figure 6.8 shows the variances of the 57 series from which we see that the variance of the ninth series is more than 10 times larger than those of the other series. This phenomenon confirms our concern that PCA depends heavily on the scales of the series involved. In some sense, the ninth series can be regarded as an outlying series. To mitigate the scaling effect, we standardized the growth rate series and repeat the analysis. The standardized

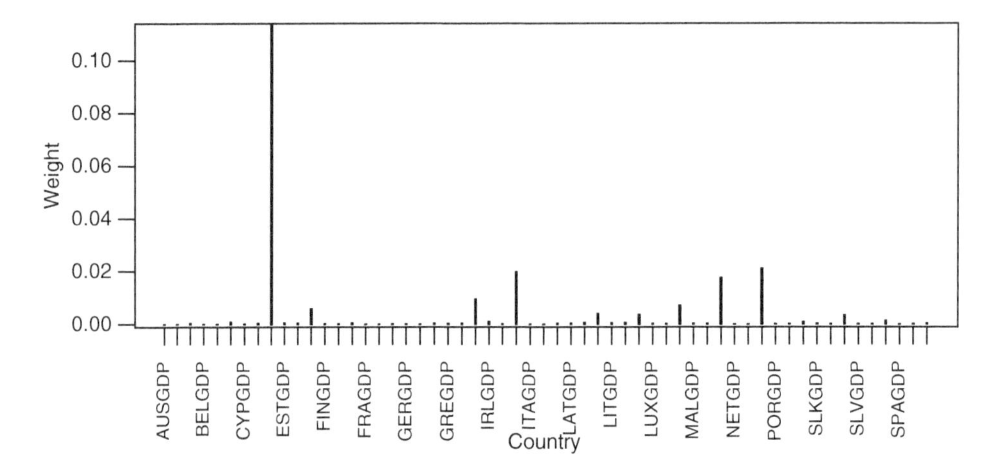

Figure 6.8 Variances of the 57 economic growth rates of the EMU countries.

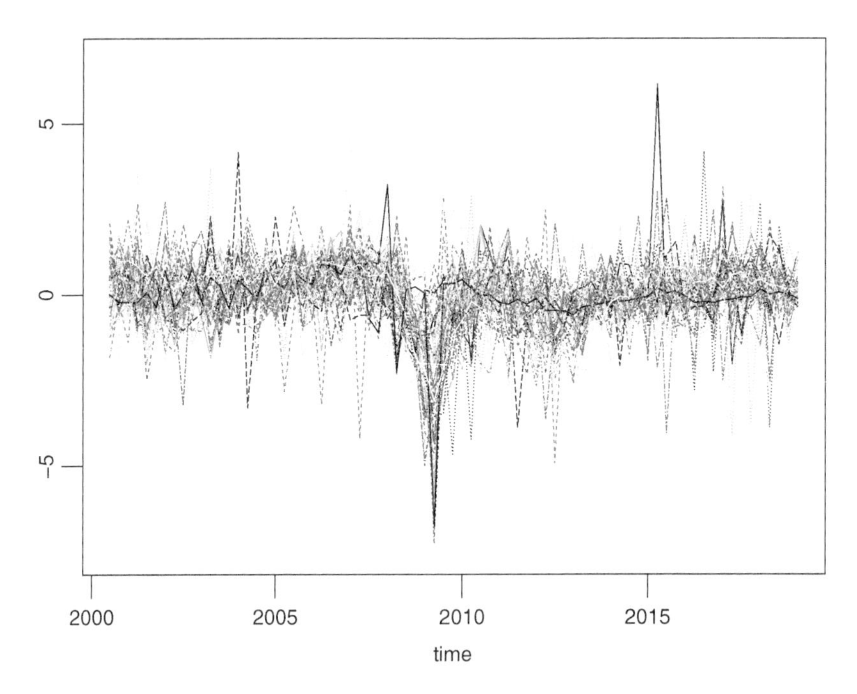

Figure 6.9 Time plots of standardized growth rates of 57 economic series of the 19 countries in European Monetary Union. Quarterly data from 2000 to 2018.

growth rates are shown in Figure 6.9. It is clear now that all growth rates are affected by the 2008 financial crisis. Analysis of the standardized growth rates also selects a single common factor, which is shown in Figure 6.10. The corresponding loadings are in Figure 6.11. The loadings indicate that the common factor is a weighted average of the standardized growth rates.

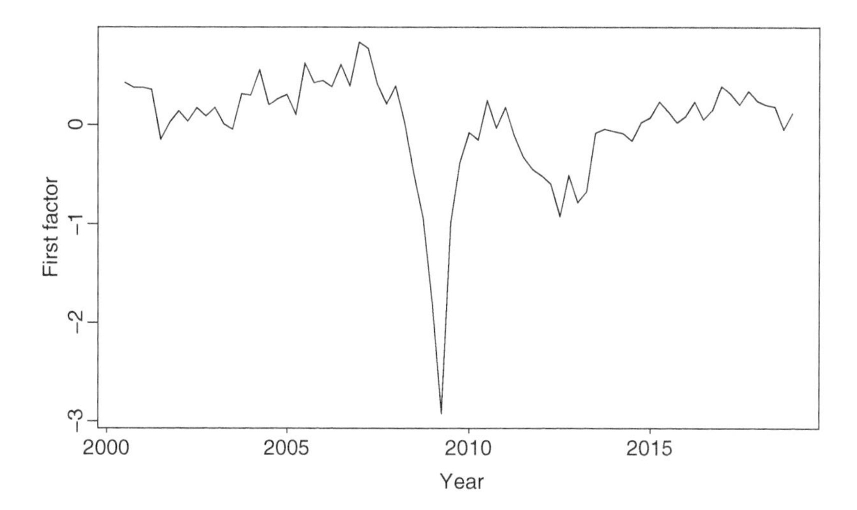

Figure 6.10 Time plot of the first common factor for the standardized growth rates of the EMU countries.

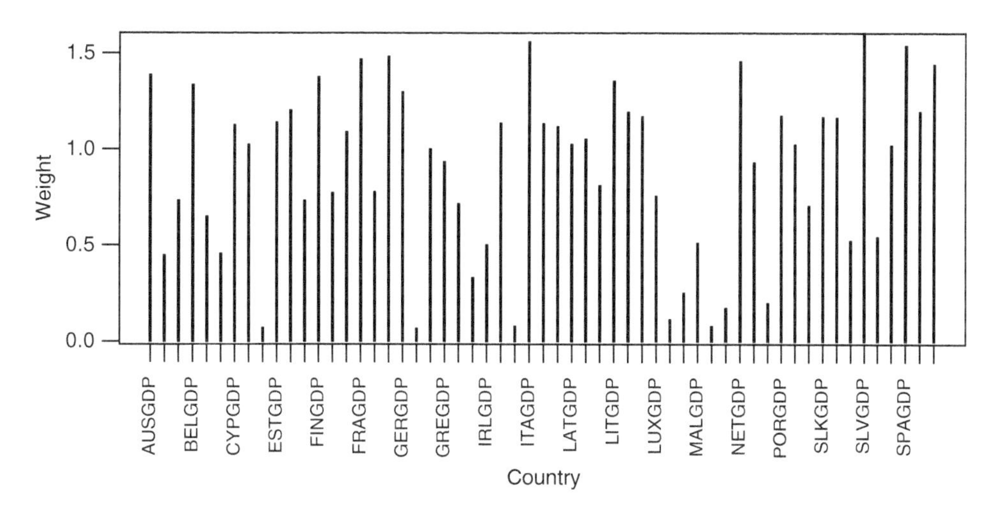

Figure 6.11 Loadings of the first common factor on the standardized growth rates of the EMU countries.

R commands for Example 6.3.

```
>library(MTS)
>x=as.matrix(UMEdata)
>y=log(x)
>y1=diffM(y)
>tdx<- c(2:76)/4+2000 # calendar time
>MTSplot(y1,tdx)
>sal1=dfmpc(y1, stand = 0, mth = 0, r = 1, lagk = 0)
>plot(tdx,sal1$F[,1],xlab='year',ylab='1st factor',type='l')
>plot(1:57,-sal1$L,xlab='country',ylab='1st factor',type='l')
```

```
>sy1=scale(y1)
>sal2=dfmpc(sy1, stand = 0, mth = 0, r = 1, lagk = 0)
>MTSplot(sy1,tdx)
>plot(tdx,-sal2$F[,1],xlab='year',ylab='1st factor',type='l')
>plot(1:57,-sal2$L,xlab='country',ylab='1st factor',type='l')
```
■

6.2.7 Alternative Formulations of the DFM

A different FM was proposed by Li and Shedden (2002). They assume that each component y_{it} of the vector time series $\boldsymbol{y}_t$ is generated by

$$y_{it} = \sum_{j=1}^{r} \lambda_{ij} Z_{ji}(t), \quad i = 1, \ldots, k, \tag{6.23}$$

where the λ_{ij} are unknown real numbers and the components, $Z_{ji}(t)$, $i = 1, \ldots, k$ are independent processes drawn from a common probability distribution. Assuming that the series are standardized, writing Eq. (6.23) as $y_{it} = \boldsymbol{\lambda}_i' \boldsymbol{Z}_i(t)$, where $\boldsymbol{\lambda}_i' = (\lambda_{i1}, \ldots, \lambda_{ir})$ and $\boldsymbol{Z}_i(t) = (Z_{1i}(t), \ldots, Z_{ri}(t))'$, and calling $\rho_i(h) = E(y_{it} y_{it+h})/E(y_{it}^2)$, we have $\rho_i(h) = \boldsymbol{\lambda}_i' \boldsymbol{\Gamma}_{Z_i}(h) \boldsymbol{\lambda}_i / \boldsymbol{\lambda}_i' \boldsymbol{\Gamma}_{Z_i}(0) \boldsymbol{\lambda}_i$, where $\boldsymbol{\Gamma}_{Z_i}(h)$ is the lag-h autocovariance matrix of $\boldsymbol{Z}_i(t)$. As these components are standardized and are uncorrelated, the matrix $\boldsymbol{\Gamma}_{Z_i}(0)$ is the identity matrix and $\boldsymbol{\Gamma}_{Z_i}(h)$ is diagonal. Calling $\phi_{ij}(h) = E(Z_{ij}(t) Z_{ij}(t+h))$ be the jth diagonal element of $\boldsymbol{\Gamma}_{Z_i}(h)$, which represents the lag-h autocorrelation of the process $Z_{ij}(t)$, we have

$$\rho_i(h) = \sum_{j=1}^{r} \lambda_{ij}^2 \phi_{ij}(h) / \sum_{j=1}^{r} \lambda_{ij}^2 = \sum_{j=1}^{r} p_{ij}^2 \phi_{ij}(h),$$

where $p_{ij}^2 = \lambda_{ij}^2 / \sum_{j=1}^{r} \lambda_{ij}^2$ are weights satisfying $\sum_{i=1}^{r} p_{ij}^2 = 1$. This is the same expression as that of the EDFM in Eq. (6.1) with independent factors that follow the same distribution. In this sense, as shown by Peña (2009), the two formulations are equivalent.

Another DFM model has been proposed by Gao and Tsay (2019, 2020) that avoids the singularity of the covariance matrix of the noises. They assume that an observed k-dimensional time series is a nonsingular linear transformation of r common factors and $k - r$ idiosyncratic noises. Thus, instead of the DFM model $\boldsymbol{z}_t = \boldsymbol{P}\boldsymbol{f}_t + \boldsymbol{n}_t$, where $\boldsymbol{n}_t$ is of dimension k, they assume that the noises have dimension $k - r$. As shown in Tiao and Tsay (1989), using canonical correlation analysis between the series and its past values, any finite-order VARMA model can be written as

$$\boldsymbol{z}_t = \boldsymbol{L}_1 \boldsymbol{f}_t + \boldsymbol{L}_2 \boldsymbol{e}_t, \tag{6.24}$$

where $\boldsymbol{L}_1$ is $k \times r$ and $\boldsymbol{L}_2$ is $k \times (k - r)$, $\boldsymbol{f}_t$ is a r-dimensional common factors with dynamic structure and satisfying $\boldsymbol{\Gamma}_f(0) = \boldsymbol{I}$, and $\boldsymbol{e}_t$ is a $(k - r)$-dimensional white noise process with mean zero and a positive-definite covariance matrix $\boldsymbol{\Sigma}_e$. In addition, no linear combination of $\boldsymbol{f}_t$ is a white noise process and $\{\boldsymbol{f}_t\}$ and $\{\boldsymbol{e}_t\}$ are uncorrelected. Without loss of generality, one can assume that $\boldsymbol{L}_1' \boldsymbol{L}_1 = \boldsymbol{I}_r$ and $\boldsymbol{L}_2' \boldsymbol{L}_2 = \boldsymbol{I}_{k-r}$, but $\boldsymbol{L}_1$ is not orthogonal to $\boldsymbol{L}_2$ in general. Then, calling $\boldsymbol{L}_1^*$ the orthogonal complement of $\boldsymbol{L}_1$, we have

$$(\boldsymbol{L}_1^*)' \boldsymbol{z}_t = (\boldsymbol{L}_1^*)' \boldsymbol{L}_2 \boldsymbol{e}_t,$$

so that we have $k - r$ linear combinations of z_t that are white noise. Therefore, the number of factors can be obtained by searching for the number of linear combinations that are white noise.

Similarly, letting L_2^* be the orthogonal complement of L_2 and pre-multiplying Eq. (6.24), we have

$$(L_2^*)' z_t = (L_2^*)' L_1 f_t,$$

which does not involve the noise component e_t. Consequently, one can derive an estimator of the common factor f_t that is free of the impact of the noise term. This is different from the estimation of common factors mentioned before via the PCA, where the estimate $\hat{f}_t$ is always affected by the noise component. Gao and Tsay (2019) refer to such an estimation procedure the *projected PCA*. As this model has some advantages in the high-dimensional case, in which $k > T$, we discuss it further in Section 6.10.

6.3 GENERALIZED DFM (GDFM) FOR STATIONARY SERIES

The DFM assumes a contemporaneous relationship between the series and the common factors. This model can be generalized, see Forni et al. (2000), by assuming that the observed time series are affected by the factors and all their past values. The resulting model is called the GDFM. Assuming a finite number of lags, the representation is:

$$z_t = \sum_{l=0}^{L-1} P_l f_{t-l} + n_t, \tag{6.25}$$

where L is a positive integer, f_t is a q-dimensional stationary vector of uncorrelated common factors, the L matrices P_l are $k \times q$ factor loading matrices and n_t is the noise or idiosyncratic part, that is stationary and assumes the conditions of the approximate DFM. We denote the number of factors by q in Eq. (6.25) whenever $L > 1$ and return to the notation r if $L = 1$. The reason for this change will be clear shortly. The common component of the series z_t is $c_t = \sum_{l=0}^{L-1} P_l f_{t-l}$ and is generated by the $q < k$ unobserved factor series. Although the GDFM was proposed by allowing L to go to infinity in this chapter we only discuss the case that L is finite, that is the most important one in practice. The conditions for identification of the loading matrices and the factors are similar to the DFM, $P_l' P_l = I_q$ for $l = 0, 1, \ldots, L-1$, and $E(f_t f_t')$ is diagonal.

Model (6.25) can also be written as

$$z_t = P(B) f_t + n_t,$$

where $P(B) = P_0 + P_1 B + \cdots + P_{L-1} B^{L-1}$ is a matrix polynomial in the lag operator. Assuming that the factors follow a vector autoregressive (VAR) model with diagonal matrices

$$A(B) f_t = a_t,$$

we have that

$$z_t = \Psi(B) a_t + n_t,$$

where $\Psi(B) = P(B) A(B)$. Note that in finite samples there is an identification problem in the decomposition of $\Psi(B)$ into $A(B)$ and $P(B)$ as none of them are observed.

In order to identify the GDFM, Forni et al. (2000) postulate conditions in the frequency domain with respect to the eigenvalues of the spectral density matrix of the observed data for each frequency. They assume that these matrices have q eigenvalues that diverge as $k \to \infty$. However, for the noise component, all the eigenvalues of these matrices are bounded. As the spectral density matrix is the Fourier transform of the covariance matrix (see, Shumway and Stoffer, 2000) this property is equivalent to say that all the lag covariance matrices have q eigenvalues that diverge when $k \to \infty$, whereas the largest eigenvalue of the covariance matrices of the idiosyncratic component are bounded for all lags. From an intuitive point of view, this means that the factors have some effect on almost all the series and only a few values of the loading parameters are zero. It is said that the factors have a pervasive effect on the observed series. However, the noises are concentrated on a finite number of series. These assumptions for the noises are similar to the ones presented for the ADFM. In other words, the noise-to-signal ratio must converge to zero as $k \to \infty$ for the model to be identifiable.

6.3.1 Some Properties of the GDFM

Model (6.25) can be written as model (6.1) by defining a $k \times r$ loading matrix $\boldsymbol{B} = [\boldsymbol{P}_0, \dots, \boldsymbol{P}_{L-1}]$, where now $r = qL$ and a $r \times 1$ vector of series $\boldsymbol{F}_t^D = [\boldsymbol{f}_t', \dots, \boldsymbol{f}_{t-L+1}']'$. Then

$$\boldsymbol{z}_t = \boldsymbol{B}\boldsymbol{F}_t^D + \boldsymbol{n}_t, \tag{6.26}$$

but this equation does not represent the standard DFM because $\boldsymbol{B}'\boldsymbol{B} \neq \boldsymbol{I}_r$, as in general $\boldsymbol{P}_i'\boldsymbol{P}_j \neq \boldsymbol{0}_r$ for $i \neq j$, and the covariance matrices $\boldsymbol{\Gamma}_F^D(h)$ are no longer diagonal. The covariance matrices of the observed series $\boldsymbol{z}_t$ satisfy

$$\boldsymbol{\Gamma}_z(h) = \boldsymbol{B}\boldsymbol{\Gamma}_F^D(h)\boldsymbol{B}' + \boldsymbol{\Gamma}_n(h),$$

where now $\boldsymbol{B}$ is $k \times qL$ and $\boldsymbol{\Gamma}_F^D(h)$ is $qL \times qL$. The matrix $\boldsymbol{B}$ can be written as $\boldsymbol{UDV}'$ for the singular value decomposition and model (6.26) is equivalent to

$$\boldsymbol{z}_t = \boldsymbol{U}\boldsymbol{F}_t^S + \boldsymbol{n}_t,$$

where the vector of static factors $\boldsymbol{F}_t^S = \boldsymbol{D}\boldsymbol{V}'\boldsymbol{F}_t^D$ is a linear transformation of the original dynamic factors. These factors can be made orthogonal with some additional transformation

$$\boldsymbol{z}_t = \boldsymbol{U}\boldsymbol{M}\boldsymbol{M}'\boldsymbol{F}_t^S + \boldsymbol{n}_t = \boldsymbol{P}\boldsymbol{f}_t + \boldsymbol{n}_t,$$

where $\boldsymbol{M}$ is an orthonormal matrix with the eigenvectors of the covariance matrix of $\boldsymbol{F}_t^S$ and $\boldsymbol{P} = \boldsymbol{UM}$ and $\boldsymbol{f}_t = \boldsymbol{M}'\boldsymbol{F}_t^S$. Note that although $\boldsymbol{P}\boldsymbol{\Gamma}_F^S(h)\boldsymbol{P}'$ could have rank qL, the space of the static factor could be well approximated by a space of lower dimension than qL. As an illustration, consider the simple model with $L = 2$

$$\boldsymbol{z}_t = \boldsymbol{P}_0\boldsymbol{f}_t + \boldsymbol{P}_1\boldsymbol{f}_{t-1} + \boldsymbol{n}_t,$$

and the factors follow the VAR(1) model $\boldsymbol{f}_t = \boldsymbol{\Phi}\boldsymbol{f}_{t-1} + \boldsymbol{u}_t$. Then, this GDFM model can also be written as

$$\boldsymbol{z}_t = (\boldsymbol{P}_0\boldsymbol{\Phi} + \boldsymbol{P}_1)\boldsymbol{f}_{t-1} + \boldsymbol{n}_t + \boldsymbol{P}_0\boldsymbol{u}_t = \boldsymbol{P}^*\boldsymbol{f}_t^* + \boldsymbol{n}_t^*,$$

This model can be seen as a DFM with $r = 2q$, as in (6.26), or as a DFM with with $r = q$ factors, $\boldsymbol{f}_t^*$, that have VAR(1) structure and a white noise, $\boldsymbol{n}_t^*$, with a non-diagonal covariance matrix. In particular, if the variances of the factor noises, $\boldsymbol{u}_t$, are smaller than those of the $\boldsymbol{n}_t$ and the off-diagonal elements of the covariance matrix are small, the model will be close to a standard EDFM with $r = q$ factors. In practice, there is strong evidence that the number of dynamic factors found, q, is usually smaller than the number of statics factors, r, but also that $r < qL$.

The estimation of GDFM can be carried out by DPC analysis in the frequency domain, as proposed by Forni et al. (2000, 2005) or by linear generalized DPC analysis in the time domain, as proposed by Peña et al. (2019). This last approach seems to work better in finite samples, see Peña and Yohai (2016) and Peña et al. (2019), and has the advantage that it can be applied to the nonstationary case, where the spectral approach cannot.

6.3.2 GDFM and VARMA Models

As mentioned in Chapter 3, to overcome the problem of using a large number of parameters in a VARMA model when k is large, Velu, Reinsel and Wichern (1986) proposed reduced rank models. In particular, the reduced rank VAR model assumes that $\boldsymbol{z}_t$ follows a VAR model with parameter matrices that can be decomposed as:

$$\boldsymbol{z}_t = \Phi(B)\boldsymbol{z}_{t-1} + \boldsymbol{e}_t = \boldsymbol{A}(B)\boldsymbol{C}(B)\boldsymbol{z}_{t-1} + \boldsymbol{e}_t, \tag{6.27}$$

where $\boldsymbol{A}(B) = \sum_{i=0}^{p_1} \boldsymbol{A}_i B^i$ and $\boldsymbol{C}(B) = \sum_{i=0}^{p_2} \boldsymbol{C}_i B^i$ are matrix polynomials of dimensions $k \times r$ and $r \times k$, where $r < k$, and with degrees p_1 and p_2, respectively. The series $\boldsymbol{z}_t$ follows a VAR($p_1 + p_2 + 1$) model. Calling $\boldsymbol{C}(B)\boldsymbol{z}_{t-1} = \boldsymbol{f}_t$, we can write the model as

$$\boldsymbol{z}_t = \boldsymbol{A}(B)\boldsymbol{f}_t + \boldsymbol{e}_t, \tag{6.28}$$

which looks like a GDFM with r factors and a white noise term. However, there is an important difference between the two models. In the reduced rank representation (6.28), the factors $\boldsymbol{f}_t$ must be linear combinations of lagged values of the series, whereas in the GDFM the factors do not have this restrictions and they include the contemporaneous information as well as the past information. See Ahn and Reinsel (1988) for generalizations of Model (6.28) to the nested reduced rank AR model, and its relationship to the scalar components model of Tsay and Tiao (1985) and Tiao and Tsay (1989).

6.4 DYNAMIC PRINCIPAL COMPONENTS

6.4.1 Dynamic Principal Components for Optimal Reconstruction

We have seen in Chapter 1 that the standard PCs provide an optimal reconstruction of the time series data using only contemporaneous information. Brillinger (1981) addressed the reconstruction problem in a more general form, allowing the use of lagged values, and defined Dynamic Principal Components, (DPCs) as linear combinations of the time series that provide an optimal reconstruction using all leads and lags of the data. Formally, consider a zero-mean k-dimensional stationary process

$\{z_t| -\infty < t < \infty\}$. Define the first DPC as a linear combination of all the values of the series

$$f_t = \sum_{h=-\infty}^{\infty} c'_h z_{t-h}, \tag{6.29}$$

where the c_h are k-dimensional vectors such that f_t provides an optimal reconstruction of the data using all of its lags and leads, that is, the first DPC minimizes:

$$E\left[\left(z_t - \sum_{j=-\infty}^{\infty} \beta_j f_{t+j}\right)' \left(z_t - \sum_{j=-\infty}^{\infty} \beta_j f_{t+j}\right)\right], \tag{6.30}$$

using some $k \times 1$ vectors $\beta_j, -\infty < j < \infty$. Brillinger elegantly solved this problem in the frequency domain by showing that c_k is the inverse Fourier transform of the PCs of the cross-spectral matrices for each frequency, and β_j is the inverse Fourier transform of the conjugates of the same PCs. See Brillinger (1981) and Shumway and Stoffer (2000) for the details. When this procedure is adapted to finite samples, the number of lags in Eq. (6.29) and in the reconstruction of the series in Eq. (6.30) are to be defined. Also, the approach proposed by Brillinger to solve the problem can only be applied to stationary time series. These DPC has been used by Forni et al. (2000, 2005) for estimation of GDFMs.

6.4.2 One-Sided DPCs

Peña et al. (2019) proposed a way to compute the DPC in the time domain that is more efficient than that in the frequency domain in finite samples. They called the estimates one-sided dynamic principal component (ODPC), because they used the lagged values of the series to differentiate them from those defined by Brillinger that use both the past and future values. The ODPCs are summarized as follows. Consider two nonnegative integers c_1 and c_2. Let $\boldsymbol{a}' = (\boldsymbol{a}'_0, \dots, \boldsymbol{a}'_{c_1})$ be a $k(c_1+1)$-dimensional vector, where $\boldsymbol{a}'_h = (a_{h,1}, \dots, a_{h,k})$. Also, let $\boldsymbol{B}$ be a $(c_2+1) \times k$ matrix of the form $\boldsymbol{B}' = [\mathbf{b}_0, \dots, \mathbf{b}_{c_2}]$, where each $\mathbf{b}_i \in R^k$. Define the first DPC as,

$$f_t(\hat{\boldsymbol{a}}) = \sum_{h=0}^{c_1} z'_{t-h}\hat{\boldsymbol{a}}_h, \quad t = c_1+1, \dots, T, \tag{6.31}$$

which is a linear combination of the lagged values of the series that has the property that the reconstruction of the original data via

$$z_t^{R,1}(\boldsymbol{a}, \boldsymbol{B}) = \sum_{h=0}^{c_2} \mathbf{b}_h f_{t-h}(\boldsymbol{a}), \quad t = c_1 + c_2 + 1, \dots, T, \tag{6.32}$$

minimizes the MSE in the reconstruction. That is, the vector $\hat{\boldsymbol{a}}$ and the matrix $\hat{\boldsymbol{B}}$ are such that

$$(\hat{\boldsymbol{a}}, \hat{\boldsymbol{B}}) = \arg \min_{\|\boldsymbol{a}\|=1, \boldsymbol{B}} \frac{1}{T^* k} \sum_{t=(c_1+c_2)+1}^{T} \|z_t - z_t^{R,1}(\boldsymbol{a}, \boldsymbol{B})\|^2, \tag{6.33}$$

where $T^* = T - (c_1 + c_2)$ is the effective number of observations we can reconstruct with the ODPC. Note that by (6.31), we can compute the component for

$t = c_1 + 1, \ldots, T$, and by (6.32), the first value we can reconstruct is $t = c_1 + c_2$. The constraint $\|\boldsymbol{a}\| = 1$ is included for identification purpose, because if we multiply $\boldsymbol{a}_h$ in Eq. (6.31) by g and divide $\mathbf{b}_h$ by g, the solution remains the same.

The second ODPC is then defined as a linear combination of the lagged series that can be used to optimally reconstruct the residuals from the first component. Higher-order components are defined similarly. Finally, the reconstruction of $\boldsymbol{z}_t$ using the first υ one-sided DPCs is given by

$$\widehat{\boldsymbol{z}}_t^{\upsilon} = \sum_{i=1}^{\upsilon} \sum_{h=0}^{c_2^i} \widehat{\mathbf{b}}_h^i \widehat{f}_{t-h}^i, \quad c_{\max}^{\upsilon} + 1 \leq t \leq T,$$

where now c_1^i and c_2^i are the values of c_1 and c_2 for the ith ODPC and $c_{\max}^{\upsilon} = \max\{c_1^i + c_2^i | i = 1, \ldots, \upsilon\}$.

We now describe an algorithm to compute the ODPC. Since their definition is sequential, it suffices to propose an algorithm to compute the first component. In what follows, to keep the notation light, we drop the superscript indicating the component number. In order to derive an algorithm to compute $\widehat{\boldsymbol{a}}$, and $\widehat{\boldsymbol{B}}$, we need first to express the objective function in Eq. (6.33) in a more manageable form. To this end, we introduce some further notation. Let $\boldsymbol{Z}_h$ be the $T^* \times k$ data matrix of observations, i.e.

$$\boldsymbol{Z}_h = \begin{bmatrix} \boldsymbol{z}'_{h+1} \\ \boldsymbol{z}'_{h+2} \\ \vdots \\ \boldsymbol{z}'_{h+T^*} \end{bmatrix}, \tag{6.34}$$

and consider the sequence of matrices $\boldsymbol{Z}_h$ for $h = 0, \ldots, (c_1 + c_2)$. The matrix $\boldsymbol{Z}_{c_1+c_2}$ includes the observations to be reconstructed, and the matrices $\boldsymbol{Z}_0, \ldots, \boldsymbol{Z}_{c_1+c_2-1}$ are used to compute the component. We also need, for $l = c_1, \ldots, (c_1 + c_2)$, the sequence of $T^* \times k(c_1 + 1)$ matrices $\boldsymbol{Z}_{l,c_1} = [\boldsymbol{Z}_l, \boldsymbol{Z}_{l-1}, \ldots, \boldsymbol{Z}_{l-c_1}]$ that include the lagged values for the computation of the component, and the $T^* \times (c_2 + 1)$ matrix $\boldsymbol{F}_{c_1,c_2} = [\boldsymbol{f}_{(c_1+c_2)}, \boldsymbol{f}_{(c_1+c_2)-1}, \ldots, \boldsymbol{f}_{c_1}]$ with the values of the component in columns $\boldsymbol{f}_j = (f_{j+1}, \ldots, f_{T-(c_1+c_2)+j})'$. These last matrices are related by

$$\boldsymbol{F}_{c_1,c_2} = (\boldsymbol{Z}_{c_1+c_2,c_1}\boldsymbol{a}, \ldots, \boldsymbol{Z}_{c_1,c_1}\boldsymbol{a}),$$

where $\boldsymbol{F}_{c_1,c_2} = \boldsymbol{F}_{c_1,c_2}(a)$, even though this dependence is not made explicit in the notation. The reconstruction of $\boldsymbol{Z}_{c_1+c_2}$ can be written as a matrix $\boldsymbol{Z}_{c_1+c_2}^R$ of the same dimension, $T^* \times k$,

$$\boldsymbol{Z}_{c_1+c_2}^R = \boldsymbol{F}_{c_1,c_2}\boldsymbol{B}.$$

Then, $(\widehat{\boldsymbol{a}}, \widehat{\boldsymbol{B}})$ can be obtained by minimizing

$$\|\boldsymbol{Z}_{c_1+c_2} - \boldsymbol{F}_{c_1,c_2}\boldsymbol{B}\|_F^2 = \|\text{vec}(\boldsymbol{Z}_{c_1+c_2}) - \text{vec}(\boldsymbol{Z}_{c_1+c_2}^R)\|^2,$$

with the restriction $\|\boldsymbol{a}\| = 1$, where $\| \cdot \|_F$ is the Frobenius norm.

Then, it can be shown (see Peña et al. 2019) that the least squares estimate of $\widehat{\boldsymbol{a}}$ is

$$\widehat{\boldsymbol{a}} = (\boldsymbol{Y}'\boldsymbol{Y})^{-1}\boldsymbol{Y}' \text{ vec}(\boldsymbol{Z}_{c_1+c_2}), \tag{6.35}$$

where the matrix $\boldsymbol{Y}$ is given by

$$\boldsymbol{Y} = (\boldsymbol{B} \otimes \boldsymbol{I}_{T^*})\boldsymbol{Z}^B,$$

and the matrix $\boldsymbol{Z}^B$ is formed by the matrices $\boldsymbol{Z}_{l,c_1}$ that have the lagged values for the computation of the component. This solution should be standardized to unit norm.

On the other hand, for a fixed $\boldsymbol{F}_{c_1,c_2}$, the optimal $\boldsymbol{B}$ can also be computed by least squares method

$$\widehat{\boldsymbol{B}} = (\boldsymbol{F}'_{c_1,c_2}\boldsymbol{F}_{c_1,c_2})^{-1}\boldsymbol{F}'_{c_1,c_2}\boldsymbol{Z}_{c_1+c_2}. \tag{6.36}$$

We can apply an alternating least squares algorithm for computing $\widehat{\boldsymbol{a}}$ and $\widehat{\boldsymbol{B}}$. Let $\boldsymbol{a}^{(i)}$, $\boldsymbol{B}^{(i)}$, and $\boldsymbol{f}^{(i)}$ be the values of $\boldsymbol{a}$, $\boldsymbol{B}$, corresponding to the ith iteration. Let $\delta \in (0, 1)$ be a tolerance parameter to stop the iterations. In our implementation, we take $\delta = 10^{-4}$. As before, the algorithm is defined by giving an initial value of the component, say $\boldsymbol{f}^{(0)}$, and describing the rule to compute $\boldsymbol{B}^{(i+1)}$, $\boldsymbol{a}^{(i+1)}$, and $\boldsymbol{f}^{(i+1)}$ from $\boldsymbol{f}^{(i)}$. This can be done as follows:

1. Given $\boldsymbol{f}^{(i)}$, define $\boldsymbol{B}^{(i+1)}$ by Eq. (6.36), where $\boldsymbol{F}_{c_1,c_2}$ corresponds to $\boldsymbol{f}^{(i)}$.
2. Compute $\boldsymbol{a}_p^{(i+1)}$ by Eq. (6.35) with $\boldsymbol{B} = \boldsymbol{B}^{(i+1)}$, and let $\boldsymbol{a}^{(i+1)} = \boldsymbol{a}_p^{(i+1)}/||\boldsymbol{a}_p^{(i+1)}||$.
3. The tth coordinate of $\boldsymbol{f}^{(i+1)}$ is given by Eq. (6.31) with $\boldsymbol{a} = \boldsymbol{a}^{(i+1)}$.

We stop the iteration when the relative decrease in the MSE is smaller than δ. Clearly in this algorithm at each step the MSE decreases and, therefore, it converges to a local minimum. To obtain a global minimum, the initial value $\boldsymbol{f}^{(0)}$ should be sufficiently close to the optimal one. We propose to take $\boldsymbol{f}^{(0)}$ as the last $T - c_1$ coordinates of the first ordinary PC of the data.

6.4.3 Model Selection and Forecasting

In practice, the number of components and the number of lags in each component need to be chosen. To simplify the notation, assume that for each component $c_1 = c_2$, that is, the number of lags of $\boldsymbol{z}_t$ used to define the ith DPC and the number of lags of $\widehat{f}_t^i$ used to reconstruct the original series are the same. One possible approach to choose the number of components and the number of lags is to minimize the cross-validated forecasting error in a stepwise fashion. Suppose we want to make h-step ahead forecasts. Fix a maximum number of lags $G_{\max}$ and a window size w. Then, given $g \in \{1, \ldots, G_{\max}\}$, we compute the first ODPC with g lags, using periods $1, \ldots, T - h - t + 1$ for $t = 1, \ldots, w$, and for each of these fits we compute an h-step ahead forecast and the corresponding MSE $E_{t,h}$. The cross-validation estimate of the forecasting error is then

$$\widehat{\text{MSE}}_{1,g} = \frac{1}{w}\sum_{t=1}^{w} E_{t,h}.$$

We choose for the first component the value $g^{*,1}$ that minimizes $\widehat{\text{MSE}}_{1,g}$. Then, we fix the first component computed with $g^{*,1}$ lags and repeat the procedure with the second component. If the optimal cross-validated forecasting error using the two components, $\widehat{\text{MSE}}_{2,g}$, is larger than the one using only one component, $\widehat{\text{MSE}}_{1,g}$, we stop and the ODPC has one component with $g^{*,1}$ lags; otherwise, we add the second component defined using $g^{*,2}$ lags and proceed as before.

The same stepwise approach could be applied to minimize an information criterion instead of the cross-validated forecasting error. This would reduce the computational burden significantly. The following criterion, inspired by the IC_{p3} criterion proposed by Bai and Ng (2002), seems to give good results in practice. Let $\hat{\sigma}^2_{1,g}$ be the reconstruction MSE for the first ODPC defined using g lags. Let $T^{*,1,g} = T - 2g$. Then we choose the value $g^{*,1}$ among $1, \ldots, G_{\max}$ that minimizes

$$\text{BNG}_{1,g} = \log(\hat{\sigma}^2_{1,g}) + (g+1)\frac{\log(\min(T^{*,1,g}, g))}{\min(T^{*,1,g}, g)}.$$

Note that $(g+1)$ is the number of static factors implied by fitting a single component with g lags. Suppose now that we have computed $q-1$ DPCs, each with $c_1 = c_2 = g^{*,i}$ lags, $i = 1, \ldots, q-1$. Let $\hat{\sigma}^2_{q,g}$ be the reconstruction MSE for the fit obtained using q components, where the first $q-1$ components are defined using $g^{*,i}$, $i = 1, \ldots, q-1$ and the last component is defined using g lags. Let $T^{*,q,g} = T - \max\{2g^{*,1}, \ldots, 2g^{*,q-1}, 2g\}$. Let $g^{*,q}$ be the value among $1, \ldots, G_{\max}$ that minimizes

$$\text{BNG}_{q,g} = \log(\hat{\sigma}^2_{q,g}) + \left(\sum_{i=1}^{q-1}(g^{*,i}+1) + g + 1\right)\frac{\log(\min(T^{*,q,g}, g))}{\min(T^{*,q,g}, g)}.$$

Note that $\sum_{i=1}^{q-1}(g^{*,i}+1)$ is the number of static factors implied by fitting $q-1$ dynamic components, each with $g^{*,i}$ lags. If $\text{BNG}_{q,g^{*,q}}$ is larger than $\text{BNG}_{q-1,g^{*,q-1}}$ we stop and the final model is the ODPC with $q-1$ components. Otherwise, we add the qth component defined using $g^{*,q}$ and continue as before.

After fitting q DPCs to the data, each with (c_1, c_2) lags, $i = 1, \ldots, q$, we can build forecasts as follows. Suppose we have decided upon a procedure to forecast each of these DPCs separately. For instance, we can fit ARMA models to the components in an automatic fashion. Let $\hat{f}^i_{T+h|T}$ for $h > 0$ be the forecast of f^i_{T+h} with information until time T. Then we can obtain a h-step ahead forecast of z_T as

$$\hat{\boldsymbol{z}}_{T+v|T} = \sum_{i=1}^{q}\sum_{h=0}^{c_2}\hat{\boldsymbol{b}}^i_h \hat{f}^i_{T+v-h|T}.$$

6.4.4 One Sided DPC and GDFM Estimation

The one-sided DPC or ODPC provides a useful way to estimate GDFM. In fact, it was proved by Peña et al. (2019) that asymptotically, when both the number of series and the sample size go to infinity, the reconstruction obtained with ODPC converges in mean square to the common part of the GDFM. Hence, the ODPC provides an alternative way to estimate the common part of DFMs with a finite number of lags. For nonstationary time series, we will present next a generalized version of these time domain DPCs to estimate nonstationary GDFM.

Example 6.4

Consider, again, the growth rates of standardized GDP series of the 19 Euro countries from 2000.II to 2018.IV. These data were introduced in Example 6.1. We demonstrate the estimation of ODPC using the R package **odpc**. Figure 6.12 shows the time plot of the first estimated ODPC and Figure 6.13 the second factor or second ODPC. These results are computed with three lags. Figure 6.14 plots the coefficients of $\boldsymbol{a}^1$ of the first DPC, whereas Figure 6.15 shows the loading in $\boldsymbol{B}$ also for the first DPC.

R commands for dynamic principal components:

```
>require(odpc) # load package
sal<- odpc(G1, ks=c(3,3))
> sal
    k1 k2   MSE
Component 1  3   3 0.006
Component 2  3   3 0.004
# 3Two components are selected with 3 lags. Now
# sal[[1]] has the information of the 1st component, sal[[2]] of the
# second component and so on.
> f1<- sal[[1]]$f
> a1<- sal[[1]]$a
> f2<- sal[[2]]$f
> a2<- sal[[2]]$a
> yl <- range(f1)
> tdx <- c(5:76)/4+2000
```

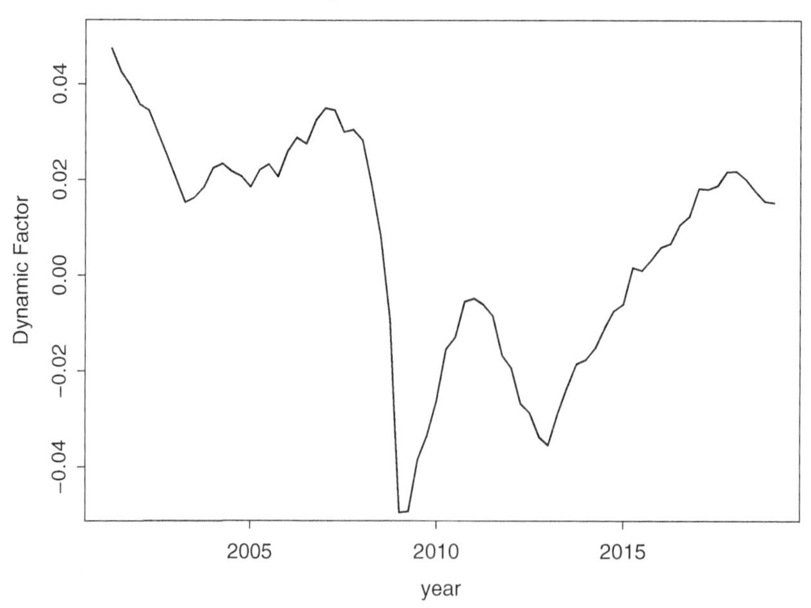

Figure 6.12 First dynamic factor for the growth rates of GDP series for the 19 countries in the European Monetary Union. The factor was built with three lags and entered with three lags.

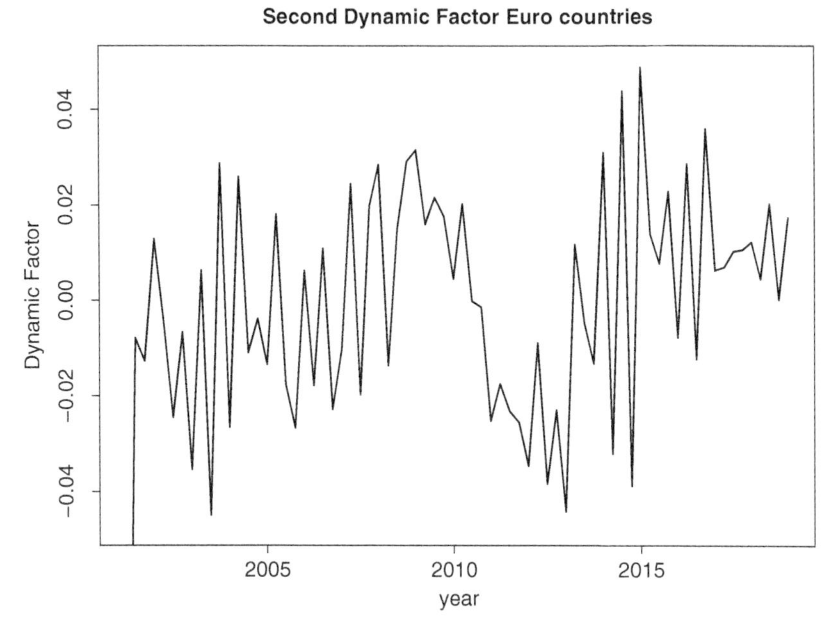

Figure 6.13 Second dynamic factor for the growth rates of GDP series for the 19 countries in the European Monetary Union. The factor was built with three lags and entered with three lags.

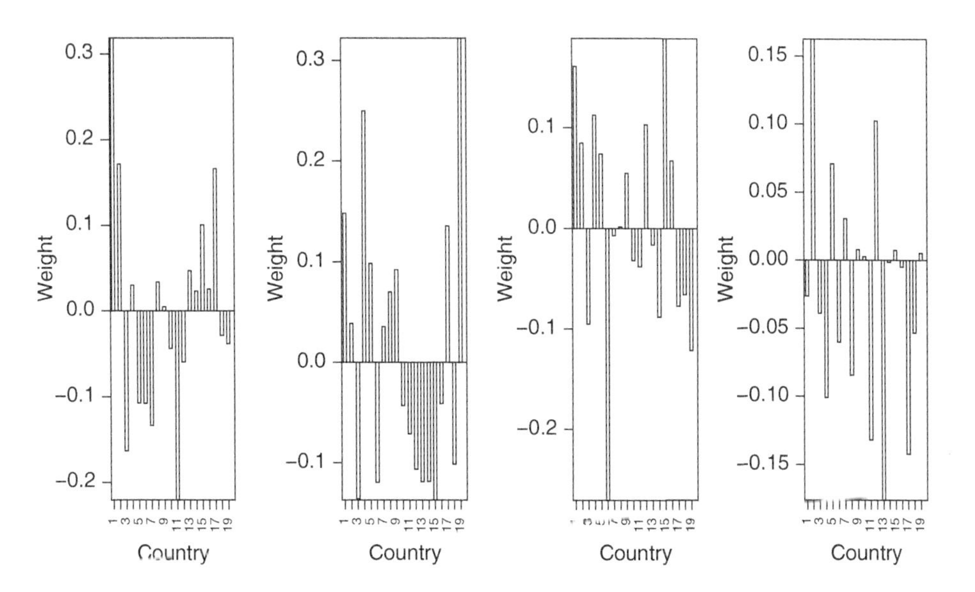

Figure 6.14 Weights of the lag series (**a** coefficients) to build the first dynamic factor of the GDP growth rates of the 19 countries of the European Monetary Union. The weights apply to lags from zero to three.

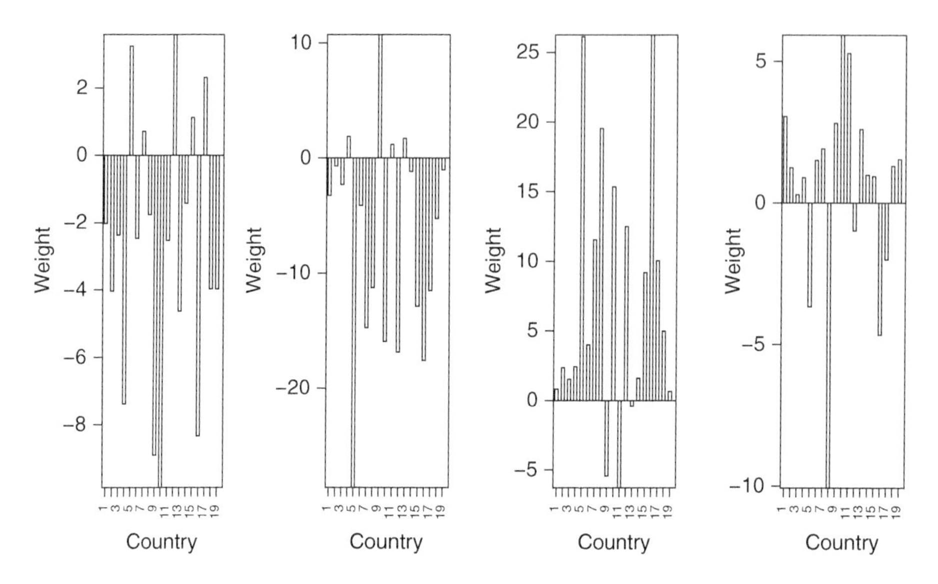

Figure 6.15 Loadings (**b** coefficients) of the lag first dynamic principal component to reconstruct the GDP growth of the European Monetary Union. The loadings apply to lags from zero to three.

```
> plot(tdx,-f1,xlab="year",ylab="Dynamic Factor", main="First Dynamic
Factor Euro countries", type="l", ylim=yl)
> plot(tdx,-f2,xlab="year",ylab="Dynamic Factor", main="Second Dynamic
Factor Euro countries", type="l", ylim=yl)
# A vector is split into its lags in order to plot them.
> a10=a1[1:19]
> a11=a1[20:38]
> a12=a1[39:57]
> a13=a1[58:76]
> par(mfrow=c(1,4))
> plot(a10,xlab="Country",ylab="Lag 0 weight ", type="l")
> plot(a11,xlab="Country",ylab="Lag 1 weight ", type="l")
> plot(a12,xlab="Country",ylab="Lag 2 weight ", type="l")
> plot(a13,xlab="Country",ylab="Lag 3 weight ", type="l")
> B1 <- sal[[1]]$B
> par(mfrow=c(1,4))
# The Loadings B are plotted
> plot(B1[1,],xlab="Countries",ylab="Loadings 1st DF lag 0", type="l")
> plot(B1[2,],xlab="Countries",ylab="Loadings 1st DF lag 1", type="l")
> plot(B1[3,],xlab="Countries",ylab="Loadings 1st DF lag 2", type="l")
> plot(B1[4,],xlab="Countries",ylab="Loadings 1st DF lag 3", type="l")
```
■

6.5 DFM FOR NONSTATIONARY SERIES

For simplicity, assume that the time series $\boldsymbol{z}_t$ satisfies $\bar{\boldsymbol{z}} = \sum_{t=1}^{T} \boldsymbol{z}_t/T = \boldsymbol{0}$ and are an $I(1)$ process. Then, some of the common factors are $I(1)$, but some of them may be stationary. This model was considered by Hu and Chou (2004) that propose to use scalar component models (Tiao and Tsay, 1989) to separate the stationary and non-stationary factors. A formal approach was proposed by Peña and Poncela (2006) with the following model:

$$z_t = P_1 f_{1t} + P_2 f_{2t} + n_t, \tag{6.37}$$

where P_1 and f_{1t} correspond to the r_1 -dimensional $I(1)$ common factors, P_2 and f_{2t} are for the r_2-dimensional stationary common factors, and n_t is a stationary noise process. For an $I(d)$ process y_t, the authors defined the generalized sample autocovariance matrices $C_y(h)$ as

$$C_y(h) = \frac{1}{T^{2d}} \sum_{t=h+1}^{T} z_{t-h} z_t', \tag{6.38}$$

and showed that (1) these generalized covariance matrices converge to some random matrices and (2) for identification of nonstationary factors these matrices play the same role as the sample autocovariance matrices of a stationary series. In fact, they showed that, under Model (6.37), the generalized sample covariance matrix, $C_y(h)$ with finite h, converges weakly to a random matrix Γ_y, that has, almost surely, r_1 eigenvalues greater than zero and $k - r_1$ eigenvalues equal to zero. Also, the eigenvectors corresponding to the r_1 eigenvalues greater than zero form a basis of the space spanned by the columns of the loading submatrix P_1. This implies that the number of common nonstationary factors can be found as the number of non-zero eigenvalues of $C_y(h)$ and the corresponding eigenvectors provide an estimate of the P_1 loading matrix. Thus, calling $\widehat{P}_1$ the matrix formed by the eigenvectors associated with the r_1 largest eigenvalues of $C_y(0)$, we estimate the integrated factors by $\widehat{f_{1t}} = \widehat{P}_1' z_t$. Next, we obtain a new series y_t free from the nonstationary factors by

$$y_t = z_t - \widehat{P}_1 \widehat{f_{1t}} = P_2 f_{2t} + n_t + \epsilon_t, \tag{6.39}$$

where $\epsilon_t = P_1 f_{1t} - \widehat{P}_1 \widehat{f_{1t}}$ is a process that converges to zero as $T \to \infty$. Then, the process y_t would approximately follow a stationary DFM in Eq. (6.1), and we can apply the procedure discussed before for stationary time series to estimate P_2 and f_{2t}. Finally, the noise is computed and the models for the factors and the noises can be built. As an alternative to the generalized covariance matrix we can use the pooled covariance matrix (6.21) to compute the eigenvectors.

This approach requires that the noises be stationary. Otherwise, the eigenstructure of $C_y(0)$ would not be informative and the series y_t in Eq. (6.39) is likely to have integrated noises, resulting in an identification problem. Also, as we saw in Chapter 3, independent $I(1)$ series may have strong correlation coefficients that, in turn, can cause the spurious regression problem. See, Section 3.5.1. Thus, a set of independent $I(1)$ series may have covariance matrices with many off diagonal non-zero coefficients and applying PCs can lead to spurious factors.

An alternative procedure proposed by Bai and Ng (2004), which does not make the assumption of stationary noises, is to take the first difference of observed series and apply the DFM to the resulting stationary series. Let $y_t = \nabla z_t$ and write:

$$y_t = P_1 \widetilde{f}_{1t} + P_2 \widetilde{f}_{2t} + \widetilde{n}_t.$$

This is a stationary DFM with factors $\widetilde{\boldsymbol{f}}_{1t} = \nabla \boldsymbol{f}_{1t}$ and $\widetilde{\boldsymbol{f}}_{2t} = \nabla \boldsymbol{f}_{2t}$ and noises that are stationary, but may not be invertible. We can apply after differencing the procedure for stationary DFM to find a set of factors $\widetilde{\boldsymbol{f}}_t$ that includes the stationary and nonstationary factors. Then, we can construct the integrated factors by $\boldsymbol{f}_1 = \widetilde{\boldsymbol{f}}_1$ and

$$\widehat{\boldsymbol{f}}_t = \sum_{s=2}^{t} \widetilde{\boldsymbol{f}}_s, \quad t \geq 2,$$

and check stationarity of the components of $\widehat{\boldsymbol{f}}_t$. This can be done using the augmented Dickey-Fuller test discussed in Chapter 2. Note that the number of integrated factors found by this procedure could be larger than the true number r_1, and a co-integration test is needed to select the number of factors. See, Chapter 3. Then, the noise series $\widehat{\boldsymbol{n}}_t = \boldsymbol{z}_t - \widehat{\boldsymbol{P}}_1 \widehat{\boldsymbol{f}}_{1t} - \widehat{\boldsymbol{P}}_2 \widehat{\boldsymbol{f}}_{2t}$ is computed and each component is tested for stationarity and used for modeling.

The first approach, that works on levels, has no identification problem and the advantage that the dimension of the integrated factor space is found directly. It is recommended if we expect that the series have common factors with stationary noises. On the other hand, as explained before, it may find spurious factors if the series are independent $I(1)$ series or if some of the noises are integrated (see, Onatski and Wang, 2019). The second approach, that works on the differenced series, is better to test for the presence of integrated factors when the noises are integrated, but it is less efficient to find the dimension of the integrated factors when the noises are stationary, as shown in the Monte Carlos study by Corona et al. (2020). One reason for this effect indicated by Choi (2017) is that differencing a stationary noise usually increases its variance and its autocorrelations, making it less efficient the estimation of the common factors. In any case, this is still an open problem and more research is needed to judge the relative advantages of both approaches. For instance, see, Corona et al. (2017) for a study of the performance of methods for selecting the number of factors after stationary univariate transformations. Therefore, our recommendation is always to try both and compare their results. Another approach is recently proposed by Gao and Tsay (2020) that also works on the levels of $\boldsymbol{z}_t$.

Example 6.5

We analyze, again, the GDP series of the 19 Euro countries, but now use the levels directly. Figure 6.16 shows the time plots of these 19 standardized GDP series. The Figure shows that the GDP series of all countries grew rapidly before the 2008 financial crisis. Then, all countries were greatly affected by the financial crisis, although most of them resumed their growth soon, except for three countries that took another drop around 2013. These three economies are Greece, Portugal, and Italy. In fact, the GDP series of two countries have not returned to their pre-crisis levels.

We analyze the series using the program `dfmpc` to estimate the factors and found two common factors via the Lam and Yao test. The two factors are displayed in Figure 6.17. The first one explains 79.72% of the variability and the second 14.95%. These results are in agreement with the main eigenvalues of the lag covariance matrices, that, as the series are standardized, are the eigenvalues of the correlation matrices. These eigenvalues are shown in Table 6.2, and two factor are suggested.

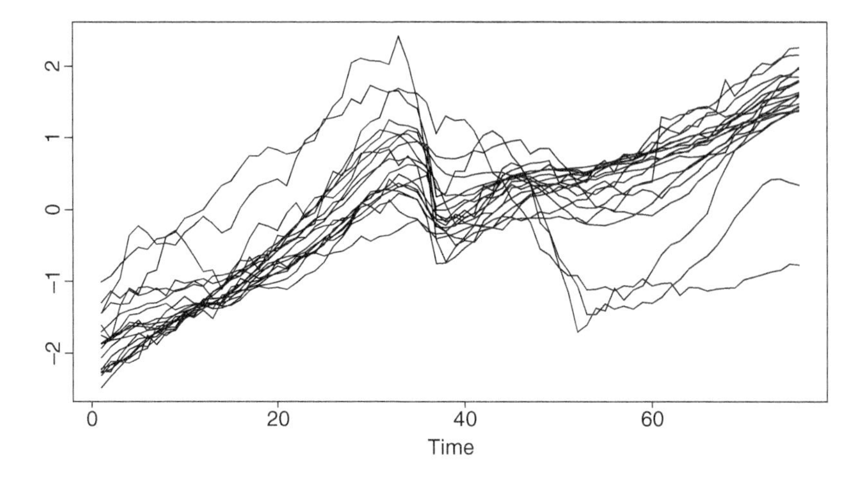

Figure 6.16 Time plots of the log GDP series for the 19 countries of the EMU.

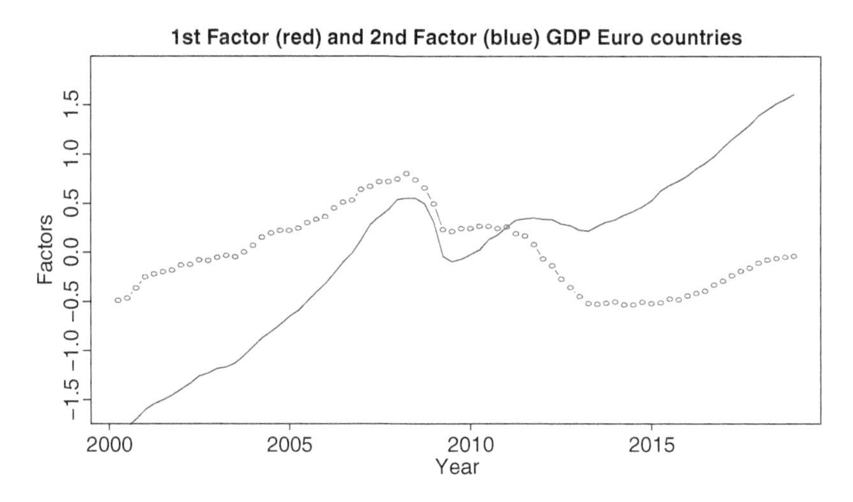

Figure 6.17 Time plot of the two factors for the standardized logGDP series for the 19 countries in the EMU. The first factor is in red and solid line and the second in blue and points line.

TABLE 6.2 Absolute Values of Eigenvalues of Lag Covariance Matrices for Standardized logGDP Series of 19 EMU Countries

	0	1	2	3	4	5
λ_1	0.298	0.282	0.266	0.250	0.234	0.218
λ_2	0.270	0.261	0.250	0.235	0.217	0.196
λ_3	0.070	0.067	0.065	0.062	0.061	0.060
λ_4	0.039	0.038	0.036	0.032	0.029	0.024
λ_5	0.008	0.007	0.006	0.004	0.003	0.003

The **R** commands to obtain the eigenvalues and eigenvectors of the lag correlation matrices are given below, where for $i = 0, \ldots, 5$ the matrix Si includes the lag-i covariance matrix, Ei the five largest eigenvalues of the matrix Si and $EVfi$ the eigenvectors linked to the first eigenvalue of the Si matrix and $EVsi$ those corresponding to the second eigenvector. Also the loadings for the two eigenvalues are stable for different lags. The estimated loadings for the factors are shown in Figure 6.18. The first factor is a weighted average of all the countries, although Greece (8), Italy (10), and Portugal (16) have smaller weights than the others. The second factor loadings differentiate countries with positive and negative weights: The countries with largest negative loading are Greece (8), Italy (10), and Malta (14).

R commands used for Example 6.5:

```
> x=as.matrix(UMEdata)
> n=nrow(x)
> G=matrix(0,n,19)
> for (i in 1:19){G[,i]=x[,1+3*(i-1)]}
> G0=log(G)
> G1=scale(G0)
> ##the data in logs and scale is in G1
> yl <- range(G1)
> tdx <- c(1:76)/4+2000
> plot(tdx,G1[,1],xlab="year",ylab="logGDP", main="GDP Euro countries",
```

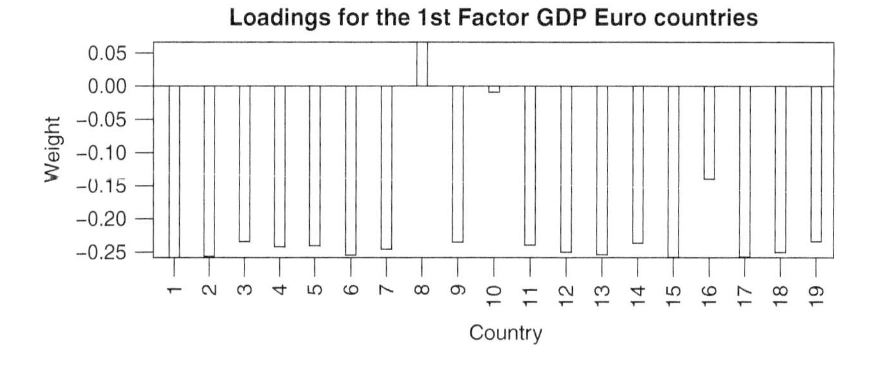

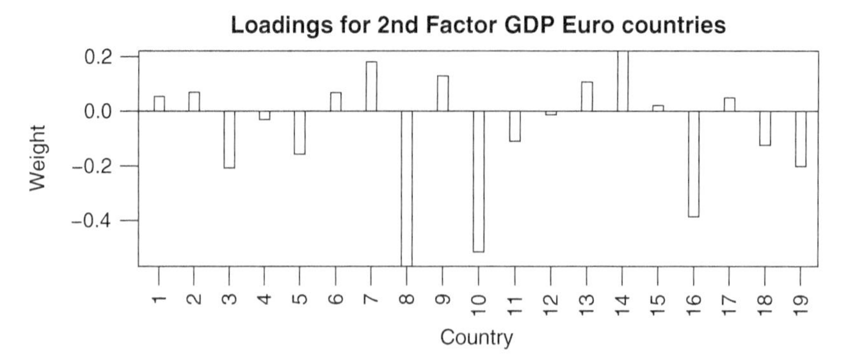

Figure 6.18 Loadings for the two factors of the logGDP series for the 19 countries of the EMU.

```
type="l", ylim=y1)
> for (i in 2:19){lines(tdx,G1[,i],col=i)}
> out=acf(G1,lag.max=5,type="covariance",plot=FALSE)
>
> S0=eigen(out$acf[1,,])
> E0=S0$values[1:5]
> EVf0=S0$vectors[,1]
> EVs0=S0$vectors[,2]
## Same commands to compute S1-S5 to EVs1-EVs5
> EVf=cbind(EVf0,EVf1,EVf2,EVf3,EVf4,EVf5)
> EVs=cbind(EVs0,EVs1,EVs2,EVs3,EVs4,EVs5)
> out=dfmpc(G1)
> out$r
[1] 2

> y1 <- range(out$F)
> tdx <- c(1:76)/4+2000
> plot(tdx,-out$F[,1],xlab="year",ylab="Factors", main="1stFactor (red)
and 2ndFactor (blue) GDP Euro countries", type="l",ylim=y1,col="red")
> lines(tdx,-out$F[,2], type="b", col="blue")
> par(mfrow=c(2,1))
> plot(out$L[,1],xlab="Countries",ylab="Loadings",
main="Loadings for the 1stFactor (red) GDP Euro countries",
type="b",col="red")
> plot(out$L[,2],xlab="Countries",ylab="Loadings",
main="Loadings for 2ndFactor (blue) GDP Euro countries",
type="o",col="blue")
```

■

6.5.1 Cointegration and DFM

Nonstationary vector of time series, as discussed in Chapter 3, can be cointegrated, and some linear combinations of the series are stationary; see, Banarjee et al. (1993) and Johansen (1995). A DFM with some integrated factors and stationary noises implies cointegration of the series $\boldsymbol{z}_t$. Letting $\boldsymbol{P}_1^{\perp}$ be the $k \times (k - r_1)$ orthogonal complement matrix of $\boldsymbol{P}_1$ so that $\boldsymbol{P}_1' \boldsymbol{P}_1^{\perp} = \boldsymbol{0}_{(r_1 \times k - r_1)}$, we have

$$\boldsymbol{P}_1^{\perp\prime} \boldsymbol{z}_t = \boldsymbol{P}_1^{\perp\prime} \boldsymbol{P}_2 \boldsymbol{f}_{2t} + \boldsymbol{P}_1^{\perp\prime} \boldsymbol{n}_t.$$

Since the right-hand side of the prior equation consists of linear combinations of stationary series, the $k - r_1$ series in $\boldsymbol{P}_1^{\perp\prime} \boldsymbol{z}_t$ are stationary. Indeed, these series represent the co-integrating series of $\boldsymbol{z}_t$. See, Escribano and Peña (1994) for further analysis. Thus, the DFM offers a straightforward approach to find the cointegration relations. See, Zhang et al. (2018) for a consistent proposal to estimate the cointegration rank via DFM. With many time series (i.e. a large k), the number of integrated factors, r_1, is expected to be relatively small and, therefore, the number of cointegrating relations, $k - r_1$, would be large. It has been argued, see Barigozzi et al. (2016, 2017), that the large number of cointegration relations implied by the DFM with stationary noises is not in agreement with empirical evidence, implying that the noises are likely to follow integrated processes in real applications. These authors also considered the possibility that the factors have a reduced rank representation or cointegration among the factors, and proposed a procedure for estimating them when some of the noises are cointegrated. The fact that the noises might be nonstationary implies that the covariance matrix of the noise in a FM may have some dominating eigenvalues, which can lead to difficulties in identifying the number of common factors by PCA or examining the ratios of eigenvalues. This problem has been addressed in Gao and Tsay (2018).

6.6 GDFM FOR NONSTATIONARY SERIES

There are few results of GDFM for nonstationary data, and most applications have been carried out by differencing the series in order to build a stationary GDFM. The estimation procedure based on DPCs in the frequency domain can only be applied to stationary series but there is a time domain approach with Generalized Dynamic Principal Components (GDPC), proposed by Peña and Yohai (2016), that does not require the stationarity assumption and can be applied to nonstationary time series. These generalized DPCs are optimal functions to reconstruct the data that are not restricted to be linear combinations of the observations. Note that both the DPC by Brillinger and the ODPC presented in Section 6.5 are linear combinations of the observed series. Dropping this restrictions the GDPC can be nonlinear functions of the observations and, therefore, can have a better adaptation to a nonstationary behavior of the series. Smucler (2019) proved that these GDPCs are consistent estimators of the common part of a GDFM.

6.6.1 Estimation by Generalized DPC

The GDPCs can be computed as follows. See, Peña and Yohai (2016) for a more detailed explanation.

Suppose that we observe series $z_{j,t}$ verifying $\sum_{t=1}^{T} z_{j,t} = 0$. Consider two positive integer numbers c_1 and c_2. We define the first GDPC with c_1 lags and c_2 leads as a vector $\boldsymbol{f} = (f_t)_{-c_1+1 \leq t \leq T+c_2}$, so that the reconstruction of series $z_{j,t}$ $(1 \leq j \leq k)$ as a linear combination of elements in $\{f_j | t - c_1 \leq j \leq t + c_2\}$ is optimal with respect to the MSE criterion. More precisely, the reconstruction of the original series $z_{j,t}$ is defined as

$$\hat{z}_{j,t} = \sum_{i=-c_1}^{c_2} \beta_{j,i} f_{t+i}.$$

We can write this expression using only lags of the component by calling $L = c_1 + c_2$ and defining as new component $f_t^* = f_{t+c_2}$, so that the values $(f_{t+c_2}, \ldots, f_{t-c_1})$ are transformed to $(f_t^*, \ldots, f_{t-L}^*)$, and calling $\beta_{j,i}^*$ the new coefficients corresponding to f_{t-i}^* that are related to the previous ones by $\beta_{j,i}^* = \beta_{j,c_2-i}$. Then, the reconstructed series can also be obtained as

$$\hat{z}_{j,t} = \sum_{i=-c_1}^{c_2} \beta_{j,i} f_{t+i} = \sum_{h=0}^{L} \beta_{j,h}^* f_{t-h}^*.$$

Thus, without loss of generality, we can use L lags of the PC to reconstruct the series. Defining the $(L+1) \times 1$ vector $\boldsymbol{f}_t = (f_t, \ldots, f_{t-L})'$ and the $k \times (L+1)$ matrix $\boldsymbol{B}$ that has rows $\boldsymbol{\beta}_j' = (\beta_{j,0}, \ldots, \beta_{j,L})$, we want to minimize

$$\text{MSE}(\boldsymbol{f}, \boldsymbol{B}) = \frac{1}{Tk} \sum_{j=1}^{k} \sum_{t=1}^{T} \left(z_{j,t} - \sum_{i=0}^{L} \beta_{j,i} f_{t-i} \right)^2 = \frac{1}{Tk} \sum_{t=1}^{T} (\boldsymbol{z}_t - \boldsymbol{B}\boldsymbol{F}_t)'(\boldsymbol{z}_t - \boldsymbol{B}\boldsymbol{F}_t). \quad (6.40)$$

Note that this loss function is well defined even in the case of nonstationary vector time series. The optimal choices of the $T+L$ vector $\boldsymbol{f} = (f_{1-L}, \ldots, f_T)'$ and $k \times (L+1)$ matrix $\boldsymbol{B}$ are defined by

$$(\widehat{\boldsymbol{f}}, \widehat{\boldsymbol{B}}) = \arg \min_{\boldsymbol{f} \in \mathbb{R}^{T+L}, \boldsymbol{B} \in \mathbb{R}^{k\times(L+1)}} \text{MSE}(\boldsymbol{f}, \boldsymbol{B}). \tag{6.41}$$

Clearly, if $\boldsymbol{f}$ is optimal, then $\gamma\boldsymbol{f} + \delta$ is optimal too. Thus, we can choose $\boldsymbol{f}$ so that $\sum_{t=1-L}^{T} f_t = 0$ and $(1/(T+L)) \sum_{t=1-L}^{T} f_t^2 = 1$. We call $\widehat{\boldsymbol{f}}$ the first (possible non stationary) GDPC of order L of the observed series $z_1, \ldots, z_t$. Note that the first GDPC of order 0 corresponds to the first regular PC of the data. As

$$\text{MSE}(\boldsymbol{f}, \boldsymbol{B}) = \frac{1}{Tk} \left[\sum_{t=1}^{T} z_t' z_t - 2 \sum_{t=1}^{T} z_t' \boldsymbol{B} \boldsymbol{F}_t + \sum_{t=1}^{T} \boldsymbol{f}_t' \boldsymbol{B}' \boldsymbol{B} \boldsymbol{F}_t \right].$$

Then

$$\frac{\partial \text{MSE}(\boldsymbol{f}, \boldsymbol{B})}{\partial \boldsymbol{B}} = 0 = -\sum_{t=1}^{T} \boldsymbol{f}_t z_t' + \sum_{t=1}^{T} \boldsymbol{f}_t \boldsymbol{f}_t' \boldsymbol{B}'. \tag{6.42}$$

Define $\boldsymbol{F}' = [\boldsymbol{f}_1, \ldots, \boldsymbol{f}_T]$ as the $(L+1) \times T$ matrix that includes values of all the lags the component needed to reconstruct the data Z. Equation (6.42) implies

$$\widehat{\boldsymbol{B}}' = (\boldsymbol{F}'\boldsymbol{F})^{-1} \boldsymbol{F}' \boldsymbol{Z}. \tag{6.43}$$

The partial derivative of Eq. (6.40) with respect to f_t takes into account that all the terms $z_{j,t}, \ldots, z_{j,t+L}$ are reconstructed using f_t so that squares and cross products of the $\beta_{j,i}$ coefficients will appear in the equation. It can be shown (see, Peña and Yohai, 2016) that the estimate of $\widehat{\boldsymbol{f}}$ can be written as

$$\widehat{\boldsymbol{f}} = \boldsymbol{D}(\boldsymbol{B})^{-1} \sum_{j=0}^{L} \boldsymbol{C}_j(\boldsymbol{Z}) \boldsymbol{\beta}_j, \tag{6.44}$$

where the squared $(T+L)$ matrix $\boldsymbol{D}(\boldsymbol{B})$ includes the cross products of the elements of the matrix $\boldsymbol{B}$, the matrix $\boldsymbol{C}_j(\boldsymbol{Z})$ includes elements of $\boldsymbol{Z}$ and has dimensions $(T+L) \times (L+1)$.

In order to define an iterative algorithm to compute $(\widehat{\boldsymbol{f}}, \widehat{\boldsymbol{B}})$, it suffices to give an initial value $\boldsymbol{f}^{(0)}$ and a rule describing how to compute $\boldsymbol{B}^{(h)}$ and $\boldsymbol{f}^{(h+1)}$ once $\boldsymbol{f}^{(h)}$ is known. For Eq. (6.43), given $\boldsymbol{f}^{(h)}$ the coefficients $\boldsymbol{B}^{(h)}$ can be obtained by the linear regression method

$$\boldsymbol{B}^{(h)} = [\boldsymbol{F}(\boldsymbol{f}^{(h)})' \boldsymbol{F}(\boldsymbol{f}^{(h)})]^{-1} \boldsymbol{F}(\boldsymbol{f}^{(h)})' \boldsymbol{Z},$$

and by Eq. (6.44), $\boldsymbol{f}^{*(h+1)}$ is computed by

$$\boldsymbol{f}^{*(h+1)} = \boldsymbol{D}(\boldsymbol{B}^{(h)})^{-1} \sum_{j=0}^{L} \boldsymbol{C}_j(\boldsymbol{Z}) \boldsymbol{\beta}_j^{(h)},$$

and the vector of the component is standardized to mean zero and variance one.

The initial value $\boldsymbol{f}^{(0)}$ can be set to the standard (nondynamic) first PC, completed with L zeros. We stop the iterations when

$$\frac{\mathrm{MSE}(\boldsymbol{f}^{(h)}, \boldsymbol{B}^{(h)}) - \mathrm{MSE}(\boldsymbol{f}^{(h+1)}, \boldsymbol{B}^{(h+1)})}{\mathrm{MSE}(\boldsymbol{f}^{(h)}, \boldsymbol{B}^{(h)})} < \varepsilon,$$

for some value ε.

Note that the dimension of the matrices to be inverted to compute $\boldsymbol{f}^{(h)}, \boldsymbol{B}^{(h)}$ is independent of the number of time series and, therefore, we can deal with a large number of series.

The second GDPC is computed using the residuals $r_{jt}(1) = z_{jt} - \hat{z}_{jt}(1)$, where $\hat{z}_{jt}(1)$ is the reconstructed value using the first GDPC. These residuals are used as observed data and the new component is computed. In general, if we have computed the hth GDPC $\hat{z}_{jt}(h)$, using the residuals $r_{jt}(h-1)$, the new data for computing the next $(h+1)$th component is $r_{jt}(h) = r_{jt}(h-1) - \hat{z}_{jt}(h)$. In this way, the components are uncorrelated for all the lags involved. Finally, the selection of the number of components and lags can be carried out as that in the ODPC case.

Example 6.6

We analyze, as in Example 6.5, the GDP series of the 19 Euro countries, using the levels directly. We apply the program `auto.gdpc` of the CRAN package `gdpc`. See Peña et al. (2020) for description of the package. We run program using the default option, where the number of GDPC or dynamic factors, and the lags required are chosen by cross-validation, as explained before. A single factor is selected with 10 lags, that explains 99.2% of the variability. Note that this factor explains by itself a larger proportion of the variability than the two factors estimated in Example 6.5, 94.67% (79.72 + 14.95% = 94.67%). Figure 6.19 shows the factor, and Figure 6.20 the loadings of the series in the factor for the 11 lags, including lag zero.

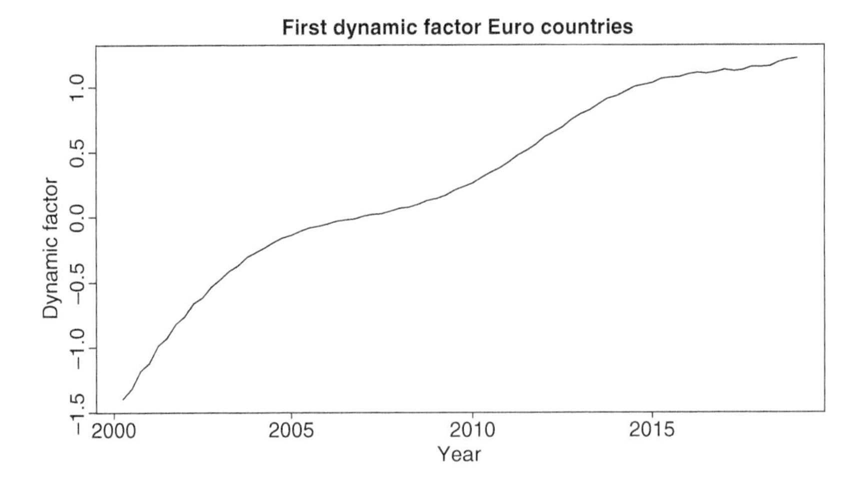

Figure 6.19 The first GDPC for the standardized logGDP series for the 19 countries in the EMU.

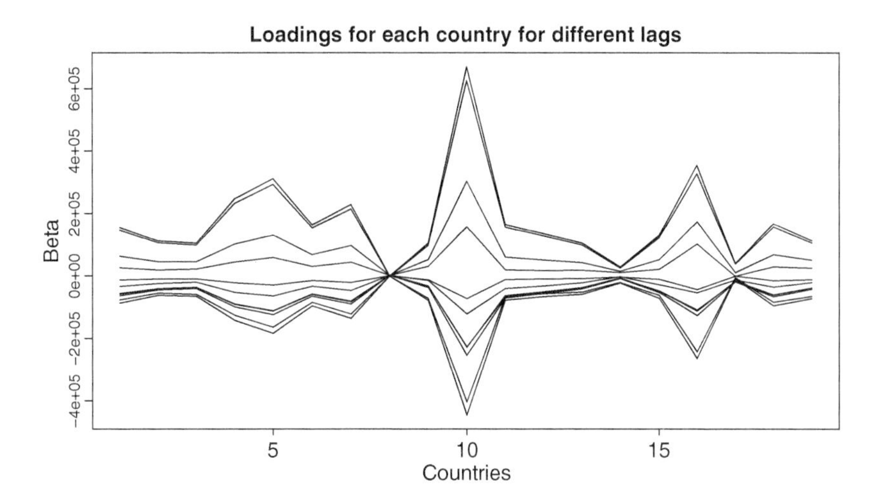

Figure 6.20 Loadings for the GDPC with 10 lags of the logGDP series for the 19 countries of the EMU.

R commands used for Example 6.6:

```
>require(gdpc)
>x=as.matrix(UMEdata)
> n=nrow(x)
> G=matrix(0,n,19)
> for (i in 1:19){G[,i]=x[,1+3*(i-1)]}
> G0=log(G)
> G1=scale(G0)
> sal<- auto.gdpc(G1, niter_max=1000)
> sal
#the summary of the fitted GDPC:
    Number.of.lags   LOO   MSE Explained.Variance
Component 1             10 0.014 0.008              0.992
#sal[[1]] has the information of the 1st component, sal[[2]] of the
#second component and so on
> f1<- sal[[1]]$f
>  beta1<- sal[[1]]$beta
> y1 <- range(f1)
> tdx <- c(1:76)/4+2000
>  plot(tdx,f1,xlab="year",ylab="Dynamic Factor",
main="First Dynamic Factor Euro countries", type="l", ylim=y1)
>  ts.plot(beta1,xlab="countries",ylab="beta",
main="Loadings for each country for different lags")
```

■

6.7 OUTLIERS IN DFMS

6.7.1 Factor and Idiosyncratic Outliers

We consider two types of outliers in DFM. They are outliers in the factors, that affect all or most of the time series, and outliers in the idiosyncratic components, that only influence one or a few time series. These two types of outliers are similar to the multivariate and univariate outliers studied in Chapter 4. However, by incorporating the structure of the DFM in the estimation, the power of the detection procedure increases.

Let $\boldsymbol{f}_t = (f_{1t}, \ldots, f_{rt})'$ be the vector of dynamic factors. A factor outlier appears in the jth dynamic factor at time index $t = o$ if, instead of $\boldsymbol{f}_t$, we have a set of latent factors $\boldsymbol{f}_t^* = (f_{1t}^*, \ldots, f_{rt}^*)'$ given by

$$\boldsymbol{f}_t^* = \boldsymbol{f}_t + \boldsymbol{w}_j^{(o)} I_t^{(o)}, \tag{6.45}$$

where $\boldsymbol{w}_j^{(o)} = (0, \ldots, 0, w_j^{(o)}, 0, \ldots, 0)'$ is the size of the outlier in the jth dynamic factor and $I_t^{(o)}$ is an indicator variable for time $t = o$; see Chapter 4. The factor outlier in Eq. (6.45) can potentially affect all the time series components, depending on the loadings. Note also that all the outliers of the same factor have effects on the series that are proportional to the loadings of this factor. The observed time series is

$$\boldsymbol{z}_t = \boldsymbol{P}\boldsymbol{f}_t^* + \boldsymbol{n}_t = \boldsymbol{P}(\boldsymbol{f}_t + \boldsymbol{w}_k^{(o)} I_t^{(o)}) + \boldsymbol{n}_t = \boldsymbol{y}_t + \boldsymbol{P}\boldsymbol{w}_j^{(o)} I_t^{(o)},$$

where $\boldsymbol{y}_t$ is the outlier-free series and $\boldsymbol{P}\boldsymbol{w}_j^{(o)} = w_j^{(o)} \boldsymbol{p}_{.j}$ is the effect of the outlier.

On the other hand, an idiosyncratic outlier is an additive outlier in a single idiosyncratic noise component. Consequently, if $\boldsymbol{n}_t = (n_{1t}, \ldots, n_{kt})'$ is a vector of idiosyncratic noises, an idiosyncratic outlier appears in the lth idiosyncratic noise component at time $t = i$ if instead of $\boldsymbol{n}_t$, we have $\boldsymbol{n}_t^* = (n_{1t}^*, \ldots, n_{kt}^*)'$ given by

$$\boldsymbol{n}_t^* = \boldsymbol{n}_t + \boldsymbol{w}_l^{(i)} I_t^{(i)}, \tag{6.46}$$

where $\boldsymbol{w}_l^{(i)} = (0, \ldots, 0, w_l^{(i)}, 0, \ldots, 0)'$ is the size of the idiosyncratic outlier and $I_t^{(i)}$ is an indicator variable for $t = i$. The idiosyncratic outlier in Eq. (6.46) only affects one of the time series because the observed time series is

$$\boldsymbol{z}_t = \boldsymbol{P}\boldsymbol{f}_t + \boldsymbol{n}_t^* = \boldsymbol{P}\boldsymbol{f}_t + (\boldsymbol{n}_t + \boldsymbol{w}_l^{(i)} I_t^{(i)}) = \boldsymbol{y}_t + \boldsymbol{w}_l^{(i)} I_t^{(i)}. \tag{6.47}$$

Therefore, an idiosyncratic outlier in the lth noise component at time $t = i$ produces an outlier of size $w_l^{(i)}$ in the lth observed time series.

In practice, the observed time series $\{\boldsymbol{z}_1, \ldots, \boldsymbol{z}_t\}$, may be affected by the presence of several factors and/or idiosyncratic outliers as follows

$$\begin{aligned}\boldsymbol{z}_t &= \boldsymbol{P}\left(\boldsymbol{f}_t + \sum_{i=1}^{F} \boldsymbol{w}_{j_i}^{(o_{j_i})} I_t^{(o_{j_i})}\right) + \left(\boldsymbol{n}_t + \sum_{j=1}^{I} \boldsymbol{w}_{l_j}^{(i_j)} I_t^{(i_j)}\right) \\ &= \boldsymbol{y}_t + \sum_{i=1}^{F} \boldsymbol{P}\boldsymbol{w}_{j_i}^{(o_{j_i})} I_t^{(o_{j_i})} + \sum_{j=1}^{I} \boldsymbol{w}_{l_j}^{(i_j)} I_t^{(i_j)},\end{aligned} \tag{6.48}$$

for $t = 1, \ldots, T$, where $\boldsymbol{w}_{j_1}^{(o_1)}, \ldots, \boldsymbol{w}_{j_F}^{(o_r)}$ are the sizes of the F factor outliers in the j_1th,...,j_Fth dynamic factors at times $o_1, \ldots, o_F$, and $\boldsymbol{w}_{l_1}^{(i_1)}, \ldots, \boldsymbol{w}_{l_I}^{(i_I)}$ are the sizes of the I idiosyncratic outliers in the l_1th,...,l_Ith idiosyncratic noise components at times $i_1, \ldots, i_I$, respectively. Note that expression in Eq. (6.48) is flexible enough to cover situations such as a set of factor outliers affecting different dynamic factors at the same time point or a factor outlier and an idiosyncratic outlier at the same time point.

The effects of these outliers are to introduce biases in the sample covariance matrix that can be severe if some of the outliers are large.

6.7.2 A Procedure to Find Outliers in DFM

Galeano et al. (2020) have proposed a procedure for outlier detection in stationary high-dimensional time series divided into five steps as follows:

1. Perform an initial cleaning of the time series $\{\boldsymbol{x}_1^*, \ldots, \boldsymbol{x}_T^*\}$, leading to the data $\{\boldsymbol{x}_1^\circ, \ldots, \boldsymbol{x}_T^\circ\}$.
2. Use the cleaned data to obtain robust estimates of the number of factors r, the factor loading matrix $\boldsymbol{P}$, and its orthogonal complement matrix $\boldsymbol{V}$, denoted by $\hat{r}$, $\hat{\boldsymbol{P}}$, and $\hat{\boldsymbol{V}}$, respectively.
3. Project the time series $\{\boldsymbol{x}_1^*, \ldots, \boldsymbol{x}_T^*\}$ with $\hat{\boldsymbol{V}}'$ to detect and clean the series from idiosyncratic outliers, leading to a time series $\{\boldsymbol{x}_1^\diamond, \ldots, \boldsymbol{x}_T^\diamond\}$.
4. Project the time series $\{\boldsymbol{x}_1^\diamond, \ldots, \boldsymbol{x}_T^\diamond\}$ with $\hat{\boldsymbol{P}}'$ to detect and clean the series from factor outliers, leading to a time series $\{\boldsymbol{x}_1^\bullet, \ldots, \boldsymbol{x}_T^\bullet\}$.
5. Use the time series $\{\boldsymbol{x}_1^\bullet, \ldots, \boldsymbol{x}_T^\bullet\}$ to estimate again the factor loading matrix $\boldsymbol{P}$ and its orthogonal complement matrix $\boldsymbol{V}$, and repeat Steps 3-5 until no more outliers are found.

These five steps can be carried out automatically. In the first step of the procedure, an initial time series cleaning of the observed data is carried out by the Galeano et al. (2006) procedure explained in Chapter 4. In the second step, the number of factors r, the loading matrix $\boldsymbol{P}$, and its orthogonal complement $\boldsymbol{V}$ are estimated with the clean data obtained in step 1. In the third, the idiosyncratic outliers are detected and their effects are removed from the observed time series $\{\boldsymbol{x}_1^*, \ldots, \boldsymbol{x}_T^*\}$. This is carried out by testing for an idiosyncratic outlier in each of the time series and in each of the time points using the information of the transformed time series $\boldsymbol{z}_t = \hat{\boldsymbol{V}}'\boldsymbol{x}_t^*$, for $t = 1, \ldots, T$. In the fourth step, we take advantage that the idiosyncratic outliers have been removed from the time series $\{\boldsymbol{x}_1^\diamond, \ldots, \boldsymbol{x}_T^\diamond\}$ to estimate the dynamic factors with $\boldsymbol{f}_t^\diamond = \hat{\boldsymbol{P}}'\boldsymbol{x}_t^\diamond$, for $t = 1, \ldots, T$, where $\hat{\boldsymbol{P}}$ is the robust estimate of $\boldsymbol{P}$ obtained in step 2, and then, we use the algorithm for univariate time series to detect the factor outliers in each dynamic factor. In the last step of the procedure, we iterate the previous steps until no more outliers are found.

Note that an important difference from the projection procedure of Chapter 4 is that here we project the series on the factor space and on the orthogonal noise space, instead of projecting them in the extreme kurtosis directions. The DFM leads the search for each type of outlier whereas in the previous model-free algorithm we just want to clean the data from serious errors. Thus, the previous procedure can be applied in step 1 of the algorithm presented in this section to make an initial cleaning of the data and compute a robust estimation of the covariance matrix.

Outliers in DFM were first studied by Baragona and Battaglia (2007). Breitung and Eickmeier (2011) considered level shits or structural breaks in these models, and

Cheng et al. (2016) proposed a shrinkage estimation for level shifts. Missing value estimation in DFM has been studied by Bai and Ng (2019).

6.8 DFM WITH CLUSTER STRUCTURE

Often, there are factors that affect all the series, whereas others are group specific and affect only some group of the series. For instance, we may have groups of countries with various degrees of economic developments, inter-connected industrial sectors, and bilateral trade. Then, the dynamic evolution of economic time series of those countries may be affected by some general factors, that reflect the evolution of the global economy, and some specific factors that are group-dependent. In this case, we have a DFM with cluster structure (DFMCS). Such models have been studied, among others, by Wang (2010), Hallin and Liška (2011), Lin and Ng (2012), Bonhomme and Manresa (2015), Su et al. (2016), Ando and Bai (2016, 2017), and Alonso et al. (2020).

Partition the k-dimensional time series as $\boldsymbol{z}_t = (\boldsymbol{z}'_{1t}, \boldsymbol{z}'_{2t}, \dots, \boldsymbol{z}'_{ct})'$, where $\boldsymbol{z}_{it}$ is a k_i-dimensional time series such that $\sum_{i=1}^{c} k_i = k$. In other words, we have c clusters of time series and the ith cluster has k_i time series. The DFMCS can be written as

$$\boldsymbol{z}_t = \boldsymbol{P}_0 \boldsymbol{f}_{0t} + \sum_{i=1}^{c} \boldsymbol{P}_i \boldsymbol{f}_{it} + \boldsymbol{n}_t, \tag{6.49}$$

where $\boldsymbol{n}_t$ is the vector of noises, $\boldsymbol{f}_{0t} = (f_{01t}, \dots, f_{0r_0t})'$ is an r_0-dimensional vector of global factors, $\boldsymbol{P}_0$ is a $k \times r_0$ global loading matrix, $\boldsymbol{f}_{it} = (f_{i,1t}, \dots, f_{i,r_it})'$ is an r_i-dimensional common factor of the ith cluster and $\boldsymbol{P}_i$ is a $k \times r_i$ loading matrix for the ith cluster such that $\boldsymbol{P}_i = [\boldsymbol{0}'_{i-1}, \boldsymbol{w}'_i, \boldsymbol{0}'_{c-i}]'$, where $\boldsymbol{0}_{i-1}$ is a zero matrix of dimension $(\sum_{j=1}^{i-1} k_i)$ by-r_i provided that $i > 1$, $\boldsymbol{w}_i$ is a $k_i \times r_i$ loading matrix for the ith cluster, and $\boldsymbol{0}_{c-i}$ is another zero matrix of dimension $(\sum_{j=i+1}^{c} k_i)$-by-r_i provided that $i < c$. In this way, $\boldsymbol{P}_i \boldsymbol{f}_{it}$ only affects time series in the ith cluster. The total number of factors is $r = r_0 + r_1 + \cdots + r_c$.

The identification of the DFMCS in Eq. (6.49) has been studied by Wang (2010) and the conditions needed are: (1) $\boldsymbol{P}'_0 \boldsymbol{P}_0 = \boldsymbol{I}_{r_0}$, where $\boldsymbol{I}_{r_0}$ is the identity matrix of order r_0; (2) $\boldsymbol{P}'_i \boldsymbol{P}_i = \boldsymbol{w}'_i \boldsymbol{w}_i = \boldsymbol{I}_{r_i}$ for $i = 1, \dots, c$; (3) $\boldsymbol{P}'_0 \boldsymbol{P}_i = \boldsymbol{0}_{r_0 \times r_i}$; and (4) the covariance matrix of the $r = \sum_{j=0}^{c} r_j$ factors is diagonal. Note that also, by definition, $\boldsymbol{P}'_i \boldsymbol{P}_j = \boldsymbol{0}_{r_i \times r_j}$, for $i \neq j$. We can write this model as a standard DFM. Letting $\boldsymbol{f}_t = (\boldsymbol{f}'_{0t}, \boldsymbol{f}'_{1t}, \dots, \boldsymbol{f}'_{ct})'$, and $\boldsymbol{P} = [\boldsymbol{P}_0 | \boldsymbol{P}_1 | \cdots | \boldsymbol{P}_c]$, we have

$$\boldsymbol{z}_t = \boldsymbol{P} \boldsymbol{f}_t + \boldsymbol{n}_t, \tag{6.50}$$

and the previous conditions implied the usual identification restriction $\boldsymbol{P}'\boldsymbol{P} = \boldsymbol{I}_r$.

The assumptions for the idiosyncratic term and the factors are the same as before. The idiosyncratic term or noise, $\boldsymbol{n}_t = (n_{1t}, \dots, n_{kt})'$, is a general sequence of stationary time series with mean $\boldsymbol{0}_k$ and weak serial dependency as in the ADFM. The global and specific factors are orthogonal to each other and follow a diagonal VARMA. Additionally, we assume that both innovation processes appearing in the FM are uncorrelated for all lags. However, the number of clusters and the allocation of the series to the clusters are unknown.

The estimation of a DFMCS requires obtaining the following parameters: (1) the number of global factors, r_0, the number of groups, c, and the number of specific factors in each of the c groups, $r_1, \ldots, r_c$; (2) the label variable $g_i \in \{1, \ldots, c\}$, indicating to which group the series belongs, and we call $\boldsymbol{G}$ the $k \times 1$ vector with components g_i, for $i = 1, \ldots, k$; (3) the loading matrices of the global and specific factors, $\boldsymbol{P}_0, \boldsymbol{P}_1, \ldots, \boldsymbol{P}_c$ and the time series of these factors, $\boldsymbol{f}_{0t}, \boldsymbol{f}_{1t}, \ldots, \boldsymbol{f}_{ct}$. Given the estimated factors and noises, where $\widehat{\boldsymbol{n}}_t = \boldsymbol{z}_t - \widehat{\boldsymbol{P}}\widehat{\boldsymbol{f}}_t$, estimators of the parameters of the scalar ARMA models for the factors and noises can be obtained.

6.8.1 Fitting DFMCS

The procedure we consider has the following three steps:

1. An initial set of factors and their loadings are estimated with all the time series and used to build the common component of each time series. Then, a cluster by dependency using the generalized cross-correlation (GCC) method (proposed in Chapter 5) is applied to these common components to find the groups.
2. A new set of factors and their loadings are estimated in each group of time series found in the previous step. All the factors found in the groups as well as those found in step 1 are compared and classified as global or specific;
3. The effect of the global factors is removed from each of the time series and the group-specific residuals obtained are used to re-estimate the specific factors. Finally, groups are checked for possible recombination.

In the first step, the initial estimation of the factors and loadings is made with PCA applied to the sample covariance matrix of the time series $\widehat{\boldsymbol{\Gamma}}_z(0)$. As we have seen, the eigenvectors associated to the r_* largest eigenvalues of this matrix provide us with an estimate of the factor loading matrix, $\widehat{\boldsymbol{P}}$, with columns $\widehat{\boldsymbol{p}}_1, \ldots, \widehat{\boldsymbol{p}}_{r_*}$. The number of factors, r_*, is determined by using the test proposed by Ahn and Horenstein (2013) based on the ratios of consecutive eigenvalues of the matrix $\widehat{\boldsymbol{\Gamma}}_z(0)$. Note that the matrix $\widehat{\boldsymbol{P}}$ is expected to include all the global factors and some (or all) of the group-specific factors, so that the number of factors in this matrix, r_*, is in general larger than the true number of global factors, r_0. Also, in practice, when the factors are of different degree of strength, the results of ratio of eigenvalues tests are usually better if they are applied twice, first to detect a few dominant factors and second to find the others, as suggested by Lam and Yao (2012). Then, the factors are estimated by $\widehat{\boldsymbol{f}}_{i,t} = \widehat{\boldsymbol{p}}_i' \boldsymbol{z}_t$, for $i = 1, \ldots, r_*$, and the common component by $\boldsymbol{f}_t = \widehat{\boldsymbol{P}}\widehat{\boldsymbol{P}}' \boldsymbol{z}_t$.

The groups are now built applying a hierarchical clustering algorithm with single linkage to the dissimilarity matrix using the GCC. The number of clusters is obtained by a modification of the Silhouette algorithm adding the restriction that the clusters must have a minimum size. We implement this restriction by omitting time series in the dendrogram analysis that have a relatively small dependency with the rest (for instance, the 90% percentile of the dendrogram's unions). Once the groups are formed, the omitted time series are assigned to the closer cluster in the single linkage sense. In this way, we obtain an estimated value of c, the number of groups, and an estimator of the vector $\boldsymbol{G}$ that gives the allocation of the series to the c groups.

In the second step, we use the series in the groups found in the previous step to estimate new sets of factors and their loadings. Let $r_1^s, \ldots, r_c^s$ be the number of factors found in each group by using the Ahn and Horenstein (2013) test applied to the eigenvalues of the sample covariance matrices of the time series in each group. The specific loading matrices $\widehat{\boldsymbol{P}}_i$ of dimension $k \times r_i^s$ and columns $\widehat{\boldsymbol{p}}_{i1}, \ldots, \widehat{\boldsymbol{p}}_{ir_i^s}$ are built by adding to the eigenvectors corresponding to the largest r_i^s eigenvalues in the ith group, a set of zero values for the observations not included in the group. The factors in each group are estimated by $\widehat{\boldsymbol{f}}^s_{ij,t} = \widehat{\boldsymbol{p}}'_{ij}\boldsymbol{z}_t$, with $j = 1, \ldots, r_i^s$. These group factors are expected to include all the specific factors and some (or all) of the global factors.

Next, in order to decide whether a factor is global or specific, we compare the set of r_* factors found in step 1 and the set of $\sum_{i=1}^{c} r_i^s$ factors found in this second step. Note that the factors contained in the first set may be a rotation of the factors contained in the second set and, therefore, it is not evident which ones should be classified as global and which as specific. Consequently, we first decide if each factor f_t in the first set of r_* factors is global or specific by applying the following three simple rules:

1. If f_t does not belong to any of the second set of factors then it is a global factor.
2. If f_t belongs to only one of the sets of the second set of factors then it is a specific factor in this group.
3. If f_t belongs to more than one of the second set of factors then it is a global factor.

We decide if a factor, f_t, belongs to a set of specific factors by computing the empirical canonical correlation between the factor, f_t, and the ones in the set, $\widehat{\boldsymbol{f}}^s_{i1,t}, \ldots, \widehat{\boldsymbol{f}}^s_{ir_i^s,t}$, with $i = 1, \ldots, c$. When the empirical canonical correlation of factor f_t with elements of the set is higher than some threshold value, ρ_0, we say that f_t belongs to this set. The threshold value of $\rho_0 = 0.9$ seems to work well in our Monte Carlo exercise. Afterwards, we check if any of the groups with $r_1^s, \ldots, r_c^s$ factors include any factor that does not belong to the set of factors found in step 2. If this is the case, the factor is classified as specific factor in the corresponding group.

In the third step, we compute the residuals $\boldsymbol{v}_t = \boldsymbol{z}_t - \widehat{\boldsymbol{P}}_0\widehat{\boldsymbol{f}}_{0t}$, where $\widehat{\boldsymbol{f}}_{0t}$ is the vector of estimated global factors obtained in step 3 and $\widehat{\boldsymbol{P}}_0$ is the loading matrix corresponding to these factors, and the specific factors are re-estimated using the series v_{it} corresponding to each group. Then, we verify that the groups obtained are due to different specific factors and not due to differences between factor loadings in a global factor. This is made by checking whether all the groups have at least one specific factor. We may face the following cases: (1) All the c groups found include at least one specific factor, and we have a DFMCS with c groups; (2) c_1 groups, $(1 \leq c_1 < c)$ contain specific factors, and $c_2 = c - c_1$ groups only contain global factors, then we have a DFMCS with $c_1 + 1$ groups; and (3) All the groups only contain global factors, then we have the standard DFM.

To justify this last verification to check that the groups are generated by different specific factors and not for different loadings of the global factors, consider the simple model

$$\begin{bmatrix} \boldsymbol{z}_{1t} \\ \boldsymbol{z}_{2t} \end{bmatrix} = \begin{bmatrix} a\mathbf{1}_{k_1} \\ b\mathbf{1}_{k_2} \end{bmatrix} f_t + \begin{bmatrix} \boldsymbol{n}_{1t} \\ \boldsymbol{n}_{2t} \end{bmatrix},$$

where $\boldsymbol{z}_{it}$ is $k_i \times 1$, with $k = k_1 + k_2$, and $\mathbf{1}_{k_i} = (1, \ldots, 1)'$ is also $k_i \times 1$. In this model, there are two groups of time series of similar dependency. For simplicity, let us assume that the noises are i.i.d. with the same variance σ^2 and let us call $s = \text{Var}(f_t)/\sigma^2$ the signal-to-noise ratio. Then, the cross-correlations of any two variables in the first and second group of series will be respectively $r_1 = a^2s/(1 + a^2s)$ and $r_2 = b^2s/(1 + b^2s)$, while the correlation between series in different groups is $r_{12} = abs/\sqrt{(1 + a^2s)(1 + b^2s)}$. Consequently, if a is very different from b, the clustering should detect two groups of time series.

Finally, given the estimated factors, loadings $(\widehat{\boldsymbol{P}}_0\widehat{\boldsymbol{f}}_{0t}, \widehat{\boldsymbol{P}}_1\widehat{\boldsymbol{f}}_{1t}, \ldots, \widehat{\boldsymbol{P}}_c\widehat{\boldsymbol{f}}_{ct})$, and groups, we can estimate AR models for the factors and compute the residuals or idiosyncratic component, $\widehat{\boldsymbol{n}}_t = z_t - \widehat{\boldsymbol{P}}_0\widehat{\boldsymbol{f}}_{0t} - \sum_{i=1}^{c} \widehat{\boldsymbol{P}}_i\widehat{\boldsymbol{f}}_{it}$, and also fit AR($p$) models to them.

Example 6.7

We analyze a dataset of hourly day-ahead demand for the ISO New England electricity market from January 2004 to December 2016. The dataset is available in the file `PElectricity1344.csv` and was presented in Example 2.1. The New England region is divided into eight load zones: Connecticut (CT), Maine (ME), New Hampshire (NH), Rhode Island (RI), Vermont (VT), Northeastern Massachusetts and Boston (NEMA), Southeastern Massachusetts (SEMA), and Western/Central Massachusetts (WCMA).

Each of the time series, D_{it}, for $1 \leq i \leq 1344$ and $1 \leq t \leq 678$, corresponds to the demand of electricity in one of the eight regions at one of the 24 hours of one of the days of the week, that is, we have $8 \times 24 \times 7 = 1344$ time series. The number of points in each series is $T = 678$ that corresponds to about 52 weeks for 13 years. The series require a seasonal difference and a logarithm transformation to become stationary (see, García-Martos and Conejo, 2013). Also, as some values of the series are negative, we add 15 to all the data before taking the log transformation, and the series analyzed are $z_{it} = \nabla_7 \log(D_{it} + 15)$.

Figure 6.21 shows a plot of the series in logarithm and with a difference. More plots of the series were presented in Example 2.1. Some clear outliers are shown in these

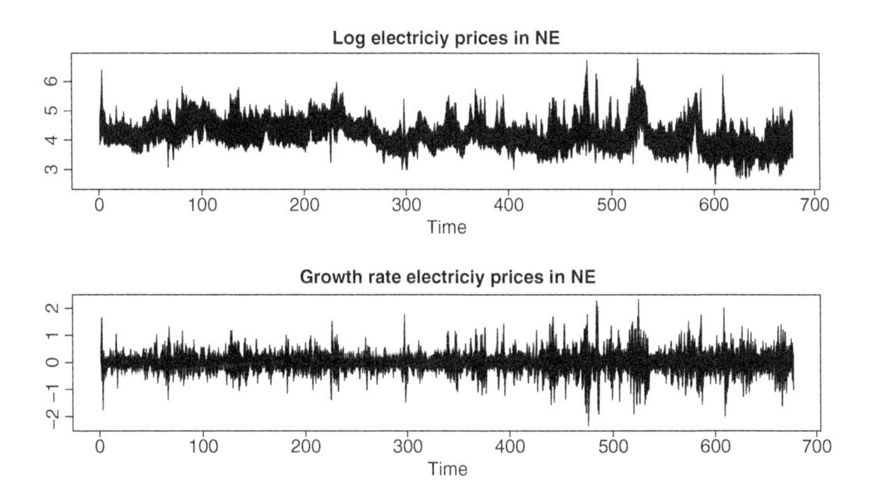

Figure 6.21 Time plots of 1344 weekly series of electricity prices in New England for 13 years.

plots, and to illustrate the relevance of outlier detection, we summarize the cluster solution with the original time series, z_{it}, to compare it with the one obtained with the outlier-corrected time series, z^c_{it}. Three factors are selected for the original series z_{it} using the two-step Ahn and Horenstein's procedure and it was clear that hour 02:00 had a different behavior from the others. Indeed, the first factor is essentially measuring the effect of this particular hour.

We then apply the cleaning procedure to z_{it}. In the first step with projections in the directions of maximum kurtosis, the procedure identifies 70 multivariate additive outliers (MAOs, see Chapter 4) in seven projected time series, an average of 10 outliers per projected time series. Once the outlier effects are removed, the second step with projections in the directions of minimum kurtosis, the procedure identifies 23 MAOs in five projected time series, i.e. 4.6 outliers in average per projected time series. After removing these outliers, the third step with random projections finds 20 MAOs in 12 random projections, i.e. 1.66 outliers in average per projected time series. Finally, the fourth step using univariate search discovers 59 additive outliers in 47 out of the 192 series, i.e. 1.25 outliers on average per series. In summary, the procedure detects 113 MAOs and 59 univariate additive outliers and cleans 2.38% of the total number of data points among all 192 time series.

The largest number of MAOs appears on Sundays and Mondays. In fact, 24 out of the 56 MAOs detected on Sundays correspond to the daylight saving time days, where the demand is set to zero at hour 02:00.

With the 192 outlier corrected series, the two-step Ahn and Horenstein test finds two factors, that explain 77.1 and 8.8% of the total variability, respectively. The loadings of these two factors are shown as curves in Figure 6.22. The first factor is essentially a weighted average of the series with *similar weights* (in the range from 0.037 to 0.100) across the 24 hours. Therefore, it reproduces the average global dynamic of the

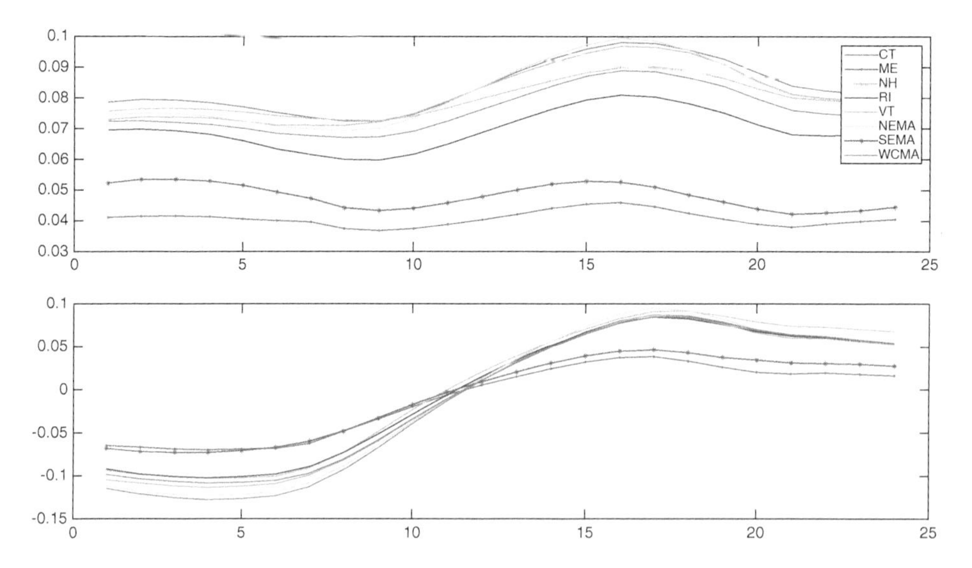

Figure 6.22 Estimated loadings for the two initial factors for the outlier corrected series. The loadings are shown as functions of the 24 hours of the day and each of the eighth curves represents one of the zones.

differentiated series. Regarding the effect of each hour, it gives more relative weights to the afternoon (13:00 to 19:00). The second factor gives negative weights to series from the 1st to the 11th hours and positive weights to series from the 12th to 24th hours. Also, it differentiates between the night (1:00 to 7:00) and the rest of the hours, with a peak in 17:00 to 19:00. Note that these factors are not the same across regions, because the loadings for the second (ME) and seventh (SEMA) zones are different from the others. Thus, if we do not consider the presence of clusters in the data, we may conclude that a DFM with two factors seems to be appropriate for the data.

We search for clusters using the GCC of the series. First, to see the effect of the outliers, we apply this measure to the original series. Figure 6.23(a) shows the resulting dendrogram. The demand series of the second hour (blue cluster) appear at the top, revealing that these series are far away from the others. No other groups are found. On the other hand, the dendrogram for the outlier corrected time series in Figure 6.23(b) shows clearly two clusters. The first cluster (in red) contains the time series of demand for hours 11:00 to 24:00 and the second one (in green) contains the ones for hour 01:00 to 10:00. The Silhouette statistics also indicates two clusters. Thus, we conclude that the series after outlier corrections form two groups: the first one broadly includes series in daylight hours and the second one in the night hours.

When the two-step Ahn and Horenstein procedure is applied to series within the two groups, seven factors are obtained in each cluster. These seven factors explain approximately 96.8 and 97.6% of the variability of the series at the first and the second cluster, respectively. As some of these factors may be global and others are specific, we compare the 2 factors found with all the series shown early and the 14 factors

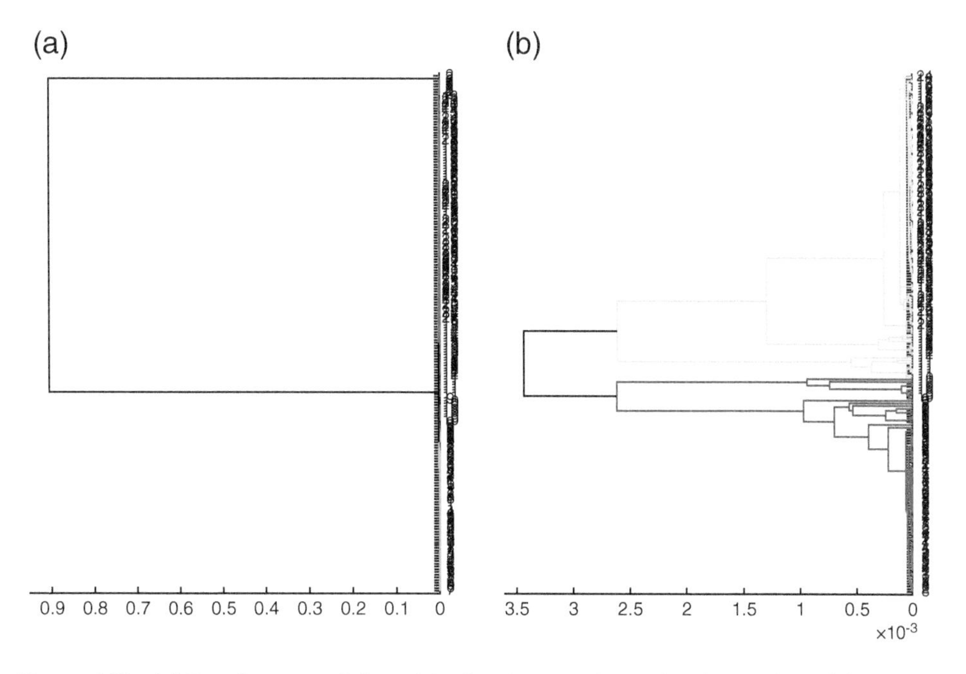

Figure 6.23 (*a*) Dendrogram of the original series, z_{it}, of the electricity prices. (*b*) Dendrogram for the outlier corrected series, z^c_{it}.

found in the two clusters using the rules presented in Section 6.8.1. The two initial factors are classified as global factors. The first one has canonical correlations with the factors in the two clusters 0.984 and 0.967, respectively. The second one has weaker correlations, 0.673 and 0.799, respectively, but its canonical correlation with the set of all the specific factors is almost one. This implies that its effect is distributed among several factors found in the groups. Now, we apply step 4 of the procedure and obtain the $R_{it} = z^c_{it} - \widehat{\boldsymbol{P}}_0\widehat{\boldsymbol{f}}_{0t}$, where $\widehat{\boldsymbol{P}}_0$ and $\widehat{\boldsymbol{f}}_{0t}$ are the estimated loadings and factors, respectively, for the two global factors. The Ahn and Horenstein's test applied to R_{it}, for series in each cluster, obtains six and five factors for the first and the second cluster, respectively. These factors are clearly specific and orthogonal to the two global factors.

Figures 6.24 and 6.25 show the loadings for these specific factors in the two groups. Note that two extreme zones from the geographical point of view, Maine (ME) and Southeastern Massachusetts (SEMA), have the largest effect in almost all the specific factors in both clusters, whereas for the global factors the situation was just the opposite: these zones have the smallest weights in the two global factors in Figure 6.22. Regarding the effect of each hour, a richer picture appears in the structure of these group factors with respect to the global ones. In group 1, the first three factors give more weights to hours from 11:00 to 18:00 than those from 19:00 to 24:00, and factors four and six account for a peak in electricity demand when most people return home, hours 17:00–18:00. In the second cluster, the first two factors have opposite peaks of demand around 1:00–2:00 and 7:00–8:00. The other three factors have small variability in the hours but they differentiate strongly among the eight zones.

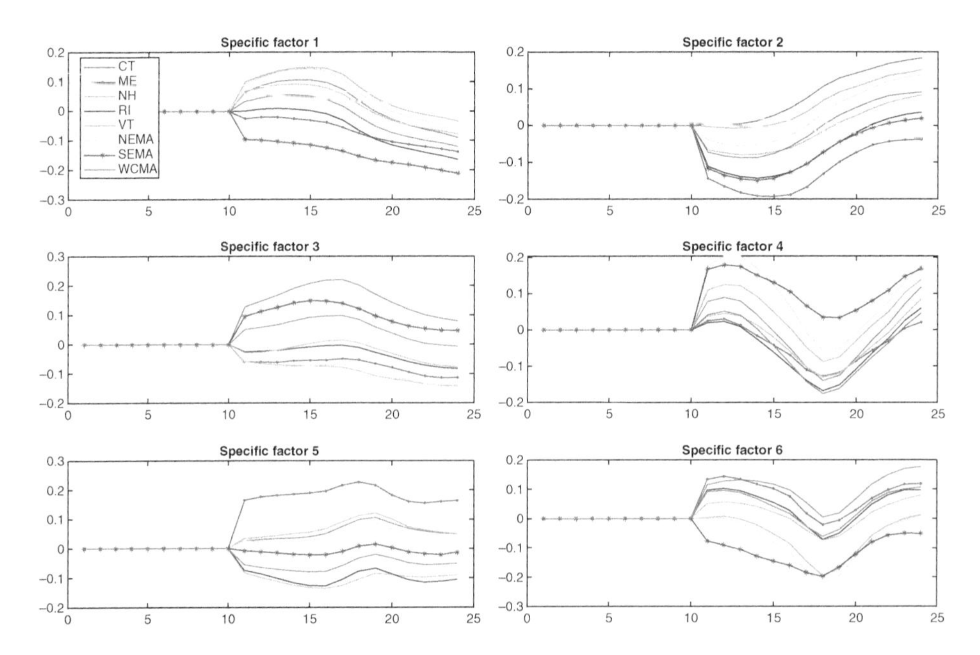

Figure 6.24 Estimated loadings for the six specific factors in the first group (hours 11:00 to 24:00) using the outlier corrected series.

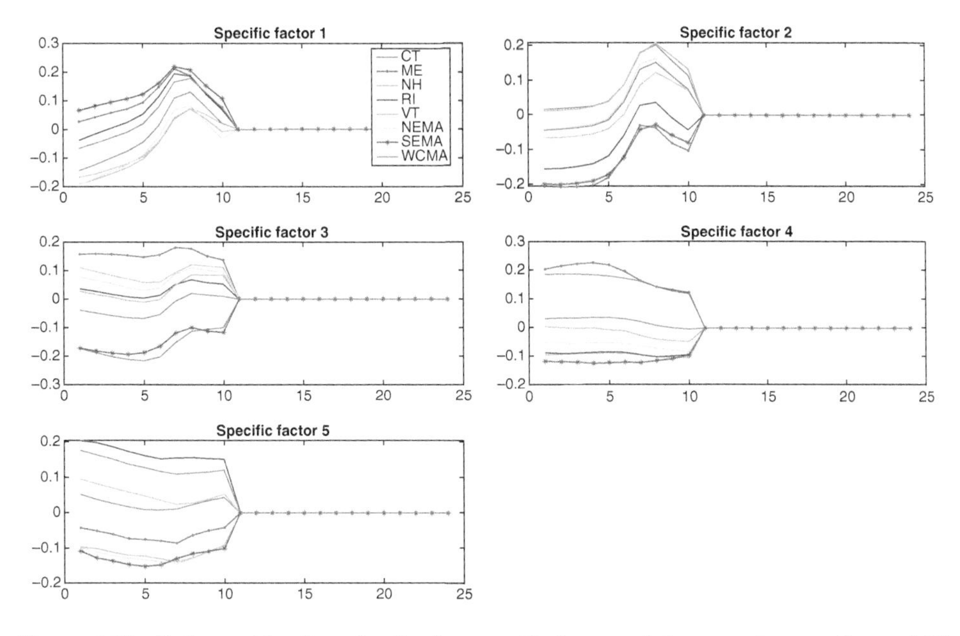

Figure 6.25 Estimated loadings for the five specific factors of the second group (hours 1:00 to 10:00) using the outlier-corrected series.

Finally, we perform an out-of-sample prediction exercise to compare the model fitted with and without clustering. Thus we consider: (1) the fitted DFM with two factors (M1); and (2) the fitted DFMCS model with two global factors and eleven specific factors (M2). The first 10 years of data were used as training period for the estimation of the models and the last 3 years (1095 days) as testing period. Seasonal autoregressive integrated moving-average (ARIMA) models (as in Alonso et al., 2006; García-Martos and Conejo, 2013) were fitted to the factors in models M1 and M2. For simplicity, in both models the idiosyncratic terms are assumed to be white noise. We perform one-day ahead predictions using a rolling window across the testing period. We calculate the mean absolute prediction errors (MAE) and the root mean squared prediction errors (RMSE) using the following expressions:

$$\mathrm{MAE} = \frac{1}{192}\frac{1}{1095}\sum_{i=1}^{192}\sum_{t=1}^{1095} |X_{t,i} - \widehat{X}_{t,i}|,$$

and

$$\mathrm{RMSE} = \left(\frac{1}{192}\frac{1}{1095}\sum_{i=1}^{192}\sum_{t=1}^{1095} (X_{t,i} - \widehat{X}_{t,i})^2\right)^{1/2},$$

where z_{it} corresponds to the ith series at day t and $\widehat{z}_{it}$ is its prediction. The MAE and RMSE for Model M1 were 0.0526 and 0.1131, while for Model M2 were 0.0487 and 0.1100, respectively. These results point out that introducing the cluster structure and the specific factors produce a moderate improvement on the overall out-of-sample prediction. The MAE and RMSE calculated at the time series of cluster 1 (cluster 2) are 0.0545 and 0.0734 (0.0501 and 0.1522) for Model M1, and 0.0502 and 0.0677

(0.0465 and 0.1505) for Model M2, an improvement of 7.88% in MAE and 7.86% in RMSE (7.04 and 1.10%), respectively. That is, the improvement is observed in both clusters and is larger in the first cluster. ■

6.9 SOME EXTENSIONS OF DFM

The models presented in this chapter have been extended in several ways. First, the loadings could not be constant but be changing over time, or the parameters of the FMs may also change in the observed period. Note that as only the common component, product of the loadings and the factors, can be identified, it makes no sense to assume changes in both structures. The most studied variation is with factor loadings. Stock and Watson (2002a, 2002b) proved that if the amplitude of the variations of these loadings is small the estimation by PC is still consistent. However, it is easy to see that if these changes are not very small the PC estimation would become inefficient. Motta et al. (2011) studied locally stationary FMs. Motta and Ombao (2012) assumed that the loadings change smoothly over time and estimate them by the spectral decomposition of the estimated time-varying covariance matrix, and Su and Wang (2017) applied a local version of PC to estimate the factors and the loadings. Other approaches have been based on using wavelets to estimate the loading evolution; see, Cataño et al. (2020). Regarding parameter changes in the model for the factor Corral and Peña (2008) assumed that the factors follow an AR threshold model and studied its estimation.

An important future development for DFM is the analysis of tensor data over time. For instance, suppose that we analyze the evolution of the demand for different products, in different markets, over time. At each time t, $t = 1, \ldots, T$, we observe a matrix with the sales of the products, $i = 1, \ldots, I$, in different markets, $j = 1, \ldots, J$. The factors may operate on both dimensions, the products and the markets. Calling $\boldsymbol{Z}_t$ the data matrix of dimension $I \times J$ at time t, Wang et al. (2019) and Chen et al. (2020) assume the FM

$$\boldsymbol{Z}_t = \boldsymbol{R}\boldsymbol{F}_t\boldsymbol{C}' + \boldsymbol{U}_t,$$

where $\boldsymbol{R}$ is an $I \times r_1$ row loading matrix, $\boldsymbol{F}_t$ a $r_1 \times r_2$ matrix of factors and $\boldsymbol{C}$ a $J \times r_2$ column loading matrix. In this example, the data is a third-order tensor but the model can be generalized to any h-order tensor. With the increasing growth of big data that can be classified on different dimensions such extensions would have increasing importance in the future.

Another extension is to consider seasonality in DFM. These model can be extended to include seasonal factors as

$$z_t = \boldsymbol{P}_1\boldsymbol{f}_{1t} + \boldsymbol{P}_2\boldsymbol{f}_{2t} + \boldsymbol{P}_3\boldsymbol{f}_{3t} + \boldsymbol{e}_t,$$

where we assume, for simplicity, a white noise, $\boldsymbol{e}_t$, and three sets of common factors in $\boldsymbol{f}_t = (\boldsymbol{f}'_{1t}, \boldsymbol{f}'_{2t}, \boldsymbol{f}'_{3t})'$. These common factors are associated with integrated, seasonal, and stationary factors, respectively. Thus, $\boldsymbol{f}_{1t}$ is a vector of nonstationary and nonseasonal factors with dimension r_1 and follows the model with $d \geq 1$:

$$\boldsymbol{\phi}_1(B)\nabla^d \boldsymbol{f}_{1t} = \boldsymbol{\theta}_1(B)\boldsymbol{a}_{1t}, \tag{6.51}$$

$\boldsymbol{f}_{2t}$ is a seasonal factor with period s and dimension r_2 such that

$$\boldsymbol{\phi}_2(B^s)\nabla_s^D \boldsymbol{f}_{2t} = \boldsymbol{\theta}_2(B^s)\boldsymbol{a}_{2t}, \tag{6.52}$$

where $\nabla_s = (1 - B^s)$ and $D \geq 0$, and $\boldsymbol{f}_{3t}$ consists of stationary factors with dimension r_3 and follows the model

$$\boldsymbol{\phi}_3(B)\boldsymbol{f}_{3t} = \boldsymbol{\theta}_3(B)\boldsymbol{a}_{3t}, \tag{6.53}$$

where the AR and MA matrix polynomials satisfy the identifiability conditions discussed in Chapter 3. In fact, we assume that, for each $i = 1, 2, 3$, $\{\boldsymbol{a}_{it}\}$ is a Gaussian white noise process with mean $\boldsymbol{0}$ and diagonal covariance matrix $\boldsymbol{\Sigma}_i$ and all parameter matrices are diagonal. Here, $r_1 + r_2 + r_3 = r$ and we write $\boldsymbol{P} = [\boldsymbol{P}_1\boldsymbol{P}_2\boldsymbol{P}_3]$, where the submatrix $\boldsymbol{P}_i$ is of dimension $k \times r_i$, and $\boldsymbol{f}_{it} = (f_{i1,t}, \ldots, f_{ir_i,t})'$, where $i = 1, 2, 3$. The innovation processes $\{\boldsymbol{a}_{1t}\}, \{\boldsymbol{a}_{2t}\}, \{\boldsymbol{a}_{3t}\}$ and $\{\boldsymbol{e}_t\}$ are mutually uncorrelated. We assume $\boldsymbol{P}'\boldsymbol{P} = \boldsymbol{I}_r$, implying that $\boldsymbol{P}_i'\boldsymbol{P}_i = \boldsymbol{I}_{r_i}$ for $i = 1, 2, 3$.

Assume, for simplicity, that $d = D = 1$. We define the sample generalized seasonal autocovariance (SGCV) matrices $C_{z,s}(h)$ as

$$\boldsymbol{C}_{z,s}(h) = \frac{s^{2d}}{T^{2d}} \sum_{t=h+1}^{T} \boldsymbol{z}_t\boldsymbol{z}_{t-h}'. \tag{6.54}$$

The normalization constant in $\boldsymbol{C}_{z,s}(h)$ is equal to $1/(\frac{T}{s})^{2d}$, where T/s is the number of seasons in the sample, whenever T is a multiple of s. It can be shown, see Nieto et al. (2016), that when h is fixed and $T \to \infty$, then, the sequence $\{\boldsymbol{C}_{z,s}(h)\}$ converges weakly to a random matrix $\boldsymbol{\Gamma}_{z,s}(h)$ that has the following properties: (a) for $h = js$ with j being a positive integer, $\boldsymbol{\Gamma}_{z,s}(h)$ almost surely has $r_1 + r_2$ positive eigenvalues and $k - r_1 - r_2$ zero eigenvalues; (b) for $k \neq js$, $\boldsymbol{\Gamma}_{z,s}(h)$ has, almost surely, r_1 positive eigenvalues and $k - r_1$ zero eigenvalues. Let $\lambda_1(h), \ldots, \lambda_k(h)$ be the eigenvalues of $\boldsymbol{\Gamma}_{z,s}(h)$. Then, if $h = js$, the eigenvectors corresponding to the $r_1 + r_2$ positive eigenvalues form a basis of the column space of the submatrix $[\boldsymbol{P}_1\boldsymbol{P}_2]$ and, if $h \neq js$, the eigenvectors corresponding to the r_1 positive eigenvalues form a basis for the column space of the submatrix $\boldsymbol{P}_1$.

Thus, we expect that the sample covariance matrices $\boldsymbol{C}_{z,s}(h)$ have two disjoint subsets of eigenvalues: one contains relatively large positive values and the other consists of small ones close to zero. For the ordered eigenvalues, the small ones would occur at $i = r_1 + 1, \ldots, k$, when $h \neq js$, and at $i = r_1 + r_2 + 1, \ldots, k$, when $k = js$. A procedure based on these results to fit the model has been proposed by Nieto et al. (2016).

6.10 HIGH-DIMENSIONAL CASE

6.10.1 Sparse PCs

In this section, we discuss the PC estimation of DFMs in high-dimensional time series with the dimension k greater than the sample size T. In these cases, the PC estimation is not consistent and may not be accurate in finite samples.

As an illustration, Tables 6.3 and 6.4 present two measures of the estimation precision by PC of the largest eigenvector of the covariance matrix in an EDFM with one factor, that is a white noise with variance 25, and $N(0,1)$ noises. Table 6.3 presents

TABLE 6.3 Absolute Value of the Correlation Coefficient Between the First Eigenvector of the Covariance Matrix and the One Estimated by PC as a Function of the Sample Size, *T*, and the Number of Series, *k* in a One Factor EDFM with Large Signal-to-Noise Ratio

T/k	10	20	50	100	200	300	400	500
100	0.992	0.983	0.960	0.924	0.868	0.813	0.769	0.726
200	0.996	0.991	0.979	0.961	0.924	0.894	0.862	0.838
300	0.997	0.994	0.986	0.972	0.948	0.924	0.903	0.883

TABLE 6.4 Absolute Value of the Angle in Grades Between the First Eigenvector of the Covariance Matrix and the One Estimated by PC as a Function of the Sample Size, *T*, and the Number of Series, *k* in a One Factor EDFM with Large Signal-to-Noise Ratio

T/k	10	20	50	100	200	300	400	500
100	3.46	5.06	8.18	11.57	16.15	19.44	22.38	25.25
200	2.39	3.52	5.82	8.08	11.59	14.03	16.13	18.03
300	1.92	2.92	4.71	6.74	9.56	11.60	13.31	14.85

the absolute value of the correlation coefficient between the true eigenvector and the one estimated by PC, and Table 6.4 the angle in grades between the two vectors. It is clear that the estimation precision decreases when k/T increases and can be bad if $k > T$. The R command to build these tables are given in the appendix and the reader can check that decreasing the signal-to-noise ratio the precision of PC can be very low. For instance, with a variance of the factor being 4 and T=100 and k=40, the correlation between the true and estimated eigenvectors is as low as 0.35.

To explain these results, consider the simple case of an EDFM with one factor. The covariance matrix of the observations is given by

$$\boldsymbol{\Gamma}_z(0) = \boldsymbol{p}\boldsymbol{p}'\gamma_0 + \sigma^2\boldsymbol{I},$$

where $\boldsymbol{p}$ is the loading vector. Let $s = \gamma_0/\sigma^2$ be the signal-to-noise ratio. Calling $\hat{\boldsymbol{p}}$ to the PC estimate of $\boldsymbol{p}$, the estimation precision can be measured by $\cos\alpha = \boldsymbol{p}'\hat{\boldsymbol{p}}$ or by $\text{corr}(\hat{\boldsymbol{p}}, \boldsymbol{p})$, and the closer to zero the angle between the two vectors, or the closer to one the correlation coefficient, the better the accuracy of the estimation. It can be proved (Johnstone and Lu, 2009) that allowing both T and k to go to infinity and calling $c = \lim_{T,k\to\infty} k/T$, we have

$$\lim_{T,k\to\infty} \cos^2\alpha = \frac{(s^2 - c)_+}{s^2 + sc} \tag{6.55}$$

where $(a)_+ = \max\{a, 0\}$. Thus, if $\lim_{T,k\to\infty} \cos^2\alpha = 1$, $\hat{\boldsymbol{p}}$ is a consistent estimate of $\boldsymbol{p}$. This requires $c = 0$, that is, k must be fixed, or has to grow very slowly with T, when T goes to infinity. This is the standard PC application when $T > k$ and fixed k. When $c > 0$, the estimation accuracy depends on the signal-to-noise ratio s. If this ratio is small and $s^2 < c$, then $\lim_{T,k\to\infty} \cos^2\alpha = 0$ and the vector $\widehat{\boldsymbol{p}}$ is orthogonal to $\boldsymbol{p}$: the estimation does not provide any information about the true value. If the signal-to-noise ratio is strong and $s^2 > c$, the estimation accuracy grows with s^2, but the estimator is never consistent and always has some bias in the limit. See, Wang and Fan (2017) for a generalization of this analysis.

Several solutions have been proposed to overcome the difficulty. Johnstone and Lu (2009) proved that, even if $k > p$, if the vector $\boldsymbol{p}$ is sparse, i.e. having many coefficients equal to zero, PC can still provide consistent estimates. They proposed a two-step procedure in which first, the variables z_{it} with larger variances are selected and then PC is applied to this small set of selected variables. The final vector $\widehat{\boldsymbol{p}}$ is built with the estimated values for the selected variables and zero for the others. An alternative solution is to add a sparsity condition to the method of computing PC, either by variance maximization or optimal reconstructions. Jolliffe et al. (2003) use the largest variance condition and Lasso, which is studied in Chapter 7, to solve the following problem:

$$\max\{\boldsymbol{v}'\boldsymbol{X}'\boldsymbol{X}\boldsymbol{v}\}, \quad \text{subject to } \|\boldsymbol{v}\|_1 \leq t, \|\boldsymbol{v}\|_2 = 1. \tag{6.56}$$

Dropping the restriction $\|\boldsymbol{v}\|_1 \leq t$, the problem is the standard way to compute PC as the main eigenvectors of the matrix $\boldsymbol{X}'\boldsymbol{X}$. Adding the restriction, $\|\boldsymbol{v}\|_1 \leq t$, that the sum of the absolute values coefficients of the PC must be small, a sparse solution is obtained. Efficient methods to solve Eq. (6.56) are discussed in Hastie et al. (2015). The optimal reconstruction property of PC was used by Zou and Hastie (2005) by finding the projection direction $\boldsymbol{v}$ and the coefficients $\boldsymbol{\beta}$ that verify

$$\min\left\{\frac{1}{T}\sum_{i=1}^{T}\|z_i - \boldsymbol{\beta}\boldsymbol{v}'z_i\|_2^2 + \lambda_1\|\boldsymbol{v}\|_1 + \lambda_2\|\boldsymbol{v}\|_2^2\right\}, \text{subject to } \|\boldsymbol{\beta}\|_2 = 1.$$

This procedure also uses Lasso restrictions and is highly related to the elastic net, which is studied in the next chapter.

In the next section, we present a structural-FM for multivariate and high-dimensional time series analysis. The model postulates that an observed k-dimensional time series is a nonsingular linear transformation of r common factors and $k - r$ white noise processes. For a finite k, the structural-FM enjoys certain advantages over the traditional exact FM as the sample covariance matrix of the noises is positive definite. For the case of k approaching infinity, the structural-FM uses different methods to specify the number of common factors, instead of ratios of eigenvectors, so that it does not encounter the difficulties of PCA mentioned here.

6.10.2 A Structural-FM Approach

Gao and Tsay (2018) proposed the following Structural-FM

$$z_t = \boldsymbol{P}\begin{bmatrix} \boldsymbol{f}_t \\ \epsilon_t \end{bmatrix} = [\boldsymbol{P}_1, \boldsymbol{P}_2]\begin{bmatrix} \boldsymbol{f}_t \\ \epsilon_t \end{bmatrix} = \boldsymbol{P}_1\boldsymbol{f}_t + \boldsymbol{P}_2\epsilon_t, \tag{6.57}$$

where $\boldsymbol{P}$ is a $k \times k$ nonsingular matrix, $\boldsymbol{f}_t = (f_{1t}, \ldots, f_{rt})'$ is the common factors, $\epsilon_t = (\epsilon_{1t}, \ldots, \epsilon_{vt})'$ is a v-dimensional white noise with $v = k - r$, $\boldsymbol{P}_1$ is a $k \times r$ loading matrix, and $\boldsymbol{P}_2$ is a $k \times v$ real matrix. In Model (6.57), we assume that $\boldsymbol{\Gamma}_f(0) = \boldsymbol{I}_r$, $\boldsymbol{\Gamma}_\epsilon(0) = \boldsymbol{I}_v$, $\text{Cov}(\boldsymbol{f}_t, \boldsymbol{\epsilon}_t) = \boldsymbol{0}$, and no linear combination of $\boldsymbol{f}_t$ is a white noise series; otherwise, we can move the white noise to $\boldsymbol{\epsilon}_t$.

Let $\boldsymbol{P}_1 = \boldsymbol{A}_1\boldsymbol{Q}_1$ and $\boldsymbol{P}_2 = \boldsymbol{A}_2\boldsymbol{Q}_2$, where $\boldsymbol{A}_1$ and $\boldsymbol{A}_2$ are half-orthonormal matrices such that $\boldsymbol{A}_1'\boldsymbol{A}_1 = \boldsymbol{I}_r$ and $\boldsymbol{A}_2'\boldsymbol{A}_2 = \boldsymbol{I}_v$. This can always be done via the QR decomposition or singular value decomposition. Under such a parameterization, the strengths of $\boldsymbol{f}_t$ and $\boldsymbol{\epsilon}_t$ are shown by $\boldsymbol{Q}_1$ and $\boldsymbol{Q}_2$, respectively. The model in Eq. (6.57) can be rewritten as

$$\boldsymbol{z}_t = \boldsymbol{A}_1\boldsymbol{x}_t + \boldsymbol{A}_2\boldsymbol{e}_t, \tag{6.58}$$

where $\boldsymbol{x}_t = \boldsymbol{Q}_1\boldsymbol{f}_t$ and $\boldsymbol{e}_t = \boldsymbol{Q}_2\boldsymbol{\epsilon}_t$. Note that the identifiability of FMs continues to apply to the model in Eq. (6.58), but the column spaces generated by $\boldsymbol{A}_1$ (hence, by $\boldsymbol{P}_1$) can be identified. The same applies to the column space of $\boldsymbol{A}_2$. The main difference between Eq. (6.57) and the traditional FMs is that the random vectors of Eq. (6.57) and, hence, of Eq. (6.58), always have a total dimension k. As such, the number of common factors r can be identified by $k - v$, where v is the number of white noise processes embedded in the observed vector $\boldsymbol{z}_t$. Several methods have recently been proposed to test high-dimensional white noise processes. Our analysis below takes advantages of those white noise test statistics. See, Section 6.10.4.

6.10.3 Estimation

In this section, we outline the estimation procedure for the model in Eq. (6.58). We begin with a brief discussion on the rationale underlying the estimation method used in Gao and Tsay (2018). Let $\boldsymbol{B}_1$ and $\boldsymbol{B}_2$ be the orthogonal complement of $\boldsymbol{A}_1$ and $\boldsymbol{A}_2$, respectively, so that $\boldsymbol{B}_1 \in R^{k\times v}$ and $\boldsymbol{B}_2 \in R^{k\times r}$ are two half-orthonormal matrices satisfying $\boldsymbol{B}_1'\boldsymbol{A}_1 = \boldsymbol{0}$ and $\boldsymbol{B}_2'\boldsymbol{A}_2 = \boldsymbol{0}$. From Eq. (6.58), we have

$$\boldsymbol{B}_1'\boldsymbol{z}_t = \boldsymbol{B}_1'\boldsymbol{A}_2\boldsymbol{e}_t, \tag{6.59}$$

so that $\boldsymbol{B}_1'\boldsymbol{z}_t$ is a v-dimensional white noise process. Also, from Eq. (6.58), we have

$$\boldsymbol{\Gamma}_z(h) = \boldsymbol{A}_1\boldsymbol{\Gamma}_x(h)\boldsymbol{A}_1' + \boldsymbol{A}_1\boldsymbol{\Sigma}_{xe}(h)\boldsymbol{A}_2', \quad h \geq 1, \tag{6.60}$$

where $\boldsymbol{\Sigma}_{xe}(h) = \text{Cov}(\boldsymbol{x}_t, \boldsymbol{e}_{t-h})$, and

$$\boldsymbol{\Gamma}_z(0) = \boldsymbol{A}_1\boldsymbol{\Gamma}_x(0)\boldsymbol{A}_1' + \boldsymbol{A}_2\boldsymbol{\Gamma}_e(0)\boldsymbol{A}_2'. \tag{6.61}$$

For a prespecified positive integer h_0, define, as in Eq. (6.21),

$$\boldsymbol{M} = \sum_{h=1}^{h_0} \boldsymbol{\Gamma}_z(h)\boldsymbol{\Gamma}_z(h)', \tag{6.62}$$

which is a $k \times k$ nonnegative definite matrix. By $\boldsymbol{B}_1'\boldsymbol{A}_1 = \boldsymbol{0}$, we have $\boldsymbol{M}\boldsymbol{B}_1 = \boldsymbol{0}$ so the columns of $\boldsymbol{B}_1$ are eigenvectors associated with zero eigenvalues of $\boldsymbol{M}$ and the factor loading space, which is the column space of $\boldsymbol{A}_1$, is spanned by the eigenvectors associated with the r non-zero eigenvalues of $\boldsymbol{M}$. This provides an estimate of $\boldsymbol{A}_1$.

Next, from Eq. (6.58), we have

$$\boldsymbol{B}_2'\boldsymbol{z}_t = \boldsymbol{B}_2'\boldsymbol{A}_1\boldsymbol{x}_t, \tag{6.63}$$

which is uncorrelated with $\boldsymbol{B}_1'\boldsymbol{z}_t$ defined in Eq. (6.59). Therefore,

$$\boldsymbol{B}_2'\boldsymbol{\Gamma}_z(0)\boldsymbol{B}_1\boldsymbol{B}_1'\boldsymbol{\Gamma}_z(0)\boldsymbol{B}_2 = \boldsymbol{0}, \tag{6.64}$$

which implies that $\boldsymbol{B}_2$ consists of r eigenvectors corresponding to the zero eigenvalues of the following matrix

$$\boldsymbol{S} = \boldsymbol{\Gamma}_z(0)\boldsymbol{B}_1\boldsymbol{B}_1'\boldsymbol{\Gamma}_z(0). \tag{6.65}$$

Gao and Tsay (2018) showed that the matrix $\boldsymbol{B}_2'\boldsymbol{A}_1$ is invertible and, hence, from Eq. (6.63), we have $\boldsymbol{x}_t = (\boldsymbol{B}_2'\boldsymbol{A}_1)^{-1}\boldsymbol{B}_2\boldsymbol{z}_t$. This provides an estimate of the common factor $\boldsymbol{x}_t$ and, hence, $\boldsymbol{f}_t$. From Eq. (6.58), $\boldsymbol{x}_t$ does not involve the white noise term $\boldsymbol{e}_t$, so that the estimated common factor $\widehat{\boldsymbol{f}}_t = \boldsymbol{Q}_1^{-1}\boldsymbol{x}_t$ would not involve the noise term ϵ_t. This approach of estimation thus improves the efficiency in estimating the common factors $\boldsymbol{f}_t$. The authors called the eigenvalue-eigenvector analysis of $\boldsymbol{S}$ in Eq. (6.65) a projected PCA, because it projects the observed series $\boldsymbol{z}_t$ into the directions of $\boldsymbol{B}_1$ before performing the PCA.

Turn to estimation in practice. Assume for now that the number of common factors r is known. The selection of r is given in the next section. First, we perform an eigenvalue-eigenvector analysis of the sample matrix

$$\widehat{\boldsymbol{M}} = \sum_{h=1}^{h_0} \widehat{\boldsymbol{\Gamma}}_z(h)\widehat{\boldsymbol{\Gamma}}_z(h)',$$

where h_0 is a small positive integer. Let $\widehat{\boldsymbol{A}}_1 = [\widehat{\boldsymbol{a}}_1, \ldots, \widehat{\boldsymbol{a}}_r]$ and $\widehat{\boldsymbol{B}}_1 = [\widehat{\boldsymbol{B}}_1, \ldots, \widehat{\boldsymbol{B}}_v]$ be the half-orthonormal matrices consisting of the eigenvectors of $\widehat{\boldsymbol{M}}$ associated with non-zero and zero eigenvalues, respectively. Next, we perform eigenvalue-eigenvector analysis of $\widehat{\boldsymbol{S}} = \widehat{\boldsymbol{\Gamma}}_z(0)\widehat{\boldsymbol{B}}_1\widehat{\boldsymbol{B}}_1'\widehat{\boldsymbol{\Gamma}}_z(0)$, which is a $k \times k$ matrix with rank at most $k - r$. Let $\widehat{\boldsymbol{B}}_2 = [\widehat{\boldsymbol{B}}_{v+1}, \ldots, \widehat{\boldsymbol{B}}_k]$, where $\widehat{\boldsymbol{B}}_i$ (for $i = v + 1, \ldots, k$) are the eigenvectors associated with the r smallest eigenvalues of $\widehat{\boldsymbol{S}}$. For large k, Gao and Tsay (2018) provide a modification to compute $\widehat{\boldsymbol{B}}_2$ if one allows for some largest eigenvalues of the noise covariance matrix $\boldsymbol{\Gamma}_e(0)$ to diverge to infinity as $k \to \infty$. Interested readers are referred to that paper for details. We can then estimate the factor process via $\widehat{\boldsymbol{x}}_t = (\widehat{\boldsymbol{B}}_2'\widehat{\boldsymbol{A}}_1)^{-1}\widehat{\boldsymbol{B}}_2'\boldsymbol{z}_t$. Finally, the white noise process can be estimated by $\widehat{\boldsymbol{e}}_t = \widehat{\boldsymbol{A}}_2'\boldsymbol{z}_t - \widehat{\boldsymbol{A}}_2'\widehat{\boldsymbol{A}}_1\widehat{\boldsymbol{x}}_t$, where $\widehat{\boldsymbol{A}}_2 = [\widehat{\boldsymbol{B}}_1, \ldots, \widehat{\boldsymbol{B}}_v]$ consists of the eigenvectors associated with the v largest eigenvalues of $\widehat{\boldsymbol{S}}$.

6.10.4 Selecting the Number of Common Factors

In the case of $k \geq T$, the canonical correlation analysis cannot be used to select the number of common factor r. However, from the discussion of the previous section, $\widehat{\boldsymbol{B}}_1'\boldsymbol{z}_t$ is a v-dimensional white noise, where $v = k - r$. Thus, by testing for the number of white noise series in the PCs of the matrix $\widehat{\boldsymbol{M}}$, we can estimate v. Denote the estimated number of white noise series by $\widehat{v}$. Then, $\widehat{r} = k - \widehat{v}$. The key issue then is how to test for high-dimensional white noise time series.

In recent years, testing for high-dimensional white noise process has been widely studied. We discuss two approaches here. Other possible approaches include Ling et al. (2020) and the reference therein. The two approaches we discuss make use of the maximum absolute cross-correlation of the lag autocorrelation matrices of a time series. For our purpose, we want to detect the number of white noise series in $\hat{\boldsymbol{u}}_t = [\hat{\boldsymbol{A}}_1, \hat{\boldsymbol{B}}_1]'\boldsymbol{z}_t$, where $[\hat{\boldsymbol{A}}_1, \hat{\boldsymbol{B}}_1]$ are the matrix of eigenvectors of the matrix $\widehat{\boldsymbol{M}}$ of Eq. (6.62). For simplicity in discussion, let $\boldsymbol{w}_t$ be a d-dimensional subseries of $\hat{\boldsymbol{u}}_t$, where $1 \leq d \leq k$. Our goal is to test the null hypothesis $H_0 : \boldsymbol{\Gamma}_w(1) = \cdots = \boldsymbol{\Gamma}_w(m) = \boldsymbol{0}$ versus $H_a : \boldsymbol{\Gamma}_w(h) \neq \boldsymbol{0}$ for some h satisfying $1 \leq h \leq m$, where m is a positive integer and, for simplicity, we assume that $\hat{\boldsymbol{u}}_t$ is standardized so that $\boldsymbol{\Gamma}_w(h)$ is the lag-h autocorrelation matrix of $\boldsymbol{w}_t$. The first approach is proposed by Chang et al. (2017) with test statistic

$$T_1(m) = \sqrt{T} \times \max_{1\leq h\leq m} \max_{1\leq i,j\leq d} |\hat{\gamma}_{w,ij}(h)|, \tag{6.66}$$

where $\hat{\gamma}_{w,ij}(h)$ is the (i,j)th element of $\hat{\boldsymbol{\Gamma}}_w(h)$. The authors showed that, under finite fourth-order moments of $\boldsymbol{w}_t$ and the null hypothesis, the limiting distribution of $T_1(m)$ can be approximately by that of the L_∞-norm of a normal random vector and, hence, simulation can be used to obtain the asymptotic critical values of the test statistic $T_1(m)$. Limited experience indicates that the test statistic works reasonably well if d is moderate. For large d, the test statistic requires intensive computing time.

The second approach to white noise testing is proposed by Tsay (2020). Instead of using $\boldsymbol{w}_t$ directly, the author proposes to use the rank series of $\boldsymbol{w}_t$. For a scalar time series y_t with a realization $\{y_1, \ldots, y_t\}$, its rank series is the time series obtained by replacing y_t by its rank statistic in the sample. Let $\hat{\boldsymbol{R}}_w(h)$ be the lag-h sample autocorrelation matrix of the rank series of $\boldsymbol{w}_t$. Define

$$T_{\max}(h) = \max_{1\leq i,j\leq d} \hat{r}_{w,ij} \quad \text{and} \quad T_{\min}(h) = -\min_{1\leq i,j\leq d} \hat{r}_{w,ij},$$

where $\hat{r}_{w,ij}$ is the (i,j)th element of $\hat{\boldsymbol{R}}_w(h)$. Note that $T_{\max}(h)$ is the maximum rank cross-correlations of $\boldsymbol{w}_t$ at lag h whereas $-T_{\min}(h)$ is the minimum rank cross-correlation of $\boldsymbol{w}_t$ at lag h. Under the hull hypothesis that $\boldsymbol{w}_t$ is a d-dimensional white noise and $\boldsymbol{\Gamma}_w(0)$ is diagonal, Tsay (2020) showed that $\{\hat{r}_{w,ij}(h)|1 \leq i,j \leq d\}$ are asymptotically independent and normally distributed as $N(0,1)$. Therefore, both $T_{\max}(h)$ and $T_{\min}(h)$ are extreme values of d^2 $N(0,1)$ random variates so that they follow asymptotically a Gumbel distribution. Moreover, $T_{\max}(h)$ and $T_{\min}(h)$ are asymptotically independent. Now, define the test statistic

$$T_2(m) = \sqrt{T} \times \max_{1\leq h\leq m} \max\{T_{\max}(h), T_{\min}(h)\}. \tag{6.67}$$

Since components of $\boldsymbol{w}_t$ in our application are uncorrelated so that one of the assumptions used by Tsay (2020) automatically satisfied. Therefore, under the assumption that $\boldsymbol{w}_t$ has a continuous distribution and the null hypothesis, $T_2(m)$ of Eq. (6.67) follows asymptotically a product of independent Gumbel distribution from which the critical values are available in closed form. Specifically, let α be the type I error, then the null hypothesis H_0 is rejected at the 5% level if

$$T_2(m) \geq c_{d,m} \times x_{1-\alpha/2} + s_{d,m},$$

where $x_{1-\alpha/2} = -\ln(-\ln(1-\alpha/2))$ is the $(1-\alpha/2)$th quantile of the standard Gumbel distribution and

$$c_{d,m} = [2\ln(d^2m)]^{-1/2} \quad \text{and} \quad s_{d,m} = \sqrt{2\ln(d^2m)} - \frac{\ln(4\pi) + \ln(\ln(d^2m))}{2(2\ln(d^2m))^{1/2}}.$$

Besides having closed-form critical values, the test statistic $T_2(m)$ of Eq. (6.67) also has some other advantages. It does not require the existence of any moment of $\boldsymbol{w}_t$, such as the Cauchy distribution, and is robust to outliers in $\boldsymbol{w}_t$.

6.10.5 Asymptotic Properties of Loading Estimates

For two $k \times r$ half orthogonal matrices $\boldsymbol{H}_1$ and $\boldsymbol{H}_2$ satisfying $\boldsymbol{H}_1'\boldsymbol{H}_1 = \boldsymbol{H}_2'\boldsymbol{H}_2 = \boldsymbol{I}_r$, the difference between the column spaces spanned by $\boldsymbol{H}_1$ and $\boldsymbol{H}_2$ can be measured by

$$D(\boldsymbol{H}_1, \boldsymbol{H}_2) = \sqrt{1 - \frac{1}{r}\,\text{tr}(\boldsymbol{H}_1\boldsymbol{H}_1'\boldsymbol{H}_2\boldsymbol{H}_2')}. \tag{6.68}$$

Here $D(\boldsymbol{H}_1, \boldsymbol{H}_2) \in [0, 1]$. It is equal to zero if and only if the two column spaces of $\boldsymbol{H}_1$ and $\boldsymbol{H}_2$ are equivalent, and it is equal to 1 if and only if the two column spaces are orthogonal to each other. We use this measure to quantify the accuracy in estimating a FM. For instance, we use $D(\widehat{\boldsymbol{A}}_1, \boldsymbol{A}_1)$ to measure the consistency of using $\widehat{\boldsymbol{A}}_1$ to estimate $\boldsymbol{A}_1$.

For a fixed k, we can derive asymptotic properties of the estimates of FMs under some mild conditions such as the existence of certain finite-order moments of $\boldsymbol{f}_t$ and ϵ_t and some α-mixing conditions of z_t and $\boldsymbol{f}_t$. For instance, as expected, we have $D(\widehat{\boldsymbol{A}}_1, \boldsymbol{A}_1) = O(T^{-1/2})$. Details can be found, for instance, in Gao and Tsay (2019). For high-dimensional case, similar results can be obtained so long as $k = o(T^{1/2})$. On the other hand, the limiting properties of the estimates become more complicated when both k and T increase to infinity. Consider the model in Eq. (6.57). We focus only on the case in which $\boldsymbol{f}_t$ and ϵ_t are independent. The case that $\boldsymbol{f}_t$ is correlated with ϵ_{t-h} for some $h > 0$ can be found in Gao and Tsay (2018).

Let $\boldsymbol{L}_1 = [\boldsymbol{c}_1, \ldots, \boldsymbol{c}_r]$ and $\boldsymbol{L}_2 = [\boldsymbol{c}_{r+1}, \ldots, \boldsymbol{c}_k]$. As k increases, we need some measurements to quantify the strengths of the loadings and noises.

1. Condition A: The columns of $\boldsymbol{L}_1$ satisfy that $\|\boldsymbol{c}_j\|_2^2 \asymp k^{1-\delta_1}$ for $j = 1, \ldots, r$ and $\delta_1 \in [0, 1)$, where $a \asymp b$ means $a = O(b)$ and $b = O(a)$. Also, for each column $\boldsymbol{c}_j$, $\min_{\theta_i \in R, i \neq j} \|\boldsymbol{c}_j - \sum_{1 \leq i \leq r, i \neq j} \theta_i \boldsymbol{c}_j\|_2^2 \asymp k^{1-\delta_1}$.
2. Condition B: $\boldsymbol{L}_2$ admits a singular value decomposition $\boldsymbol{L}_2 = \boldsymbol{A}_2\boldsymbol{D}_2\boldsymbol{V}_2'$, where $\boldsymbol{A}_2 \in R^{k \times v}$ is given in Eq. (6.58), $\boldsymbol{D}_2 = \text{diag}(d_1, \ldots, d_v)$ and $\boldsymbol{V}_2 \in R^{v \times v}$ satisfying $\boldsymbol{V}_2'\boldsymbol{V}_2 = \boldsymbol{I}_v$. Also, there exists a finite integer $0 < g < v$ such that $d_1 \asymp \cdots \asymp d_g \asymp k^{(1-\delta_2)/2}$ for some $\delta_2 \in [0, 1)$ and $d_{g+1} \asymp \cdots \asymp d_v \asymp 1$.

The quantity δ_1 of Condition A is used to quantify the strength of the factors and the eigenvalues of $\boldsymbol{L}_1\boldsymbol{L}_1'$ are all of order $k^{1-\delta_1}$. If $\delta_1 = 0$, the corresponding factors are called strong factors, since they include the case where each element of $\boldsymbol{c}_j$ is $O(1)$. If $\delta_1 > 0$, the corresponding factors are weak factors and the smaller the δ_1 is, the stronger the factors are. An advantage of using index δ_1 is to link the convergence rates of the estimated factors explicitly to the strengths of the factors. Condition A ensures that all common factors in $\boldsymbol{x}_t$ are of equal strength δ_1. In practice, the factors may have multiple levels of strength. In this case, the consistency of the loading matrix would then depend on the strength of the weakest factors. We do not consider this issue here to save space. There are many sufficient conditions for Condition B to hold. For example, it holds if we allow $(\boldsymbol{c}_{r+1}, \ldots, \boldsymbol{c}_{r+g})$ to satisfy Condition A for some $\delta_2 \in [0, 1)$, and the L_1- and L_∞-norms of $(\boldsymbol{c}_{r+g+1}, \ldots, \boldsymbol{c}_k)$ are all finite. A special case is to let $\boldsymbol{c}_{r+g+j}$ be a standard unit vector. The constraint between δ_1 and δ_2 is used later under different scenarios to guarantee the consistency in estimation.

If r is known and fixed, Gao and Tsay (2018) showed that, under certain conditions including Conditions A and B and $\boldsymbol{f}_t$ and ϵ_t being independent, if $kT^{-1/2} = o(1)$, then $D(\widehat{\boldsymbol{A}}_1, \boldsymbol{A}_1) = O_p(k^{\delta_1} T^{-1/2})$ and $D(\widehat{\boldsymbol{B}}_1, \boldsymbol{B}_1) = O_p(k^{\delta_1} T^{-1/2})$. Therefore, the convergence rate is slower than that of the case with fixed k if $\delta_1 > 0$. The strength of the loading thus affects the convergence rate of the estimates in FMs. The estimation of $\boldsymbol{B}_2$ is much harder. In some cases, no consistent estimate is available for large k. To mitigate this difficulty, we estimate $\boldsymbol{B}_2^*$, which is the subspace spanned by the eigenvectors of the matrix $\boldsymbol{S}$ in Eq. (6.65) associated with the smallest $k - g$ eigenvalues. Let $\widehat{\boldsymbol{B}}_2^*$ consist of the eigenvectors of $\widehat{\boldsymbol{S}}$ associated with the $k - g$ smallest eigenvalues. The choice of $\widehat{\boldsymbol{B}}_2$ is then a subspace of $\widehat{\boldsymbol{B}}_2^*$. Under the aforementioned conditions, if $kT^{-1/2} = o(1)$, we also have

$$D(\widehat{\boldsymbol{B}}_2^*, \boldsymbol{B}_2^*) = O_p[k^{2\delta_2-\delta_1} T^{-1/2} + k^{\delta_2} T^{-1/2} + (1 + k^{2(\delta_2-\delta_1)}) D(\widehat{\boldsymbol{B}}_1, \boldsymbol{B}_1)].$$

Finally, the case of high-dimensional FMs with unit roots has been studied in Gao and Tsay (2020).

APPENDIX 6.A: SOME R COMMANDS

The commands to compute Tables 6.3 and 6.4 with two precision measures for the estimation of the largest eigenvector of the covariance matrix via PC are given below.

```
t0 <- proc.time() # start the clock
N=250
R0=rep(0,N)
ANG=rep(0,N)
COS=rep(0,N)
##define the signal to noise ratio with the sd of the factor
fsd=5
vT=c(100,200,300)
vk=c(10,20,50,100,200,300,400,500)
dT=length(vT)
dk=length(vk)
TAB1=matrix(0,dT,dk)
TAB2=matrix(0,dT,dk)
for (j in 1:dk){
for (t in 1:dT){
T=vT[t]
```

```
k=vk[j]
for(i in 1:N){
a=rnorm(k*T)
A=matrix(a,T,k)
f=rnorm(T)*fsd
## a general arima model for the factor can be used with
##f=arima.sim(n=T,list(ar=c(.7)))
F=matrix(f,T,1)
p=runif(k)
pe=p/sqrt(sum(p^2))
P=matrix(pe,1,k)
YC=F%*%P
Y=YC+A
S=cov(Y)
ES=eigen(S)
EV=ES$values
EVV=ES$vectors
pv=ES$vectors[,1]
pv=as.matrix(pv,1,k)
ro=abs(cor(pv,pe))
R0[i]=ro
cos=abs(sum(pv*pe))
angr=acos(cos)
ang=angr*180/pi
ANG[i]=ang}
TAB1[t,j]=mean(R0)
TAB2[t,j]=mean(ANG)}}
tf=proc.time()-t0
```

EXERCISES

1. In the DFM $\boldsymbol{z}_t = \boldsymbol{P}\boldsymbol{f}_t + \boldsymbol{a}_t$ with $\boldsymbol{P} = \frac{1}{\sqrt{k}}\mathbf{1}$ is a $k \times 1$ vector, $\mathbf{1} = (1, \ldots, 1)'$ and $\boldsymbol{a}_t$ is white noise with $\boldsymbol{\Sigma}_a = \sigma^2\boldsymbol{I}$, (1) explain the form of the linear combinations of the series $\boldsymbol{c}_j'\boldsymbol{z}_t$ for $j = 1, \ldots, k-1$ that are white noise; (2) find the variance of these linear combinations.
2. Suppose the DFM $\boldsymbol{z}_t = \boldsymbol{P}\boldsymbol{f}_t + \boldsymbol{n}_t$, where $\boldsymbol{n}_t$ follows a diagonal VAR model. Under what conditions are the factors $\boldsymbol{f}_t$ linear combinations of the data? What should be the relationship between the variance of the factors and those of the noises so that we can obtain by PCA consistent estimates of the factors?
3. Consider the GDFM with one factor and two lags, $\boldsymbol{z}_t = \boldsymbol{P}_0 f_t + \boldsymbol{P}_1 f_{t-1} + \boldsymbol{P}_2 f_{t-3} + \boldsymbol{n}_t$, where the factor follows $f_t = \phi f_{t-1} + u_t$. Prove that this model can be written as a DFM with one factor, $\boldsymbol{z}_t = \boldsymbol{P}^* f_t^* + \boldsymbol{n}_t^*$ with $f_t^* = \phi f_{t-1}^* + u_t^*$.
4. Suppose that the GDFM $\boldsymbol{z}_t = \boldsymbol{P}_0 f_t + \boldsymbol{P}_1 f_{t-1} + \boldsymbol{a}_t$, where $\boldsymbol{a}_t$ is white noise, is estimated by the ODPC with one lag, and $c_1 = c_2 = 1$. Show that the estimation of $\boldsymbol{z}_t$ can be written as $\hat{\boldsymbol{z}}_t = \boldsymbol{A}_0\boldsymbol{z}_t + \boldsymbol{A}_1\boldsymbol{z}_{t-1} + \boldsymbol{A}_2\boldsymbol{z}_{t-2}$.

5. Fit a DFM to the EUUS (CPI) price indexes in file `CPIEurope2000-15.csv`. Apply the transformation $\nabla\nabla_{12} \log z_t$ and the command `dfmpc` to fit the model. Analyze the properties of the three factors found and the models for the residuals.
6. Fit a DFM to the data in levels of price indexes EUUS (CPI) in file `CPIEurope2000-15.csv`. Apply the transformation $\log(z_t)$ and the command `dfmpc` to fit the model. Analyze the properties of the first factor found. Take the residuals from this factor and apply again the `dfmpc` command. Compare the results with those obtained in Exercise 5.
7. Use the data in file `gdpsimple6c8010.txt` of the GDP of six countries to fit a FM. Compare the results with those of Example 3.2 where a VAR model was fitted.
8. Modify the commands given in the appendix to check the precision of PC in the estimation of the EDFM for different values of the signal-to-noise ratio and see the decrease in the precision when this ratio decreases. Also, check the effect of more factors in the precision of the estimation.

REFERENCES

Ahn, S. C. and Horenstein, A. R. (2013). Eigenvalue ratio test for the number of factors. *Econometrica*, **83**: 1203–1227.

Ahn, S. K. and Reinsel, G. C. (1988). Nested reduced-rank autoregressive models for multiple time series. *Journal of the American Statistical Association*, **83**: 849–856.

Alonso, A. M., Berrendero, J. R., Hernández, A., and Justel, A. (2006). Time series clustering based on forecast densities. *Computational Statistics and Data Analysis*, **51**: 762–766.

Alonso, A. M., Galeano, P., and Peña, D. (2020). A robust procedure to build dynamic factor models with cluster structure. *Journal of Econometrics*, **216**: 35–52.

Anderson, T. W. (1963). The use of factor analysis in the statistical analysis of multiple time series. *Psychometrika*, **28**: 1–25.

Ando, T. and Bai, J. (2016). Panel data models with grouped factor structure under unknown group membership. *Journal of Applied Econometrics*, **31**: 163–191.

Ando, T. and Bai, J. (2017). Clustering huge number of financial time series: A panel data approach with high-dimensional predictors and factor structures. *Journal of the American Statistical Association*, **519**: 1182–1198.

Bai, J. and Li, K. (2012). Statistical analysis of factor models of high dimension. *The Annals of Statistics*, **40**: 436–465.

Bai, J. and Li, K. (2016). Maximum likelihood estimation and inference for approximate factor models of high dimension. *Review of Economics and Statistics*, **98**: 298–309.

Bai, J. and Ng, S. (2002). Determining the number of factors in approximate factor models. *Econometrica*, **70**: 191–221.

Bai, J. and Ng, S. (2004). A PANIC attack on unit roots and cointegration. *Econometrica*, **72**: 1127–1177.

Bai, J. and Ng, S. (2013). Principal components estimation and identification of static factors. *Journal of Econometrics*, **176**: 18–29.

Bai, J. and Ng, S. (2019). Matrix completion, counterfactuals, and factor analysis of missing data. arXiv preprint arXiv:1910.06677.

Banarjee, A., Dolado, J., Galbraith, J. W., and Hendry, D. (1993). *Cointegration, Error Correction and the Econometric Analysis of Non-stationary Data*. Oxford University Press, Oxford, UK.

Baragona, R. and Battaglia, F. (2007). Outliers in dynamic factor models. *Electronic Journal of Statistics*, **1**: 392–432.

Barigozzi, M., Lippi, M., and Luciani, M. (2016). Non-stationary dynamic factor models for large datasets. *Finance and Economics Discussion Series Divisions of Research & Statistics and Monetary*. Federal Reserve Board, Washington, DC. ArXiv preprint arXiv:1602.02398.024.

Barigozzi, M., Lippi, M., and Luciani, M. (2017). Dynamic factor models, cointegration, and error correction mechanisms. Working paper, arXiv:1510.02399v3.

Boivin, J. and Ng, S. (2006). Are more data always better for factor analysis?. *Journal of Econometrics*, **132**: 169–194.

Bolivar, S., Nieto, F., and Peña, D. (2020). On a new procedure for identifying a dynamic factor model. *Revista Colombiana de Estadística* (in press).

Bonhomme, S. and Manresa, E. (2015). Grouped patterns of heterogeneity in panel data. *Econometrica*, **83**: 1147–1184.

Box, G. E. P. and Tiao, G. (1977). A canonical analysis of multiple time series. *Biometrika*, **64**: 355–365.

Breitung, J. and Eickmeier, S. (2011). Testing for structural breaks in dynamic factor models. *Journal of Econometrics*, **163**: 71–84.

Breitung, J. and Tenhofen, J. (2011). GLS estimation of dynamic factor models. *Journal of the American Statistical Association*, **106**: 1150–1166.

Brillinger, D. R. (1964). The generalization of the techniques of factor analysis, canonical correlation adn principal components to stationary time series. *Royal Statistical Society Conference Cardiff*, Wales, UK.

Brillinger, D. R. (1981). *Time Series Data Analysis and Theory*, Expanded Edition. Holden-Day, San Francisco, CA.

Caro, A. and Peña, D. (2020). Estimating dynamic factor models with lag covariance matrices. Working paper, Universidad Carlos III de Madrid.

Cataño, D. H., Rodríguez-Caballero, C. V., and Peña, D. (2020). Wavelet estimation for dynamic factor models with time-varying loadings. Working paper no. 2019-23, Department of Economics and Business Economics, Aarhus University.

Chamberlain, G. (1983). Funds, factors, and diversification in arbitrage pricing models. *Econometrica*, **51**: 1305–1323.

Chamberlain, G. and Rothschild, M. (1983). Arbitrage, factor structure in arbitrage pricing models. *Econometrica*, **51**:1281–1304.

Chan, N. H., Lu, Y., and Yau, C. Y. (2017). Factor modelling for high-dimensional time series: Inference and model selection. *Journal of Time Series Analysis*, **38**: 285–307.

Chang, J., Yao, Q., and Zhou, W. (2017). Testing for high-dimensional white noises using maximum cross correlations. *Biometrika*, **104**: 111–127.

Chen, E. Y., Tsay, R. S., and Chen, R. (2020). Constrained factor models for high-dimensional matrix-variate time series. *Journal of the American Statistical Association*, **115**(530), 775–793.

Cheng, X., Liao, Z., and Schorfheide, F. (2016). Shrinkage estimation of highdimensional factor models with structural instabilities. *Review of Economic Studies*, **83**: 1511–1543.

Choi, I. (2012). Efficient estimation of factor models. *Econometric Theory*, **28**: 274–308.

Choi, I. (2017). Efficient estimation of nonstationary factor models. *Journal of Statistical Planning and Inference*, **183**: 18–43.

Connor, G. and Korajczyk, R. A. (1986). Performance measurement with the arbitrage pricing theory: A new framework for analysis. *Journal of Financial Economics*, **15**: 373–394.

Cook, R. D. (2018). *An Introduction to Envelopes: Dimension Reduction for Efficient Estimation in Multivariate Statistics*. John Wiley & Sons, Hoboken, NJ.

Corona, F., Poncela, P., and Ruiz, E. (2017). Determining the number of factors after stationary univariate transformations. *Empirical Economics*, **53**: 351–372.

Corona, F., Poncela, P., and Ruiz, E. (2020). Estimating non-stationary common factors: implications for risk sharing. *Computational Economics*, **55**: 37–60.

Corral, M. and Peña, D. (2008). The factorial dynamic model threshold. *Revista Colombiana de Estadística*, **31**: 183–190.

Creal, D. and Tsay, R. S. (2015). High dimensional dynamic stochastic copula models. *Journal of Econometrics*, **189**: 335–345.

Doz, C., Giannone, D., and Reichlin, L. (2012). A quasi–maximum likelihood approach for large, approximate dynamic factor models. *Review of Economics and Statistics*, **94**: 1014–1024.

Engle, R. and Watson, M. (1981). A one-factor multivariate time series model of metropolitan wage rates. *Journal of the American Statistical Association*, **76**: 774–781.

Escribano, A. and Peña, D. (1994). Cointegration and common factors. *Journal of Time Series Analysis*, **15**: 577–586.

Fan, J., Xue, L., and Yao, J. (2017). Sufficient forecasting using factor models. *The Journal of Econometrics*, **201**: 292–306.

Forni, M., Giannone, D., Lippi, M., and Reichlin, L. (2009). Opening the black box: Structural factor models with large cross sections. *Econometric Theory*, **25**: 1319–1347.

Forni, M., Hallin, M., Lippi, M., and Reichlin, L. (2000). The generalized dynamic factor model: Identification and estimation. *The Review of Economic and Statistics*, **82**: 540–554.

Forni, M., Hallin, M., Lippi, M., and Reichlin, L. (2005). The generalized dynamic factor model: One sided estimation and forecasting. *Journal of the American Statistical Association*, **100**: 830–840.

Forni, M., Hallin, M., Lippi, M., and Zaffaroni, P. (2015). Dynamic factor models with infinite-dimensional factor spaces: One-sided representations. *Journal of Econometrics*, **185**: 359–371.

Galeano, P., Peña, D., and Tsay, R. S. (2006). Outlier detection in multivariate time series by projection pursuit. *Journal of the American Statistical Association*, **101**: 654–669.

Galeano, P., Peña, D., and Tsay, R. S. (2020). Outlier detection in high dimensional time series. Working paper. Universidad Carlos III de Madrid.

Gao, Z. and Tsay, R. S. (2018). Modeling of high-dimensional time series: Another look at factor models with diverging eigenvalues. arXiv preprint arXiv:1808.07932.

Gao, Z. and Tsay, R. S. (2019). A structural-factor approach to modeling high-dimensional time series and space-time data. *Journal of Time Series Analysis*, **40**: 343–362.

Gao, Z. and Tsay, R. S. (2020). Modeling high-dimensional unit-root time series. *International Journal of Forecasting* (to appear).

García-Martos, C. and Conejo, A. J. (2013). Price forecasting techniques in power system. In J. Webster (ed.). *Encyclopedia of Electrical and Electronics Engineering*. John Wiley & Sons, Hoboken, NJ.

Geweke, J. F. (1977). The dynamic factor analysis of economic time series. In D. Aigner and A. Goldberger (eds.). *Latent Variables in Socioeconomic Models*. North Holland, Amsterdam, NL.

Geweke, J. F. and Singleton, K. J. (1981). Maximum likelihood confirmatory analysis of economic time series. *International Economic Review*, **22**: 37–54.

Hallin, M. and Liška, R. (2011). Dynamic factors in the presence of blocks. *Journal of Econometrics*, **163**: 29–41.

Harman, H. H. (1976). *Modern Fctor Analysis*. University of Chicago Press, Chicago, IL.

Hastie, T., Tibshirani, R., and Wainwright, M. (2015). *Statistical Learning with Sparsity: The Lasso and Generalizations*. CRC Press.

Hu, Y. (2005). Identifying the time-effect factors of multiple time series. *Journal of Forecasting*, **24**: 379–387.

Hu, Y. and Chou, R. (2004). On the Peña-Box model. *Journal of Time Series Analysis*, **25**: 811–830.

Johansen, S. (1995). *Likelihood-Based Inference in Cointegrated Vector Autoregressive Models*. Oxford University Press, Oxford, UK.

Johnstone, I. M. and Lu, A. Y. (2009). On consistency and sparsity for principal components analysis in high dimensions. *Journal of the American Statistical Association*, **104**: 682–693.

Jolliffe, I. T., Trendafilov, N. T., and Uddin, M. (2003). A modified principal component technique based on the LASSO. *Journal of computational and Graphical Statistics*, **12**(3), 531–547.

Jöreskog, K. G. (1967). Some contributions to maximum likelihood factor analysis. *Psychometrika*, **32**: 443–482.

Lam, C. and Yao, Q. (2012). Factor modeling for high dimensional time series: Inference for the number of factors. *The Annals of Statistics*, **40**: 694–726.

Lam, C., Yao, Q., and Bathia, N. (2011). Estimation of latent factors for high-dimensional time series. *Biometrika*, **98**: 901–918.

Li, K. C. (1991). Sliced inverse regression for dimension reduction. *Journal of the American Statistical Association*, **86**(414), 316–327.

Li, K. C. and Shedden, K. (2002). Identification of shared components in large ensembles of time series using dimension reduction. *Journal of the American Statistical Association*, **97**: 759–765.

Lin, C. and Ng, S. (2012). Estimation of panel data models with parameter heterogeneity when group membership is unknown. *Journal of Econometric Methods*, **1**: 42–55.

Ling, S., Tsay, R. S., and Yang, Y. (2020). Testing serial correlation and ARCH effect of high-dimensional time series data. *Journal of Business and Economic Statistics*, (to appear).

McAlinn, K., Rockova, V., and Saha, E. (2018). Dynamic sparse factor analysis. arXiv preprint arXiv:1812.04187.

Molenaar, P. C. M., De Gooijer, J. G., and Schmitz, B. (1992). Dynamic factor analysis of nonstationary multivariate time series. *Psychometrika*, **57**: 333–349.

Motta, G., Hafner, C. M., and von Sachs, R. (2011). Locally stationary factor models: Identification an nonparametric estimation. *Econometric Theory*, **27**: 1279–1319.

Motta, G. and Ombao, H. (2012). Evolutionary factor analysis of replicated time series. *Biometrics*, **68**: 825–836.

Nieto, F. H., Peña, D., and Saboya, D. (2016). Common seasonality in multivariate time series. *Statistica Sinica*, **26**: 1389–1410.

Onatski, A. (2009). Testing hypotheses about the number of factors in large factor models. *Econometrica*, **77**: 1447–1479.

Onatski, A. (2010). Determining the number of factors from empirical distribution of eigenvalues. *The Review of Economics and Statistics*, **92**: 1004–1016.

Onatski, A. and Wang, C. (2019). Spurious factor analysis. Manuscript. University of Cambridge.

Peña, D. (2009). Dimension reduction in time series and the dynamic factor model. *Biometrika*, **96**: 494–496.

Peña, D. and Box, G. (1987). Identifying a simplifying structure in time series. *Journal of the American Statistical Association*, **82**: 836–843.

Peña, D. and Poncela, P. (2004). Forecasting with nonstationary dynamic factor models. *Journal of Econometrics*, **119**: 291–321.

Peña, D. and Poncela, P. (2006). Nonstationary dynamic factor analysis. *Journal of Statistical Planning and Inference*, **136**: 1237–1256.

Peña, D., Smucler, E., and Yohai, V. J. (2019). Forecasting multiple time series with one-sided dynamic principal components. *Journal of the American Statistical Association*, **114**: 1683–1694.

Peña, D., Smucler, E., and Yohai, V. J. (2020). gdpc: An R Package for generalized dynamic principal components. *Journal of Statistical Software*, **92**: 1–23.

Peña, D. and Yohai, V. J. (2016). Generalized dynamic principal components. *Journal of the American Statistical Association*, **111**: 1121–1131.

Sargent, T. J. and Sims, C. A. (1977). Business cycle modeling without pretending to have too much a priori economic theory. *New Methods in Business Cycle Research*, **1**: 145–168.

Seber, G. A. F. (1984). *Multivariate Observations*. Wiley, Hoboken, NJ.

Shumway, R. H. and Stoffer, D. S. (2000). *Time Series Analysis and Its Applications*. Springer, New York, NY.

Smucler, E. (2019). Consistency of generalized dynamic principal components in dynamic factor models. *Statistics & Probability Letters*, **154**: 1–10.

Stock, J. H. and Watson, M. W. (1988). Testing for common trends. *Journal of the American Statistical Association*, **83**: 1097–1107.

Stock, J. H. and Watson, M. W. (2002a). Forecasting using principal components from a large number of predictors. *Journal of the American Statistical Association*, **97**: 1167–1179.

Stock, J. H. and Watson, M. W. (2002b). Macroeconomic forecasting using diffusion indexes. *Journal of Business and Economic Statistics*, **20**, 147–162.

Su, L., Shi, Z., and Phillips, P. C. (2016). Identifying latent structures in panel data. *Econometrica*, **84**: 2215–2264.

Su, L. and Wang, X. (2017). On time-varying factor models: Estimation and testing. *Journal of of Econometrics*, **198**: 84–101.

Tiao, G. C. and Tsay, R. S. (1989). Model specification in multivariate time series. *Journal of the Royal Statistical Society, Series B*, **51**: 157–213.

Tsay, R. S. (2014). *Multivariate Time Series Analysis: With R and Financial Applications*. John Wiley & Sons, Hoboken, NJ.

Tsay, R. S. (2020). Testing serial correlations in high-dimensional time series via extreme value theory. *Journal of Econometrics*, **216**: 106–117.

Tsay, R. S. and Tiao, G. C. (1985). Use of canonical analysis in time series model identification. *Biometrika*, **72**: 299–315.

Velu, R. P., Reinsel, G. C., and Wichern, D. W. (1986). Reduced rank models for multiple time series. *Biometrika*, **73**: 105–118.

Wang, P. (2010). Large dimensional factor models with a multi-level factor structure. Working paper, Department of Economics, HKUST.

Wang, W. and Fan, J. (2017). Asymptotics of empirical eigenstructure for ultra-high dimensional spiked covariance model. *Annals of Statistics*, **45**: 1342–1374.

Wang, D., Liu, X., and Chen, R. (2019). Factor models for matrix-valued high-dimensional time series. *Journal of of Econometrics*, **208**: 231–248.

Zhang, R., Robinson, P., and Yao, Q. (2018). Identifying cointegration by eigenanalysis. *Journal of the American Statistical Association*, **114**: 916–927.

Zou, H. and Hastie, T. (2005). Regularization and variable selection via the elastic net. *Journal of the Royal Statistical Society: Series B (Statistical Methodology)*, **67**(2), 301–320.

CHAPTER 7

FORECASTING WITH BIG DEPENDENT DATA

Let y_t be a scalar variable of interest and $\boldsymbol{x}_t = (x_{1t}, \dots, x_{kt})'$ be a collection of k explanatory variables, which may contain some lagged values of y_t. In this chapter, we focus on the problem of predicting y_{n+h} given $\{\boldsymbol{x}_1, \dots, \boldsymbol{x}_n\}$, where n is the forecast origin and $h > 0$ is the forecast horizon. For ease in presentation, let F_n be the available information at time $t = n$. Mathematically, F_n denotes the σ-field generated by $\{\boldsymbol{x}_1, \dots, \boldsymbol{x}_n\}$. Our goal is to study the conditional distribution $p(y_{n+h}|F_n)$. This is an old prediction problem with many statistical methods available in the literature. However, in the current data-rich environment, several interesting methods have been developed in recent years. We introduced some of those methods based on the dynamic factor model in Chapter 6. In this chapter, we discuss and demonstrate several other new developments. The methods considered include (a) regularized linear models, (b) diffusion index (DI) or principal component regression (PCR), (c) partial least squares, and (d) boosting algorithms. For regularized methods, we discuss Least Absolute Shrinkage and Selection Operator (Lasso) and its various extensions. We also introduce some recent developments concerning Lasso methods for dependent data. For boosting, we discuss ℓ_2 boosting for forecasting and some boosting algorithms for classification. Finally, we introduce the mixed-data sampling (MIDAS) approach to forecasting and the idea of nowcasting for updating prediction. The latter uses high-frequency data to update the prediction of a lower-frequency variable of interest. For instance, the gross domestic product (GDP) of an economy is typically available quarterly. Using nowcasting, one can revise the GDP forecast once data of monthly variables become available when the first month into the quarter has passed. Neural networks, tree-based methods such as random forest, and machine learning are discussed in the next chapter.

Statistical Learning for Big Dependent Data, First Edition. Daniel Peña and Ruey S. Tsay.
© 2021 John Wiley & Sons, Inc. Published 2021 by John Wiley & Sons, Inc.

The discussion of this chapter contains certain deviations from the traditional methods available in the literature, because we focus on dependent data and emphasize on the case when the number of predictors k is large, even greater than the sample size T or the forecast origin n. Keep in mind, however, that most of the modern methods discussed in the chapter were developed for independent data so that in some cases we may start with this assumption. Extensions or justifications of those methods to dependent data will be given later, if available.

Care must be exercised, however, in applying methods developed for independent data to dependent data. It is well known that serial dependence and high multicollinearity may lead to erroneous inference if they are overlooked. See the discussions in Chapter 1 and Section 7.2. Also, some regularizations are needed to build a forecasting model when $k > \min\{n, T\}$. We start our discussion with short-range dependence data, i.e. serial dependence of the variables involved decays to zero exponentially. The case of strong serial dependence will be briefly addressed in a later section.

7.1 REGULARIZED LINEAR MODELS

One of the important recent developments in statistics is the regularized estimation under the *sparsity* assumption, especially the Lasso (Tibshirani, 1996). The linear regression model for the problem of interest of this chapter can be written as

$$y_{t+h} = \boldsymbol{x}_t'\boldsymbol{\beta} + e_t = \sum_{i=1}^{k} \beta_i x_{it} + e_t, \quad t = 1, \dots, n \tag{7.1}$$

where the forecast horizon h is given and fixed and $\boldsymbol{\beta}$ is the vector of parameters. For simplicity in notation, it is understood that $\boldsymbol{x}_t$ may contain the vector of 1. Often in applications, each predictor x_{it} is mean-adjusted and standardized to have unit variance. This normalization works well for independent data, but may encounter difficulties when the data have strong serial correlations. We give some details in Section 7.2. For regularized linear model one estimates the unknown parameters and uses the estimates along with values of the predictors to obtain prediction.

In matrix- and vector-notation, the linear regression model of Eq. (7.1) becomes

$$\boldsymbol{Y} = \boldsymbol{X}\boldsymbol{\beta} + \boldsymbol{E}, \tag{7.2}$$

where $\boldsymbol{Y}$ is the n-dimensional response vector, $\boldsymbol{X}$ is the design matrix with ith row being $\boldsymbol{x}_i'$, and $\boldsymbol{E}$ is the error vector.

Under the traditional linear regression analysis, one assumes that (a) the errors $e_1, \dots, e_n$ are independent and identically distributed (i.i.d.) with mean zero and variance σ_e^2, (b) e_t is independent of $\boldsymbol{x}_t$, and (c) the predictor $\boldsymbol{x}_t$ satisfies the condition that $E(\boldsymbol{x}_t\boldsymbol{x}_t')$ is positive definite. We shall relax these conditions later.

If k is sufficiently less than n, then the ordinary least squares (LS) method can be used to estimate the model in Eq. (7.1). The estimates are $\widehat{\boldsymbol{\beta}} = (\sum_{t=1}^{n} \boldsymbol{x}_t\boldsymbol{x}_t')^{-1} \sum_{t=1}^{n} \boldsymbol{x}_t y_{t+h}$. Furthermore, we have

$$\sqrt{n}(\widehat{\boldsymbol{\beta}} - \boldsymbol{\beta}) \rightarrow_d N\left[\boldsymbol{0}, \sigma_e^2\left(\sum_{t=1}^{n} \boldsymbol{x}_t\boldsymbol{x}_t'\right)^{-1}\right],$$

where $\to_d$ denotes convergence in distribution. For details of properties of $\widehat{\boldsymbol{\beta}}$, readers are referred to, for instance, Seber and Lee (2003).

On the other hand, if $k \geq n$, then the LS estimates are not unique and will over-fit the data. Some form of regularization is needed. The Lasso estimator of Tibshirani (1996) uses the ℓ_1-penalty with this objective. For a given penalty parameter (or tuning parameter) $\lambda > 0$, the Lasso estimator $\widehat{\boldsymbol{\beta}}(\lambda)$ is given by

$$\widehat{\boldsymbol{\beta}}(\lambda) = \arg\min_{\boldsymbol{\beta}} \left(\frac{\|\boldsymbol{Y} - \boldsymbol{X}\boldsymbol{\beta}\|_2^2}{n} + \lambda\|\boldsymbol{\beta}\|_1 \right), \tag{7.3}$$

where $\|\boldsymbol{Y} - \boldsymbol{X}\boldsymbol{\beta}\|_2^2 = \sum_{t=1}^{n} (y_{t+h} - \boldsymbol{x}_t'\boldsymbol{\beta})^2$ and $\|\boldsymbol{\beta}\|_1 = \sum_{i=1}^{k} |\beta_i|$. Note that, for convenience, in the Lasso literature, the predictors are often standardized so that $\sum_{t=1}^{n} x_{it} = 0$ and $\sum_{t=1}^{n} x_{it}^2 = 1$ for $i = 1, \ldots, k$. Also, one further assumes the dependent variable is mean-adjusted, i.e. $\bar{Y} = \sum_{t=1}^{n} y_{t+h}/n = 0$.

The ℓ_1-penalty of Lasso has several nice properties. First, the objective function of Eq. (7.3) is convex so that efficient computation methods can be used to obtain the Lasso estimator, including the coordinate descent method and the least angle regression of Efron et al. (2004). See, for instance, Giraud (2015, chapter 4) for further details. Second, the ℓ_1-geometry, that implies linear restrictions, enables Lasso to perform variable selection in the sense that some estimates $\hat{\beta}_i$ are often exactly zero. It can be shown that the optimization problem of Eq. (7.3) is equivalent to

$$\widehat{\boldsymbol{\beta}}_{\text{primal}}(R) = \arg \min_{\boldsymbol{\beta}:\|\boldsymbol{\beta}\|_1 \leq R} \frac{\|\boldsymbol{Y} - \boldsymbol{X}\boldsymbol{\beta}\|_2^2}{n}, \tag{7.4}$$

where there is a one-to-one correspondence between the penalty parameter λ in (7.3) and R in (7.4). The correspondence may depend on the data. Details can be found, for instance, in Bertsekas (1995, chapter 5). In linear programming with linear objective function and linear constraints, it is well known that the solution must occur at the vertexes or the simplex. With many variables several of them would be highly correlated and the contours of the quadratic function in Eq. (7.4) for these variables will be approximately linear, and the solution would often be on the vertexes. In this way Lasso could perform variable selection. The assumption that only a few coefficients β_i are non-zero is referred to as the *sparsity* assumption in the literature. The sparsity condition plays an important role in data analysis when $k \geq n$.

To illustrate further the variable selection property of the Lasso, suppose that all the variables are standardized to zero mean and unit variance and that all the β_i coefficients are positive. Taking the derivative in Eq. (7.3) with respect to every coefficient β_i, we obtain

$$\frac{\partial\widehat{\boldsymbol{\beta}}(\lambda)}{\partial\boldsymbol{\beta}} = 0 \quad \Rightarrow \boldsymbol{\beta} = (\boldsymbol{X}'\boldsymbol{X})^{-1}[\text{Cov}(\boldsymbol{X}, y) - \mathbf{1}\lambda^*],$$

where $\lambda^* = \lambda n/2$ and $\text{Cov}(\boldsymbol{X}, y) = \boldsymbol{X}'\mathbf{y}$. Thus, making $\lambda^* = \text{Cov}(x_i, y)$ the estimated value of β_i would be zero. Therefore, if we have a set of correlated predictors, say $\boldsymbol{x}$, with similar values in $\text{Cov}(\boldsymbol{x}, y)$, $\lambda^* = \text{Cov}(\boldsymbol{x}, y)$ would make most of the corresponding β_i coefficients equal to zero.

There are other penalty functions available for the regularization. For instance, the well-known ridge-regression is obtained when the ℓ_2-penalty is used, i.e.

$$\hat{\beta}_{\text{ridge}}(\lambda) = \arg\min_{\beta} \left(\frac{\|\boldsymbol{Y} - \boldsymbol{X}\boldsymbol{\beta}\|_2^2}{n} + \lambda \|\boldsymbol{\beta}\|_2^2 \right). \tag{7.5}$$

See Hoerl (1962). The ℓ_2 penalty is also known as the *Tikhonov* regularization.

It can be shown that these regularizations can be interpreted as Bayes estimators with different priors; see, for instance, Hastie et al. (2009, 2015). See also Taddy (2017) for related work.

An important problem in application of regularized regression is the selection of the penalty parameter λ. A small penalty would result in over-fitting whereas a large penalty, such as $\lambda = \infty$, would lead to $\beta_i = 0$ for all i. In the literature, for independent data, as explained in Chapter 5, one can use cross-validation (CV) to select λ, such as the 10-fold CV. Roughly speaking, one randomly divides the data set $S = \{(y_{t+h}, \boldsymbol{x}_t) | t = 1, \ldots, n\}$ into 10 subsamples, say $S_1, \ldots, S_{10}$. Let $S_{(i)}$ be all the data but the subsample S_i. That is, $S_{(i)} = \cup_{j \neq i}^{10} S_j$. Let $\{\lambda_j | j = 1, \ldots, m\}$ be a set of possible values for λ and let $\text{SSE}(\lambda_j)$ be the sum of squared forecast errors defined below for penalty parameter λ_j. The CV procedure works as follows:

1. For each λ_j $(j = 1, \ldots, m)$
   (a) Initialize $\text{SSE}(\lambda_j) = 0$
   (b) For $i = 1, \ldots, 10$:
      - Fit the Lasso model with penalty λ_j using data $S_{(i)}$
      - Predict the response using the fitted parameters and the predictors in S_i
      - Compute the prediction errors for the subsample S_i
      - Update $\text{SSE}(\lambda_j)$ by adding the sum of squared prediction errors of subsample S_i
2. Select λ by $\lambda = \arg\min_{\lambda_j} \text{SSE}(\lambda_j)$

This procedure is commonly used in most software packages of Lasso regression. However, the validity of such a CV is questionable when the data are serially dependent. We discuss some methods later for choosing the penalty parameter λ when the data are serially dependent.

7.1.1 Properties of Lasso Estimator

In this section, we briefly outline some important properties of Lasso estimator when the errors e_t of Eq. (7.3) satisfy the usual linear regression assumptions. Here the number of predictors k may be greater than the forecast origin n (or sample size T) so that one can think of k as a function of n, and the asymptotic properties discussed are for $n \to \infty$. Readers are referred to Bühlmann and van de Geer (2011) for further details and additional theoretical properties of the Lasso estimator.

We start with some notation. For two real sequences $\{a_n\}$ and $\{b_n\}$, $a_n = o(b_n)$ if $a_n / b_n \to 0$ as $n \to \infty$ and $a_n \asymp b_n$ if there exist constants c and C such that $c \leq a_n / b_n \leq C$ for all sufficiently large n. Also, define the qth norm of $\boldsymbol{\beta}$ as $\|\boldsymbol{\beta}\|_q = \left(\sum_{i=j}^{k} |\beta_j|^q \right)^{1/q}$.

To introduce the basic properties of Lasso estimator, consider the linear model

$$y_i = \sum_{j=1}^{k} \beta_j x_{i,j} + \epsilon_i, \quad i = 1, \ldots, n, \tag{7.6}$$

where the design matrix $\boldsymbol{X}$ is fixed and the error vector $\boldsymbol{\epsilon} = (\epsilon_1, \ldots, \epsilon_n)' \sim N(0, \sigma_e^2 \boldsymbol{I})$. Assume that the linear model in Eq. (7.6) is the true model and denote the true parameter vector by $\boldsymbol{\beta}^0$. If $n > k$, $\boldsymbol{X}$ is full rank and the LS estimator $\hat{\boldsymbol{\beta}} = (\boldsymbol{X}'\boldsymbol{X})^{-1}(\boldsymbol{X}'\boldsymbol{Y})$ enjoys various nice properties. Furthermore,

$$\|\boldsymbol{X}(\hat{\boldsymbol{\beta}} - \boldsymbol{\beta}^0)\|_2^2 / \sigma_e^2 \sim \chi_k^2,$$

where χ_k^2 denotes the chi-square distribution with k degrees of freedom. Therefore,

$$\frac{E\|\boldsymbol{X}(\hat{\boldsymbol{\beta}} - \boldsymbol{\beta}^0)\|_2^2}{n} = \frac{\sigma_e^2}{n} k.$$

This means that, after re-parameterization so that the columns of $\boldsymbol{X}$ are orthogonal, each parameter β_j^0 is estimated with squared accuracy $\sigma_e^2/n, j = 1, \ldots, k$.

Turn to the case of $k > n$. Here certain restriction is needed. It can be shown (Bühlmann and van de Geer, 2011) that for the Lasso estimate $\hat{\boldsymbol{\beta}}$ we have

$$\frac{\|\boldsymbol{X}(\hat{\boldsymbol{\beta}} - \boldsymbol{\beta}^0)\|_2^2}{n} \leq C \|\boldsymbol{\beta}^0\|_1 \sqrt{\log k / n},$$

so that if $\|\boldsymbol{\beta}^0\|_1$ has a smaller number of non-zero coefficients than $\sqrt{n/\log k}$, say $\|\boldsymbol{\beta}^0\|_1 = o(\sqrt{n/\log k})$, the Lasso estimator is consistent when both k and n go to infinity. Let

$$S_0 = \{j | \beta_j^0 \neq 0\}$$

be the set of indexes with non-zero true coefficients and $s_0 = |S_0|$ be the number of non-zero coefficients of $\boldsymbol{\beta}^0$. In the literature, s_0 is the *sparsity index* of $\boldsymbol{\beta}^0$ and S_0 can be regarded as the *active set* of the linear model (7.6). If S_0 were known, then we can simply select those columns of $\boldsymbol{X}$ associated with non-zero parameters to perform the analysis. But S_0 is unknown in practice. Under regularized estimation with a properly chosen penalty parameter λ (of order $\sigma_e \sqrt{\log(k)n}$), we have the *oracle inequality*, see Candes (2006) for an illustration of this concept, the Lasso estimator satisfying

$$\frac{\|\boldsymbol{X}(\hat{\boldsymbol{\beta}} - \boldsymbol{\beta})\|_2^2}{n} \leq c \frac{\sigma_e^2 \log(k)}{n} s_0, \tag{7.7}$$

with a large probability, where the constant c depends on k or n through the Gram matrix $\hat{\boldsymbol{\Sigma}} = \boldsymbol{X}'\boldsymbol{X}/n$. Compared with the case of $k < n$, we have added the factors $\log(k)$ to the inequality.

Next, consider the linear model of Eq. (7.6). By the definition of Lasso estimator, we have

$$\|\boldsymbol{Y} - \boldsymbol{X}\hat{\boldsymbol{\beta}}\|_2^2/n + \lambda \|\hat{\boldsymbol{\beta}}\|_1 \leq \|\boldsymbol{Y} - \boldsymbol{X}\boldsymbol{\beta}^0\|_2^2/n + \lambda \|\boldsymbol{\beta}^0\|_1.$$

Consequently, by rewriting $\boldsymbol{Y} - \boldsymbol{X}\hat{\boldsymbol{\beta}} = (\boldsymbol{Y} - \boldsymbol{X}\boldsymbol{\beta}^0) - \boldsymbol{X}(\hat{\boldsymbol{\beta}} - \boldsymbol{\beta}^0)$, we have the following basic inequality

$$\|\boldsymbol{X}(\hat{\boldsymbol{\beta}} - \boldsymbol{\beta}^0)\|_2^2/n + \lambda \|\hat{\boldsymbol{\beta}}\|_1 \leq 2\boldsymbol{\epsilon}'\boldsymbol{X}(\hat{\boldsymbol{\beta}} - \boldsymbol{\beta}^0)/n + \lambda \|\boldsymbol{\beta}^0\|_1. \tag{7.8}$$

The term $2\epsilon'\boldsymbol{X}(\hat{\boldsymbol{\beta}} - \boldsymbol{\beta}^0)$ is referred to as the *empirical process* of the problem. That is, it is the random part of the problem and it can be bounded by the ℓ_1 norm of the parameters involved as below:

$$2|\epsilon'\boldsymbol{X}(\hat{\boldsymbol{\beta}} - \boldsymbol{\beta}^0)| \leq \left(\max_{1\leq j\leq k} 2|\epsilon\boldsymbol{X}_{.j}|\right) ||\hat{\boldsymbol{\beta}} - \boldsymbol{\beta}^0||_1,$$

where $\boldsymbol{X}_{.j}$ denotes the jth column of $\boldsymbol{X}$. The idea of the penalty is then to dominate the empirical process of the problem. Therefore, define

$$F = \left\{\max_{1\leq j\leq k} 2|\epsilon'\boldsymbol{X}_{.j}|/n \leq \lambda_0\right\}, \tag{7.9}$$

where we assume that $\lambda > 2\lambda_0$ to ensure that on the set F we can get rid of the random part of the problem. Let $\hat{\sigma}_j^2$ be the (j, j)th element of $\hat{\boldsymbol{\Sigma}} = \boldsymbol{X}'\boldsymbol{X}/n$, which is the estimate of the variance of the jth predictor. The following result confirms that the set F is not small.

Lemma 7.1 Assume that $\hat{\sigma}_j = 1$ for $j = 1, \ldots, k$. Then, for all $t > 0$ and for $\lambda_0 = 2\sigma_e\sqrt{\frac{t^2+2\log(k)}{n}}$, we have

$$P(F) \geq 1 - 2\exp(-t^2/2).$$

Proof: Under the assumption $\hat{\sigma}_j = 1$, the variable $\boldsymbol{U}_j = \epsilon'\boldsymbol{X}_{.j}/(\sqrt{n\sigma_e^2})$ follows a $N(0, \boldsymbol{I})$ distribution. Therefore,

$$P\left(\max_{1\leq j\leq k} |U_j| > \sqrt{t^2 + 2\log(k)}\right) \leq 2k\exp\left[-\frac{t^2 + 2\log(k)}{2}\right] = 2\exp\left(\frac{-t^2}{2}\right).$$

Using the basic inequality and Lemma 7.1, we can obtain the consistency of the Lasso estimator.

Theorem 7.1 Assume that $\hat{\sigma}_j = 1$ for all j. For $t > 0$, let the penalty parameter be

$$\lambda = 4\hat{\sigma}_e\sqrt{\frac{t^2 + 2\log(k)}{n}},$$

where $\hat{\sigma}_e$ is an estimate of the standard error σ_e of the error ϵ_i. Then, with probability at least $1 - \alpha$, where $\alpha = 2\exp[-t^2/2] + P(\hat{\sigma}_e \leq \sigma_e)$,

$$2||\boldsymbol{X}(\hat{\boldsymbol{\beta}} - \boldsymbol{\beta}^0)||_2^2 \leq 3\lambda||\boldsymbol{\beta}^0||_1.$$

See corollary 6.1 of Bühlmann and van de Geer (2011). It says that, for the linear model in Eq. (7.6), by taking the penalty (or regularization) parameter λ of the order $\sqrt{\log(k)/n}$ and assuming that the ℓ_1-norm of the true parameter $\boldsymbol{\beta}^0$ is of smaller order than $\sqrt{n/\log(k)}$, we obtain the consistency result of the Lasso estimator.

The same result can be obtained when the design matrix is random, i.e. $\boldsymbol{X}$ consists of random variables. In general, let

$$\boldsymbol{\Sigma}_X = \begin{cases} \boldsymbol{X}'\boldsymbol{X}/n, & \text{if the design is fixed,} \\ \text{Cov}(\boldsymbol{x}_t), & \text{if the design is random.} \end{cases}$$

Under some mild regularity condition on the error distribution and a properly chosen range of the penalty parameter λ, say $\lambda \asymp \sqrt{\log(k)/n}$, the Lasso estimator is consistent as follows:

$$[\hat{\beta}(\lambda) - \beta^0]'\Sigma_X[\hat{\beta}(\lambda) - \beta^0] = o_p(1), \quad n \to \infty, \tag{7.10}$$

if the design is fixed and the basic sparsity assumption

$$s_0 = \|\beta^0\|_0 = o\left(\sqrt{\frac{n}{\log(k)}}\right) \tag{7.11}$$

holds. The consistency result of Eq. (7.10) continues to hold if the design is random provided that the true parameters satisfy

$$\|\beta^0\|_1 = o((n/\log(n))^{1/4}). \tag{7.12}$$

See Greenshtein and Ritov (2004) and Bartlett et al. (2012).

Let S be a subset of $\{1, \dots, k\}$, i.e. $S \subset \{1, \dots, k\}$. Let β_S be a k-dimensional vector such that its jth element is β_j if $j \in S$ and is 0, otherwise. Similarly, β_{S^c} be a k-dimensional vector such that its jth element is β_j if $j \ni S$ and is 0, otherwise, where S^c denotes the complement of S with respect to $\{1, \dots, k\}$. Clearly, $\beta = \beta_S + \beta_{S^c}$. We can then consider further properties of the Lasso estimator.

Lemma 7.2 On the set F of Eq. (7.9) with $\lambda > 2\lambda_0$, we have

$$2\|X(\hat{\beta} - \beta^0)\|_2^2/n + \lambda\|\hat{\beta}_{S_0^c}\|_1 \le 3\|\hat{\beta}_{S_0} - \beta_{S_0}\|_1.$$

The lemma can be shown by the basic inequality and triangular inequality. See Lemma 6.3 of Bühlmann and van de Geer (2011). By Lemma 7.2, we have

$$\|\hat{\beta}_{S_0^c}\|_1 \le 3\|\hat{\beta}_{S_0} - \beta_{S_0}\|_1.$$

For the linear model of Eq. (7.6), we say that the *compatibility condition* is met for the set S_0 if for some $\phi_0 > 0$ and for all β satisfying $\|\beta_{S_0^c}\|_1 \le 3\|\beta_{S_0}\|_1$, the inequality condition

$$\|\beta_{S_0}\|_1^2 \le (\beta'\hat{\Sigma}\beta)s_0/\phi_0^2 \tag{7.13}$$

holds. Note that the coefficient 3 of the inequality is rather arbitrary, but commonly used in the literature. It can be replaced by any constant greater than 1. This compatibility condition is similar to a condition on the smallest eigenvalue of $\hat{\Sigma}$ commonly seen in the literature if one replaces $\|\beta_{S_0}\|_1^2$ by its upper bound $s_0\|\beta_{S_0}\|_2^2$. In practice, S_0 is unknown so that the compatibility condition cannot be verified. If s_0, the cardinality of S_0 is known, then one can verify the condition by checking all subsets of $\{1, \dots, k\}$ with cardinality s_0, resulting in the *restricted eigenvalue assumption* of Bickel et al. (2009).

Theorem 7.2 Suppose that the compatibility condition holds for S_0. Then, on the set F and for $\lambda \ge 2\lambda_0$, we have

$$\|X(\hat{\beta} - \beta^0)\|_2^2/n + \lambda\|\hat{\beta} - \beta^0\|_1 \le 4\lambda^2 s_0/\phi_0^2.$$

See theorem 6.1 of Bühlmann and van de Geer (2011). The result of Theorem 7.2 shows two error bounds. First, $\|\boldsymbol{X}(\widehat{\boldsymbol{\beta}} - \boldsymbol{\beta}^0)\|_2^2/n \le 4\lambda^2 s_0/\phi_0^2$ for the linear regression prediction error. Second, $\|\widehat{\boldsymbol{\beta}} - \boldsymbol{\beta}^0\|_1 \le 4\lambda s_0/\phi_0^2$ for the ℓ_1 estimation error.

Finally, we state two further properties of the Lasso estimator available in Bühlmann and van de Geer (2011). First, under the compatibility condition of the design matrix $\boldsymbol{X}$ and on the sparsity s_0 of the linear model, it can be shown that for λ in a suitable range of order $\lambda \asymp \sqrt{\log(k)/n}$, we have

$$\|\widehat{\boldsymbol{\beta}}(\lambda) - \boldsymbol{\beta}^0\|_q \to_p \mathbf{0}, \quad n \to \infty,$$

where $\to_p$ denotes convergence in probability and $q \in \{1, 2\}$. Second, to use Lasso estimator for variable selection, one needs some rather restrictive conditions. Define

$$\widehat{S}(\lambda) = \{j | \widehat{\beta}_j(\lambda) \neq 0, j = 1, \ldots, k\} \tag{7.14}$$

as the set of indexes with non-zero elements in $\widehat{\boldsymbol{\beta}}(\lambda)$. For a proper variable selection, we wanted to have $P(\widehat{S}(\lambda) = S_0) \to 1$ as the sample size (or forecast origin) n increases. To this end, certain conditions on the design matrix $\boldsymbol{X}$ is needed. Again, let $\hat{\boldsymbol{\Sigma}} = \boldsymbol{X}'\boldsymbol{X}/n$ and, without loss of generality, assume that the first s_0 elements of $\boldsymbol{\beta}$ is non-zero. Partition $\hat{\boldsymbol{\Sigma}}$ as follows

$$\hat{\boldsymbol{\Sigma}} = \begin{bmatrix} \hat{\boldsymbol{\Sigma}}_{11} & \hat{\boldsymbol{\Sigma}}_{12} \\ \hat{\boldsymbol{\Sigma}}_{21} & \hat{\boldsymbol{\Sigma}}_{22} \end{bmatrix},$$

where $\hat{\boldsymbol{\Sigma}}_{11}$ is a $s_0 \times s_0$ matrix, $\hat{\boldsymbol{\Sigma}}_{22}$ is a $(k - s_0) \times (k - s_0)$ matrix, and $\hat{\boldsymbol{\Sigma}}_{12}$ is of dimension $(s_0, k - s_0)$. The *irrepresentable condition* is

$$\|\hat{\boldsymbol{\Sigma}}_{21}\hat{\boldsymbol{\Sigma}}_{11}^{-1}\text{sign}(\beta_1^0, \beta_2^0, \ldots, \beta_{s_0}^0)\|_\infty \le \theta, \quad 0 < \theta < 1, \tag{7.15}$$

where $\|\boldsymbol{x}\|_\infty = \max_j |x_j|$ and $\text{sign}(\beta_1^0, \beta_2^0, \ldots, \beta_{s_0}^0) = [\text{sign}(\beta_1^0), \text{sign}(\beta_2^0), \ldots, \text{sign}(\beta_{s_0}^0)]'$. See Zhao and Yu (2006). Assuming the irrepresentable condition and the non-zero coefficients satisfying

$$\inf_{j \in S_0} |\beta_j^0| \gg \sqrt{s_0 \log(k)/n},$$

Meinshausen and Bühlmann (2006) show that, for a suitably chosen $\lambda \gg \sqrt{\log(k)/n}$,

$$P(\widehat{S}(\lambda) = S_0) \to 1, \quad n \to \infty.$$

The derivations of this section are brief and might be hard to follow. The message is clear, however. Under some regularity conditions on the properties of the design matrix, the error term, and the cross-dependence between predictors and the error, one can obtain good results of the Lasso regression for a properly chosen penalty parameter λ. This is particularly so for the case of strong sparsity, i.e. the number of non-zero true coefficients in $\boldsymbol{\beta}^0$ is finite.

7.1.2 Some Extensions of Lasso Regression

In finite samples, the Lasso estimator may contain biases. Also, in some applications, domain knowledge or past experience may provide certain information on the coefficient vector $\boldsymbol{\beta}$. For instance, the problem of interest may contain functions or a subset of the parameters β_j ($j = 1, \dots, k$). To expand the applicability of Lasso methodology various extensions have been proposed and in this section we introduce some of them.

7.1.2.1 *Adaptive Lasso*

Selecting the penalty parameter λ is important in Lasso application. The commonly used CV method is based on prediction. For variable selection, experience indicates that Lasso tends to select too many variables. To mitigate this over-selection problem, Zou (2006) proposes an adaptive Lasso, which is a two-stage procedure that assigns different penalties to each coefficient based on weight as follows. Let

$$\widehat{\boldsymbol{\beta}}_{\text{adapt}}(\lambda) = \arg\min_{\boldsymbol{\beta}} \left(\|\boldsymbol{Y} - \boldsymbol{X}\boldsymbol{\beta}\|_2^2/n + \lambda \sum_{j=1}^{k} \frac{|\beta_j|}{|\widehat{\beta}_{\text{ini},j}|} \right), \tag{7.16}$$

where $\widehat{\beta}_{\text{ini},j}$ denotes an initial estimator of β_j. In practice, one often runs the usual Lasso estimation with CV to select the penalty parameter. Denote the selected penalty as $\lambda_{\text{ini},cv}$ and one can use $\widehat{\beta}_j(\lambda_{\text{ini},cv})$ as the initial estimate $\widehat{\beta}_{\text{ini},j}$ and run the adaptive Lasso. The penalty parameter of the second-stage Lasso is also selected by CV. Of course, in the second stage, if $\widehat{\beta}_{\text{ini},j} = 0$, then $\widehat{\beta}_{\text{adapt},j} = 0$. A nice feature of the adaptive Lasso is that different initial estimates result in using different penalties at the second-stage analysis. A large $\widehat{\beta}_{\text{ini},j}$ implies a smaller penalty for $\widehat{\beta}_{\text{adapt},j}$, whereas a small $\widehat{\beta}_{\text{ini},j}$ implies a heavier penalty, making it easier that this parameter is estimated as zero and, hence, reducing the number of variables selected in the second stage of estimation.

7.1.2.2 *Group Lasso*

In some applications, the data might contain some factor structure and the Group Lasso is designed for this situation. It was introduced in Chapter 4 for detecting parameter changes in autoregressive (AR) processes. Now, suppose that the k predictors are divided into g groups. Specifically, suppose $G_1, \dots, G_g$ form a partition of $\{1, \dots, k\}$. That is, $\cup_{i=1}^{g} G_i = \{1, \dots, k\}$ and $G_i \cap G_j = \emptyset$ for $i \neq j$. Also, let $\boldsymbol{\beta}_{G_i}$ be the vector of parameters for variables in G_i. Let $\|\boldsymbol{\beta}_{G_i}\|_2$ be the usual Euclidean norm of $\boldsymbol{\beta}_{G_i}$, where $i = 1, \dots, g$. The Group Lasso estimator is defined as

$$\widehat{\boldsymbol{\beta}}_g(\lambda) = \arg\min_{\boldsymbol{\beta}} \left(\|\boldsymbol{Y} - \boldsymbol{X}\boldsymbol{\beta}\|_2^2/n + \lambda \sum_{j=1}^{g} m_j \|\boldsymbol{\beta}_{G_j}\|_2 \right), \tag{7.17}$$

where m_j is a constant, e.g. $m_j = \sqrt{n_j}$ with n_j being the cardinality of the group G_j. Clearly, Group Lasso considers simultaneously all parameters in each group. Depending on λ, either the vector of parameters for variables in G_i are zero or all of its elements are non-zero. Note that if $g = k$ so that all groups contain a single variable, then, by letting $\|\boldsymbol{\beta}_{G_i}\|_2 = \|\boldsymbol{\beta}_{G_i}\|_1$, the Group Lasso reduces to the standard Lasso. The Group Lasso was proposed by Yuan and Lin (2006). It can be used in modeling high-dimensional time series (HDTS) if one is interested in the coefficient matrix jointly at a given lag of a vector autoregressive (VAR) model. Note that the L2 penalty function is needed instead of the L1 in order to consider jointly the parameters in each group. Several procedures for estimating sparse additive models are related to this estimation method (see Hastie et al. 2015).

7.1.2.3 *Elastic Net* Zou and Hastie (2005) use a combination of ℓ_1- and ℓ_2-penalties to improve the interpretation of Lasso estimator and propose the elastic net defined as

$$\widehat{\boldsymbol{\beta}}_{\mathrm{EN}}(\alpha) = \arg\min_{\boldsymbol{\beta}}(\|\boldsymbol{Y} - \boldsymbol{X}\boldsymbol{\beta}\|_2^2/n + \lambda\alpha\|\boldsymbol{\beta}\|_1 + \lambda(1-\alpha)\|\boldsymbol{\beta}\|_2), \tag{7.18}$$

where $0 \le \alpha \le 1$ and the subscript "EN" is used to denote elastic net. Clearly, the elastic net estimator reduces to the Lasso estimator if $\alpha = 1$ and it is the ridge regression estimator if $\alpha = 0$. In Eq. (7.18), a simple linear combination is used to define the penalty, which can be selected using CV. In Zou and Hastie (2005), two penalties λ_1 and λ_2 are used, resulting in the need to select two tuning parameters. In the literature, there exists no clear guidelines for selecting the parameter α. One can choose a few possible values for α and compare their out of sample performance to select a proper value.

7.1.2.4 *Fused Lasso* To improve the parsimony (or smoothness) in parameterization of a linear regression model, Tibshirani et al. (2005) proposed the fused Lasso, which is defined as

$$\widehat{\boldsymbol{\beta}}_{\mathrm{fused}}(\lambda_1, \lambda_2) = \arg\min_{\boldsymbol{\beta}}\left(\|\boldsymbol{Y} - \boldsymbol{X}\boldsymbol{\beta}\|_2^2/n + \lambda_1\|\boldsymbol{\beta}\|_1 + \lambda_2\sum_{j=2}^{k}|\beta_j - \beta_{j-1}|\right), \tag{7.19}$$

where $\lambda_1 \ge 0$ and $\lambda_2 \ge 0$. The second penalty λ_2 focuses on the smoothness of the parameters from one variable to another variable. A large λ_2 would force all parameters to be the same. As a special case, suppose the true model is $Y_i = \beta\sum_{j=1}^{k} x_{ji} + \epsilon_i$, which is parsimonious but not sparse. The fused Lasso can easily handle such a parsimonious model, yet the Lasso may encounter some difficulties.

7.1.2.5 *SCAD Penalty* To reduce the finite sample biases of Lasso estimator in variable selection, Fan and Li (2001) proposed the Smoothly Clipped Absolute Deviation (SCAD) penalty for the linear regression model, where each coefficient is penalized depending on its absolute size. The estimator is defined as

$$\widehat{\boldsymbol{\beta}}_{\mathrm{scad}}(\lambda, a) = \arg\min_{\boldsymbol{\beta}}\left(\|\boldsymbol{Y} - \boldsymbol{X}\boldsymbol{\beta}\|_2^2/n + \sum_{j=1}^{k}\mathrm{pen}_{\lambda,a}(\beta_j)\right), \tag{7.20}$$

where the penalty is given by

$$\mathrm{pen}_{\lambda,a}(u) = \begin{cases} \lambda|u|, & \text{if } |u| \le \lambda, \\ -(u^2 - 2a\lambda|u| + \lambda^2)/[2(a-1)], & \text{if } \lambda < |u| \le a, \\ (a+1)\lambda^2/2, & \text{if } |u| > a\lambda, \end{cases}$$

where u is a real number, $\lambda \ge 0$, and a is a positive real number. Fan and Li (2001) recommend $a = 3.7$. From the definition, it is clear that the penalty of SCAD depends on the absolute value of the regression coefficient β_j: it is the standard Lasso penalty if the coefficient is small, it depends on both the absolute value and the squares in an intermediate interval, and it is bounded by $(a+1)\lambda^2/2$ when the coefficient is large in magnitude. This is different from that of the Lasso for which the penalty is unbounded. The SCAD penalty is not differentiable at zero and is non-convex. The derivative of the penalty does exist except for the zero point provided that $a > 2$.

Example 7.1

To demonstrate the application of regularized linear regression, consider the US monthly macroeconomic data set described in McCracken and Ng (2016) and available at https://research.stlouisfed.org/econ/mccracken/fred-databases/. The data set is updated monthly and contains 128 macroeconomic variables. Details of the variables are given in McCracken and Ng (2016). Due to missing values, we remove six variables (numbered 58, 60, 95, 104, 123, and 128) and use the data from January 1960 to April 2019 in our analysis. Furthermore, we use the differencing and transformation given in McCracken and Ng (2016) to create stationary series. Consequently, our analysis employs 710 observations and 122 variables. Figure 7.1 shows the time plots of the 122 variables (properly scaled so that the sample variance of each variable is 1).

For illustration, we select the monthly growth rates of the US Industrial Production Index (IPI) as the dependent variable with forecast horizon $h = 1$, and the first 6 lagged values of all 122 variables as the predictors. Consequently, we have

$$\boldsymbol{Y} = \boldsymbol{X}\beta + \epsilon,$$

where $\boldsymbol{Y}$ is a 704-dimensional vector of response (for $t = 7, \ldots, 710$) and $\boldsymbol{X}$ is a 704×732 ($122 \times 6 = 732$) design matrix. In this particular instance, $k = 732 > n = 704$.

Figure 7.2 shows a coefficient profile plot of the Lasso regression, obtained by running the `glmnet` command of the **glmnet** package with $\alpha = 1$. See Eq. (7.18). The lines in the plot show the estimated value of each regression coefficient as a function of the ℓ_1 norm of the estimated coefficient vector. When the penalty is high, the number of zero coefficients will be large and the ℓ_1 norm of the regression coefficient vector will be small. Consequently, the ℓ_1 norm of the x-axis serves as a proxy of the penalty. As the plot only shows the non-zero estimated coefficients the number of

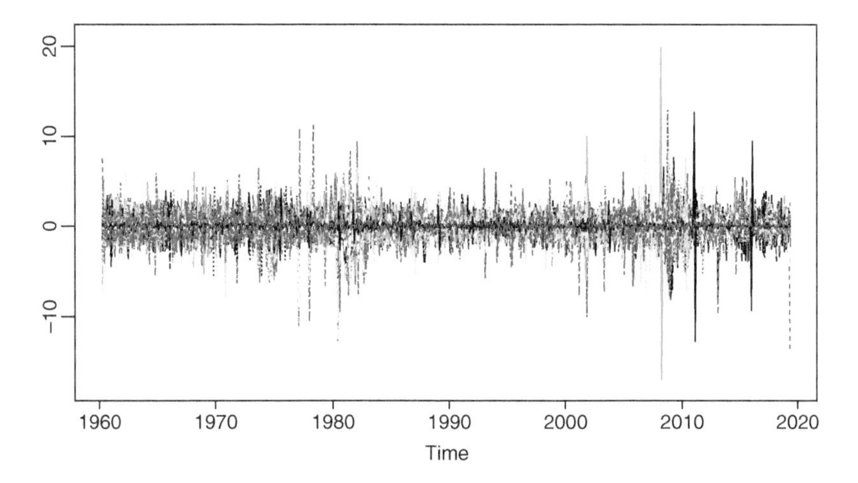

Figure 7.1 Time plots of 122 monthly macroeconomic variables. The series are scaled so that sample variance of each series is 1 for ease in visualization.

lines for each value of the ℓ_1 norm is not the same and increases with the ℓ_1 norm. The numbers shown on top of the plot are how many non-zero coefficient are estimated for a given ℓ_1 norm. For instance, Figure 7.2 shows that there are 189 non-zero coefficient estimates when ℓ_1 norm is 5. Note that the maximum number of coefficients is 732 and for $\ell_1 = 15$ only 351 coefficients, less than half of the possible, are shown. The package **glmnet** allows for different types of regression coefficients profile plot. Figure 7.3 shows another coefficient profile plot, where the x-axis is the logarithm of the penalty parameter λ, and it provides the same information but in a different way. Again each coefficient is represented by a line, but now as heavier penalty reduces the number of non-zero coefficient estimates, the number of lines in

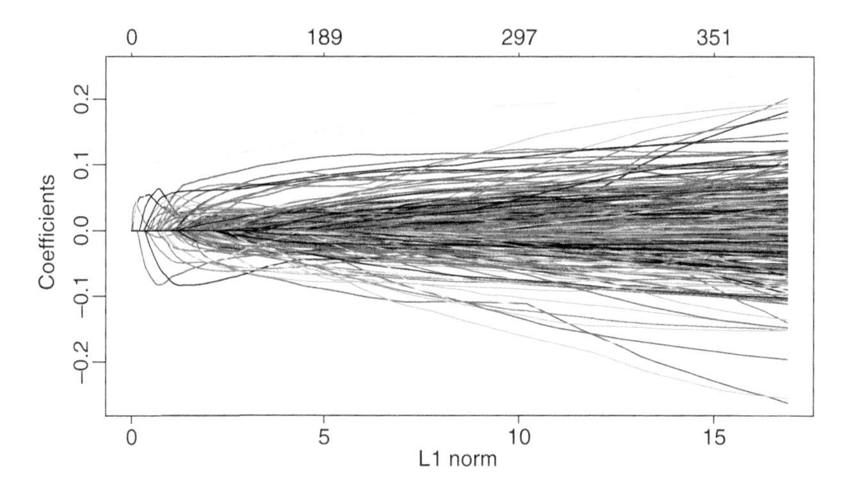

Figure 7.2 Coefficients versus ℓ_1-norm of Lasso linear regression for the monthly growth rates of US IPI from 1960 to 2019.

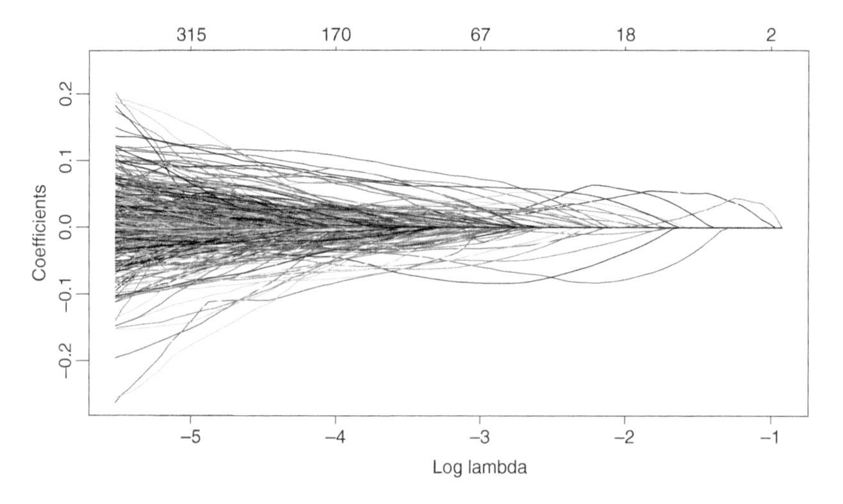

Figure 7.3 Coefficients versus log penalty, $\ln(\lambda)$, of Lasso linear regression for the growth rates of US IPI from 1960 to 2019.

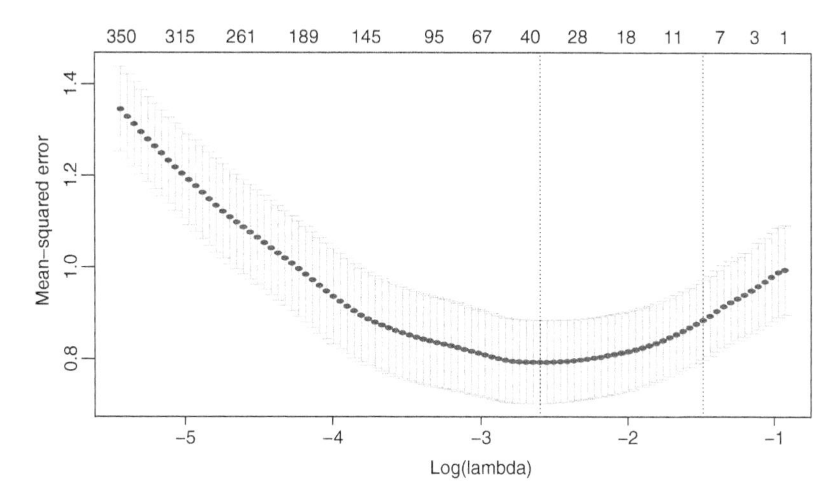

Figure 7.4 Results of 10-fold CV of Lasso linear regression for the growth rates of US IPI from 1960 to 2019. The left vertical dashed line locates the penalty for minimum MSE and the right vertical dashed line marks the one standard error limit of λ.

the plot decreases as λ increases. The numbers on top of the plot are also counts of non-zero coefficient estimates.

CV is often used to select the penalty parameter in linear regression analysis with independent data. For time series data, out-of-sample forecasts are preferred. However, for illustrative purpose of using Lasso, we use CV in this example. Figure 7.4 shows the results of a 10-fold CV (CV). The out of sample mean squared error (MSE) is plotted versus the logarithm of penalty parameter λ and the two vertical dashed lines locate the penalty corresponding to minimum MSE and a possible range of the penalty parameter. From the plot, a choice of $\log(\lambda) \approx -2.6$ seems reasonable. As a matter of the fact, the CV selects $\lambda \approx 0.0743$, which is close to $\exp(-2.6) = 0.07427$. Figure 7.5 shows estimated coefficients $\hat{\beta}_j$ when $\lambda = 0.07429$. Based on the plot, we find that there are nine predictors with $|\hat{\beta}_j| > 0.04$. These predictors are

1. IPDMAT: Industrial production: durable materials (lag-1),
2. CLAIMSx: Initial claims (lag-1),
3. NDMANEMP: All employees: nondurable goods (lag-1),
4. T1YFFM: 1-Year treasury constant minus Fed Funds (lag-1),
5. CES2000000008: Average hourly earnings: construction (lag-1),
6. ISRATIOx: Total business: inventories to sales ratio (lag-2)
7. S&P div. yield: S&P composite common stock: dividend yield (lag-2),
8. S&P indust: S&P's common stock price index: industrials (lag-3),
9. TB3SMFFM: 3-Month treasury constant minus Fed Funds (lag-5)

Therefore, the Lasso regression suggests that the growth rates of US IPI depend on various sectors, including output, labor market, inventories, interest rates, and stock market. This seems reasonable. Note that the choice of threshold 0.04 is arbitrary. Since the predictors are standardized, we can directly compare the magnitude of

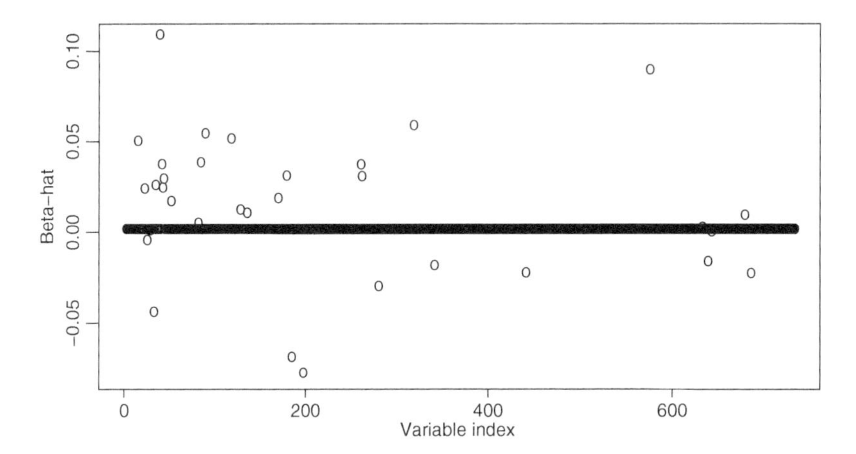

Figure 7.5 Coefficient estimates of Lasso linear regression for the growth rates of US IPI from 1960 to 2019. The penalty parameter is selected by the 10-fold CV.

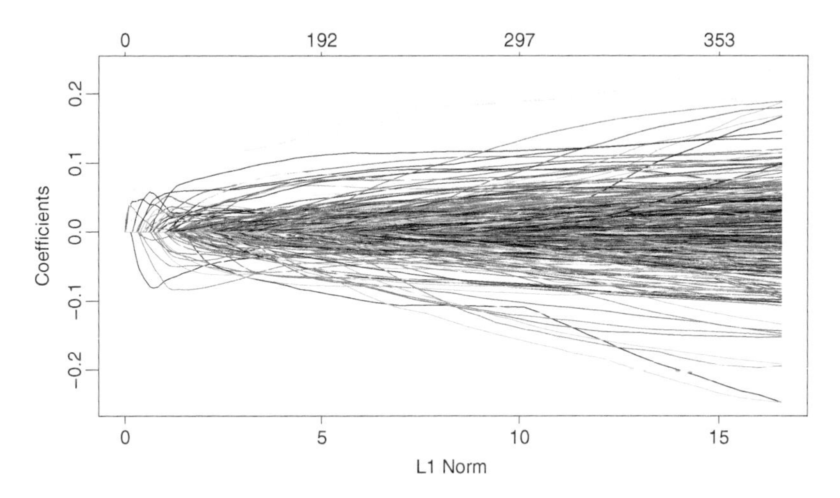

Figure 7.6 Coefficients versus ℓ_1-norm of elastic net linear regression for the monthly growth rates of U.S. IPI from 1960 to 2019. The penalty is $0.8\ell_1 + 0.2\ell_2$.

estimated coefficient. If a higher threshold, around 0.05, is chosen, then we would select about seven predictors, instead of nine mentioned earlier.

Next, turn to the elastic net with a linear combination defined by the α parameter of the ℓ_1- and ℓ_2-penalties. Figure 7.6 shows the coefficient estimates versus ℓ_1 norm of elastic net regression with $\alpha = 0.8$. This coefficient profile plot works in the same manner as that of Figure 7.2 and, as expected, it provides similar results. Figure 7.7 shows the results of a 10-fold CV versus the logarithm of penalty parameter λ. From the plot, a choice of $\log(\lambda) \approx -2.5$ seems reasonable. Specifically, the CV selects $\lambda \approx 0.0846$. Figure 7.8 shows the estimated coefficients $\hat{\beta}_j$ when λ is selected by the CV. Based on the plot, we find that there are 10 predictors with $|\hat{\beta}_j| > 0.04$. These predictors contain the same selections as those of the Lasso regression shown

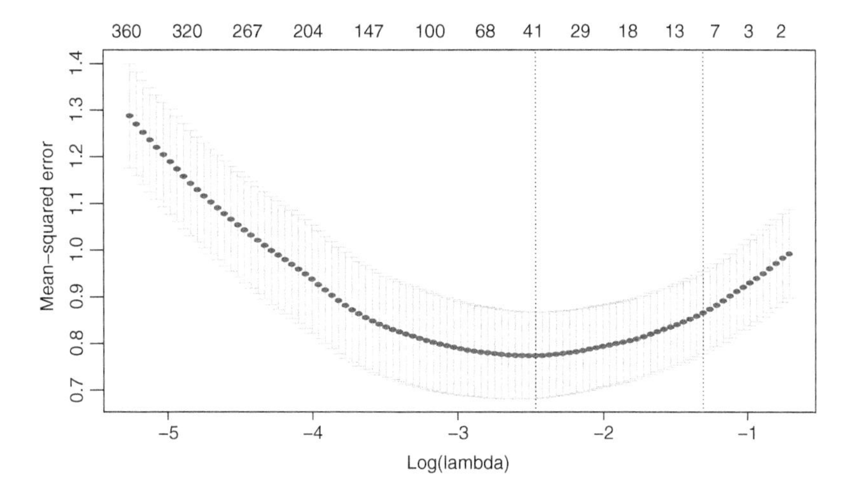

Figure 7.7 Results of 10-fold CV of elastic net linear regression for the growth rates of US IPI from 1960 to 2019. $\alpha = 0.8$ is used to define the penalty. The left vertical dashed line locates the minimum MSE and the right vertical dashed line provides one-standard error limit.

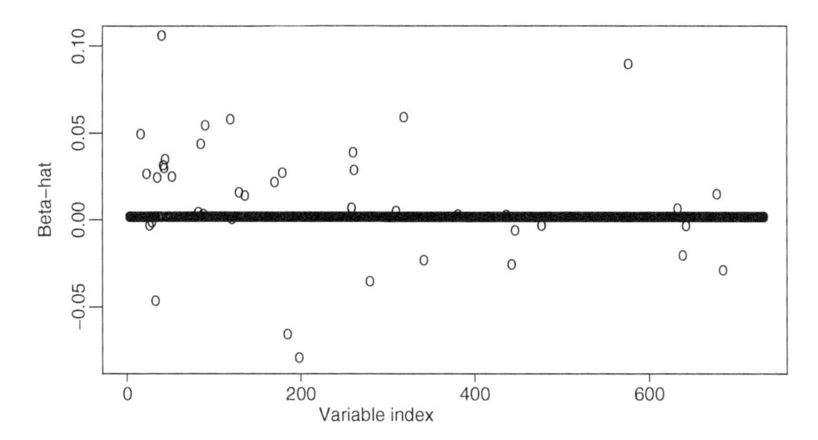

Figure 7.8 Coefficient estimates of elastic net linear regression for the monthly growth rates of US IPI from 1960 to 2019. The penalty parameter is selected by a 10-fold CV and $\alpha = 0.8$.

before with an additional predictor, which is the Moody's seasoned Aaa corporate bond yield (lag-1). Therefore, the elastic net suggests that the US monthly growth rates of IP index depend on the same sectors as the Lasso regression.

R commands used: Lasso and elastic net regressions.

```
da <- read.csv('FREDMDApril19.csv',header=TRUE)
x <- as.matrix(da)
require(MTS)
y <- scale(x,scale=TRUE) # standardize the data for plotting
tdx <- c(1:710)/12+1960
MTSplot(y,tdx)
```

```
set.seed(311)
require(glmnet)
Y <- y[7:710,6]
X <- cbind(y[6:709,],y[5:708,],y[4:707,],y[3:706,],y[2:705,],y[1:704,])
m2 <- glmnet(X,Y,family=c("gaussian"),alpha=1)
plot(m2)  #coefficient profile plot (x-axis of ell-1 norm)
plot(m2,xvar=''lambda'') #coefficient profile plot (x-axis is log penalty)
cv.m2 <- cv.glmnet(X,Y,family=c("gaussian"),alpha=1)
plot(cv.m2)
names(cv.m2)
cv.m2$lambda.min
beta <- coef(m2,s=cv.m2$lambda.min)
beta <- c(beta[,1])[-1]
plot(1:732,beta,xlab="variable index",ylab="beta-hat",pch="o",cex=0.6)
idx <- c(1:732)[abs(beta) >0.04]
idx
colnames(X)[idx]
######### glmnet with alpha = 0.8
m3 <- glmnet(X,Y,family=c("gaussian"),alpha=0.8)
plot(m3)
cv.m3 <- cv.glmnet(X,Y,family="gaussian",alpha=0.8)
plot(cv.m3)
cv.m3$lambda.min
beta1 <- coef(m3,s=cv.m3$lambda.min)
beta1 <- c(beta1[,1])[-1]
plot(1:732,beta,xlab="variable index",ylab="beta-hat",pch="o",cex=0.6)
idx <- c(1:732)[abs(beta1) > 0.04]
idx
colnames(X)[idx]
```

■

Example 7.2

For HDTS analysis Group Lasso could be useful. In this section, we demonstrate how to use it. We expect that the serial dependence in stationary time series data decays to zero as lag increases. Group Lasso enables us to implement this domain knowledge by grouping predictors of lagged variables together. Consider the quarterly GDP growth rates of six countries used in Chapter 3. The six countries are US, UK, France (FR), Australia (AU), Germany (GE), and Canada (CA). The sample period is from 1980 to 2018. Suppose that we are interested in predicting the GDP growth rate of US. and employ the lagged values of GDP growth rates of all countries as predictor. Here the dependent variable is y_t, the GDP growth rate of US. The predictors used consist of lagged values of GDP growth rates of all countries. Specifically, we have $\boldsymbol{x}_t = (y_{t-1}, \ldots, y_{t-8}, \boldsymbol{z}'_{t-1}, \ldots, \boldsymbol{z}'_{t-8})'$, where $\boldsymbol{z}_t$ is the quarterly GDP growth rates of all countries but US. To make use of properties of Group Lasso, we consider the following groups:

1. Group 1: The lagged values of US GDP growth rates, i.e. $y_{t-1}, \ldots, y_{t-8}$.
2. Group $i + 1$: The lag-i values of GDP growth rates of other countries, i.e. $\boldsymbol{z}_{t-i}$, where $i = 1, \ldots, 8$.

Therefore, we have divided the 48 predictors into 9 groups. The basic idea of our predictor grouping is that (a) the lagged values of US GDP growth rate are important and (b) for a given lag, we treat the growth rates of other countries equally. There are other ways to group the predictors, our grouping is mainly for illustrative purpose.

There are several R packages available for Group Lasso, e.g. **gglasso** and **grplassp**. We use **gglasso** in our analysis. The R commands used are given below, including grouping. Figure 7.9 shows the coefficient profile of Group Lasso for the GDP example. This plot can be interpreted in a similar way as that of Lasso. As expected, the profile shows that lagged predictors are useful in predicting US GDP growth rate. Figure 7.10 shows the result of 10-fold CV in selecting the penalty parameter λ. The plot shows that the minimum of LS loss is achieved approximately at $\ln(\lambda) = -3.4$. Figure 7.11 shows the plot of $\hat{\beta}_i$ (including the constant term) when λ is at the minimum of CV. From the plot, we see that lag-1 predictors are more important

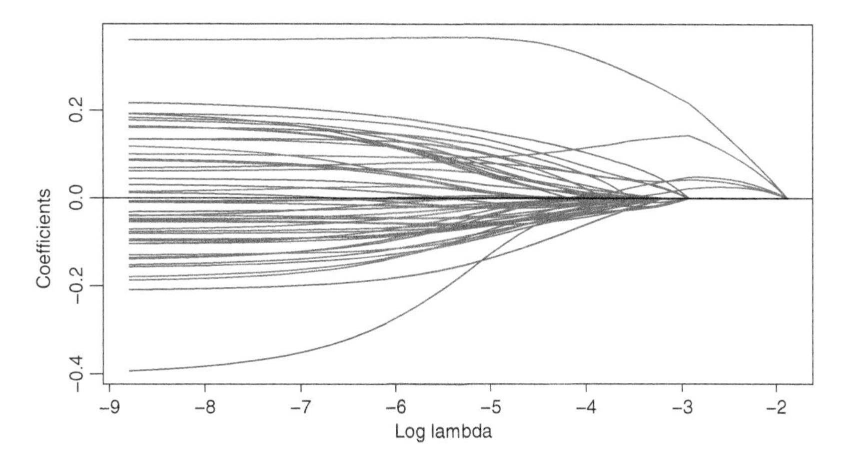

Figure 7.9 Coefficients versus $\ln(\lambda)$ of Group Lasso for the quarterly GDP growth rates of six countries from 1980 to 2018. The dependent variable is US GDP growth rate.

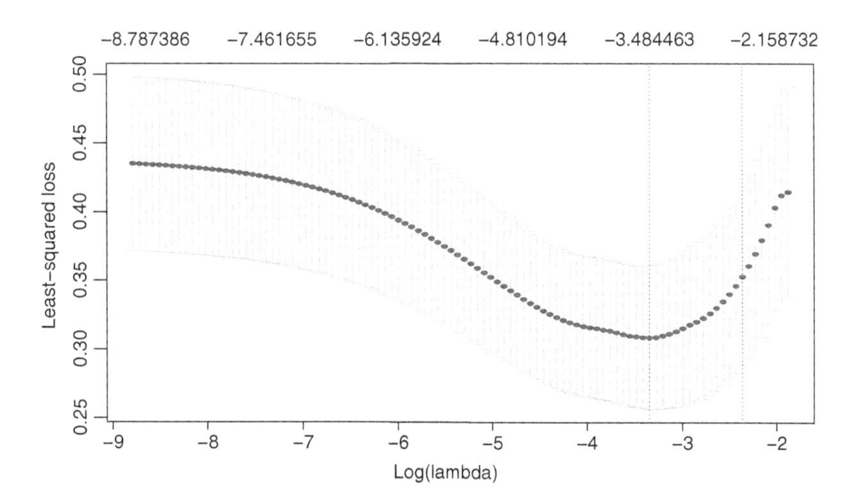

Figure 7.10 Results of 10-fold CV of Group Lasso for the GDP growth series. The dependent variable is the US GDP growth rate.

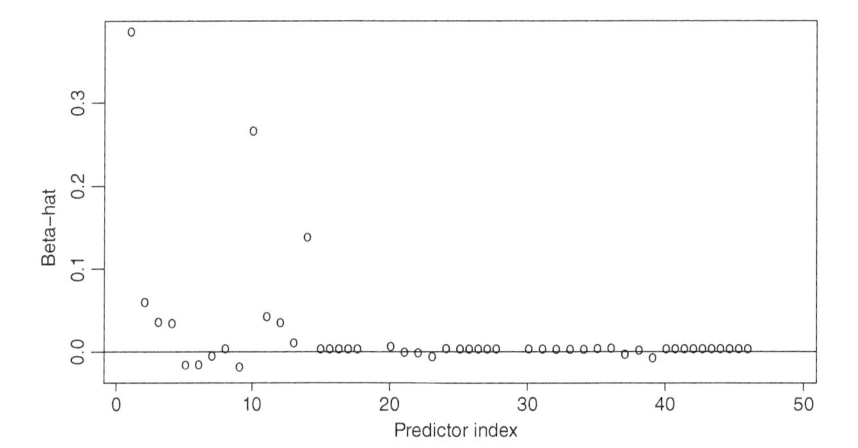

Figure 7.11 Coefficient estimates of Group Lasso for the GDP data set. The dependent variable is the US GDP growth rate and the penalty is $\lambda \approx 0.0353$.

than higher-order lagged predictors. As a matter of fact, for this particular instance, the important predictors have predictor indexes 1, 10, 14, and 2 that correspond to the intercept and the lag-1 GDP growth rates of UK, CA, and US. This is reasonable given the close tie between these three economies.

R Demonstration: Group Lasso for GDP growth rates.

```
> da <- read.table("gdpsimple6c8018.txt",header=TRUE)
> fix(da)
> dim(da)
[1] 153   6
> da[1:2,]
US          UK          FR         AU          GE          CA
1  0.3145153 -0.9719569  1.1197030 0.4478638  0.8012350  0.70678280
2 -2.0604770 -1.9796670 -0.5731772 0.2604569 -0.6438022 -0.03293659
> y <- da[9:153,1] ## Dependent variable
> X <- NULL ## setup predictors
> grp <- NULL  ## Grouping
> for (i in 1:8){
+ X <- cbind(X,da[(9-i):(153-i),1])
+ grp <- c(grp,1)
+ }
> for (i in 1:8){
+ X <- cbind(X,da[(9-i):(153-i),-1])
+ grp <- c(grp,rep(i+1,5))
+ }
> dim(X)
[1] 145  48
> n1 <- paste("US",1:8,sep="") ## name predictors
> n2 <- colnames(da)[-1]
> n2
[1] "UK" "FR" "AU" "GE" "CA"
> nn1 <- paste(n2,1,sep="")
> nn2 <- paste(n2,2,sep="")
> nn3 <- paste(n2,3,sep="")
```

```
> nn4 <- paste(n2,4,sep="")
> nn5 <- paste(n2,5,sep="")
> nn6 <- paste(n2,6,sep="")
> nn7 <- paste(n2,7,sep="")
> nn8 <- paste(n2,8,sep="")
> colnames(X) <- c(n1,nn1,nn2,nn3,nn4,nn5,nn6,nn7,nn8)
> require(gglasso)
> X <- as.matrix(X)
> m1 <- gglasso(X,y,group=grp,loss=c("ls"))
> plot(m1)
> cv.m1 <- cv.gglasso(X,y,group=grp,loss=c("ls"))
> names(cv.m1)
[1] "lambda"      "cvm"          "cvsd"         "cvupper"     "cvlo"
[6] "name"        "gglasso.fit"  "lambda.min"   "lambda.1se"
> plot(cv.m1)
> cv.m1$lambda.min
[1] 0.03526331
> beta <- coef(m1,s=cv.m1$lambda.min)
> length(beta[,1]) ## include the constant term
[1] 49
> plot(1:49,beta[,1],xlab="predictor index",ylab="beta-hat",pch="o",cex=0.6)
> abline(h=c(0))
> idx <- c(1:49)[abs(beta[,1]) > 0.04]
> idx
[1]  1  2 10 14
> pre <- c("cnst",colnames(X))
> pre[idx]
[1] "cnst" "US1"  "UK1"  "CA1"
> grp
[1] 1 1 1 1 1 1 1 1 2 2 2 2 2 2 3 3 3 3 3 4 4 4 4 4 5 5 5 5 5
6 6 6 6 6 7 7 7 7 7 8 8 8 8 8 9 9 9 9 9
```

■

7.2 IMPACTS OF DYNAMIC DEPENDENCE ON LASSO

As seen in the previous section, most properties of penalized regression models were derived under the independence assumption. On the other hand, in most applications, the data are likely to be dynamically dependent either in time or in space or both. In this section, we first summarize the consequences of dynamic dependence in regression analysis and then discuss the situation of Lasso regression. It is well known, see, for instance, Greene (2003) or Griffiths et al. (1985), that when applying regression analysis to time series data, the noises usually have autocorrelations. Consequently, using the LS estimation instead of the generalized LS has the following shortcomings:

1. The LS estimator is unbiased, but inefficient;
2. The LS variances of the coefficient estimates are biased, and generally underestimate the true variances of the coefficients;
3. The residual variance is underestimated.

These properties of the LS estimation can produce some undesirable effect such as the *spurious regression*, where explanatory variables unrelated to the response may appear as significant with the standard t-test because of the underestimation of their regression coefficient variance.

For instance, consider a simple regression model $y_t = \beta x_t + e_t$ with T data points, where the explanatory variable x_t follows an AR(1) model, $x_t = \phi x_{t-1} + a_t$, and the errors e_t also follows an AR(1) model, $e_t = \alpha e_{t-1} + \epsilon_t$. Then, the usual LS estimate of β is $\widehat{\beta}_{\text{dep}} = \sum_{t=1}^{T} x_t y_t / \sum_{t=1}^{T} x_t^2$ and its variance can be approximated (see Greene, 2003) by

$$\text{Var}(\widehat{\beta}_{\text{dep}}) = \text{Var}(\widehat{\beta}_{\text{ind}})\frac{(1+\phi\alpha)}{(1-\phi\alpha)}, \tag{7.21}$$

where $\text{Var}(\widehat{\beta}_{\text{ind}}) = \sigma^2/T\sigma_x^2$ is the variance of the LS estimate applied to independent data, that is, when $\phi = \alpha = 0$. Assuming the usual situation with $\phi > 0$ and $\alpha > 0$, we see that the variance of $\widehat{\beta}_{\text{dep}}$ can be much larger than the one computed assuming independence. In particular, if both x_t and e_t processes are serially dependent with large AR coefficients, the variance of $\text{Var}(\widehat{\beta}_{\text{dep}})$ can be very large and, hence, even if $\beta = 0$, we can obtain with high probability a large estimate $\widehat{\beta}_{\text{dep}}$, which would be seen as significant with a t statistic computed by the usual regression formula. This is the well-known spurious regression problem. The phenomenon also occurs often in the presence of strong serial correlations. See the discussion in Section 3.5.1.

With Lasso linear regression the parameter estimates are usually biased and we do not have exact formulas to compute their variances but due to the autocorrelations in the data some irrelevant variable may appear with large coefficients and a similar problem of spurious regression is likely to occur. We saw such an impact in Chapter 1 when the temporal dependence is strong. In this section, we use simulation to demonstrate further some impacts of serial dependence on Lasso linear regression. Our objective is to show that the dynamic dependence of the data should not be overlooked when it exists.

Example 7.3

Consider the ARIMA(1,1,0) model that we wrote as a scalar AR(2) model

$$y_t = 1.4y_{t-1} - 0.4y_{t-2} + a_t, \quad t = 1, \dots, 400, \tag{7.22}$$

where $\{a_t\}$ are i.i.d. $N(0,1)$ and $y_t = 0$ if $t < 1$. We generated independently 11 series from the model. The first series is called y_t and the remaining 10 series is denoted by $\boldsymbol{x}_t$. Figure 7.12 shows the time plots of these 11 realizations, where y_t is in bold line. Then, we consider the following linear regression

$$y_{t+2} = \boldsymbol{\beta}_1'\boldsymbol{x}_{t+1} + \boldsymbol{\beta}_2'\boldsymbol{x}_t + e_{t+2}, \quad t = 1, \dots, 398. \tag{7.23}$$

This model can be written in matrix form as $\boldsymbol{Y} = \boldsymbol{X}\boldsymbol{\beta} + \boldsymbol{E}$, where $\boldsymbol{X}$ is the design matrix its ith row is $(\boldsymbol{x}_{i+1}', \boldsymbol{x}_i')$ and $\boldsymbol{Y} = (y_3, \dots, y_{400})'$. Since y_t and $\boldsymbol{x}_t$ are independent, $\boldsymbol{\beta} = \boldsymbol{0}$. However, we apply the Lasso estimation to model (7.23) pretending that the data are serially independent. Note that in model fitting the components of the predictor $\boldsymbol{x}_t$ are normalized so that the sample variance of each predictor series is one.

Figure 7.13 shows the coefficient profile plot of the Lasso fitting of model (7.23). From the plot, several of the coefficient estimates are relatively large. For instance, there are 12 non-zero coefficient estimates, when the ℓ_1 norm is 10. Figure 7.14 shows

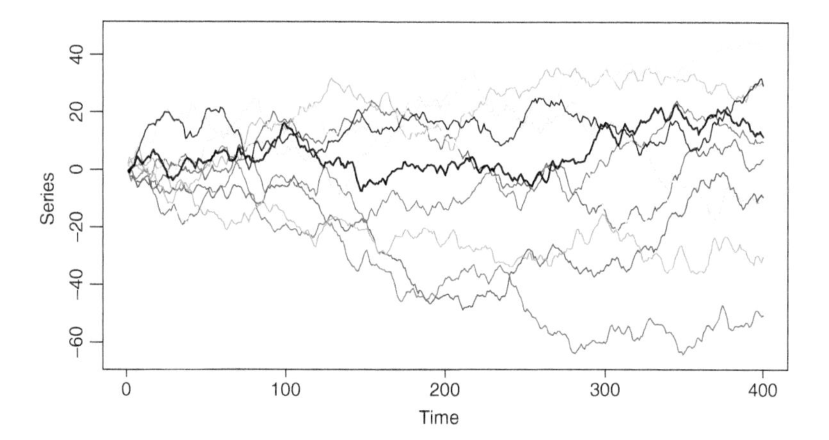

Figure 7.12 Time series plots of the 11 scalar time series used in Example 7.3. The dependent variable y_t is in bold line.

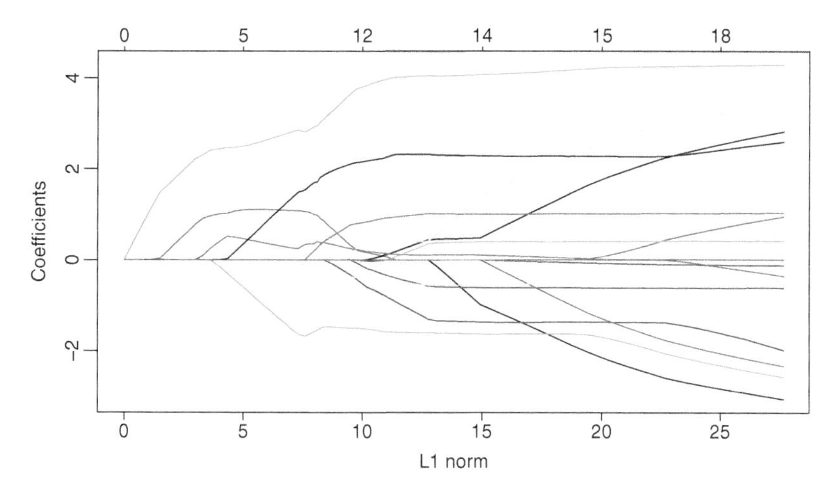

Figure 7.13 Coefficient profile plot of the Lasso fitting of the model in Eq. (7.23). The x-axis is the ℓ_1 norm of the fitted coefficient vectors.

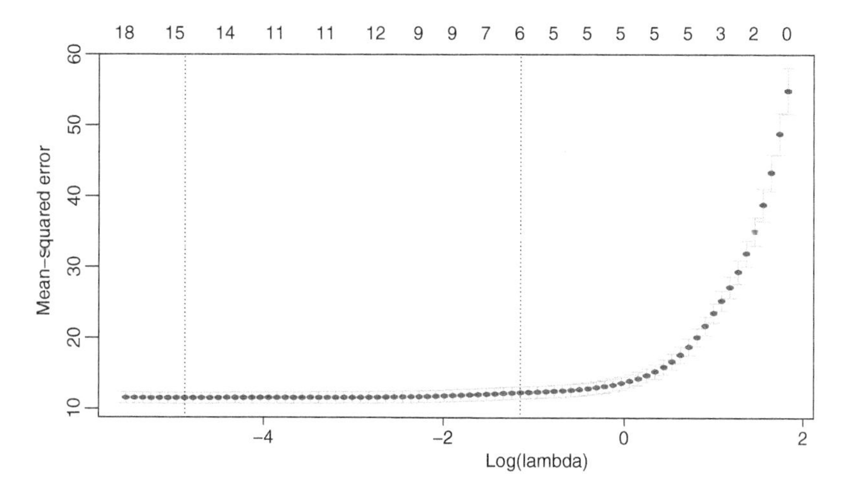

Figure 7.14 Result of 10-fold CV for the Lasso fit of the model in Eq. (7.23).

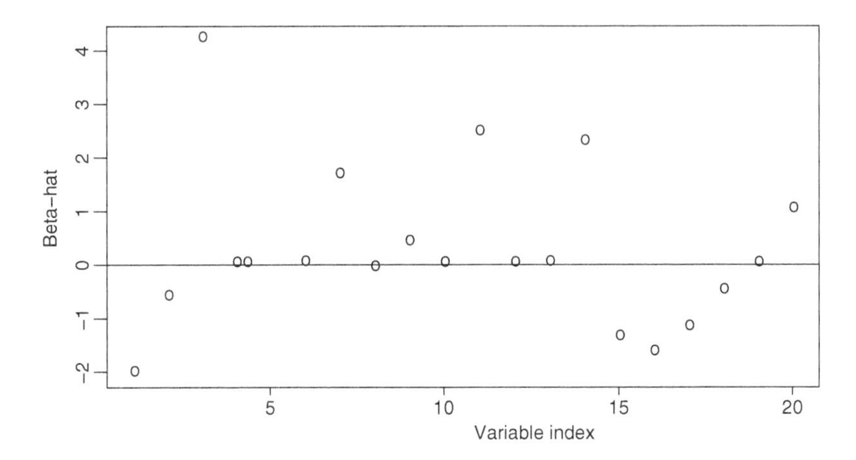

Figure 7.15 Estimated coefficients $\hat{\beta}_i$ of Lasso fit to Eq. (7.23) when the penalty λ is selected by a 10-fold CV.

the result of 10-fold CV of Lasso fit to model (7.23). The plot is not informative as there is no clear minimum, indicating that there exists a wide range of choice for the penalty parameter λ. The `cv.glmnet` command selects $\lambda \approx 0.00765$. Figure 7.15 shows $\hat{\beta}_i$ when λ is selected by the 10-fold CV. From the plot, we see that several estimates $\hat{\beta}_i$ are relatively large even though $\beta_i = 0$ for all i. As a matter of fact, the plot shows nine $\hat{\beta}_i$ greater than 1 in absolute value. Theoretically speaking, the phenomenon found in our simulation can be briefly explained as follows. First, since the series employed are unit-root nonstationary, the limiting distributions of the efficient estimates $\hat{\beta}_i$ are non-standard. They are functions of the standard Brownian motion so that the magnitude of $\hat{\beta}_i$ can be large, even though they should not be statistically significant. Second, the normalization of each predictor in $\boldsymbol{x}_{t+1}$ and $\boldsymbol{x}_t$ exacerbates the problem of multicollinearity. Using the unit-root properties, e.g. Tsay (2014, chapter 5) and the references therein, it can be shown that $\sum_{t=i+1}^T y_t y_{t-i} / \sum_{t=1}^T y_t^2$ converges to 1 at the rate of T^{-1} for any fixed i as $T \to \infty$. To illustrate, rewrite the model in Eq. (7.22) as $y_t = y_{t-1} + 0.4\nabla y_{t-1} + a_t$. Then,

$$\sum_t y_t y_{t-1} = \sum_t y_{t-1}^2 + 0.4 \sum_t y_{t-1} \nabla y_{t-1} + \sum_t a_t y_{t-1}.$$

Since $\sum_t y_{t-1} \nabla y_{t-1} = O_p(T)$ and $\sum_t a_t y_{t-1} = O_p(T)$, because of stationarity, but $\sum_t y_t^2 = O_p(T^2)$ as $T \to \infty$, we see that $\sum_t^T y_t y_{t-1} / \sum_t y_t^2 = 1 + O_p(T^{-1})$. This implies that, for large T, the impact of the stationary part of the model vanishes in comparison with $\sum_t y_t^2$. Consequently, normalizing the predictors increases the multicollinearity of Lasso regression in the presence of unit roots.

This simple example shows that overlooking the strong serial dependence in the data may lead to an erroneous fitted model in Lasso regression. Even in the case of weak dependency as in stationary processes, the same problem may appear, although with smaller magnitude. If we repeat the example with series generated by an AR(1) process with coefficient 0.8, four coefficients are found with absolute values between 0.1 and 0.2. If a standard linear regression is fitted using as explanatory variables these four variables, two of them are significant with absolute t values larger than 3.5.

R commands used in Example 7.3

```
require(SLBDD} ## Package for the book
require(glmnet)
set.seed(123)
Yt <- sim.uarima(T=400,ar=c(0.4),ma=NULL,D=0)
Xt <- NULL
for (i in 1:10){
xt <- sim.uarima(T=400,ar=c(0.4),ma=NULL,D=0)
Xt <- cbind(Xt,xt)
}
y <- Yt[3:400]
xt <- scale(Xt,scale=TRUE)
colnames(xt) <- paste("xt",1:10,sep="")
X <- cbind(xt[2:399,],xt[1:398,])
m4 <- glmnet(X,y,family="gaussian",alpha=1)
plot(m4)
cv.m4 <- cv.glmnet(X,y,family="gaussian",alpha=1)
plot(cv.m4)
cv.m4$lambda.min
beta <- coef(m4,s=cv.m4$lambda.min)
beta <- c(beta[,1])[-1]
plot(1:20,beta,xlab="variable index",ylab="beta-hat",pch="o",cex=0.6)
abline(h=0)
idx <- c(1:20)[abs(beta) > 1]
idx
```

■

Example 7.4

Consider the linear regression

$$y_i = \boldsymbol{x}_i'\boldsymbol{\beta} + e_i, \quad i = 1, \ldots, 100, \tag{7.24}$$

where $\boldsymbol{x}_i$ is a 100-dimensional random vector with elements drawn independently from $N(0, 1)$, the error term e_i follows a scalar AR(1) model given by $e_i = \phi e_{i-1} + \epsilon_i$, with ϵ_i being i.i.d. $N(0, 1)$ and the coefficient vector $\boldsymbol{\beta}$ is defined below

$$\beta_i = \begin{cases} U[-6, 6] & \text{for } i = 1, \ldots, 10, \\ 0 & \text{for } i > 10, \end{cases}$$

where $U[a, b]$ denotes a random draw from the uniform distribution on $[a, b]$. The goal of this example is to study the impact of serial correlations in the error term e_i on the confidence intervals of the Lasso estimates $\hat{\beta}_i$ of non-zero coefficient. That is, we focus on the Lasso estimates $\hat{\beta}_i$ for $i = 1, \ldots, 10$. To this end, we conduct the following simulation:

1. Draw randomly the true coefficients β_i for $i = 1, \ldots, 10$.
2. The true AR(1) coefficient ϕ assumes three values, namely $\phi = 0, 0.6$, or -0.6.
3. For a given ϕ, do the following procedure 3000 iterations:
   (a) Generate the data using the model of Eq. (7.24) with true parameters.
   (b) Apply the Lasso to the data set.
   (c) Use 10-fold CV to select the penalty parameter λ.
   (d) Store the estimates $\hat{\beta}_i$ for $i = 1, \ldots, 10$. Denote the resulting $\hat{\beta}_i$ as $\widehat{\boldsymbol{\beta}}(\phi)$, which is a 3000-by-10 matrix for a given ϕ.
4. For each ϕ, compute some column summary statistics of $\widehat{\boldsymbol{\beta}}(\phi)$.

The column summary statistics computed include (a) sample mean, (b) sample standard deviation, and (c) empirical coverage probabilities that the conventional 95% confidence interval of β_i contains the true β_i ($i = 1, \ldots, 10$). Figure 7.16 shows the box plots of the $\hat{\beta}_i(\phi)$ for $\phi = 0, 0.6, -0.6$ and $i = 1, \ldots, 4$. Also given in the plot are the true values of β_i. From the plots, it is clear that serial correlations in the error e_i of Eq. (7.24), if overlooked, would substantially affect the variability of the estimates $\hat{\beta}_i$. This is true whether the serial correlations are positive or negative. Table 7.1 provides some summary statistics of selected $\hat{\beta}_i(\phi)$. Again, $\phi = 0$ is the case that the assumptions for Lasso hold. From the table, it is seen that (a) the Lasso estimates work

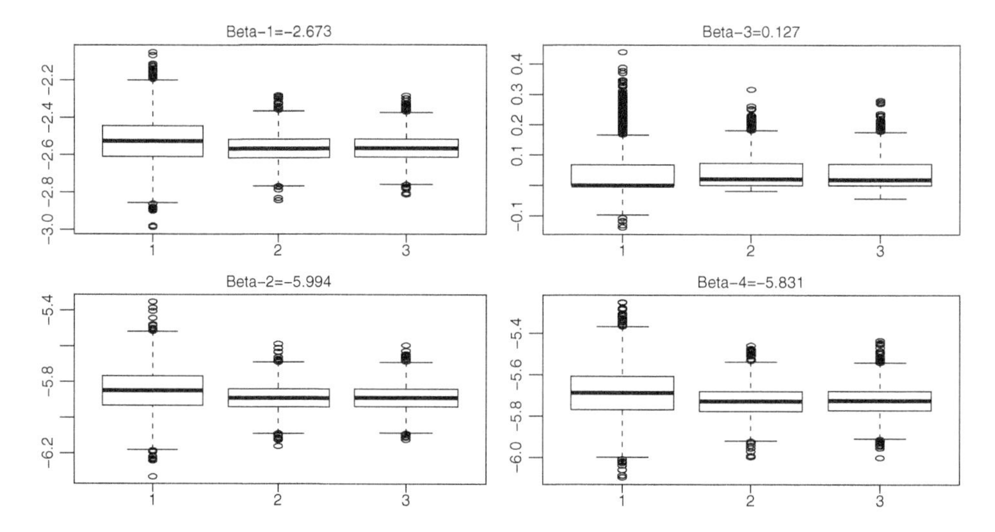

Figure 7.16 Boxplots of $\hat{\beta}_i(\phi)$ for $i = 1, \ldots, 4$. The left box is $\phi = 0$ and the middle box is for $\phi = 0.6$. The estimates are based on 3000 iterations and the model is in Eq. (7.24).

TABLE 7.1 Selected Summary Statistics of Lasso Estimates $\hat{\beta}_i(\phi)$ for the Model in Eq. (7.24)

ϕ	Statistics	$\hat{\beta}_1$	$\hat{\beta}_2$	$\hat{\beta}_3$	$\hat{\beta}_6$	$\hat{\beta}_7$	$\hat{\beta}_9$
	Trueϕ	−2.673	−5.994	0.127	5.458	−4.962	4.568
0	Mean	−2.526	−5.849	0.040	5.310	−4.816	4.422
	SE	0.124	0.124	0.067	0.125	0.124	0.123
	C.Prob	0.999	0.999	0.987	0.788	0.998	0.784
0.6	Mean	−2.566	−5.889	0.042	5.351	−4.857	4.462
	SE	0.075	0.074	0.051	0.075	0.074	0.075
	C.Prob	0.999	1.000	**0.465**	0.716	1.000	0.716
−0.6	Mean	−2.565	−5.887	0.041	5.353	−4.854	4.461
	SE	0.075	0.075	0.051	0.076	0.075	0.074
	C.Prob	1.000	1.000	**0.452**	0.727	0.999	0.700

[a] "C.Prob" denotes the coverage probability of the conventional 95% confidence interval of β_i. The results are based on 3000 iterations and $\phi = 0, 0.6$, and -0.6, respectively.

well when the errors of the linear regression model are uncorrelated, (b) the Lasso substantially underestimates the standard errors of $\hat{\beta}_i(\phi)$ when $\phi \neq 0$, implying that overlooking the serial dependence in the error term can affect statistical inference in using Lasso, and (c) the true coverage probability of a 95% confidence interval can be much lower than 0.95 if the serial dependence is ignored. The lower coverage probability is caused by two factors. First, the serial correlations in the residuals lead to underestimation of the variability of coefficient estimates (i.e. smaller empirical standard errors). Second, the serial correlations introduce downward biases in coefficient estimates, and this negative bias can be substantial if the true β_i is not large. ■

7.3 LASSO FOR DEPENDENT DATA

We demonstrated via simulation in the previous section that serial dependence can affect statistical inference of Lasso methodology. In the literature, some researchers have provided certain theoretical justifications for Lasso when the serial dependence in the data is not strong. Wang et al. (2007) proposed a Lasso estimator for linear regression model with AR errors by applying different penalties to the regression and AR coefficients in the linear regression model

$$y_t = \boldsymbol{x}_t' \boldsymbol{\beta} + e_t, \tag{7.25}$$

where e_t denotes the error term that is assumed to follow the AR(p) model $\phi_p(B)e_t = a_t$. As before, $\boldsymbol{x}_t = (x_{1t}, \ldots, x_{kt})'$ is a k-dimensional random vectors of predictors. Here the subscript is changed to t to signify the time series data. They proposed the estimation method

$$\widehat{\boldsymbol{\beta}}_{\text{adapt}}(\lambda) = \arg\min_{\beta} \left\{ \|\boldsymbol{Y} - \boldsymbol{X}\boldsymbol{\beta}\|_2^2/n + \sum_{j=1}^{k} \lambda_j |\beta_j| + \sum_{i=1}^{p} \gamma_i |\phi_i| \right\}, \tag{7.26}$$

and showed that this modified Lasso is more efficient for the problem under study than the traditional one. They also proposed an algorithm to estimate this modified Lasso estimator. Gupta (2012) investigated Lasso estimator in model (7.25), where e_t is a weakly dependent process. A weakly dependent process has the property that the autocorrelations go to zero as lag increases. This condition is a generalization of independence and is useful in asymptotic analysis. ARMA processes and, in general, stationary and ergodic processes are weakly dependent. See Dedecker et al. (2007).

The aforementioned two papers focus on the case of sample size n is greater than the number of predictors k. Basu and Michailidis (2015) investigated theoretical properties of Lasso estimator with a random design for high-dimensional Gaussian processes. Kock and Callot (2015) established oracle inequalities for the Lasso regression with Gaussian errors in stationary VAR models. Wu and Wu (2016) analyzed Lasso estimator with a fixed design matrix and assumed that a restricted eigenvalue condition is satisfied. Medeiros and Mendes (2016) studied asymptotic properties of adaptive Lasso when the errors are non-Gaussian and may be conditionally heteroskedastic. Han and Tsay (2020) established the rate of convergence of Lasso estimator under weakly sparsity condition and provided sign consistency of Lasso regression. The paper extends Lasso results beyond fixed design and exact sparsity in time series analysis, where the number of true non-zero coefficients is finite, without assuming restricted eigenvalue condition on the sample covariance matrix or population covariance matrix. In what follows, for a k-dimensional random vector z, the qth norm is defined as $\|z\|_q = (E|z|^q)^{1/q}$ for $1 \leq q \leq k$.

Consider the linear regression model (7.25). Let $\{\epsilon_t\}$ be an i.i.d. sequence of random vectors and $F_t = \sigma\{\epsilon_i | i \leq t\}$ be the σ-field generated by ϵ_i with $i \leq t$. We assume that the components of $\boldsymbol{x}_t$ and the error e_t are well-defined real-valued measurable functions of F_t, that is,

$$\boldsymbol{x}_t = (g_1(F_t), \ldots, g_k(F_t))' \quad \text{and} \quad e_t = g_e(F_t).$$

For simplicity, one can think of x_{it} and e_t are linear functions of F_t. To quantify the serial dependence of time series data, we adopt the functional dependence measure of Wu (2005). Specifically, let

$$\delta_{t,q,i} = \|x_{it} - x_{it}^*\|_q = \|g_i(F_t) - g_i(F_t^*)\|_q, \tag{7.27}$$

$$\delta_{t,q,e} = \|e_i - e_i^*\|_q = \|g_e(F_t) - g_e(F_t^*)\|_q, \tag{7.28}$$

where $\boldsymbol{x}_t^* = (g_1(F_t^*), \ldots, g_k(F_t^*))'$ and $e_t^* = g_e(F_t^*)$ with $F_t^* = \sigma\{[\epsilon_i | i \leq t \cap i \neq 0] \cup \epsilon_0^*]\}$, where ϵ_0^* is a random vector independent of $\{\epsilon_i\}$ but follows the same distribution as ϵ_0. Roughly speaking, for a given time index t and positive integer q, $\delta_{t,q,i}$ of Eq. (7.27) quantifies the impact of changing ϵ_0 to ϵ_0^* on x_{it} as measured by the qth norm. Similarly, $\delta_{t,q,e}$ of Eq. (7.28) quantifies the impact of changing ϵ_0 to ϵ_0^* on the error term e_t. These two dependence measures play a similar role as the impulse response weight ψ_i of the moving-average representation of a linear time series model.

For a positive integer m, define the cumulative dependence measures as

$$\Delta_{m,q,i} = \sum_{t=m}^{\infty} \delta_{t,q,i}, \tag{7.29}$$

$$\Delta_{m,q,e} = \sum_{t=m}^{\infty} \delta_{t,q,e}. \tag{7.30}$$

These two measures quantify the cumulative impacts on x_{it} and e_t, respectively, when one change ϵ_0 to ϵ_0^*. Again, one can think of these two measures as cumulative sums of the ψ-weights of a linear ARMA model. We say that the series x_{it} and e_t are *short-range dependence* if $\Delta_{m,q,i} < \infty$ and $\Delta_{m,q,e} < \infty$ for some fixed m. From the definition, it is seen that the impact of changing ϵ_0 to ϵ_0^* on a time series is bounded if the series is short-range dependence, because $\delta_{t,q,i}$ and $\delta_{t,q,e}$ should approach zero under short-range dependence.

The cumulative dependence measures can be used, under the short-range dependence, to define dependence-adjusted norm (DAN) as follows:

$$\|x_{i.}\|_{q,\alpha} = \sup_{m \geq 0} (m+1)^{\alpha} \Delta_{m,q,i}, \quad \alpha \geq 0,$$

$$\|e_.\|_{q,\alpha} = \sum_{m \geq 0} (m+1)^{\alpha} \Delta_{m,q,e}, \quad \alpha \geq 0.$$

These DANs are used to derive consistency of Lasso estimator for dependent data.

Next, turn to the cross-sectional dependence of $\boldsymbol{x}_t$. Let $\boldsymbol{\Sigma} = [\sigma_{ij}]$ be the covariance matrix of $\boldsymbol{x}_t$, assuming that the series is stationary. Define the L^∞ functional dependence measure and its corresponding DAN as

$$w_{t,q} = \| \max_{1\le j\le k} |x_{jt} - x^*_{jt}| \|_q,$$

$$\| |\boldsymbol{x}_.|_\infty \|_{q,\alpha} = \sum_{m\ge 0} (m+1)^\alpha \sum_{t=m}^{\infty} w_{t,q}, \quad \alpha \ge 0.$$

In addition, define

$$\Psi_{q,\alpha} = \max_{1\le j\le k} \|\boldsymbol{x}_{j.}\|_{q,\alpha} \quad \text{and} \quad \Upsilon_{q,\alpha} = \left(\sum_{j=1}^{k} \|\boldsymbol{x}_{j.}\|^q_{q,\alpha}\right)^{1/q}.$$

Here $\Psi_{q,\alpha}$ and $\Upsilon_{q,\alpha}$ can be regarded as the uniform and the overall DANs of $\boldsymbol{x}_t$.

Consider the model in Eq. (7.25). We impose the following weak sparsity condition: there exist some constant θ satisfying $0 \le \theta < 1$ and a uniform radius K_θ such that

$$\sum_{j=1}^{k} |\beta_j|^\theta \le K_\theta. \tag{7.31}$$

Assume the weak sparsity condition holds and both x_{it} $(i = 1, \ldots, k)$ and e_t satisfy the short-range dependence. More specifically, assume $\Psi_{\gamma,\alpha_X} = M_X < \infty$ with $\gamma > 4$ and $\alpha_X > 0$, and $\|e_.\|_{q,\alpha_e} = M_e < \infty$ with $q > 2$ and $\alpha_e > 0$. Define

$$v = \begin{cases} 1, & \text{if } \alpha_X \ge 1/2 - 2/\gamma, \\ \gamma/4 - \alpha_X\gamma/2, & \text{if } \alpha_X < 1/2 - 2/\gamma. \end{cases}$$

Let $\tau = q\gamma/(q+\gamma) > 2$ and $\alpha = \min(\alpha_X, \alpha_e)$. Define

$$\rho = \begin{cases} 1, & \text{if } \alpha \ge 1/2 - 1/\tau, \\ \tau/2 - \alpha\tau, & \text{if } \alpha < 1/2 - 1/\tau. \end{cases}$$

Let $w = \sqrt{\log(k)/n}M_X^2 + n^{2v/\gamma-1}(\log(k))^{3/2}\| |\boldsymbol{x}_.|_\infty \|^2_{\gamma,\alpha_X}$. In addition, assume the minimum eigenvalue of $\boldsymbol{\Sigma}$ is bounded away from zero, say $\lambda_{\min}(\boldsymbol{\Sigma}) \ge \kappa > 0$. Under these assumptions, Han and Tsay (2020) derive the consistency of Lasso estimates of the linear regression in Eq. (7.25) provided that the penalty parameter λ satisfies

$$\lambda \ge \sqrt{\log(k)/n}M_e M_X + n^{\rho/\tau-1}(\log(k))^{3/2} M_e \| |\boldsymbol{x}_.|_\infty \|_{\gamma,\alpha_X}$$

and the quantity w satisfies $K_\theta w \lambda^{-\theta} < \infty$. Specifically, with high probability, we have that, as $n \to \infty$,

$$|\hat{\boldsymbol{\beta}} - \boldsymbol{\beta}|_2 \le \sqrt{K_\theta}\left(\frac{\lambda}{\kappa}\right)^{1-\theta/2}, \tag{7.32}$$

$$|\hat{\boldsymbol{\beta}} - \boldsymbol{\beta}|_1 \le \sqrt{K_\theta}\left(\frac{\lambda}{\kappa}\right)^{1-\theta}. \tag{7.33}$$

Details are given in theorem 1 of Han and Tsay (2020). The consistency results can be improved under strong sparsity, namely $|\boldsymbol{\beta}|_0 = s < \infty$. The conditions for the consistency of Lasso estimates under the short-range dependence look complicated and lack intuition, but the message is clear. The convergence rate of the Lasso estimates is slower than that of the i.i.d. case. One can think of the effective sample size to understand the convergence rate for dependent data. In the presence of serial dependence, the effective sample size of T data points is smaller than T. The stronger the serial dependence, the smaller the effective sample size. Consequently, the convergence rate is slower in the presence of serial dependence. Furthermore, for model selection consistency, Han and Tsay (2020) show that the number of predictors k can only grow at a polynomial rate instead of an exponential rate. See theorem 3 of Han and Tsay (2020). Again, one can use effective sample size to understand the issue. Since the effective sample size of dependent data is smaller than that of the i.i.d. case, the number of allowed predictors k should be smaller than that of the i.i.d. case. Other applications under the weak dependence assumption of Lasso estimation in time series analysis are Nicholson et al. (2017a), who study structural regularization for large vector autoregressions and Nicholson et al. (2017b) for modeling sparse HDTS.

Another important issue of applying Lasso to dependent data is the choice of the penalty parameter λ. The serial dependence can easily invalidate the CV method. To overcome this difficulty, one can divide the data into estimation and forecasting subsamples and use out-of-sample prediction errors to select the penalty parameter, as we have seen in Chapter 6 with the rolling forecasting. Of course, how to divide the data into estimation and forecasting subsamples requires a careful study. In general, the sample size of the forecasting subsample should not be too small so that the summary statistics of the forecast errors are stable. See Peña et al. (2021) for an application of the selection of λ in regularizing Dynamical Principal Components. An alternative approach to select λ is to employ some information criteria such as the Bayesian information criterion (BIC). For the degrees of freedom of Lasso regression, readers are referred to Zou et al. (2007) and Tibshirani and Taylor (2012).

Example 7.5

Consider, again, the monthly macroeconomic variables used in Example 7.1. The dependent variable is the monthly grow rates of US IPI and the predictors consist of the first six lagged values of all variables. Our goal here is to illustrate the use of out-of-sample prediction to select the penalty parameter. To this end, we divide the data into estimation and forecasting subsamples with the former consisting of the first 670 data point and the latter having 34 observations. Figure 7.17 shows the result of 10-fold CV of the Lasso linear regression via the command `cv.glmnet`. The selected value is $\lambda \approx 5.956 \times 10^{-4}$. Figure 7.18 shows the out-of-sample mean squared errors versus the λ. The selected value is $\lambda \approx 1.777 \times 10^{-4}$. The root mean squared errors of the out-of-sample predictions are 0.00503 and 0.00467, respectively, for $\lambda = 5.956 \times 10^{-4}$ and 1.777×10^{-4}. The difference is not surprising because CV does not use data in the forecasting subsample. Figure 7.19 plots the estimates $\hat{\beta}_j$ for both selection methods. From the plots, it is clear that, as expected, Lasso selects fewer non-zero estimates than the out-of-sample prediction does.

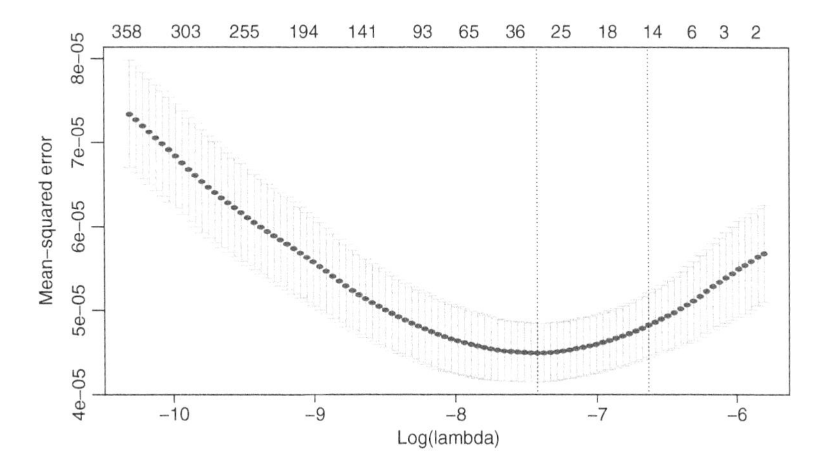

Figure 7.17 Results of 10-fold CV for the monthly macroeconomic data using the first 670 data points.

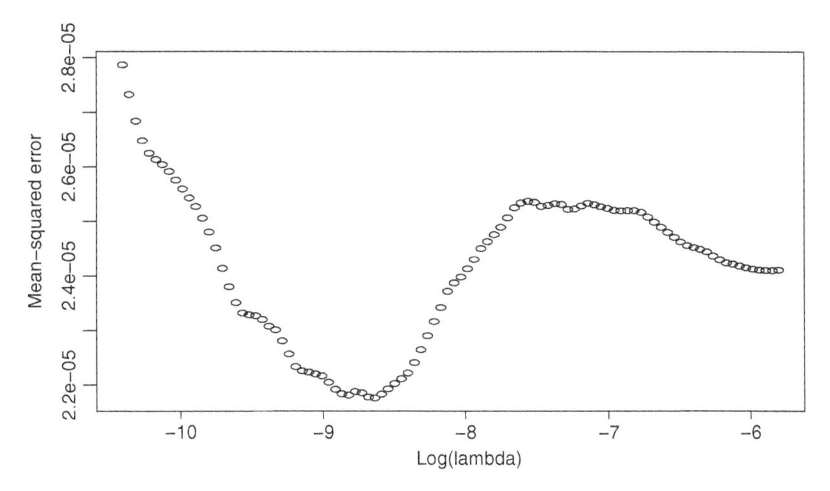

Figure 7.18 Results of out-of-sample prediction for the monthly macroeconomic data. The estimation subsample contains the first 670 data points.

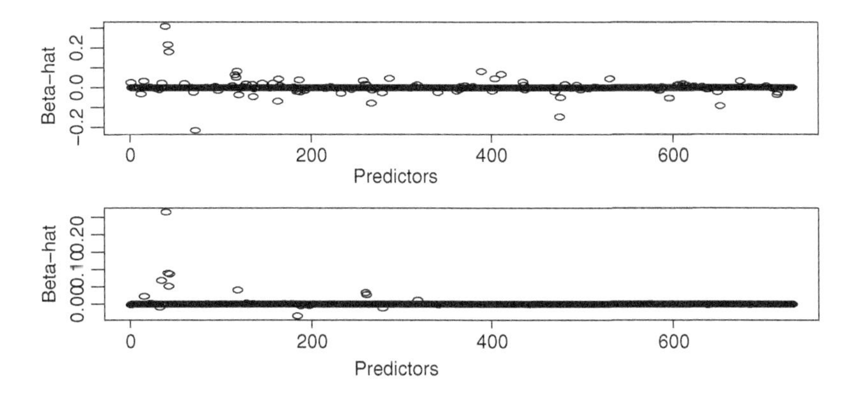

Figure 7.19 Scatterplots of coefficient estimates $\hat{\beta}_j$ versus j of the monthly macroeconomic data. (*a*) Out-of-sample prediction and (*b*) CV.

R commands used:

```
require(SLBDD}; require(glmnet)
da <- read.csv("FREDMDApril19.csv'',header=TRUE)
x <- as.matrix(da)
dim(da)
[1] 710 122
Y <- x[7:710,6]
X <- cbind(x[6:709,],x[5:708,],x[4:707,],x[3:706,],x[2:705,],x[1:704,])
dim(X)
m2 <- Lambda.sel(X[1:670,],Y[1:670],X[671:704,],Y[671:704])
m2$lambda.min
[1] 0.0001777053
m3 <- glmnet(X[1:670,],Y[1:670],family="gaussian",alpha=1)
cv.m3 <- cv.glmnet(X[1:670,],Y[1:670],family='gaussian',alpha=1)
plot(cv.m3)
coef <- coef(m3,s=cv.m3$lambda.min)
cv.m3$lambda.min
[1] 0.0005955974
par(mfcol=c(2,1))
plot(1:733,m2$beta,xlab="predictors",ylab="beta-hat",
main="(a) Out-sample")
plot(1:733,coef[,1],xlab="predictors",ylab="beta-hat",
main="(b) Cross-validation")
```

■

7.4 PRINCIPAL COMPONENT REGRESSION AND DIFFUSION INDEX

Consider the linear regression model of Eq. (7.2) with n data points and k predictors in $\boldsymbol{X}$, where it is understood that $\boldsymbol{X}$ may contain a column of ones or both $\boldsymbol{Y}$ and $\boldsymbol{X}$ are mean-adjusted. In some applications, the matrix $\boldsymbol{X}'\boldsymbol{X}$ is singular, such as when $n < k$ or when columns of $\boldsymbol{X}$ are highly correlated (multicollinearity), so that the LS estimate of $\boldsymbol{\beta}$ is not unique. Many methods have been proposed in the literature to overcome this difficulty. Principal component regression (PCR) and diffusion index (DI) are two of such methods. The idea of these two methods is simple. They perform the conventional principal component analysis (PCA) on $\boldsymbol{X}$ and then, use a set of the first few principal components to predict $\boldsymbol{Y}$. Since the principal components are orthogonal to each other, the multicollinearity issue is overcome. It remains the problem of selecting the number of principal components to use, which is often addressed by either CV or out-of-sample prediction.

Mathematically, PCR can be formulated as follows:

$$\boldsymbol{X} = \boldsymbol{O}\boldsymbol{P}', \tag{7.34}$$

$$\boldsymbol{Y} = \boldsymbol{F}\boldsymbol{\beta} + \boldsymbol{E}, \tag{7.35}$$

where $\boldsymbol{O}$ is the $n \times k$ *score matrix*, $\boldsymbol{P}$ is the $k \times k$ *loading matrix*, $\boldsymbol{Y}$ is a $n \times 1$ response vector, $\boldsymbol{F}$ is a $n \times r$ matrix consisting of the first few r columns of $\boldsymbol{O}$ and, for ease in notation, $\boldsymbol{E}$ continues to denote the error term. The loading $k \times k$ matrix $\boldsymbol{P}$ is an orthonormal matrix such that $\boldsymbol{P}'\boldsymbol{P} = \boldsymbol{P}\boldsymbol{P}' = \boldsymbol{I}$. Equation (7.34) can be obtained by the singular value decomposition of $\boldsymbol{X}$, namely $\boldsymbol{X} = \boldsymbol{UDP}'$, where $\boldsymbol{U}$ is the $n \times k$ matrix of eigenvectors of $\boldsymbol{XX}'$ corresponding to non-null eigenvalues, $\boldsymbol{D}$ is a diagonal $k \times k$ matrix with elements, the squared roots of the eigenvalues of $\boldsymbol{XX}'$, and $\boldsymbol{P}$ is the $k \times k$ matrix with the eigenvectors of $\boldsymbol{X}'\boldsymbol{X}$. Clearly, $\boldsymbol{O} = \boldsymbol{UD}$.

The DI of Stock and Watson (2002) can be regarded as a generalization of PCR. It uses the concept of factor models of Chapter 6 and postulates

$$\boldsymbol{X} = \boldsymbol{FL}' + \boldsymbol{\eta}, \tag{7.36}$$

$$\boldsymbol{Y} = \boldsymbol{F\beta} + \boldsymbol{Z\gamma} + \boldsymbol{E}, \tag{7.37}$$

where $\boldsymbol{F}$ consists of common factors (latent variables) of dimension $n \times r$ with $r \ll k$, $\boldsymbol{\eta}$ denotes the matrix of idiosyncratic term of $\boldsymbol{X}$, $\boldsymbol{Z}$ consists of some lagged values of the dependent variable y_{t+h} with h being the forecast horizon, and $\boldsymbol{\gamma}$ is the (partial) dependence of y_{t+h} on its lagged values. In the econometric literature, Eq. (7.36) is often referred to as an approximate factor model. Again, see Chapter 6 for further details. In practice, PCA of $\boldsymbol{X}$ is used to estimate the common factors in $\boldsymbol{F}$. If one expands $\boldsymbol{X}$ to include $\boldsymbol{Z}$, then Eq. (7.37) can be replaced by Eq. (7.35). In this sense, DI approach to forecasting can be regarded as an extension of PCR.

Some discussions of PCR or DI method are in order. First, PCR seeks orthogonal linear combinations of the k predictors to explain as much as possible the variance of $\boldsymbol{x}_t$ and uses the first few of these linear combinations to predict the dependent variable y_{t+h}. Yet there are no good reasons to suggest that y_{t+h} depends indeed only on the first few principal components. A way to select the number of factors is to use Lasso regression in fitting the model, see Bai and Ng (2008). Second, since PCA depends critically on the scales of the variables involved, it is important to standardize the predictors before applying PCR or DI. Third, there are methods proposed in the literature to select the number of common factor r for DI method. See Chapter 6 for details. Fourth, there are several R packages available to perform PCR, e.g. **pls** and **plsdof**. The DI approach to forecasting is available in **MTS** via the command `SWfore`. Finally, for the high-dimensional case in which $k > T$, one can apply sparse PCA to generalize PCR. The sparse PCA is available in Zou et al. (2006).

Example 7.6

Consider, again, the monthly macroeconomic data set used in Example 7.5. Our goal here is to demonstrate the use of PCR and we have standardized each of the 732 predictors. For simplicity, we set the number of principal components used to 30. If CV is used to select the number of principal components, one selects 4 components. See Figure 7.20. On the other hand, if one treats principal component scores as variable and uses AIC to select the number of principal components, then one selects 24 PCs. See Figure 7.21, which shows the scatterplot of AIC criterion versus the number of PCs. In this particular instance, the PCR with 30 components explains about 33% of the variability of the IPI growth rates. Figure 7.22 shows the cross-validated prediction of the PCR with 30 components. The straight line indicates the target. From the plot, the fitted model seems reasonable.

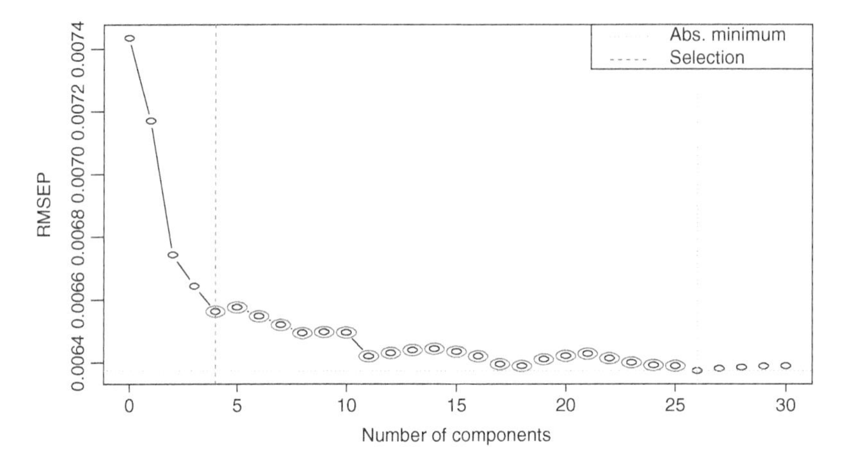

Figure 7.20 Residual mean squared errors of CV for selecting the number of principal components of monthly macroeconomic data set. The dependent variable is the growth rate of IPI and predictors consists of the first six lagged values of all variables.

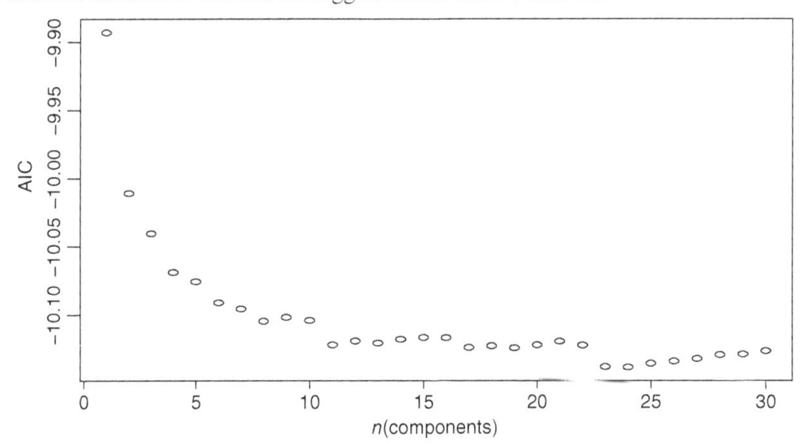

Figure 7.21 Akaike's information criterion (AIC) versus the number of principal components of the monthly macroeconomic data set. The dependent variable is the growth rate of IPI and predictors consists of the first six lagged values of all variables.

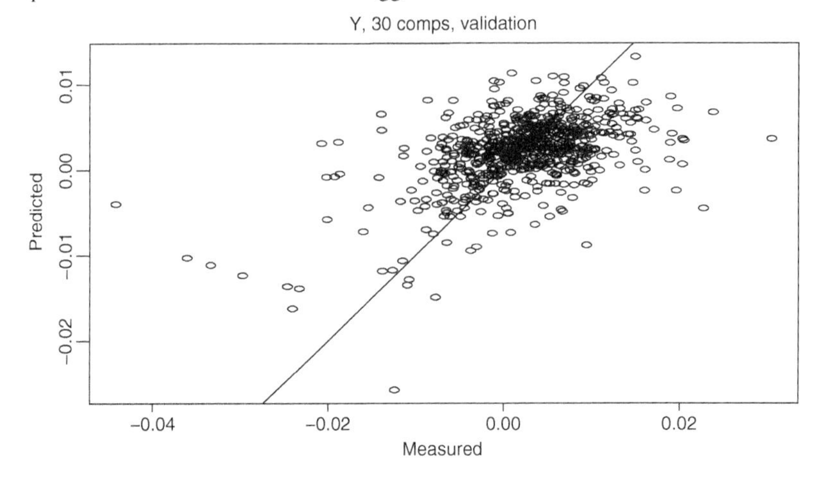

Figure 7.22 Cross-validated predictions of PCR regression with 30 components for the macroeconomic data set. The dependent variable is the growth rate of IPI and predictors consists of the first six lagged values of all variables.

R output of PCR with package **pls**: output edited.

```
> m1 <- read.csv("FREDMSApril19.csv",header=TRUE)
> x <- as.matrix(m1)
> Y <- x[7:710,6]
> X <- cbind(x[6:709,],x[5:708,],x[4:707,],x[3:706,],x[2:705,],x[1:704,])
> name <- paste("V",1:122,sep="") ### create names for predictors
> n1 <- paste(name,"L1",sep=""); n2 <- paste(name,"L2",sep="")
> n3 <- paste(name,"L3",sep=""); n4 <- paste(name,"L4",sep="")
> n5 <- paste(name,"L5",sep=""); n6 <- paste(name,"L6",sep="")
> colnames(X) <- c(n1,n2,n3,n4,n5,n6)
> X <- scale(X,center=TRUE,scale=TRUE) ### Standardize the predictors.
> require(pls)
> m2 <- pcr(Y~.,data=data.frame(X),ncomp=30,validation=''CV'')
> names(m2)
[1] "coefficients"  "scores"        "loadings"      "Yloadings"
[5] "projection"    "Xmeans"        "Ymeans"        "fitted.values"
[9] "residuals"     "Xvar"          "Xtotvar"       "fit.time"
[13] "ncomp"         "method"        "call"          "terms"
[17] "model"
> dim(m2$coefficients)
[1] 732   1  30
> dim(m2$scores)
[1] 704  30
> dim(m2$loadings)
[1] 732  30
> residuals <- m2$residuals[,1,]
> dim(residuals)
[1] 704  30
> mse <- apply(residuals^2,2,mean)
> npar <- c(1:30)
> aic <- log(mse)+2*npar/704
> which.min(aic)
24 comps
24
> plot(1:30,aic,xlab="n(components)",ylab="aic")
> selectNcomp(m2,"randomization",plot=TRUE)
[1] 4
> summary(m2)
Data:   X dimension: 704 732
Y dimension: 704 1
Fit method: svdpc
Number of components considered: 30

VALIDATION: RMSEP
Cross-validated using 10 random segments.
(Intercept)  1 comps   2 comps   3 comps   4 comps   5 comps   6 comps
CV      0.007435  0.007125  0.006725  0.006635  0.006553  0.006572  0.006528
adjCV   0.007435  0.007124  0.006722  0.006632  0.006548  0.006569  0.006522
7 comps   8 comps   9 comps  10 comps  11 comps  12 comps  13 comps
CV      0.006541  0.006532  0.006551  0.006549  0.006480  0.006447  0.006443
adjCV   0.006547  0.006522  0.006546  0.006545  0.006455  0.006436  0.006430
14 comps  15 comps  16 comps  17 comps  18 comps  19 comps  20 comps
CV      0.006448  0.006455  0.006453  0.006417  0.006424  0.006413  0.006412
adjCV   0.006437  0.006448  0.006446  0.006406  0.006413  0.006401  0.006399
21 comps  22 comps  23 comps  24 comps  25 comps  26 comps  27 comps
CV      0.006421  0.006409  0.006373  0.006384  0.006397  0.006410  0.006418
```

```
adjCV  0.006414  0.006407  0.006345  0.006360  0.006375  0.006393  0.006400
28 comps  29 comps  30 comps
CV      0.006428  0.006441  0.006449
adjCV  0.006413  0.006423  0.006431

TRAINING: % variance explained
1 comps  2 comps  3 comps  4 comps  5 comps  6 comps  7 comps  8 comps
X   11.157    17.35    21.52    24.02    26.09    28.00    29.80    31.51
Y    8.589    18.96    21.57    23.98    24.71    26.07    26.61    27.48
9 comps  10 comps  11 comps  12 comps  13 comps  14 comps  15 comps
X    33.09     34.62     36.04     37.34     38.57     39.76     40.91
Y    27.48     27.85     29.32     29.33     29.64     29.64     29.75
16 comps  17 comps  18 comps  19 comps  20 comps  21 comps  22 comps
X      42.03     43.10     44.01     44.91     45.74     46.57     47.32
Y      29.97     30.67     30.79     31.08     31.13     31.14     31.53
23 comps  24 comps  25 comps  26 comps  27 comps  28 comps  29 comps
X      48.06     48.77     49.45     50.12     50.77     51.41     52.04
Y      32.80     33.01     33.02     33.10     33.17     33.17     33.32
30 comps
X      52.67
Y      33.36
> plot(m2,line=TRUE)
```

■

7.5 PARTIAL LEAST SQUARES

Partial least squares (PLS) regression was proposed by H. Wold (1966) and has been widely used in many scientific fields, especially in chemometrics (i.e. computational chemistry). Geladi and Kowalski (1986) provide a tutorial and Abdi (2003) gives a nice introduction of PLS. Ng (2013) considers PLS from an algorithmic point of view. Like PCR, PLS also addresses the singularity difficulty of the $\boldsymbol{X}'\boldsymbol{X}$ matrix, but it can handle multivariate dependent variables. Therefore, we treat $\boldsymbol{Y}$ as a $n \times m$ data matrix in this section. The goal of PLS regression is to predict $\boldsymbol{Y}$ from $\boldsymbol{X}$ and to describe their common structure. For readers familiar with multivariate statistical analysis, PLS can be regarded as a method between canonical correlation analysis and PCR of $\boldsymbol{Y}$ and $\boldsymbol{X}$. In contrast with PCR, which seeks orthogonal transformations to explain as much as possible the variance of $\boldsymbol{X}$, PLS finds linear combinations of $\boldsymbol{X}$ that are also relevant to $\boldsymbol{Y}$. Specifically, PLS seeks a set of linear transformations that performs a simultaneous decomposition of $\boldsymbol{X}$ and $\boldsymbol{Y}$ such that those transformed components explain as much as possible the covariance between $\boldsymbol{X}$ and $\boldsymbol{Y}$.

Mathematically, PLS can be formulated as follows:

$$\boldsymbol{X} = \boldsymbol{O}\boldsymbol{P}' + \boldsymbol{\eta}, \tag{7.38}$$

$$\boldsymbol{Y} = \boldsymbol{U}\boldsymbol{Q}' + \boldsymbol{E}, \tag{7.39}$$

where $\boldsymbol{\eta}$ and $\boldsymbol{E}$ denote the error terms and it is understood that they can be zero. Similarly to the PCR case, $\boldsymbol{O}$ and $\boldsymbol{U}$ are the score matrices and $\boldsymbol{P}$ and $\boldsymbol{Q}$ are the loading matrices. PLS then considers the linear regression between $\boldsymbol{U}$ and $\boldsymbol{O}$. The idea of PLS is to perform the decompositions in Eqs. (7.38) and (7.39) so that the covariance between $\boldsymbol{X}$ and $\boldsymbol{Y}$ is maximized. Intuitively, one must take information from each other into account in performing the decompositions. To gain further insight, it pays to consider the spectral decomposition of PCR in Eq. (7.34). The decomposition can be achieved by an iterative method called the *Nonlinear Iterative Partial Least Squares* (NIPALS) algorithm, which is similar to the power method for obtaining eigenvalue and eigenvector of a positive-definite matrix. Let $\boldsymbol{p}$ be a column of the loading matrix $\boldsymbol{P}$ of Eq. (7.34) and $\boldsymbol{o}$ be its corresponding score vector. From the eigenvalue-eigenvector analysis, we have

$$\mathbf{X}'\boldsymbol{X}\boldsymbol{p} = \lambda\boldsymbol{p}, \tag{7.40}$$

Letting

$$\boldsymbol{o} = \boldsymbol{X}\boldsymbol{p}, \tag{7.41}$$

we have

$$\boldsymbol{p} = \frac{1}{\lambda}\boldsymbol{X}'\boldsymbol{o}. \tag{7.42}$$

NIPALS Algorithm for $\boldsymbol{X}$:

I. Initialization: Let $\boldsymbol{o}$ be an arbitrary n-dimensional vector, e.g. an arbitrary column of $\boldsymbol{X}$.
II. Let $\boldsymbol{p} = \boldsymbol{X}'\boldsymbol{o}/\|\boldsymbol{X}'\boldsymbol{o}\|$,
III. Update $\boldsymbol{o} = \boldsymbol{X}\boldsymbol{p}$,

where $\|\boldsymbol{x}\|$ denotes the Euclidean norm of $\boldsymbol{x}$. Iterate step II and step III until $\boldsymbol{o}$ converges. In other words, the algorithm starts with an initial value of $\boldsymbol{o}$, then applies Eq. (7.42) to update $\boldsymbol{p}$ followed by updating $\boldsymbol{o}$ using Eq. (7.41). When convergence occurs, one obtains the first principal component $\boldsymbol{p}_1$ and the associated scores $\boldsymbol{o}_1$. To find the subsequent components, one set $\boldsymbol{X} = \boldsymbol{X} - \boldsymbol{o}_1\boldsymbol{p}_1'$ and repeat the algorithm.

The algorithm also applies to PCA of $\boldsymbol{Y}$ to obtain the score matrix $\boldsymbol{U}$ and loading matrix $\boldsymbol{Q}$ of Eq. (7.39). The question then is how to take information from each other between $\boldsymbol{X}$ and $\boldsymbol{Y}$ to achieve the goal of PLS. A simple idea is to swap $\boldsymbol{o}$ and $\boldsymbol{u}$ in the updating. The resulting NIPALS algorithm is as follows:

NIPALS Algorithm:

1. Initialization: Let $\boldsymbol{u}$ be an arbitrary starting vector, e.g. $\boldsymbol{u}$ being an arbitrary column of $\boldsymbol{Y}$.
2. Loop:
   - $\boldsymbol{p} = \boldsymbol{X}'\boldsymbol{u}/\|\boldsymbol{X}'\boldsymbol{u}\|$,
   - $\boldsymbol{o} = \boldsymbol{X}\boldsymbol{p}$,
   - $\boldsymbol{q} = \boldsymbol{Y}'\boldsymbol{o}/\|\boldsymbol{Y}'\boldsymbol{o}\|$,
   - $\boldsymbol{u} = \boldsymbol{Y}\boldsymbol{q}$.
3. Until $\boldsymbol{o}$ converges.

If $\boldsymbol{Y}$ is a vector, one can set $\boldsymbol{q} = 1$ and omits the last two steps of the loop. The algorithm gives the first set of PLS components and loadings. For subsequent components and loadings, we define

$$\boldsymbol{X} = \boldsymbol{X} - \boldsymbol{o}\boldsymbol{p}', \quad \boldsymbol{Y} = \boldsymbol{Y} - \boldsymbol{u}\boldsymbol{q}',$$

and repeat the algorithm.

After r iterations, we obtain two $n \times r$ matrices $\boldsymbol{O}$ and $\boldsymbol{U}$ and two loading matrices $\boldsymbol{P}$ and $\boldsymbol{Q}$. We then relate $\boldsymbol{Y}$ to $\boldsymbol{X}$ by the linear regression

$$\boldsymbol{U} = \boldsymbol{O}\boldsymbol{\beta} + \epsilon,$$

where ϵ denotes the error term. The fitted value of $\boldsymbol{Y}$ is then

$$\hat{\boldsymbol{Y}} = \hat{\boldsymbol{U}}\boldsymbol{Q}' = \boldsymbol{O}\hat{\boldsymbol{\beta}}\boldsymbol{Q}' = \boldsymbol{X}\boldsymbol{P}\hat{\boldsymbol{\beta}}\boldsymbol{Q}', \tag{7.43}$$

where $\boldsymbol{O} = \boldsymbol{X}\boldsymbol{P}$ of PCA is used.

To gain insight into the NIPALS algorithm, consider

$$\begin{aligned}
\boldsymbol{p} &= \boldsymbol{X}'\boldsymbol{u}/\|\boldsymbol{X}'\boldsymbol{u}\| \\
&= \boldsymbol{X}'\boldsymbol{Y}\boldsymbol{q}/\|\boldsymbol{X}'\boldsymbol{Y}\boldsymbol{q}\| \\
&= \boldsymbol{X}'\boldsymbol{Y}\boldsymbol{Y}'\boldsymbol{o}/\|\boldsymbol{X}'\boldsymbol{Y}\boldsymbol{Y}'\boldsymbol{o}\| \\
&= \boldsymbol{X}'\boldsymbol{Y}\boldsymbol{Y}'\boldsymbol{X}\boldsymbol{p}/\|\boldsymbol{X}'\boldsymbol{Y}\boldsymbol{Y}'\boldsymbol{X}\boldsymbol{p}\| \\
&= \frac{1}{\lambda}(\boldsymbol{Y}'\boldsymbol{X})'(\boldsymbol{Y}'\boldsymbol{X})\boldsymbol{p}.
\end{aligned} \tag{7.44}$$

Clearly, $\boldsymbol{p}$ is an eigenvector of the sample covariance matrix of $\boldsymbol{Y}'\boldsymbol{X}$. The algorithm in Eq. (7.44) is exactly the updating rule in the Power method for computing the largest eigenvalue-eigenvector of the symmetric matrix $\boldsymbol{X}'\boldsymbol{Y}\boldsymbol{Y}'\boldsymbol{X}$.

Besides the NIPALS algorithm, there are alternative methods available to compute the PLS regression. See, for instance, de Jong (1993). Helland (1990) provides an explanation of PLS from a prediction point of view, which is easy to follow when $\boldsymbol{Y}$ is a scalar dependent variable.

PLS algorithm 1: (PLS1)

1. Define the starting values of the $\boldsymbol{X}$ residuals, $\boldsymbol{\eta}_0$, and y residuals $\boldsymbol{f}_0$:

$$\boldsymbol{\eta}_0 = \boldsymbol{X} - \boldsymbol{\mu}_x,$$
$$\boldsymbol{f}_0 = \boldsymbol{Y} - \boldsymbol{\mu}_y,$$

   where $\boldsymbol{\mu}_X$ denotes the matrix consisting of mean vector of $\boldsymbol{x}_t$ and $\boldsymbol{\mu}_y$ is the vector consisting of the mean of dependent variable y_{t+h}. In other words, $\boldsymbol{\eta}_0$ and $\boldsymbol{f}_0$ are the mean-adjusted $\boldsymbol{X}$ and $\boldsymbol{Y}$.

   For $i = 1, 2, \ldots$, do the steps 2 to 4 below:

2. Introduce the X-scores, $\boldsymbol{o}_i$, as linear combinations of the $\boldsymbol{X}$ residuals $\boldsymbol{\eta}_{i-1}$ using the covariances between $\boldsymbol{\eta}_{i-1}$ and $\boldsymbol{f}_{i-1}$ as weights:

$$\boldsymbol{o}_i = \boldsymbol{\eta}'_{i-1}\boldsymbol{w}_i, \quad \text{with} \quad \boldsymbol{w}_i = \mathrm{Cov}(\boldsymbol{\eta}_{i-1}, \boldsymbol{f}_{i-1}).$$

3. Determine $\boldsymbol{X}$ loading $\boldsymbol{p}_i$ and $\boldsymbol{Y}$ loading $\boldsymbol{q}_i$ by least squares:

$$\boldsymbol{p}_i = \mathrm{Cov}(\boldsymbol{\eta}_{i-1}, \boldsymbol{o}_i)/\mathrm{Var}(\boldsymbol{o}_i),$$
$$\boldsymbol{q}_i = \mathrm{Cov}(\boldsymbol{f}_{i-1}, \boldsymbol{o}_i)/\mathrm{Var}(\boldsymbol{o}_i).$$

   That is, regress $\boldsymbol{\eta}_{i-1}$ and $\boldsymbol{f}_{i-1}$ on $\boldsymbol{o}_i$ to obtain the LS estimates.

4. Find the new residuals:

$$\boldsymbol{\eta}_i = \boldsymbol{\eta}_{i-1} - \boldsymbol{o}_i\boldsymbol{p}_i', \quad \boldsymbol{f}_i = \boldsymbol{f}_{i-1} - \boldsymbol{o}_i\boldsymbol{q}_i'.$$

From the Steps 1 and 4, it is clear that the algorithm at each step m gives two representations

$$\boldsymbol{X} = \boldsymbol{\mu}_X + \boldsymbol{p}_1\boldsymbol{o}_1' + \cdots + \boldsymbol{p}_m\boldsymbol{o}_m' + \boldsymbol{\eta}_m, \quad \boldsymbol{Y} = \boldsymbol{\mu}_y + \boldsymbol{q}_1\boldsymbol{o}_1' + \cdots + \boldsymbol{q}_m\boldsymbol{o}_m' + \boldsymbol{f}_m. \tag{7.45}$$

From the properties of LS regression, it is clear from steps 3 and 4 that the residuals $\boldsymbol{\eta}_i$ is uncorrelated with $\boldsymbol{o}_i$ so that the scores $\boldsymbol{o}_1, \ldots, \boldsymbol{o}_m$ are uncorrelated. Furthermore, it can be shown that the weights $\boldsymbol{w}_1, \ldots, \boldsymbol{w}_m$ are also uncorrelated. From the algorithm, it is also clear that the linear dependence between $\boldsymbol{Y}$ and the residuals $\boldsymbol{\eta}_i$ is used to construct the scores of $\boldsymbol{X}$, which is then used to predict $\boldsymbol{Y}$.

Finally, several R packages are available for PLS regression. The package **pls** and **plsdepot** provide PLS analysis, **plsdof** considers the degrees of freedom of PLS, **plsRglm** gives PLS analysis of generalized linear models, and **plsVarSel** provides variable selection of PLS.

Example 7.7

Again, consider the monthly macroeconomic data set with the monthly growth rates of U.S. IPI as the dependent variables and the first six lagged values of all variables as predictors as we have done in Examples 7.5 and 7.6. Setting the maximum number of PLS components to 30 and using CV, the PLS regression selects $m = 5$ for the data if the observed predictors are used. See Figure 7.23. The summary of the PLS regression indicates that with 30 components the model explains about 54% of the variance of IPI growth rate. Figure 7.24 shows the cross-validated prediction of PLS with 30 components. The straight line of the plot indicates the target line. The plot shows that the PLS regression with 30 components provides a reasonable fit. If the 732 predictors are standardized, the PLS regression with CV selects a single PLS component.

R demonstration of PLS: with package **pls**.

```
> da <- read.csv("FREDMDApril19.csv",header=TRUE)
> x <- as.matrix(da)
> Y <- x[7:710,6]
> X <- cbind(x[6:709,],x[5:708,],x[4:707,],x[3:706,],x[2:705,],x[1:704,])
>  name <- paste("V",1:122,sep="")
>  n1 <- paste(name,"L1",sep="")
>  n2 <- paste(name,"L2",sep="")
>  n3 <- paste(name,"L3",sep="")
>  n4 <- paste(name,"L4",sep="")
>  n5 <- paste(name,"L5",sep="")
>  n6 <- paste(name,"L6",sep="")
>  colnames(X) <- c(n1,n2,n3,n4,n5,n6)
> require(pls)
> m.plsr <- plsr(Y~.,data=data.frame(X),ncomp=30,validation="CV")
```

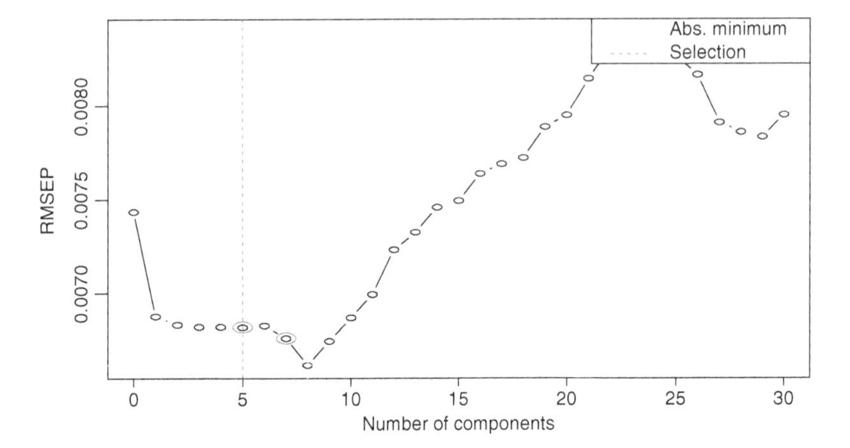

Figure 7.23 Results of CV of PLS regression. The dependent variable is the growth rate of IPI and predictors consists of the first six lagged values of all variables.

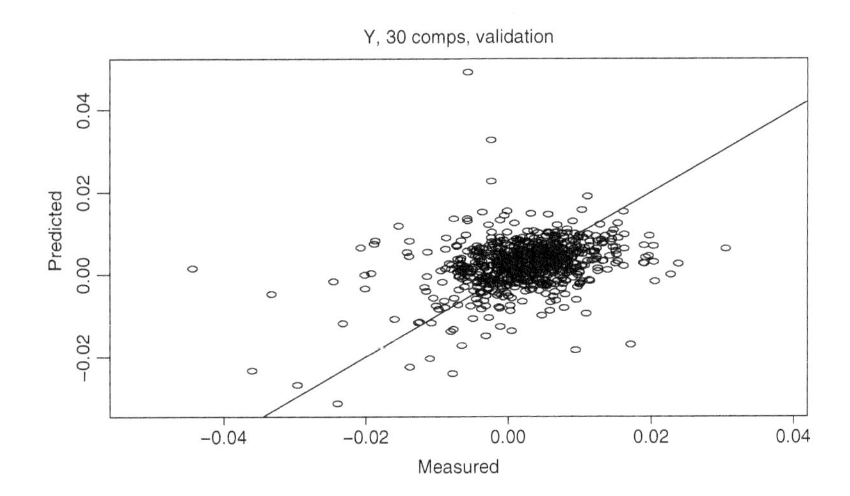

Figure 7.24 Cross-validated predictions of PLS regression with 30 components for the macroeconomic data set. The dependent variable is the growth rate of IPI and predictors consists of the first six lagged values of all variables.

```
> summary(m.plsr)
Data:    X dimension: 704 732
Y dimension: 704 1
Fit method: kernelpls
Number of components considered: 30

VALIDATION: RMSEP
Cross-validated using 10 random segments.
(Intercept) 1 comps   2 comps   3 comps   4 comps   5 comps   6 comps
CV      0.007435  0.006880  0.006851  0.006842  0.006843  0.006837  0.006839
adjCV   0.007435  0.006874  0.006846  0.006837  0.006837  0.006832  0.006839
7 comps   8 comps   9 comps  10 comps  11 comps  12 comps  13 comps
```

```
CV       0.006778  0.006642  0.006754  0.006878  0.006935  0.007024  0.007057
adjCV  0.006775  0.006633  0.006724  0.006824  0.006873  0.006945  0.006978
14 comps  15 comps  16 comps  17 comps  18 comps  19 comps  20 comps
CV       0.007126  0.007145  0.007226  0.007299  0.007423  0.007653  0.007956
adjCV  0.007031  0.007041  0.007123  0.007195  0.007301  0.007506  0.007779
21 comps  22 comps  23 comps  24 comps  25 comps  26 comps  27 comps
CV       0.008149  0.008298  0.008255  0.008169  0.008145  0.008042  0.007974
adjCV  0.007962  0.008099  0.008053  0.007982  0.007963  0.007866  0.007796
28 comps  29 comps  30 comps
CV       0.007825  0.007891  0.007988
adjCV  0.007658  0.007724  0.007806

TRAINING: % variance explained
1 comps  2 comps  3 comps  4 comps  5 comps  6 comps  7 comps  8 comps
X     15.26    40.45    49.27    72.16    82.47    98.50    99.97    99.98
Y     16.32    17.08    17.43    17.44    17.48    17.69    19.77    26.04
9 comps  10 comps  11 comps  12 comps  13 comps  14 comps  15 comps
X     99.98     99.98     99.98     99.98     99.99     99.99     99.99
Y     31.61     35.26     36.85     38.88     39.81     41.43     42.81
16 comps  17 comps  18 comps  19 comps  20 comps  21 comps  22 comps
X      99.99     99.99     99.99     99.99     99.99     99.99     99.99
Y      43.50     44.18     45.48     46.61     47.43     48.05     48.63
23 comps  24 comps  25 comps  26 comps  27 comps  28 comps  29 comps
X      99.99     99.99     99.99     99.99     99.99     99.99     99.99
Y      49.38     49.94     50.65     51.37     52.14     52.80     53.28
30 comps
X      99.99
Y      53.97
plot(m.plsr,ncomp=30,asp=1,line=TRUE)
> selectNcomp(m.plsr,"randomization",plot=TRUE)
[1] 5
```

■

7.6 BOOSTING

Boosting algorithms are proposed originally as ensemble methods, that is, methods that combine forecasts of several models (or approaches) to obtain an optimal prediction. The basic idea of boosting is to generate multiple predictions from adjusted data which are then aggregated, using linear combination or majority voting, to build the final estimator or prediction. It is similar to model averaging. Good references for boosting are Bühlmann and van de Geer (2011, chapter 12) and Hastie et al. (2009). Friedman et al. (2000) and Friedman (2001) develop a more general statistical framework which gives a direct interpretation of boosting as a statistical method for function estimation.

Given the data $\{(\boldsymbol{x}_t, y_t)|t = 1, \ldots, n\}$, boosting consists of two key ingredients. The first ingredient is called a *weak learner*, which is an easy to compute statistical prediction method. Examples of weak learner include a simple regression tree or a simple linear regression. The second ingredient is a method for data adjustment, where the variables to be fit are adjusted at each iteration in a way that often depends on the weak learner. Consider the model

$$y_t = f(\boldsymbol{x}_t) + \epsilon_t, \tag{7.46}$$

where ϵ_t is the error term and $f(.)$ is the unknown real-valued function of interest. A boosting algorithm works as follows:

$$\begin{aligned} \text{Adjusted data 1} &\rightarrow_{\text{weak learner}} \hat{f}^{(1)}(.) \\ \text{Adjusted data 2} &\rightarrow_{\text{weak learner}} \hat{f}^{(2)}(.) \\ \vdots &\quad \vdots \\ \text{Adjusted data } r &\rightarrow_{\text{weak learner}} \hat{f}^{(r)}(.) \end{aligned}$$

where r is a prespecified positive integer. Then, the aggregated result is

$$\text{aggregation} : \hat{f}_A = \sum_{i=1}^{r} \alpha_i \hat{f}^{(i)}(.), \tag{7.47}$$

where α_i are suitable coefficients or weights. Obviously, the choice of weak learner, the number of iterations, and the selection of weights α_i are important issues for boosting to be useful in application.

Mathematically speaking, to estimate the function $f(.)$ of Eq. (7.46), we often employ a proper loss function and minimize it over some space of measurable functions. That is,

$$f_o(.) = \arg\min_{f \in S} E[L(f(\boldsymbol{X}), \boldsymbol{y})] \tag{7.48}$$

where S denotes a space of measurable functions and $L(f(.), y)$ is a loss function which is assumed to be differentiable and convex with respect to $f(.)$. For the squared error loss, $L(f(.), y) = (y - f(.))^2$ and the associated population minimizer is $f_o(\boldsymbol{x}) = E[Y|\boldsymbol{X} = \boldsymbol{x}]$.

Friedman (2001) gives a functional gradient descent (FGD) algorithm to estimate $f_o(.)$. The algorithm is as follows:

FGD Algorithm:

1. Initialize $\hat{f}^{(0)}(.)$ such as $\hat{f}^{(0)}(.) \equiv 0$ or
$$\hat{f}^{(0)} = \arg\min_{c} \frac{1}{n} \sum_{t=1}^{n} L(c, y_t).$$
Set $r = 0$.
2. Increase r by 1, i.e. $r < -r + 1$. Generate a new set of response variables by computing the negative gradient $-\frac{\partial L(f,y)}{\partial f}$ and evaluating it at $\hat{f}^{(r-1)}(\boldsymbol{x}_t)$. That is, compute
$$v_t = -\frac{\partial L(f, y_t)}{\partial f}|_{f=\hat{f}^{(r-1)}(\boldsymbol{x}_t)}, \quad t = 1, \ldots, n.$$
3. Fit the new response variables $v_1, \ldots, v_n$ by the explanatory variables $\boldsymbol{x}_1, \ldots, \boldsymbol{x}_n$ by the weak learner, i.e.
$$(\boldsymbol{x}_t, v_t) \rightarrow_{\text{weak learner}} g^{(r)}(.).$$
4. Update $\hat{f}^{(r)}(.) = \hat{f}^{(r-1)}(.) + \delta \times \hat{g}^{(r)}(.)$, where $0 < \delta \leq 1$ is a step-size parameter.
5. Iterate Steps 2 to 4 until $r = r_o$ for a prespecified positive integer r_o.

For independent data, r_o can be selected by CV. It turns out the choice of r_o is important in the above FGD algorithm. On the other hand, the step-size δ is not critical so long as it is small. In what follows, we focus on the ℓ_2 boosting as our main focus is prediction.

7.6.1 ℓ_2 Boosting

Assume that the function $f(\boldsymbol{x}_t)$ of Eq. (7.46) is linear so that we have a linear regression problem. The ℓ_2 boosting is then a useful algorithm, especially if the dimension of $\boldsymbol{x}_t$ is large. Here it is convenient to modify the squared error loss function as $L(f, y) = \frac{1}{2}(y - f)^2$ for which the negative gradient is the simply regression residuals. Applying the FGD algorithm to this high-dimensional linear regression problem, we obtain the following ℓ_2 boosting algorithm:

ℓ_2 **Boosting Algorithm**:

1. Initialize $\hat{f}^{(0)}(.) = \bar{y}$ and set $r = 0$.
2. Increase r by 1. That is, $r < -r + 1$. Compute the residuals

$$v_t = y_t - \hat{f}^{(r-1)}(\boldsymbol{x}_t), \quad t = 1, \ldots, n.$$

3. Fit the residuals $v_1, \ldots, v_n$ to $\boldsymbol{x}_1, \ldots, \boldsymbol{x}_n$ by the weak learner:

$$(\boldsymbol{x}_t, v_t) \rightarrow_{\text{weak learner}} g^{(r)}(.).$$

4. Update $\hat{f}^{(r)}(.) = \hat{f}^{(r-1)}(.) + \delta \times \hat{g}^{(r)}(.)$, where $0 < \delta \leq 1$.
5. Iterate Steps 2 to 4 until $r = r_o$, a pre-specified positive number of iterations.

The resulting estimate is

$$\hat{f}^{(r_o)}(.) = \delta \sum_{r=1}^{r_o} \hat{g}^{(r)}(.) + \hat{f}^{(0)}(.). \tag{7.49}$$

Therefore, the properties of the boosting estimate $\hat{f}^{(r_o)}(.)$ is determined by those the $\hat{g}^{(r)}(.)$, which is the estimate of the weak learner. It pays to discuss choices of the weak learner and to study properties of $\hat{g}^{(r)}(.)$.

7.6.2 Choices of Weak Learner

An important factor to consider in choosing a weak learner is ease in computation. Several commonly used weak learners have been proposed in the literature, including component-wise linear LS for (generalized) linear models, component-wise smoothing spline for additive models, and regression or classification trees. See Bühlmann and van de Geer (2011, chapter 12). Here we use componentwise linear LS. Our data set is $\{(\boldsymbol{x}_t, y_t) | t = 1, \ldots, n\}$, which is assumed to satisfy the usual assumptions, e.g. independence, homoscedastic, and $E(y_t|\boldsymbol{x}_t) = \sum_{i=1}^{k} \beta_i x_{it}$. For each boosting iteration, the data set is $\{(\boldsymbol{x}_t, v_t)|t = 1, \ldots, n\}$ and $\hat{g}^{(r)}(.)$ is given by

$$\hat{g}^{(r)}(\boldsymbol{x}_t) = \hat{\gamma}_{\hat{j}_r} x_{\hat{j}_r t},$$

$$\hat{\gamma}_j = \frac{\sum_{t=1}^n x_{jt} v_t}{\sum_{t=1}^n x_{jt}^2}, \quad \hat{j}_r = \arg\min_{1 \le j \le k} \sum_{t=1}^n (v_t - \hat{\gamma}_j x_{jt})^2. \tag{7.50}$$

In other words, in each iteration, we select the best predictor in a simple linear regression framework and use its coefficient estimate $\hat{\gamma}_j$. This is similar to the conventional forward procedure in variable selection of linear regression analysis.

The model fit of the above weak learner is

$$\hat{f}^{(r)}(\boldsymbol{x}_t) = \hat{f}^{(r-1)}(\boldsymbol{x}_t) + v \times \hat{\gamma}_{\hat{j}_r} x_{\hat{j}_r, t},$$

where $\hat{j}_r$ is defined in Eq. (7.50). In terms of coefficient estimates, we have

$$\hat{\beta}_j^{(r)} = \begin{cases} \hat{\beta}_j^{(r-1)} + \delta \times \hat{\gamma}_j, & \text{if } j = \hat{j}_r, \\ \hat{\beta}_j^{(r-1)}, & \text{if } j \neq \hat{j}_r. \end{cases}$$

Consequently, only the coefficient estimate of the $\hat{j}_r$th predictor is updated in the rth iteration. Clearly, the ℓ_2 boosting with componentwise linear LS weak leaner produces a linear model fit for every boosting iteration r. Consistency of such a ℓ_2 boosting algorithm for the high-dimensional linear regression has been derived in the literature. See, for instance, Theorem 12.2 of Bühlmann and van de Geer (2011).

Bai and Ng (2009) applies this algorithm to select the factors to be included in the model of PCR and DIs of Section 7.4. Two ways to compute the final predictor by boosting are compared. In the first case they considered the lags of the variables as any other variable and use simple regression to evaluate its prediction. In the second case they considered all the lags together and use multiple regression to compute its prediction. Their results show that the best of these two ways depends on the data generating process that is unknown.

Example 7.8

There are several packages available in R to perform boosting. We use the **gbm** package in our demonstration. The package **caret** is also available. These packages use tree-based methods, which is discussed in the next chapter. Here it suffices to say that a regression tree consists of certain partitioning of the predictor space. The partition is carried out sequentially, and each partition uses a selected predictor and a threshold. The variable and the threshold are chosen to minimize the sum of squared errors of residuals in each subregion of the predictor space.

To illustrate boosting, we apply the ℓ_2 boosting to the monthly macroeconomic data of Examples 7.5-7.7. Since this is a regression problem, we use the subcommand `distribution=''gaussian''` in the `gbm` command. First, we use 10 000 trees with `interaction.depth=1`, which limits the tree to one split only, to model the growth rate of the IPI. In addition, we employ a small $\delta = 0.001$. With fivefold CV, the algorithm took 9994 iterations to achieve the minimum CV error of 6.50×10^{-3}.

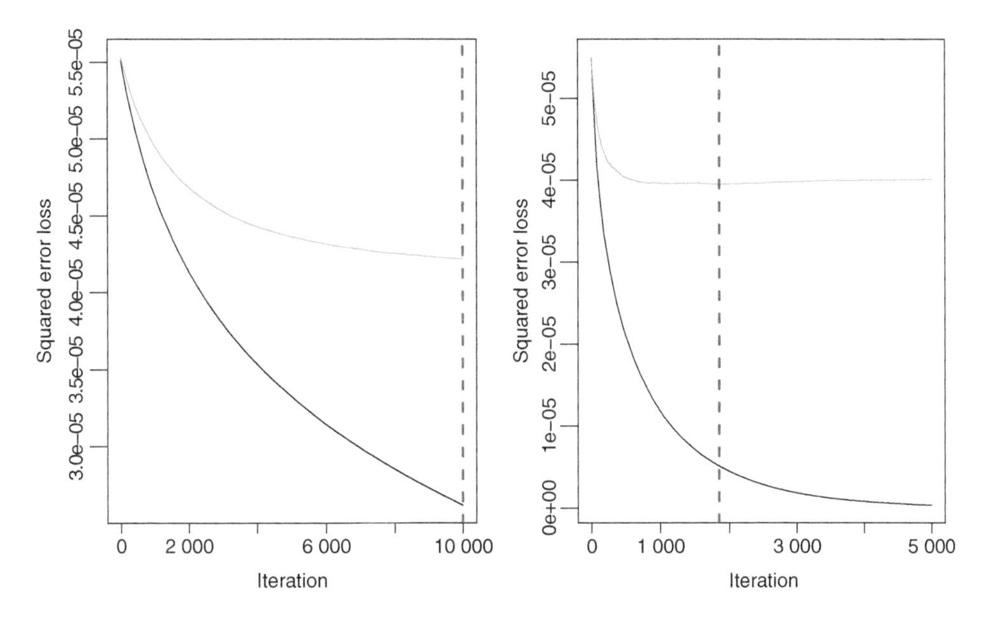

Figure 7.25 Results of fivefold CV for ℓ_2 boosting algorithm. The data consist of monthly macroeconomic variables. The dependent variable is the monthly growth rate of IPI and the predictors are lag-1 to lag-6 of all 122 variables available. The left plot is for 10 000 trees with interaction depth 1 and the right plot is for 5000 trees with depth 3.

See the left plot of Figure 7.25. Among 732 predictors, the boosting algorithm identifies 485 variables which have non-zero influence. Second, we use 5000 trees with `interaction.depth=3`, which again limits the tree size, to model the IP growth rate. The shrinkage rate used is 0.01. See the right plot of Figure 7.25. In this particular instance, the CV selects 1848 iterations with CV error 6.29×10^{-3}. Now, 682 variables out of 732 predictors had non-zero influence.

Finally, we use the command `tsBoost` of the **SLBDD** package that uses simple linear regression as weak learner. With $\delta = 0.01$ and 5000 iterations, the program selects 124 predictors. Figure 7.26 shows the time plots of the monthly growth rates of the IPI and the fitted values of the ℓ_2 boosting. From the plots, the boosting algorithm works well as it is able to describe nicely the behavior of IP growth rates. Figure 7.27 shows the sample autocorrelations of the IP growth rates and the residuals of the boosting fit. Clearly, the residuals show no sign of significant serial correlations. In this particular application, as expected, the simple linear regression seems to be a good choice of weak learner.

R Demonstration of Boosting:

```
> da <- read.csv("FREDMDApril19.csv",header=TRUE)
> x <- as.matrix(da)
> Y <- x[7:710,6]
> X <- cbind(x[6:709,],x[5:708,],x[4:707,],x[3:706,],x[2:705,],x[1:704,])
> X1 <- scale(X,center=TRUE)
> require(gbm)
> gbm1 <- gbm(Y~.,data=data.frame(X1),distribution="gaussian",
```

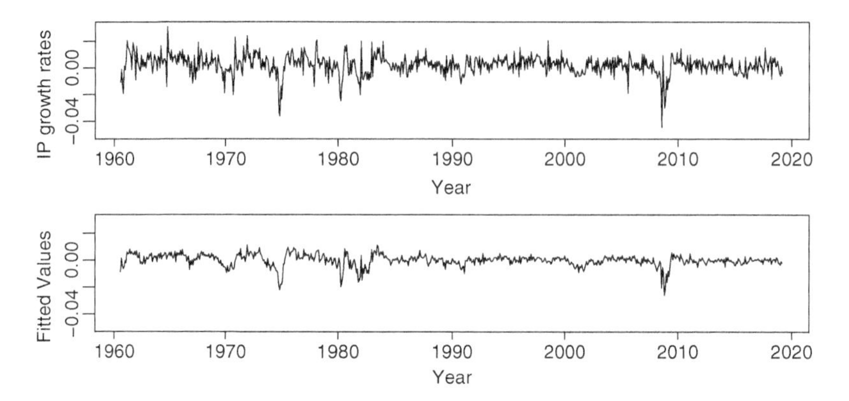

Figure 7.26 Time plots of the monthly growth rates of U.S. IPI: The top panel is the observed rates and the bottom panel is the fitted values of ℓ_2 boosting with simple linear regression, $\nu = 0.01$ and 5000 iterations.

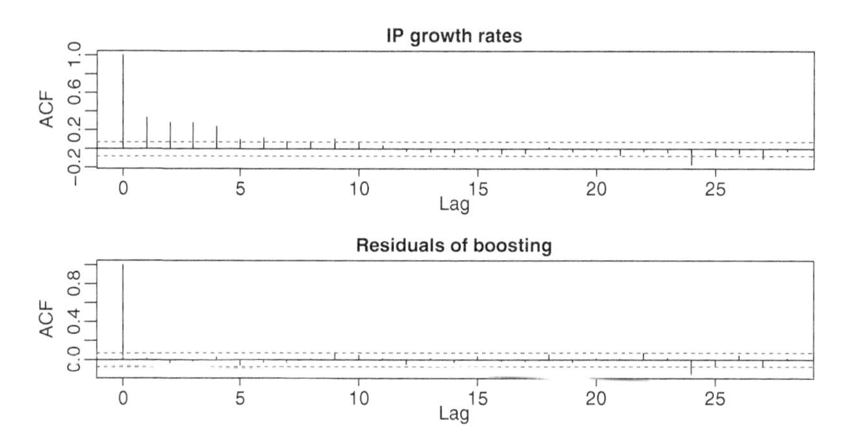

Figure 7.27 Autocorrelation functions (ACFs): the top panel is for the monthly growth rate of US IPI and the bottom panel is for the residuals of ℓ_2 boosting with $\nu = 0.01$ and 5000 iterations.

```
n.trees=10000,shrinkage=0.001,cv.folds=5)
> print(gbm1)
gbm(formula = Y ~., distribution = "gaussian", data = data.frame(X1),
n.trees = 10000, shrinkage = 0.001, cv.folds = 5)
A gradient boosted model with gaussian loss function.
10000 iterations were performed.
The best CV iteration was 9994.
There were 732 predictors of which 485 had non-zero influence.
> sqrt(min(gbm1$cv.error))
[1] 0.006497476
> names(gbm1)
[1] "initF"              "fit"                "train.error"
[4] "valid.error"        "oobag.improve"      "trees"
[7] "c.splits"           "bag.fraction"       "distribution"
[10] "interaction.depth" "n.minobsinnode"     "num.classes"
```

```
[13] "n.trees"          "nTrain"          "train.fraction"
[16] "response.name"    "shrinkage"       "var.levels"
[19] "var.monotone"     "var.names"       "var.type"
[22] "verbose"          "data"            "Terms"
[25] "cv.error"         "cv.folds"        "call"
[28] "m"                "cv.fitted"
>
> gbm2 <- gbm(Y~.,data=data.frame(X1),distribution="gaussian",
n.trees=5000,interaction.depth=3,shrinkage=0.01,cv.folds=5)
> print(gbm2)
gbm(formula = Y ~., distribution = "gaussian", data = data.frame(X1),
n.trees = 5000, interaction.depth = 3, shrinkage = 0.01,
cv.folds = 5)
A gradient boosted model with gaussian loss function.
5000 iterations were performed.
The best CV iteration was 1848.
There were 732 predictors of which 682 had non-zero influence.
> sqrt(min(gbm2$cv.error))
[1] 0.006286221
#sBoot command
> m1 <- tsBoost(Y,X,v=0.01,m=5000) #Original X.
ell-2 boosting via simple linear regression:
Both Y and X are mean-adjusted.
v and m:  0.01 5000
number of predictors selected and number of predictors:  124 732
> names(m1)
[1] "beta"      "residuals" "m"      "v"      "selection" "count"
[7] "yhat"
> tdx <- c(1:710)/12+1960
> tdx <- tdx[7:710]
> par(mfcol=c(2,1))
> range(Y)
> plot(tdx,Y,xlab="year",ylab="IP growth rates",type="l",
ylim=c(-0.05,0.031))
> plot(tdx,m1$yhat,xlab="year",ylab="Fitted Values",type="l",
ylim=c(-0.05,0.031))
> acf(Y,main="IP growth rates")
> acf(m1$residuals,main="Residuals of Boosting")
```
■

7.6.3 Boosting for Classification

The first application for Boosting was for binary classification with the AdaBoost algorithm of Freund and Schapire (1996, 1997). In the statistical literature, Breiman (1998, 1999) showed that the AdaBoost algorithm can be represented as a steepest descent algorithm in function space. As discussed in Chapter 5, given a set of n elements and a vector of classification variables, $\boldsymbol{x}$, for each element, a classifier is a function $C(\boldsymbol{x})$ that assigns to each element a value $C(\boldsymbol{x}_i) \in \{-1, 1\}$ for $1 \le i \le n$. Assuming we have a vector $\boldsymbol{y} = (y_1, \dots, y_n)'$ with values $y_i \in \{-1, 1\}$ with the correct classification, the error rate in the set of the classifier $C(\boldsymbol{x})$ is given by

$$\text{Err} = \frac{1}{n} \sum_{i=1}^{n} I(C(\boldsymbol{x}_i) \neq y_i) \tag{7.51}$$

where $I(x)$ is the indicator function, i.e. $I(x) = 1$ if x is true and 0 otherwise. Suppose we choose a very simple classifier, for instance m nearest neighbors or, better, a classification tree that is studied in the next chapter. Let $C(\boldsymbol{x})$ be the chosen weak classifier. Then, the AdaBoost algorithm works as follows:

1. Set the weights $w_i^{(1)} = 1/n$.
2. For $l = 1, \ldots, L$
   (a) Fit the classifier $C_l(\boldsymbol{x})$ to the data using weights $w_i^{(l)}$.
   (b) Compute the error rate with weights $w_i^{(l)}$,
$$\text{Err}_l = \frac{\sum_{i=1}^{n} w_i^{(l)} I(C_l(\boldsymbol{x}_i) \neq y_i)}{\sum_{i=1}^{n} w_i^{(l)}}. \tag{7.52}$$
   (c) Compute
$$\alpha_l = \log((1 - \text{Err}_l)/\text{Err}_l). \tag{7.53}$$
   (d) Modify the weights as
$$w_i^{(l+1)} = w_i^{(l)} \exp\{\alpha_l I(C_l(\boldsymbol{x}_i) \neq y_i)\}. \tag{7.54}$$
3. Make the final classification as
$$C(\boldsymbol{x}_i) = \text{sign}\left[\sum_{l=1}^{L} \alpha_l C_l(\boldsymbol{x}_i)\right]. \tag{7.55}$$

Note that the weights of each observation are modified in each iteration, according to Eq. (7.54). If the element is well classified, the weight remains the same, $w_i^{(l+1)} = w_i^{(l)}$. If it is wrongly classified, its weight is increased by $w_i^{(l+1)} = w_i^{(l)} \exp\{\alpha_l\}$. Therefore, the relative weights of the well-classified observations decrease and those of wrongly classified increase. The change of weights depends on α_l, the logit of the ratio between the success rate and error rate, see Eq. (7.53), which is the weight of the classifier in the final solution. Figure 7.28 shows the relationship between the error rate of a classifier and its weight. Finally, the final classifier is a linear combination of all the classifiers. In Chapter 8, we see an example of AdaBoost using classification tree as a weak classifier.

7.7 MIXED-FREQUENCY DATA AND NOWCASTING

The data-rich environment also gives rise to the case in which data observed at different frequencies are widely available. Consider, for example, the problem of predicting the quarterly growth rates of US GDP. There are many monthly or weekly economic data available that may contribute to the US GDP growth. For example, monthly unemployment rate and weekly (or even daily) crude oil price are likely to affect the US economy and, hence, its GDP growth. The issue then is how to make use of the information embedded in the higher-frequency data to aid the prediction of GDP. One can aggregate monthly or weekly economic data into quarterly

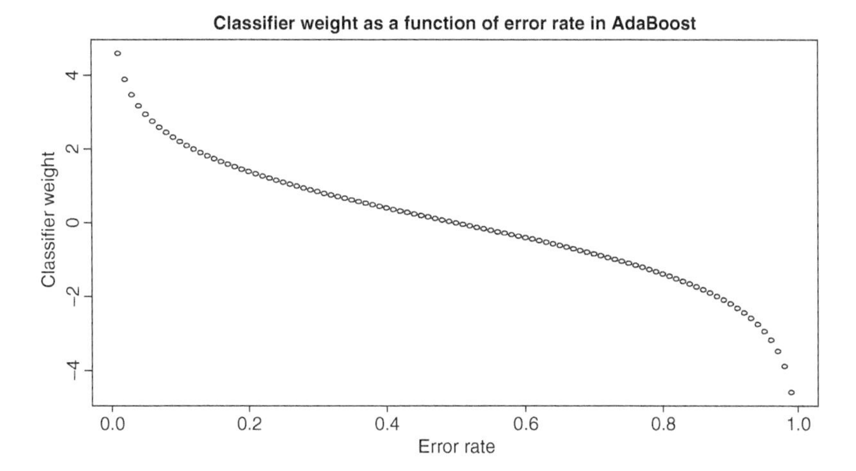

Figure 7.28 Relationship between the error rate of a classifier and its weight in the final classifier with ADABoost.

ones and uses the quarterly aggregates to predict GDP. But aggregations tend to encounter information loss owing to the effect of central limit theorem. Indeed, quarterly data tend to follow more closely the normal distribution than their weekly or daily counterparts do. Therefore, it seems advisable to employ directly the available monthly or weekly data to aid the prediction of quarterly GDP. Such a forecasting problem is often referred to as a MIDAS approach in the econometric literature. See, for instance, Andreou et al. (2010). Of particular interest in the MIDAS approach to forecasting is to use the newly available high-frequency data to update the prediction of a lower-frequency dependent variable of interest. For instance, at the quarterly forecast origin $t = n$, one can use a suitable econometric model to predict the GDP growth rate of quarter $t = n + 1$. This is similar to the conventional 1-step ahead forecast. However, as time passes, say one month into the quarter $t = n + 1$, monthly predictors become available. One can then use the mixed-frequency model to update the GDP prediction. Such an updating procedure is referred to as *nowcasting* in the literature. In what follows, we briefly introduce the MIDAS regression and econometric models available for nowcasting.

7.7.1 MIDAS Regression

To introduce the MIDAS regression, we consider the simple case of a lower-frequency dependent variable y_t and a single high-frequency predictor $x_{t,j}^{(g)}$, where $g > 1$ is a positive integer denoting the number of observations of the predictor between t and $t - 1$ and $j = 1, \ldots, g$. For example, if the time index t denotes quarter, then $g = 3$ for monthly variables, and $g = 91$ or 66 for daily variables, assuming an average of 30.4 days or 22 trading days in a month. Figure 7.29 sketches a rough time stamps for the observed data, where time flows from left to right, $y(t)$ is the dependent variable and $x(t, j)$ is the predictor.

For simplicity in notation, in this section we treat the subscript i of a time series as a pure index of the sequence of observations. It is understood that for the

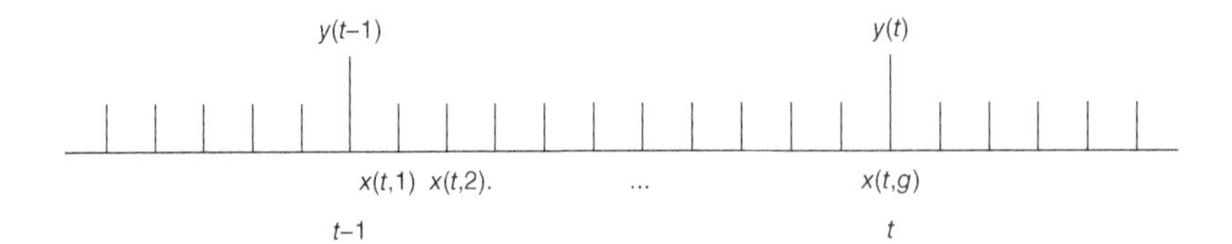

Figure 7.29 A brief sketch of time stamps of mixed-frequency data.

low-frequency dependent variable y_t, the index i is identical to time t, but for the high-frequency predictor $x_{t,j}$, the index i is given by $i = (t-1)g + j = tg - (g-j)$, which is the jth observation in time period t. To illustrate, suppose y_t is quarterly, $x_{t,j}$ is monthly, and the data available started in 1980. Then, y_1 and y_2 are the data of y_t for the first and second quarters of 1980. On the other hand, x_1 and x_4 are, respectively, the data of $x_{t,j}$ for the first and fourth months of 1980 with $4 = (t-1) \times 3 + 1$ with $t = 2$.

At the forecast origin $t-1$, a naive model for the mixed-frequency data may assume the form

$$y_i = \phi_0 + \sum_{j=1}^{p} \phi_j y_{i-j} + \sum_{v=1}^{q} \beta_v \left[\sum_{j=1}^{g} w_j x_{(i-v)g-(g-j)} \right] + e_i, \qquad (7.56)$$

where p and q are positive integers and e_i denotes the error term. In Eq. (7.56), ϕ_j are AR parameters, β_v measures the impact of the aggregate of $\{x_{i-v,j} | j = 1, \ldots, g\}$ on y_i, and w_j are weights, which are assumed to satisfy the conditions $0 \le w_j < 1$ and $\sum_{j=1}^{g} w_j = 1$. In particular, if $w_j = w$, a non-zero constant, then a simple aggregation is used to process the predictor. When there are many high-frequency predictors with large g, the model in Eq. (7.56) would employ many weight coefficients w_j making the model hard to estimate and inefficient. Andreou et al. (2010) postulates that the weights w_j follow some well-known functions driven by only a small number of parameters such as the exponential Almon lag polynomial. For instance,

$$w_j \equiv w_j(\theta_1, \theta_2) = \frac{\exp[\theta_1 j + \theta_2 j^2]}{\sum_{i=1}^{g} \exp[\theta_1 i + \theta_2 i^2]},$$

where θ_1 and θ_2 are the underlying unknown parameters. Here the weights w_j are defined so that it decays smoothly as the predictor moves away from y_i. Other weight functions can be used. Ghysels et al. (2016) provides some choices of weight functions. The key idea of MIDAS regression is to employ a weight function that results in a parsimonious regression model for y_i.

With the weight function w_j specified, the MIDAS regression model in Eq. (7.56) can be estimated by the nonlinear LS method. Some properties of the model are given in Andreou et al. (2010).

7.7.2 Nowcasting

The MIDAS regression model in Eq. (7.56) follows the tradition of forecasting y_t using the information available at time $t-1$. Such a prediction can be thought of as an initial forecast. With the availability of high-frequency data $\{x_{t,j}\}$, one can make use of the newly available information to update the forecast of y_t. Such an updating approach to forecasting is referred to as *nowcasting*, because some information in the current time period t is used in the prediction.

Again, we use a single dependent variable and a single high-frequency predictor $\{x_{t,j}|j = 1, \ldots, g\}$ to introduce nowcasting. Consider the situation of j_o into the quarter t, where $1 \leq j_o < g$. The nowcasting regression may assume the form

$$y_i = \phi_0 + \sum_{j=1}^{p} \phi_j y_{i-j} + \sum_{v=1}^{q} \beta_v \left[\sum_{j=1}^{g} w_j x_{(i-v)g-(g-j)} \right] + \sum_{j=1}^{j_o} \gamma_j x_{ig-(g-j)} + e_i, \tag{7.57}$$

where all parameters, but those of the newly added predictors $x_{i,j}$, are defined in Eq. (7.56) and γ_j denotes the conditional contribution of $x_{i,j}$ on y_i in the presence of all other predictors. If j_o is large, then γ_j may follow some weight functions similar to those used in Eq. (7.56). In other words, an alternative model for nowcasting is

$$y_i = \phi_0 + \sum_{j=1}^{p} \phi_j y_{i-j} + \sum_{v=1}^{q} \beta_v \left[\sum_{j=1}^{g} w_j x_{(i-v)g-(g-j)} \right] + \gamma \sum_{j=1}^{j_o} v_j x_{ig-(g-j)} + e_i, \tag{7.58}$$

where v_j are weights for aggregating the predictors $\{x_{t,j}|j = 1, \ldots, j_o\}$ in time period t.

Obviously, the idea of nowcasting applies to most statistical forecasting models when high-frequency predictors are available. For instance, instead of using the weight functions of the MIDAS regression, one can cast the model as a high-dimensional AR model with exogenous variables and estimate the model by some regularized estimation method discussed before, e.g. the Lasso method.

MIDAS regression is available in R via the package **midasr**. The package can also be used to compute the weights of Almon lag polynomials. Details of the package can be found in Ghysels et al. (2016).

Example 7.9

To illustrate the MIDAS regression, we consider the growth rate of US quarterly GDP from the first quarter of 1986 to the second quarter of 2019. The two high-frequency predictors used are (a) changes in the monthly unemployment rates, $(x_{t,i}|i = 1, 2, 3)$, and (b) changes in the weekly crude oil price, Western Texas Intermediate, Cushing, Oklahoma, $(z_{t,i}|i = 1, \ldots, 12)$. Thus, we have one monthly and one weekly exogenous predictor. For the forecasting model of Eq. (7.56), we use $p = 2$ for the AR part, $q = 2$ for unemployment rate and $q = 4$ for crude oil price. In addition, we use exponential Almon lag polynomial (`nealmon` command in **midasr**) as the weight function. For the monthly unemployment rate, the lags used are 3-5 and for the weekly crude oil price, the lags used are 12-23. These lags cover the high-frequency predictors in a quarter. The R commands used for such a MIDAS regression and the associated output are given below in the R demonstration. In this particular case, the output shows that the conditional contributions of the high-frequency predictor are not strong.

Turn to nowcasting. Two cases are considered. In the first case, we use high-frequency data one month into the quarter. Here we use $q = 2$ for the unemployment rate and $q = 3$ for the crude oil price. Details of the commands and the associated output are also given in the R demonstration below. It is interesting to see that for both high-frequency predictors the concurrent values (i.e. `unrate11` and `coilchg11`) are significant at the conventional 5% level. In the second case, we use high-frequency data two months into the quarter. The results are similar to those of the one-month case as the concurrent high-frequency data are helpful in explaining the quarterly growth rate of US GDP.

R Demonstration: MIDAS regression. The R scripts are provided by Dr. Yuefeng Han of the Rutgers University.

```
(A) MIDAS regression for forecasting:
(1) R scripts

gdp8619<-read.csv("q-gdp8619.csv", header = TRUE)
gdp<-gdp8619$A191RP1Q027SBEA
unrate8619d.csv<-read.csv("m-unrate8619d.csv", header = TRUE)
unrate<-unrate8619d.csv$ratechg
coilchg.csv<-read.csv("w-coilchg.csv", header = TRUE)
coilchg<-coilchg.csv$coilchg
library(midasr)
t=2 # time lag for gdp
t1=3 # time lag for unrate
t2=12 # time lag for crude oil price

fit<-midas_r(gdp~mls(gdp,1:t,1)+mls(unrate,3:(t1+2),3,nealmon)+
mls(coilchg,12:(t2+11),12,nealmon),
start=list(unrate=c(1,-1),coilchg=rep(0,4)),
#aggregation within a quarter 2 and 4
method="Nelder-Mead")
#print(fit); print(summary(fit))

(2) Output
> names(fit)
[1] "coefficients"       "midas_coefficients" "model"
[4] "unrestricted"       "term_info"          "fn0"
[7] "rhs"                "gen_midas_coef"     "opt"
[10] "argmap_opt"         "start_opt"          "start_list"
[13] "call"               "terms"              "gradient"
[16] "hessian"            "gradD"              "Zenv"
[19] "use_gradient"       "nobs"               "tau"
[22] "lhs"                "lhs_start"          "lhs_end"
[25] "convergence"        "fitted.values"      "residuals"
> summary(fit)
MIDAS regression model with "numeric" data:
Start = 3, End = 133

Formula gdp ~ mls(gdp,1:t,1)+mls(unrate,3:(t1+2),3,nealmon) +
mls(coilchg, 12:(t2 + 11), 12, nealmon)

Parameters:
Estimate Std. Error t value Pr(>|t|)
(Intercept)  2.03136    0.51532    3.942 0.000135 ***
gdp1         0.34660    0.10876    3.187 0.001826 **
gdp2         0.23341    0.10295    2.267 0.025131 *
unrate1     -0.96387    3.29969   -0.292 0.770700
```

```
unrate2      1.18138    7.07705   0.167 0.867700
coilchg1     0.52039    0.34126   1.525 0.129871
coilchg2     2.07281    5.64456   0.367 0.714089
coilchg3    -0.29348    0.93787  -0.313 0.754873
coilchg4     0.01082    0.05004   0.216 0.829154
—
Residual standard error: 2.316 on 122 degrees of freedom

(B) Nowcasting: one-month
(1) R scripts
t=2 # time lag for gdp
t1=3+1 # time lag for unrate
t2=12+4 # time lag for crude oil price
gdp1=c(NA,gdp)
unrate1=c(0,0,unrate,0)
coilchg1=c(rep(0,8),coilchg,rep(0,4))

fit1<-midas_r(gdp1~mls(gdp1,1:t,1)+mls(unrate1,3:(t1+2),3,nealmon)+
mls(coilchg1,12:(t2+11),12,nealmon),
start=list(unrate1=c(1,-1),coilchg1=rep(0,3)),
method="Nelder-Mead")
#print(fit1); print(summary(fit1))
(2) Output
> fit1
MIDAS regression model with "numeric" data:
Start = 4, End = 134
model: gdp1 ~ mls(gdp1,1:t,1)+mls(unrate1,3:(t1+2),3,nealmon) +
mls(coilchg1, 12:(t2 + 11), 12, nealmon)
(Intercept) gdp11 gdp12 unrate11 unrate12 coilchg11 coilchg12 coilchg13
2.4596 0.3081 0.184  -4.2635  -3.6072    0.7236    0.7873   -0.2933

Function optim was used for fitting
> summary(fit1)

MIDAS regression model with "numeric" data:
Start = 4, End = 134

Formula gdp1 ~ mls(gdp1,1:t,1)+mls(unrate1,3:(t1+2),3,nealmon) +
mls(coilchg1, 12:(t2 + 11), 12, nealmon)

Parameters:
Estimate Std. Error t value Pr(>|t|)
(Intercept)  2.45959    0.32542   7.558 8.12e-12 ***
gdp11        0.30808    0.08904   3.460 0.000743 ***
gdp12        0.18353    0.09137   2.009 0.046755 *
unrate11    -4.26354    1.75713  -2.426 0.016699 *
unrate12    -3.60720   11.36851  -0.317 0.751556
coilchg11    0.72356    0.09164   7.895 1.37e-12 ***
coilchg12    0.78727    1.47175   0.535 0.593670
coilchg13   -0.29329    0.31563  -0.929 0.354588
—
Residual standard error: 1.963 on 123 degrees of freedom

(C) Nowcasting: 2 months
(1) R scripts
t=2 # time lag for gdp
t1=3+2 # time lag for unrate
```

```
t2=12+8 # time lag for crude oil price

gdp2=c(NA,gdp)
unrate2=c(0,unrate,0,0)
coilchg2=c(rep(0,4),coilchg,rep(0,8))

fit2<-midas_r(gdp1~mls(gdp2,1:t,1)+mls(unrate2,3:(t1+2),3,nealmon)+
mls(coilchg2,12:(t2+11),12,nealmon),
start=list(unrate2=c(1,-1),coilchg2=rep(0,3)),
method="Nelder-Mead")
#print(fit2);  print(summary(fit2))
(2) Output
> fit2
MIDAS regression model with "numeric" data:
Start = 4, End = 134
model: gdp1 ~ mls(gdp2, 1:t, 1)+mls(unrate2, 3:(t1+2),3,nealmon) +
mls(coilchg2, 12:(t2 + 11), 12, nealmon)
(Intercept)  gdp21  gdp22 unrate21 unrate22 coilchg21 coilchg22 coilchg23
2.8584 0.2575 0.1835  -8.3487  -0.7293    0.7542    3.3481   -0.2915

Function optim was used for fitting
> summary(fit2)
MIDAS regression model with "numeric" data:
Start = 4, End = 134

Formula gdp1 ~ mls(gdp2, 1:t, 1)+mls(unrate2,3:(t1+2),3,nealmon) +
mls(coilchg2, 12:(t2 + 11), 12, nealmon)

Parameters:
Estimate Std. Error t value Pr(>|t|)
(Intercept)  2.85839    0.44802   6.380 3.26e-09 ***
gdp21        0.25755    0.08463   3.043 0.002863 **
gdp22        0.18347    0.08676   2.115 0.036469 *
unrate21     8.34869    2.28870  -3.648 0.000389 ***
unrate22    -0.72935    0.41003  -1.779 0.077747.
coilchg21    0.75425    0.16080   4.691 7.12e-06 ***
coilchg22    3.34811    2.95422   1.133 0.259280
coilchg23   -0.29153    0.25830  -1.129 0.261242
—
Residual standard error: 2.019 on 123 degrees of freedom
```

■

Example 7.10

To demonstrate the power of using high-frequency data to improve the prediction of a low-frequency variable of interest, we consider a simple application. Figure 7.30 shows the time plot of the square-root transform of daily maximum $PM_{2.5}$ measurement of a monitoring station in the Kaohsiung City of Taiwan from 1 January 2006 to 31 December 2015. The data are available in the file `TaiwanPM25.csv` (Station 1) and, for simplicity, the data of 29 February were removed. From the plot, there is, as expected, a clear annual cycle. We took the square-root transform as the daily maximum $PM_{2.5}$ is non-Gaussian. We reserve the last two years as the forecasting subsample and our goal is to demonstrate that using the first few hourly $PM_{2.5}$ measurements can improve markedly the prediction of y_t compared with AR models that only employ lagged values of y_t as predictors.

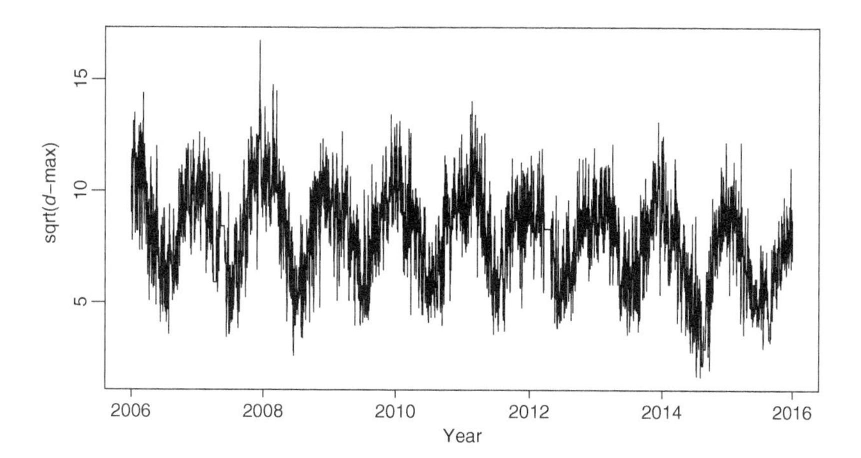

Figure 7.30 Time plot of the square-root transform of daily maximum $PM_{2.5}$ measurements of a monitoring station in the Kaohsiung City of Taiwan from 1 January 2006 to 31 December 2015. The data for 29 February were removed.

Specifically, we divide the data into estimation and forecasting subsamples with the forecasting subsample consisting of the last 730 observations. The estimation subsample has 2920 data points. The AIC selects an AR(22) model for y_t and the model indeed fits the data well. See the model checking statistics shown in Figure 7.31. For instance, the residual Q-statistics fail to reject the null hypothesis of no serial correlations in the residuals.

For nowcasting, we employ the model

$$y_t = \phi_0 + \sum_{i=1}^{22} \phi_i y_{t-i} + \sum_{j=1}^{q} \beta_j x_{t,j} + \epsilon_t, \tag{7.59}$$

where $q = 1, \ldots, 6$ and $x_{t,j}$ denotes the square-root transform of the hourly observed $PM_{2.5}$ at the jth hour of each day. The maximum value of q is 6, indicating that one can update the prediction of y_t at the 6'o clock in the morning. The model is Eq. (7.59) is a simple ARX model, which is very easy to estimate. Table 7.2 summarizes the results of out-of-sample predictions, where AR(22) is the benchmark for comparison and AR(22)+q denotes the first q hourly data are used in nowcasting. From the table, it is clear that using the hourly data in nowcasting indeed provides improvements over the benchmark in out-of-sample forecasting. The R script for this example is given below.

R script for Example 7.10:

```
da <- read.csv("TaiwanPM25.csv",header=TRUE)
st1 <- sqrt(da[,4]) # Square-root transformation
y <- NULL; X <- NULL# Find daily maximum and first 6 hours data
for (i in 1:3650){
ist <- (i-1)*24+1
iend <- i*24
y <- c(y,max(st1[ist:iend]))
```

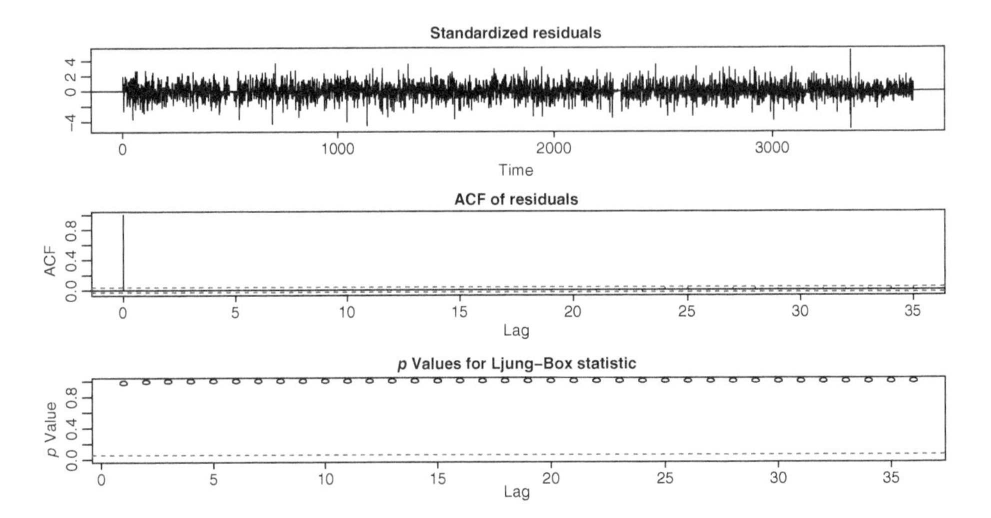

Figure 7.31 Model checking of a scalar AR(22) model for the square-root transform of daily maximum $PM_{2.5}$ measurements of a monitoring station in the Kaohsiung City of Taiwan from 1 January 2006 to 31 December 2015. The data for 29 February were removed.

TABLE 7.2 Summary Statistics of Out-of-Sample Forecasts of the Square-Root Transform of Daily Maximum $PM_{2.5}$ of a Station in Kaohsiung City

Model	Bias	RMSE	MAE
AR(22)+0	−0.0991	1.145	0.886
AR(22)+1	0.0778	1.015	0.783
AR(22)+2	0.0135	0.985	0.760
AR(22)+3	−0.0319	0.949	0.731
AR(22)+4	−0.0045	0.917	0.697
AR(22)+5	0.0096	0.913	0.693
AR(22)+6	−0.0050	0.884	0.683

In the Table AR(22)+q denotes the first q hourly data are used in Nowcasting, and Bias, RMSE, and MAE are the average bias, root mean squared forecast error, and mean absolute forecast error, respectively.

```
X <- rbind(X,st1[ist:(ist+5)])
}
m1 <- ar(y)
p <- m1$order
p
ist <- p+1
nT <- 3650
Y <- y[ist:nT]
Z <- NULL
for (i in 1:p){
Z <- cbind(Z,y[(ist-i):(nT-i)])
}
```

```
nobe <- nT-p
nfit <- nobe-730
Zfit <- Z[1:nfit,]
Yfit <- Y[1:nfit]
Ztest <- Z[(nfit+1):nobe,]
Ytest <- Y[(nfit+1):nobe]
m2 <- lm(Yfit~.,data=data.frame(Zfit)) #AR(22) model
pm2 <- predict(m2,newdata=data.frame(Ztest))
s1 <- mean(Ytest-pm2)
s2 <- sqrt(mean((Ytest-pm2)^2))
s3 <- mean(abs(Ytest-pm2))
M <- c(s1,s2,s3)
## select the corresponding time period.
X1 <- X[ist:nT,] # Nowcasting
for (i in 1:6){
Zfitnow <- cbind(Zfit,X1[1:nfit,1:i])
Ztestnow <- cbind(Ztest,X1[(nfit+1):nobe,1:i])
m3 <- lm(Yfit~.,data=data.frame(Zfitnow))
pm3 <- predict(m3,newdata=data.frame(Ztestnow))
s1 <- mean(Ytest-pm3)
s2 <- sqrt(mean((Ytest-pm3)^2))
s3 <- mean(abs(Ytest-pm3))
M <- rbind(M,c(s1,s2,s3))
}
print(M)
```
■

7.8 STRONG SERIAL DEPENDENCE

Most of the methods discussed in this chapter may fail in the presence of strong serial correlations in the data. See, for instance, the discussion of Lasso in Section 7.1 and the conditions in Section 7.3. Care must be exercised when strong serial dependence exists. But there is no unique way to handle the strong serial correlations if the dimension of the data is high. In fact, many important questions concerning the properties of regularized estimation remain open when the data have strong temporal dependence. Ideally, based on the available theoretical justifications, one would like to transform the observed data with strong serial dependence into observations with short-range dependence for which the methods discussed in the chapter apply. For multivariate time series, the method of co-integration discussed in Chapter 3 and the dynamic factor models in Chapter 6 can be useful. However, how to analyze co-integrated processes in HDTS analysis is still under investigation. In this section, we consider two approaches to handle HDTS when the serial dependence is strong. The first approach is to fit a VAR(1) model with some regularization to the observed series, then employ the residuals in the subsequent analysis. This approach should work reasonably well so long as the multiplicity of any unit root is one. Indeed, most macroeconomic time series are I(1) processes, that is, they have a single unit root. We note that it is in general not a good idea to differencing individual time series to achieve marginal stationarity because, as discussed in Chapter 3, this can lead to over-differencing which, in turn, introduces non-invertible time series. The second approach is to perform residual analysis of any statistical model used. Limited experience indicates that both under-differencing and over-differencing can easily result in serial correlations in the residuals. Thus, proper residual analysis can detect mistreatment of strong serial dependence.

Finally, some recent investigations on modeling high-dimensional unit-root time series include Zhang et al. (2019), who investigate identifying cointegration via eigenanalysis, and Gao and Tsay (2020), who study modeling high-dimensional unit-root time series.

EXERCISES

1. Consider the monthly macroeconomic data set used in Section 7.3. The dependent variable of interest is the inflation, consumer price index all items, which is `CPIAUCSL`. The predictors consist of the first 6 lagged values of all 122 variables available. Perform a Lasso linear regression analysis on the data, including coefficient profile plot and the CV plot. Use CV to select the optimal penalty parameter. Plot the resulting estimated coefficients $\hat{\beta}_i$ of the Lasso regression.
2. Repeat the analysis of Problem 1 but using `glmnet` with $\alpha = 0.75$.
3. Consider, again, the monthly macroeconomic data set of Problem 1. Apply Group Lasso by letting (a) lagged predictors of CPIAUCSL as group 1, and (b) predictors of other variables with the same lag form a group.
4. Again, consider the monthly macroeconomic data set of Problem 1. Apply boosting to the problem with subcommands `n.trees` = 10000 and shrinkage = 0.001.
5. Consider the hourly $PM_{2.5}$ measurements of Station 2 (Column 5) in the data file `TaiwanPM25.csv`. Obtain the series y_t of the square-root transform of daily maximum $PM_{2.5}$. Reserve the last two years as the forecasting subsample. (a) Entertain a scalar AR model for the y_t series. Compute the root mean squared errors of the 1-step ahead predictions of the AR model. (b) Augment the predictors with the first six hourly $PM_{2.5}$ of each day. Compute the root mean squared errors of the 1-step ahead prediction using nowcasting. Is nowcasting helpful in this particular instance? Why?

REFERENCES

Abdi, H. (2003). Partial least squares (pls) regression. In M. Lewis-Beck, A. Bryman, and T. Futing (eds.). *Encyclopedia of Social Sciences Research Methods*. Sage, Thousand Oaks, CA.

Andreou, E., Ghysels, E., and Kourtellos, A. (2010). Regression models with mixed sampling frequencies. *Journal of Econometrics*, **158**: 246–261.

Bai, J. and Ng, S. (2008). Forecasting economic time series using targeted predictors. *Journal of Econometrics*, **146**: 304–317.

Bai, J. and Ng, S. (2009). Boosting diffusion indices. *Journal of Applied Econometrics*, **24**: 607–629.

Bartlett, P., Mendelson, S., and Neeman, J. (2012). ℓ_1 regularized linear regression: Persistence and oracle inequalities. *Probability Theory and Related Fields*, **152**: 193–224.

Basu, S. and Michailidis, G. (2015). Regularized estimation in sparse high-dimensional time series models. *Annals of Statistics*, **43**: 1535–1567.

Bertsekas, D. (1995). *Nonlinear Programming*. Athena Scientific, Belmont, MA.

Bickel, P., Ritov, Y., and Tsybakov, A. (2009). Simultaneous analysis of Lasso and Dantzig selector. *Annals of Statistics*, **37**: 1705–1732.

Breiman, L. (1998). Arcing classifiers (with discussions). *Annals of Statistics*, **26**: 801–849.

Breiman, L. (1999). Prediction games and arcing algorithms. *Neural Computation*, **11**: 1493–1517.

Bühlmann, P. and van de Geer, S. (2011). *Statistics for High-Dimensional Data*. Springer, New York, NY.

Candes, E. J. (2006). Modern statistical estimation via oracle inequalities. *Acta Numerica*, **15**: 257–325.

de Jong, S. (1993). IMPLS: An alternative approach to partial least squares regression. *Chemometrics and Intelligent Laboratory Systems*, **18**: 251–263.

Dedecker, J., Doukhan, P., Lang, G., Rafael, L. R. J., Louhichi, S., and Prieur, C. (2007). *Weak dependence: With examples and applications*. Springer, New York, NY.

Efron, B., Hastie, T., Johnstone, I., and Tibshirani, R. (2004). Least angle regression (with discussion). *Annals of Statistics*, **32**: 407–499.

Fan, J. and Li, R. (2001). Variable selection via nonconcave penalized likelihood and its oracle properties. *Journal of the American Statistical Association*, **96**: 1348–1360.

Freund, Y. and Schapire, R. (1996). Experiments with a new boosting algorithm. In *Proceedings of the Thirteenth International Conference on Machine Learning*. Morgan Kaufmann Publishers Inc., San Francisco, CA.

Freund, Y. and Schapire, R. (1997). A decision-theoretic generalization of on-line learning and an application to boosting. *Journal of Computer and System Sciences*, **55**: 119–139.

Friedman, J. (2001). Greedy function approximation: A gradient boosting machine. *Annals of Statistics*, **29**: 1189–1232.

Friedman, J., Hastie, T., and Tibshirani, R. (2000). Additive logistic regression: A statistical view of boosting (with discussion). *Annals of Statistics*, **28**: 337–407.

Gao, Z. and Tsay, R. S. (2020). Modeling high-dimensional unit-root time series. *International Journal of Forecasting* (to appear).

Geladi, P. and Kowalski, B. (1986). Partial least squares regression: A tutorial. *Analytica Chemica Acta*, **35**: 1–17.

Ghysels, E., Kvedaras, V., and Zemlys, V. (2016). Mixed frequency data sampling regression models: The R package midasr. *Journal of Statistical Software*, **72**(4). doi: 10.18637/jss.v072.i04.

Giraud, C. (2015). *Introduction to High-Dimensional Statistics*. CRC Press, Boca Raton, FL.

Greene, W. H. (2003). *Econometric Analysis*. Pearson Education, India.

Greenshtein, E. and Ritov, Y. (2004). Persistence in high-dimensional predictor selection and the virtue of over-parameterization. *Bernoulli*, **10**: 971–988.

Griffiths, W. E., Judge, G. G., Hill, R. C., Lütkepohl, H., and Lee, T. C. (1985). *The Theory and Practice of Econometrics*. Wiley, Hoboken, NJ.

Gupta, S. (2012). A note on the asymptotic distribution of lasso estimator for correlated data. *Sankhya: The Indian Journal of Statistics, Series A*, **74**: 10–28.

Han, Y. and Tsay, R. S. (2020). High-dimensional linear regression for dependent data with applications to nowcasting. *Statistica Sinica*, **30**: 1797–1827.

Hastie, T., Tibshirani, R., and Friedman, J. (2009). *The elements of statistical learning: Data mining, inference, and prediction*. Springer Science & Business Media. New York, NY.

Hastie, T., Tibshirani, R., and Wainwright, M. (2015). *Statistical Learning with Sparsity: The Lasso and Generalizations*. CRC press, Boca Raton, FL.

Helland, I. S. (1990). Partial least squares regression and statistical methods. *Scandinavian Journal of Statistics*, **17**: 97–114.

Hoerl, A. E. (1962). Application of ridge analysis to regression problems. *Chemical Engineering Progress*, **58**: 54–59.

Kock, A. B. and Callot, L. (2015). Oracle inequalities for high dimensional vector autoregressions. *Journal of Econometrics*, **186**: 325–344.

McCracken, M. W. and Ng, S. (2016). FRED-MD: A monthly database for macroeconomic research. *Journal of Business & Economic Statistics*, **34**: 574–589.

Medeiros, M. C. and Mendes, E. F. (2016). L1-regularization of high-dimensional time-series models with non-Gaussian and heteroskedastic errors. *Journal of Econometrics*, **191**: 255–271.

Meinshausen, N. and Bühlmann, P. (2006). High-dimensional graphs and variable selection with the Lasso. *Annals of Statistics*, **34**: 1436–1462.

Ng, K. S. (2013). A simple explanation of partial least squares. Penn State University. http://citeseerx.ist.psu.edu/viewdoc/summary?doi=10.1.1.352.4447

Nicholson, W. B., Matteson, D. S., and Bien, J. (2017a). VARX-L: Structural regularization for large autoregressions with exogenous variables. *International Journal of Forecasting*, **33**: 627–651.

Nicholson, W. B., Matteson, D. S., and Bien, J. (2017b). Bigvar: Tools for modeling sparse high-dimensional multivariate time series. arXiv: 1702.07094.

Peña, D., Smucler, E., and Yohai, V. (2021). Sparse estimation of dynamic principal components for forecasting high-dimensional time series. *International Journal of Forecasting*. (to appear).

Seber, G. A. F. and Lee, A. J. (2003). *Linear Regression Analysis*, 2nd Edition. Wiley, Hoboken, NJ.

Stock, J. H. and Watson, M. W. (2002). Macroeconomic forecasting using diffusion indexes. *Journal of Business & Economic Statistics*, **20**: 147–162.

Taddy, M. (2017). One-step estimator paths for concave regularization. *Journal of Computational and Graphical Statistics*, **26**: 525–536.

Tibshirani, R. (1996). Regression shrinkage and selection via Lasso. *Journal of the Royal Statistical Society, Series B*, **58**, 267–288.

Tibshirani, R., Saunders, M., Rosset, S., Zhu, J., and Knight, K. (2005). Sparsity and smoothness via fused lasso. *Journal of the Royal Statistical Society, Series B*, **67**: 91–108.

Tibshirani, R. J. and Taylor, J. (2012). Degrees of freedom in lasso problems. *Annals of Statistics*, **40**: 1198–1232.

Tsay, R. S. (2014). *Multivariate Time Series Analysis: With R and Financial Applications*. Wiley, Hoboken, NJ.

Wang, H., Li, G., and Tsai, C. L. (2007). Regression coefficient and autoregressive order shrinkage and selection via the lasso. *Journal of the Royal Statistical Society, Series B*, **69**: 63–78.

Wold, H. (1966). Estimation of principal components and related models by iterative least squares. In P.R. Krishnaiah (ed.). *Multivariate Analysis*, pp. 391–420. Academic Press, New York.

Wu, W. B. (2005). Nonlinear system theory: another look at dependence. *Proceedings of the National Academy of Sciences of the United States of America*, **102**: 14150–14154.

Wu, W. B. and Wu, Y. N. (2016). Performance bounds for parameter estimates of high-dimensional linear models with correlated errors. *Electronic Journal of Statistics*, **10**: 352–379.

Yuan, M. and Lin, Y. (2006). Model selection and estimation in regression with grouped variables. *Journal of the Royal Statistical Society, Series B*, **68**: 49–67.

Zhang, R., Robinson, P., and Yao, Q. (2019). Identifying cointegration by eigenanalysis. *Journal of the American Statistical Association*, **114**: 916–927.

Zhao, P. and Yu, B. (2006). On model selection consistency of Lasso. *Journal of Machine Learning Research*, **7**: 2541–2563.

Zou, H. (2006). The adaptive Lasso and its oracle properties. *Journal of the American Statistical Association*, **101**: 1418–1429.

Zou, H. and Hastie, T. (2005). Regularization and variable selection via the Elastic Net. *Journal of the Royal Statistical Society, Series B*, **67**: 301–320.

Zou, H., Hastie, T., and Tibshirani, R. (2006). Sparse principal component analysis. *Journal of Computational and Graphical Statistics*, **15**: 265–286.

Zou, H., Hastie, T., and Tibshirani, R. (2007). On the degrees of freedom of the lasso. *Annals of Statistics*, **35**: 2173–2192.

CHAPTER 8

MACHINE LEARNING OF BIG DEPENDENT DATA

Classification and discriminant analysis are useful statistical tools with many applications, as shown in Chapter 5. They are widely used in both the traditional statistical analysis and modern machine learning. In some cases, they serve as simple statistical tools to leverage the modern computing power to process efficiently large datasets. In this chapter, we focus on statistical and machine learning methods that are useful not only in prediction, but also in classification and discriminant analysis. In the first part, tree-based methods are introduced including classification and regression tree (CART) and random forest (RF). These statistical methods are nonparametric in nature. They do not postulate any specific statistical model for the data under study. Instead, they use recursive partitioning to explore the structure hidden in the dataset. The CART has a long history in the statistical literature. See Breiman et al. (1984). With the advances in ensemble learnings, CART is further extended to RF by Breiman (2001). In recent years, there are many articles concerning Bayesian CART and Bayesian additive regression trees (BART). See, for instance, Chipman et al. (2010) and a recent improvement over BART by He et al. (2018).

Machine learning or artificial intelligence (AI) is popular nowadays. It leverages the availability of big data (mainly from Internet of Things), advances in optimization methods, and powerful computers to process information to aid decision-making. Its goal is to let a machine (or a collection of computer algorithms) to think and act like a human brain. Deep learning (DL) is part of the machine learning. It often employs a sophisticated neural network (NN) with multiple layers to extract useful information embedded in the data. Basically, a DL machine consists of four parts: (1) a big dataset for training and validation, (2) a well-defined loss function (or cost function or objective function), (3) an efficient or feasible optimization procedure, and (4) a model

Statistical Learning for Big Dependent Data, First Edition. Daniel Peña and Ruey S. Tsay.
© 2021 John Wiley & Sons, Inc. Published 2021 by John Wiley & Sons, Inc.

(or a deep network structure). In the second part of this chapter, we discuss some basic concepts of NN and DL, and demonstrate their applications using dependent data. These methods are semi-parametric statistical models because a NN typically consists of some sophisticated model, formed by layers of activation functions, with many unknown parameters, referred to as biases and weights in the machine learning literature. For further details of DL, readers are referred to Schmidhuber (2015) and Goodfellow et al. (2016), among many others.

8.1 REGRESSION TREES AND RANDOM FORESTS

Tree-based methods are widely used in machine learning and in decision-making. They use recursive partitions and simple models to explore the structure of the data. These trees are usually called CART (classification and regression trees) and were introduced by Breiman et al. (1984). We explain first their use in regression and, second, for classification. For prediction, the methods essentially partition the predictor space into non-overlapping subregions and use the sample mean (or the majority voting) of the dependent variable in each subregion as its prediction. One can think of a regression tree with multiple predictors as a multivariate step function. The key question in applying regression tree is to find efficient ways to perform the partition of the predictor space so as to minimize the sum of squared forecast errors (or misclassification rates) in the training sample.

We start with binary trees by discussing ways to grow and prune a regression tree. In the literature, the size of a tree is often referred to as the number of partitions, or the number of resulting subregions of the predictor space. The starting point of a tree denotes the whole predictor space, i.e. no partition, and the largest tree for a given data set corresponds to the tree in which every data point becomes a leaf. For prediction, the smallest tree means using the sample mean in prediction and the largest tree means using a single data point in prediction. Therefore, neither the smallest nor the largest tree is useful in practice. For tree-based methods to be useful, one needs to select a proper tree, where prediction is made by using the conditional mean of the response when the explanatory variables take values in a defined region of the predictor space. To this end, we need to discuss ways to grow and to prune a regression tree.

8.1.1 Growing Tree

Consider the data $\{(\boldsymbol{x}_t, y_t)|t = 1, \ldots, n\}$, where $\boldsymbol{x}_t = (x_{1t}, \ldots, x_{kt})'$ is a realization of the predictor $\boldsymbol{X}_t$ and y_t is a realization of the dependent variable Y_t. In this chapter, we follow the conventional statistical literature by using n to denote the sample size. A *regression tree* is typically a binary tree constructed by growing and pruning branches to better describe the relationship between Y_t and $\boldsymbol{X}_t$. See, for instance, Figure 8.1. It contains a stem (or trunk), several branches (or nodes), and many leaves. The leaves are called *terminal* nodes and each branch gives a split on the predictor space. At the stem, the predictor space is R^k. That is, there is no splitting in the space. For the ith predictor x_{it}, let $x_{i(1)} \leq x_{i(2)} \leq \cdots \leq x_{i(n)}$ be its order statistics. Any real number $\eta \in [x_{i(1)}, x_{i(n)}]$ partitions R^k into the following two subregions:

$$R_{i,1} = \{\boldsymbol{x}_t | x_{it} \leq \eta\}, \quad R_{i,2} = \{\boldsymbol{x}_t | x_{it} > \eta\}.$$

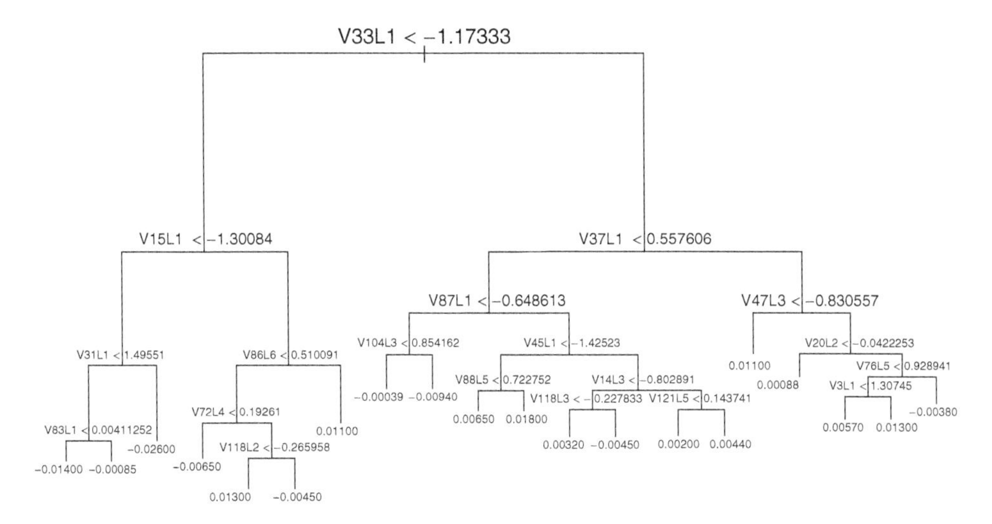

Figure 8.1 Regression tree for the monthly growth rate of US IP index. There are 732 predictors.

One can compute the sum of squares of residuals of y_t of the partition as

$$SS(i, \eta) = \sum_{x_t \in R_{i,1}} (y_t - \bar{y}_1)^2 + \sum_{x_t \in R_{i,2}} (y_t - \bar{y}_2)^2,$$

where $\bar{y}_1$ and $\bar{y}_2$ are the sample means of y_t in $R_{i,1}$ and $R_{i,2}$, respectively. In practice, η assumes the order statistics of x_{it} and the best partition is selected by

$$\hat{\eta}_i = \arg \min_{\eta \in \{x_{i(1)}, \ldots, x_{i(n)}\}} SS(i, \eta),$$

where the subscript i is added to signify that the partition is based on the ith predictor. The resulting sum of squared forecast errors is $SS(i, \hat{\eta}_i)$. The first branch of the tree is then determined by the i_1th predictor with threshold $\hat{\eta}_{i_1}$, where i_1 is given by

$$(i_1, \hat{\eta}_{i_1}) = \arg \min_{1 \le i \le k} SS(i, \hat{\eta}_i),$$

and the associated two subregions are $R_{i_1,1}$ and $R_{i_1,2}$. Therefore, the first split ends with two leaves, namely $R_{i_1,1}$ and $R_{i_1,2}$.

Next, the growing process is repeated for $R_{i_1,1}$ and $R_{i_1,2}$ separately. The only modification for $R_{i_1,1}$ is that $x_{i_1,t}$ only assumes values less than or equal to $\hat{\eta}_{i_1}$. For $R_{i_1,2}$, $x_{i_1,t}$ assumes values greater than $\hat{\eta}_{i_1}$. In theory, this growing process can continue until every subregion contains a single observation (or a small prespecified number of data points). But such a fully grown tree would result in overfitting and is not useful in practice. The method used to avoid overfitting of a regression tree is called *pruning*, which we discuss next. Note that there exist some variants in growing trees. For instance, instead of using order statistics to perform the partition, one can randomly select some values of each predictor to do the partition.

8.1.2 Pruning

The number of splits of a binary regression tree is called the *depth* of the tree. For depths greater than zero, the number of leaves (or terminal nodes) is equal to the depth of the tree plus one and is often referred to as the size of the tree. Denote a given tree with depth m by T_m. It splits the predictor space R^k into $m+1$ subregions, denoted by $\{R_j | j = 1, \ldots, m+1\}$, where, for simplicity, the predictors used in the splitting are dropped. The sum of squared of forecast errors of T_m is given by

$$\text{SS}(T_m) = \sum_{j=1}^{m+1} \sum_{\boldsymbol{x}_t \in R_j} (y_t - \bar{y}_j)^2, \tag{8.1}$$

where $\bar{y}_j$ denotes the sample mean of y_t in R_j, that is, the observed conditional mean of the response when the explanatory variables belong to R_j. To prune the tree, one often employs the cost of complexity criterion as

$$C(T_m, \lambda) = \text{SS}(T_m) + \lambda |T_m|, \tag{8.2}$$

where $|T_m|$ denotes the size of the tree and $\lambda > 0$ is a penalty parameter, which governs the trade-off between tree complexity and the goodness of fit of the tree. In practice, λ can be chosen by cross-validation or out-of-sample forecast. For a given λ, one selects the tree by

$$T_{\hat{m}} = \arg\min_m C(T_m, \lambda).$$

Clearly, a large λ prefers a smaller tree and $\lambda = 0$ corresponds the largest tree allowed.

The R packages **tree** and **rpart** can be used to perform tree-based methods for regression and classification. We use **tree** in the following demonstration.

8.1.3 Classification Trees

When the objective is the classification of new observations into G possible classes, the splitting criterion is modified as follows. In the first split, as before, for each predictor x_{it} we find a real number $\eta \in [x_{i(1)}, x_{i(n)}]$ that splits the data in an optimal classification, according to some criterion. In the next partitions, we continue splitting each subsample, as in regression trees, but now using the criterion chosen for optimal classification. This optimality criterion can be defined in several ways. The first one is by minimizing the proportion of observations wrongly classified. Thus, at each partition R_j the value of η is chosen to produce the smallest proportion of classification errors:

$$p_{j,\hat{g}} = \arg\max_g p_{j,g},$$

where $p_{j,g}$, for $g = 1, \ldots, G$, is the proportion of observations in the partition that belong to class g and the split is made classifying all the observations in the more frequent class. Then, the proportion of classification errors for partition j will be:

$$\text{CE}_j = 1 - p_{j,\hat{g}}.$$

Another criterion is to minimize the average proportion of classification errors. The relative errors in each class, $1 - p_{j,g}$, are weighted by the proportion of observations in the class, $p_{j,g}$, where $\sum_{g=1}^{G} p_{j,g} = 1$. This average value is called the Gini index of the partition, and is defined by

$$\text{Gini}_j = \sum_{g=1}^{G} p_{j,g}(1 - p_{j,g}).$$

A similar criterion is to minimize the cross-entropy or deviance, or impurity of the node, defined by

$$\text{Ent}_j = -\sum_{g=1}^{G} p_{j,g} \log p_{j,g}. \tag{8.3}$$

It is easy to see that the entropy assumes the maximum at a node when the split leads to equal proportion of observations in the groups and $p_{j,g} = 1/G$. It attains the minimum if the split has no classification error and, for each group g, either $p_{j,g} = 1$ or $p_{j,g} = 0$. Of course, this minimum value can only be reached if the subsample to be split only contains observations from two groups. Note that, for small x, $\log(1 - x) \approx -x$, and the entropy criterion would be similar to the Gini index. Both criteria are differentiable, and the most commonly used ones in practice. The pruning of the tree is as before, but using in Eq. (8.2) one of these criteria of optimal classification, instead of $SS(T_m)$.

Example 8.1

Consider, again, the US monthly macroeconomic data of Example 7.1, but use the industrial production (IP) growth rate as the dependent variable. The predictors consist of lag-1 to lag-6 values of all 122 variables. We standardize the predictors for ease in graphical display. The tree-based method selects a tree with 20 leaves (terminal nodes) and the residual mean deviance (i.e., squared errors) is 2.52×10^{-5}. The 19 predictors selected for splitting are given in the R demonstration below. The first split is based on lag-1 value of the 33th predictor with threshold −1.173. The output shows that, at the root node, there are 704 observations with deviance 3.88×10^{-2} and sample mean 0.002163. Also, the first leaf has five observations with deviance 1.128×10^{-4} and sample mean −0.01439. Figure 8.1 displays the selected regression tree.

R demonstration: Regression tree

```
> require(tree)
> FREDMDApril19 <- read.csv("FREDMDApril19.csv", header=TRUE)
> x=as.matrix(FREDMDApril19)
> Y=x[7:710,6] # dependent variable
> X=cbind(x[6:709,],x[5:708,],x[4:707,],x[3:706,],x[2:705,],x[1:704,])
> X1 <- scale(X,center=TRUE,scale=TRUE)
> tree.fit <- tree(Y~.,data=data.frame(X1),split="deviance")
> tree.fit
node), split, n, deviance, yval
      * denotes terminal node

  1) root 704 3.880e-02  0.0021630
    2) V33L1 < -1.17333 72 1.144e-02 -0.0056860
      4) V15L1 < -1.30084 20 3.590e-03 -0.0157700
        8) V31L1 < 1.49551 11 8.355e-04 -0.0070080
```

```
          16) V83L1 < 0.00411252 5 1.128e-04 -0.0143900 *
          17) V83L1 > 0.00411252 6 2.230e-04 -0.0008548 *
        9) V31L1 > 1.49551 9 8.754e-04 -0.0264900 *
      5) V15L1 > -1.30084 52 5.033e-03 -0.0018060
       10) V86L6 < 0.510091 44 3.044e-03 -0.0040430
         20) V72L4 < 0.19261 34 8.944e-04 -0.0064640 *
         21) V72L4 > 0.19261 10 1.273e-03  0.0041860
           42) V118L2 < -0.265958 5 4.892e-04  0.0128700 *
           43) V118L2 > -0.265958 5 2.915e-05 -0.0044990 *
       11) V86L6 > 0.510091 8 5.579e-04  0.0105000 *
    3) V33L1 > -1.17333 632 2.242e-02  0.0030570
      6) V37L1 < 0.557606 481 1.526e-02  0.0021890
       12) V87L1 < -0.648613 67 2.445e-03 -0.0014600
         24) V104L3 < 0.854162 59 1.499e-03 -0.0003903 *
         25) V104L3 > 0.854162 8 3.799e-04 -0.0093500 *
       13) V87L1 > -0.648613 414 1.178e-02  0.0027790
         26) V45L1 < -1.42523 14 6.244e-04  0.0106000
           52) V88L5 < 0.722752 9 1.797e-04  0.0064910 *
           53) V88L5 > 0.722752 5 1.840e-05  0.0180100 *
         27) V45L1 > -1.42523 400 1.027e-02  0.0025050
           54) V14L3 < -0.802891 34 1.297e-03 -0.0010940
            108) V118L3 < -0.227833 15 3.127e-04  0.0032070 *
            109) V118L3 > -0.227833 19 4.880e-04 -0.0044890 *
           55) V14L3 > -0.802891 366 8.490e-03  0.0028390
            110) V121L5 < 0.143741 239 5.161e-03  0.0020070 *
            111) V121L5 > 0.143741 127 2.852e-03  0.0044060 *
      7) V37L1 > 0.557606 151 5.643e-03  0.0058230
       14) V47L3 < -0.830557 30 8.874e-04  0.0110000 *
       15) V47L3 > -0.830557 121 3.753e-03  0.0045400
         30) V20L2 < -0.0422253 32 8.761e-04  0.0008838 *
         31) V20L2 > -0.0422253 89 2.296e-03  0.0058540
           62) V76L5 < 0.928941 84 1.580e-03  0.0064280
            124) V3L1 < 1.30745 75 1.012e-03  0.0056830 *
            125) V3L1 > 1.30745 9 1.804e-04  0.0126400 *
           63) V76L5 > 0.928941 5 2.232e-04 -0.0037850 *
> summary(tree.fit)

Regression tree:
tree(formula = Y ~., data = data.frame(X1), split = "deviance")
Variables actually used in tree construction:
 [1] "V33L1" "V15L1" "V31L1" "V83L1" "V86L6" "V72L4" "V118L2" "V37L1"
 [9] "V87L1" "V104L3" "V45L1" "V88L5" "V14L3" "V118L3" "V121L5"
[16] "V47L3" "V20L2" "V76L5" "V3L1"
Number of terminal nodes:  20
Residual mean deviance:  2.522e-05 = 0.01725 / 684
Distribution of residuals:
      Min.    1st Qu.     Median       Mean    3rd Qu.       Max.
-2.085e-02 -3.167e-03 -1.751e-05  0.000e+00  3.237e-03  1.993e-02
> plot(tree.fit)
> text(tree.fit,cex=0.5)
```

■

8.1.4 Random Forests

The RF is another ensemble method for classification and regression. As seen in Chapter 7 with boosting, ensemble methods combine different procedures or algorithms to obtain better predictions than those obtained from any of the constituent algorithms, as for instance, in model averaging. RF is an extension of bagging (bootstrap aggregating) to avoid overfitting in tree-based methods. The method was introduced by Ho (1995) and developed into a proper statistical tool by Breiman (2001). Our discussion below focuses on forecasting. Basically, RFs employ many simple trees to produce predictions, then combine those predictions to provide a consensus forecast. The method is nonparametric as it does not assume any specific model for the underlying data.

Again, consider the data set $\{(\boldsymbol{x}_t, y_t)|t = 1, \ldots, n\}$, where $\boldsymbol{x}_t$ is a realization of the predictor $\boldsymbol{X}_t = (X_{1t}, \ldots, X_{kt})'$ and y_t is a realization of the dependent variable Y_t. The goal is to predict y_{t+h} with h being the forecast horizon. Before discussing RFs, we state the *bagging* methodology below:

1. Draw a random sample with replacement from the data set. Denoted the bootstrap sample as $\{(\boldsymbol{x}_t^*, y_t^*)|t = 1, \ldots, n\}$.
2. Build a regression tree using the bootstrap sample to obtain a prediction for y_{t+h} of interest.

By repeating the aforementioned bootstrap procedure many times, one obtains a collection of bootstrap forecasts, which, in turn, can be used to obtain a consensus forecast of y_{t+h}.

RFs extend the bagging procedure by introducing a novel idea to explore fully the relationship between y_{t+h} and $\boldsymbol{x}_t$. The idea is that, for each tree split, one selects a random sample of predictors as candidates for splitting. Let g be a positive integer satisfying $1 \leq g \leq k$. Often $g = [k/3]$ is used. To build a regression tree from a bootstrap sample for RFs, one takes a random sample of g predictors from $\{1, \ldots, k\}$ to serve as candidates for splitting at each split. This novel step serves multiple purposes. First, it reduces the computation intensity, especially when k is large. Second, it mitigates the multicollinearity between the predictors, because highly correlated predictors may not be jointly selected. Finally, it avoids the chance of using similar trees in the ensemble. A simple average of bootstrap forecasts is often used to produce a consensus prediction. Weighted average can also be used under a given loss function.

Discussion: Some discussions of RFs are in order. First, the R package **randomForest** can be used to perform RF prediction. See the demonstration below. Second, some modifications are needed to apply effectively RFs to dependent data. For instance, in time series analysis, the lag-1 value of the dependent variable is often more important than the lag-2 value. Similarly, seasonal lags of the dependent variable are often important in modeling seasonal time series. Consequently, one might use a weighted random sample to select the g predictors as candidates for splitting at each split instead of using a pure random sample. Third, care must be exercised in drawing bootstrap samples for dependent data and some form of block bootstrap should be used; see, for instance, Alonso et al. (2004) and Lahiri (2013).

Example 8.2

To demonstrate the use of RFs, we also employ the monthly macroeconomic data used in Example 8.1. Again, the dependent variable is the monthly growth rate of US

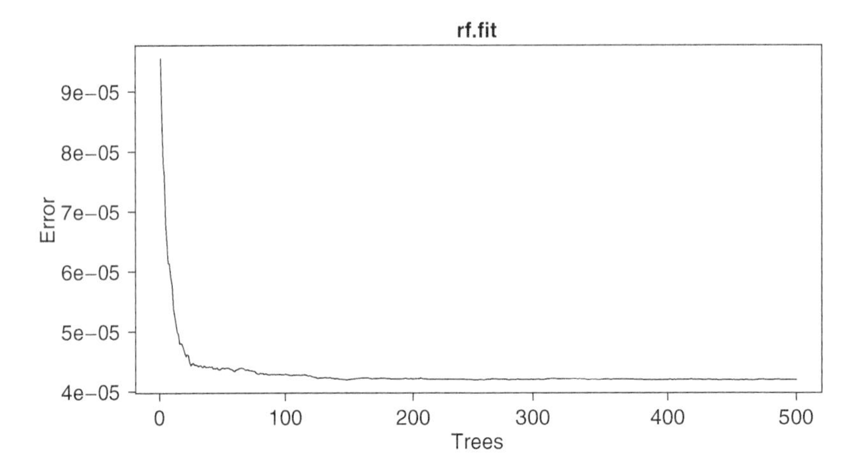

Figure 8.2 Result of RFs applied to the monthly macroeconomic data with the growth rate of IP index as the dependent variable. 500 trees and mtry = 20 are used.

IP index and the predictors are lag-1 to lag-6 values of all available variables. Thus, we have $n = 704$ and $k = 732$. The predictors are standardized. We run the **randomForest** command twice to show the impact of selecting the number of predictors in each tree split. The subcommand `ntree` is used to specify the number of trees used and the subcommand `mtry` is the number of predictors used in each split. That is, g in our notation. We use the default option so that the tree is grown with the constraint that the number of data points in each leaf cannot be less than 5. An alternative approach is to specify the maximum number of leaves a tree can have. See the subcommands `nodesize` and `maxnodes` of **randomForest**, respectively.

With 500 trees and `mtry = 20`, the RFs explain 23.92% of the variability of dependent variable. Figure 8.2 shows the MSE versus the number of trees used whereas Figure 8.3 displays the mean decrease in MSE of each predictor. From

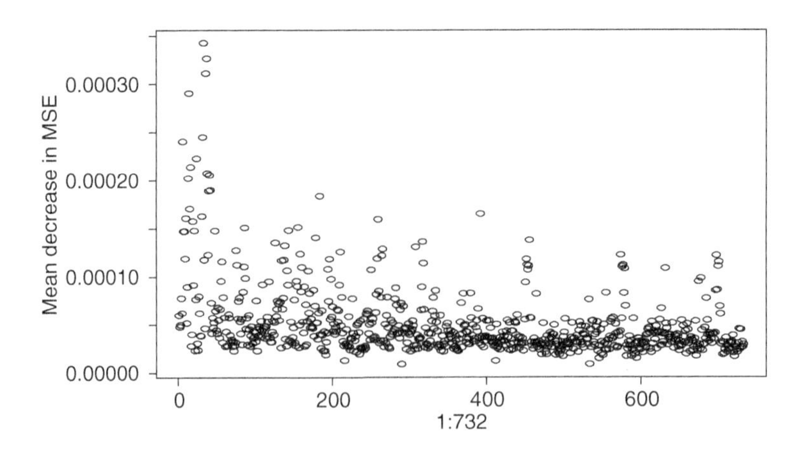

Figure 8.3 Mean decrease in MSE of each predictor for the monthly macroeconomic data with the growth rate of IP index as the dependent variable. 500 trees and mtry = 20 are used.

Figure 8.2, we see that the MSE decreases quickly with the number of trees used. If one increases `mtry = 30`, then the RFs explain about 24.47% of the variability of dependent variable. In this particular example, the choice of `mtry` appears to be not sensitive.

R demonstration: Random forests

```
> require(randomForest)
> set.seed(117)
> rf.fit <- randomForest(X1,Y,ntree=500,mtry=20,importance=TRUE)
> rf.fit
Call: randomForest(x=X1,y=Y,ntree=500,mtry=20,importance=TRUE)
               Type of random forest: regression
                     Number of trees: 500
No. of variables tried at each split: 20

          Mean of squared residuals: 4.19353e-05
                    % Var explained: 23.92
> plot(rf.fit)
> dim(rf.fit$importance)
[1] 732   2
> plot(1:732,rf.fit$importance[,2],ylab="mean decrease in MSE")
> set.seed(117)
> rf.fit <- randomForest(X1,Y,ntree=500,mtry=30,importance=TRUE)
> rf.fit
Call: randomForest(x=X1,y=Y,ntree=500,mtry=30,importance=TRUE)
               Type of random forest: regression
                     Number of trees: 500
No. of variables tried at each split: 30

          Mean of squared residuals: 4.163187e-05
                    % Var explained: 24.47
```

■

8.2 NEURAL NETWORKS

NNs (or artificial neural networks) are a semi-parametric learning method that mimics human brain in processing information. A network consists of multiple layers of nodes, which are also called neurons. The first layer of a network is the *input* layer with each predictor serving as a node. The last layer is the *output* layer that contains as many nodes as the dimension of the dependent variable. All the middle layers of neurons are called the *hidden* layers. In a feedforward network, information flows from the input layer to the first hidden layer, then to the second hidden layer, and so on, until it reaches the output layer. Thus, information passes through each neuron only once. For recurrent networks, feedback from latter layers to former ones or from one neuron back to itself is allowed. Alternatively, a recurrent network means that the information processing may go over one layer multiple times. Feedforward NNs are also known as *perceptrons*. See Figure 8.4, which is a $P - m - k$ NN.

A NN tries to approximate a response function using a set of explanatory variables. It relies on universal approximation theory stating that, under certain smoothness conditions of the response function, one can provide adequate

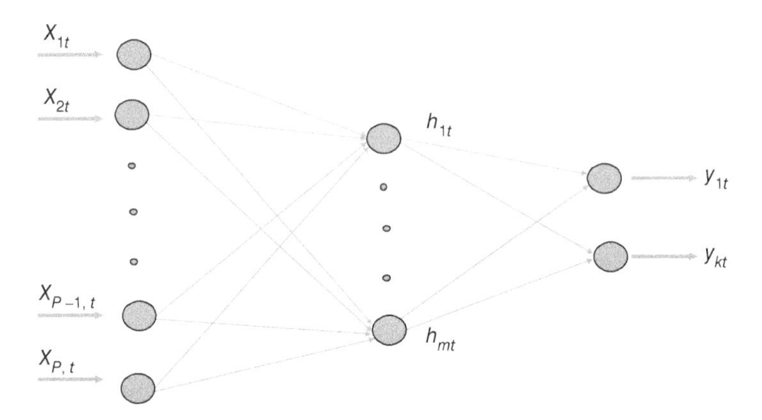

Figure 8.4 A perceptron with P input variables, a layer with m nodes, and an output with k response variables.

approximations with some activation function and network structure. Indeed, there have been many approaches to obtain universal approximation functions. For instance, we can approximate any continuous function in an interval by using polynomials, by Taylor theorem, or sinusoidal functions, as in Fourier analysis. Kolmogorov (1957) showed that all continuous functions of m variables can be well approximated in terms of finite superpositions and compositions of a small number of functions of one variable. Instead of functions of each variable, NNs search for approximations of functions of linear combinations of all input variables. Cybenko (1989) and Hornik et al. (1989) proved that any continuous function of m real variables can be well approximated by compositions of finite linear combinations of the variables, which implies that a feedforward NN with a single hidden layer of arbitrary width, that is, with a large number of neurons or nodes, can approximate, with any given precision, any continuous function.

The function used to process information from one layer to the next of a network is referred to an *activation* function. Different layers may use different functions even though the same activation function is commonly used except for the output layer. The property of universal approximation for a single layer NN is valid for any activation function that it is not polynomial. The activation function for the output layer is determined by whether the dependent variable is continuous or discrete. Details are given later. Many activation functions have been proposed in the literature, but the following ones are commonly used in the traditional NN. Section 8.3.3 provides some others that are used in the deep learning or machine learning.

1. Logistic function (or Sigmoid function):

$$g(\boldsymbol{x}) = \frac{1}{1+\exp(-\boldsymbol{x})} = \frac{\exp(b+\sum_{i=1}^{m} w_i x_i)}{1+\exp(b+\sum_{i=1}^{m} w_i x_i)},$$

   where $\boldsymbol{x} = (x_1, \ldots, x_m)'$ with m being the number of nodes in the previous layer, b is called the *bias*, and w_i are the weights. This function is the one used in logistic regression, and it transforms a continuous variable in the real line into a variable between zero and one.

2. Hyperbolic tangent (tanh):

$$g(\boldsymbol{x}) = \frac{\sinh(\boldsymbol{x})}{\cosh(\boldsymbol{x})} = \frac{\exp(\boldsymbol{x}) - \exp(-\boldsymbol{x})}{\exp(\boldsymbol{x}) + \exp(-\boldsymbol{x})},$$

where $\exp(\boldsymbol{x}) = \exp(b + \sum_{i=1}^{m} w_i x_i)$ as before with bias b and weights w_i.

3. Heaviside function:

$$g(\boldsymbol{x}) = \begin{cases} 0, & \text{if } b + \sum_{i=1}^{m} w_i x_i \leq 0, \\ 1, & \text{if } b + \sum_{i=1}^{m} w_i x_i > 0. \end{cases}$$

The Heaviside function is for the output layer when the dependent variable is binary. On the other hand, the usual linear function is used for the output layer if the dependent variable is continuous. We refer to NNs as a semi-parametric learning method because once the activation functions and the structure of the network are given the network becomes a nonlinear statistical model with unknown parameters consisting of biases and weights. Figure 8.5 compares the sigmoid and tanh activation functions. The first one produces an output number that can be positive or negative, whereas the second one always gives a positive number.

8.2.1 Network Training

Network training is to estimate the biases and weights of a given network. This is often carried out by optimizing certain objective function (or minimizing some loss function) using the available data. In practice, one often divides the data set $\{(\boldsymbol{x}_t, y_t)|t = 1, \ldots, n\}$ into training and validation subsamples. For independent data, the training subsample can be selected by a random draw without replacement from the data set and the remaining observations form the validation subsample. For instance, let n_1 be a positive integer less than n. One can draw n_1 observations from $\{1, \ldots, n\}$ uniformly without replacement, say $\{i_1, \ldots, i_{n_1}\}$.

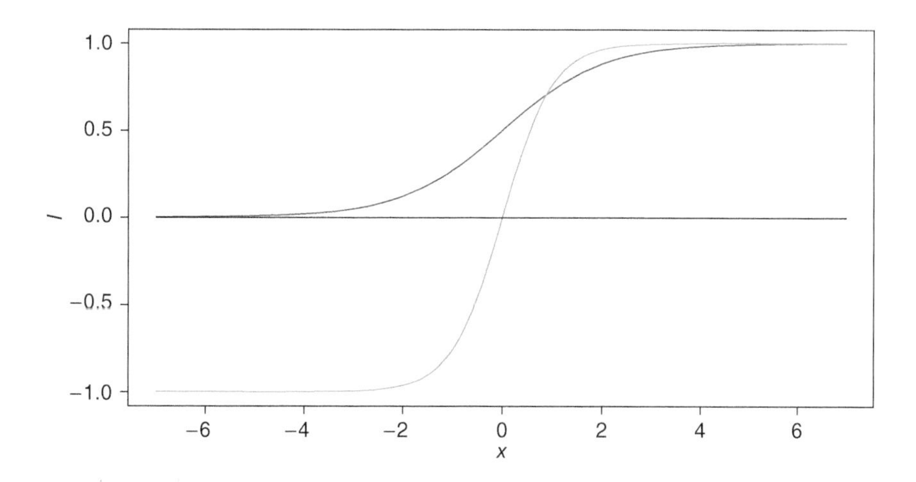

Figure 8.5 Logistic (green) and tanh (red) activation functions.

Then, use $\{(\boldsymbol{x}_t, y_t)|t = i_1, \ldots, i_{n_1}\}$ as the training subsample. On the other hand, for serially dependent data, one typically uses the first n_1 observations as the training subsample.

In what follows, we briefly discuss the training of a feedforward $P - M - K$ network, which has P input nodes, a single hidden layer with M nodes, and K output nodes. See Hastie et al. (2001). This network is also referred to as a $P - M - K$ perceptron. Denote the input as $\boldsymbol{x}_t = (x_{1t}, \ldots, x_{Pt})'$, the output as $\boldsymbol{y}_t = (y_{1t}, \ldots, y_{Kt})'$, and the vector of outputs of the hidden layer as $\boldsymbol{h}_t = (h_{1t}, \ldots, h_{Mt})'$. We can write down the model of the network as follows:

$$h_{mt} = g(w_{m0} + \boldsymbol{w}'_m \boldsymbol{x}_t), \quad m = 1, \ldots, M, \tag{8.4}$$

$$q_{vt} = \beta_{v0} + \boldsymbol{\beta}'_v \boldsymbol{h}_t, \quad v = 1, \ldots, K, \tag{8.5}$$

$$y_{vt} \equiv f_v(\boldsymbol{x}_t) = g_v(\boldsymbol{q}_t), \quad v = 1, \ldots, K, \tag{8.6}$$

where $g(.)$ and $g_v(.)$ are activation functions, w_{m0} and $\boldsymbol{w}_m = (w_{m1}, \ldots, w_{mP})'$ are the bias and weights associated with the mth neuron of the hidden layer, β_{v0} and $\boldsymbol{\beta}_v = (\beta_{v1}, \ldots, \beta_{vM})'$ are, respectively, the bias and weights of the vth output neuron, and $\boldsymbol{q}_t = (q_{1t}, \ldots, q_{Kt})$. In statistical terms, the model can be explained as a nonlinear factor model, where the original P explanatory variables are combined to create M linear combinations or factors. Each factor is then transformed in a nonlinear way to create new explanatory variables, h_{mt}, and the K response variables are functions of the linear combinations of these new variables h_{mt}. Also the bias can be incorporated into the weights by redefining the input variables as $\boldsymbol{x}_t = (1, x_{1t}, \ldots, x_{Pt})'$ and the hidden nodes as $\boldsymbol{h}_t = (1, h_{1t}, \ldots, h_{Mt})'$. The function $f_v(.)$ of Eq. (8.6) is used to denote the underlying function for the vth output variable as a function of $\boldsymbol{x}_t$. For continuous dependent variables $\boldsymbol{y}_t$, $g_v(.)$ often assumes the linear function $g_v(\boldsymbol{q}_t) = q_{vt} = \beta_{v0} + \boldsymbol{\beta}'_v \boldsymbol{h}_t$. For discrete variables such as classification, $g_v(.)$ typically assumes a multi-logit model, i.e.

$$f(\boldsymbol{x}_t) = \arg \max_{1 \le v \le K} \hat{f}_v(\boldsymbol{x}_t), \tag{8.7}$$

where $\hat{f}_v(.)$ denotes the fitted value of $f_v(.)$ in the training subsample and $f(.)$ denotes the underlying model. In this particular instance, given $g(.)$ and $g_v(.)$, the network contains the following parameters:

$$\{(w_{m0}, \boldsymbol{w}_m | m = 1, \ldots, M\} : \ M(P+1) \quad \text{parameters}, \tag{8.8}$$

$$\{(\beta_{v0}, \boldsymbol{\beta}_v | v = 1, \ldots, K\} : \ K(M+1) \quad \text{parameters}. \tag{8.9}$$

Let $\boldsymbol{\theta}$ be the collection of all parameters (biases and weights). For continuous dependent variables $\boldsymbol{y}_t$, one often uses the squared error loss function

$$L(\boldsymbol{\theta}) = \sum_{v=1}^{K} \sum_{t=1}^{n_1} [y_{vt} - f_v(\boldsymbol{x}_t)]^2, \tag{8.10}$$

where n_1 is the sample size of the training subsample. For classification, one often uses the cross-entropy (or deviance), defined in Eq. (8.3), to measure the goodness of fit

$$L(\boldsymbol{\theta}) = \sum_{v=1}^{K} \sum_{t=1}^{n_1} y_{vt} \log[f_v(\boldsymbol{x}_t)], \tag{8.11}$$

and the classification is based on Eq. (8.7).

Network training is then amount to minimizing the loss function in the training subsample over the parameter space. This is a nonlinear optimization problem that can be solved by many procedures, from the simple gradient method to Newton and quasi-Newton algorithms including genetic algorithms. For a large number of variables, the gradient algorithm is fast to compute and requires only to keep the first derivatives, instead of the second as in the Newton method. In the literature of NN, the gradient algorithm is commonly known as *backpropagation*. It is called *forward-propagation* to the process of starting with the input variables and applying weights and the activation function to compute the intermediate variables $\boldsymbol{h}_t$. Applying new weights to this intermediate variables the output is obtained. Backpropagation is the inverse process of comparing the output with the observed data, compute the forecast errors and use these errors to modify the weights so that the forecast error is reduced. We present next the simplest backpropagation algorithm. One of its variants is the resilient backpropagation with or without weight backtracking. See Riedmiller and Braun (1993). A gradient decent method requires calculating the partial derivatives of the activation function so that the activation functions used in practice all have simple derivatives.

In what follows, we provide some details of the backpropagation for the squared error loss function of Eq. (8.10). We assume a continuous dependent variable with output activation function equal to the identity, $g_v(\boldsymbol{q}_t) = q_{vt}$ and sigmoid as the activating function of the hidden layer. Also, without loss of generality and for simplicity, we omit the bias terms. The problem can be rewritten as

$$L(\boldsymbol{\theta}) = \sum_{t=1}^{n_1} L_t \quad \text{with} \quad L_t = \sum_{v=1}^{K} [y_{vt} - \boldsymbol{\beta}_v' \boldsymbol{h}_t]^2. \tag{8.12}$$

Taking partial derivatives and using chain rules, we have

$$\frac{\partial L_t}{\partial \boldsymbol{\beta}_v} = -2\boldsymbol{h}_t(y_{vt} - \boldsymbol{h}_t'\boldsymbol{\beta}_v), \tag{8.13}$$

$$\frac{\partial L_t}{\partial \boldsymbol{\omega}_m} = -\sum_{v=1}^{K} 2(y_{vt} - \boldsymbol{h}_t'\boldsymbol{\beta}_v)\beta_{vm} g'(\boldsymbol{\omega}_m'\boldsymbol{x}_t)\boldsymbol{x}_t, \tag{8.14}$$

It is easy to see that the derivative of the sigmoid function $g(x)$ is $g'(x) = g(x)[1 - g(x)]$. Therefore,

$$\frac{\partial L_t}{\partial \boldsymbol{\omega}_m} = -\sum_{v=1}^{K} 2(y_{vt} - \boldsymbol{h}_t'\boldsymbol{\beta}_v)\beta_{vm}[g(\boldsymbol{\omega}_m'\boldsymbol{x}_t)(1 - g(\boldsymbol{\omega}_m'\boldsymbol{x}_t))]\boldsymbol{x}_t. \tag{8.15}$$

Equation (8.13) leads, summing over t, to the LS estimate of $\boldsymbol{\beta}_v$ when the $\boldsymbol{h}_t$ variables are known, showing that the residuals must be orthogonal to the predictors $\boldsymbol{h}_t$. Equation (8.15), summing over t, gives a way to estimate $\boldsymbol{\omega}_m$ given the $\boldsymbol{\beta}_v$. Calling

$$p_m = g'(\boldsymbol{\omega}_m'\boldsymbol{x}_t) = [g(\boldsymbol{\omega}_m'\boldsymbol{x}_t)(1 - g(\boldsymbol{\omega}_m'\boldsymbol{x}_t))], \tag{8.16}$$

the estimation of $\boldsymbol{\omega}_m$ must verify

$$\sum_{t=1}^{n}\sum_{v=1}^{K}(y_{vt} - \boldsymbol{h}_t'\boldsymbol{\beta}_v)\beta_{vm}p_m\mathbf{x}_t = \mathbf{0},$$

and the weighted residuals must also be orthogonal to the input variables.

Given the derivatives, a gradient descent update from the ith to the $(i+1)$th iteration has the form

$$\beta_{vm}^{(i+1)} = \beta_{vm}^{(i)} - \gamma_i \sum_{t=1}^{n_1} \frac{\partial L_t}{\partial \beta_{vm}^{(i)}}, \tag{8.17}$$

$$w_{m\ell}^{(i+1)} = w_{m\ell}^{(i)} - \gamma_i \sum_{t=1}^{n_1} \frac{\partial L_t}{\partial w_{m\ell}^{(i)}}, \tag{8.18}$$

where γ_i is the *learning rate*, which must satisfy certain condition to ensure the convergence of the estimates. For instance, $\gamma_i \to 0$ as $i \to \infty$. Some discussions on γ_i can be found in Hastie et al. (2001). In practice, one may try different values of the learning rate and choose the one that attains the minimum loss function. To implement the update, one can rewrite Eqs. (8.13) and (8.14) for each weight as the *backpropagation equations*:

$$\frac{\partial L_t}{\partial \beta_{km}} \equiv \delta_{vt} h_{mt}, \tag{8.19}$$

$$\frac{\partial L_t}{\partial w_{m\ell}} \equiv s_{mt} x_{\ell t}, \tag{8.20}$$

where

$$\delta_{vt} = -2(y_{vt} - \boldsymbol{h}_t'\boldsymbol{\beta}_v),$$

and

$$s_{mt} = p_m \sum_{v=1}^{K} \beta_{vm}\delta_{vt}. \tag{8.21}$$

For instance, suppose $K = 1$ so that we have only an output variable and call $e_t^{(i)} = y_t - \boldsymbol{h}_t^{(i)\prime}\boldsymbol{\beta}_v^{(i)}$. The updating equations are

$$\boldsymbol{\beta}^{(i+1)} = \boldsymbol{\beta}^{(i)} - \gamma_i \sum_{t=1}^{n_1} e_t^{(i)}\boldsymbol{h}_t^{(i)}, \tag{8.22}$$

$$\boldsymbol{w}_m^{(i+1)} = \boldsymbol{w}_m^{(i)} - \gamma_i\beta_m^{(i)}p_m^{(i)} \sum_{t=1}^{n_1} e_t^{(i)}\boldsymbol{x}_t, \tag{8.23}$$

and have a straightforward interpretation: the $\boldsymbol{\beta}^{(i)}$ weights are updated depending on orthogonality of the forecast errors and the intermediate explanatory variables, $\boldsymbol{h}_t^{(i)}$; the $\boldsymbol{w}_{m\ell}^{(i)}$ weights in neuron m are updated depending on the orthogonality of the prediction errors and the input variables, the importance of this neuron to the response, as measured by $\beta_m^{(i)}$, and the effect of changes in the $\boldsymbol{w}_m^{(i)}$ on the $\boldsymbol{h}_t^{(i)}$, as measured by the derivative of the activation function, $p_m^{(i)}$.

The training of a NN is made as follows:

1. Initialization: The initial values for the parameters are chosen and some constants of the training algorithm are set, such as (a) the learning rate; (b) the batch size, B, which is the number of observations used to update the NN parameters or size of the block of observations to be processed jointly; and (c) the number of epochs, or iterations with the n_1 data through the updating algorithm.
2. Feed forward: With the current weights (parameters) input the next block of B input variables and compute the output, or predictions, of NN $\hat{f}_v(\boldsymbol{x}_t)$ with Eqs. (8.4)–(8.6).
3. Compute prediction errors: The output errors, $y_{vt} = y_{vt} - \hat{f}_v(\mathbf{x}_t)$, are computed as well as the value of the loss function for the observations processed so far.
4. Backpropagation: The quantities δ_{vt} are computed, and back propagated via Eq. (8.21) to obtain s_{mt}. Both δ_{vt} and s_{mt} are then used to compute the gradient update in Eqs. (8.17) and (8.18).

Steps 2–4 are repeated until the n_1 observations of the training set have been analyzed and an epoch is completed. Then, the process is iterated until the prespecified number of epochs is reached.

Some discussions of network training are in order. First, the scales of predictors may affect the usefulness of a network. As such, one often standardizes the predictors (or input variables) so that they have mean zero and variance one. Second, the training requires some initial estimates of biases and weights. With standardized predictors, one can employ random draws of $U[-w, w]$, where $0 < w < 1$, as initial parameter values. As the function we are minimizing may have many local minimums, it is important to start the algorithm with different starting values and compare the results in the validation subsample. Third, the batch size, B, is determined depending on the sample size. If the subsample n_1 is not large we can take $B = 1$, updating the parameters after each input $\boldsymbol{x}_t$, as is made in online recursive estimation (see Section 2.8.2). For large n_1, to speed up the estimation, B is chosen as some fraction of the data. Processing the B observations in a batch to update the parameters is called an iteration of the algorithm, and the number of iterations required to process the n_1 observations once is an epoch. (For instance, if $n_1 = 50000$, and $B = 100$, 500 iterations are needed to complete an epoch.) Fourth, it is possible that overfitting may occur if a network has too many layers or nodes. To avoid overfitting, the validation subsample becomes valuable. One can compare the performance of fitted networks in the validation subsample and selects the one with best performance as the preferred network. Finally, several R packages are available for NN analysis, including **nnet** and **neuralnet**. **nnet** only allows a single hidden layer, whereas **neuralnet** is more flexible. We use the latter in our demonstration below.

The advantages of the backpropagation method include simplicity and local nature. Each neuron of the hidden layer receives and passes information only from and to neurons that share a connection. As such, the algorithm can be carried out on a parallel manner. On the other hand, the backpropagation could be slow, and some other methods such as those require second-order derivatives are often used.

Obviously, the training idea can be generalized to networks with multiple hidden layers. However, some computational difficulties may occur when there are too many hidden layers. We return to this issue in the next section of deep learning.

Example 8.3

A widely used data set in the statistical literature is the Boston housing data set, which is available from the R package **MASS**. See Belsley et al. (2005) for a careful analysis of the data set. Here we use the data set only as an illustration. The dependent variable of interest is the median value of owner-occupied homes (in thousands) of Boston suburbs. There are 13 predictors including `rm` (average number of rooms per dwelling), `age` (proportion of owner-occupied units built prior to 1940), and `lstat` (lower status of the population in percent). Our goal here is to demonstrate the use of NN so we only consider the aforementioned three predictors in the following analysis.

To begin, we take the log transformation of the dependent variable `medv` and standardize the input variables so that each of the predictor has mean 0 and variance 1. We then employ two NN models. The first model is a 3-2-1 network, which has 3 input variables, 1 single hidden layer with 2 nodes, and 1 output variable. Figure 8.6 displays the resulting network, where each numerical value associated with an arrow denotes the weight and the value associated with the arrow coming out of circle 1 denotes the bias. We use the default option so that the activation function is the logistic function. In this particular instance, the first node of the hidden layer is given by

$$h_1(\boldsymbol{x}_t) = \frac{\exp(2.923 - 0.404\text{rm}_t - 1.373\text{age}_t - 2.08\text{lstat}_t)}{1 + \exp(2.923 - 0.404\text{rm}_t - 1.373\text{age}_t - 2.08\text{lstat}_t)},$$

and the output node is

$$\widehat{\text{medv}}_t = 2.326 + 0.686h_1(\boldsymbol{x}_t) + 0.871h_2(\boldsymbol{x}_t).$$

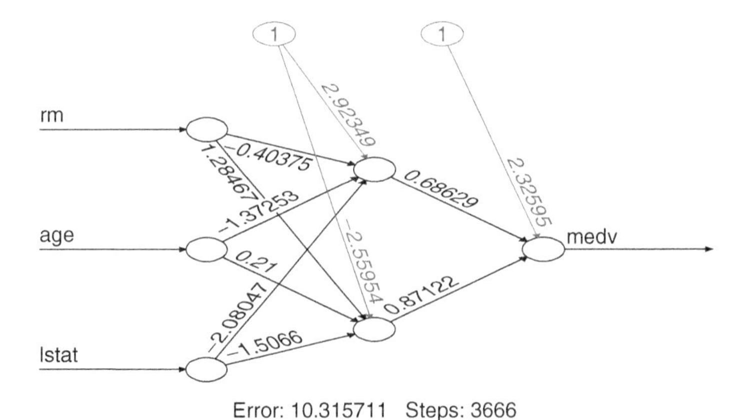

Figure 8.6 A fitted 3-2-1 NN for the log median home value with `rm`, `age`, and `lstat` as predictors. The data set is Boston housing.

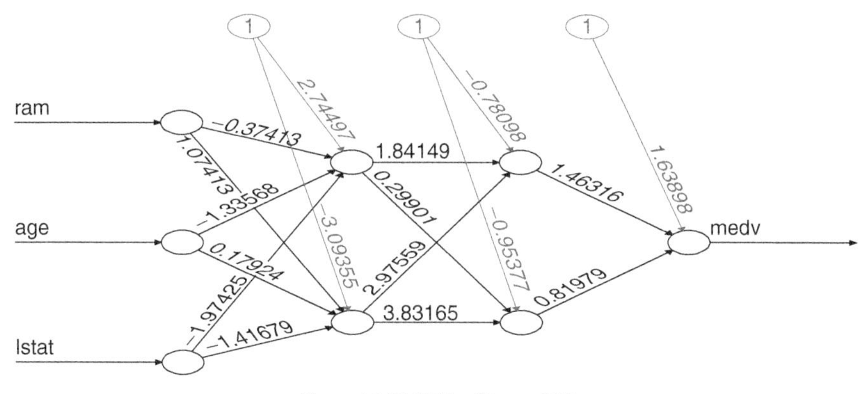

Figure 8.7 A fitted 3-2-2-1 NN for the log median home value with `rm`, `age`, and `lstat` as predictors. The data set is Boston housing.

The residual mean squared error of this network is 0.2019 whereas the sample standard error of the log(medv) is 0.4088.

Figure 8.7 displays the fitted 3-2-2-1 network, which has 2 hidden layers each with 2 nodes. The network can be interpreted in a similar manner. The residual mean squared error of this network is also close to 0.2019. In this particular instance, the additional hidden layer does not contribute much.

R demonstration: NN with **neuralnet** package.

```
> require(MASS)
> require(neuralnet)
> fix(Boston)
> names(Boston)
 [1] "crim"  "zn"    "indus"  "chas"    "nox"     "rm"     "age"
 [8] "dis"   "rad"   "tax"    "ptratio" "black"   "lstat"  "medv"
> X = cbind(Boston$rm,Boston$age,Boston$lstat) #Input nodes
> X = scale(X,center=TRUE,scale=TRUE) #Standardization
> colnames(X) <- c("rm","age","lstat")
> medv <- log(Boston$medv)
> netw <- neuralnet(medv~.,data=data.frame(X),hidden=c(2),
             algorithm="rprop+")
> names(netw)
 [1] "call"              "response"            "covariate"
 [4] "model.list"        "err.fct"             "act.fct"
 [7] "linear.output"     "data"                "exclude"
[10] "net.result"        "weights"             "generalized.weights"
[13] "startweights"      "result.matrix"
> plot(netw)
>
> fitted.values <- as.numeric(netw$net.result[[1]])
> rmse <- sqrt(mean((medv-fitted.values)^2))
> rmse
[1] 0.2019246
> sqrt(var(medv))
[1] 0.4087569
> netw1 <- neuralnet(medv~.,data=data.frame(X),hidden=c(2,2),
```

```
        algorithm="rprop+")
> fit1 <- as.numeric(netw1$net.result[[1]])
> rmse1 <- sqrt(mean((medv-fit1)^2))
> rmse1
[1] 0.2019169
```

■

8.3 DEEP LEARNING

Deep learning is part of a broader family of *machine learning* based on NNs. It uses multiple layers in a NN to progressively extract higher level features from the data. The word *deep* in DL is referred to the number of layers used to process the data, and DL typically means that the network used has two or more hidden layers. In this section, we briefly introduce DL for forecasting and classification.

Similar to the traditional NNs, the basic theory supporting deep NNs is the *universal approximation theorem*. This property holds for a single layer NN of arbitrary width but also for a multilayer NN with fixed width but arbitrary depth. Also some results indicate that NNs with several layers provide a more efficient (less number of parameters) approximation to any function than the single layer NN. See Hornik (1991), Hanin and Sellke (2017), and Lu et al. (2017).

8.3.1 Types of Deep Networks

Besides the feedforward design, various types of NNs have been proposed in the literature for DL. Here we briefly discuss some of the commonly used networks. In practice, these networks can be mixed together to form a sophisticate one. Typically, NNs are fully connected in the sense that each node in a layer is connected to all nodes of the next layer. Such a network would encounter various difficulties in training when it is deep, e.g. the vanishing gradient problem, and may require extensive computational cost. Some details are given in the next section. The network structures discussed below are designed to increase the power and applicability of NNs and to simplify the network training.

Convolutional neural network (CNN): They are regularized multilayer perceptrons that mimic the function of an animal visual cortex and are mostly used in computer vision. Here cortical neurons respond to stimuli only in a restricted region of the visual field known as the receptive field. The receptive fields of neurons are relatively small but they may overlap so that the whole image is properly connected. Usually the input of CNN are the standardized pixel values of one or more matrices [three in a red–green–blue (RGB) image]. Some filters (convolutions) are applied to small groups of pixels in the image, and repeating the process until the entire image is traversed. These filters are designed to extract features, as edges, colors, straight lines, and so on. This process is repeated several times with different filters and the final features found are the input of a feedforward NN that makes the classification of the image. A commonly used activation function of CNN is the ramp function, i.e. the rectifier. CNN is widely used in classification. For example, Lawrence et al. (1997) use a CNN for face recognition. It employs a self-organizing map, which is an unsupervised learning process that learns the distribution of a set of patterns without any class of information, and an updating algorithm, which employs a neighborhood function similar to a kernel function in the statistical literature, to perform network training. Kalchbrenner et al. (2014) apply CNN for modeling sentences.

Recursive NN: A recursive network is a deep NN created by applying the same set of weights recursively over a structured input to produce a structured prediction. It has been applied for speech recognition as a sentence is the sum of the words and, therefore, has structure. Each word is represented by a vector and a binary tree is formed. Recursive NNs operate on any hierarchical structure, combining the information in two nodes (children) into a new one (parent) and employ tree-type structure to process information.

Recurrent neural network (RNN): These networks can be considered a particular case of recursive NN representations operating on the linear progression of time, combining the previous time step and a hidden representation to obtain the output of the current time step. RNNs are designed to process sequential data and are especially useful for time series analysis. They were first created for data that arrive with some kind of sequential structure, as handwriting and speech recognition. See, for instance, Mikolov et al. (2010). RNN have been applied to forecasting vectors of time series assuming a nonlinear autoregressive model with exogenous variables (NARX). See Menezes and Barreto (2008). The issue of regularization of RNN is studied by Zaremba et al. (2015), and properties of RNN have been investigated in the literature. See, for instance, Hammer et al. (2005) and Hammer (2007). Since RNN focuses on dynamic dependence, we provide further details of them in the next section.

8.3.2 Recurrent NN

RNNs are a family of NNs designed to process sequential data. They are similar to the dynamic system

$$\boldsymbol{s}_t = f(\boldsymbol{s}_{t-1}, \boldsymbol{x}_t, \boldsymbol{\theta}),$$

where $\boldsymbol{s}_t \in \mathbb{R}^d$ denotes a state vector at time t, $\boldsymbol{x}_t \in \mathbb{R}^k$ is the input at time t, $\boldsymbol{\theta} \in \mathbb{R}^p$ is a set of parameters, and $f(.)$ denotes the transition from state $\boldsymbol{s}_{t-1}$ to $\boldsymbol{s}_t$ with new data $\boldsymbol{x}_t$. A special feature of the dynamic system is that the parameter vector $\boldsymbol{\theta}$ is time-invariant. That is, the same values are shared by all time indexes t. Given the input data, $\{\boldsymbol{x}_1, \ldots, \boldsymbol{x}_n\}$, let $\boldsymbol{h}_t \in \mathbb{R}^d$ be the output of the hidden layer, where $\boldsymbol{h}_0 = \boldsymbol{0}$. Then, a simple RNN can be symbolically described as

$$\mathbf{s}_t = \boldsymbol{b} + \boldsymbol{W}\boldsymbol{h}_{t-1} + \boldsymbol{U}\boldsymbol{x}_t, \tag{8.24}$$

$$\boldsymbol{h}_t = g(\boldsymbol{s}_t), \tag{8.25}$$

$$\hat{\boldsymbol{y}}_t = g_y(\boldsymbol{c} + \boldsymbol{V}\boldsymbol{h}_t), \tag{8.26}$$

where $g(.)$ and $g_y(.)$ are activation functions such as tanh, $\boldsymbol{b}$ and $\boldsymbol{c}$ are the bias terms, and $\boldsymbol{W}, \boldsymbol{U}$ and $\boldsymbol{V}$ are parameter matrices and $\hat{\boldsymbol{y}}_t \in \mathbb{R}^m$. In the literature, the RNN in Eqs. (8.24)–(8.26) is referred to as the Elman network; see Elman (1990). Figure 8.8 is a representation of the RNN. At each time t, the available information is the input $\boldsymbol{x}_t$ and the vector of memory, $\boldsymbol{h}_{t-1}$, from $t-1$. These two vectors are combined linearly to form the state vector $\boldsymbol{s}_t$ at time t. The new vector of memory is a nonlinear function

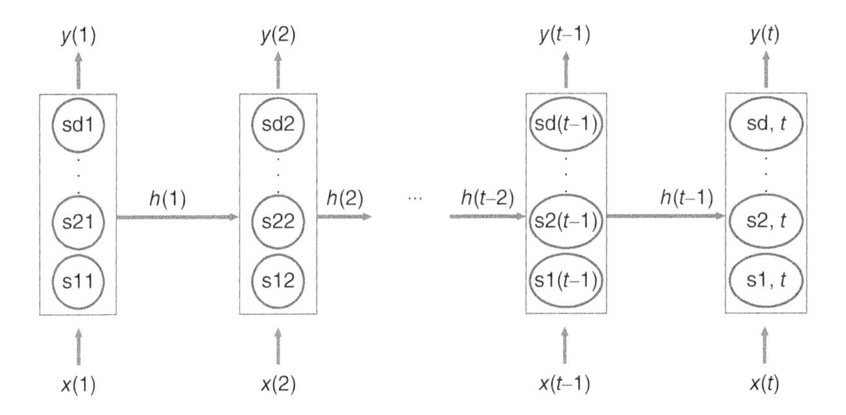

Figure 8.8 A representation of a simple RNN.

of the state vector and the forecast (or output) is built as a linear combination of the components of the new vector of memory $\boldsymbol{h}_t$, that is also feed into the next period.

For instance, consider the simplest case of a stationary zero-mean univariate time series, x_t, and $k = m = 1$, with linear activation functions and $d = 1$. Then, $s_t = wh_{t-1} + ux_t = h_t$ and the state is also the memory function, and $\hat{y}_t = \hat{x}_{t+1} = vh_t$. Assuming $(w, u, w) < 1$, we have that $s_t = u(1 + wB + w^2B^2 + \cdots + w^tB^t)x_t$ and, for large t, $(1 - wB)s_t = ux_t$ and $\hat{x}_{t+1} = w\hat{x}_t + uvx_t$. Thus, the forecast generated by this simple RNN corresponds to that of an ARMA(1,1) process. Increasing the dimension d and for $k = 1$, other univariate autoregressive integrated moving-average (ARIMA) forecasts can be generated. For $k > 1$, the output is a vector time series. An RNN can generate forecasts for nonlinear autoregressive with exogenous variables (NARX) models

$$\hat{\boldsymbol{y}}_t = F(\boldsymbol{y}_{t-1}, \ldots, \boldsymbol{y}_{t-p}, \boldsymbol{x}_t, \ldots, \boldsymbol{x}_{t-h}).$$

See Siegelmann et al. (1997).

Another type of RNN is the Jordan (1997) network given by

$$\boldsymbol{h}_t = g_h(\boldsymbol{W}y_{t-1} + \boldsymbol{U}\boldsymbol{x}_t + \boldsymbol{b}), \tag{8.27}$$

$$y_t = g_y(\boldsymbol{V}\boldsymbol{h}_t + c), \tag{8.28}$$

where $g_h(.)$ and $g_y(.)$ are activation functions. Here the state $\boldsymbol{s}_t$ is omitted.

In general, the output y_t of an RNN follows certain probability distribution and the maximum likelihood function is used to estimate the parameters. Also, a deep RNN may assume feedback in many ways. It can be from a hidden layer to a hidden layer or from nodes to nodes. See Goodfellow et al. (2016) for some examples. One can think of RNNs as networks that allow information to pass through certain neurons multiple times. Training an RNN can also be done by similar methods to those of a feedforward network. The only change needed is to compute the gradient recursively.

8.3.3 Activation Functions for Deep Learning

In addition to the linear, sigmoid, and tanh functions mentioned before, there are other types of activation functions available for DL, as the following.

1. Rectified Linear Unit (ReLU): $f(x) = \max(0, x)$. This activation function is computationally efficient with a simple gradient function. A disadvantage of the function is that its gradient is zero if $x < 0$, which could deactivate the neurons when $x < 0$.
2. Leaky ReLU:
$$f(x) = \begin{cases} 0.1x, & \text{if} x < 0, \\ x, & \text{if} x \geq 0. \end{cases}$$
This function avoids the deactivation issue associated with ReLU.
3. Parameterized ReLU:
$$f(x, a) = \begin{cases} x, & \text{if} x \geq 0, \\ ax, & \text{if} x < 0, \end{cases}$$
where a is a constant, representing the slope when $x < 0$. Often a is a small positive number such as 0.01. In practice, a can also be a parameter included in estimation.
4. Exponential Linear Unit (ELU):
$$f(x, a) = \begin{cases} x, & \text{if} x \geq 0, \\ a[\exp(x) - 1], & \text{if} x < 0. \end{cases}$$
This function is a modification of the leaky ReLU as it allows for nonlinear derivative when $x < 0$.
5. Swish function:
$$f(x) = x \times \text{sigmoid}(x) = \frac{x}{1 + e^{-x}}.$$
6. Softmax: for $\boldsymbol{x} = (x_1, \ldots, x_m)'$,
$$f(\boldsymbol{x}) = \frac{\exp(\boldsymbol{x})}{\sum_{j=1}^{m} \exp(x_j)}.$$

With many choices of activation functions, there are some general guidelines for selecting one in applications. We list some of the guidelines below:

- The ReLU function is commonly used in DL as it mitigates the vanishing gradient problem.
- If dead neurons are likely to exist, then ReLU function would work better.
- The ReLU function should only be used in hidden layers. The output layer is either the linear function for continuous output variables or the softmax function for classification.

- Sigmoid and softmax functions and their combinations work well for classifiers in general.

Of course, experience may dictate the choice of activation function or one may experiment with different types of activation function for a given application.

8.3.4 Training Deep Networks

The conventional backpropagation method would encounter various difficulties in training a deep network. For instance, the chain rule is used multiple times to obtain the gradients. Yet for most activation functions, the derivatives are less than 1. Consequently, the gradients can approach zero quickly and, hence, no clear directions are given for the next optimization step. This is referred to as the vanishing gradient problem. As another example, consider the RNN in Eqs. (8.24) and (8.25). The recursive nature of the dynamic implies that the eigenvalues of $\boldsymbol{W}$ play an important role in gradient calculation. Let λ be an eigenvalue of $\boldsymbol{W}$. If $|\lambda| < 1$, then $\lambda^t \to 0$ as $t \to \infty$. On the other hand, if $|\lambda| > 1$, then $|\lambda|^t \to \infty$ as t increases. Thus, we have the vanishing and exploding gradient problems. The exploding gradient can make the DL unstable. As a matter of fact, a large gradient itself may cause difficulty in learning. Finally, local minimums, saddle points, and flat parameter regions can also happen when the network is deep.

8.3.4.1 ***Long Short-Term Memory Model*** The vanishing or exploding gradient problem is referred to as the long-term dependencies in machine learning. This difficulty is particularly relevant for RNNs. Many methods have been proposed in the literature to overcome the difficulty, including adding skip connections through time and organizing the states of RNN at multiple time scales with information flowing more easily through long distances at the slower time scales. The long short-term memory (LSTM) model of Hochreiter and Schmidhuber (1997) introduces self-loops that produce paths where the gradient can flow for long durations. The LSTM model can mitigate the vanishing gradient problem, but not the exploding gradient one. Consider the RNN in Eqs. (8.24)–(8.25). The common LSTM recurrent networks employ *LSTM cells* that have an internal loop in addition to the usual outer recurrence. Each LSTM cell has the same inputs and outputs as an ordinary RNN, but employs three regulators, often referred to as *gates*, to govern the flow of information inside the cell. Consider Eq. (8.24). The state node s_{jt} of $\boldsymbol{s}_t$ now is accompanied by three gates. The first gate is called the *forget gate*, which is defined as

$$f_{jt} = \sigma\left(b_j^f + \sum_v U_{jv}^f x_{vt} + \sum_v W_{jv}^f h_{v,t-1}\right), \tag{8.29}$$

where $\sigma(.)$ denotes the sigmoid (logistic) function, $\boldsymbol{x}_t$ is the input vector and $\boldsymbol{h}_t$ is the vector of current hidden layer containing the outputs of all the LSTM cells, and $\boldsymbol{b}^f$, $\boldsymbol{U}^f$, and $\boldsymbol{W}^f$ are, respectively, biases, input weights, and recurrent weights for the forget gate. The second gate is the *external input gate* i_{jt}, which also uses a sigmoid activation function but with its own parameters, namely,

$$i_{jt} = \sigma\left(b_j^i + \sum_v U_{jv}^i x_{vt} + \sum_v W_{jv}^i h_{v,t-1}\right). \tag{8.30}$$

The third gate is the *output gate*, through which the output h_{it} of the LSTM cell can be shut off. Again, the sigmoid activation function is often used and we have

$$o_{jt} = \sigma\left(b_j^o + \sum_v U_{jv}^o x_{vt} + \sum_v W_{jv}^o h_{v,t-1}\right), \tag{8.31}$$

where, as before, $\boldsymbol{b}^o, \boldsymbol{U}^o$, and $\boldsymbol{W}^o$ are, respectively, the biases, input weights, and recurrent weights of the output gate. Next, the state cell is given by

$$s_{jt} = f_{jt}s_{j,t-1} + i_{jt}\sigma\left(b_j + \sum_v U_{jv}x_{vt} + \sum_v W_{jv}h_{v,t-1}\right), \tag{8.32}$$

where the activation $\sigma(.)$ often assumes the tanh function and $\boldsymbol{b}$, $\boldsymbol{U}$, and $\boldsymbol{W}$ denote, respectively, the biases, input weights, and recurrent weights of the state node. Clearly, the forget gate and the external input gate govern the evolution of the state nodes. Finally, we have

$$h_{jt} = \tanh(s_{jt})o_{jt}, \tag{8.33}$$

from which h_{jt} can be dropped if o_{jt} is close to zero.

From the description, we see that the LSTM uses more parameters but the three gates can control the information flow so that the length of memory becomes flexible. It is also clear that the gates can mitigate the vanishing gradient problem because they can break the chain rule in computing the gradients. There are some variants to the aforementioned LSTM cell, but the differences are minor.

8.3.4.2 *Training Algorithm* Deep networks often contain many parameters, which may be highly correlated. Thus, some regularizations are often needed. The concept of ℓ_1 or ℓ_2 penalty discussed before for high-dimensional linear regression continues to apply here.

Stochastic gradient decent (SGD) and its variants are probably the most used optimization algorithms of machine learning and deep learning. The basic idea of SGD is that it is possible to obtain an unbiased estimate of the gradient by taking the average gradient on a minibatch of ℓ data points drawn randomly from the data-generating distribution. In other words, instead of calculating gradient from the whole available data, one can do the following:

SGD at the training iteration j:

1. Specify a learning rate, say ϵ_j.
2. Let the initial parameter be $\boldsymbol{\theta}$.
3. While stopping criterion is not met, do
   - Sample a minibatch of ℓ data points, say $\{(\boldsymbol{x}^{(i)}, y^{(i)})|i = 1, \ldots, \ell\}$.
   - Compute the gradient estimate:

$$\hat{\boldsymbol{g}} = \frac{1}{\ell}\nabla_{\theta}\sum_i L(f(\boldsymbol{x}^{(i)}, \boldsymbol{\theta}), y^{(i)}).$$

   - Update: $\boldsymbol{\theta} = \boldsymbol{\theta} - \epsilon_j\hat{\boldsymbol{g}}$.

In the algorithm, ∇_θ denotes the partial derivative with respect to θ and $L(f(.), y)$ denotes a loss function between $f(.)$ and y with $f(.)$ representing the output of a deep network. A sufficient condition for the convergence of SGD algorithm is that

$$\sum_{j=1}^{\infty} \epsilon_j = \infty \quad \text{and} \quad \sum_{j=1}^{\infty} \epsilon_j^2 < \infty.$$

In practice, one can start with an initial learning rate ϵ_0 and decrease the rate linearly until iteration τ as

$$\epsilon_j = (1-\alpha)\epsilon_0 + \alpha\epsilon_\tau,$$

with $\alpha = j/\tau$. After iteration τ, the learning rate becomes a constant. In some application, one can even use $\ell = 1$, i.e. a single data point.

Another useful training method is the **momentum** of Polyak (1964), which can be summarized as follows:

SGD with momentum:

1. Specify a learning rate ϵ and a momentum parameter α.
2. Let the initial parameter be $\boldsymbol{\theta}$ and initial velocity be $\boldsymbol{v}$.
3. While stopping criterion is not met, do
   - Sample a minibatch of ℓ data points, say $\{(\boldsymbol{x}^{(i)}, y^{(i)})|i = 1, \ldots, \ell\}$.
   - Compute the gradient estimate:

$$\hat{\boldsymbol{g}} = \frac{1}{\ell}\nabla_\theta \sum_i L(f(\boldsymbol{x}^{(i)}, \boldsymbol{\theta}), y^{(i)}).$$

   - Compute the velocity update: $\boldsymbol{v} = \alpha\boldsymbol{v} - \epsilon\hat{\boldsymbol{g}}$.
   - Update: $\boldsymbol{\theta} = \boldsymbol{\theta} + \boldsymbol{v}$.

There are many other methods available to train a deep network, including the coordinate decent, adaptive learning rates (the AdaGrad algorithm), the RMSProp algorithm of Hinton (2012), and the Adam method of Kingma and Ba (2014). Again, interested readers are referred to Goodfellow et al. (2016).

8.4 SOME APPLICATIONS

We demonstrate some applications of RNN and DL using several real data sets. For the RNN example, we use the R package **rnn** and for other examples, we use the package **keras**, which is available from the RStudio at https://keras.rstudio.com. The web also contains information on installation and several examples for running machine learning in R. Interested readers are referred to Chollet and Allaire (2020) for further information. Also, the vignettes of **keras** on R CRAN include several tutorials on using the package such as basic classification and basic regression.

8.4.1 The Package: keras

The package **keras** available in RStudio is a useful package to learn DL in R. It has some built-in functions for applications. Users can build a NN model by using one of three methods: (a) sequential models, (b) functional application programming interface (API), and (c) custom models. For beginners, the sequential models are easy to learn. This is the method we use in this book. Once a reader is comfortable with the package, more sophisticated NN models can be built using the other two methods. The package contains many different types of network layer, including `layer_dense`, which is a densely connected layer of the traditional NN, `layer_dropout`, which governs randomly dropping of some input nodes, `layer_lstm`, which is a LSTM layer, among many others. The package also contains various activation functions of Section 8.3.3. The sequential method enables users to specify one layer after another for a NN and to choose the activation function of each layer. Readers are recommended to experiment with it to learn more about DL in R.

Example 8.4

In this example, we consider US monthly unemployment rate from January 1948 to February 2020. The data are seasonally adjustment and can be downloaded directly from the FRED, which is the Economic Database of the Federal Reserve Banks at St. Louis. In fact, one can download the data over the Internet via the R package **quantmod**. Details are given in the attached R commands and output. Alternatively, the data are also in the file `unrate4820.txt`. Figure 8.9 shows the time plot of the US monthly unemployment rate data.

The unemployment rate is measured to 1 digit after the decimal point, making it suitable for the R package **rnn**, which is a package for recurrent NN that converts integers into binary codes. Let u_t be the observed unemployment rate. We use $z_t = 10u_t$ so that z_t assumes an integer value. However, we use u_t in forecasting comparison with other statistical methods. The sample size is 866, and we reserve the last

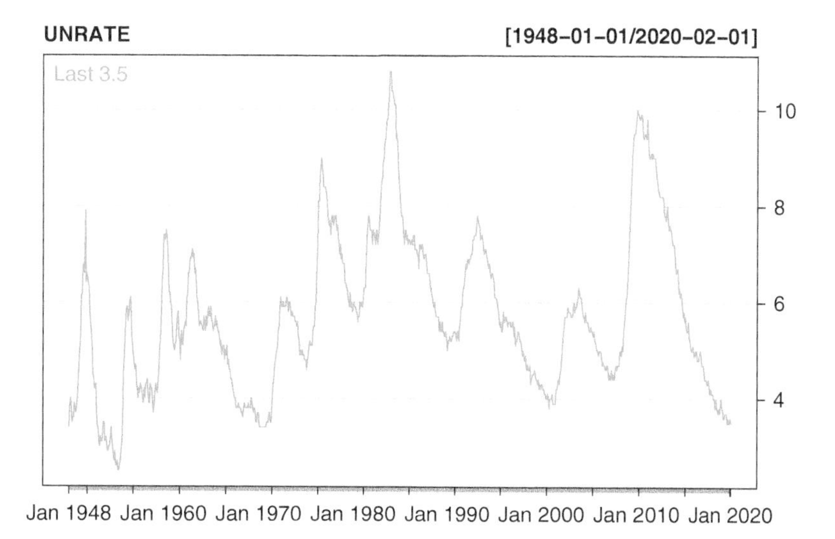

Figure 8.9 Monthly US unemployment rate from January 1948 to February 2020. The plot is obtained by the command `chartSeries` of the R package **`quantmod`**.

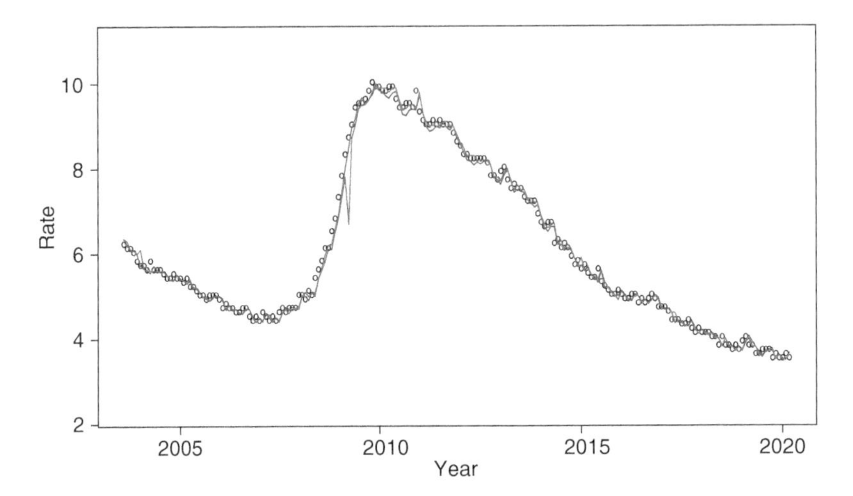

Figure 8.10 Time plots of observed data (in circle), predictions of RNN (in blue) and predictions of an AR(13) model (in red) in the forecasting subsample of the US monthly unemployment rate.

200 data points for out-of-sample comparison. The forecasting subsample is from July 2003 to February 2020.

As a baseline model, we use a scalar AR(13) model. Therefore, we use $z_{t-1}, \dots, z_{t-13}$ as the input variables for the RNN network and, in addition, we use three hidden layers of sizes (5,5,5). The activation function chosen is the logistic (i.e. the sigmoid) function with bias term included. The training was carried out with a learning rate of 0.05 and 20 epochs. Figure 8.10 shows the time plots of the observed data (in circle), predictions of RNN (in blue) and the predictions of AR(13) (in red) in the forecasting subsample. From the plots, we see that both RNN and scalar AR(13) model track the observed data reasonably well, but the RNN has a relatively large deviation around 2010. The root mean squared error (RMSE), the mean absolute error, and the median absolute error (MedAE) of the scalar AR(13) model are 0.15, 0.12, and 0.11, respectively. On the other hand, the three measures for the RNN are 0.22, 0.13, and 0.1, respectively. The relatively large RMSE of the RNN is caused by the large error around 2010 mentioned before.

R demonstration: RNN package

```
> require(quantmod)
> require(rnn)
> getSymbols("UNRATE",src="FRED")
> chartSeries(UNRATE,theme="white")
> rate <- as.numeric(UNRATE[,1])
> x <- rate*10 ### change rate to integers
> require(SLBDD)
> mm <- rnnStream(x,h=13,nfore=200) #A R command to setup the input and
## output for the RNN neural network. See R scripts below.
> names(mm)
[1] "Xfit" "Yfit" "yp"   "Xp"   "X"    "yfit" "newX"
## Xfit, Yfit, and newX are arrays in binary of length 8.
> Xo <- mm$X
```

```
> dim(Xo)
[1] 653  13
> yfit <- mm$yfit
> Xo=data.frame(Xo)
> newX=data.frame(mm$newX)
> m1 <- lm(yfit~.,data=Xo)
> m1
Call:
lm(formula = yfit ~., data = Xo) ## output truncated

Coefficients:
(Intercept)        zz      zz.1       zz.2        zz.3     zz.4       zz.5
   0.880015  0.984881 0.223721 -0.053961  -0.068886 0.008282 -0.116379
       zz.8      zz.9     zz.10       zz.11       zz.12
  -0.027701 -0.108302 0.131595  -0.140236   0.112020

> y1 <- predict(m1,newdata=newX)
> yp <- mm$yp  # observed rates
> mean(abs(yp-y1)) # should be divided by 10
[1] 1.195978
> sqrt(mean((yp-y1)^2))
[1] 1.499084
> median(abs(y1-yp))
[1] 1.076162
#
> m2 <- trainr(Y=mm$Yfit,X=mm$Xfit,learningrate = 0.05,
hidden_dim = c(5,5,5),use_bias=TRUE,numepochs = 20)
Trained epoch: 1 - Learning rate: 0.05
Epoch error: 3.8143465630104
...
...
Trained epoch: 20 - Learning rate: 0.05
Epoch error: 2.12295450909187
> y2 <- predictr(m2,mm$Xp)
> y3 <- bin2int(y2) # convert back to integers
> mean(abs(yp-y3))
[1] 1.305
> sqrt(mean((yp-y3)^2))
[1] 2.164486
> median(abs(y3-yp))
[1] 1
> tdx <- c(6+c(1:200))/12+2003 ## x-axis for plotting
> plot(tdx,yp/10,xlab='year',ylab='rate',ylim=c(2.3,11),pch='o',cex=0.6)
> lines(tdx,y1/10,col="red")
> lines(tdx,y3/10,col="blue")

## Below is the R script of rnnStream
"rnnStream" <- function(z,h=25,nfore=200){
# z: input in integer values
# h: number of lags used as input
# nfore: data points in the testing subsample.
if(!is.integer(z))z=as.integer(z)
ist <- h+1
nT <- length(z)
y <- z[ist:nT]
X <- NULL
for (i in 1:h){
```

```
 zz <- z[(ist-i):(nT-i)]
 X <- cbind(X,zz)
}
## convert to binary
nobe <- length(y)
orig <- nobe-nfore
yp<- y[(orig+1):nobe]
yfit <- y[1:orig]
s <- NULL
for (i in 1:h){
 x1 <- int2bin(X[1:orig,i],length=8)
 s <- c(s,c(x1))
}
Xfit <- array(s,dim=c(orig,8,h))
Yfit <- int2bin(yfit,length=8)
Yfit <- array(Yfit,dim=c(orig,8,1))
##
s <- NULL
for (i in 1:h){
 x2 <- int2bin(X[(orig+1):nobe,i],length=8)
 s <- c(s,c(x2))
}
Xp <- array(s,dim=c(nfore,8,h))
rnnStream <- list(Xfit=Xfit,Yfit=Yfit,yp=yp,Xp=Xp,X=X[1:orig,],
   yfit=yfit,newX=X[(orig+1):nobe,])
}
```

■

Example 8.5

In this example, we also use the US monthly unemployment rate series. The only difference is that we add the newly available 1 March 2020 data to the series. As before, we keep the last 200 observations for out-of-sample forecasting comparison and use the AR(13) model as the baseline model in comparison. The train sample thus has 654 observations. In this particular instance, we experiment with several NN models, including (a) 13-64-32-1 NN, (b) 13-32-16-1 NN, (c) 13-16-4-1 NN, and (d) 13-8-2-1 NN, where N1-N2-N3-1 means the network has N1 input variables, N2 neurons in the first hidden layer, N3 neurons in the second hidden layer, and a single neuron at the output layer. In this study, we use the ReLU activation for the hidden layers and the linear activation function $f(x) = x$ for the output layer. Also, we use the mean squared error as the loss function. For each NN model, the training is set for 500 epochs, but we also employed a stopping function to stop early if the value of the loss function does not decrease much. Figure 8.11 shows the mean absolute error versus the epoch for the 13-8-2-1 NN. Both training and validation are given. Here validation uses 20% of the training data in each epoch. The plot shows that the convergence of the training was achieved rather quickly.

Table 8.1 summarizes the out-of-sample performance of various NNs of the **keras** package, where the AR(13) model serves as the baseline model for comparison. From the table, we see that (a) the NN indeed can produce similar forecasts as that of the AR(13) model, and (b) overfitting in NN fares poorly in this particular instance. Finally, the R commands used for the 13-8-2-1 NN are given in the attached R demonstration.

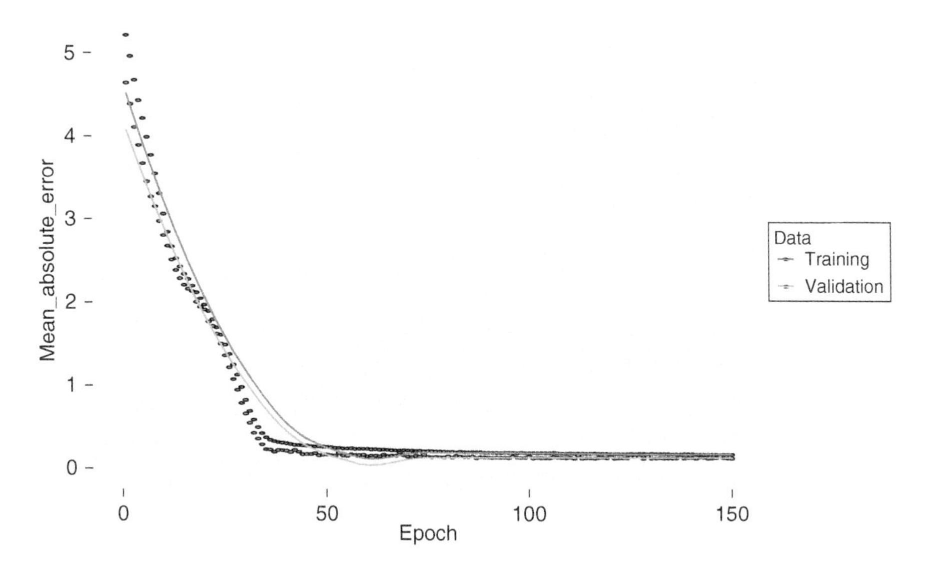

Figure 8.11 Convergence of training an 13-8-2-1 NN for the US monthly unemployment rate via the **keras** package.

TABLE 8.1 Out-of-Sample Performance of Various NNs for the Monthly US Unemployment Rate

Model	RMSE	MAE	MedAE	NP
13-64-32-1	0.314	0.248	0.224	3009
13-32-16-1	0.361	0.255	0.149	993
13-16-4-1	0.258	0.197	0.158	369
13-8-2-1	0.169	0.130	0.108	133
AR(13)	0.161	0.123	0.107	14

[a] RMSE, MAE, MedAE, and NP stand for root mean squared error, mean absolute error, median absolute error, and number of parameters, respectively.

R demonstration for DL with keras package: Readers are referred to https://keras.rstudio.com for further information.

```
rate <- scan("unrate4820.txt") # load data
nT <- length(rate)
nfore <- 200 # number of data in the forecasting subsample
p <- 13
ist=p+1
Y <- rate[ist:nT]
X <- NULL
for (i in 1.p){
X <- cbind(X,rate[(ist-i):(nT-i)])
}
nobe <- nT-p
nfit <- nobe-nfore
x_train <- X[1:nfit,]
```

```
y_train <- Y[1:nfit]
x_pred <- X[(nfit+1):nobe,]
y_pred <- Y[(nfit+1):nobe]
### Linear regression prediction
m1 <- lm(y_train~.,data=data.frame(x_train))
pm1 <- predict(m1,newdata=data.frame(x_pred))
rmse_lm <- sqrt(mean((y_pred-pm1)^2))
mae_lm <- mean(abs(y_pred-pm1))
med_lm <- median(abs(y_pred-pm1))
cat("lm results: (rmse,mae,med): ",c(rmse_lm,mae_lm,med_lm),"\n")

## standardize predictors for neural networks
x_train <- scale(x_train)
col_means_train <- attr(x_train,"scaled:center")
col_stddevs_train <- attr(x_train,"scaled:scale")
## standardize the predictor accordingly
x_pred <- scale(x_pred,center=col_means_train,scale=col_stddevs_train)
##
library(keras)

build_mod <- function(){
model <- keras_model_sequential() %>%
layer_dense(units=8,activation="relu",input_shape=dim(x_train)[2]) %>%
layer_dense(units=2,activation="relu") %>%
layer_dense(units=1)

model %>% compile(
   loss="mse",
   optimizer=optimizer_rmsprop(),
   metrics=list("mean_absolute_error")
)

 model
}
summary(model) # print out the model
# display training progress by printing a single dot for each
# completed epoch
model <- build_mod()
model %>% summary()
print_dot_callback <- callback_lambda(
   on_epoch_end=function(epoch,logs){
     if(epoch %% 80==0) cat("\n")
     cat(".")
    }
)
epochs <- 500 ## set number of training epochs

# fit the model and store training stats
historyO <- model %>% fit(
    x_train,
    y_train,
    epochs = epochs,
    validation_split=0.2,
    verbose = 0,
    callbacks = list(print_dot_callback)
```

```
)

### visualize the result
library(ggplot2)

plot(history0,metrics="mean_absolute_error",smooth=FALSE) +
  coord_cartesian(ylim=c(0,5))

## The patience is the amount of epochs to check for improvement
early_stop <- callback_early_stopping(monitor="val_loss",patience=20)
model <- build_mod()
history <- model %>% fit(
   x_train,
   y_train,
   epochs=epochs,
   validation_split=0.2,
   verbose=0,
   callbacks = list(early_stop,print_dot_callback)
)
plot(history,metrics="mean_absolute_error",smooth=FALSE)+
  coord_cartesian(xlim=c(0,150),ylim=c(0,5))

### performance at the test
c(loss,mae) %<-% (model %>% evaluate(x_pred,y_pred,verbose=0))

print(paste0("Mean absolute error on test set: ",sprintf("%.2f",mae)))
##
predTest <- model %>% predict(x_pred)
#cat("predictions are in predTest","\n")
yp <- as.numeric(predTest[,1])
rmse_dl <- sqrt(mean((y_pred-yp)^2))
mae_dl <- mean(abs(y_pred-yp))
med_dl <- median(abs(y_pred-yp))
cat("DL model results: (rmse,mae,med): ",c(rmse_dl,mae_dl,med_dl),"\n")
```

■

8.4.2 Dropout Layer

As shown by Example 8.5, a network with too many neurons may not fare well for the simple application considered. On the other hand, one may want to use as many input variables as possible because of no prior knowledge concerning the usefulness of the individual variables. As a compromise, one can build in some dropout layers to mitigate the impact of noisy input. Typically, a dropout layer contains a rate by which certain input variables are randomly dropped in the network training. To illustrate, we consider the 13-32-16-1 NN of Example 8.5. For the second hidden layer, we add a dropout layer with rate 0.5 and retrain the network. In this particular instance, we train the network for 500 epochs without early stopping. The convergence is shown in Figure 8.12. The RMSE, MAE, and MedAE of this new network in the forecasting subsample are 0.229, 0.184, and 0.150, respectively. While the results are still not as good as those of the simple 13-8-2-1 network, but they mark substantial improvements over the original network in RMSE and MAE. See Table 8.1.

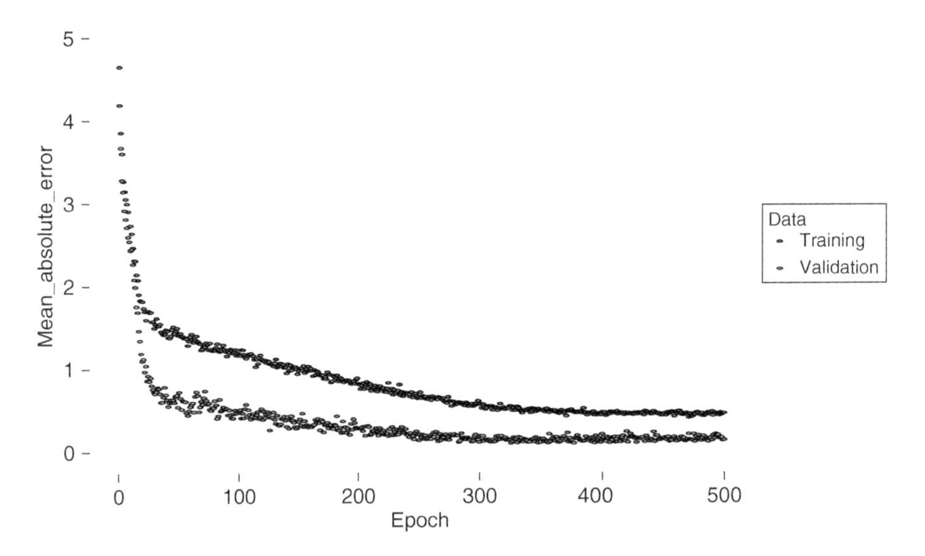

Figure 8.12 Convergence of training an 13-32-16-1 NN with dropout for the US monthly unemployment rate via the **keras** package.

R commands for network with dropout layer: The same training and forecasting subsamples as those of Example 8.5.

```
build_mod <- function(){
model <- keras_model_sequential() %>%
layer_dense(units=32,activation="relu",input_shape=dim(x_train)[2]) %>%
layer_dense(units=16,activation="relu") %>%
layer_dropout(rate=0.5) %>%  ## newly added dropout layer
layer_dense(units=1)

model %>% compile(
   loss="mse",
   optimizer=optimizer_rmsprop(),
   metrics=list("mean_absolute_error")
)
 model
}
summary(model) # print out the model
```

8.4.3 Application of Convolution Networks

To illustrate another application of DL and to consider convolution NNs, we consider the commonly used data set `MNIST`, which is concerned with reading handwritten digits $\{0, 1, \ldots, 9\}$. See the documents on the R packages **keras** and **kerasR**. The data set is available on the R package **keras** and contains a training and a testing subsample. The training data contain an array $\boldsymbol{X}$ of dimension (60000,28,28) as the predictors and a 60000-dimensional vector $\boldsymbol{y}$ of true digits as the response variable. Each face $X[i,,]$ of $\boldsymbol{X}$ is a 28×28 matrix of gray levels in [0,255], each of which represents a handwritten digit. We analyze this classification problem in two ways to assess the efficacy of the convolution network.

First, we follow the approach shown in the vignette of **keras** by vectorization each 28×28 matrix into as 784-dimensional vector and scale the gray level to be in [0,1]. Therefore, in our first analysis of the MNIST data, the predictors are in a 60000×784 data matrix and the response is a 60000-dimensional vector of digits from $\{0, \ldots, 9\}$. The response is treated as a categorical variable with 10 categories. In statistical term, the response is a factor with 10 categories. Thus, in our first analysis, we treat the problem as a one-dimensional classification problem, to which DL has been proven to be useful.

We specify a network with five layers as our first model. The first layer is dense with 256 neurons and employs the ReLU as its activation function. The subcommand `input_shape` specifies the input dimension. Note that the first dimension of the input data matrix is identified from the training sample so that the `input_shape` parameter starts with 784, the second dimension, in our application. The second layer is a dropout layer with rate 0.4. The third layer is also a dense layer with 128 neurons and employs the ReLU as its activation function. This is followed yet by another dropout layer with rate 0.3. Finally, the output layer has 10 neurons for the 10 digits and uses the `softmax` activation function. The resulting NN model has 235,146 parameters.

The model is compiled using categorical cross-entropy as the loss function and we use `rmsprop` as the criterion for training. We then run the training for 30 epochs with a batch size 128 and a validation split with probability 0.2, which roughly corresponding to a fivefold cross-validation. The default initial parameters are used in the training. Figure 8.13 plots the convergence of the 30 training epochs. Once the network is trained, we apply it to the testing subsample. In this particular case, the

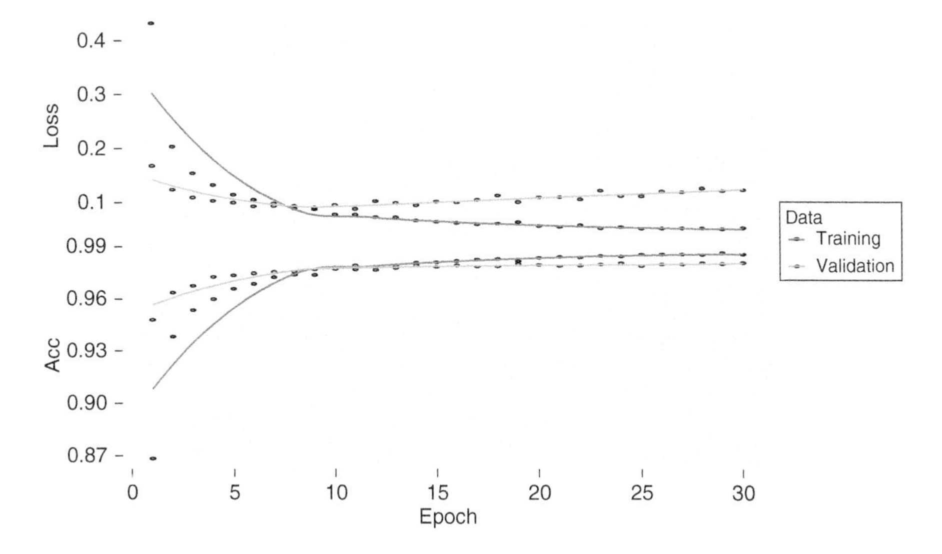

Figure 8.13 The convergence plot of a dense NN for the MNIST data set.

successful rate is 0.9811, which is very good. The details of the classification results of the network is given below:

```
> table(mnist$test$y,yhat)
   yhat
       0    1    2    3    4    5    6    7    8    9
  0  971    1    1    2    1    0    1    1    2    0
  1    0 1125    3    1    0    0    2    1    3    0
  2    3    0 1016    2    1    0    1    6    3    0
  3    0    0    2  996    0    3    0    6    2    1
  4    1    2    4    0  961    0    3    3    3    5
  5    2    1    0    8    1  868    5    2    4    1
  6    7    2    0    1    4    5  937    0    2    0
  7    1    4    8    5    0    0    0 1006    2    2
  8    1    1    3    4    4    3    0    4  951    3
  9    1    2    0    7   10    2    0    6    1  980
> mean(mnist$test$y==yhat)
[1] 0.9811
```

From the output, we see that 9 and 4 can occasionally be mixed up. The same is also true for 7 and 2, and for 5 and 3. These misclassification seems reasonable in hand writing. Finally, the R commands and some selected output are given in the attached R demonstration.

R demonstration: A five-layer dense network for MNIST data set:

```
> library(keras)
> mnist <- dataset_mnist()
> x_train <- mnist$train$x
> y_train <- mnist$train$y
> x_test <- mnist$test$x
> y_test <- mnist$test$y
> yobserved <- y_test
> ## reshape the training from 28-by-28 images into 784 vectors
> x_train <- array_reshape(x_train,c(nrow(x_train),784))
> x_test <- array_reshape(x_test,c(nrow(x_test),784))
> ## rescale integers into between 0 and 1.
> x_train <- x_train/255
> x_test <- x_test/255
>
> y_train <- to_categorical(y_train,10)
> y_test <- to_categorical(y_test,10)
> ## define a model
> mod <- keras_model_sequential()
> mod %>%
+   layer_dense(units=256,activation="relu",input_shape=c(784)) %>%
+   layer_dropout(rate=0.4) %>%
+   layer_dense(units=128,activation="relu") %>%
+   layer_dropout(rate=0.3) %>%
+   layer_dense(units=10,activation="softmax")
```

```
>
> ## print out mod
> summary(mod)
Model: "sequential_1"
________________________________________________________________
Layer (type)                  Output Shape              Param #
================================================================
dense_2 (Dense)               (None, 256)               200960
________________________________________________________________
dropout_2 (Dropout)           (None, 256)               0
________________________________________________________________
dense_3 (Dense)               (None, 128)               32896
________________________________________________________________
dropout_3 (Dropout)           (None, 128)               0
________________________________________________________________
dense_4 (Dense)               (None, 10)                1290
================================================================
Total params: 235,146
Trainable params: 235,146
Non-trainable params: 0
______________________________________________________
> mod %>% compile(
+     loss="categorical_crossentropy",
+     optimizer = optimizer_rmsprop(),
+     metrics=c("accuracy")
+)
> ### training and evaluation
> history1 <- mod %>% fit(
+     x_train, y_train,
+     epochs = 30, batch_size = 128,
+     validation_split = 0.2
+)
Train on 48000 samples, validate on 12000 samples
Epoch 1/30
48000/48000 [==============================] - 3s 58us/sample -
loss: 0.4323 - acc: 0.8685 - val_loss: 0.1680 - val_acc: 0.9481
Epoch 2/30 ## output shortened.
> plot(history1)
`geom_smooth()` using formula 'y ~ x'
>
> mod %>% evaluate(x_test,y_test)
$loss
[1] 0.1094975
$acc
[1] 0.9811

> ### prediction
> yhat <- mod %>% predict_classes(x_test)
> table(mnist$test$y,yhat)
> mean(mnist$test$y==yhat)
[1] 0.9811
```

Next, turn to the convolution NN. Following the vignettes of **kerasR**, we keep the 28×28 matrices of gray levels as a two-dimensional (2D) image so that we can define a kernel size (or neighborhood) for convolution network. Specifically, we use an eight-layer convolution network in our second analysis. The first and second layers are convolution layers with filters 32, kernel size (3,3) and activation function `relu` (rectified linear unit). Since we kept the 28×28 matrix so that we use `input_shape=c(28,28,1)`. The third layer pools the first two layers together with pooling size (2,2). The fourth layer is a dropout with rate 0.25, and the fifth layer is a flatten layer, which flattens the input. The sixth layer is dense with 128 neurons and activation function `relu`. The seventh layer is a dropout with rate 0.25, and the final layer is a dense output with 10 neurons and activation function `softmax` as before. The resulting network has 600,810 parameters most of them are from the sixth layer of dense network. We also use the categorical cross entropy as the loss function with criterion rmsprop as before. The training of the convolution network is carried in five epochs and validation split 0.1. Figure 8.14 shows the convergence of the five training epochs. Even though we only use five epochs in training, the resulting network performs well. In the testing subsample, the accuracy of classification is 0.9874, which is slightly higher than the dense network used in the first analysis. The details of classification results are

```
> table(Y_test, Yhat) # See R commands below
      Yhat
Y_test    0    1    2    3    4    5    6    7    8    9
     0  972    1    0    0    0    0    5    1    1    0
     1    0 1133    1    0    0    1    0    0    0    0
     2    3    2 1019    0    3    0    2    3    0    0
     3    0    0    2 1003    0    2    0    1    2    0
     4    0    0    0    0  975    0    3    0    2    2
     5    3    1    0    5    0  877    4    0    0    2
     6    4    3    0    0    2    2  945    0    2    0
     7    1    2   10    2    0    1    0 1007    1    4
     8    3    1    2    1    1    0    2    1  962    1
     9    2    0    0    0   14    2    0    6    4  981
> mean(Y_test == Yhat)
[1] 0.9874
```

From the results, we see that, again, 9 and 4 are occasionally mixed up. The same also applies to 7 and 2. Such mistakes are reasonable for handwritten digits. Again, the R commands used and some selected results are given below in the R demonstration.

R demonstration: 2D convolution network for MNIST data set:

```
> require(keras)
> mnist <- dataset_mnist()
> names(mnist)
[1] "train" "test"
> X_train <- array(mnist$train$x, dim = c(dim(mnist$train$x), 1))/255
```

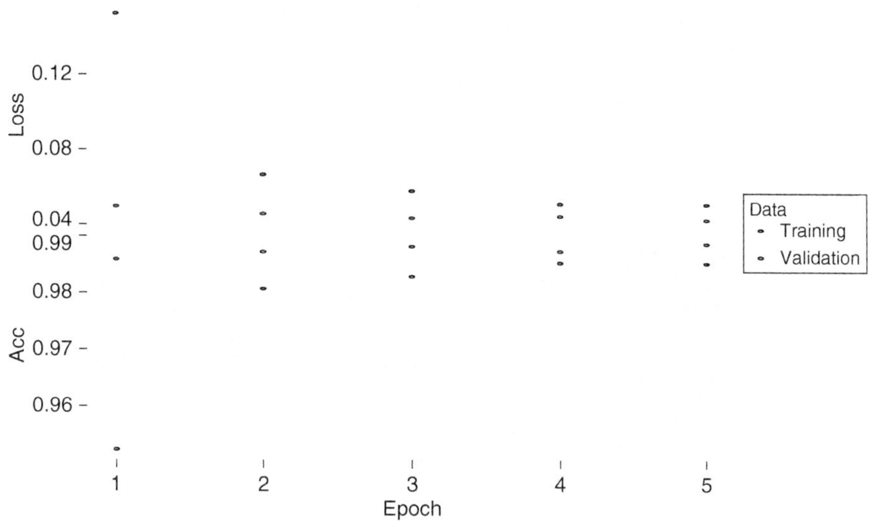

Figure 8.14 The convergence plot of a convolution NN for the MNIST data set.

```
> Y_train <- to_categorical(mnist$train$y, 10)
> X_test <- array(mnist$test$x, dim = c(dim(mnist$test$x), 1))/255
> Y_test <- mnist$test$y
> mod <- keras_model_sequential()
> mod %>%
+      layer_conv_2d(filters=32,kernel_size=c(3,3),activation="relu",
                 input_shape=c(28,28,1)) %>%
+
+      layer_conv_2d(filters=32,kernel_size=c(3,3),activation="relu",
                 input_shape=c(28,28,1)) %>%
+
+      layer_max_pooling_2d(pool_size=c(2,2)) %>%
+      layer_dropout(rate=0.25) %>%
+      layer_flatten() %>%
+      layer_dense(units=128,activation="relu") %>%
+      layer_dropout(rate=0.25) %>%
+      layer_dense(units=10,activation="softmax")
>
> summary(mod)
Model: "sequential"
________________________________________________________________________
Layer (type)                      Output Shape                  Param #
========================================================================
conv2d (Conv2D)                   (None, 26, 26, 32)            320
________________________________________________________________________
conv2d_1 (Conv2D)                 (None, 24, 24, 32)            9248
________________________________________________________________________
max_pooling2d (MaxPooling2D)      (None, 12, 12, 32)            0
________________________________________________________________________
dropout (Dropout)                 (None, 12, 12, 32)            0
________________________________________________________________________
```

```
flatten (Flatten)              (None, 4608)              0
________________________________________________________________________
dense (Dense)                  (None, 128)               589952
________________________________________________________________________
dropout_1 (Dropout)            (None, 128)               0
________________________________________________________________________
dense_1 (Dense)                (None, 10)                1290
========================================================================
Total params: 600,810
Trainable params: 600,810
Non-trainable params: 0
________________________________________________________________________
> mod %>% compile(
+     loss = "categorical_crossentropy",
+     optimizer= optimizer_rmsprop(),
+     metrics = list("accuracy")
+)
> history1 <- mod %>% fit(
+     X_train,
+     Y_train,
+     epochs = 5,
+     batch_size=32,
+     verbose=1,
+     validation_split=0.1
+)
Train on 54000 samples, validate on 6000 samples
Epoch 1/5
54000/54000 [==============================] - 102s 2ms/sample -
loss: 0.1526 - acc: 0.9524 - val_loss: 0.0497 - val_acc: 0.9858
....
Epoch 5/5
54000/54000 [==============================] - 113s 2ms/sample -
loss: 0.0511 - acc: 0.9852 - val_loss: 0.0427 - val_acc: 0.9887
> plot(history1)
> mod %>% evaluate(X_train,Y_train,batch_size=32)
$loss
[1] 0.03316241
$acc
[1] 0.99025
> Yhat <- mod %>% predict_classes(X_test,batch_size=32)
> table(Y_test, Yhat)
> mean(Y_test == Yhat)
```

By comparing the two networks used in this section, we see that the convolution networks can be useful in classification. This seems reasonable in this particular application because the neighboring cells of the 28×28 matrices contain information concerning the handwritten digits. In fact, a majority of the cells in the matrix would be empty (or zero in the gray level), which carries no information.

Remark. The package **kerasR** is an interface between **keras** and users. It simplifies model specification in **keras**. Unfortunately, there exist some problems in reading integers with the subcommand `input_shape` in layer specification. We use **keras** in this chapter.

8.4.4 Application of LSTM

To demonstrate the application of LSTM, we consider the well-known movie review data set **IMDB**, which is also available from the **keras** package. The dependent variable is binary "1" or "0" (approval or disapproval) and the predictors consist of comments from the reviewers. Following the approach commonly used for the data set, we use exactly 100 words of every review. If a review is less than 100 words, we pad with a special word coded as zeros. This preprocessing of the predictors can be carried out using the command `pad_sequences` of the package. Since words in a review are connected RNN and LSTM appear to be useful. We start our analysis with a RNN. The analysis is then followed by a LSTM network so that the contribution of LSTM can be assessed. The data set has 5736 reviews, and we use the first 4000 as the training subsample and the last 1736 as the testing subsample. The commands used to preprocessing of predictors are given in the attached R demonstration.

Recall that **keras** has many built-in network layers and we use sequential approach to build our network. Again, readers are referred to the document and vignettes of the package for further details. For the RNN, we start with an embedding layer to formulate the data into an array with shape (100,32). We then use a reshape layer to ensure that the data are properly processed through the network. This is followed by a dropout layer with rate 0.25 to simplify the training. We use a flatten layer to embedded data. Next, we apply a dense layer with 128 neurons with activation function `relu` (rectified linear unit). This is also followed by a dropout layer with rate 0.25 to simplify the computation. Finally, the output layer is a dense layer with a single neuron and sigmoid activation function. This is so because our response variable is binary. The resulting RNN has 425,857 parameters.

We train the RNN using binary cross-entropy as the loss function and residual mean squared error as criterion. Since the wording varies from reviewer to reviewer, we use a small learning rate of 0.00025. This learning rate was used by the **kerasR** package, even though we cannot run the package. We train the network for 30 epochs with batch size 32 and validation split of 0.1. Figure 8.15 shows the convergence of the 30 epochs. From the plot, the error is a decreasing function of the number of epochs, but the cross-validation suggests that we may stop earlier. After the training, the loss is about 0.075. For the testing subsample, we obtain the following result:

```
 > table(R_test,Rhat)
       Rhat
R_test    0   1
     0 608 193
     1 180 755
> mean(R_test == Rhat)
[1] 0.7851382
```

indicating that the success rate of movie review by the employed RNN is about 78.5%.

Turn to the LSTM network. We modify the RNN with several changes. (a) We replace the flatten layer by a LSTM layer with 32 neurons. (b) We increase the dense layer to 256 neurons. (c) We remove the dropout layer before the output layer. The

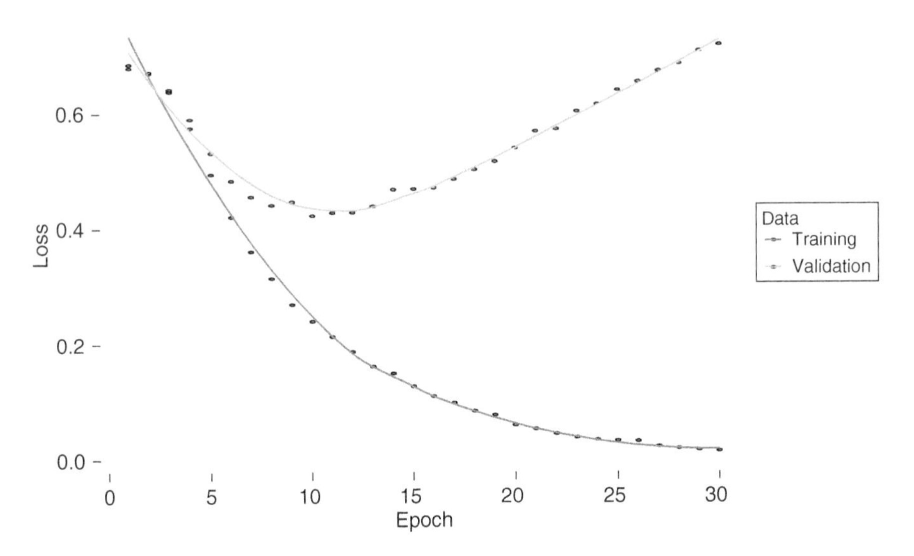

Figure 8.15 Convergence of the RNN for the IMDB data set in the training subsample, which consists of the first 4000 reviews.

resulting LSTM network has 33,025 parameters, which is much smaller than that of the RNN. We use the same loss function and criterion, but only train the LSTM network for 10 epochs. Figure 8.16 shows the convergence of the LSTM network in the 10 epochs. In this case, both the in sample and cross-validation curves continue to decrease, indicating that we can either train the network a bit longer or use a larger cross validation rate. We experiment with a few runs, but the out-of-sample

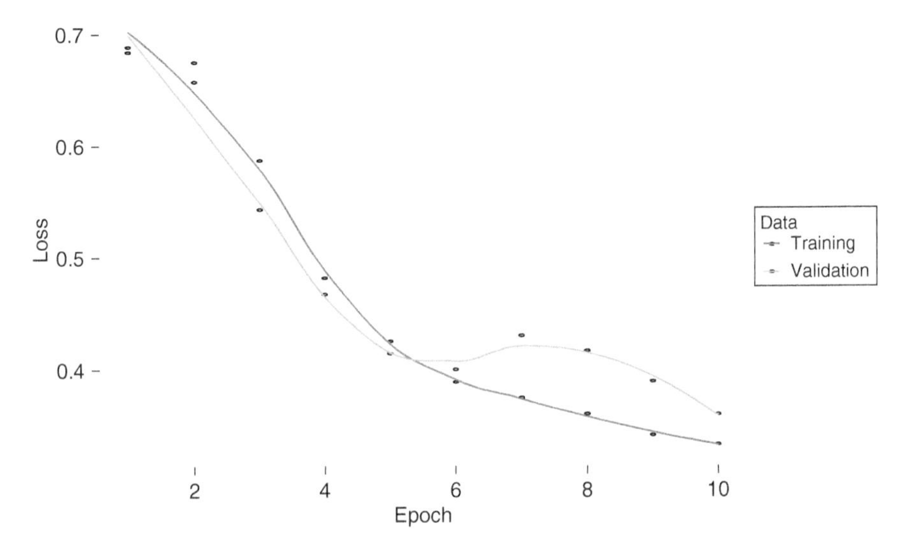

Figure 8.16 Convergence of the RNN for the IMDB data set in the training subsample, which consists of the first 4000 reviews.

prediction seems similar. For the testing subsample, the LSTM network shows the following results:

```
> table(R_test,Yhat2)
      Yhat2
R_test   0   1
     0 651 150
     1 135 800
> mean(R_test==Yhat2)
[1] 0.8358295
```

indicating a success rate of 83.6%, which is higher than that of the RNN network. Thus, in this particular instance, the LSTM seems to be helpful.

R commands and output of RNN and LSTM networks:

```
>library(keras)
## Load data and split into training and testing
>imdb <- dataset_imdb(num_words=500,maxlen=100)
> Z_train <- pad_sequences(imdb$train$x[1:4000],maxlen=100)
> R_train <- imdb$train$y[1:4000]
> Z_test <- pad_sequences(imdb$train$x[4001:5736],maxlen=100)
> R_test <- imdb$train$y[4001:5736]
## Network specification: RNN
> mod1 <- keras_model_sequential()
> mod1 %>%
+     layer_embedding(500,32,input_length=100) %>%
+     layer_reshape(target_shape=c(100,32),input_shape=c(100)) %>%
+     layer_dropout(rate=0.25) %>%
+     layer_flatten() %>%
+     layer_dense(units=128,activation="relu") %>%
+     layer_dropout(rate=0.25) %>%
+     layer_dense(units=1,activation="sigmoid")
> ## print the model
> summary(mod1)
Model: "sequential_9"
____________________________________________________________________
Layer (type)                 Output Shape                Param #
====================================================================
embedding (Embedding)       (None, 100, 32)             16000
____________________________________________________________________
reshape (Reshape)           (None, 100, 32)             0
____________________________________________________________________
dropout (Dropout)           (None, 100, 32)             0
____________________________________________________________________
flatten (Flatten)           (None, 3200)                0
____________________________________________________________________
dense (Dense)               (None, 128)                 409728
____________________________________________________________________
dropout (Dropout)          (None, 128)                 0
____________________________________________________________________
dense (Dense)              (None, 1)                   129
====================================================================
Total params: 425,857
Trainable params: 425,857
```

```
Non-trainable params: 0
________________________________________________________________________
> ## compile the model
> mod1 %>% compile(
+     loss="binary_crossentropy",
+     optimizer=optimizer_rmsprop(lr=0.00025),
+)
> ## fit
> history2 <- mod1 %>% fit(
+     Z_train, R_train,batch_size=32,epochs=30,verbose=1,
+     validation_split=0.1
+)
Train on 3600 samples, validate on 400 samples
Epoch 1/30
3600/3600 [==============================] - 2s 529us/sample -
loss: 0.6861 - val_loss: 0.6797
.....
Epoch 30/30
3600/3600 [==============================] - 1s 337us/sample -
loss: 0.0144 - val_loss: 0.7202
>
> plot(history2)
'geom_smooth()' using formula 'y ~ x'
> mod1 %>% evaluate(Z_train,R_train,batch_size=32)
4000/4000 [==============================] - 0s 58us/sample -
loss: 0.0748
      loss
0.07476683

> Rhat <- mod1 %>% predict_classes(Z_test,batch_size=32)
> table(R_test,Rhat)
> mean(R_test == Rhat)

## Network specification:  LSTM  network
> mod2 <- keras_model_sequential()
> mod2 %>%
+     layer_embedding(500,32,input_length=100) %>%
+     layer_reshape(target_shape=c(100,32),input_shape=c(100)) %>%
+     layer_dropout(rate=0.25) %>%
+     layer_lstm(units=32) %>%
+     layer_dense(units=256,activation="relu") %>%
+     layer_dense(units=1,activation="sigmoid")
>
> summary(mod2)
Model: "sequential_10"
________________________________________________________________________
Layer (type)                   Output Shape                 Param #
========================================================================
embedding_10 (Embedding)       (None, 100, 32)              16000
________________________________________________________________________
reshape_9 (Reshape)            (None, 100, 32)              0
________________________________________________________________________
dropout_18 (Dropout)           (None, 100, 32)              0
________________________________________________________________________
lstm (LSTM)                    (None, 32)                   8320
________________________________________________________________________
dense_19 (Dense)               (None, 256)                  8448
```

```
dense_20 (Dense)                 (None, 1)                        257
=====================================================================
Total params: 33,025
Trainable params: 33,025
Non-trainable params: 0
_____________________________________________________________________
> mod2 %>% compile(
+     loss="binary_crossentropy",
+     optimizer=optimizer_rmsprop(lr=0.00025),
+)
>
> history3 <- mod2 %>% fit(
+     Z_train, R_train,batch_size=32,epochs=10,
+     verbose=1, validation_split=0.1
+)
Train on 3600 samples, validate on 400 samples
Epoch 1/10
3600/3600 [==============================] - 7s 2ms/sample -
loss: 0.6885 - val_loss: 0.6839
> plot(history3)
'geom_smooth()' using formula 'y ~ x'
>
> mod2 %>% evaluate(Z_train,R_train,batch_size=32)
     loss
0.3006215
>
> Yhat2 <- mod2 %>% predict_classes(Z_test,batch_size=32)
> table(R_test,Yhat2)
> mean(R_test==Yhat2)
```

Example 8.6

To demonstrate the application of DL in economic data analysis, we revisit the monthly macroeconomic data set from FRED. The data span is from January 1959 to April 2019. Due to missing values and various transformations used to achieve stationarity, as explained before, we work with 710 observations and 122 variables. Again, we use the growth rate of the IP index as the dependent variables and the lag-1 to lag-6 of all variables to form the predictors. Therefore, we have an effective sample size of 704 and 732 predictors. We further divide the data into (a) a training subsample with 504 observations and (b) a forecasting subsample of 200 observations.

First, as a baseline model, we use a scalar AR(12) model. The fitted model is

$$\begin{aligned} y_t &= 0.0014 + 0.264y_{t-1} + 0.136y_{t-2} + 0.133y_{t-3} + 0.080y_{t-4} - 0.12y_{t-5} \\ &\quad - 0.011y_{t-6} + 0.018y_{t-7} + 0.01y_{t-8} + 0.11y_{t-9} - 0.01y_{t-10} \\ &\quad - 0.002y_{t-11} - 0.089y_{t-12} + a_t, \quad \hat{\sigma}_a = 0.0068. \end{aligned}$$

This model fits the data reasonably well. For instance, the Ljung–Box statistics of the residuals show $Q(24) = 22.55$ with p-value 0.55. For the testing subsample, this AR model gives the RMSE, mean absolute error (MAE), and median absolute error (MedAE) of 0.0065, 0.0046, and 0.0037, respectively.

Second, consider DL. We standardize the predictors of the training subsample so that each predictor has mean zero and variance 1. We then transformed the predictors of the testing subsample based on the means and standard errors of the predictors in the training subsample. Details of the transformation are given in the attached R demonstration. We have tried some DL models and the best model we obtained is as follows: It is a NN with six layers. The first layer is dense with 64 neurons, which has 46,912 parameters. The second layer is a dropout with rate 0.4, and the third layer is also dense with 16 neurons. This is followed by another dropout with rate 0.25. The fifth layer is a dense layer with four neurons, and the final output layer has a single neuron with simple linear activation function. All other dense layers employ the ReLU as the activation function. Overall, the network has 48,025 parameters. Figure 8.17 shows the convergence plot of the network in the training subsample over 100 epochs. Both the loss function and mean absolute error are shown with a cross-validation rate of 0.2. The plots indicate that the training has converged. For the testing subsample, the network gives RMSE, MAE, and MedAE of 0.0070, 0.0047, and 0.0036. These results are essentially the same as those of the AR(12) model. The results of other DL models we tried are given in Table 8.2. From the table, we see that the performance of DL seems relatively stable, except for the first model that has 103,137 parameters.

R commands and some output of DL: Monthly macroeconomic data set.

```
> da <- read.csv("FREDMDApril19.csv") #load data
> source("DLdata.R") # R scripts to formulate data, given below
> m1 <- DLdata(da,forerate=c(200/nrow(da)),lag=6,locY=6)
> Xtrain <- m1$Xtrain
> Ytrain <- m1$Ytrain
> Xtest <- m1$Xtest
> Ytest <- m1$Ytest
```

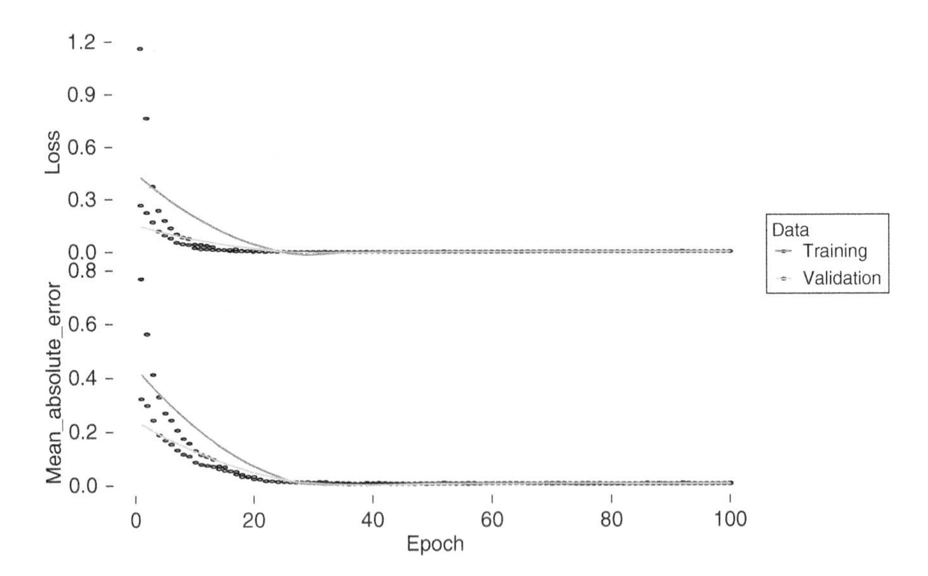

Figure 8.17 Convergence of the a six-layer DL network for Example 8.6. The data used are the US monthly macroeconomic data.

TABLE 8.2 Forecasting Performance of Various Six-Layer NNs for the Monthly Macroeconomic Data Set

M	Nod	Rate	Nod	rate	Nod	Nod	RMSE	MAE	MedAE
1	128	0.3	64	0.25	16	1	0.0224	0.0086	0.0040
2	64	0.3	32	0.25	4	1	0.0072	0.0049	0.0037
3	32	0.3	16	0.25	4	1	0.0073	0.0050	0.0038
4	64	0.4	16	0.25	4	1	0.0070	0.0047	0.0036
AR							0.0065	0.0046	0.0037

In the Table M is the model, Nod denotes the number of neurons, rate is the dropout rate, and RMSE, MAE, MedAE are the root mean squared error, mean absolute error and median absolute error.

```
> k <- ncol(Xtrain)
> k
[1] 732
> require(keras)
### First DL model used.
> ## model specification
> mod <- keras_model_sequential()
> mod %>%
+      layer_dense(units=128,activation="relu",input_shape=c(k)) %>%
+      layer_dropout(rate=0.3) %>%
+      layer_dense(units=64,activation="relu") %>%
+      layer_dropout(rate=0.25) %>%
+      layer_dense(units=16,activation="relu") %>%
+      layer_dense(units=1)
> summary(mod)
Model: "sequential"
____________________________________________________________________
Layer (type)                  Output Shape               Param #
====================================================================
dense (Dense)                 (None, 128)                93824
____________________________________________________________________
dropout (Dropout)             (None, 128)                0
____________________________________________________________________
dense_1 (Dense)               (None, 64)                 8256
____________________________________________________________________
dropout_1 (Dropout)           (None, 64)                 0
____________________________________________________________________
dense_2 (Dense)               (None, 16)                 1040
____________________________________________________________________
dense_3 (Dense)               (None, 1)                  17
====================================================================
Total params: 103,137
Trainable params: 103,137
Non-trainable params: 0
____________________________________________________________________
> mod %>% compile(
+      loss="mse",
+      optimizer=optimizer_rmsprop(),
+      metrics=list("mean_absolute_error")
+)
```

```
> print_dot_callback <- callback_lambda(
+     on_epoch_end=function(epoch,logs){
+         if(epoch %% 80==0) cat("\n")
+         cat(".")
+     }
+)
> ## train the model
> history <- mod %>% fit(
+     Xtrain,
+     Ytrain,
+     epochs = 100,
+     validation_split=0.2,
+     verbose = 0,
+     callbacks = list(print_dot_callback)
+)
> ## visualize the training result
> plot(history)
'geom_smooth()' using formula 'y ~ x'
> ## Model evaluation (in sample)
> mod %>% evaluate(Xtrain,Ytrain,verbose=0)
$loss
[1] 2.478281e-05
$mean_absolute_error
[1] 0.003508422
> ## prediction
> Yhat <- mod %>% predict(Xtest)
> RMSE <- sqrt(mean((Ytest-Yhat)^2))
> MAE <- mean(abs(Ytest-Yhat))
> MedAE <- median(abs(Ytest-Yhat))
> cat("RMSE, MAE, MedAE: ",c(RMSE,MAE,MedAE),"\n")
RMSE, MAE, MedAE:  0.02237929 0.008638669 0.004014824
### 4th model used
> mod <- keras_model_sequential()
> mod %>%
+  layer_dense(units=64,activation="relu",input_shape=c(k)) %>%
+  layer_dropout(rate=0.4) %>%
+  layer_dense(units=16,activation="relu") %>%
+  layer_dropout(rate=0.25) %>%
+  layer_dense(units=4,activation="relu") %>%
+  layer_dense(units=1)
> ## print out the model
> summary(mod)
Model: "sequential_3"
```

Layer (type)	Output Shape	Param #
dense_12 (Dense)	(None, 64)	46912
dropout_6 (Dropout)	(None, 64)	0
dense_13 (Dense)	(None, 16)	1040
dropout_7 (Dropout)	(None, 16)	0
dense_14 (Dense)	(None, 4)	68
dense_15 (Dense)	(None, 1)	5

```
=====================================================================
Total params: 48,025
Trainable params: 48,025
Non-trainable params: 0
_____________________________________________________________________
> ## compile the model
> mod %>% compile(
+        loss="mse",
+        optimizer=optimizer_rmsprop(),
+        metrics=list("mean_absolute_error")
+ )
> ## command to show training status
> print_dot_callback <- callback_lambda(
+    on_epoch_end=function(epoch,logs){
+      if(epoch %% 80==0) cat("\n")
+      cat(".")
+     }
+)
> ## train the model
> history <- mod %>% fit(
+     Xtrain,
+     Ytrain,
+     epochs = 100,
+     validation_split=0.2,
+     verbose = 0,
+     callbacks = list(print_dot_callback)
+)
> plot(history)
> ## Model evaluation (in sample)
> mod %>% evaluate(Xtrain,Ytrain,verbose=0)
$loss
[1] 7.85964e-05
$mean_absolute_error
[1] 0.005820423
> ## prediction
> Yhat <- mod %>% predict(Xtest)
> RMSE <- sqrt(mean((Ytest-Yhat)^2))
> MAE <- mean(abs(Ytest-Yhat))
> MedAE <- median(abs(Ytest-Yhat))
> cat("RMSE, MAE, MedAE: ",c(RMSE,MAE,MedAE),"\n")
RMSE, MAE, MedAE:  0.007001048 0.004705776 0.003581634
###  DLdata R script
"DLdata" <- function(da,forerate=0.2,locY=1,lag=1){
# Input:
# da: data matrix T-by-k matrix
# forerate: fraction of sample size to form testing sample
# locY: locator for the dependent variable.
# lag: number of lags to be used to form predictors
# output:
# Xtrain: standardized predictors matrix
# Ytrain: dependent variable in training sample
# Xtest: predictor in testing sample, standardized according to X_train
# Ytest: dependent variable in the testing sample
#
if(!is.matrix(da))da <- as.matrix(da)
k <- ncol(da)
nT <- nrow(da)
```

```
name <- colnames(da)
if(is.null(name))name <- paste("V",1:k,sep="")
nfore <- floor(forerate*nT)
nobe <- nT-lag
ist <- lag+1
if(nfore < nobe){
 ntrain <- nobe-nfore
 } else {nfore=0
       ntrain=nobe
       cat("Not enough data for testing","\n")
      }
Xtrain = Ytrain = Xtest = Ytest <- NULL
if(nfore > 0){
 Y <- da[ist:nT,locY]
 X <- NULL
 na <- NULL
 nb <- paste(name,"L",sep="")
 for (i in 1:lag){
 X <- cbind(X,da[(ist-i):(nT-i),])
 na <- c(na,paste(nb,1:k,sep=""))
 }
colnames(X) <- na
Xtrain <- X[1:ntrain,]
Ytrain <- Y[1:ntrain]
Xtest <- X[(ntrain+1):nobe,]
Ytest <- Y[(ntrain+1):nobe]
# standardization of predictors in training sample
#  Also, the same standardization of predictors in the test sample.
Xtrain <- scale(Xtrain,center=TRUE,scale=TRUE)
col_means_train <- attr(Xtrain,"scaled:center")
col_sds_train <- attr(Xtrain,"scaled:scale")
Xtest <- scale(Xtest,center=col_means_train,scale=col_sds_train)
}
list(Xtrain=Xtrain,Ytrain=Ytrain,Xtest=Xtest,Ytest=Ytest,nfore=nfore)
}
```

■

8.5 DEEP GENERATIVE MODELS

In the literature, various generate models have been proposed for DL. They include Boltzmann machine, restricted Boltzmann machine, deep belief network, and deep Boltzmann machine. Different models serve different purposes. For instance, Boltzmann machine is a type of stochastic recurrent NN and has been used in cognitive science. See Hinton (2007). Typically, the network is symmetrically connected and each neuron (state) is binary. An energy function is associated with the network and each connection between visible and hidden nodes is assigned a probability. The probability is determined by the weights and biases of the network in the training sample. Thus, the energy function is the objective function in statistical term. We shall not discuss these generate models, as our focus is on forecasting dependent data. Interested readers are referred to Goodfellow et al. (2016).

8.6 REINFORCEMENT LEARNING

Reinforcement learning (RL) is another area of machine learning concerned with how the program (or machine or agent) should take actions (or decisions) in an environment in order to maximize some well-defined cumulative reward. It is one of the three basic machine learnings, alongside supervised learning and unsupervised learning. A good reference for RL is Sutton and Barto (2018).

RL is a computational approach to understanding and automating goal-directed learning and decision-making. It distinguishes itself from other machine learning approaches by emphasizing on learning by an agent from direct interaction with its environment and by focusing on achieving long-term goals. It uses a Markov decision process to define the interaction between an agent and its environment using the concepts of states, actions, and rewards. It is an extension of the multi-armed bandit problem in the statistical literature. As an example, consider a Go player. She makes a move by (a) planning and anticipating possible replies and counter-replies and (b) the desirability of particular positions and moves available to her at that particular instance (environment). The objective is to win the game and she may explore the environment by taking some suboptimal moves with certain small probability in the process, because her moves would lead to changes in the environment and, hence, the choices of her subsequence moves.

One can think of RL as a decision maker (or agent) who interacts with an environment over a sequence of observations and chooses actions to maximize the reward over time. Thus, an RL model consists of a finite set of environment states (denoted by $\boldsymbol{S}$), a finite set of agent actions (denoted by $\boldsymbol{A}$), and a set of scalar rewards (denoted by R). At time (or step) t, the agent observes some representation of the environment, say $s_t \in \boldsymbol{S}$. The agent then selects an action $a_t \in \boldsymbol{A}(s_t)$, where $\boldsymbol{A}(s_t) \subset \boldsymbol{A}$ is the collection of available actions under state s_t. With the choice of action a_t, the agent receives a reward r_{t+1} and observes a new state s_{t+1}. From the description, the sequence of states $\{s_t\}$ is dynamically dependent.

The goal of RL is to take the sequence of actions to maximize the total reward

$$R = \sum_{t=0}^{\infty} \gamma^t r(s_t, a_t),$$

where γ is called a discounting factor, which assumes a value between 0 and 1. In the business and economics literature, discounting factor is well known and associated with the present values of the future rewards.

RL has found applications in many engineering fields, but less so in business and economics. A relevant application is in display advertising in Jin et al. (2018). For a good reference of RL algorithms, readers are referred to Szepesvari (2010). There exists a recent package **reinforcementlearning** in R for running RL. However, since RL is not yet an important area for the analysis of big dependent data in business and economics, we defer a detailed discussion on the topic for the future.

EXERCISES

1. Consider the US monthly macroeconomic data used in Example 8.4. Use the same training and testing subsamples, and apply a DL network below

```
mod <- keras_model_sequential()
mod %>%
 layer_dense(units=32,activation="relu",input_shape=c(k)) %>%
 layer_dropout(rate=0.4) %>%
```

```
layer_dense(units=16,activation="relu") %>%
layer_dropout(rate=0.25) %>%
layer_dense(units=16,activation="relu") %>%
layer_dropout(rate=0.25) %>%
layer_dense(units=4,activation="relu") %>%
layer_dense(units=1)
```

What are the root mean squared error, mean absolute error, and median absolute error in the testing subsample? Compare the results with those in Example 8.4.

2. Again, consider the US monthly macroeconomic data set used in Example 8.4, but use the unemployment rate (UNRATE) as the dependent variable. Apply a DL network to obtain forecasts in the testing subsample. Compute the root mean squared error, mean absolute error, and median absolute error of the forecasts. Compare the results with those of Example 8.3. Do the additional predictors helpful in predicting the unemployment rate?
3. Apply the random forest to the monthly macroeconomic data set as in Example 8.2, but using the unemployment rate as the dependent variable. You may try different choices of `mtry` and `ntree`.
4. Consider the US monthly unemployment rate series from 1948 to 2020 in the file `unrate4820.txt`. Use lag-1 to lag-24 of the series as predictors to build a tree model for the US unemployment rate.
5. Consider the IMDB data set used in Section 8.4.4. Use the first 4500 observations as the training subsample and the last 1236 observations as the testing sample. Repeat the analyses of Section 8.4.4. Is the LSTM network helpful in the movie rating?

REFERENCES

Alonso, A. M., Peña, D., and Romo, J. (2004). Introducing model uncertainty in time series bootstrap. *Statistica Sinica*, **14**: 155–174.

Belsley, D. A., Kuh, E., and Welsch, R. E. (2005). *Regression Diagnostics: Identifying Influential Data and Sources of Collinearity*. Wiley, New York.

Breiman, L. (2001). Random forests. *Machine Learning*, **45**: 5–32.

Breiman, L., Friedman, J. H., Olshen, R. A., and Stone, C. J. (1984). *Classification and Regression Trees*. Chapman and Hall/CRC, New York.

Chipman, H. A., George, E. I., and McCulloch, R. E. (2010). BART: Bayesian additive regression trees. *The Annals of Applied Statistics*, **4**: 266–298.

Chollet, F. and Allaire, J. J. (2020). *Deep Learning with R*. Manning Shelter Island, NY.

Cybenko, G. (1989). Approximation by superpositions of a sigmoidal function. *Mathematics of Control, Signals, and Systems*, **2**: 303–314.

Elman, J. L. (1990). Finding structure in time. *Cognitive Science: A Multidisciplinary Journal*, **14**: 179–211.

Goodfellow, I., Dengio, Y., and Courville, A. (2016). *Deep Learning*. MIT Press, Boston, MA.

Hammer, B. (2007). *Learning with Recurrent Neural Networks*. Lecture Notes in Control and Information Sciences 254. Springer, London, UK.

Hammer, B., Micheli, A., and Sperduti, A. (2005). Universal approximation capability of cascade correlation for structure. *Neural Computation*, **17**: 1109–1159.

Hanin, B. and Sellke, M. (2017). Approximating continuous functions by ReLU nets of minimal width. ArXiv:1710.11278.

Hastie, T., Tibshirani, R., and Friedman, J. (2001). *The Elements of Statistical Learning*. Springer, New York.

He, J., Yalov, S., and Hahn, P. R. (2018). XBART: Accelerated Bayesian additive regression trees. arXiv: 1810.02215.

Hinton, G. (2007). Boltzmann machine. *Scholarpedia*, **2**: 1668.

Hinton, G. (2012). *Neural Networks for Machine Learning*. Course, video lectures.

Ho, T. K. (1995). Random decision forests. In *Proceeding of the 3rd International Conference on Document Analysis and Recognition*. Montreal, QC, 14–16 August 1995. pp: 278–282.

Hochreiter, S. and Schmidhuber, J. (1997). Long short-term memory. *Neural Computation*, **9**: 1735–1780.

Hornik, K. (1991). Approximation capabilities of multilayer feedforward networks. *Neural Networks*, **4**: 251–257.

Hornik, K., Stinchcombe, M., and White, H. (1989). Multilayer feedforward networks are universal approximators. *Neural Networks*, **2**: 359–366.

Jin, J., Song, C., Li, H., Gai, K., Wang, J., and Zhang, W. (2018). Real-time bidding with multi-agent reinforcement learning in display advertising. ARxiv: 1802.09756.

Jordan, M. I. (1997). Serial order: A parallel distributed processing approach. *Advances in Psychology*, **121**: 471–495.

Kalchbrenner, N., Grefenstette, E., and Blunsom, P. (2014). A convolutional neural network for modeling sentences. ArXiv: 1404.2188v1.

Kingma, D. and Ba, J. (2014). Adam: A method for stochastic optimization. ArXiv: 1412.6980.

Kolmogorov, A. N. (1957). On the representation of continuous functions of many variables by superposition of continuous functions of one variable and addition. *Doklady Akademii. Nauk SSSR*, **114**: 953–956.

Lahiri, S. N. (2013). *Resampling Methods for Dependent Data*. Springer Science and Business Media, New York, NY.

Lawrence, S., Giles, C. L., Tsoi, A. C., and Back, A. D. (1997). Face recognition: A convolutional neural network approach. *IEEE Transactions on Neural Networks*, **8**: 98–113.

Lu, Z., Pu, H., Wang, F., Hu, Z., and Wang, L. (2017). The expressive power of neural networks: A view from the width. *In Advances in Neural Information Processing Systems*, **30**: 6231–6239. Curran Associates, Inc.

Menezes Jr, J. M. P. and Barreto, G. A. (2008). Long-term time series prediction with the NARX network: An empirical evaluation. *Neurocomputing*, **71**: 3335–3343.

Mikolov, T., Karafiat, M., Burget, L., Cernocky, J., and Khudanpur, S. (2010). Recurrent neural network based language model. In *Eleventh Annual Conference of the International Speech Communication Association*, pp. 1045–1048. Makuhari, Chiba, Japan.

Polyak, B. T. (1964). Some methods of speeding up the convergence of iteration methods. *USSR Computational Mathematics and Mathematical Physics*, **4**: 1–17.

Riedmiller, M. and Braun, H. (1993). A direct adaptive method for faster backpropagation learning: The RPROP algorithm. *IEEE International Conference on Neural Networks*, **1**: 586–591.

Schmidhuber, J. (2015). Deep learning in neural networks: An overview. *Neural Networks*, **61**: 85–117.

Siegelmann, H. T., Horne, B. G., and Giles, C. L. (1997). Computational capabilities of recurrent NARX neural networks. *IEEE Transactions on Systems, Man, and Cybernetics, Part B (Cybernetics)*, **27**: 208–215.

Sutton, R. S. and Barto, A. G. (2018). *Reinforcement Learning: An Introduction*, 2nd Edition. MIT Press. Cambridge, MA.

Szepesvari, C. (2010). *Algorithms for Reinforcement Learning*. Morgan & Claypool, San Rafael, CA.

Zaremba, W., Sutskever, I., and Vinyals, O. (2015). Recurrent neural network regularization. ArXiv: 1409.23329v5.

CHAPTER 9

SPATIO-TEMPORAL DEPENDENT DATA

Dependent data occur in many ways with the most common ones being observations taken over time, and these data have been the main focus of the previous chapters. Here we turn to another type of commonly observed dependent data, observations taken at different locations in space. The space can be geographical space, socioeconomic space, or network space. Consider, for instance, a given geographical region, such as the state of California or the Midwest of the United States. The daily average temperatures of the counties in the state of California on 1 October 2019, and the $PM_{2.5}$ measurements of all monitoring stations in the Midwest at 8:00 a.m. on 1 January 2020 are two examples of dependent data in space. These data are referred to as the *spatial data*. Let $\boldsymbol{s}_i$ denote the ith location in the space such as $\boldsymbol{s}_i = (\text{Latitude}_i, \text{Longitude}_i)$ and t_{ij} be the time of the jth data point taken at location $\boldsymbol{s}_i$. We can denote the data as $z(\boldsymbol{s}_i, t_{ij})$.

The simplest spatial data are those in which t_{ij} is the same for all locations $\boldsymbol{s}_i$ and they can be written as $\{z(\boldsymbol{s}_i)|i = 1, \ldots, m\}$ with m being the number of locations under study. Such data can be thought of as a snap shot of the variables of interest over the space. On the other hand, suppose that, at each location $\boldsymbol{s}_i$, we have more than one observation, then the time points when the observations were taken become important and we can denote the data collection as $\{z(\boldsymbol{s}_i, t_{ij})|i = 1, \ldots, m; j = 1, \ldots, n_i\}$, where n_i is a positive integer denoting the sample size of location $\boldsymbol{s}_i$. Such data are referred to as the *spatio-temporal (S-T) data*. If $t_{ij} = t_j$ for all i and the sequence $\{t_j\}$ are equally spaced time points, then the data $\{\boldsymbol{z}_{t_j} = [z(\boldsymbol{s}_1, t_j), z(\boldsymbol{s}_2, t_j), \ldots, z(\boldsymbol{s}_m, t_j)]'\}$ are similar to an m-dimensional vector time series discussed in Chapter 3. In general, for most S-T data, $\boldsymbol{s}_i$ can be irregularly located over the space and $\{t_{ij}\}$ can be irregularly spaced over time.

Statistical Learning for Big Dependent Data, First Edition. Daniel Peña and Ruey S. Tsay.
© 2021 John Wiley & Sons, Inc. Published 2021 by John Wiley & Sons, Inc.

Analysis of S-T data has a long history in the literature and there are several excellent books available on the topic. Interested readers are referred to Cressie (1993), Cressie and Wikle (2011), Diggle (2013), and Wikle et al. (2019). The S-T data $\{z(\boldsymbol{s}_i, t_{ij})\}$ are realizations of a S-T stochastic process $\{Y(\boldsymbol{s}, t)\}$ at location $\boldsymbol{s} = \boldsymbol{s}_i$ and $t = t_{ij}$. Typically, $\boldsymbol{s} \in R^d$ with $d = 2$ or $d = 3$. In this chapter, we briefly study properties of these processes and the analysis of S-T data. Our goal is to introduce readers to this important topic, to explore the features of spatial and S-T data, and to demonstrate their applications. To this end, we make good use of the recent book by Wikle et al. (2019) and its companying R package **STRbook**. We also take advantages of several useful R packages for spatial-temporal data analysis such as **sp**, **gstat**, **SpatPCA**, **autoFRK**, and **SpatioTemporal**.

Similarly to the multivariate or high-dimensional time series analysis, the objectives of studying S-T data include (a) spatial prediction, including smoothing and filtering, (b) temporal forecasting, and (c) understanding the relationships between various S-T processes. There are, however, some differences between S-T and multivariate time series analyses. In S-T analysis one often wants to make a prediction on a variable of interest at a location where no data are available. For instance, one might be interested in assessing the $PM_{2.5}$ measurement near an elementary school, which does not have any monitoring station nearby. On the other hand, the two analyses share the same basic concepts and statistical foundation.

9.1 EXAMPLES AND VISUALIZATION OF SPATIO TEMPORAL DATA

To begin, we consider some examples of S-T data, including two from Wikle et al. (2019) and available in the R package **STRbook** of Zammit-Mangion (2018a). These data sets will be used in the chapter to demonstrate S-T analysis.

1. NOAA daily weather data: The original data are from the US National Oceanic and Atmospheric Administration (NOAA) National Climatic Data Center and can be obtained from the IRI/LDEO Climate Data Library at Columbia University.[1] Following Wikle et al. (2019), we consider four variables: daily maximum temperature (Tmax), minimum temperature (Tmin), and dew point temperature (TDP), all in degrees of Fahrenheit (°F), and precipitation in inches at 138 weather stations in the central United States (between 32 °N–46 °N and 80 °W–100 °W), recorded between the years 1990 and 1993 (inclusive). We refer to this data set as the *NOAA data set*, which is directly available from the STRbook package.
2. Sea-surface temperature anomalies: These sea-surface temperature (SST) anomaly data are from the NOAA Climate Prediction Center and are also available from the IRI/LDEO Climate Data Library at Columbia University.[2] The data are gridded at 2° by 2° resolution from 124 °E–70 °W to 30 °S–30 °N, and they represent monthly anomalies from a January 1970 to December 2003 climatology (averaged over time). Again, this data set is from the STRbook package.

[1] http://iridl.ldeo.columbia.edu/SOURCES/.NOAA/.NCDC/.DAILY/.FSOD
[2] http://iridl.ldeo.columbia.edu/SOURCES/.CAC/

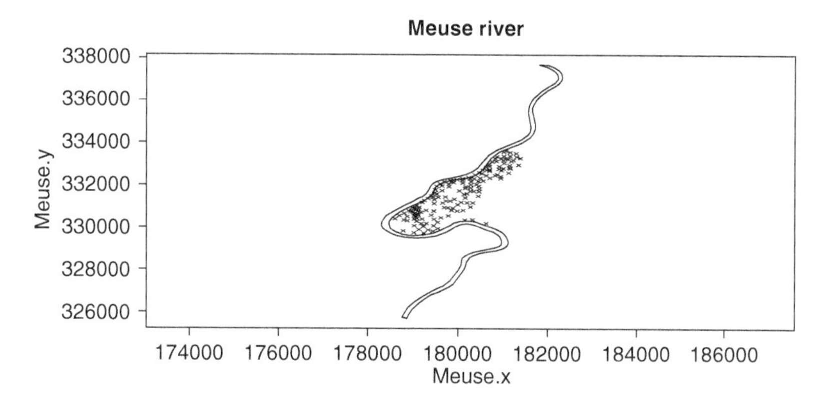

Figure 9.1 Meuse river and observation stations near the village Stein of the Netherlands.

3. Ozone data in Midwestern United States: The daily 8-hour average (from 9 a.m. to 4 p.m.) of surface ozone measurements in parts per billion (PPB) for 153 sites in the Midwestern United States from 3 June to 31 August 1987 for 89 days. The data set is available from the R package **fields**. See also Nychka et al. (1998).
4. Meuse river data set: This is a good example of spatial data. Topsoil heavy metal concentrations, along with a number of soil and landscape variables at the observation stations, are collected in a flood plain of the River Meuse, near the village of Stein, The Netherlands. Heavy metal concentrations are from composite samples of an area of approximately 15 m by 15 m. They include cadmium, copper, lead, and zinc. The coordinates of the observation locations are also given with x denoting easting (m) in Rijksdeiehoek (RDH, The Netherlands topographical) map coordinates and y northing (m) in RDH coordinates. The data set is available in several R packages such as **sp** and **gstat**. See Burrough and McDonnell (1998) for more details of the data. There are 155 observations for 14 variables. Figure 9.1 plots part of river Meuse and the locations of observation stations.
5. Hourly measurements of air pollutants: The data are hourly measurements of various air pollutants for 71 monitoring stations in Taiwan from 1 January 2006 to 31 December 2015. The data are available from the Environmental Protection Administration, Executive Yuan, Taiwan.[3] The air pollutants considered include $PM_{2.5}$, PM_{10}, NO_2, SO_2, and CO. We focus on the $PM_{2.5}$ measurements ($\mu g/m^3$) in this chapter. Figure 9.2 shows the snap shot of the monitoring stations at 9:00 a.m. on 26 September 2019. In the figure, different symbols and colors are used to indicate the air quality index at each monitoring station with green (circle), yellow (square), and red (triangle) indicating good, moderate, and poor air quality, respectively.

Figure 9.3 shows the locations of the Midwestern US sites used in the ozone data set (in circle) and the ozone measurements (PPB) on 18 June 1987 (color-coded). The state map is imposed to help readers understand better the measurements. From

[3]https://taqm.epa.gov.tw/taqm/en/YearlyDataDownload.aspx

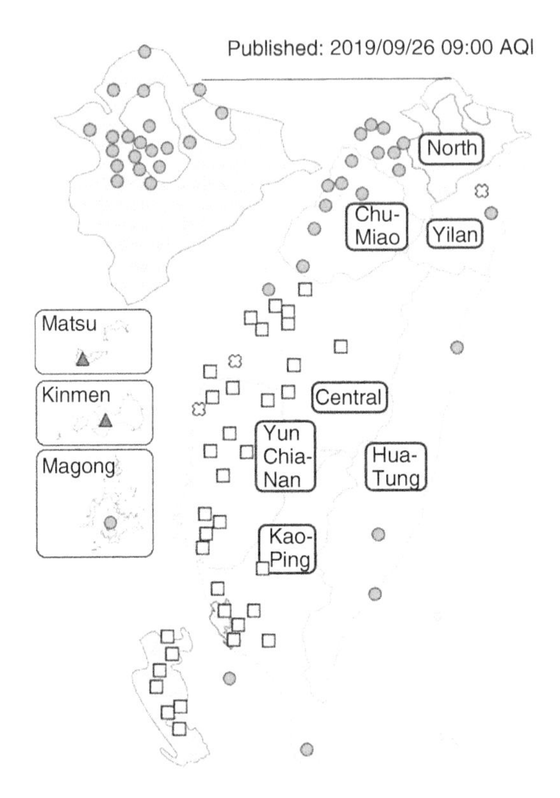

Figure 9.2 Air quality indexes at various monitoring stations in Taiwan at 9:00 a.m. on 26 September 2019. Green circle, yellow square, and red triangle indicate good, moderate, and poor air quality index, respectively.

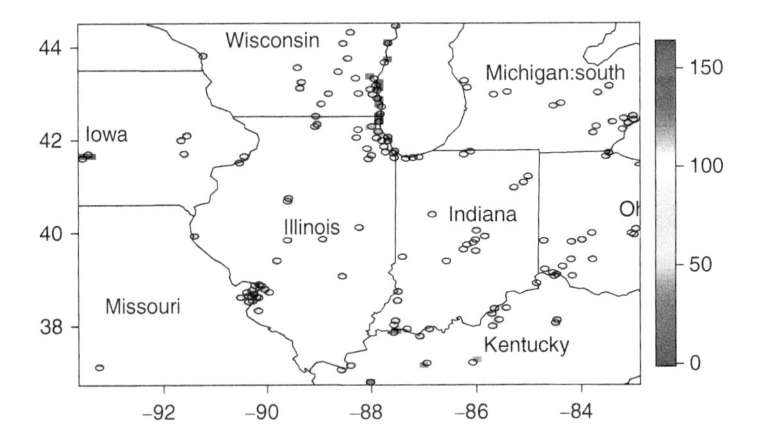

Figure 9.3 Daily 8-hour averages (9 a.m. to 4 p.m.) of ozone measurement in PPB for 153 sites in the Midwestern United States on 18 June 1987. The circles denote the sites and the ozone measurements are in color squares.

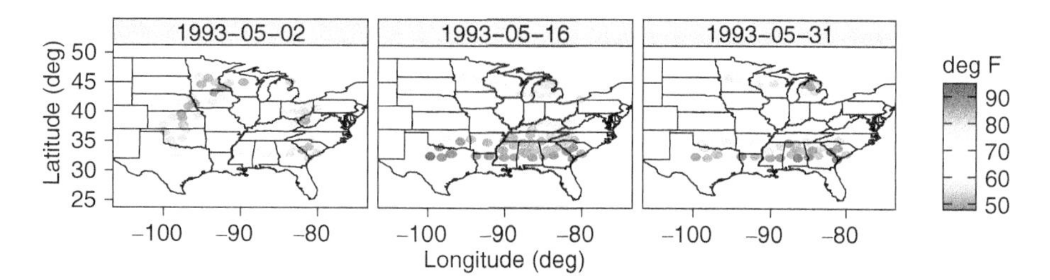

Figure 9.4 Daily maximum temperatures of 138 weather stations in the central United States. The temperatures are for 2 May (left), 16 May (center), and 31 May (right) 1993.

the plot, it is easily seen that high ozone measurements were concentrated near the cities along the west shoreline of Lake Michigan in northern Illinois and southern Wisconsin.

Figure 9.4 provides the spatial plots of daily maximum temperature for the 138 weather stations in the central United States for three selected days in May 1993. The plots show that, as expected, the daily maximum temperatures were higher in the south and cooler in the north. On the other hand, the pattern of daily maximum temperatures is less clear between the west and the east. The plots also indicate there existed marked temporal variations in the daily maximum temperature.

Figures 9.2–9.4 are examples of spatial plots. They represent a snap shot of the spatial information of interest. Figure 9.5 shows the time series plots of daily maximum temperatures of 10 randomly selected weather stations. These time series plots show the time evolution of daily maximum temperature (°F) from 1 May to 30 September 1993. The number in the gray heading of each plot denotes the station ID. Figure 9.6 shows the time plots of hourly $PM_{2.5}$ measurement (μg/m^3) of three randomly selected monitoring stations in the Kao-Ping area of Taiwan. See Figure 9.2. These plots clearly exhibit the seasonal patterns (or annual cycles) of the $PM_{2.5}$ measurements. They also demonstrate that missing values exist in empirical applications with some data missing in patches. Figures 9.5 and 9.6 are examples of time series plots which show the temporal dependence of the data and are extensively used in the previous chapters.

Figure 9.7 shows the Hovmöller plots for both the longitude (left) and latitude (right) coordinates for the daily maximum temperature (°F) of the NOAA data set from 1 May to 30 September 1993. In the plots, the data are interpolated so that the longitude or latitude has 25 equally spaced points whereas the time has 120 equally spaced points. A Hovmöller plot (Hovmöller, 1949) is a two-dimensional space-time visualization in which space is collapsed onto one dimension and where the second dimension denotes time. From the plots, we see that the temporal trend is rather constant with longitude (left panel), but it decreases considerably with increasing latitude (right panel). This is understandable because overall daily maximum temperature decreases with increasing latitude in the conterminous United States. See, also, the patterns in Figure 9.4.

Finally, R packages and commands used to create the S-T plots of this section are given in Appendix 9.A. See also the Lab work in Wikle et al. (2019).

The chapter is organized as follows. Section 9.2 introduces spatial processes and spatial data analysis. Section 9.3 focuses on geostatistical processes and their

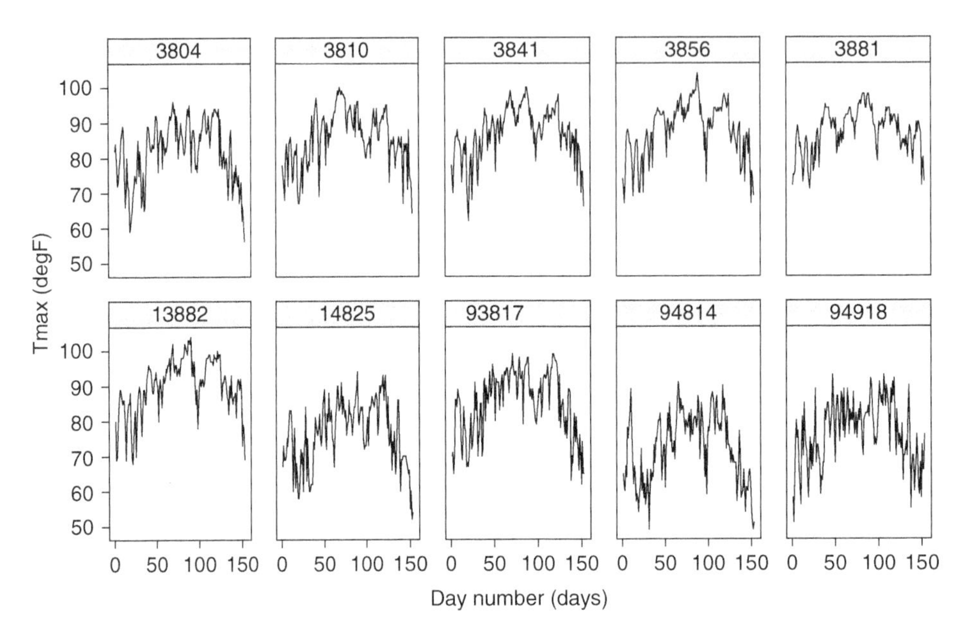

Figure 9.5 Daily maximum temperature (°F) of 10 randomly selected stations from the NOAA data set. The x-axis denotes the time sequence starting from 1 May to 30 September 1993. The number in the gray heading of each plot denotes the station ID.

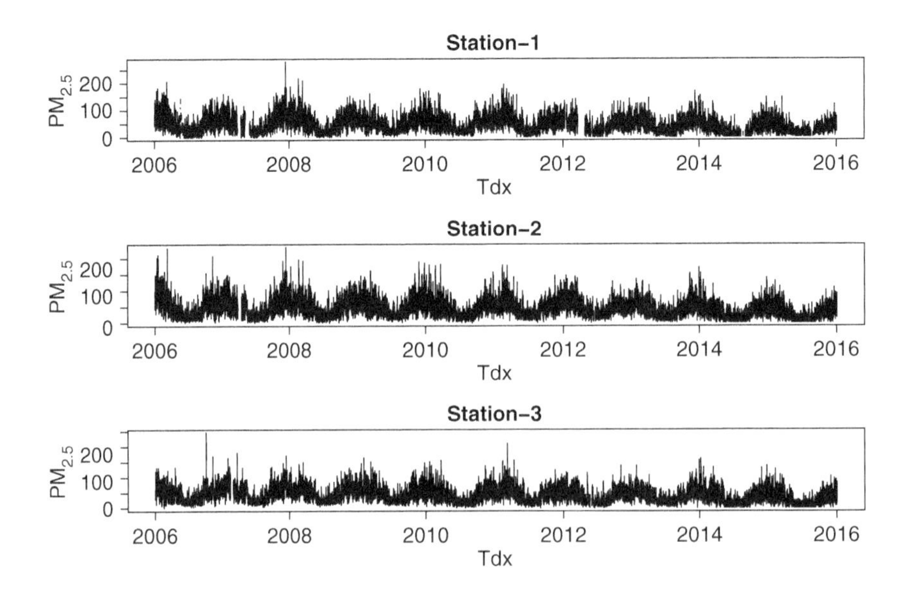

Figure 9.6 Hourly $PM_{2.5}$ measurements of three randomly selected stations in the Kao-Ping area of Taiwan from 1 January 2006 to 31 December 2015 for 87 648 observations.

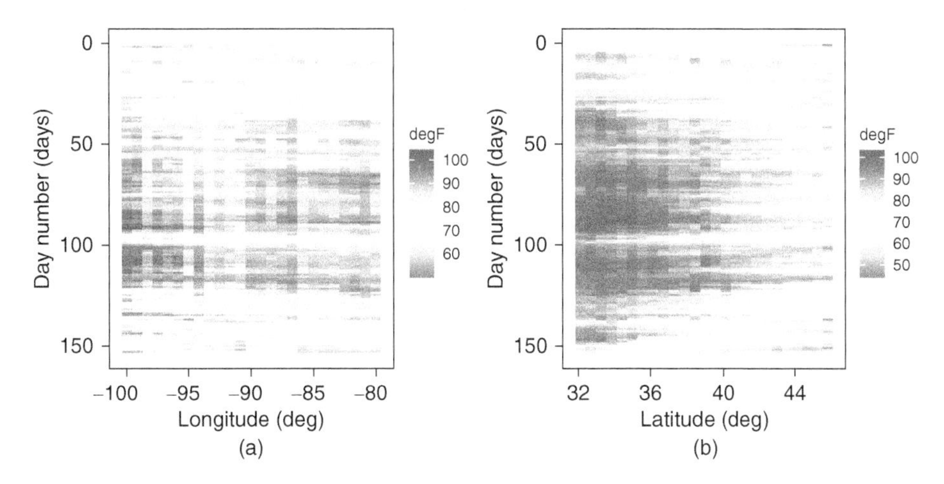

Figure 9.7 Hovmöller plots for both the longitude (left) and latitude (right) coordinates for the daily maximum temperature (°F) of the NOAA data set from 1 May to 30 September 1993. Source: NOAA (National Oceanic and Atmospheric Administration) data set.

properties. It also introduces variogram and three types of kriging. Section 9.4 discusses lattice processes whereas Section 9.5 introduces spatial point processes. Section 9.6 studies S-T processes and extends the spatial methods to S-T methods. Section 9.8 provides various descriptive models for S-T data, including use of S-T basis functions and spatial principal component analysis. Finally, Section 9.9 briefly discusses dynamic S-T models. Some R commands for plotting spatial or S-T data are given in Appendix 9.A.

9.2 SPATIAL PROCESSES AND DATA ANALYSIS

A useful statistical framework for spatial data analysis is the hierarchical model (HM). See Cressie and Wikle (2011). Under the HM framework, the data are observed from the underlying spatial process with noises. The spatial process then depends on some covariates with some unknown parameters and follows a well-specified distribution function. Statistical inference can then be made by Bayesian analysis or the conventional maximum likelihood approach. See below for more details.

Let $D \subset R^d$ be the region in R^d of interest and the spatial process is $\{Y(\boldsymbol{s})|\boldsymbol{s} \in D\}$. The region D is referred to as the domain of $Y(\boldsymbol{s})$. Assume that noisy observations are observed at known locations $\{\boldsymbol{s}_1, \ldots, \boldsymbol{s}_m\}$ and the data model is

$$Z(\boldsymbol{s}_i) = Y(\boldsymbol{s}_i) + \epsilon(\boldsymbol{s}_i), \quad i = 1, \ldots, m, \tag{9.1}$$

where $\epsilon(\boldsymbol{s})$ is a white noise process with $\boldsymbol{s} \in D$ and is independent of $Y(\boldsymbol{s})$. As usual, by white noise, we have $E[\epsilon(.)] = 0$ and $\text{Var}[\epsilon(.)] = \sigma_\epsilon^2 > 0$ and $\text{Cov}[\epsilon(\boldsymbol{s}_1), \epsilon(\boldsymbol{s}_2)] = 0$ if $\boldsymbol{s}_1 \neq \boldsymbol{s}_2$. The $\epsilon(\boldsymbol{s})$ process can be considered as a measurement error. Let $\boldsymbol{Z} = \{z(\boldsymbol{s}_i)|i = 1, \ldots, m\}$ be the data collection. In geostatistics, an important problem is to obtain an optimal spatial predictor of $Y(\boldsymbol{s}_0)$ at a known location $\boldsymbol{s}_0$ based on the squared error loss and data $\boldsymbol{Z}$. The optimal predictor is $E[Y(\boldsymbol{s}_0)|\boldsymbol{Z}]$. In some special situations, one might be interested in $E[Z(\boldsymbol{s}_0)|\boldsymbol{Z}]$. Note that $E[Y(\boldsymbol{s}_0)|\boldsymbol{Z}] = E[Z(\boldsymbol{s}_0)|\boldsymbol{Z}]$, but these two conditional expectations have different properties and they serve different purposes.

Before discussing data analysis, we start with some basic concepts of the spatial process $Y(\boldsymbol{s})$ with $\boldsymbol{s} \in D \subset R^d$, where D is assumed to have positive volume in R^d. The process $Y(\boldsymbol{s})$ is *second-order stationary* if

$$E[Y(\boldsymbol{s})] = \mu, \quad \text{for all } \boldsymbol{s} \in D, \tag{9.2}$$

$$\text{Cov}[Y(\boldsymbol{s}_1), Y(\boldsymbol{s}_2)] = C(\boldsymbol{s}_1 - \boldsymbol{s}_2), \quad \text{for all } \boldsymbol{s}_1, \boldsymbol{s}_2 \in D. \tag{9.3}$$

Eq. (9.2) states that the expectation of $Y(\boldsymbol{s})$ is invariant in D, and the function $C(.)$ of Eq. (9.3) is called a *covariogram* or a stationary covariance function. Furthermore, if $C(\boldsymbol{s}_1 - \boldsymbol{s}_2)$ is a function only of $\|\boldsymbol{s}_1 - \boldsymbol{s}_2\|$, then $C(.)$ is called *isotropic*. From the definition, readers immediately know that the concept of second-order (or weak) stationarity of a spatial process is similar to that of a time series. The only difference is that for spatial processes the location $\boldsymbol{s}$ has no well-defined direction if $d > 1$. The concept of *ergodicity* of $Y(\boldsymbol{s})$ is also defined in a similar way as that of a time series. Roughly speaking, for an ergodic process $Y(\boldsymbol{s})$, one can estimate $E[Y(\boldsymbol{s})]$ by spatial averages.

For a second-order stationary spatial process $Y(\boldsymbol{s})$, define

$$C_y(\boldsymbol{h}) = \text{Cov}[Y(\boldsymbol{s}), Y(\boldsymbol{s}+\boldsymbol{h})], \quad \text{for all } \boldsymbol{s}, \boldsymbol{s}+\boldsymbol{h} \in D. \tag{9.4}$$

For simplicity, we further assume that $Y(\boldsymbol{s})$ and $\epsilon(\boldsymbol{s})$ are independent Gaussian processes so that the first two moments are of interest. Define $\sigma_y^2 = C_y(\boldsymbol{0})$ and

$$\sigma_0^2 = \lim_{\boldsymbol{h}\to\boldsymbol{0}}[C_y(\boldsymbol{0}) - C_y(\boldsymbol{h})], \tag{9.5}$$

which is nonnegative. Note that σ_0^2 denotes the jump in the covariogram of $Y(\boldsymbol{s})$ process at $\boldsymbol{h} = \boldsymbol{0}$. Then, we have

$$\text{Cov}[Z(\boldsymbol{s}), Z(\boldsymbol{s}+\boldsymbol{h})] \equiv C_z(\boldsymbol{h}) = \begin{cases} \sigma_y^2 + \sigma_\epsilon^2, & \boldsymbol{h} = \boldsymbol{0} \\ C_y(\boldsymbol{h}), & \boldsymbol{h} \neq \boldsymbol{0}. \end{cases}$$

The quantity

$$\lim_{\boldsymbol{h}\to\boldsymbol{0}}[C_z(\boldsymbol{0}) - C_z(\boldsymbol{h})] = \sigma_0^2 + \sigma_\epsilon^2$$

is referred to the *nugget effect* in the geostatistics literature. See, for instance, Journel and Huijbregts (1978). The name *nugget* is used to signify the jump in $C_z(\boldsymbol{h})$ at $\boldsymbol{h} = \boldsymbol{0}$. From its definition, the nugget is the sum of the jump in covariogram of $Y(\boldsymbol{s})$ at $\boldsymbol{h} = \boldsymbol{0}$ and the variance of observed noise. In time series analysis, if there exist measurement errors, then an observed stationary time series becomes $z_t = y_t + \epsilon_t$, where y_t is the underlying time series and ϵ_t is the measurement error, which is assumed to be independent of y_t. In this case, it is easy to see that (a) $\text{Var}(z_t) = \gamma_z(0) = \gamma_y(0) + \sigma_\epsilon^2$, where σ_ϵ^2 is the variance of ϵ_t; and (b) $\gamma_z(\ell) = \gamma_y(\ell)$, for $\ell \neq 0$ is an integer. If t is continuous, then $\gamma_z(0) = \gamma_y(0) + \sigma_\epsilon^2$ and $\gamma_z(h) = \gamma_y(h)$ for $h > 0$. Therefore, nugget effect also occurs in time series analysis in the presence of measurement error. In the spatial statistics, the nugget effect further includes any intrinsic jump σ_0^2 of the covariance function of the spatial process $Y(\boldsymbol{s})$ at $\boldsymbol{h} = \boldsymbol{0}$.

With nugget effect, a simple geostatistical model on the domain $D \subset R^d$ can be written as a HM below:

- Data model: Conditional on σ_ϵ^2, for $i = 1, \ldots, m$,

$$Z(\boldsymbol{s}_i)|\,Y(\boldsymbol{s}_i), \sigma_\epsilon^2 \sim_{\text{ind}} N[Y(\boldsymbol{s}_i), \sigma_\epsilon^2].$$

- Process model: Conditional on μ and $C_y(.)$, $Y(.)$ is a stationary Gaussian process with mean μ and covariance function $C_y(\boldsymbol{h})$ with $\boldsymbol{h} \in D$.

Based on the nature of D, spatial processes $Y(\boldsymbol{s})$ are often classified into geostatistical processes, lattice processes, spatial point processes, and random-set processes. For instance, when D consists of two-dimensional grid points, $Y(\boldsymbol{s})$ is a lattice process, and when D consists of random points in R^2, $Y(\boldsymbol{s})$ is a point process. See Cressie and Wikle (2011, chapter 4) for further details. In what follows, we provide some further details concerning statistical inference of the first three types of spatial process. The random-set processes are less employed in applications.

9.3 GEOSTATISTICAL PROCESSES

In applications, some covariates are often available and we postulate that the spatial process $Y(\boldsymbol{s})$ follows the linear model

$$Y(\boldsymbol{s}) = \boldsymbol{x}(\boldsymbol{s})'\boldsymbol{\beta} + \eta(\boldsymbol{s}), \quad \boldsymbol{s} \in D, \tag{9.6}$$

where $\eta(\boldsymbol{s})$ is a mean-zero Gaussian process with stationary covariance function $C_y(.)$, $\boldsymbol{x}(.)$ is a k-dimensional vector of covariates, which are available at every location in D, and $\boldsymbol{\beta}$ is a k-dimensional vector of parameters. At the observation locations $\boldsymbol{s}_1, \ldots, \boldsymbol{s}_m$, we then have a linear mixed model

$$\boldsymbol{Y} = \boldsymbol{X}\boldsymbol{\beta} + \boldsymbol{\eta}, \tag{9.7}$$

where $\boldsymbol{Y} = [Y(\boldsymbol{s}_1), \ldots, Y(\boldsymbol{s}_m)]'$, the jth row of $\boldsymbol{X}$ is $\boldsymbol{x}(\boldsymbol{s}_j)'$, $\boldsymbol{\eta} = [\eta(\boldsymbol{s}_i), \ldots, \eta(\boldsymbol{s}_m)]'$, which denotes a spatial *random* effect and $\text{Var}(\boldsymbol{\eta}) = [C_y(\boldsymbol{s}_i - \boldsymbol{s}_j)]$ is an $m \times m$ positive-definite matrix. In Eq. (9.7), $\boldsymbol{\beta}$ is the *fixed* effect of the covariates and we assume that $k < m$. The data model becomes

$$Z(\boldsymbol{s}_i) = Y(\boldsymbol{s}_i) + \epsilon(\boldsymbol{s}_i), \quad i = 1, \ldots, m, \tag{9.8}$$

where $\{\epsilon(\boldsymbol{s}_i)\}$ are *iid* $N(0, \sigma_\epsilon^2)$. Then, for $\boldsymbol{s} \in D$, we have $\text{Var}[Z(\boldsymbol{s})] = \sigma_y^2 + \sigma_\epsilon^2$, which is referred to as the *sill* in the geostatistics literature. The nugget effect of the model is

$$c_0 = \sigma_0^2 + \sigma_\epsilon^2,$$

where $\sigma_0^2 = \lim_{\boldsymbol{h} \to \boldsymbol{0}}[C_y(\boldsymbol{0}) - C_y(\boldsymbol{h})]$. See Eq. (9.5). The quantity $\sigma_z^2 - c_0 = \sigma_y^2 - \sigma_0^2$ is called the *partial sill*.

9.3.1 Stationary Variogram

For a spatial process $Y(\boldsymbol{s})$ in $D \subset R^d$, assume that the mean is constant and differences of variables that are $\boldsymbol{h}$-apart vary in a way that depends only on $\boldsymbol{h}$. Then, the function

$$2\gamma_y(\boldsymbol{h}) = \text{Var}[Y(\boldsymbol{s}+\boldsymbol{h}) - Y(\boldsymbol{s})], \quad \text{for all } \boldsymbol{s}, \boldsymbol{s}+\boldsymbol{h} \in D, \tag{9.9}$$

is called the stationary *variogram*. If $\text{Var}[Y(\boldsymbol{s})]$ exists, then we have

$$2\gamma_y(\boldsymbol{h}) = 2C_y(\boldsymbol{0})(1 - C_y(\boldsymbol{h})/C_y(\boldsymbol{0})) = 2C_y(\boldsymbol{0})[1 - \rho_y(\boldsymbol{h})], \tag{9.10}$$

where $\rho_y(\boldsymbol{h})$ serves as the autocorrelation of the spatial process if the variogram is isotropic, which is defined below.

The variogram must satisfy the conditional nonpositive-definiteness condition,

$$\sum_{i=1}^{\ell}\sum_{j=1}^{\ell} \alpha_i\bar{\alpha}_j 2\gamma_y(\boldsymbol{s}_i - \boldsymbol{s}_j) \le 0, \tag{9.11}$$

for any positive integer ℓ, any set of spatial locations $\{\boldsymbol{s}_1, \ldots, \boldsymbol{s}_\ell\}$, and any set of complex numbers $\{\alpha_1, \ldots, \alpha_\ell\}$ satisfying $\sum_{i=1}^{\ell} \alpha_i = 0$, where $\bar{\alpha}_i$ is the complex conjugate of α_i. For a second-order stationary process $Y(\boldsymbol{s})$, it is easy to show that the condition (9.11) holds. Basically, because $\sum_{i=1}^{\ell} \alpha_i = 0$, the left side of Eq. (9.11) becomes $-2\text{Var}[\sum_{i=1}^{\ell} \alpha_i Y(\boldsymbol{s}_i)]$. This condition is important, because it guarantees that all model-based variances are nonnegative.

If the variogram can be written as a function of $\|\boldsymbol{h}\| = (h_1^2 + \cdots + h_d^2)^{1/2}$, the variogram is said to be *isotropic*; otherwise it is said to be *anisotropic*. Sometimes, we write $\|\boldsymbol{h}\|$ as $\|\boldsymbol{h}\|_2$ to emphasize the ℓ_2 norm.

In the previous chapters, we use the autocovariance function to describe the temporal dependence of a second-order stationary time series. In a similar vein, one often uses variogram to capture the spatial dependence of a second-order stationary spatial process. However, one should note that variogram is more general than the covariance function. For instance, consider the scalar case. The variogram is well defined for a random walk process, for which the covariance function is not defined. In practice, variogram models that depend on only a few parameters $\boldsymbol{\theta}$ are often used to summarize the spatial dependence in the literature. Some examples are given next.

9.3.2 Examples of Semivariogram

The function $\gamma_y(\boldsymbol{h})$ of Eq. (9.9) is referred as the semivariogram of $Y(\boldsymbol{s})$, which can be written as $\gamma_y(\boldsymbol{h}) = C_y(\boldsymbol{0}) - C_y(\boldsymbol{h})$. If it depends on a finite-dimensional parameter $\boldsymbol{\theta}$, then we have

$$\gamma_y(\boldsymbol{h}, \boldsymbol{\theta}) = C_y(\boldsymbol{0}, \boldsymbol{\theta}) - C_y(\boldsymbol{h}, \boldsymbol{\theta}). \tag{9.12}$$

In this section, we consider some specific examples of semivariogram. Parametric models can be useful in spatial prediction such as kriging. The first example of semivariogram is the *Matérn semivariogram* with

$$C_y(\boldsymbol{h}, \boldsymbol{\theta}) = \sigma_0^2(\|\boldsymbol{h}\| = 0) + \sigma_1^2\{2^{\theta_2-1}\Gamma(\theta_2)\}^{-1}\{\|\boldsymbol{h}\|/\theta_1\}^{\theta_2} K_{\theta_2}(\|\boldsymbol{h}\|/\theta_1), \tag{9.13}$$

where $K_{\theta_2}(.)$ is a modified Bessel function of the second kind of order θ_2 (e.g. Abramowitz and Stegun, 1964, pp. 374–379), $\Gamma(.)$ is the gamma function, and $\boldsymbol{\theta} = (\theta_1, \theta_2, \sigma_0^2, \sigma_1^2)'$ with $\theta_1 > 0, \theta_2 > 0, \sigma_0^2 \ge 0, \sigma_1^2 \ge 0$. Figure 9.8(*a*) shows a plot

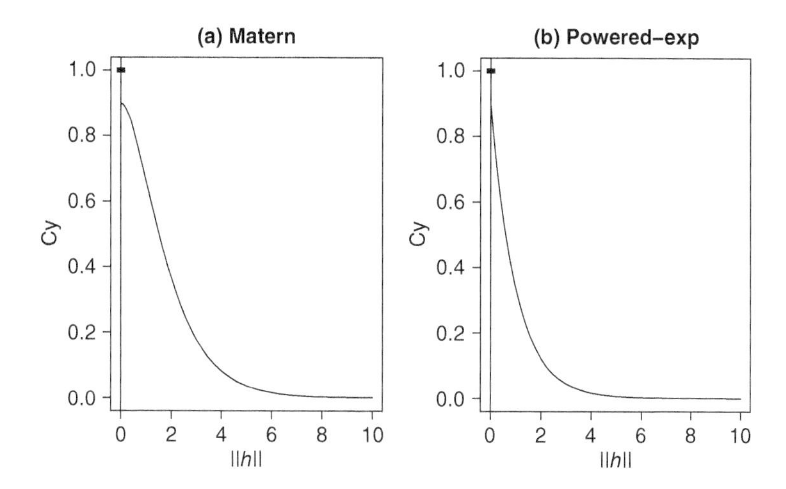

Figure 9.8 Scatterplots of $C_y(\boldsymbol{h}, \boldsymbol{\theta})$ for selected semivariograms. (*a*) Matérn semivariogram with $\boldsymbol{\theta} = (1, 1.5, 0.1, 0.9)'$. (*b*) Powered-exponential semivariogram with $\boldsymbol{\theta} = (1, 1, 0.1, 0.9)'$. Both plots show a 10% nugget effect.

of $C_y(\boldsymbol{h}, \boldsymbol{\theta})$ of Eq. (9.13) with $\boldsymbol{\theta} = (1, 1.5, 0.1, 0.9)'$ versus $||\boldsymbol{h}||$, indicating that the covariance function decays smoothly to zero as $||\boldsymbol{h}||$ increases. The plot also shows a 10% nugget effect. The values of $\boldsymbol{\theta}$ used are from Cressie and Wikle (2011). For the model in Eq. (9.13), σ_0^2 is the nugget, σ_1^2 is the variance, θ_1 is called the *scale* or *range* parameter, and θ_2 is the *shape* or *smoothness* parameter. This is an isotropic semivariogram.

Another example of commonly used isotropic semivariogram is the powered exponential given by

$$C_y(\boldsymbol{h}, \boldsymbol{\theta}) = \sigma_0^2 I(||\boldsymbol{h}|| = 0) + \sigma_1^2 \exp[-(||\boldsymbol{h}||/\theta_1)^{\theta_2}], \tag{9.14}$$

where $\boldsymbol{\theta} = (\theta_1, \theta_2, \sigma_0^2, \sigma_1^2)'$ with $\theta_1 > 0$, $0 < \theta_2 \leq 2$, $\sigma_0 \geq 0$, and $\sigma_1^2 \geq 0$. Figure 9.8(*b*) shows a plot of $C_y(\boldsymbol{h}, \boldsymbol{\theta})$ of Eq. (9.14) versus $||\boldsymbol{h}||$ with $\boldsymbol{\theta} = (1, 1, 0.1, 0.9)'$. From the plot, the covariance function decays quickly and smoothly to zero as $||\boldsymbol{h}||$ increases. A 10% nugget effect is also shown in the plot. The powered-exponential semivariogram is discussed in Diggle and Ribeiro (2007). In practice, $\theta_2 = 1$ is often used.

The following isotropic semivariogram models without nugget effect are also available in some R packages:

1. Linear model:
$$\gamma_y(\boldsymbol{h}, \boldsymbol{\theta}) = \begin{cases} 0, & \text{if } \boldsymbol{h} = \boldsymbol{0}, \\ b_0 + b_1||\boldsymbol{h}||, & \text{if } \boldsymbol{h} \neq \boldsymbol{0}, \end{cases}$$
where $\boldsymbol{\theta} = (b_0, b_1)'$ with $b_0 \geq 0$ and $b_1 \geq 0$.
2. Spherical model: valid for R^d with $1 \leq d \leq 3$
$$\gamma_y(\boldsymbol{h}, \boldsymbol{\theta}) = \begin{cases} 0, & \boldsymbol{h} = \boldsymbol{0}, \\ b_0 + b_1[1.5(||\boldsymbol{h}||/b_2) - 0.5(||\boldsymbol{h}||/b_2)^3], & 0 < ||\boldsymbol{h}|| \leq b_2, \\ b_0 + b_1, & ||\boldsymbol{h}|| \geq b_2, \end{cases}$$
where $\boldsymbol{\theta} = (b_0, b_1, b_2)'$ with $b_i \geq 0$ for i=1, 2, 3.

3. Exponential model:

$$\gamma_y(\boldsymbol{h}, \boldsymbol{\theta}) = \begin{cases} 0, & \boldsymbol{h} = \boldsymbol{0}, \\ b_0 + b_1[1 - \exp(-\|\boldsymbol{h}\|/b_2)], & \boldsymbol{h} \neq \boldsymbol{0}, \end{cases}$$

where $\boldsymbol{\theta} = (b_0, b_1, b_2)'$ with $b_i \geq 0$ for $i = 1, 2, 3$.

In some applications, anisotropic variograms are needed if the underlying physical process evolves differently in space, that is, different dynamic dependences of the process occur in different directions in space. For instance, the spatial dependence of air pollutants is likely affected by the geographical features surrounding the monitoring stations so that it can be different in different directions. In some cases, the anisotropy can be corrected by an invertible linear transformation of the lag vector $\boldsymbol{h}$. If $2\gamma_y(.)$ can be written as

$$2\gamma_y(\boldsymbol{h}) = 2\gamma_y^*(\|\boldsymbol{A}\boldsymbol{h}\|), \quad \boldsymbol{h} \in R^d, \tag{9.15}$$

where $\boldsymbol{A}$ is an invertible $d \times d$ matrix and $\gamma_y^*(.)$ has domain R, it is said to be *geometrically anisotropic*.

9.3.3 Stationary Covariance Function

Instead of using variogram in Eq. (9.9), if the spatial process $Y(\boldsymbol{s})$ satisfies the following two conditions:

$$C_y(\boldsymbol{h}) = \text{Cov}[Y(\boldsymbol{s} + \boldsymbol{h}), Y(\boldsymbol{s})], \quad \text{for all } \boldsymbol{s}, \boldsymbol{s} + \boldsymbol{h} \in D, \tag{9.16}$$

and

$$E[Y(\boldsymbol{s})] = \mu, \quad \text{for all } \boldsymbol{s} \in D, \tag{9.17}$$

then it is said to be a second-order stationary process in D with stationary covariance function $C_y(.)$. Similarly to time series analysis, one can also use the correlation function

$$\rho_y(.) = C_y(.)/C_y(\boldsymbol{0})$$

in studying a covariance stationary spatial process. The covariance function $C_y(.)$ must satisfy the nonnegative-definiteness condition

$$\sum_{i=1}^{\ell} \sum_{j=1}^{\ell} \alpha_i \bar{\alpha}_j C_y(\boldsymbol{s}_i - \boldsymbol{s}_j) \geq 0, \tag{9.18}$$

for any positive integer ℓ, any set of spatial locations $\{\boldsymbol{s}_1, \dots, \boldsymbol{s}_\ell\}$, and any set of complex numbers $\{\alpha_1, \dots, \alpha_\ell\}$. Note that the nonnegative-definiteness condition of Eq. (9.18) must hold for all complex numbers $\{\alpha_i\}$, not only those that sum to zero. Consequently, stationary covariance function is stronger than stationary variogram. If $C_y(\boldsymbol{h}) = C_y^*(\|\boldsymbol{h}\|)$, then it is said to be *isotropic*, otherwise it is *anisotropic*, where $C_y^*(.)$ has domain R. Note that the condition in Eq. (9.16), which implies that the process has a finite and constant variance, is not needed for a process to have a stationary variogram.

9.3.4 Estimation of Variogram

For a second-order stationary spatial process $Z(\boldsymbol{s}_i) = Y(\boldsymbol{s}_i) + \epsilon(\boldsymbol{s}_i)$ with data $\boldsymbol{Z}$, one often uses method-of-moments to estimate covariogram or variogram. Matheron (1962) proposed the estimator

$$2\hat{\gamma}(\boldsymbol{h}) = \frac{1}{|N(\boldsymbol{h})|} \sum_{N(\boldsymbol{h})} [Z(\boldsymbol{s}_i) - Z(\boldsymbol{s}_j)]^2, \quad \boldsymbol{h} \in R^d, \tag{9.19}$$

where

$$N(\boldsymbol{h}) = \{(\boldsymbol{s}_i, \boldsymbol{s}_j) | \boldsymbol{s}_i - \boldsymbol{s}_j = \boldsymbol{h}; i, j = 1, \ldots, m\},$$

and $|N(\boldsymbol{h})|$ is the number of distinct pairs in $N(\boldsymbol{h})$.

When data are irregularly spaced in R^d, the variogram estimator in Eq. (9.19) is often smoothed by using instead

$$2\gamma^+(\boldsymbol{h}) = \text{ave}\{[Z(\boldsymbol{s}_i) - Z(\boldsymbol{s}_j)]^2 | (i,j) \in N(\boldsymbol{h}), \boldsymbol{h} \in U(\boldsymbol{h}(\ell))\}, \tag{9.20}$$

where the region $U(\boldsymbol{h}(\ell))$ is some specified *tolerance* region in R^d around $\boldsymbol{h}(\ell)$ for $\ell = 1, \ldots, g$ with g being a positive integer. The choice of tolerance region is discussed in Journel and Huijbregts (1978). Following the time series literature, one can also employ the estimator

$$\hat{C}(\boldsymbol{h}) = \frac{1}{|N(\boldsymbol{h})|} \sum_{N(\boldsymbol{h})} [Z(\boldsymbol{s}_i) - \bar{Z}][Z(\boldsymbol{s}_j) - \bar{Z}], \tag{9.21}$$

where $\bar{Z} = \sum_{i=1}^m Z(\boldsymbol{s}_i)/m$ and $N(\boldsymbol{h})$ is defined as before in Eq. (9.19). For the case of $d = 1$, asymptotic properties of the estimators $\hat{C}(h)$ and $2\hat{\gamma}(h)$ have been studied in the time series literature.

9.3.5 Testing Spatial Dependence

Given the spatial data $\boldsymbol{Z} = [Z(\boldsymbol{s}_1), \ldots, Z(\boldsymbol{s}_m)]'$, one important question to ask is whether the data have spatial dependence. That is, we are interested in testing $H_0 : Y(\boldsymbol{s})$ has no spatial dependence versus $H_a : Y(\boldsymbol{s})$ is spatially dependent. Assume that the variogram is isotropic, i.e. $2\gamma_y(\boldsymbol{h}) = 2\gamma_y^*(\|\boldsymbol{h}\|)$, and that the mean of $Z(.)$ (and of $Y(.)$) is constant, Cressie and Wikle (2011) consider an estimate of the variogram

$$2\hat{\gamma}_z^*(h) = \text{ave}\{[Z(\boldsymbol{s}_i) - Z(\boldsymbol{s}_j)]^2 | \|\boldsymbol{s}_i - \boldsymbol{s}_j\| \in N(h); i, j = 1, \ldots, m\}, \tag{9.22}$$

where $N(h)$ is a tolerance region around h such as $h \pm \Delta$ for some small $\Delta > 0$. This estimate is referred to as the *empirical variogram*. It is similar to the smoothed version in Eq. (9.20). These authors propose the test statistic

$$Q = \frac{\hat{\gamma}_z^*(h_1)}{\hat{\sigma}_z^2}, \tag{9.23}$$

where $\hat{\sigma}_z^2 = \sum_{i=1}^m [Z(\boldsymbol{s}_i) - \hat{\mu}_z]^2/(m-1)$, $\hat{\mu}_z = \sum_{i=1}^m Z(\boldsymbol{s}_i)/m$, and h_1 is the smallest lag from all possible lags $h_1, \ldots, h_L$. One rejects H_0 if $|Q - 1|$ is large. A permutation-based test of H_0 can be carried out without making any distributional assumptions about $\boldsymbol{Z}$; the data locations are permuted and Q is then computed for all possible permutations. Empirical quantiles of the permuted Q values are then used to make inference, e.g. if the observed Q of Eq. (9.23) is greater than the 97.5 percentile or lower than the 2.5 percentile of the permuted Q values, then H_0 is rejected.

9.3.6 Kriging

Kriging is a spatial prediction method; see Matheron (1963). Following Cressie and Wikle (2011), we introduce kriging using a HM framework. The spatial process of interest is $\{Y(\boldsymbol{s})|\boldsymbol{s} \in D\}$ with $0 < \text{Var}[Y(\boldsymbol{s})] < \infty$ for all $\boldsymbol{s} \in D$. Note that $Y(\boldsymbol{s})$ may not be second-order stationary. The observed data $Z(\boldsymbol{s})$ are noisy and follow the model in Eq. (9.8). Assume, for now, that the first two moment equations of $Y(\boldsymbol{s})$ are known and given by

$$\mu_y(\boldsymbol{s}) = E[Y(\boldsymbol{s})], \quad \boldsymbol{s} \in D, \tag{9.24}$$

$$C_y(\boldsymbol{s}_1, \boldsymbol{s}_2) = \text{Cov}[Y(\boldsymbol{s}_1), Y(\boldsymbol{s}_2)], \quad \boldsymbol{s}_1, \boldsymbol{s}_2 \in D. \tag{9.25}$$

In addition, the variance of the measurement error σ_ϵ^2 is also known. In Eq. (9.25), we do not assume $C_y(\boldsymbol{s}_1, \boldsymbol{s}_2)$ is a function of $\boldsymbol{s}_1 - \boldsymbol{s}_2$. The variogram of $Y(\boldsymbol{s})$ is then given by

$$2\gamma_y(\boldsymbol{s}_1, \boldsymbol{s}_2) = \text{Var}[Y(\boldsymbol{s}_1) - Y(\boldsymbol{s}_2)] = C_y(\boldsymbol{s}_1, \boldsymbol{s}_1) + C_y(\boldsymbol{s}_2, \boldsymbol{s}_2) - 2C_y(\boldsymbol{s}_1, \boldsymbol{s}_2).$$

Kriging is to obtain an optimal spatial prediction of $Y(\boldsymbol{s}_0)$ conditional on $\boldsymbol{Z} = \{Z(\boldsymbol{s}_1), \dots, Z(\boldsymbol{s}_m)\}$, where $\boldsymbol{s}_0 \in D$ is the location of interest. Three types of kriging are considered in the literature, depending on the assumption imposed on the mean equation of $Y(\boldsymbol{s})$. The first kriging is simple kriging, in which $E[Y(\boldsymbol{s})]$ is known. The second type of kriging is ordinary kriging, in which $E[Y(\boldsymbol{s})] = \mu$ is an unknown constant, and the final type of kriging is universal kriging, in which $E[Y(\boldsymbol{s})]$ depends on some covariates with unknown parameters.

9.3.6.1 *Simple Kriging*

The method known as *simple kriging* is to search over all heterogeneously linear predictors, $\{\boldsymbol{a}'\boldsymbol{Z} + u|\boldsymbol{a} \in R^m, u \in R\}$, to obtain an optimal spatial predictor of $Y(\boldsymbol{s}_0)$. Like minimizing the mean squared prediction error (MSPE) in time series analysis, one seeks $\boldsymbol{a}$ and u by taking partial derivatives of

$$\text{MSPE}(\boldsymbol{a}, u) = E[Y(\boldsymbol{s}_0) - \boldsymbol{a}'\boldsymbol{Z} - u]^2, \tag{9.26}$$

with respect to $\boldsymbol{a}$ and u. Equation (9.26) can be written as

$$\text{MSPE}(\boldsymbol{a}, u) = \text{Var}[Y(\boldsymbol{s}_0) - \boldsymbol{a}'\boldsymbol{Z} - u] + \{E[Y(\boldsymbol{s}_0) - \boldsymbol{a}'\boldsymbol{Z} - u]\}^2.$$

The second term is $[\mu_y(\boldsymbol{s}_0) - \boldsymbol{a}'\boldsymbol{\mu}_y - u]^2$, where $\boldsymbol{\mu}_y = [\mu_y(\boldsymbol{s}_1), \dots, \mu_y(\boldsymbol{s}_m)]'$, which is zero if u satisfies

$$u = \mu_y(\boldsymbol{s}_0) - \boldsymbol{a}'\boldsymbol{\mu}_y. \tag{9.27}$$

With u given in Eq. (9.27), the first term of MSPE is

$$C_y(\boldsymbol{s}_0, \boldsymbol{s}_0) - 2\boldsymbol{a}' \text{Cov}[Y(\boldsymbol{s}_0), \boldsymbol{Z}] + \boldsymbol{a}'\boldsymbol{\Sigma}_z\boldsymbol{a}, \tag{9.28}$$

where the $m \times m$ covariance matrix $\boldsymbol{\Sigma}_z = [C_z(\boldsymbol{s}_i, \boldsymbol{s}_j)]$ is given by

$$C_z(\boldsymbol{u}, \boldsymbol{v}) = \begin{cases} C_y(\boldsymbol{v}, \boldsymbol{v}) + \sigma_\epsilon^2, & \text{if } \boldsymbol{u} = \boldsymbol{v}, \boldsymbol{v} \in D, \\ C_y(\boldsymbol{u}, \boldsymbol{v}), & \text{if } \boldsymbol{u} \neq \boldsymbol{v}, \end{cases}$$

and

$$\text{Cov}[Y(\boldsymbol{s}_0), \boldsymbol{Z}] = [C_y(\boldsymbol{s}_0, \boldsymbol{s}_1), \ldots, C_y(\boldsymbol{s}_0, \boldsymbol{s}_m)]' \equiv \boldsymbol{c}_y(\boldsymbol{s}_0). \tag{9.29}$$

Taking partial derivatives of Eq. (9.28) with respect to $\boldsymbol{a}$ and letting the result equal to $\boldsymbol{0}$, we obtain

$$-2\boldsymbol{c}_y(\boldsymbol{s}_0) + 2\boldsymbol{\Sigma}_z \boldsymbol{a} = \boldsymbol{0}.$$

Consequently, the optimal coefficient vector is $\boldsymbol{a}_* = \boldsymbol{\Sigma}_z^{-1}\boldsymbol{c}_y(\boldsymbol{s}_0)$ and, hence, $u_* = \mu_y(\boldsymbol{s}_0) - \boldsymbol{c}_y(\boldsymbol{s}_0)'\boldsymbol{\Sigma}_z^{-1}\boldsymbol{\mu}_y$. Based on the prior derivation, the simple-kriging predictor is

$$Y^*(\boldsymbol{s}_0) = \mu_y(\boldsymbol{s}_0) + \boldsymbol{c}_y(\boldsymbol{s}_0)'\boldsymbol{\Sigma}_z^{-1}(\boldsymbol{Z} - \boldsymbol{\mu}_y), \tag{9.30}$$

where $\boldsymbol{c}_y(\boldsymbol{s}_0)$ is given in Eq. (9.29) and $\boldsymbol{\mu}_y$ is defined in Eq. (9.27). Plugging $\boldsymbol{u}_*$ and $\boldsymbol{a}_*$ into Eq. (9.26), the minimized MSPE is

$$\sigma_{y,sk}^2(\boldsymbol{s}_0) = \text{MSPE}(\boldsymbol{a}_*, u_*) = C_y(\boldsymbol{s}_0, \boldsymbol{s}_0) - \boldsymbol{c}_y(\boldsymbol{s}_0)'\boldsymbol{\Sigma}_z^{-1}\boldsymbol{c}_y(\boldsymbol{s}_0), \tag{9.31}$$

where the subscript sk is used to denote simple kriging. The quantity $\sigma_{y,sk}^2(\boldsymbol{s}_0)$ of Eq. (9.31) is often referred to as simple-kriging variance.

When covariates are available and the model follows Eqs. (9.6) to (9.8), we have $\mu_y(\boldsymbol{s}) = \boldsymbol{x}(\boldsymbol{s})'\boldsymbol{\beta}$ and the simple-kriging predictor becomes

$$Y^*(\boldsymbol{s}_0) = \boldsymbol{x}(\boldsymbol{s}_0)'\boldsymbol{\beta} + \boldsymbol{c}_y(\boldsymbol{s}_0)'\boldsymbol{\Sigma}_z^{-1}(\boldsymbol{Z} - \boldsymbol{X}\boldsymbol{\beta}). \tag{9.32}$$

The simple-kriging variance of Eq. (9.31) remains unchanged.

Note that the simple-kriging solution and variance can also be obtained under the following HM:

- Data model: Conditional on σ_ϵ^2,

$$Z(\boldsymbol{s}_i)|Y(\boldsymbol{s}_i), \sigma_\epsilon^2 \sim_{\text{ind}} N[Y(\boldsymbol{s}_i), \sigma_\epsilon^2], \quad \boldsymbol{s}_i \in D, \quad i = 1, \ldots, m.$$

- Process model: Conditional on $\boldsymbol{\beta}$ and $C_y(\cdot, \cdot)$, $Y(\cdot)$ is a Gaussian process satisfying

$$E[Y(\boldsymbol{s})] = \boldsymbol{x}(\boldsymbol{s})'\boldsymbol{\beta}, \quad \text{Cov}[Y(\boldsymbol{s}_i, \boldsymbol{s}_j)] = C_y(\boldsymbol{s}_i, \boldsymbol{s}_j), \quad \boldsymbol{s}_1, \boldsymbol{s}_2 \in D.$$

From the data model, we have $\boldsymbol{Z}|\boldsymbol{Y} \sim N(\boldsymbol{Y}, \sigma_\epsilon^2\boldsymbol{I})$, where $\boldsymbol{Y}$ is given in Eq. (9.7) and $\boldsymbol{I}$ is the $m \times m$ identity matrix. From the process model, we have

$$\begin{bmatrix} Y(\boldsymbol{s}_0) \\ \boldsymbol{Y} \end{bmatrix} \sim N\left(\begin{bmatrix} \boldsymbol{x}(\boldsymbol{s}_0)' \\ \boldsymbol{X} \end{bmatrix}\boldsymbol{\beta}, \begin{bmatrix} C_y(\boldsymbol{s}_0, \boldsymbol{s}_0) & \boldsymbol{c}_y(\boldsymbol{s}_0)' \\ \boldsymbol{c}_y(\boldsymbol{s}_0) & \boldsymbol{\Sigma}_y \end{bmatrix}\right), \tag{9.33}$$

where $\Sigma_y = \text{Var}(Y)$. Therefore, by properties of multivariate normal distribution,

$$Y(s_0)|Y \sim N[x(s_0)'\beta + c_y(s_0)'\Sigma_y^{-1}(Y - X\beta), C_y(s_0, s_0) - c_y(s_0)'\Sigma_y^{-1}c_y(s_0)]. \quad (9.34)$$

Next, by Bayes' Theorem, $[Y|Z] \propto [Z|Y][Y]$, where $[A|C]$ denotes the conditional distribution of A given C, and $[Y]$ denotes the marginal distribution of Y. By the conjugate property of Gaussian distribution, we can obtain the posterior distribution $[Y|Z]$ as

$$Y|Z \sim N[D^{-1}(\sigma_\epsilon^{-2}Z + \Sigma_y^{-1}X\beta), D^{-1}], \quad (9.35)$$

where $D = \sigma_\epsilon^{-2}I + \Sigma_y^{-1}$ is the sum of two precision matrices. Using well-known Sherman-Morrison-Woodbury matrix identities, we can rewrite the conditional distribution as

$$Y|Z \sim N[X\beta + K(Z - X\beta), (I - K)\Sigma_y], \quad (9.36)$$

where $K = \Sigma_y\Sigma_z^{-1}$ with $\Sigma_z = \Sigma_y + \sigma_\epsilon^2 I$. Now, using iterated expectation and the fact that $[Y(s_0)|Y, Z] = [Y(s_0)|Y]$, we have

$$\begin{aligned} E[Y(s_0)|Z] &= E\{E[Y(s_0)|Y, Z]|Z\} \\ &= E[x(s_0)'\beta + c_y(s_0)'\Sigma_y^{-1}(Y - X\beta)|Z] \\ &= x(s_0)'\beta + c_y(s_0)'\Sigma_y^{-1}K(Z - X\beta) \quad \text{[Equation (9.36)]} \\ &= x(s_0)'\beta + c_y(s_0)'\Sigma_z^{-1}(Z - X\beta) \\ &= Y^*(s_0). \end{aligned} \quad (9.37)$$

For the posterior variance, we have

$$\begin{aligned} \text{Var}[Y(s_0)|Z] &= E\{\text{Var}[Y(s_0)|Y]|Z\} + \text{Var}\{E[Y(s_0)|Y]|Z\} \\ &= C_y(s_0, s_0) - c_y(s_0)'\Sigma_y^{-1}c_y(s_0) + \text{Var}[c_y(s_0)'\Sigma_y^{-1}Y|Z] \\ &= C_y(s_0, s_0) - c_y(s_0)'\Sigma_y^{-1}Kc_y(s_0) \quad \text{[Equation (9.36)]} \\ &= C_y(s_0, s_0) - c_y(s_0)'\Sigma_z^{-1}c_y(s_0) \\ &= \sigma_{y,sk}^2(s_0). \end{aligned} \quad (9.38)$$

This completes the derivation under the HM framework.

9.3.6.2 *Ordinary Kriging* The derivation of simple kriging assumes the mean and covariance function are known. Here we relax the assumption by allowing the mean function to be unknown, but impose certain unbiasedness, leading to the use of homogeneously linear combination of the data in spatial prediction. Ordinary kriging is referred to the case that the mean of $Y(s)$ is constant, but unknown.

The ordinary-kriging predictor is then a linear combination of the data $\boldsymbol{Z}$, say $\boldsymbol{b}'\boldsymbol{Z}$, that minimizes the MSPE

$$\text{MSPE}(\boldsymbol{b}) = E[Y(\boldsymbol{s}_0) - \boldsymbol{b}'\boldsymbol{Z}]^2, \tag{9.39}$$

subject to the unbiasedness condition $E(\boldsymbol{b}'\boldsymbol{Z}) = E[Y(\boldsymbol{s}_0)] = \mu$ for all $\mu \in R$. Since $E[Z(\boldsymbol{s})] = E[Y(\boldsymbol{s})] = \mu$, we have $E(\boldsymbol{Z}) = \mu\mathbf{1}$, where $\mathbf{1}$ is the m-dimensional vector of 1s. The unbiasedness condition then implies

$$\boldsymbol{b}'\mathbf{1} = 1, \tag{9.40}$$

that is, the sum of the ordinary-kriging coefficients is 1. The method of Lagrange multipliers is then used to minimize the objective function

$$\ell(\boldsymbol{b}) = \text{MSPE}(\boldsymbol{b}) - 2\lambda(\boldsymbol{b}'\mathbf{1} - 1).$$

More specifically, the objective function is

$$\ell(\boldsymbol{b}, \lambda) = C_y(\boldsymbol{s}_0, \boldsymbol{s}_0) - 2\boldsymbol{b}'\boldsymbol{c}_y(\boldsymbol{s}_0) + \boldsymbol{b}'\boldsymbol{\Sigma}_z\boldsymbol{b} - 2\lambda(\boldsymbol{b}'\mathbf{1} - 1), \tag{9.41}$$

where, again, $\boldsymbol{\Sigma}_z = \text{Var}(\boldsymbol{Z})$. Taking partial derivatives with respect to $\boldsymbol{b}$ and λ, and setting them to zero, we have

$$\begin{aligned} -\boldsymbol{c}_y(\boldsymbol{s}_0) + \boldsymbol{\Sigma}_z\boldsymbol{b} &= \lambda\mathbf{1}, \\ \boldsymbol{b}'\mathbf{1} &= 1. \end{aligned} \tag{9.42}$$

Solving the above equations, we obtain

$$\boldsymbol{b}_* = \boldsymbol{\Sigma}_z^{-1}[\boldsymbol{c}_y(\boldsymbol{s}_0) + k_*\mathbf{1}],$$

and $k_* = [1 - \mathbf{1}'\boldsymbol{\Sigma}_z^{-1}\boldsymbol{c}_y(\boldsymbol{s}_0)]/(\mathbf{1}'\boldsymbol{\Sigma}_z^{-1}\mathbf{1})$. Consequently, the ordinary-kriging predictor is

$$\widehat{Y}(\boldsymbol{s}_0) = \boldsymbol{b}_*'\boldsymbol{Z} = \{\boldsymbol{c}_y(\boldsymbol{s}_0) + \mathbf{1}(1 - \mathbf{1}'\boldsymbol{\Sigma}_z^{-1}\mathbf{1})/(\mathbf{1}'\boldsymbol{\Sigma}_z^{-1}\mathbf{1})\}'\boldsymbol{\Sigma}_z^{-1}\boldsymbol{Z}, \tag{9.43}$$

and the minimized MSPE is

$$\sigma_{y,ok}^2 = C_y(\boldsymbol{s}_0, \boldsymbol{s}_0) - \boldsymbol{c}_y(\boldsymbol{s}_0)'\boldsymbol{\Sigma}_z^{-1}\boldsymbol{c}_y(\boldsymbol{s}_0) + [1 - \mathbf{1}'\boldsymbol{\Sigma}_z^{-1}\boldsymbol{c}_y(\boldsymbol{s}_0)]^2/(\mathbf{1}'\boldsymbol{\Sigma}_z^{-1}\mathbf{1}), \tag{9.44}$$

where the subscript ok denotes ordinary kriging. $\sigma_{y,ok}^2$ is referred to as ordinary-kriging variance.

9.3.6.3 Universal Kriging Universal kriging is referred to the case that $E[Y(\boldsymbol{s})] = \boldsymbol{x}(\boldsymbol{s})'\boldsymbol{\beta}$ with unknown $\boldsymbol{\beta}$. Here $\boldsymbol{\beta}$ is estimated by the generalized least squares (gls) method and it can be shown that the universal-kriging predictor is

$$\widehat{Y}(\boldsymbol{s}_0) = \boldsymbol{x}(\boldsymbol{s}_0)'\widehat{\boldsymbol{\beta}}_{\text{gls}} + \boldsymbol{c}_y(\boldsymbol{s}_0)'\boldsymbol{\Sigma}_z^{-1}(\boldsymbol{Z} - \boldsymbol{X}\widehat{\boldsymbol{\beta}}_{\text{gls}}), \tag{9.45}$$

where $\widehat{\boldsymbol{\beta}}_{\text{gls}} = (\boldsymbol{X}'\boldsymbol{\Sigma}_z^{-1}\boldsymbol{X})^{-1}\boldsymbol{X}'\boldsymbol{\Sigma}_z^{-1}\boldsymbol{Z}$ is the generalized least squares estimator of $\boldsymbol{\beta}$. The universal-kriging variance is

$$\begin{aligned}\sigma^2_{y,uk}(\boldsymbol{s}_0) &= \boldsymbol{c}_y(\boldsymbol{s}_0)'\boldsymbol{\Sigma}_z^{-1}\boldsymbol{c}_y(\boldsymbol{s}_0)\\ &\quad + [\boldsymbol{x}(\boldsymbol{s}_0) - \boldsymbol{X}\boldsymbol{\Sigma}_z^{-1}\boldsymbol{c}_y(\boldsymbol{s}_0)]'(\boldsymbol{X}'\boldsymbol{\Sigma}_z^{-1}\boldsymbol{X})^{-1}[\boldsymbol{x}(\boldsymbol{s}_0) - \boldsymbol{X}\boldsymbol{\Sigma}_z^{-1}\boldsymbol{c}_y(\boldsymbol{s}_0)],\end{aligned} \tag{9.46}$$

where the subscript uk signifies universal kriging.

In real applications, one uses the universal kriging when some covariates are available. The ordinary kriging can be used in the absence of covariates. Indeed, most spatial packages use universal kriging.

9.4 LATTICE PROCESSES

Lattice processes $\{Y(\boldsymbol{s})|\boldsymbol{s} \in D\}$ are defined on a finite or countable subset $D \subset R^d$. They are used to study properties of and to model lattice data $\boldsymbol{Z} = \{Z(\boldsymbol{s}_1), \ldots, Z(\boldsymbol{s}_m)\}$, where $\{\boldsymbol{s}_1, \ldots, \boldsymbol{s}_m\}$ are data locations defined on discrete spatial features such as grid nodes or pixels. Often an $m \times m$ spatial neighborhood matrix $\boldsymbol{W} = [w_{ij}]$ is defined with zero diagonal elements and $w_{ij} = 1$ if location $\boldsymbol{s}_i$ is considered a neighbor of location $\boldsymbol{s}_j$. The word *neighbor* is defined by the data analyst based on her domain knowledge. For instance, one may define $\boldsymbol{s}_i$ and $\boldsymbol{s}_j$ as neighbors if $\|\boldsymbol{s}_i - \boldsymbol{s}_j\| \le r$ for some positive real number r.

Moran (1950) considered the simple case of real line with $s_i = i$ and $i = 1, \ldots, m$, and defined $w_{ij} = w_{ji} = 1$ if $0 < |s_i - s_j| \le 1$. For detrended data $\boldsymbol{Z}$ with $E(\boldsymbol{Z}) = \boldsymbol{0}$, he proposed the statistic

$$I = \frac{\boldsymbol{Z}'\boldsymbol{W}\boldsymbol{Z}}{\boldsymbol{Z}'\boldsymbol{Z}}, \tag{9.47}$$

for testing the null hypothesis $H_0 : Z(\boldsymbol{s}_1), \ldots, Z(\boldsymbol{s}_m)$ are *iid*. Since the data are mean-adjusted, the numerator and denominator of Eq. (9.47) are the neighboring covariance and variance of the process, respectively. Thus, the test belongs to a correlation test. Cliff and Ord (1981) proposed to use more general $\boldsymbol{W}$, where $\boldsymbol{W}$ may not be symmetric and its off-diagonal elements do not have to be 1, and coined the test statistic *Moran's I*. This statistic has been widely used in the literature for testing the null hypothesis of no spatial autocorrelation in $\boldsymbol{Z}$. Similarly to the F-statistic of empirical variogram in Eq. (9.23), permutation-based reference distribution can be used for the Moran's I test.

9.4.1 Markov-Type Models

A key difference between spatial and time-series processes is that there is no natural ordering in R^d (with $d > 1$) for the spatial processes. In the literature, various statistical models have been developed for spatial processes in R^d. Many of them make use of the Markov property with certain neighboring system. See, for instance,

Besag (1974). In this section, we briefly introduce some Markov-type models for spatial data.

The simplest Markov model for lattice processes is the first-order unilateral spatial process in R^1 with $D = \{1, 2, \ldots, n\}$, where n is the sample size. The model is

$$Y(s) = \phi Y(s-1) + \epsilon(s), \quad s = 2, \ldots, n, \tag{9.48}$$

where $|\phi| < 1$, $\{\epsilon(s)\} \sim_{iid} N(0, \sigma^2)$ and $\epsilon(s)$ is independent of $Y(s-1)$, which is similar to the autoregressive model of order 1, i.e. AR(1) model of Chapter 2. Since the sample size n is finite, one needs an initial condition for the spatial process to be a stationary Gaussian process. The initial condition is

$$Y(1) \sim N[0, \sigma^2/(1-\phi^2)]. \tag{9.49}$$

It is easy to see that $\text{cor}(Y(s), Y(s-1)) = \phi$ and, for $\boldsymbol{Y} = [Y(1), \ldots, Y(n)]'$, we have $E(\boldsymbol{Y}) = \boldsymbol{0}$ and

$$\text{Var}(\boldsymbol{Y}) = \frac{\sigma^2}{1-\phi^2} \begin{bmatrix} 1 & \phi & \cdots & \phi^{n-1} \\ \phi & 1 & \cdots & \phi^{n-2} \\ \vdots & \vdots & \ddots & \vdots \\ \phi^{n-1} & \phi^{n-2} & \cdots & 1 \end{bmatrix}_{n \times n}. \tag{9.50}$$

The joint distribution of $\boldsymbol{Y}$ is Gaussian and it is easy to see that

$$[\text{Var}(\boldsymbol{Y})]^{-1} = \frac{1}{\sigma^2} \begin{bmatrix} 1 & -\phi & 0 & 0 & \cdots & 0 & 0 & 0 \\ -\phi & 1+\phi^2 & -\phi & 0 & \cdots & 0 & 0 & 0 \\ 0 & -\phi & 1+\phi^2 & -\phi & \cdots & 0 & 0 & 0 \\ \vdots & \vdots & \vdots & \vdots & & \vdots & \vdots & \vdots \\ 0 & 0 & 0 & 0 & \cdots & -\phi & 1+\phi^2 & -\phi \\ 0 & 0 & 0 & 0 & \cdots & 0 & -\phi & 1 \end{bmatrix}.$$

In spatial statistics, this model belongs to the class of conditional autoregressive (CAR) models.

In general, suppose that $D = \{\boldsymbol{s}_1, \ldots, \boldsymbol{s}_n\} \subset R^d$. Let $\boldsymbol{Y} = [Y(\boldsymbol{s}_1), \ldots, Y(\boldsymbol{s}_n)]'$ and $\boldsymbol{Y}_{-i} = [Y(\boldsymbol{s}_1), \ldots, Y(\boldsymbol{s}_{i-1}), Y(\boldsymbol{s}_{i+1}), \ldots, Y(\boldsymbol{s}_n)]'$ for $i = 1, \ldots, n$. That is, $\boldsymbol{Y}_{-i}$ consists of all data in $\boldsymbol{Y}$ but $Y(\boldsymbol{s}_i)$. In addition, let $N(\boldsymbol{s}_i)$ be a set of prespecified locations that define the *neighborhood* of $\boldsymbol{s}_i$, $i = 1, \ldots, n$. Then, a CAR model on D can be specified as follows: the sequence of conditional distributions $\{[Y(\boldsymbol{s}_i)|\boldsymbol{Y}_{-i}]|i = 1, \ldots, n\}$ each of which follows a Gaussian distribution with mean and variance

$$E[Y(\boldsymbol{s}_i)|\boldsymbol{Y}_{-i}] = \sum_{\boldsymbol{s}_j \in N(\boldsymbol{s}_i)} c_{ij} Y(\boldsymbol{s}_j), \tag{9.51}$$

$$\text{Var}[Y(\boldsymbol{s}_i)|\boldsymbol{Y}_{-i}] = \tau_i > 0, \tag{9.52}$$

where the coefficients c_{ij} must satisfy certain conditions to guarantee the existence of the joint distribution of $\boldsymbol{Y}$. See, for instance, the condition below.

Since $\boldsymbol{s}_i \ni N(\boldsymbol{s}_i)$, we define $c_{ii} = 0$. Here we use the notation $\ni$ to denote *not in*. Furthermore, let $c_{ij} = 0$ if $\boldsymbol{s}_j \ni N(\boldsymbol{s}_i)$. Then, Eq. (9.51) can be written as

$$E[Y(\boldsymbol{s}_i)|\boldsymbol{Y}_{-i}] = \sum_{j=1}^{n} c_{ij} Y(\boldsymbol{s}_j), \quad i = 1, \ldots, n. \tag{9.53}$$

Let $\boldsymbol{C} = [c_{ij}]$ be the $n \times n$ coefficient matrix and $\boldsymbol{M} = \text{diag}\{\tau_1^2, \ldots, \tau_n^2\}$. Besag (1974) showed that if $\boldsymbol{M}^{-1}(\boldsymbol{I} - \boldsymbol{C})$ is positive-definite, then the joint distribution of the CAR model is

$$\boldsymbol{Y} \sim N[\boldsymbol{0}, (\boldsymbol{I} - \boldsymbol{C})^{-1}\boldsymbol{M}]. \tag{9.54}$$

An example of this type of Markov model is given below under the HM framework:

- Data model: For $i = 1, \ldots, n$,

$$Z(\boldsymbol{s}_i)|Y(\boldsymbol{s}_i) \sim_{\text{ind}} \text{Po}[\exp\{Y(\boldsymbol{s}_i)\}],$$

  where Po(u) denotes a Poisson distribution with mean u.
- Process model: Conditional on $\boldsymbol{\beta}, \tau^2, \phi$ and for $\boldsymbol{Y} = [Y(\boldsymbol{s}_1), \ldots, Y(\boldsymbol{s}_n)]'$,

$$\boldsymbol{Y}|\boldsymbol{\beta}, \tau^2, \phi \sim N[\boldsymbol{X}\boldsymbol{\beta}, \tau^2(\boldsymbol{I} - \phi\boldsymbol{H})^{-1}],$$

  where $\boldsymbol{X} = [x_j(\boldsymbol{s}_i)]$ is an $n \times k$ matrix obtained from the k-dimensional covariate $\boldsymbol{x}(\boldsymbol{s}_i)$, $\boldsymbol{H}$ is a known neighborhood matrix with zero diagonal elements and $h_{ij} = 1$ if $\boldsymbol{s}_j \in N(\boldsymbol{s}_i)$, and the parameter space of ϕ ensures that the matrix inverse exists.

The neighborhood $N(\boldsymbol{s}_i)$ of a lattice process $Y(\boldsymbol{s}_i)$ is often defined as the collection of all other locations $\boldsymbol{s}_j$ $(j \neq i)$ such that

$$[Y(\boldsymbol{s}_i)|\boldsymbol{Y}_{-i}] = [Y(\boldsymbol{s}_i)|\boldsymbol{Y}[N(\boldsymbol{s}_i)]], \quad i = 1, \ldots, n, \tag{9.55}$$

where $\boldsymbol{Y}[N(\boldsymbol{s}_i)] = [Y(\boldsymbol{s}_j) : \boldsymbol{s}_j \in N(\boldsymbol{s}_i)]$. If the conditional distributions of Eq. (9.55) define the joint distribution of $\boldsymbol{Y}$, then the process $\{Y(\boldsymbol{s})|\boldsymbol{s} \in D\}$ is called a *Markov random field* (MRF). If one further assumes that

$$Y(\boldsymbol{s}_i)|Y[N(\boldsymbol{s}_i)] \sim \text{Po}\left(\exp\left\{\alpha_i + \sum_{\boldsymbol{s}_j \in N(\boldsymbol{s}_i)} \theta_{ij} Y(\boldsymbol{s}_j)\right\}\right), \quad i = 1, \ldots, n,$$

where $\theta_{ij} = \theta_{ji}$, then the model is referred to as the *auto Poisson model*; see Besag (1974). In the literature, graphs are used to represent the conditional dependence of Eq. (9.55). Interested readers are referred to Cressie and Wikle (2011, chapter 4).

If the data are binary, Besag (1974) shows that, under certain conditions,

$$[Y(\boldsymbol{s}_i)|\boldsymbol{Y}[N(\boldsymbol{s}_i)]] = \frac{\exp\{\alpha_i Y(\boldsymbol{s}_i) + \sum_{\boldsymbol{s}_j \in N(\boldsymbol{s}_i)} \theta_{ij} Y(\boldsymbol{s}_i) Y(\boldsymbol{s}_j)\}}{1 + \exp\{\alpha_i Y(\boldsymbol{s}_i) + \sum_{\boldsymbol{s}_j \in N(\boldsymbol{s}_i)} \theta_{ij} Y(\boldsymbol{s}_i)\}}, \quad Y(\boldsymbol{s}_i) \in \{0, 1\}, \tag{9.56}$$

$\theta_{ij} = \theta_{ji}$. This is called an *auto logistic model*.

Finally, consider $D = \{(u, v)|u, v = 0, \pm 1, \pm 2, \dots\} \subset R^2$, that is, $\boldsymbol{s} = (u, v)$. The *Ising model* is defined by Eq. (9.56) under the assumption that (a) the parameter $\alpha(u, v) = \alpha$ (a constant) and (b) $N(u, v) = \{(u-1, v), (u+1, v), (u, v-1), (u, v+1)\}$, which is the *nearest-neighbor* structure. In this particular case, the numerator of Eq. (9.56) becomes

$$\exp[Y(u, v)\{\alpha + \theta_1[Y(u-1, v) + Y(u+1, v)] + \theta_2[Y(u, v-1) + Y(u, v+1)]\}],$$

where θ_1 and θ_2 are two parameters. See Ising (1925) with $\theta_1 = \theta_2$.

9.5 SPATIAL POINT PROCESSES

Another type of spatial processes is the spatial point processes, which are concerned with the random locations $\boldsymbol{s}_i$ in the domain $D \subset R^d$ of an event of interest. In addition, the number of such events is also random. For instance, consider a given area such as a forest reserve in the Cook county of Illinois. Let n be the number of ash trees in the forest reserve and $\boldsymbol{s}_i$ be the location of the ith tree. For any given area $A \subset D$ with d-dimensional area $|A|$. In our particular example, $|A|$ is the two-dimensional area of A in the forest reserve. Let $Z(A)$ be the number of ash trees in A. Then, $Z(.)$ is a stochastic process that is defined on the set of Lebesgue measurable subsets of D. Since D is bounded so that $Z(.)$ is finite for all $A \subset D$. Such a process $Z(.)$ is a counting point process.

A special, but important, example of point processes is the *Poisson point process* $Z(.)$, for which

$$Z(A)|\lambda(\boldsymbol{s}) \sim \text{Poi}(\lambda(\boldsymbol{s})|A|), \quad A \subset D, \tag{9.57}$$

where $\lambda(\boldsymbol{s})$ is an intensity function with units of per d-dimensional volume, which measures the potential for an event to appear at any location $\boldsymbol{s} \in D$. Let $\boldsymbol{ds}$ denote a small region located at $\boldsymbol{s}$ with volume $|\boldsymbol{ds}|$. Then, the first-order intensity function of the point process $Z(.)$ is defined as

$$\lambda(\boldsymbol{s}) = \lim_{|\boldsymbol{ds}| \to 0} \frac{E[Z(\boldsymbol{ds})]}{|\boldsymbol{ds}|}, \quad \boldsymbol{s} \in D, \tag{9.58}$$

provided that the limit exists. Hence,

$$E[Z(A)] = \int_A \lambda(\boldsymbol{s}) d\boldsymbol{s}, \quad A \subset D. \tag{9.59}$$

Statistical inference of spatial point process is concerned with the evolution of $\lambda(\boldsymbol{s})$ over $\boldsymbol{s} \in D$. If $\lambda(\boldsymbol{s}) = \lambda_0$, a constant, then the point process is homogeneous. In this case, λ can be estimated by some nonparametric kernel method. See, for instance, Diggle (1985). On the other hand, if $\lambda(\boldsymbol{s})$ is a function of some covariate $\boldsymbol{x}(\boldsymbol{s})$ that are available at $\boldsymbol{s}$, then we have an *inhomogeneous Poisson point process*.

Inference of spatial point processes such a point Poisson process can be done in a similar way as other spatial processes discussed before under HM.

- Data model: Conditional on $\lambda(.)$, $Z(.)$ is an inhomogeneous Poisson point process with intensity $\lambda(.)$.

- Process model: Conditional on $\boldsymbol{\beta}$ and $C_y(\cdot,\cdot)$, $Y(.) = \log(\lambda(.))$ is a Gaussian process such that

$$E[Y(\boldsymbol{s})] = \boldsymbol{x}(\boldsymbol{s})'\boldsymbol{\beta}, \quad C_y(\boldsymbol{s}_1, \boldsymbol{s}_2) = \text{Cov}[Y(\boldsymbol{s}_1), Y(\boldsymbol{s}_2)],$$

where, as before, $\boldsymbol{x}(\boldsymbol{s})$ denotes the vector of covariates at $\boldsymbol{s}$.

9.5.1 Second-Order Intensity

To model spatial interactions of spatial point processes, the second-order intensity function is often used. See Cressie (1993). For $\boldsymbol{u}, \boldsymbol{v} \in D$, the *second-order intensity* is defined as

$$\lambda_2(\boldsymbol{u}, \boldsymbol{v}) = \lim_{|d\boldsymbol{u}|\to 0, |d\boldsymbol{v}|\to 0} \frac{E[Z(d\boldsymbol{u})Z(d\boldsymbol{v})]}{|d\boldsymbol{u}||d\boldsymbol{v}|}, \quad \boldsymbol{u}, \boldsymbol{v} \in D, \tag{9.60}$$

provided the limit exists. When $\lambda(\boldsymbol{u}, \boldsymbol{v})$ exists, the *pair-correlation function* defined by

$$g(\boldsymbol{u}, \boldsymbol{v}) = \frac{\lambda_2(\boldsymbol{u}, \boldsymbol{v})}{\lambda(\boldsymbol{u})\lambda(\boldsymbol{v})}, \tag{9.61}$$

is often used to describe the spatial dependence of a spatial point process.

Suppose $Z(.)$ is stationary and isotropic, then $\lambda(\boldsymbol{s}) = \lambda_*$, a constant, and

$$\lambda_2(\boldsymbol{u}, \boldsymbol{v}) = \lambda_2^*(\|\boldsymbol{u} - \boldsymbol{v}\|), \quad \boldsymbol{u}, \boldsymbol{v} \in D.$$

In this particular case, the pair-correlation function becomes

$$g^*(h) = \frac{\lambda_2^*(h)}{\lambda_*^2}, \quad h > 0.$$

Stoyan and Stoyan (1994) demonstrate that small values of $g^*(h)$ occur if events at distance h-apart are rare and large values occur if events at distance h-apart appear frequently. For a homogeneous point Poisson process, $g^*(h) = 1$.

For stationary and isotropic spatial point processes, there is another measure of spatial dependence, which is called the K function. See Bartlett (1964) and Ripley (1976). Interested readers are referred to Cressie and Wikle (2011) for more discussions.

Example 9.1

We use the R package **gstat** in this section. The package can be used to perform variogram modeling, various kriging methods discussed in the chapter, and spatial plotting. The commands used include `variogram` (to compute sample variogram from the data or from the residuals of a fitted linear model), `fit.variogram` (to fit a parametric variogram to the sample variogram), and `krige` (to perform kriging). See the attached R commands for details, where we omit some output to save space. A good reference for the package is the R vignette (Pebesma, 2019). Consider the Meuse river data set in the package **sp**. See examples in Section 9.1. Figure 9.9 shows the bubble plot of the zinc concentrations. Comparing with the Meuse river plot in Figure 9.1, it is clear that the zinc concentrations are higher when the locations are closer to the river. Figure 9.10(a) shows the sample semivariogram of the log(zinc),

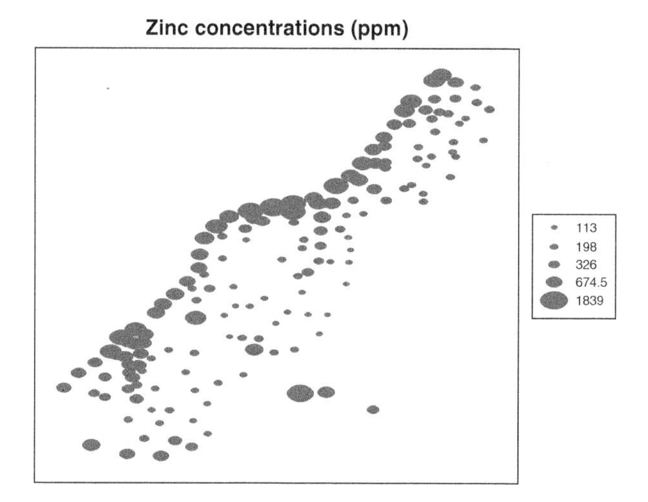

Figure 9.9 Bubble plot of the zinc concentrations of the Meuse data set.

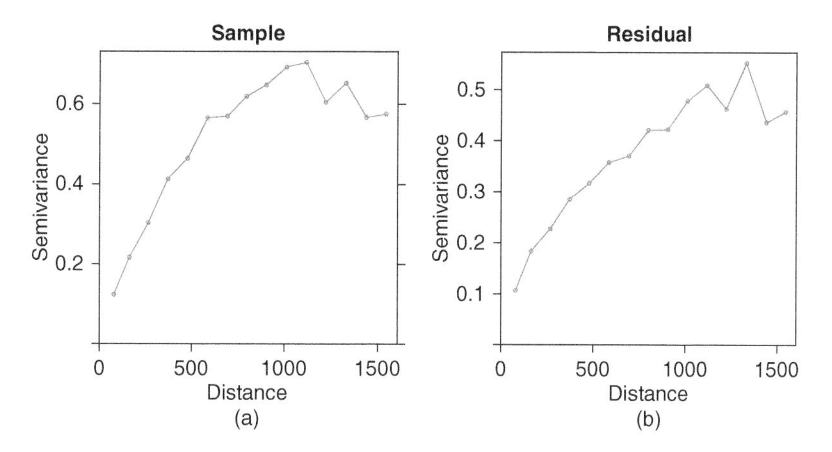

Figure 9.10 Semivariograms of log(zinc) of the Meuse data set. Part (a) is the sample semivariogram and part (b) is the residual semivariogram after fitting a linear trend.

whereas Figure 9.10(b) shows the semivariogram of the residuals after fitting a linear model with coordinates. The variances of log(zinc) and its linear-model residuals are 0.521 and 0.382, respectively, indicating that the linear model is effective in explaining the variability in zinc deposit. This is not surprising as the zinc concentrations are inversely related to the distance between the river and the location as clearly shown in the bubble plot of Figure 9.9. From the two semivariogram plots, we see that, as expected, the log(zinc) exhibits certain spatial dependence and the dependence seems to decay at a faster pace for the residuals.

If preferred, one can fit a parametric semivariogram model to the empirical semivariogram. Figure 9.11 shows the residual semivariogram of the residuals of log(zinc) with the solid line denoting the fitted curve using the Matèrn semivariogram model. The fit seems to be reasonable.

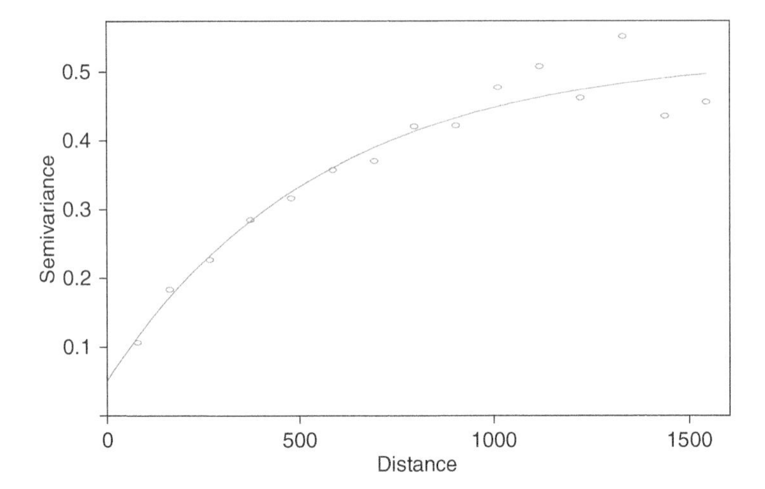

Figure 9.11 Semivariogram of linear-model residuals of the log(zinc) data of Meuse. The solid line denotes the fitted values of Matèrn semivariogram model.

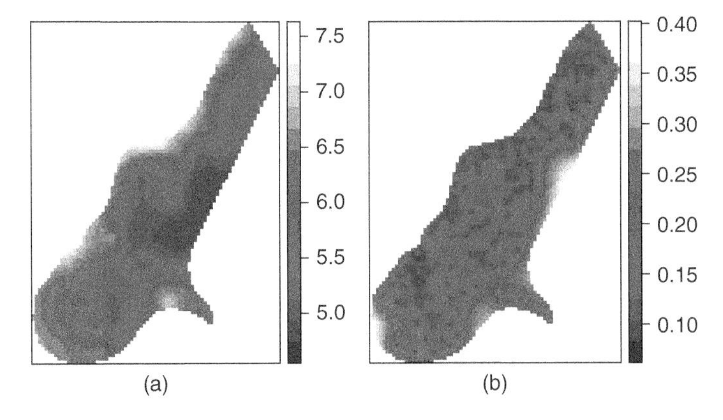

Figure 9.12 Results of universal kriging based on the fitted Matèrn model for the linear-model residuals of log(zinc) data. The new locations of interest are given in meuse.grid. Part (*a*) is the predictions of zinc concentration and part (*b*) gives the corresponding variance.

Turn to kriging. Figure 9.12(*a*) shows the result of universal kriging based on the residuals of fitted linear model. Here the locations of interest are given by the coordinates in `meuse.grid` and the model used is the fitted Matèrn model. The plot shows that the universal kriging does a good job in this particular instance, because it exhibits the general pattern of the observed zinc concentrations. Figure 9.12(*b*) shows the variance of the universal kriging. It is interesting to see that the variances are smaller when the observation locations are dense and the variance is higher when the locations are farer away from the river.

R packages and commands used: Spatial data analysis.

```
> require(gstat)
> require(sp)
```

```
> data(meuse)
> names(meuse)
> var(log(meuse$zinc))  ## 0.521
> coordinates(meuse)= ~x+y  # x- and y-coordinate are in meuse data set.
> bubble(meuse,"zinc",col=c("blue","blue"),main='Zinc concentrations(ppm)')
> v1 <- variogram(log(zinc)~1,meuse) # sample variogram
> plot(v1,type="b",main="(a) Sample")
> v2 <- variogram(log(zinc)~x+y,meuse) #variogram after fitting
##          a linear model with coordinates as explanatory variables.
> plot(v2,type="b",main="(b) Residual")
## Fit a spherical semivariogram
> v2.fit <- fit.variogram(v2,vgm(1,"Sph",700,1))
> v2.fit
  model      psill    range
1   Nug 0.08234213    0.000
2   Sph 0.38866509 1098.571
## psill is the partial sill and range is the range parameter
##   of the variogram model.
## fit a Matern semi-variogram
> v2.fit <- fit.variogram(v2,vgm(1,"Mat",900,1))
> v2.fit
  model      psill    range kappa
1   Nug 0.05029822   0.0000   0.0
2   Mat 0.47614480 555.3679   0.5
## psill is the partial sill (variance), range is the range parameter of
## the variogram model, and kappa is the smoothness parameter.
## See Equation (9.12).
> plot(v2,v2.fit)
### Universal kriging
> data(meuse.grid)
> coordinates(meuse.grid)=~x+y
> gridded(meuse.grid)=TRUE
> lz.krig <- krige(log(zinc)~x+y,meuse,meuse.grid,model=v2.fit)
[using universal kriging]
> spplot(lz.krig["var1.pred"])
> spplot(lz.krig["var1.var"])
```

■

9.6 S-T PROCESSES AND ANALYSIS

Turn to the S-T data. Consider the Sea Surface Temperature (SST) anomalies data available in the **STRbook** package. These monthly anomalies were collected from 124 °E to 290 °E and 29 °S to 29 °N on a 2 ° by 2 ° resolution so that there are 2520 (= 84×30) observations each month. Let $Z(\boldsymbol{s}, t)$ be the observed anomaly at location $\boldsymbol{s} = (\text{lon,lat})'$ and time t. Then $\boldsymbol{Z}_t = \{Z(\boldsymbol{s}, t)|\boldsymbol{s} \in D\}$ is a spatial data point discussed in the last section. The time period available is from January 1970 to December 2002 so that the data set we have is $\boldsymbol{Z} = \{Z(\boldsymbol{s}, t)|\boldsymbol{s} \in D; t = 1, \ldots, 396\}$, where $t = 1$ corresponds to January 1970. This is an example of S-T data. Due to missing values, we use data from January 1996 to December 2002. The data used are in the file `SSTdf1.csv`. Figure 9.13 shows the Hovmöller plot of a subset of the SST anomalies with Y-axis being calendar time from January 1996 to December 2002 and the X-axis being longitude from 124° to 280°. The plot shows the time revolution (from top to bottom) of the average anomalies (over latitude) along longitude. While the

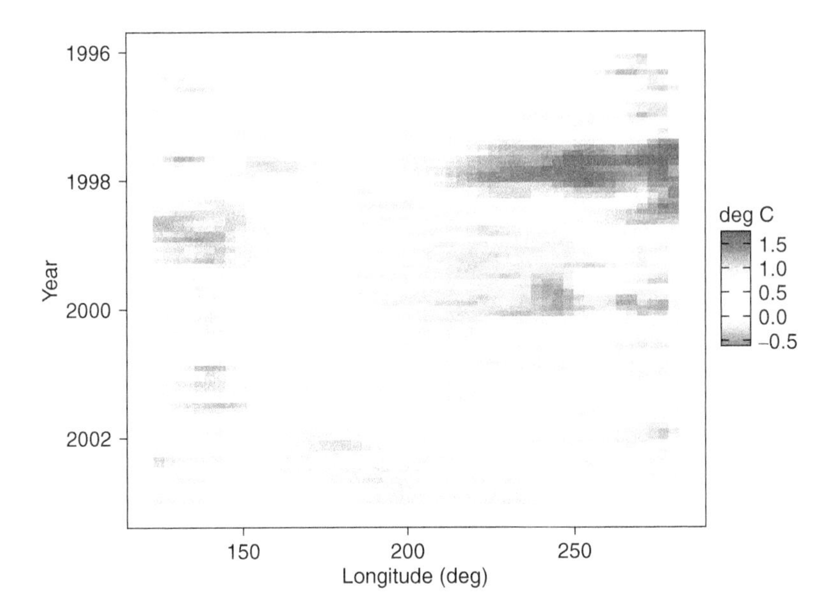

Figure 9.13 Hovmöller plot for the longitude of the sea surface temperature anomalies (°C) from January 1996 to December 2002.

plot provides certain information of the data, such as the anomalies seem to move from east to west for a given year, it fails to reveal many other features of the data. In fact, there are many ways to explore the data. First, they can be regarded as a 2520-dimensional time series with 84 observations (7 years). Second, they can also be regarded as a 84-by-30 matrix-variate time series with 84 observations. Our goal of this section is to study such S-T data.

9.6.1 Basic Properties

Consider the S-T process $\{Z(\boldsymbol{s},t)|\boldsymbol{s} \in D; t \in D_t\}$, where $D \subset R^d$ with $d > 0$ and $D_t \subset R$. In practice, we only observe data at a finite number of locations, say $\{\boldsymbol{s}_i | i = 1, \ldots, m\}$, and a finite number of time points, say $\{t_j | j = 1, \ldots, T\}$, where m and T are positive integers. For convenience, we write $\boldsymbol{Z}_{t_j} = [Z(\boldsymbol{s}_1, t_j), \ldots, Z(\boldsymbol{s}_m, t_j)]'$ and $\boldsymbol{Z}(\boldsymbol{s}_i) = [Z(\boldsymbol{s}_i, t_1), \ldots, Z(\boldsymbol{s}_i, t_T)]'$. That is, $\boldsymbol{Z}_{t_j}$ is the m-dimensional vector of observations at time t_j and $\boldsymbol{Z}(\boldsymbol{s}_i)$ is the T-dimensional vector of observations at location $\boldsymbol{s}_i$. We start with some basic concepts. Associated with the observed process $Z(\boldsymbol{s},t)$, there is an underlying S-T process $Y(\boldsymbol{s},t)$, i.e. $Z(\boldsymbol{s},t) = Y(\boldsymbol{s},t) + \epsilon(\boldsymbol{s},t)$ with $\epsilon(\boldsymbol{s},t)$ being *iid* measurement errors.

Let D be a domain of interest. For spatial processes, $D \subset R^d$. For time series, D is the set of integers if time is discrete and $D \subset R$ if time is continuous. For S-T processes, $D \subset R^d \times R$ or $D \subset R^d \times \{\ldots, -1, 0, 1, \ldots\}$. A function $\{g(\boldsymbol{u},\boldsymbol{v})|\boldsymbol{u},\boldsymbol{v} \in D\}$ defined on $D \times D$ is said to be *nonnegative-definite* if for any complex numbers $\{b_i | i = 1, \ldots, m\}$, and any locations $\{\boldsymbol{u}_i | i = 1, \ldots, m\}$ in D, and any positive integer m, we have

$$\sum_{i=1}^{m} \sum_{j=1}^{m} b_i \bar{b}_j g(\boldsymbol{u}_i, \boldsymbol{u}_j) \geq 0, \tag{9.62}$$

where $\bar{b}_i$ is the complex conjugate of b_i. If the inequality holds in Eq. (9.62) for any non-zero $\boldsymbol{b} = (b_1, \ldots, b_m)'$, then $g(\cdot, \cdot)$ is said to be *positive-definite*. We shall focus on S-T processes with covariance function (to be defined shortly) that is positive-definite.

A function $g(\boldsymbol{u}, \boldsymbol{v})$ is a *stationary S-T covariance function* on $R^d \times R$ if it satisfies the condition in Eq. (9.62) and can be written as

$$g((\boldsymbol{s}_1, t_1), (\boldsymbol{s}_2, t_2)) = C(\boldsymbol{s}_1 - \boldsymbol{s}_2, t_1 - t_2), \quad \boldsymbol{s}_i \in R^d, t_i \in R. \tag{9.63}$$

A S-T process $Y(\boldsymbol{s}, t)$ is said to be *second-order stationary* if $E[Y(\boldsymbol{s}, t)] = \mu$, a constant, and it has a stationary covariance function. More precisely, $Y(\boldsymbol{s}, t)$ is second-order stationary if $E[Y(\boldsymbol{s}, t)] = \mu$ and

$$\sum_{i=1}^{m} \sum_{j=1}^{m} b_i \bar{b}_j C_y(\boldsymbol{s}_i - \boldsymbol{s}_j, t_i - t_j) \geq 0, \tag{9.64}$$

for any $\boldsymbol{b} = (b_1, \ldots, b_m)'$, any $\{(\boldsymbol{s}_i, t_i) | i = 1, \ldots, m\}$ and any m, where $C_y(\cdot, \cdot)$ is the covariance function of $Y(\boldsymbol{s}, t)$.

The stationary S-T correlation function of $Y(\boldsymbol{s}, t)$ is defined as

$$\rho_y(\boldsymbol{h}, \tau) = \frac{C_y(\boldsymbol{h}, \tau)}{C_y(\boldsymbol{0}, 0)}, \quad \boldsymbol{h} \in R^d, \tau \in R. \tag{9.65}$$

The concept continues to apply for continuous-space and discrete-time models.

Some related concepts of covariance function are as follows:

- Spatial stationarity:

$$\text{Cov}[Y(\boldsymbol{s}_1, t_1), Y(\boldsymbol{s}_2, t_2)] = C_y(\boldsymbol{s}_1 - \boldsymbol{s}_2, t_1, t_2).$$

- Temporal stationarity:

$$\text{Cov}[Y(\boldsymbol{s}_1, t_1), Y(\boldsymbol{s}_2, t_2)] = C_y(\boldsymbol{s}_1, \boldsymbol{s}_2, t_1 - t_2).$$

- Spatial isotropy:

$$\text{Cov}[Y(\boldsymbol{s}_1, t_1), Y(\boldsymbol{s}_2, t_2)] = C_y(||\boldsymbol{s}_1 - \boldsymbol{s}_2||, t_1, t_2).$$

- Geometric spatial anisotropy:

$$\text{Cov}[Y(\boldsymbol{s}_1, t_1), Y(\boldsymbol{s}_2, t_2)] = C_y(\{h_1(s_{11} - s_{21})^2 + \cdots + h_d(s_{1d} - s_{2d})^2\}^{1/2}, t_1, t_2),$$

  where $\boldsymbol{s}_i = (s_{i1}, \ldots, s_{id})'$ and $h_i > 0$ for $i = 1, \ldots, d$.
- Geometric S-T anisotropy:

$$\text{Cov}[Y(\boldsymbol{s}_1, t_1), Y(\boldsymbol{s}_2, t_2)] = C_y(\{h_1(s_{11} - s_{21})^2 + \cdots + h_d(s_{1d} - s_{2d})^2 + h(t_1 - t_2)^2\}^{1/2}),$$

  where $\boldsymbol{s}_i = (s_{i1}, \ldots, s_{id})'$, $h_i > 0$ for $i = 1, \ldots, d$, and $h > 0$.

Another important concept is the separability of S-T correlations. A random process $Y(\boldsymbol{s}, t)$ is said to have a *separable S-T covariance function* if, for all $\boldsymbol{s}_i \in R^d$ and

$t_i \in R$ $(i = 1, 2)$, we have

$$\text{Cov}[Y(\boldsymbol{s}_1, t_1), Y(\boldsymbol{s}_2, t_2)] = C^{(s)}(\boldsymbol{s}_1, \boldsymbol{s}_2) \times C^{(t)}(t_1, t_2), \tag{9.66}$$

where $C^{(s)}$ and $C^{(t)}$ are spatial and temporal covariance functions, respectively. For stationary functions, Eq. (9.66) becomes

$$C_y(\boldsymbol{h}, \tau) = C^{(s)}(\boldsymbol{h}) \times C^{(t)}(\tau), \quad \boldsymbol{h} \in R^d, \tau \in R. \tag{9.67}$$

The separability also implies that the S-T correlation function can be written as

$$\rho(\boldsymbol{h}, \tau) = \rho^{(s)}(\boldsymbol{h}, 0) \times \rho^{(t)}(\boldsymbol{0}, \tau), \quad \boldsymbol{h} \in R^d, \tau \in R. \tag{9.68}$$

In practice, separability says that the spatial and temporal effects of the process $Y(\boldsymbol{s}, t)$ are orthogonal to each other, and they can be parameterized by separate functions. Conversely, if the S-T correlation is the product of a spatial correlation function and a temporal correlation function, then the process is separable. Thus, Eq. (9.68) is a characterization of separability in second-order stationary processes. Therefore, the equation provides a way to check the separability assumption by comparing $C_y(\boldsymbol{h}, \tau)$ with $C_y(\boldsymbol{h}, 0) \times C_y(\boldsymbol{0}, \tau)$.

Figure 9.14 shows a contour plot of a nonseparable S-T correlation given by Eq. (6.3) of Cressie and Wikle (2011). From the plot, it is seen that the correlations are nonlinear functions of spatial distance h and temporal lag τ. In the literature, many models have been proposed for the S-T processes. These models have various features in their S-T covariance functions. See, for instance, the regional-effect model of Reinsel et al. (1981) and the mixture model of Oehlert (1993), among others.

9.6.2 Some Nonseparable Covariance Functions

In what follows, we mention two specific families of S-T covariance functions that are nonseparable and flexible. The first family is proposed by Cressie and Huang (1999)

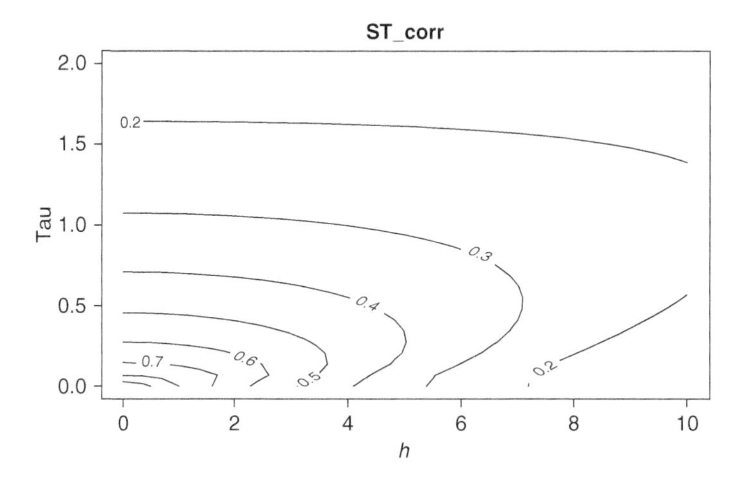

Figure 9.14 Contour plot of a nonseparable S-T correlation function given in eq. (6.3) of Cressie and Wikle (2011). The parameters are $\alpha = 1$ and $\beta = 20$. Source: Cressie and Wikle (2011). ©2011 John Wiley & Sons.

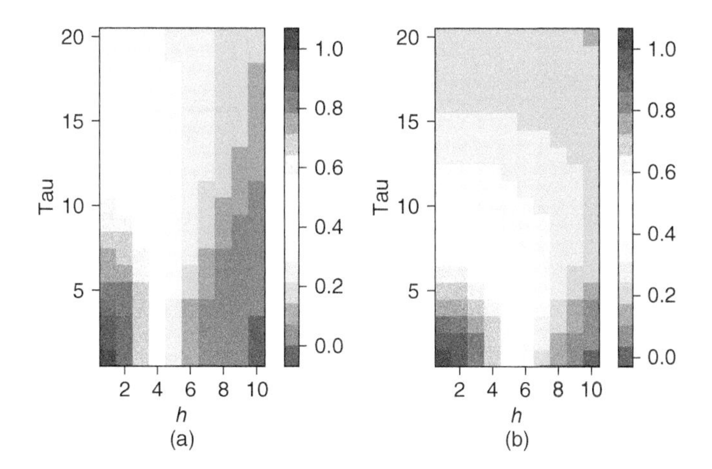

Figure 9.15 Levelplots of the nonseparable S-T covariance function of Eq. (9.69) with $d = 1$. Part (a) with $\sigma = 1, a = 1.5, b = 0.3$ and part (b) with $\sigma = 1, a = 2.3, b = 0.2$.

and given by

$$C_y(\boldsymbol{h}, \tau) = \frac{\sigma^2 \exp\{-b^2\|\boldsymbol{h}\|^2/(a^2\tau^2 + 1)\}}{(a^2\tau^2 + 1)^{d/2}}, \quad \boldsymbol{h} \in R^d, \tau \in R, \tag{9.69}$$

where $\sigma^2 = C_y(\boldsymbol{0}, 0)$ and $a \geq 0$ and $b \geq 0$ are scaling parameters for time and space, respectively. Figure 9.15 shows two examples of the nonseparable covariance function in Eq. (9.69), where $d = 1$, $\sigma = 1$, and $(a, b) = (1.5, 0.3)$ and $(2.5,0.2)$, respectively, for part (a) and (b). The effects of (a, b) on the covariance functions are clearly seen. This family of covariance function seems flexible.

The second family is proposed by Gneiting (2002) and given by

$$C_y(\boldsymbol{h}, \tau) = \frac{\sigma^2 \phi[\|\boldsymbol{h}\|^2/\psi(\tau^2)]}{\{\psi(\tau^2)\}^{d/2}}, \quad \boldsymbol{h} \in R^2, \tau \in R, \tag{9.70}$$

where $\{\phi(x)|x \geq 0\}$ is a completely monotone function and $\{\psi(x)|x \geq 0\}$ is a positive function with completely monotone derivative. For instance, assuming

$$\phi(x) = \exp(-px^{\upsilon}), \quad x \geq 0,\ p > 0,\ 0 < \upsilon \leq 1,$$

$$\psi(x) = (qx^{\alpha} + 1)^{\beta}, \quad x \geq 0,\ q > 0,\ 0 < \alpha, \beta \leq 1,$$

we have

$$C_y(\boldsymbol{h}, \tau) = \frac{\sigma^2 \exp\{-p\|\boldsymbol{h}\|^{2\upsilon}/(q|\tau|^{2\alpha} + 1)^{\beta\upsilon}\}}{(q|\tau|^{2\alpha} + 1)^{\beta d/2}}. \tag{9.71}$$

Separable covariance functions can be obtained by using the product of marginal models of nonseparable functions.

9.6.3 S-T Variogram

The *S-T variogram* of the process $Y(\boldsymbol{s}, t)$ is defined as

$$\text{Var}[Y(\boldsymbol{s}_1, t_1) - Y(\boldsymbol{s}_2, t_2)] = 2\gamma(\boldsymbol{s}_1, \boldsymbol{s}_2, t_1, t_2), \tag{9.72}$$

where $s_i \in D$ and $t_i \in R$. If $Y(s, t)$ is stationary, then Eq. (9.72) becomes $\text{Var}[Y(s_1, t1) - Y(s_2, t_2)] = 2\gamma(h, \tau)$, where $h = s_1 - s_2$ and $\tau = t_1 - t_2$. The function $\gamma(\cdot, \cdot)$ is called the *semivariogram*. The process $Y(s, t)$ is said to be *intrinsically stationary* if $E[Y(s, t)] = \mu$ and its variogram is stationary. If $Y(s, t)$ is second-order stationary with covariance function $C_y(h, \tau)$, then

$$\gamma(h, \tau) = C_y(\mathbf{0}, 0) - C_y(h, \tau), \quad h \in D, \tau \in R. \tag{9.73}$$

9.6.4 S-T Kriging

As in the spatial case, the goal of kriging is to predict $Y(s_o, t_0)$ from observed noisy data $\{Z(s_i, t_{ij})|i = 1, \ldots, m; j = 1, \ldots, T_i\}$, where

$$Z(s_i, t_{ij}) = Y(s_i, t_{ij}) + \epsilon(s_i, t_{ij}), \quad i = 1, \ldots, m; j = 1, \ldots, T_i, \tag{9.74}$$

where $T_i > 0$ and $\{\epsilon(s_i, t_{ij})\}$ is independent of $Y(s, t)$ and represents the measurement error that is assumed to be *iid* with mean zero and variance σ_ϵ^2. As before, s_0 is a new location of interest and t_0 is the time point of interest.

Let $Z_i = [Z(s_i, t_{i1}), \ldots, Z(s_i, t_{i,T_i})]'$ be the T_i-dimensional data vector at location s_i and $Z = (Z_1', \ldots, Z_m')'$ be the observed data vector. Kriging is amount to find the optimal linear combination of Z, say $Y^*(s_0, t_0)$, that minimizes the expected squared error $E[Y^*(s_0, t_0) - Y(s_0, t_0)]^2$. Similar to the spatial kriging discussed in the previous section, different types of S-T kriging can be derived under different assumptions concerning the mean $\mu(s, t) = E[Y(s, t)]$. For simple kriging, $\mu(s, t)$ is assumed to be known. For ordinary-kriging, $\mu(s, t) = \mu$, an unknown constant, and for universal-kriging, $\mu(s, t)$ is a linear function of some covariates with unknown coefficient vector β.

Assume that $Y(s, t)$ and $\epsilon(s, t)$ are both normal and write $\Sigma_z = \text{Var}(Z)$, $c_0 = \text{Cov}[Y(s_0, t_0), Z]$, and $C_{0,0} = \text{Var}[Y(s_0, t_0)]$. Also, write $Z = Y + \epsilon$ with $\text{Cov}(Y) = \Sigma_y$ and $\text{Cov}(\epsilon) = \Sigma_\epsilon$. We have

$$\begin{bmatrix} Y(s_0, t_0) \\ Z \end{bmatrix} \sim N\left(\begin{bmatrix} \mu(s_0, t_0) \\ \mu \end{bmatrix}, \begin{bmatrix} C_{0,0} & c_0' \\ c_0 & \Sigma_z \end{bmatrix} \right), \tag{9.75}$$

where $\mu = E(Z)$ and $\Sigma_z = \Sigma_y + \Sigma_\epsilon$. By Bayes' Theorem and properties of Gaussian distribution, we have

$$Y(s_0, t_0)|Z \sim N[\mu(s_0, t_0) + c_0'\Sigma_z^{-1}(Z - \mu), C_{0,0} - c_0'\Sigma_z^{-1}c_0]. \tag{9.76}$$

Since the simple-kriging predictor of $Y(s_0, t_0)$ assumes the form $Y^*(s_0, t_0) = a'Z + c$, where a and c are chosen to minimize the MSPE, $E[Y(s_0, t_0) - a'Z - c]^2$, we can use joint-Gaussian distribution in Eq. (9.75) to see that

$$Y^*(s_0, t_0) = E[Y(s_0, t_0)|Z] = \mu(s_0, t_0) + c_0'\Sigma_z^{-1}(Z - \mu). \tag{9.77}$$

Furthermore, the minimized MSPE is the posterior variance of Eq. (9.76), i.e.

$$\sigma_{sk}^2(s_0, t_0) = E[Y(s_0, t_0) - Y^*(s_0, t_0)]^2 = C_{0,0} - c_0'\Sigma_z^{-1}c_0. \tag{9.78}$$

Note that the joint Gaussian assumption is not critical for the above results of simple kriging to hold. In practice, calculating the S-T simple kriging could be time-consuming when the dimension of $\boldsymbol{Z}$ is large as it involves computing the inverse of $\boldsymbol{\Sigma}_z$. Some simplifications can be achieved with further assumptions. For instance, separability implies that $\boldsymbol{\Sigma}_z$ has block structure which can be used to simplify the computation. More specifically, under separability and assuming $T_i = T$ for all i,

$$\boldsymbol{\Sigma}_z = \boldsymbol{\Sigma}_z^{(s)} \otimes \boldsymbol{\Sigma}_z^{(t)}$$

where $\otimes$ is the Kronecker product of two matrices, $\boldsymbol{\Sigma}_z^{(s)}$ is an $m \times m$ purely spatial covariance matrix, and $\boldsymbol{\Sigma}_z^{(t)}$ is a $T \times T$ purely temporal covariance matrix. In this case, instead of inverting an $mT \times mT$ matrix, we only invert $\boldsymbol{\Sigma}_z^{(s)}$ and $\boldsymbol{\Sigma}_z^{(t)}$ because $\boldsymbol{\Sigma}_z^{-1} = [\boldsymbol{\Sigma}_z^{(s)}]^{-1} \otimes [\boldsymbol{\Sigma}_z^{(t)}]^{-1}$.

Turn to ordinary kriging for which $E[Y(\boldsymbol{s}, t)] = \mu$, a unknown constant. Let $\mathbf{1}$ be a T-dimensional vector of ones, where $T = \sum_{i=1}^{m} T_i$, which is the dimension of $\boldsymbol{Z}$. As the pure spatial case, the generalized least squares estimator of μ is

$$\hat{\mu}_{\text{gls}} = \frac{\mathbf{1}'\boldsymbol{\Sigma}_z^{-1}\boldsymbol{Z}}{\mathbf{1}'\boldsymbol{\Sigma}_z^{-1}\mathbf{1}}. \tag{9.79}$$

Therefore, the ordinary-kriging predictor is

$$\begin{aligned}\hat{Y}(\boldsymbol{s}_0, t_0) &= \hat{\mu}_{\text{gls}} + \boldsymbol{c}_0'\boldsymbol{\Sigma}_z^{-1}(\boldsymbol{Z} - \hat{\mu}_{\text{gls}}\mathbf{1}) \\ &= \{\boldsymbol{c}_0 + \mathbf{1}(1 - \mathbf{1}'\boldsymbol{\Sigma}_z^{-1}\boldsymbol{c}_0)/(\mathbf{1}'\boldsymbol{\Sigma}_z^{-1}\mathbf{1})\}'\boldsymbol{\Sigma}_z^{-1}\boldsymbol{Z},\end{aligned} \tag{9.80}$$

and the associated minimized MSPE is the ordinary-kriging variance

$$\sigma_{ok}^2(\boldsymbol{s}_0, t_0) = C_{0,0} - \boldsymbol{c}_0'\boldsymbol{\Sigma}_z^{-1}\boldsymbol{c}_0 + (1 - \mathbf{1}'\boldsymbol{\Sigma}_z^{-1}\boldsymbol{c}_0)^2/(\mathbf{1}'\boldsymbol{\Sigma}_z^{-1}\mathbf{1}). \tag{9.81}$$

The results of universal kriging also extends easily to the S-T setting.

Example 9.2

Consider the daily maximum temperature of the NOAA data set mentioned in Section 9.1. Here we only employ that data of August 1993, which are available from the package **STRbook**. The data contains 10, 168 observations and 12 variables. The variables include longitude, latitude, year, month, day, location ID, and maximum daily temperature, among others. There are 328 locations and each location has 31 observations. Our goal here is to use the August 1993 data to demonstrate S-T universal kriging using the semivariogram with the package **gstat**. Following Wikle et al. (2019), we also use the package **RColorBrewer** to color some of the surfaces in plotting. The empirical semivariogram of the August maximum temperature is given in Figure 9.16, which is produced by the command `variogram` with a fixed latitude effect (i.e. using latitude as an explanatory variable), spatial bin of 80 km, points less than 1000 km apart, and time lags from 0 to 6 days. Details of the R commands are given below.

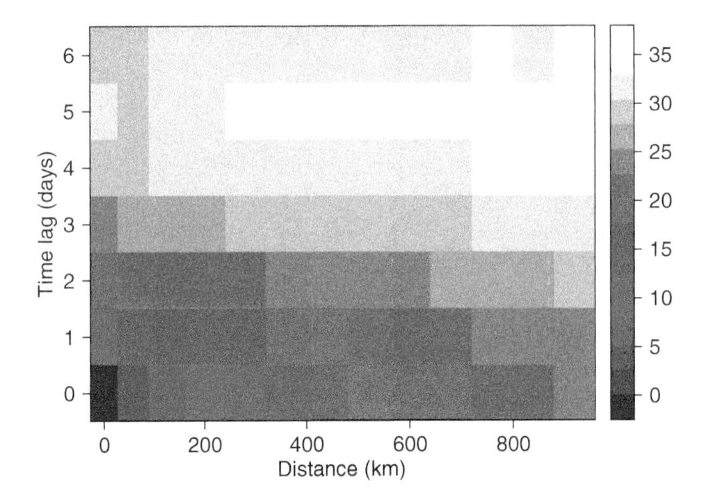

Figure 9.16 Empirical S-T semivariogram of daily maximum temperature from the NOAA data set during August 1993, computed using the command `variogram` in **gstat**. Source: NOAA (National Oceanic and Atmospheric Administration) data set.

For S-T kriging, we need to fit a parametric model to the empirical semivariogram. To this end, we employ a separable covariance function of the form

$$\gamma(\boldsymbol{h}, \tau) = s \times (\bar{\gamma}^{(s)}(\|\boldsymbol{h}\|) + \bar{\gamma}^{(t)}(|\tau|) - \bar{\gamma}^{(s)}(\|\boldsymbol{h}\|)\bar{\gamma}^{(t)}(|\tau|)),$$

where the standardized semivariograms $\bar{\gamma}^{(s)}$ and $\bar{\gamma}^{(t)}$ have separate nugget effects and $s = 1$ denoting the sill. As a second model, we also employ the covariance function

$$c_{\text{st}}(|\boldsymbol{v}_a\|) = b_1 \exp(-\phi\|\boldsymbol{v}_a\|) + b_2 I(\|\boldsymbol{v}_a\| = 0),$$

where $\boldsymbol{v}_a = (\boldsymbol{h}', a\tau)'$ with the parameter a denoting the scaling factor used for generating the space-time anisotropy. It turns out that, for this particular instance, the two models provide similar fits. The mean squared error of the separable model is 1.72, whereas that of the anisotropic model is 1.69. Figure 9.17 shows the fitted semivariograms of the two models.

Since we treat the latitude as a covariate, we use universal kriging in the following demonstration. First, we create a space-time grid for prediction. For the spatial grid, we consider 20 spatial locations between 100 °W and 80 °W and 20 spatial locations between 32 °N and 46 °N. For the temporal grid, we consider six equally spaced days in August 1993. The spatial grid and temporal grid are then used to construct an `STF` object in **gstat** for our space-time prediction grid. Finally, we omit the data on 9 August 1993 to show the capability of S-T kriging to predict across time.

Figure 9.18 shows the predictions and associated standard errors of universal kriging with the fitted separable model. From the plots, the high-standard errors of 9 August are understandable as the data were omitted.

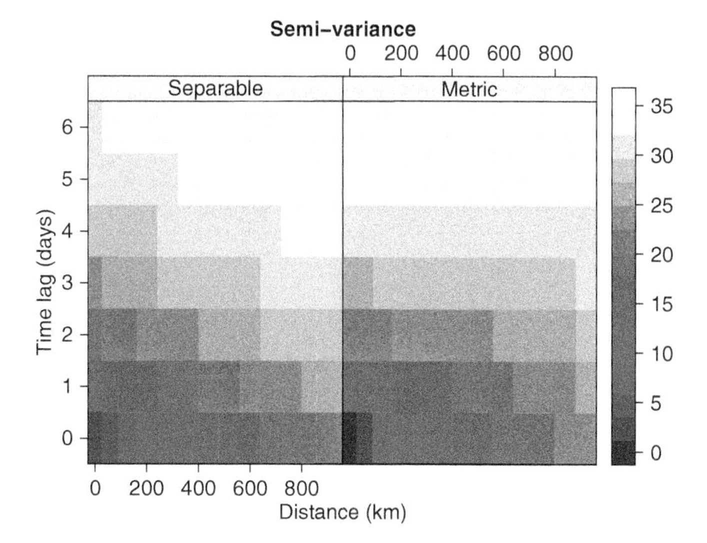

Figure 9.17 Fitted semivariograms of parametric models for the daily maximum temperature of NOAA data set during August 1993. The models used are (*a*) separable model and (*b*) metric model with anisotropy. Source: NOAA (National Oceanic and Atmospheric Administration) data set.

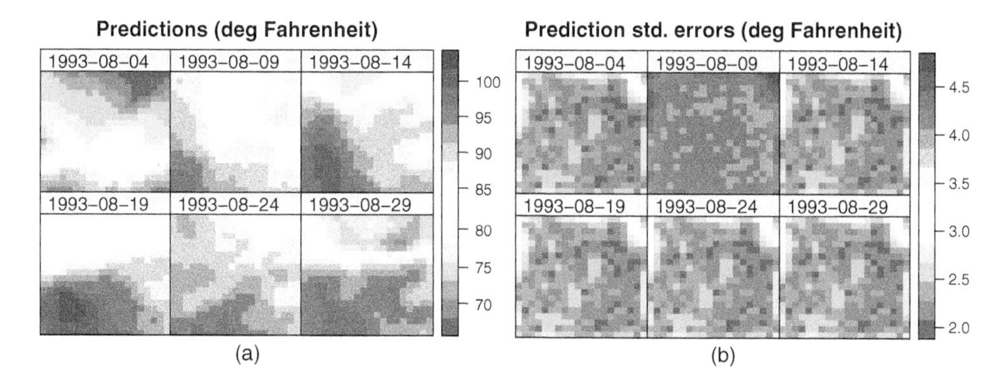

Figure 9.18 S-T kriging predictions of daily maximum temperature using a separable covariance function for August 1993. Part (*a*) is prediction and part (*b*) is the associated standard errors. The data of 9 August are omitted.

R Demonstration: S-T Kriging.

```
>require(sp)
>require(spacetime)
>require(ggplot2)
>require(dplyr)
>require(gstat)
>require(RColorBrewer)
>require(tidyr)
```

```
>require(STRbook)
##
>data("STObj3",package="STRbook")
> STObj4 <- STObj3[,"1993-08-01::1993-08-31"]
> da <- as.data.frame(STObj4)
> names(da) ## show lon, lat, sp.ID, time, …, z.
> dim(da) ## show 10168-by-12.
> length(unique(da$id)) ##show 328.
> v <- variogram(object=z~1+lat,data= STObj4,
          width=80,cutoff=1000,tlags=0.01:6.01)
> plot(v)
> print(v)
> sepVGM <- vgmST(stModel="separable",space=vgm(10,"Exp",400,nugget=0.1),
 time=vgm(10,"Exp",1,nugget=0.1),sill=30)
> sepVGM <- fit.StVariogram(v,sepVGM)
> metricVGM <- vgmST(stModel="metric",
      joint=vgm(100,"Exp",400,nugget=0.1),sill=20,stAni=100)
> metricVGM <- fit.StVariogram(v,metricVGM)
> metricMSE <- attr(metricVGM,"optim")$value
> sepMSE <- attr(sepVGM,"optim")$value
> metricMSE
#[1] 1.697554
> sepMSE
#[1] 1.723087
> plot(v,list(sepVGM,metricVGM),main="Semi-variance")

> spat_pred_grid <- expand.grid(lon=seq(-100,-80,length=20),
    lat=seq(32,46,length=20)) %>%
          SpatialPoints(proj4string=CRS(proj4string(STObj3)))
> gridded(spat_pred_grid) <- TRUE
> temp_pred_grid <- as.Date("1993-08-01")+seq(3,28,length=6)
> DE_pred <- STF(sp=spat_pred_grid,time=temp_pred_grid)
> STObj5 <- as(STObj4[,-9],"STIDF")
> STObj5 <- subset(STObj5,!is.na(STObj5$z))
> pred_kriged <- krigeST(z~1+lat,data=STObj5,newdata=DE_pred,
      modelList=sepVGM,computeVar=TRUE)
> color_pal<- rev(colorRampPalette(brewer.pal(11,"Spectral"))(16))
> stplot(pred_kriged,main="Predictions (deg Fahrenheit)",
        layout=c(3,2),col.regions=color_pal)
> pred_kriged$se <- sqrt(pred_kriged$var1.var)
> stplot(pred_kriged[,,"se"],main="Prediction std. errors(deg Fahren.)",
  layout=c(3,2),col.regions=color_pal)
```

■

9.7 DESCRIPTIVE S-T MODELS

A major difficulty in analyzing S-T data is that direct modeling the observed series $\boldsymbol{Z}_t = [Z(\boldsymbol{s}_1, t), \ldots, Z(\boldsymbol{s}_m, t)]'$ would encounter a large number of parameters, especially when the number of locations m is large. To overcome the difficulty, dimension reduction and/or parameter constraint is often used. It is important, however, in either dimension reduction or parameter constraint to make use of the knowledge on the spatial and temporal structures. The neighborhood system is an example of spatial structure, and seasonal pattern is an example of temporal structure. In this section, we discuss some approaches available in the literature to provide good descriptions of the data.

9.7.1 Random Effects with S-T Basis Functions

A widely used approach to spatial-temporal modeling is to employ additive models with random effects. When covariates $\boldsymbol{x}(\boldsymbol{s}, t)$ are available, we can write the model as

$$Y(\boldsymbol{s}, t) = \boldsymbol{x}(\boldsymbol{s}, t)'\boldsymbol{\beta} + \sum_{i=1}^{n_a} \phi_i(\boldsymbol{s}, t)\alpha_i + \nu(\boldsymbol{s}, t), \tag{9.82}$$

where $\boldsymbol{\beta}$ denotes the fixed effect of covariates, $\{\phi_i(\boldsymbol{s}, t)|i = 1, \ldots, n_a\}$ are prespecified *S-T* basis functions evaluated at location $\boldsymbol{s}$ and time t, n_a is a positive integer, $\{\alpha_i\}$ are *random effects*, and $\nu(\boldsymbol{s}, t)$ is a S-T (residual) process needed to represent small-scale S-T random effects not captured by the basis functions.

Suppose that data are available at m locations, say $\{\boldsymbol{s}_1, \ldots, \boldsymbol{s}_m\}$. Then, at each time point t, we can express the model in a matrix form as

$$\boldsymbol{Y}_t = \boldsymbol{X}_t\boldsymbol{\beta} + \boldsymbol{\Phi}_t\boldsymbol{\alpha} + \boldsymbol{\nu}_t, \tag{9.83}$$

where $\boldsymbol{Y}_t = [Y(\boldsymbol{s}_1, t), \ldots, Y(\boldsymbol{s}_m, t)]'$, $\boldsymbol{\Phi}$ is an $m \times n_a$ matrix with ith row being $[\phi_1(\boldsymbol{s}_i, t), \ldots, \phi_{n_a}(\boldsymbol{s}_i, t)]$, $\boldsymbol{\alpha} = (\alpha_1, \ldots, \alpha_{n_a})'$, and $\boldsymbol{\nu}_t$ is defined in the same location ordering as $\boldsymbol{Y}_t$. A special feature of the model is that both $\boldsymbol{\beta}$ and $\boldsymbol{\alpha}$ are both space and time invariant. Typically, we assume that $E(Y(\boldsymbol{s}, t)|\boldsymbol{x}_t) = \boldsymbol{x}_t'\boldsymbol{\beta}$ and the random effects $\boldsymbol{\alpha}$ follows the distribution $N(0, \boldsymbol{\Sigma}_a)$. Furthermore, under the assumption that (a) $\nu(\boldsymbol{s}, t)$ has no temporal dependence and (b) $\boldsymbol{\nu}_t \sim N(0, \boldsymbol{\Sigma}_\nu)$. Then, we have $\boldsymbol{Y}_t \sim N(\boldsymbol{X}_t\boldsymbol{\beta}, \boldsymbol{\Sigma}_{y,t})$, where $\boldsymbol{\Sigma}_{y,t} = \boldsymbol{\Phi}_t\boldsymbol{\Sigma}_a\boldsymbol{\Phi}_t' + \boldsymbol{\Sigma}_\nu$.

For the observed data with measurement errors $Z(\boldsymbol{s}, t) = Y(\boldsymbol{s}, t) + \epsilon(\boldsymbol{s}, t)$, we have $\boldsymbol{Z}_t \sim N(\boldsymbol{X}_t\boldsymbol{\beta}, \boldsymbol{\Sigma}_{z,t})$, where $\boldsymbol{\Sigma}_{z,t} = \boldsymbol{\Sigma}_{y,t} + \boldsymbol{\Sigma}_\epsilon = \boldsymbol{\Phi}_t\boldsymbol{\Sigma}_a\boldsymbol{\Phi}_t' + \boldsymbol{\Sigma}_\nu + \boldsymbol{\Sigma}_\epsilon$. If $n_a \ll m$, which is referred to as a *row-rank representation*, the model in Eq. (9.82) is relatively easy to estimate as its likelihood function can be evaluated effectively with the matrix inversion formula

$$\boldsymbol{\Sigma}_{z,t}^{-1} = \boldsymbol{V}^{-1} - \boldsymbol{V}^{-1}\boldsymbol{\Phi}_t(\boldsymbol{\Phi}_t'\boldsymbol{V}^{-1}\boldsymbol{\Phi}_t + \boldsymbol{\Sigma}_a^{-1})^{-1}\boldsymbol{\Phi}_t'\boldsymbol{V}^{-1},$$

where $\boldsymbol{V} = \boldsymbol{\Sigma}_\nu + \boldsymbol{\Sigma}_\epsilon$. This is particularly so if both $\boldsymbol{\Sigma}_\nu$ and $\boldsymbol{\Sigma}_\epsilon$ follow some simple structures, e.g. diagonal matrices.

Suppose that the data are observed at $t = 1, \ldots, T$. Then, we can put all the data in a matrix form

$$\boldsymbol{Z} = \boldsymbol{X}\boldsymbol{\beta} + \boldsymbol{\Phi}\boldsymbol{\alpha} + \boldsymbol{\nu} + \boldsymbol{\epsilon},$$

where $\boldsymbol{Z} = [\boldsymbol{Z}_1', \ldots, \boldsymbol{Z}_T']'$ is a mT-dimensional vector, and $\boldsymbol{X}$ and $\boldsymbol{\Phi}$ are stacked matrices of $\boldsymbol{X}_t$ and $\boldsymbol{\Phi}_t$ for $t = 1, \ldots, T$ with the same ordering as $\boldsymbol{Z}$.

For the $m = n_a$ (*full-rank*) and $n_a > m$ (*over-complete*) cases, the model can still enjoy some computational benefits if one introduces sparsity in the covariance matrix $\boldsymbol{\Sigma}_a$ of the random effects.

The choices of basis functions and the number of basis functions play a role in applying the model in Eq. (9.83). There are many basis functions available in the literature. See, for instance, the basis functions available in the R packages **FRK** of

Zammit-Mangion (2018b) and **autoFRM** of Tseng et al. (2019). See, also, the discussion of next section. Typically, S-T basis functions are tensor products of selected spatial basis functions and temporal basis functions. Also, one often assumes that, in applications, $\boldsymbol{\Sigma}_a$ depends on a small number of unknown parameters. For instance, the covariances between α_i decay exponentially with the distance between the centers of the basis functions.

9.7.2 Random Effects with Spatial Basis Functions

By simplifying the basis functions, but increasing the flexibility of the random effects, one can postulate the S-T model as

$$Y(\boldsymbol{s}, t) = \boldsymbol{x}(\boldsymbol{s}, t)'\boldsymbol{\beta} + \sum_{i=1}^{n_a} \phi_i(\boldsymbol{s})\alpha_i(t) + v(\boldsymbol{s}, t), \tag{9.84}$$

where $\{\phi_i(\boldsymbol{s})|i = 1, \ldots, n_a; \boldsymbol{s} \in D\}$ are prespecified spatial basis functions, $\{\alpha_i(t)|i = 1, \ldots, n_a\}$ are temporal random processes, and other components are defined as before in Eq. (9.82). There are many spatial basis functions available in the literature such as the bisquare function,

$$b(\boldsymbol{s}_1, \boldsymbol{s}_2) = \begin{cases} [1 - (\|\boldsymbol{s}_1 - \boldsymbol{s}_2\|/a)^2]^2, & \|\boldsymbol{s}_1 - \boldsymbol{s}_2\| \leq a, \\ 0, \text{otherwise}, & \end{cases} \tag{9.85}$$

where $a > 0$ and $\boldsymbol{s}_i \in D$. Other choices of spatial basis functions include wavelets, Gaussian functions, Fourier functions, and Wendland functions. Figure 9.19 shows the one-dimensional and two-dimensional bisquare functions with $a = 0.5$. In applications, the choice of basis functions is not critical so long as the type and number of basis functions are sufficiently flexible and large enough to capture the dependence in $Y(\boldsymbol{s}, t)$.

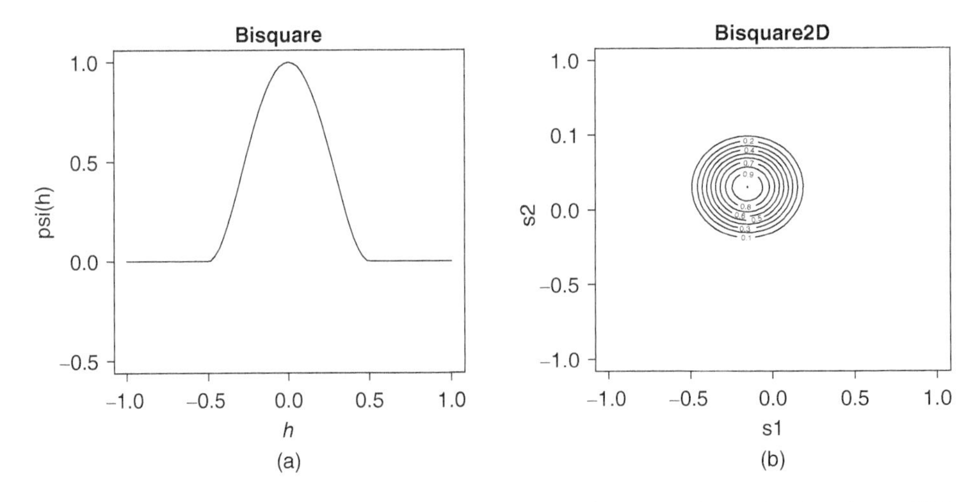

Figure 9.19 Bisquare basis functions with $a = 0.5$. Part (a) is one-dimensional and part (b) is a contour plot of two-dimensional function.

Suppose that data are observed at locations $\{s_i|i = 1, \dots, m\}$ and at time points $\{t_1, \dots, t_T\}$, where $m > 0$ and $T > 0$. Then, at time $t = t_j$, the model becomes

$$Y_{t_j} = X_{t_j}\beta + \Phi\alpha_{t_j} + v_{t_j}, \tag{9.86}$$

where, as before, $Y_{t_j} = [Y(s_1, t_j), \dots, Y(s_m, t_j)]'$ is an m-dimensional vector process, $v_{t_j} \sim N(0, \Sigma_v)$, $\Phi = [\phi(s_1), \dots, \phi(s_m)]'$, $\phi(s_j) = [\phi_1(s_j), \dots, \phi_{n_a}(s_j)]'$ for $j = 1, \dots, m$, and $\alpha_{t_j} = [\alpha_1(t_j), \dots, \alpha_{n_a}(t_j)]'$. For this model, the specification of α_{t_j} deserves some careful considerations.

If $\{\alpha_{t_j}|j = 1, \dots, T\}$ are temporally independent and $\alpha_{t_j} \sim N(\mathbf{0}, \Sigma_a)$, then the marginal distribution of Y_{t_j} is $N(X_{t_j}\beta, \Phi\Sigma_a\Phi' + \Sigma_v)$ and $Y_{t_1}, \dots, Y_{t_T}$ are temporally independent. Hence, the joint distribution of $Y = [Y'_{t_1}, \dots, Y'_{t_T}]'$ is multivariate Gaussian with covariance matrix $\Sigma_Y = I_T \otimes (\Phi\Sigma_a\Phi' + \Sigma_v)$, where I_T is a $T \times T$ identity matrix. In general, α_{t_j} is temporally dependent so that the model can better model the S-T dependence in $Y(s, t)$. For instance, if n_a is not large and time points $\{t_j\}$ are equally spaced, then a vector autoregressive moving-average (VARMA) or vector autoregressive (VAR) model of Chapter 3 can be used.

9.7.3 Fixed Rank Kriging

To keep the random-effects models applicable in practice, especially when the number of locations m is large, care is exercised to simplify the computation of the inverse covariance matrix Σ_z^{-1} of the data. One approach of model simplifications is to employ a small number of spatial basis functions, resulting in the approach of *fixed rank kriging*. See Cressie and Johannesson (2008). It is available in the R package **FRK** of Zammit-Mangion (2018b). We briefly describe it next.

Consider Eq. (9.86). For simplicity, assume that the time points are equally spaced so that $t = 1, \dots, T$ and there are no covariates. In addition, assume that Y_t is serially uncorrelated and the S-T noise v_t is Gaussian with covariance matrix $\Sigma_v = \sigma_v^2 I$. In this case, the covariance matrix of observed data Z_t is $\Sigma_z = \Phi\Sigma_a\Phi' + (\sigma_v^2 + \sigma_\epsilon^2)I$ with $\Phi\Sigma_a\Phi'$ having rank at most n_a. The resulting kriging is called *fixed rank kriging* (Cressie and Johannesson, 2008). In this particular case, letting $\sigma^2 = \sigma_v^2 + \sigma_\epsilon^2$, we have

$$\Sigma_z^{-1} = \frac{1}{\sigma^2}I - \frac{1}{\sigma^2}\Phi\left(\Sigma_a^{-1} + \frac{1}{\sigma^2}\Phi'\Phi\right)^{-1}\Phi',$$

which is manageable so long as Σ_a^{-1} is relatively easy to compute such as when the dimension n_a is not large.

One potential weakness of the FRK is that the choice of spatial basis functions does not take into consideration the characteristics of observed data. Consequently, for some applications, one may need a large n_a to provide good approximation of the sample covariance matrix, resulting in time-consuming estimation. To overcome this difficulty, Tseng and Huang (2018) propose a new class of basis functions which are resolution adaptive, and use Akaike's information criterion (AIC) to select the number of basis functions. Of particular interest is that, based on the proposed new basis functions, closed-form expressions are available for maximum likelihood estimate of Σ_a and AIC. The new basis functions are extracted from thin-plate splines and they are ordered in terms of their degrees of smoothness with higher-order

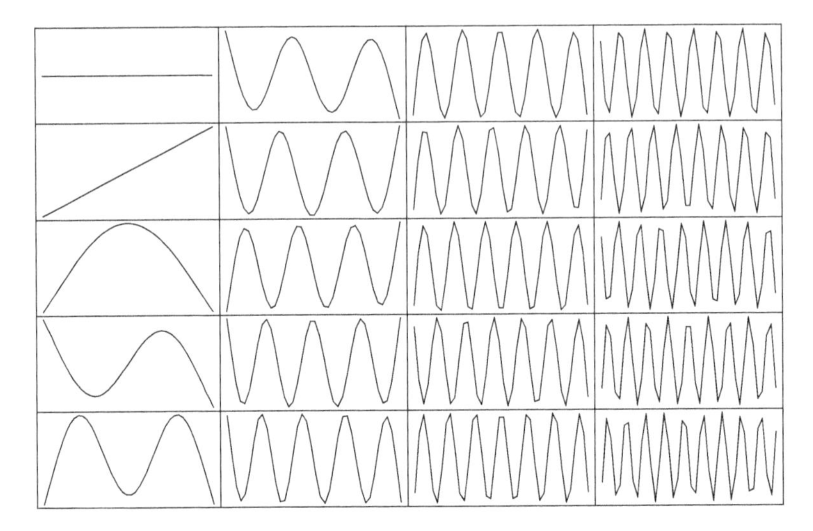

Figure 9.20 Plots of the first 20 basis functions on the unit interval [0,1] with 40 equally spaced knots. The basis functions are the multi-resolution thin-plate splines of Tseng and Huang (2018). Source: Based on Tzeng and Huang (2018).

functions corresponding to larger-scale features and lower-order ones corresponding to smaller-scale details. Consequently, the number of basis functions used determines the spatial resolution. The new basis functions also avoid the need of knot allocation. Interested readers are referred to Tseng and Huang (2018) for further details. Figure 9.20 plots the first 20 basis functions generated by the proposed method of Tseng and Huang (2018) for one-dimensional case on the unit interval [0,1] with 40 equally spaced knots. From the plots, it is clearly that the first basis function is simply a constant, the second basis function is linear, the third basis function is quadratic, and so on.

Example 9.3

Consider the ozone data set of the Midwestern United States. See Figure 9.3. Here we have $T = 89$ and $m = 153$. To remove the possibility of time trends, we use $\boldsymbol{x}_t = (1, t)'$ as covariates and focus on analyzing the trend-adjusted data. Also, the data have some missing values so that we only employed stations that have no missing values. As a result, we have $m = 67$ in this example. Our analysis is carried out via the **autoFRK** package of Tseng et al. (2019). For ease in reading, the random-effects model considered by autoFRM is

$$\boldsymbol{Z}_t = \boldsymbol{\mu} + \boldsymbol{G}\boldsymbol{w}_t + \boldsymbol{\eta}_t + \boldsymbol{\epsilon}_t, \quad t = 1, \ldots, T, \tag{9.87}$$

where $\boldsymbol{Z}_t$ is an n-vector of observed data at n locations, $\boldsymbol{\mu}$ is an n-vector of deterministic mean values, $\boldsymbol{\epsilon}_t \sim N(\boldsymbol{0}, s\boldsymbol{D})$ with $\boldsymbol{D}$ being a given $n \times n$ matrix and s being the scale parameter, $\boldsymbol{G}$ is a given $n \times K$ matrix of basis functions with K being the number of basis functions, $\boldsymbol{w}_t$ is a K-vector of unobservable random weights such that $\boldsymbol{w}_t \sim N(\boldsymbol{0}, \boldsymbol{M})$, and $\boldsymbol{\eta}_t$ is an n-vector of a spatial stationary process. Here the

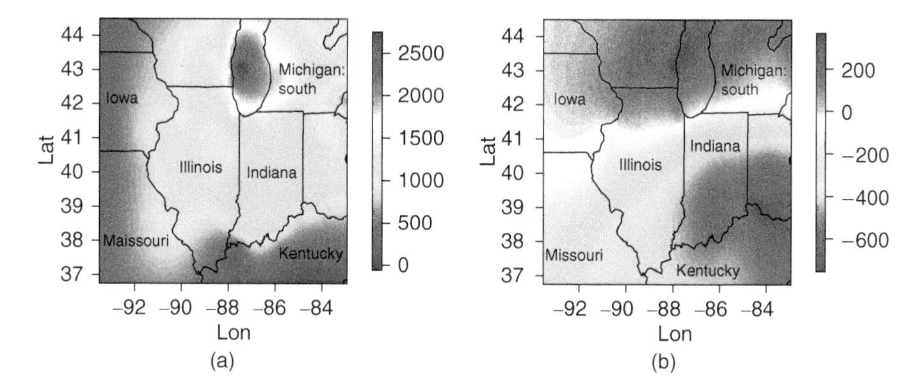

Figure 9.21 Eigenfunctions of ozone predictions at a 100×100 grid points, equally spaced on the longitude and latitude of Midwestern United States. Part (a) is the first eigenfunction and part (b) is the second eigenfunction.

columns of $\boldsymbol{G}$ are an ordered class of thin-plate spline functions and η_t is referred to as finer-scale spatial effects.

We use the default options in our demonstration so that $\boldsymbol{\mu} = \boldsymbol{0}$, $\boldsymbol{\eta}_t = \boldsymbol{0}$, and $\boldsymbol{D} = \boldsymbol{I}$. The number of basis functions K is selected by the AIC. In this particular instance, $K = 60$ is selected, which is relatively large as we have $n = 67$, and $\hat{s} = 7.21$. The program provides $\boldsymbol{G}$, which is a 67×60 matrix, and an estimate of $\boldsymbol{M}$, which is a 60×60 covariance matrix of $\boldsymbol{w}_t$. Thus, it is not an easy task to present the results. However, we can calculate the two eigenvectors associated with the two largest eigenvalues of $\boldsymbol{G}\widehat{\boldsymbol{M}}\boldsymbol{G}'$ matrix, which, for the model entertained, represents the covariance matrix of $Y(\boldsymbol{s}, t)$. These two eigenvectors can be used along the predictions of $Y(\boldsymbol{s}, t)$ at new locations to demonstrate the first two eigenfunctions of predictions based on the entertained random-effects model. Figure 9.21 plots the first two eigenfunctions of predictions with 100×100 equally spaced grid points on the longitude and latitude of Midwestern United States. From the plots, the first eigenfunction confirms the high ozone readings located at the west shoreline of Lake Michigan near southern Wisconsin and northern Illinois. The second eigenfunction marks the difference between north-west and south-east regions under study.

R demonstration of fixed rank kriging: The **autoFRK** package.

```
>require(fields)
>require(pracma)
> require(autoFRK)
> require(maps)
>data(ozone2)
>names(ozone2)
>loc <- ozone2$lon.lat
>Y <- ozone2$y
>date <- as.Date(ozone2$date,format='%y%m%d')
>good <- !colSums(is.na(Y))
>YY <- matrix(Y[,good],nrow=nrow(Y))
>YY <- detrend(YY,"linear")
>loc1 <- loc[good,]
>A1 <- autoFRK(t(YY),loc1)
```

```
>names(A1)
>A1$s
#[1] 7.210554
>dim(A1$G)
#[1] 67 60
>dim(A1$M)
#[1] 60 60

>new <- 100
>x_lon <- seq(min(loc1[,1]),max(loc1[,1]),length=new)
>x_lat <- seq(min(loc1[,2]),max(loc1[,2]),length=new)
>newloc <- as.matrix(expand.grid(x=x_lon,y=x_lat))
>Gpred <- predict(A1$G,newx=newloc)
>G <- A1$G
>Mhat <- A1$M
>dd <- eigen(G%*%Mhat%*%t(G))
>fhat <- Gpred%*%Mhat%*%t(G)%*%dd$vector[,1:2]
>quilt.plot(newloc,-fhat[,1],nx=new,ny=new,xlab="lon",ylab="lat")
>map("state",xlim=range(loc1[,1]),ylim=range(loc1[,2]),add=T)
>map.text("state",xlim=range(x_lon),ylim=range(x_lat),add=T)
#
>quilt.plot(newloc,-fhat[,2],nx=new,ny=new,xlab="lon",ylab="lat")
>map("state",xlim=range(loc1[,1]),ylim=range(loc1[,2]),add=T)
>map.text("state",xlim=range(x_lon),ylim=range(x_lat),add=T)
```

■

9.7.4 Spatial Principal Component Analysis

Instead of using prespecified spatial basis functions, one can use empirical principal component analysis (PCA) to select suitable basis functions for random-effects modeling of spatial data. Wang and Huang (2017) propose a regularized PCA approach, which is relatively easy to use and data adaptive.

For simplicity, assume that there are no covariates, and $Y(\boldsymbol{s}, t)$ is zero mean with a common spatial covariance function $C_y(\boldsymbol{s}_1, \boldsymbol{s}_2) = \text{Cov}[Y(\boldsymbol{s}_1, t), Y(\boldsymbol{s}_2, t)]$ for all $t = 1, \ldots, T$. As before, we assume that the data are observed at locations $\boldsymbol{s}_1, \ldots, \boldsymbol{s}_m$ on the spatial domain $D \subset R^d$ with independent measurement errors. Furthermore, assume that $\{Y(\boldsymbol{s}, t)\}$ are mutually temporally uncorrelated. Consider a rank K spatial random-effects model for $Y(\boldsymbol{s}, t)$:

$$Y(\boldsymbol{s}, t) = [\psi_1(\boldsymbol{s}), \ldots, \psi_K(\boldsymbol{s})]\boldsymbol{\xi}_t = \sum_{i=1}^{K} \xi_{ti}\psi_i(\boldsymbol{s}), \quad \boldsymbol{s} \in D; t = 1, \ldots, T, \tag{9.88}$$

where $\{\psi_i(\boldsymbol{s})\}$ are unknown orthonormal basis functions, $\boldsymbol{\xi}_t = (\xi_{t1}, \ldots, \xi_{tK})' \sim (\boldsymbol{0}, \boldsymbol{\Lambda})$ $(t = 1, \ldots, T)$ are uncorrelated random variables and $\boldsymbol{\Lambda}$ is an unknown symmetric nonnegative-definite matrix.

Write $\boldsymbol{\Lambda} = [\lambda_{ij}]_{K\times K}$. The spatial covariance function of $Y(\boldsymbol{s}, t)$ is

$$C_y(\boldsymbol{s}_1, \boldsymbol{s}_2) = \sum_{i=1}^{K}\sum_{j=1}^{K} \lambda_{ij}\psi_i(\boldsymbol{s}_1)\psi_j(\boldsymbol{s}_2). \tag{9.89}$$

Let $\boldsymbol{\Lambda} = \boldsymbol{V}\boldsymbol{\Lambda}^*\boldsymbol{V}'$ be the eigen-decomposition of $\boldsymbol{\Lambda}$, where $\boldsymbol{V}$ consists of the K orthonormal eigenvectors and $\boldsymbol{\Lambda}^* = \text{diag}\{\lambda_1^*, \ldots, \lambda_K^*\}$, where $\lambda_1^* \geq \cdots \geq \lambda_K^*$ are the corresponding eigenvalues. Let $\boldsymbol{\xi}_t^* = \boldsymbol{V}'\boldsymbol{\xi}_t$ and $[\psi_1^*(\boldsymbol{s}), \ldots, \psi_K^*(\boldsymbol{s})] =$

$[\psi_1(\boldsymbol{s}), \ldots, \psi_K(\boldsymbol{s})]\boldsymbol{V}$. It is easy to see that $\{\psi_i^*(\boldsymbol{s})\}$ are also orthonormal and $\xi_{ti}^* \sim (0, \lambda_i^*)$ for $t = 1, \ldots, T$ and $i = 1, \ldots, K$, are mutually uncorrelated. Consequently, we have

$$Y(\boldsymbol{s}, t) = [\psi_1^*(\boldsymbol{s}), \ldots, \psi_K^*(\boldsymbol{s})]\boldsymbol{\xi}_t^* = \sum_{i=1}^{K} \xi_{ti}\psi_i^*(\boldsymbol{s}), \quad \boldsymbol{s} \in D. \tag{9.90}$$

This expansion is known as the *Karhunen-Loève expansion* of $Y(\boldsymbol{s}, t)$ with K non-zero eigenvalues, where $\psi_i^*(\boldsymbol{s})$ is the ith eigenfunction of $C_y(\cdot, \cdot)$ with λ_i^* being the corresponding eigenvalue. See Karhunen (1947) and Loève (1978).

The observed data $\boldsymbol{Z}_t = [Z(\boldsymbol{s}_i, t), \ldots, Z(\boldsymbol{s}_m, t)]'$ can then be written as

$$\boldsymbol{Z}_t = \boldsymbol{Y}_t + \boldsymbol{\epsilon}_t = \boldsymbol{\Psi}\boldsymbol{\xi}_t + \boldsymbol{\epsilon}_t, \quad t = 1, \ldots, T, \tag{9.91}$$

where $\boldsymbol{\epsilon}_t$ denotes the vector of independent measurement errors with $\boldsymbol{\epsilon}_t \sim (\boldsymbol{0}, \sigma^2\boldsymbol{I})$, $\boldsymbol{Y}_t$ consists of the latent S-T process $[Y(\boldsymbol{s}_1, t), \ldots, Y(\boldsymbol{s}_m, t)]'$, $\boldsymbol{\Psi} = [\boldsymbol{\psi}_1, \ldots, \boldsymbol{\psi}_K]$ is an $m \times K$ matrix with (i, j)th element being $\psi_j(\boldsymbol{s}_i)$, and $\boldsymbol{\xi}_t$ and $\boldsymbol{\epsilon}_t$ are uncorrelated. The goal here is to find the first $p \le K$ dominant patterns $\boldsymbol{\psi}_1, \ldots, \boldsymbol{\psi}_p$ with large $\lambda_1^*, \ldots, \lambda_p^*$ and to estimate the covariance function $C_y(\cdot, \cdot)$ for spatial prediction. Equation (9.91) is in the form of a latent factor model, and PCA provides a way to achieve our goal. However, if the sample size T is not large compared with the dimension m, the PCA estimates of the latent factor and loadings would vary substantially. To overcome the difficulty, Wang and Huang (2017) propose to use regularization and impose certain smoothness constraints in estimation. It turns out from the theory of smoothing splines (Green and Silverman, 1994) that the estimates $\widehat{\psi}(\boldsymbol{s})$ is a natural cubic spline when $d = 1$ and a thin-plate spline when $d = 2$ or 3 with nodes at $\{\boldsymbol{s}_1, \ldots, \boldsymbol{s}_m\}$. More specifically, let $\boldsymbol{Z} = (\boldsymbol{Z}_1, \ldots, \boldsymbol{Z}_T)'$ be the $T \times m$ data matrix with the understanding that the mean of $\boldsymbol{Z}_t$ is zero. Define the sample covariance matrix as $\boldsymbol{S} = \boldsymbol{Y}'\boldsymbol{Y}/T$. Wang and Huang (2017) estimate $\boldsymbol{\Psi}$ as follows

$$\widehat{\boldsymbol{\Psi}}_{\tau_1,\tau_2} = \arg \min_{\boldsymbol{\Psi}|\boldsymbol{\Psi}'\boldsymbol{\Psi}=\boldsymbol{I}_K} \|\boldsymbol{Z} - \boldsymbol{Z}\boldsymbol{\Psi}\boldsymbol{\Psi}'\|_F^2 + \tau_1 \sum_{i=1}^{K} \boldsymbol{\psi}_i'\boldsymbol{\Omega}\boldsymbol{\psi}_i + \tau_2 \sum_{i=1}^{K}\sum_{j=1}^{m} |\psi_{ji}|, \tag{9.92}$$

subject to the constraints $\boldsymbol{\psi}_1'\boldsymbol{S}\boldsymbol{\psi}_1 \ge \boldsymbol{\psi}_2'\boldsymbol{S}\boldsymbol{\psi}_2 \ge \cdots \ge \boldsymbol{\psi}_K'\boldsymbol{S}\boldsymbol{\psi}_k$, where τ_1 and τ_2 are penalty parameters with values determined by cross-validation, and $\|\boldsymbol{M}\|_F = (\sum_{i,j} m_{ij}^2)^{1/2}$ is the Frobenius norm of the matrix $\boldsymbol{M} = [m_{ij}]$. Furthermore, to estimate the covariance function $C_y(\cdot, \cdots)$, we also need to estimate σ^2 and $\boldsymbol{\Lambda}$. To this end, Wang and Huang apply a regularized least squares method of Tzeng and Huang (2015)

$$(\widehat{\sigma}^2, \widehat{\boldsymbol{\Lambda}}) = \arg \min_{\sigma^2, \boldsymbol{\Lambda}|\sigma^2 \ge 0, \boldsymbol{\Lambda} \ge 0} \left\{ \frac{1}{2}\|\boldsymbol{S} - \widehat{\Psi}\boldsymbol{\Lambda}\widehat{\Psi}' - \sigma^2\boldsymbol{I}\|_F^2 + \gamma\|\widehat{\boldsymbol{\Psi}}\boldsymbol{\Lambda}\widehat{\boldsymbol{\Psi}}'\|_* \right\}, \tag{9.93}$$

where $\gamma \ge 0$ is a tuning parameter and $\|\boldsymbol{A}\|_* = \text{tr}[(\boldsymbol{A}'\boldsymbol{A})^{1/2}]$ is the nuclear norm of $\boldsymbol{A}$. It turns out closed-from solutions are available for both $\widehat{\boldsymbol{\Lambda}}$ and $\widehat{\sigma}^2$. See Wang and Huang (2017, proposition 1) for details. Finally, the analysis discussed in this section can be carried out via the R package **SpatPCA**.

Example 9.4

Consider again the ozone measurements of Midwestern United States. In this particular instance, we have $T = 89$ and $m = 153$ so that the sample covariance matrix

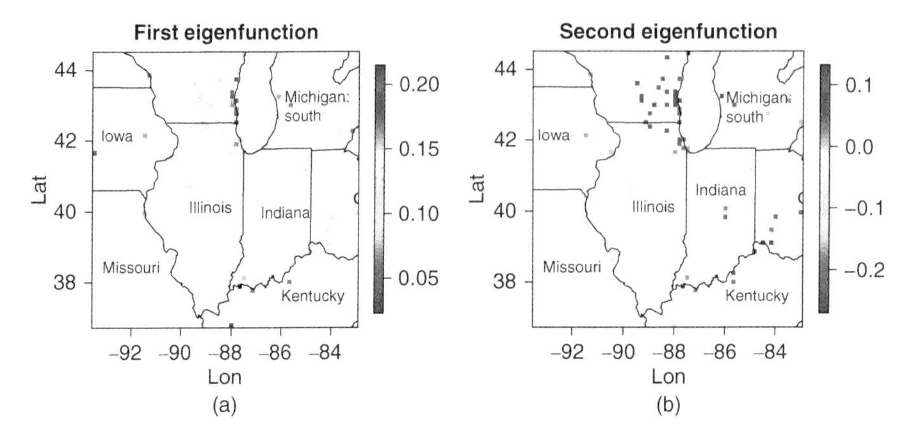

Figure 9.22 Eigenfunctions of the trend-adjusted ozone data of Midwestern United States. Part (*a*) is the first eigenfunction and part (*b*) is the second eigenfunction.

is singular and the conventional PCA is not useful. As in Example 9.2, we focus on the trend-adjusted data. Figure 9.22 shows the first two eigenfunctions of the trend-adjusted ozone data. From the plots, it is not surprising to see that the first eigenfunction has large weights along the west shoreline of Lake Michigan in northern Illinois and southern Wisconsin. On the other hand, the second eigenfunction shows certain geographical effect as it shows heavier weights on northern states (Wisconsin, Michigan, Northern Illinois) and lighter weights on southern states (Ohio and Kentucky). Figure 9.23 shows the time plots of the first two principal components. These plots indicates that the components appear to be temporally stationary.

Finally, applying the estimated covariance function, we can perform spatial prediction of the ozone data. To demonstrate, we create a new grid points by using 200

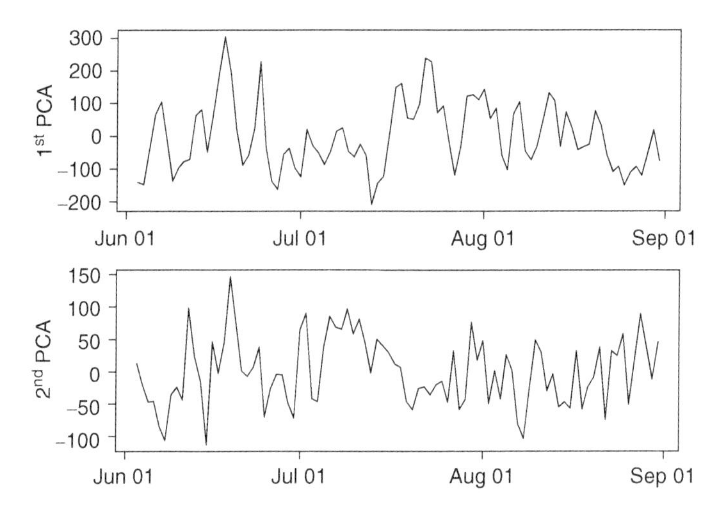

Figure 9.23 Time plots of the first two principal components of the trend-adjusted ozone data.

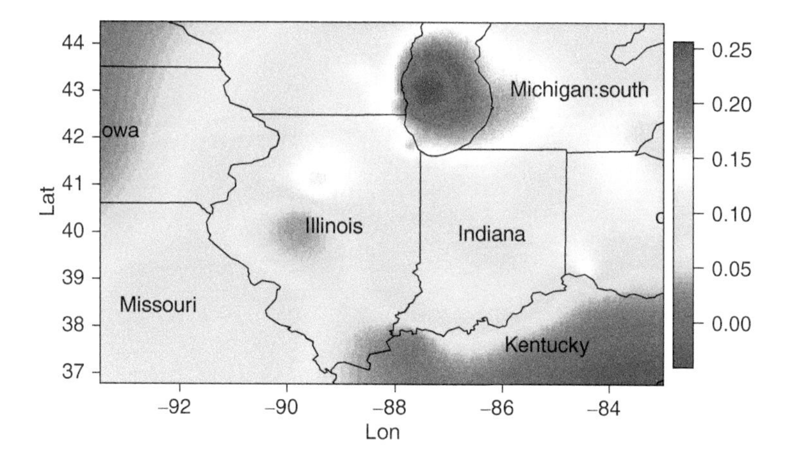

Figure 9.24 The first eigenfunction of the spatial prediction of ozone measurements of the Midwestern United States based on regularized spatial PCA.

equally spaced points along the longitude used and 200 equally spaced points along the latitude used. In other words, we have a new grid of 200×200 for the Midwestern United States. Figure 9.24 plots the first eigenfunction of the predicted ozone measurements (trend-adjusted). It is reassuring to see that the major concentration of ozone is located near the west shoreline of Lake Michigan near northern Illinois and southern Wisconsin.

Spatial PCA: **SpatPCA** package of R.

```
>require(fields)
>require(SpatPCA)
>require(pracma)
>data(ozone2)
>x <- ozone2$lon.lat
>Y <- ozone2$y
>date <- as.Date(ozone2$date,format='%y%m%d')
>rmna <- !colSums(is.na(Y))
>YY <- matrix(Y[,rmna],nrow=nrow(Y))
>YY <- detrend(YY,"linear")
>xx <- x[rmna,]
>cv <- spatpca(x=xx,Y=YY)
>quilt.plot(xx,cv$eigenfn[,1])
>map("state",xlim=range(xx[,1]),ylim=range(xx[,2]),add=T)
>map.text("state",xlim=range(xx[,1]),ylim=range(xx[,2]),cex=1,add=T)
>title(main="First Eigen Function")
###
>quilt.plot(xx,cv$eigenfn[,2])
>map("state",xlim=range(xx[,1]),ylim=range(xx[,2]),add=T)
>map.text("state",xlim=range(xx[,1]),ylim=range(xx[,2]),cex=1,add=T)
>title(main="Second Eigen Function")
## Time plots
>par(mfcol=c(2,1))
>plot(date,YY%*%cv$eigenfn[,1],type="l",ylab="1st PCA")
>plot(date,YY%*%cv$eigenfn[,2],type="l",ylab="2nd PCA")
>par(mfcol=c(1,1))
```

```
### Prediction
>new_loc <- 200
>x_lon <- seq(min(xx[,1]),max(xx[,1]),length=new_loc)
>x_lat <- seq(min(xx[,2]),max(xx[,2]),length=new_loc)
>xx_new <- as.matrix(expand.grid(x=x_lon,y=x_lat))
>eof <- spatpca(x=xx,Y=YY,K=cv$Khat,tau1=cv$stau1,tau2=cv$stau2,
           x_new=xx_new)
>quilt.plot(xx_new,eof$eigenfn[,1],nx=new_loc,ny=new_loc,xlab="lon",
            ylab="lat")
>map("state",xlim=range(x_lon),ylim=range(x_lat),add=T)
>map.text("state",xlim=range(x_lon),ylim=range(x_lat),cex=1,add=T)
```

■

9.7.5 Random Effects with Temporal Basis Functions

Alternatively, if one employs temporal basis functions and expands the random effects to be spatial, then the S-T model in Eq. (9.82) becomes

$$Y(\boldsymbol{s},t) = \boldsymbol{x}(\boldsymbol{s},t)'\boldsymbol{\beta} + \sum_{i=1}^{n_a} \phi_i(t)\alpha_i(\boldsymbol{s}) + \nu(\boldsymbol{s},t), \tag{9.94}$$

where $\{\phi_i(t)|i = 1, \ldots, n_a\}$ are temporal basis functions and $\{\alpha_i(\boldsymbol{s})\}$ are spatially indexed random effects. Let $\boldsymbol{\alpha}(\boldsymbol{s}) = [\alpha_1(\boldsymbol{s}), \ldots, \alpha_{n_a}(\boldsymbol{s})]'$ denote the n_a-dimensional spatial process. One can specify a spatial model discussed before for $\boldsymbol{\alpha}(\boldsymbol{s})$ with $\boldsymbol{s} \in D$. There are many temporal basis functions available including Fourier functions and other orthogonal functions. One can also use data-driven temporal basis functions.

Example 9.5

Consider, again, the daily maximum temperatures of August 1993 from the NOAA data set. Here we entertain a random-effect model with temporal basis functions. Specifically, the model entertained is in the form

$$Y(\boldsymbol{s},t) = \sum_{i=1}^{3} \phi_i(t)\alpha_i(\boldsymbol{s}) + \nu(\boldsymbol{s},t), \tag{9.95}$$

where $\{\phi_i(t)|i = 1,2,3\}$ are data-driven temporal basis functions, $\{\alpha_i(\boldsymbol{s})\}$ are coefficients of the temporal basis functions that are treated as multivariate spatial random fields, and $\nu(\boldsymbol{s},t)$ is a spatially correlated, but temporally independent random process. In this demonstration, we do not use any covariates so that the term $\boldsymbol{x}(\boldsymbol{s},t)'\boldsymbol{\beta}$ is omitted.

The empirical analysis of the model in Eq. (9.95) can be carried out by the R package **SpatioTemporal**. Details of R commands used are given below. See also Wikle et al. (2019, chapter 4). The space-time object used by **SpatialTemporal** is of class `STdata` and is created by the function `createSTdata`. Furthermore, the package assumes that $\nu(\boldsymbol{s},t)$ is temporally uncorrelated. The temporal dependence thus should be taken care of by either the covariates or the temporal basis functions. Since the model in Eq. (9.95) does not use any covariates, the temporal dependence in the daily maximum temperatures must be captured by the basis function. We perform some preliminary data analysis to check the validity of such assumption.

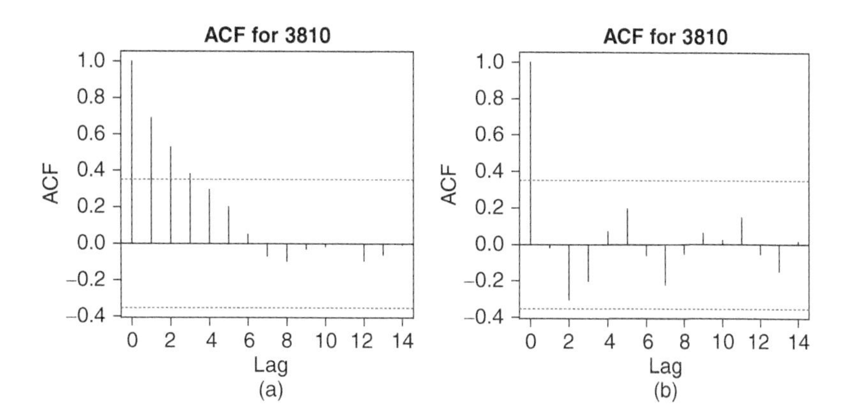

Figure 9.25 Sample autocorrelation functions (ACFs) of a selected series of daily maximum temperatures in August 1993. Part (*a*) is the ACF of the observed series and part (*b*) is the ACF of residuals after adjusting for the effect of temporal basis functions.

Figure 9.25(*a*) shows the sample autocorrelation function (ACF) of a selected series (ID = 3810). From the plot, the observed data series has significant serial correlations. Figure 9.25(*b*) provides the sample ACF of the same series after adjusting the effect of temporal basis functions. Now, the plot indicates that there is no significant serial dependence. The three temporal basis functions used are shown in Figure 9.26. The first basis function is simply a constant.

With the temporal basis functions specified, the spatial coefficients $\alpha_i(\boldsymbol{s})$ of Eq. (9.95) can be estimated by the command `estimateBetaFields`. The estimation results provide two quantities. The first quantity contains the parameter estimates, called beta in the program, and the second are the standard errors of the estimates, called beta.sd. Figure 9.27 shows the empirical estimates of $\alpha_1(\boldsymbol{s})$, $\alpha_2(\boldsymbol{s})$ and $\alpha_3(\boldsymbol{s})$ at

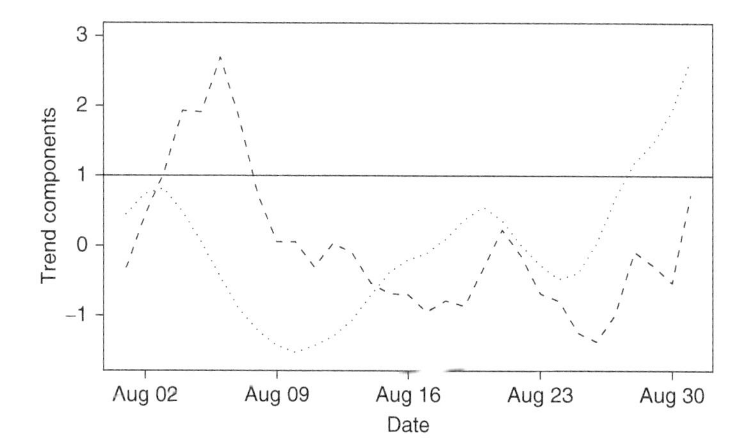

Figure 9.26 Time plots of the first three temporal basis functions for the daily maximum temperatures in August 1993. The second and third basis functions are shown in dashed and dotted lines, respectively.

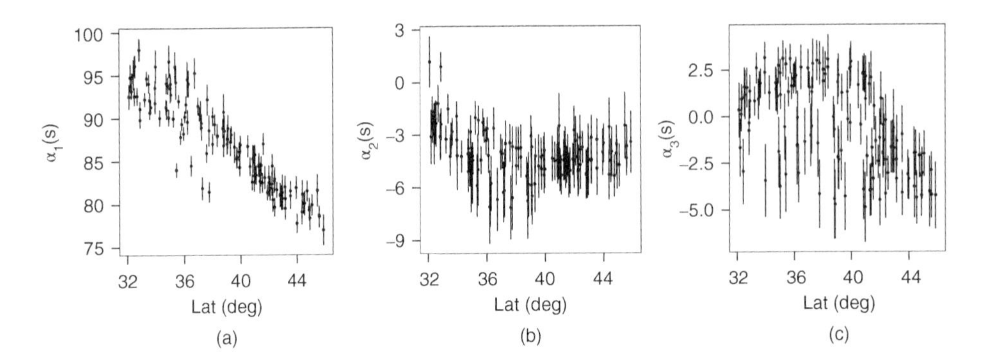

Figure 9.27 Empirical estimates of $\alpha_1(s)$, $\alpha_2(s)$ and $\alpha_3(s)$ at each location with 95% confidence intervals, plotted versus the latitude.

each location along with 95% confidence intervals, plotted against the latitude. From the plots, it is easily seen that (a), as expected, $\widehat{\alpha}_1(\boldsymbol{s})$ shows a decreasing trend with the latitude, (b) $\widehat{\alpha}_2(\boldsymbol{s})$ exhibits some weaker trend when the latitude is low, but the trend taps off as the latitude increases, and (c) $\widehat{\alpha}_3(\boldsymbol{s})$ does not have a clear dependence on the latitude except when the latter is large. Based on these findings, it is reasonable to consider the $\alpha_i(\boldsymbol{s})$ as random fields with exponential covariance functions. Specifically, we postulate the models:

$$E[\alpha_1(\boldsymbol{s})] = \alpha_{11} + \alpha_{12}s_2, \ \mathrm{Cov}[\alpha_1(\boldsymbol{s}), \alpha_1(\boldsymbol{s}+\boldsymbol{h})] = \sigma_1^2 \exp\left(\frac{-\|\boldsymbol{h}\|}{r_1}\right), \tag{9.96}$$

$$E[\alpha_2(\boldsymbol{s})] = \alpha_{21} + \alpha_{22}s_2, \ \mathrm{Cov}[\alpha_2(\boldsymbol{s}), \alpha_2(\boldsymbol{s}+\boldsymbol{h})] = \sigma_2^2 \exp\left(\frac{-\|\boldsymbol{h}\|}{r_2}\right), \tag{9.97}$$

$$E[\alpha_2(\boldsymbol{s})] = \alpha_{31}, \ \mathrm{Cov}[\alpha_3(\boldsymbol{s}), \alpha_3(\boldsymbol{s}+\boldsymbol{h})] = \sigma_3^2 \exp\left(\frac{-\|\boldsymbol{h}\|}{r_3}\right), \tag{9.98}$$

where s_2 is the latitude of the location $\boldsymbol{s} = (\mathrm{lon}, \mathrm{lat})' = (s_1, s_2)'$, r_i $(i = 1,2,3)$ are scale parameters, and σ_i^2 $(i = 1,2,3)$ are stationary variances. We further assume that $\mathrm{Cov}[\alpha_i(\boldsymbol{s}), \alpha_j(\boldsymbol{s})] = 0$ for $i \neq j$. This latter assumption is relatively strong and we use it for ease in demonstration. Finally, it remains to specify the spatial covariance function of the temporally independent residual process $\nu(\boldsymbol{s}, t)$. Following Wikle et al. (2019), we use an exponential covariance function with a nugget effect to account for measurement error. The specified overall model can be created by the command `createSTmodel` of the package **SpatioTemporal**. The model parameters that require estimation can be found using the command `loglikeSTnames`.

The fitted model can then be used to predict the daily maximum temperature. To demonstrate, we create a 20-by-20 grid with 20 equally spaced longitude from −100° to −80° and 20 equally spaced latitude from 32° to 46°. We also create six temporal days starting from 1 August 1993. The predictions can be obtained by the command `predict`. Figure 9.28(a) shows the point predictions in degrees of Fahrenheit on the new grid for 14 August whereas Figure 9.28(b) shows the corresponding standard errors of prediction.

R Demonstration and Commands used: The package **SpatioTemporal**

```
>require(dplyr)
>require(ggplot2)
```

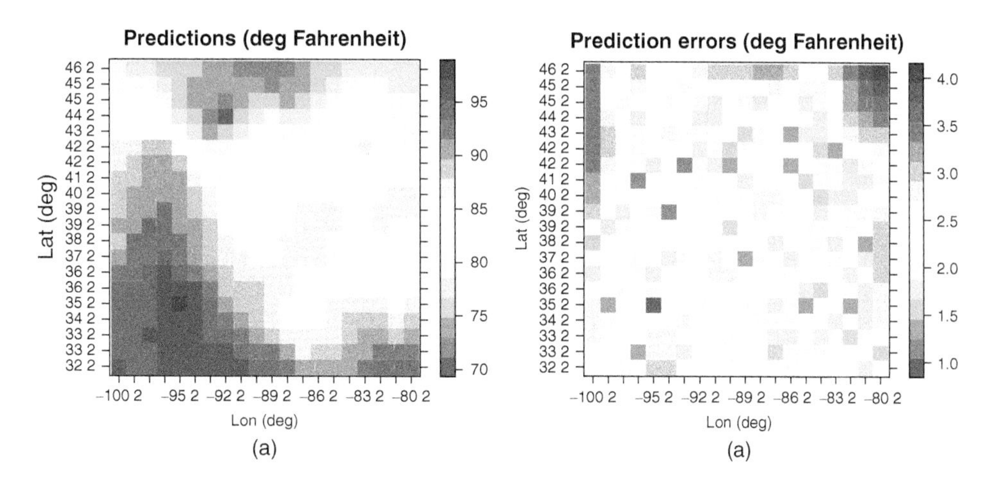

Figure 9.28 Predictions of daily maximum temperatures for 14 August 1993 based on the fitted S-T model in Eq. (9.95). Part (*a*) denotes the point predictions and part (*b*) the corresponding standard errors of prediction.

```
>require(gstat)
>require(RColorBrewer)
>require(sp)
>require(spacetime)
>require(SpatioTemporal)
>require(STRbook)
>require(tidyr)
>data("NOAA_df_1990",package="STRbook")  # load NOAA data
### Filter the data for those in August 1993 and just Tmax
>NOAA_aug <- filter(NOAA_df_1990,year==1993&month==8 & proc=="Tmax")
>NOAA_aug_for_STdata <- NOAA_aug %>%   ## Create ID as character var.
       transmute(ID = as.character(id), obs = z, date=date)
>covars <- dplyr::select(NOAA_aug, id, lat, lon) %>%
                unique() %>%
               dplyr::rename(ID = id)  # createSTdata expects "ID''
>STdata <- createSTdata(NOAA_aug_for_STdata, covars=covars)

>plot(STdata, "acf", ID="3810")
## found temporal basis functions
>STdata <- updateTrend(STdata, n.basis=2)
>names(STdata)
#[1] "obs"            "covars"          "SpatioTemporal" "trend"
#[5] "trend.fnc"
>range(STdata$trend[,1:2])
#[1] -1.523513  2.692456
>trend <- STdata$trend
>plot(trend[,3],trend[,1],type="l",lty=2,lwd=2,ylim=c(-1.6,3.0),
    xlab="date",ylab="trend components")
>  abline(h=1)
>  lines(trend[,3],trend[,2],lty=3,lwd=2)
##
>plot(STdata,"acf",ID="3810")

>alpha.lm<- estimateBetaFields(STdata)#estimate alpha (called beta in R)
>names(alpha.lm)
```

```
#[1] "beta"    "beta.sd"

### Prepare for plotting using the plotrix package.
>require(plotrix)
>head(row.names(alpha.lm$beta))
#[1] "13865" "13866" "13871" "13873" "13874" "13876"
>head(covars$ID)
#[1] 3804 3810 3811 3812 3813 3816
### matching data properly
>alpha.lm$beta <- data.frame(alpha.lm$beta)
>alpha.lm$beta.sd <- data.frame(alpha.lm$beta.sd)
>alpha.lm$beta$ID <- as.integer(row.names(alpha.lm$beta))

>ALPHA <- cbind(alpha.lm$beta,alpha.lm$beta.sd)
>colnames(ALPHA) <- c("alpha1","alpha2","alpha3","ID",
          "alpha1_CI","alpha2_CI","alpha3_CI")
>ALPHA <- left_join(ALPHA,covars,by="ID")

>ggplot(ALPHA)+geom_point(aes(x=lat,y=alpha1))+
   geom_errorbar(aes(x=lat,ymin=alpha1-1.96*alpha1_CI,
           ymax=alpha1+1.96*alpha1_CI))+
            ylab(expression(alpha[1](s)))+
               xlab("lat (deg)")+theme_bw()
### Commands for alpha2 and alpha3 are omitted.
### Sigma_a
>cov.beta <- list(covf = "exp", nugget=FALSE)
### Sigma_nu
>cov.nu <- list(covf = "exp", nugget=~1,random.effect=FALSE)
>locations <- list(coords=c("lon","lat"))
>LUR <- list(~lat, ~lat, ~1)
>STmodel <- createSTmodel(STdata, LUR=LUR,
         cov.beta=cov.beta,cov.nu=cov.nu,locations=locations)
>parnames <- loglikeSTnames(STmodel, all=FALSE)
>print(parnames)
#[1] "log.range.const.exp"          "log.sill.const.exp"
#[3] "log.range.V1.exp"             "log.sill.V1.exp"
#[5] "log.range.V2.exp"             "log.sill.V2.exp"
#[7] "nu.log.range.exp"             "nu.log.sill.exp"
#[9] "nu.log.nugget.(Intercept).exp"
>x.init <- matrix(3,9,1)
>rownames(x.init) <- loglikeSTnames(STmodel, all=FALSE)
>SpatioTemporalfit1 <- estimate(STmodel, x.init)
#Optimisation using starting value 1/1
#N = 9, M = 5 machine precision = 2.22045e-16
#At X0, 0 variables are exactly at the bounds
#At iterate     0  f=       7480.2  |proj g|=          18
#At iterate    10  f =       5615.8  |proj g|=       16.617
#At iterate    20  f =       5611.3  |proj g|=       8.3644
#At iterate    30  f =       5603.3  |proj g|=       4.4537

#iterations 36
#function evaluations 58
#segments explored during Cauchy searches 43
#BFGS updates skipped 0
```

```
#active bounds at final generalized Cauchy point 0
#norm of the final projected gradient 1.1482
#final function value 5602.96

#F = 5602.96
#final  value 5602.958497
#converged

### Prediction
>x.final <- coef(SpatioTemporalfit1,pars="cov")$par
>spat_pred_grid <- expand.grid(lon=seq(-100,-80,length=20),
        lat=seq(32,46,length=20))
>spat_pred_grid$id <- 1:nrow(spat_pred_grid)
>temp_pred_grid <- as.Date("1993-08-01")+seq(3,28,length=6)
>obs_pred_wide <- matrix(0,6,400)
>rownames(obs_pred_wide) <- as.character(temp_pred_grid)
>colnames(obs_pred_wide) <- spat_pred_grid$id
>covars_pred <- spat_pred_grid
>STdata_pred <- createSTdata(obs=obs_pred_wide,covars=covars_pred)
>E <- predict(STmodel,x.final,STdata=STdata_pred)
>names(E)
### Level plot for predictions and prediction errors of August 14, 1993
>t1 <- E$EX
>t2 <- E$VX.pred
>E1 <- matrix(t1[3,],20,20)
>E2 <- matrix(sqrt(t2[3,],20,20))
>k1 <- seq(32,46,length=20)
>k2 <- seq(-100,-80,length=20)
>colnames(E2) <- colnames(E1) <- paste(round(k1,2))
>rownames(E2) <- rownames(E2) <- paste(round(k2,2))
>require(lattice)
### define color palette
>color_pal <- rev(colorRampPalette(brewer.pal(11,"Spectral"))(16))
>levelplot(E1,col.regions=color_pal,xlab="lon(deg)",ylab="lat(deg)",
   main="Predictions (deg Fahrenheit)")
>levelplot(E2,col.regions=color_pal,xlab="lon(deg)",ylab="lat(deg)",
  main="Prediction errors (deg Fahrenheit)")
```

Our discussion focuses on Gaussian distributions. There are many non-Gaussian S-T data exist in empirical applications. For instance, S-T count data, such as the number of cases of a particular disease in a given region within a specified time period, are common in epidemiological studies. The analysis we discussed for Gaussian data can be generalized to model non-Gaussian S-T data. Interested readers are referred to Cressie and Wikle (2011) and Diggle (2013). ■

9.8 DYNAMIC S-T MODELS

In this section, we consider dynamic S-T statistical models. Given the complexity involved, there are many such models available in the literature and it would not be possible to discuss them all here. Instead, we only provide some general models that can be tailored to specific needs for a given application.

9.8.1 Space-Time Autoregressive Moving-Average Models

Assume that the locations are $\boldsymbol{s}_1, \ldots, \boldsymbol{s}_m$ and denote the process as $\boldsymbol{Y}_t = [Y(\boldsymbol{s}_1, t), \ldots, Y(\boldsymbol{s}_m, t)]'$, where the time index is assumed to be discrete and equally-spaced. Then, the model

$$\boldsymbol{Y}_t = \sum_{i=0}^{p} \left(\sum_{j=1}^{p_i} f_{ij} \boldsymbol{U}_{ij} \right) \boldsymbol{Y}_{t-i} + \sum_{\ell=0}^{q} \left(\sum_{j=1}^{q_\ell} g_{\ell j} \boldsymbol{V}_{\ell j} \right) \boldsymbol{W}_{t-\ell} \tag{9.99}$$

where $\{\boldsymbol{U}_{ij}\}$ and $\{\boldsymbol{V}_{\ell j}\}$ are known weight matrices, p and q are nonnegative integers denoting the orders of the AR part and moving-average part, respectively, $\{p_i\}$ and $\{q_\ell\}$ are nonnegative integers, $\{f_{ij}\}$ and $\{g_{\ell j}\}$ are parameters of the model, and $\{\boldsymbol{W}_t\}$ are *iid* random vectors with mean $\boldsymbol{0}$ and covariance matrix $\boldsymbol{\Sigma}_w$, which is assumed to be positive-definite. The model in Eq. (9.99) is called the *S-T autoregressive moving-average (STARMA) model*. See, for instance, Ali (1979), Pfeifer and Deutsch (1980), and Cressie (1993, p. 450). The model requires some additional assumptions for it to be well defined. The use of weight matrices $\{\boldsymbol{U}_{ij}\}$ and $\{\boldsymbol{V}_{\ell j}\}$ enables one to specify the substructures of the S-T process $\boldsymbol{Y}_t$, but it comes with the price of using potentially many parameters in $\{f_{ij}\}$ and $\{g_{\ell j}\}$, especially when p_i and q_ℓ are large.

The STARMA model in Eq. (9.99) can be rewritten as

$$\boldsymbol{Y}_t = \sum_{i=0}^{p} \boldsymbol{\Phi}_i \boldsymbol{Y}_{t-i} + \sum_{j=0}^{q} \boldsymbol{\Theta}_j \boldsymbol{W}_{t-j}, \tag{9.100}$$

where, without loss of generality, we assume that $\text{Cov}(\boldsymbol{W}_t) = \sigma_w^2 \boldsymbol{I}$, and, for identifiability, the diagonal elements of $\boldsymbol{\Phi}_0$ are zero. In addition, we assume that $(\boldsymbol{I} - \boldsymbol{\Phi}_0)$ is invertible. Similarly to the VARMA models discussed in Chapter 3, the STARMA model in Eq. (9.100) might be hard to apply because it may contain too many parameters, especially when m, p or q is large. The weak stationarity condition of $\boldsymbol{Y}_t$ in Eq. (9.100) can be obtained in a similar manner as that of VARMA models in Chapter 3.

If $q = 0$, the model reduces to the spatial-temporal AR models. If $p = q = 0$, then the model becomes

$$\boldsymbol{Y}_t = \boldsymbol{\Phi}_0 \boldsymbol{Y}_t + \boldsymbol{\Theta} \boldsymbol{W}_t,$$

which can be written as

$$\boldsymbol{Y}_t = (\boldsymbol{I} - \boldsymbol{\Phi}_0)^{-1} \boldsymbol{\Theta}_0 \boldsymbol{W}_t.$$

Since $\boldsymbol{W}_t$ is temporally uncorrelated, $\boldsymbol{Y}_t$ then becomes a spatial process. If $\boldsymbol{\Theta}_0 = \boldsymbol{I}$, then the prior model is known as a *simultaneous autoregressive* (SAR) model.

If $p = 1$ and $q = 0$, the model in Eq. (9.100) becomes

$$\boldsymbol{Y}_t = (\boldsymbol{I} - \boldsymbol{\Phi}_0)^{-1} \boldsymbol{\Phi}_1 \boldsymbol{Y}_{t-1} + (\boldsymbol{I} - \boldsymbol{\Phi}_0)^{-1} \boldsymbol{\Theta}_0 \boldsymbol{W}_t,$$

which can be written as

$$\boldsymbol{Y}_t = \boldsymbol{M} \boldsymbol{Y}_{t-1} + \boldsymbol{\eta}_t,$$

where $\boldsymbol{M} = (\boldsymbol{I} - \boldsymbol{\Phi}_0)^{-1} \boldsymbol{\Phi}_1$ and $\{\boldsymbol{\eta}_t\}$ is an *iid* sequence of random variables with mean zero and covariance matrix $\text{Cov}(\boldsymbol{\eta}_t) = \sigma_w^2 (\boldsymbol{I} - \boldsymbol{\Phi}_0)^{-1} \boldsymbol{\Theta}_0 \boldsymbol{\Theta}_0' (\boldsymbol{I} - \boldsymbol{\Phi}_0)^{-1}$.

Some specific S-T AR models have been proposed in the literature. For instance, Gao et al. (2019) consider models with banded AR coefficient matrices.

9.8.2 S-T Component Models

A general additive component model for S-T process $\{Y(\boldsymbol{s},t)|\boldsymbol{s} \in D \subset R^d, t \in D_t \subset R\}$ is

$$Y(\boldsymbol{s},t) = \mu(\boldsymbol{s},\boldsymbol{\theta}_t) + \gamma(t,\boldsymbol{\theta}_s) + \kappa(\boldsymbol{s},t,\boldsymbol{\theta}_{s,t}) + \eta(\boldsymbol{s},t), \tag{9.101}$$

where $\mu(\boldsymbol{s},\boldsymbol{\theta}_t)$ denotes a *spatial* trend surface with parameter vector $\boldsymbol{\theta}_t$ which may vary with time t, $\gamma(t,\boldsymbol{\theta}_s)$ represents a *temporal* trend with parameter vector $\boldsymbol{\theta}_s$ which may vary with location $\boldsymbol{s}$, $\kappa(\boldsymbol{s},t,\boldsymbol{\theta}_{s,t})$ denotes a S-T random process with parameter vector $\boldsymbol{\theta}_{s,t}$ which may depend on both location $\boldsymbol{s}$ and time t, and $\eta(\boldsymbol{s},t)$ represents micro-scale S-T random process which follows a zero-mean uncorrelated process in $D \times D_t$. This model is considered by Wikle (2003). The first component $\mu(\boldsymbol{s},\boldsymbol{\theta}_t)$ may contain the effects of covariates such as $\boldsymbol{x}(\boldsymbol{s},t)'\boldsymbol{\beta}$. The second component $\gamma(t,\boldsymbol{\theta}_s)$ may contain the impact of seasonality such as trigonometric series for annual cycles with parameters depend on the location $\boldsymbol{s}$. The third and fourth components of Eq. (9.101) may be random so that the model for $Y(\boldsymbol{s},t)$ is a random-effect model. In general, the random components are assumed to be independent for ease in model identification.

The component models in Eq. (9.101) are flexible and applicable in many situations. The estimation, however, can be computationally intensive, especially when the number of locations m and/or the sample size T are large. Deb and Tsay (2019) employ such a model for $PM_{2.5}$ data observed at 13 monitoring stations located in the southern part of Taiwan.

9.8.3 S-T Factor Models

Similarly to other statistical analyses, one can also introduce factor structure to S-T modeling to reduce the dimension and to simplify the analysis. However, care must be exercised as the spatial structure needs to be respected in postulating factor models. A natural approach is to let the common factors describe the temporal dependence whereas the loading matrix be spatially dependent only. See, for instance, the generalized spatial dynamic factor (GSDF) model of Lopes et al. (2011). Let $\boldsymbol{Y}_t = [Y(\boldsymbol{s}_1,t), \ldots, Y(\boldsymbol{s}_m,t)]'$ as before with locations $\{\boldsymbol{s}_i \in D \subset R^d$ for $i = 1, \ldots, m$. Under the GSDF model, $Y(\boldsymbol{s}_i,t)$ follows a one-parameter natural exponential distribution as

$$p(Y(\boldsymbol{s}_i,t)|\eta_{ti},\psi) = \exp\{\psi[Y(\boldsymbol{s}_i,t)\eta_{ti} - b(\eta_{ti})] + c[Y(\boldsymbol{s}_i,t),\psi]\}, \tag{9.102}$$

where η_{ti} is the natural parameter and ψ is a known dispersion parameter. The mean and variance of $Y(\boldsymbol{s}_i,t)$ are the first and second derivatives, $b'(\eta_{ti})$ and $b''(\eta_{ti})/\psi$, respectively. This natural parameter is defined as a linear combination of spatial and temporal components through the link function $\eta_{ti} = v(\theta_{ti})$, where θ_{ti} is a S-T process.

Let $\boldsymbol{\theta}_t = (\theta_{t1}, \ldots, \theta_{tm})'$. The GSDF model uses the following two-level hierarchy to describe the temporal behavior of $\boldsymbol{Y}_t$:

$$\boldsymbol{\theta}_t = \boldsymbol{\mu}_t + \boldsymbol{L}\boldsymbol{f}_t, \tag{9.103}$$

$$\boldsymbol{f}_t = \boldsymbol{\Phi}\boldsymbol{f}_{t-1} + \boldsymbol{\epsilon}_t, \tag{9.104}$$

where $\boldsymbol{\mu}_t$ is the mean of the space-time process, $\boldsymbol{f}_t = (f_{t1}, \ldots, f_{tr})'$ is the vector of common factors with $r \ll m$, $\boldsymbol{L}$ is a loading matrix to be defined later, $\boldsymbol{\Phi}$ is the transition matrix of the common factors, ϵ_t is an $r \times r$ *iid* Gaussian process with mean zero and covariance $\boldsymbol{\Sigma}_\epsilon > 0$, and $\boldsymbol{f}_0 \sim N(\boldsymbol{m}_0, \boldsymbol{C}_0)$ with known hyperparameters $\boldsymbol{m}_0$ and $\boldsymbol{C}_0$. In applications, both $\boldsymbol{\mu}_t$ and $\boldsymbol{\Phi}$ may account for seasonality and $\boldsymbol{\mu}_t$ may also include the effect of covariates.

The spatial variation of $\boldsymbol{Y}_t$ is modeled via the columns of the loading matrix $\boldsymbol{L}$. Write the jth column of $\boldsymbol{L}$ as $\boldsymbol{L}_j = (L_{1j}, \ldots, L_{mj})'$ for $j = 1, \ldots, r$. The GSDF model assumes that $\boldsymbol{L}_j$ is a conditionally independent, distance-based Gaussian process (or Gaussian random field). That is,

$$\boldsymbol{L}_j \sim N(\boldsymbol{\kappa}_j, \tau_j^2 \boldsymbol{R}_j), \quad j = 1, \ldots, r, \tag{9.105}$$

where $\boldsymbol{\kappa}_t$ is an m-dimensional mean vector, τ_j is a scaling parameter, and $\boldsymbol{R}_j = [r_{uv}(\phi_j)]$ with $r_{uv}(\phi_j) = \rho_{\phi_j}(|\boldsymbol{s}_u - \boldsymbol{s}_v|)$, $u, v = 1, \ldots, m$ for suitably defined correlation functions $\rho_{\phi_j}(.)$ with $j = 1, \ldots, r$. For simplicity, the parameters ϕ_j are typically a scalar or a low-dimensional vector. For example, ϕ_j is a scalar if ρ_ϕ is the exponential function $\rho_{\phi_j}(h) = \exp(-h/\phi_j)$ or the spherical function $\rho_{\phi_j}(h) = [1 - 1.5(h/\phi_j) + 0.5(h/\phi_j)^3] I_{h/\phi_j \le 1}$ with h denoting distance.

Estimation of the GSDF model in Eqs. (9.103)-(9.105) can be carried out by Markov chain Monte Carlo methods in conjunction with techniques of extended Kalman filter and Metropolis-Hasting algorithm. Of course, important issues to consider in using the GSDF model include the selection of the number of common factors r and the estimation of factor process $\boldsymbol{f}_t$. These issues can be addressed with the use of information criteria or under a Bayesian framework. See Lopes et al. (2011) for further details and applications.

9.8.4 S-T HMs

Assume that the locations $\{\boldsymbol{s}_1, \ldots, \boldsymbol{s}_m\}$ are fixed, where $\boldsymbol{s}_i \in D \subset R^d$, and that the time points are discrete and equally spaced, say $t = 1, \ldots, T$. A general framework for analyzing the S-T data is the hierarchical dynamic S-T model (DSTM) discussed in Cressie and Wikle (2011) and Wikle et al. (2019). See also the review article Harvill (2010).

Under this framework, the data model is

$$\boldsymbol{Z}_t = \boldsymbol{H}_t(\boldsymbol{Y}_t, \boldsymbol{\theta}_{h,t}, \epsilon_t), \quad t = 1, \ldots, T, \tag{9.106}$$

where $\boldsymbol{Z}_t$ denotes the observed data at time t, $\boldsymbol{Y}_t$ is the latent S-T process of interest, $\boldsymbol{H}_t$ is a linear or nonlinear mapping that connects the data to the latent process, and ϵ_t is the data error, which represents measurement error and/or certain small-scale temporal variability. The model parameters are denoted by $\boldsymbol{\theta}_{h,t}$, which may vary with space (signified by the distance h) and/or time.

Under the model, one assumes that the data $\{\boldsymbol{Z}_t\}$ are temporally independent conditioned on the latent process $\boldsymbol{Y}_t$ and the parameters $\boldsymbol{\theta}_{h,t}$. Therefore, the joint distribution of the data can be written in product form

$$p[\{\boldsymbol{Z}_t\}_{t=1}^T | \{\boldsymbol{Y}_t\}_{t=1}^T, \{\boldsymbol{\theta}_{h,t}\}_{t=1}^T] = \prod_{t=1}^T p[\boldsymbol{Z}_t | \boldsymbol{Y}_t, \boldsymbol{\theta}_{h,t}]. \tag{9.107}$$

Typically, the conditional distribution of $\boldsymbol{Z}_t$ is assumed to be normal, but other members of the exponential family can also be used. See the previous section.

Another important issue of applying the DSTM is the decomposition of the joint distribution of the latent S-T process $\{\boldsymbol{Y}_t\}$. To this end, the technique of time series analysis can be used and we have

$$p[\boldsymbol{Y}_1, \ldots, \boldsymbol{Y}_T] = \prod_{t=2}^{T} p[\boldsymbol{Y}_t|\boldsymbol{Y}_{t-1}, \ldots, \boldsymbol{Y}_1] \times p(\boldsymbol{Y}_1), \tag{9.108}$$

where, for simplicity, the parameters are omitted from the distribution functions. Equation (9.108) provides a natural way to introduce assumptions under which the joint distribution of $\{\boldsymbol{Y}_t\}$ can easily be calculated. For instance, if one assumes that the first-order Markov property holds, then

$$p[\boldsymbol{Y}_t|\boldsymbol{Y}_{t-1}, \ldots, \boldsymbol{Y}_1, \{\boldsymbol{\theta}_{h,t}\}_{t=1}^{T}] = p[\boldsymbol{Y}_t|\boldsymbol{Y}_{t-1}, \boldsymbol{\theta}_{h,t}], \tag{9.109}$$

where, again, for simplicity, we put all parameters (including those for the time evolution of $\boldsymbol{Y}_t$) into the parameter vector $\boldsymbol{\theta}_{h,t}$. Consequently, assume that $\boldsymbol{Y}_t$ is a first-order Markov process, we have

$$p[\boldsymbol{Y}_T, \ldots, \boldsymbol{Y}_1|\{\boldsymbol{\theta}_{h,t}\}] = \left(\prod_{t=2}^{T} p[\boldsymbol{Y}_t|\boldsymbol{Y}_{t-1}, \boldsymbol{\theta}_{h,t}]\right) p[\boldsymbol{Y}_1|\boldsymbol{\theta}_{h,1}]. \tag{9.110}$$

Under the first-order Markov property, we can express the model of $\boldsymbol{Y}_t$ as

$$\boldsymbol{Y}_t = \boldsymbol{M}(\boldsymbol{Y}_{t-1}, \boldsymbol{\theta}_{h,t}, \boldsymbol{\eta}_t), \tag{9.111}$$

where $\boldsymbol{M}$ denotes the temporal transition and $\boldsymbol{\eta}_t$ is a spatial noise process that is temporally uncorrelated. The transition $\boldsymbol{M}$ of Eq. (9.111) can be linear or nonlinear. For instance, $\boldsymbol{Y}_t$ may follow a VAR(1) model or a threshold VAR(1) model with multiple regimes. Interested readers are referred to Tsay (2014) and Tsay and Chen (2019), for properties of VAR(1) and threshold VAR models, respectively, among other multivariate time series books. The above model can be generalized to any finite-order Markov process so long as the order is small so that the computation remains manageable.

There are two general approaches to estimate DSTM. The first approach is to use the expectation-maximization (EM) algorithm of Dempster et al. (1977) and the second approach is the Markov chain Monte Carlo (MCMC) method. In some nonlinear cases, sequential Monte Carlo methods or particle filtering can also be used, especially when the dimension m is small. Some details can be found in Cressie and Wikle (2011).

APPENDIX 9.A: SOME R PACKAGES AND COMMANDS

In this appendix, we demonstrate R commands used to produce some of the plots used in the chapter, especially those in Section 9.1. Many R commands used are also available at http://spacetimewithr.org.

```
### S-T plots
### Figure 9.1
>library(sp)
>library(gstat)
>data(meuse)
>data(meuse.riv)
>plot(meuse.riv,type="l",asp=1,xlab="meuse.x",ylab='meuse.y',
    main="Meuse river")
>points(meuse$x,meuse$y,pch="x",cex=0.6)
## Figure 9.3
>require(fields)
>data(ozone2)
> good <- !is.na(ozone2$y[16,]) # day 16 and remove missing values
> out <- as.image(ozone2$y[16,good],x=ozone2$lon.lat[good,])
> image.plot(out)
> map("state",xlim=range(ozone2$lon.lat[good,1]),
    ylim=range(ozone2$lon.lat[good,2]),add=T)
> map.text("state",xlim=range(ozone2$lon.lat[good,1]),
   ylim=range(ozone2$lon.lat[good,2]),cex=1,add=T)
> points(ozone2$lon.lat[good,])
##
>library("animation")
>require("dplyr")
>require("ggplot2")
>require("maps")
>require("STRbook")
>set.seed(1)
>data("NOAA_df_1990",package="STRbook") ## Load data
>Tmax=filter(NOAA_df_1990,proc=="Tmax" & month %in% 5:9 & year==1993)
>Tmax %>% select(lon,lat,date,julian,z) %>% head()

>Tmax$t <- Tmax$julian - 728049
>Tmax_1 <- subset(Tmax, t %in% c(2,16,31))

>NOAA_plot<-ggplot(Tmax_1)+geom_point(aes(x=lon,y=lat,colour=z),size=2)+
   col_scale(name = "degF")+xlab("Longitude (deg)")+
   ylab("Latitude (deg)") +
   geom_path(data = map_data("state"),aes(x=long,y=lat,group=group)) +
   facet_grid(~date)+coord_fixed(xlim=c(-105,-75),ylim=c(25,50)) +
   theme_bw()
>print(NOAA_plot)

### Time Series plot
>set.seed(11)
>UIDs <- unique(Tmax$id)
>UIDs_sub <- sample(UIDs,10)
>Tmax_sub <- filter(Tmax, id %in% UIDs_sub)
>TmaxTS <- ggplot(Tmax_sub) +
   geom_line(aes(x = t,y=z))+
   facet_wrap(~id,ncol=5) +
   xlab("Day number (days)") +
   ylab("Tmax (degF)") +
   theme_bw() +
   theme(panel.spacing = unit(1,"lines"))
>print(TmaxTS)

#### Hovmoller plot
```

```
>lim_lat <- range(Tmax$lat)
>lim_t <- range(Tmax$t)
>lat_axis <- seq(lim_lat[1],lim_lat[2],length=25)
>t_axis <- seq(lim_t[1],lim_t[2],length=120)
>lat_t_grid <- expand.grid(lat=lat_axis,t=t_axis)
>Tmax_grid <- Tmax
>dists <- abs(outer(Tmax$lat,lat_axis,"-"))
>Tmax_grid$lat <- lat_axis[apply(dists,1,which.min)]
>Tmax_lat_hov <- group_by(Tmax_grid,lat,t) %>% summarise(z=mean(z))
>Hovller_lat <- ggplot(Tmax_lat_hov) +
  geom_tile(aes(x=lat,y=t,fill=z)) +
  fill_scale(name = "degF") +
  scale_y_reverse() +
  ylab("Day number (days)") +
  xlab("Latitude (deg)") +
  theme_bw()
>print(Hovller_lat)

>lim_long <- range(Tmax$lon)
>lim_t <- range(Tmax$t)
>long_axis <- seq(lim_long[1],lim_long[2],length=25)
>t_axis <- seq(lim_t[1],lim_t[2],length=120)
>long_t_grid <- expand.grid(lon=long_axis,t=t_axis)
>Tmax_grid <- Tmax
>dists <- abs(outer(Tmax$lon,long_axis,"-"))
>Tmax_grid$lon <- long_axis[apply(dists,1,which.min)]
>Tmax_long_hov <- group_by(Tmax_grid,lon,t) %>% summarize(z=mean(z))
>Hovller_long <- ggplot(Tmax_long_hov) +
   geom_tile(aes(x=lon,y=t,fill=z)) +
   fill_scale(name = "degF") +
   scale_y_reverse() +
   ylab("Day number (days)") +
   xlab("Longitude (deg)") +
   theme_bw()
>print(Hovller_long)
```

EXERCISES

1. Consider the ozone data of Midwestern United States. Obtain a similar plot as Figure 9.3 for the data of 25 June 1987.
2. Consider, again, the ozone data of Midwestern United States. Obtain the counterpart plot of Exercise 1 for the trend-adjusted data.
3. Perform the analysis of Example 9.4 but on the daily maximum temperature of July 1993.
4. Perform the universal S-T kriging as that of Example 9.2 but using the daily maximum temperature of July 1993.

REFERENCES

Abramowitz, M. and Stegun, I. A. (1964). *Handbook of Mathematical Functions with Formulas, Graphs, and Mathematical Tables*. National Bureau of Standards, Washington, DC.

Ali, M. M. (1979). Analysis of stationary spatial-temporal processes: Estimation and prediction. *Biometrika*, **66**: 513–518.

Bartlett, M. S. (1964). The spectral analysis of two-dimensional point processes. *Biometrika*, **51**: 299–311.

Besag, J. E. (1974). Spatial interaction and the statistical analysis of lattice systems. *Journal of the Royal Statistical Society, Series B*, **36**: 192–225.

Burrough, P. A. and McDonnell, R. A. (1998). *Principles of Geographical Information Systems*. Oxford University Press, Oxford, UK.

Cliff, A. D. and Ord, J. K. (1981). *Spatial Processes: Models and Applicaions*. Pion, London, UK.

Cressie, N. (1993). *Statistics for Spatial Data*, Revised Edition. Wiley, New York.

Cressie, N. and Huang, H. C. (1999). Classes of nonseparable, spatio-temporal stationary covariance functions. *Journal of the American Statistical Association*, **94**: 1330–1340.

Cressie, N. and Johannesson, G. (2008). Fixed rank kriging for very large spatial data sets. *Journal of the Royal Statistical Society, Series B*, **70**: 208–226.

Cressie, N. and Wikle, C. K. (2011). *Statistics for Spatio-Temporal Data*. Wiley, Hoboken, NJ.

Deb, S. and Tsay, R. S. (2019). Spatio-temporal models with space-time interaction and their applications to air pollution data. *Statistical Sinica* **29**: 1181–1207.

Dempster, A. P., Laird, N. M., and Rubin, D. B. (1977). Maximum likelihood from incomplete data with the EM algorithm. *Journal of the Royal Statistical Society, Series B*, **39**: 1–38.

Diggle, P. J. (1985). A kernel method for smoothing point process data. *Applied Statistics*, **34**: 138–147.

Diggle, P. J. (2013). *Statistical Analysis of Spatial and Spatio-Temporal Point Patterns*. Chapman & Hall/CRC, Boca Raton, FL.

Diggle, P. J. and Ribeiro, P. J. (2007). *Model-Based Geostatistics*. Springer, New York, NY.

Gao, Z., Ma, Y., Wang, H., and Yao, Q. W. (2019). Banded spatio-temporal autoregressions. *Journal of Econometrics*, **208**: 211–230.

Gneiting, T. (2002). Nonseparable, stationary covariance functions for space-time data. *Journal of the American Statistical Association*, **97**: 590–600.

Green, P. J. and Silverman, B. W. (1994). *Nonparametric Regression and Generalized Linear Model: A Roughness Penalty Approach*. Chapman & Hall: Boca Raton, FL.

Harvill, J. L. (2010). Spatio-temporal processes. *WIREs: Computational Statistics*, **2**: 375–382.

Hovmöller, E. (1949). The trough-and-ridge diagram. *Tellus*, **1**: 62–66.

Ising, E. (1925). Beitrag zur Theorie des Ferromagnetismus. *Zeitschrift für Physik*, **31**: 253–258.

Journel, A. G. and Huijbregts, C. J. (1978). *Mining Geostatistics*. Academic Press, London, UK.

Karhunen, K. (1947). Über lineare methoden in der wahrscheinlichkeitsrechnung. *Annales Academia Scientiarum Fennica, Series A*, **37**: 1–79.

Loève, M. (1978). *Probability Theory*. Springer-Verlag, NY.

Lopes, H. F., Gamerman, D., and Salazar, E. (2011). Generalized spatial dynamic factor models. *Computational Statistics and Data Analysis*, **55**: 1319–1330.

Matheron, G. (1962). *Traite de Geostatistique Appliquèe, Tome I*. Memoires du Bureau de Recherches Geologiques et Minieres, No. **14**, Technip, Paris.

Matheron, G. (1963). *Traitè de Geostatistique Appliquèe, Tome II: le Krigeage*. Memoires du Bureau de Recherches Geologiques et Minieres, No. **24**, Paris.

Moran, P. A. P. (1950). Notes on continuous stochastic phenomena. *Biometrika*, **37**: 17–23.

Nychka, D., Cox, L., and Piegorsch, W. (1998). Case studies in environmental statistics. In *Lecture Notes in Statistics*. Springer-Verlag, New York, NY.

Oehlert, G. W. (1993). Regional trends in sulfate wet deposition. *Journal of the American Statistical Association*, **88**: 390–399.

Pebesma, E. (2019). The meuse data set: A brief tutorial for the `gstat` R package. www.r-project.org (accessed 26 September 2019).

Pfeifer, P. E. and Deutsch, S. J. (1980). A three-stage iterative procedure for space-time modeling. *Technometrics*, **22**: 35–47.

Reinsel, G, Tiao, G. C., Wang, N. W., Lewis, R., and Nychka, D. (1981). Statistical analysis of stratospheric ozone data for the detection of trend. *Atmospheric Environment*, **15**: 1569–1577.

Ripley, B. D. (1976).The second-order analysis of stationary point processes. *The Journal of Applied Statistics*, **13**: 255–266.

Stoyan, D. and Stoyan, H. (1994). *Fractals, Random Shapes and Point Fields: Methods of Geometrical Statistics*. Wiley, Chichester, UK.

Tsay, R. S. (2014). *Multivariate Time Series Analysis with R and Financial Applications*. Wiley, Hoboken, New Jersey.

Tsay, R. S. and Chen, R. (2019). *Nonlinear Time Series Analysis*. Wiley, Hoboken, NJ.

Tseng, S. and Huang, H. C. (2015). Non-stationary multivariate spatial covariance estimation via low-rank regularization. *Statistica Sinica*, **26**: 151–172.

Tseng, S. and Huang, H. C. (2018). Resolution adaptive fixed rank kriging. *Technometrics*, **60**: 198–208.

Tseng, S. L., Huang, H. C., Wang, W. T., Nychka, D., and Gillespie, C. (2019). *autoFRK*, R package. www.r-project.org (accessed 29 March 2019.).

Wang, W. T. and Huang, H. C. (2017). Regularized principal component analysis for spatial data. *Journal of Computational and Graphical Statistics*, **26**:14–25.

Wikle, C. K. (2003). Hierarchical models in environmental science. *International Statistical Review*, **71**: 181–199.

Wikle, C. K., Zammit-Mangion, A., and Cressie, N. (2019). Spatio-Temporal Statistics with R. Chapman & Hall/CRC, Boca Raton, FL.

Zammit-Mangion, A. (2018a). *STRbook: Supplementary Package for Book on ST Modeling with R*, R Package. https://github.com/andrewzm/STRbook (accessed 25 September 2019).

Zammit-Mangion, A. (2018b). *Fixed Rank Kriging: The FRK package* for modeling spatio-temporal data with spatio-temporal basis functions. See R Core Web.

INDEX

Statistical Learning for Big Dependent Data, First Edition. Daniel Peña and Ruey S. Tsay.
© 2021 John Wiley & Sons, Inc. Published 2021 by John Wiley & Sons, Inc.